Handbook of Experimental Pharmacology

Continuation of Handbuch der experimentellen Pharmakologie

Vol. 56/I

Cardiac Glycosides

Part I: Experimental Pharmacology

Contributors

T. Akera · H. Bahrmann · G. G. Belz · H. F. Benthe · B. Dénes
W. Diembeck · E. Erdmann · H. Flasch · A. L. Fox · T. Godfraind
K. Greeff · Th. W. Güntert · K. Güttler · U. Gundert-Remy
D. Hafner · O. Heidenreich · P. H. Joubert · W. Klaus · H. H. A. Linde
H. Lüllmann · D. T. Mason · W. G. Nayler · E. A. Noack
H. Osswald · Th. Peters · M. Reiter · H. Schadewaldt · M. C. Schaub
K. Stellner · P. G. Waser · E. Weber · R. Weingart

Editor

K. Greeff

Springer-Verlag Berlin Heidelberg New York 1981

Professor Dr. med. KURT GREEFF
Direktor des Instituts für Pharmakologie der Universität Düsseldorf,
Moorenstrasse 5, D-4000 Düsseldorf

With 164 Figures

ISBN-13:978-3-642-68165-3 e-ISBN-13:978-3-642-68163-9
DOI: 10.1007/978-3-642-68163-9

Softcover reprint of the hardcover 1st edition 1981

2122/3130-543210

List of Contributors

Professor Dr. T. AKERA, Department of Pharmacology and Toxicology, Michigan State University, East Lansing, Michigan 48824, USA

Dr. H. BAHRMANN, Abt. Forschung und Entwicklung, Fa. Gödecke AG, Postfach 569, D-7800 Freiburg i. Br.

Professor Dr. G. G. BELZ, Institut für Kardiovaskuläre Therapie, Kardiologische Praxis, Wilhelmstrasse 16, D-6200 Wiesbaden

Professor Dr. H. F. BENTHE, Pharmakologisches Institut des Universitäts-Krankenhauses Eppendorf, Martinistrasse 52, D-2000 Hamburg 20

Dr. B. DÉNES, Institut für Pharmakologie der Universität Düsseldorf, Moorenstrasse 5, D-4000 Düsseldorf

DR. W. DIEMBECK, c/o Beiersdorf AG, Unnastrasse 48, D-2000 Hamburg 20

Priv.-Doz. Dr. med. E. ERDMANN, Klinikum Großhadern, Medizinische Klinik I, Ludwig-Maximilians-Universität, Marchioninistrasse 15, D-8000 München 70

Dr. H. FLASCH, c/o Beiersdorf AG, Unnastrasse 48, D-2000 Hamburg 20

Dr. ALANNA L. FOX, Institut für Pharmakologie der Universität Düsseldorf, Moorenstrasse 5, D-4000 Düsseldorf

Professor Dr. T. GODFRAIND, Faculté de Médicine, Laboratoire de Pharmacodynamie, Générale et du Pharmacologie, Université de Louvain, Av. E. Mounier, 73-U.C.L. 7350, B-1200 Bruxelles

Professor Dr. med. K. GREEFF, Direktor des Instituts für Pharmakologie der Universität Düsseldorf, Moorenstrasse 5, D-4000 Düsseldorf

Dr. TH. W. GÜNTERT, Pharmazeutisches Institut der Universität Basel, Totengässlein 3, CH-4051 Basel

Dr. med. K. GÜTTLER, Pharmakologisches Institut der Universität Köln, Gleueler Strasse 24, D-5000 Köln 41

Dr. URSULA GUNDERT-REMY, Medizinische Universitätsklinik, Bergheimer Strasse 58, D-6900 Heidelberg

Dipl.-Math. D. HAFNER, Institut für Pharmakologie der Universität Düsseldorf, Moorenstrasse 5, D-4000 Düsseldorf

Professor Dr. O. HEIDENREICH, Abt. Pharmakologie der Medizinischen Fakultät der TH Aachen, Schneebergweg, D-5100 Aachen

Professor Dr. P. H. JOUBERT, Department of Pharmacology, Medical University of Southern Africa, PO Medunsa 0202, Pretoria, South Afrika

Professor Dr. W. KLAUS, Direktor des Pharmakologischen Institutes der Universität Köln, Gleueler Strasse 24, D-5000 Köln 41

Professor Dr. H. H. A. LINDE, Pharmazeutisches Institut der Universität Basel, Totengässlein 3, CH-4051 Basel

Professor Dr. H. LÜLLMANN, Leiter der Abteilung Pharmakologie, Klinikum der Christian-Albrechts-Universität Kiel, Hospitalstrasse 4–6, D-2300 Kiel

Dr. D. T. MASON, Professor of Medicine, Chief of Cardiovascular Medicine, School of Medicine, University of California, Davis, California 95616/USA

Dr. WINIFRED G. NAYLER, Dept. of Medicine, University of Melbourne, Austin Hospital, Heidelberg, 3084 Victoria, Australia

Professor Dr. med. E. A. NOACK, Institut für Pharmakologie der Universität Düsseldorf, Moorenstrasse 5, D-4000 Düsseldorf

Priv.-Doz. Dr. H. OSSWALD, Abteilung Pharmakologie der Medizinischen Fakultät der TH Aachen, Schneebergweg, D-5100 Aachen

Priv.-Doz. Dr. med. TH. PETERS, Institut für Pharmakologie, Christian-Albrechts-Universität, Hospitalstrasse 4–6, D-2300 Kiel

Professor Dr. M. REITER, Direktor des Instituts für Pharmakologie und Toxikologie der Technischen Universität, Biedersteiner Strasse 29, D-8000 München 40

Professor Dr. med. H. SCHADEWALDT, Direktor des Instituts für Geschichte der Medizin, Universität Düsseldorf, Moorenstrasse 5, D-4000 Düsseldorf 1

Professor Dr. M. C. SCHAUB, Pharmakologisches Institut der Universität Zürich, Gloriastrasse 32, CH-8006 Zürich

Dr. K. STELLNER, c/o Boehringer Mannheim GmbH, Biochemica-Werk, Bahnhofstrasse, D-8132 Tutzing

Professor Dr. P. G. WASER, Pharmakologisches Institut der Universität Zürich, Gloriastrasse 32, CH-8006 Zürich

Professor Dr. ELLEN WEBER, Abt. Klinische Pharmakologie der Medizinischen Klinik der Universität Heidelberg, Bergheimer Strasse 58, D-6900 Heidelberg 1

Dr. R. WEINGART, Physiologisches Institut der Universität Bern, Bühlplatz 5, CH-3000 Bern

Preface

Following the monographs by STRAUB (1924) and LENDLE (1935), this is the third contribution to the "Pharmacology of Cardiac Glycosides" within the *Handbook of Experimental Pharmacology*, which was founded by ARTHUR HEFFTER and continued by WOLFGANG HEUBNER. Because of the need created by the length of time that had elapsed since LENDLE's work, the editorial board requested the rapid appearance of this 56th volume, which represents current knowledge of the pharmacology and clinical pharmacology of cardiac glycosides. In order to avoid any delay, numerous authors were invited to contribute because shorter contributions take less time to prepare and are consequently more up-to-date. The disadvantage is that some overlap between certain chapters could not be avoided, despite the editor's efforts. Overlapping can, however, actually be useful, in that differing opinions may be provided and topical issues discussed from varying viewpoints. This reminds the reader that scientific horizons in medicine should often be widened or revised.

I would like to thank DR. ALANNA FOX and DR. K. ANANTHARAMAN for their help and advice in the revision of certain chapters. I am also grateful to Springer-Verlag, and particularly to MR. WINSTANLEY and MR. EMERSON, for their contribution to the completion of this volume through translation and corrections. In conclusion I would like to thank MRS. WALKER, MR. BISCHOFF, MRS. SEEKER, and MR. BERGSTEDT of Springer-Verlag for their helpful support.

When ARTHUR HEFFTER, the founder of this handbook, invited colleagues in 1913 both at home and abroad to contribute, many years had to pass before the first two volumes appeared between 1919 and 1923. Bearing this in mind, I would like to thank all authors for their efforts and cooperation, which have made it possible to complete this volume within a reasonable time.

K. GREEFF

Contents

CHAPTER 6

ATPase for the Determination of Cardiac Glycosides
URSULA GUNDERT-REMY and ELLEN WEBER. With 4 Figures

CHAPTER 7

Rubidium Uptake in Erythrocytes. G. G. BELZ. With 7 Figures

Biological Methods for the Evaluation of Cardiac Glycosides

CHAPTER 8

Evaluation of Cardiac Glycosides in the Intact Animal

H. BAHRMANN and K. GREEFF. With 6 Figures

Mode of Action of Cardiac Glycosides

CHAPTER 11

The Positive Inotropic Action of Cardiac Glycosides on Cardiac Ventricular Muscle. M. REITER. With 15 Figures

CHAPTER 12

Influence of Cardiac Glycosides on Electrophysiologic Processes
R. WEINGART. With 11 Figures

CHAPTER 19

Effects of Cardiac Glycosides on Myofibrils
P. G. WASER and M. C. SCHAUB. With 7 Figures

CHAPTER 20

Substances Possessing Inotropic Properties Similar to Cardiac Glycosides
T. AKERA, A. L. FOX and K. GREEFF. With 8 Figures

Non-Cardiac Effects of Cardiac Glycosides

CHAPTER 21

Effects of Cardiac Glycosides on Central Nervous System. H. F. BENTHE

CHAPTER 22

Effects of Cardiac Glycosides on Vascular System. D. T. MASON
With 11 Figures

Contents

Part II: Pharmacokinetics and Clinical Pharmacology

Clinical Pharmacology

CHAPTER 1

Introduction and Remarks on the History of Cardiac Glycosides

K. GREEFF and H. SCHADEWALDT

If cardiac glycosides were to be introduced into therapy today the state authorities would most likely refuse permission, due to the extreme toxicity and narrow therapeutic range of these compounds. Today, however, digitalis glycosides are indispensable, as there are no other substances known which can increase the contractility of the heart, without increasing but lowering beat frequency, a characteristic of cardiac glycosides which ensures an increase in cardiac efficiency.

Due to the narrow therapeutic range of digitalis glycosides their prescription requires special medical experience, as, unlike other drugs, the optimal dosage has to be determined for each individual. This has been an obvious problem ever since W. WITHERING introduced digitalis therapy and runs like a thread throughout the history of the use of digitalis. WITHERING had already given clear details as to the required dosage and brought attention to the fact that potency depends not only on the season of picking of the digitalis leaves but also on their preparation as infusions, decoctions or pills; all this had to be taken into account in digitalis therapy. In order to check the potency of various preparations, pharmacological methods were later developed to standardize these drugs, using frogs, cats, guinea pigs or other animals. More recently chemical and radioimmunological methods have been introduced in order to test the drug content and potency of various preparations. The questions of enteral absorption and bioavailability must still be taken into account when recommending dosage.

Just as important when administering cardiac glycosides are the individual differences in sensitivity, which can be caused by age, constitution, hormonal factors, disturbances in kidney or liver function or interactions with other drugs; this point will be discussed in more detail in the appropriate chapters of this book. We are therefore still occupied with the problem – as in the time of WITHERING – of estimating an individual and optimal dosage in digitalis glycoside therapy.

A Chronological History of Cardiac Glycosides

ca. 1600 B.C. The healing effects of the sea onion are described in the Egyptian EBERS PAPYRUS.

ca. 400 B.C. Extracts of sea onion mixed with honey and wine recommended in the Greek CORPUS HIPPOCRATICUM to enforce diuresis.

ca. 200 B.C. THEOPHRAST OF ERESOS (380-286 BC) introduces the term "scilla."

ca. 500 A.D. The leaves of digitalis are used for the external treatment of inflammation or in the form of an ointment for abscesses.

1250 The digitalis plant was mentioned in the Welsh pharmaceutical book "Meddygon myddmai" under the name "foxes glofa," as a component of ointments and as a treatment for headaches and spasms.

1542 LEONHARD FUCHS (1501–1566), botanist and physician in Tübingen, in his book of herbs, describes the digitalis plant in great detail and produces the first illustrations of digitalis purpurea, lutea and lanata. He also mentioned the external and internal application as a diuretic and laxative and for the first time gives this plant the scientific name: "digitalis": "Diss gewechss würdt von unsern Teutschen Fingerhut geheyssen, darumb das seine blumen einen fingerhut, so man zu dem näen braucht, gantz und gar ähnlich seind. Man mags in mittler zeit, bis man einen bessern namen findt, wie wir in unserem Lateinischen Kreuterbuch gethan haben, Digitalem zu Latein, dem Teutschen namen nach nennen."

1546 HIERONYMUS BOCK (1498–1554) also mentions digitalis ointments and for the first time digitalis pills for oral use in his book of herbs.

1573 Digitalis appears in the pharmaceutical book of JODOCUS HARCH.

1608 REMBERT DODOENS (1517–1585) in his book of herbs, which appeared posthumously, describes digitalis as a herb with healing power which could be administered externally or internally. He warned, however, against the careless use of this plant.

1640 JOHN PARKINSON (1567–1650) confirms the centuries-old popular use of digitalis in epileptic fits and its external application in goitre.

1650 A digitalis ointment is mentioned for the first time in the official PHARMACOPOEIA LONDONIENSIS.

1710 WILLIAM SALMON (1644–1713) mentions a further indication for the internal application of an infusion or syrup made of digitalis leaves in the treatment of phthisis.

1744 GERARD VAN SWIETEN (1700–1772), personal physician to the empress Maria Theresia of Austria, rediscovers the diuretic effect of scilla maritima.

1748 FRANCOIS SALERNE (†1760) before the Académie française in Paris, describes the first animal experiments in which digitalis leaves had

[1] This is the first time that the foxglove is mentioned in German literature as "Fingerhut"

been administered to turkeys; the animals died in convulsions, which was a lesson to the French doctors to be cautious.

1775 The Scottish physician WILLIAM WITHERING (1741–1799) is informed on the diuretic effect of digitalis infusions by a female herbalist and begins gaining his own experience with this new therapy.

1780 ERASMUS DARWIN (1731–1802), encouraged by WITHERING, tries this form of therapy and is the first physician to publish what appear to be extraordinary experiences with this drug.
FRANCIS HOME (1719–1813), professor of materia medica in Scotland, reports that scilla influences cardiac function.

1785 WITHERING publishes the first classical monograph on the action of digitalis, summarizing 10 years of experience with this drug: "An Account of the Foxglove, and some of its Medical Uses: With practical Remarks on Dropsy, and other Diseases." He describes indirectly the use of this drug in the treatment of cardiac insufficiency in the following words: "if the pulse be feeble or intermitting, the countenance pale, the lips livid, the skin cold, the swollen belly soft, a fluctuating or the anasarcous limbs readily pitting under the pressure of the finger, we may expect the diuretic effects to follow in a kindly manner." The first preliminary report has already been given by WITHERING in 1779 before the "Medical Society" of Edinburgh. Starting in 1785 an exchange of correspondance developed with the American physician HALL JACKSON (1739–1797) concerning WITHERING's observations on the actions of digitalis. JACKSON received seeds from WITHERING and carried out digitalis therapy in North America against oedema.

1786 CARL CHRISTIAN SCHIEMANN (1763–1835) mentions a further indication for digitalis in the treatment of scrophulosis.

1790 JOSEPH MASON COX (1762–1822), English psychiatrist, introduces the use of digitalis in the treatment of the lunatics. The one time mental patient Hölderlin was treated in Tübingen with digitalis infusions, the same treatment being used 90 years later for the painter van Gogh.

1796 SAMUEL HAHNEMANN (1766–1843), the founder of homeopathy, condemns the use of digitalis because of unpleasant side-effects. These include among others excessive appetite, moodiness, head- and stomach-ache.

1799 JOHN FERRIAR (1763–1815), psychiatrist in Manchester, publishes the second comprehensive monograph on digitalis. This mentions, apart from the diuretic action of digitalis, a number of other indications as to the use of this substance.

1807 WILLIAM HAMILTON (1764–1847) recommends the use of digitalis against fever, which in those days was treated as an independent illness.

FRIEDRICH JAHN (1766–1813) concurrently confirms several indications for the use of digitalis, such as hydrops, scrofulosis, goitre, and phthisis with haemoptysis.

1809 P.G. VASSAL defines a digitalis-sensitive and a digitalis-resistant form of hydrops.

1814 FRIEDRICH LUDWIG KREYSIG (1770–1839), personal physician to the electorate of Saxony, is the first to postulate a direct effect of digitalis on the heart and blood vessels and presumes that the slowing of the pulse originates from the brain.

1824 AUGUSTE LE ROYER uses for the first time the term "digitaline" for the ether extract of the leaves of the foxglove. This extract was found to be so poisonous to rabbits, cats, and dogs, that he is afraid to use it in therapy.

1825 JOHANN EVANGELISTA PURKINJE (1787–1869), physiologist in Breslau, describes arrhythmias and optical sensations upon the self-administration of digitalis leaves, these effects having also been mentioned by WITHERING.

1836 JOSEPH FRIEDRICH SOBERNHEIM (1803–1846), practitioner in Berlin, widens the therapeutic spectrum for digitalis by using it against inflammation and infectious illnesses of all kinds, in particular pneumonia. He claims to have had success in the treatment of delirium tremens.

1845 AUGUSTIN-EUGÈNE HOMOLLE (1808–1883), physician and chemist in Paris, isolates a fraction (Digitaline Homolle) of the leaves of digitalis purpurea, which contains mainly digitoxin and gitoxin. This is used for the first time as a pharmaceutical preparation and is included in the French Pharmacopoeia of 1866.

1850 LUDWIG TRAUBE (1818–1876), clinician in Berlin, recognizes the increase in efficiency of cardiac muscle under the influence of digitalis. He suggests that the bradycardia is a result of vagal stimulation. The encouraging results in the treatment of feverish illnesses are said by TRAUBE to be a result of the direct antipyretic influence of digitalis.

1852 The Strasbourg pharmacist CHARLES FRÉDERIC GERHARDT (1816–1856) gives the name glycoside to a group of sugar-containing substances, extracted from plants.

HERMANN FRIEDRICH STANNIUS (1808–1883), professor of anatomy in Rostock, in experiments with frog hearts, establishes the bradycardian effect of digitalis to be a result of vagal stimulation. He describes the systolic cardiac standstill at higher digitalis concentrations.

1859 BERNHARD BÄHR publishes the first German monograph concerning digitalis.

1859 RUDOLF BUCHHEIM (1820–1879), professor of materia medica in Dorpat, describes in his "Lehrbuch der Arzneimittellehre" the cumulative effect of digitalin, recognised by the continual decrease in pulse frequency.

1865 DAVID LIVINGSTONE (1813–1873), African explorer, reports that the botanist JOHN KIRK (1832–1922) noticed a slowing of the pulse upon the self-administration of strophanthus seeds. CHARLES HILTON FAGGE (1838–1883) and his colleague THOMAS STEVENSON (1838–1908) report on the cardiac action of an African arrow poison, also mentioning the unpublished works of WILLIAM SHARPEY (1802–1880). In the same year the Russian scientist EUGEN WENCELALOWITSCH PELIKAN (1824–1884) carries out relevant experiments in France. Both groups of research workers compare the effects of the substances in strophanthus seeds with those of digitalis.

1869 CLAUDE-ADOLPHE NATIVELLE (1812–1889) isolates an insoluble fraction from digitalis leaves with alcohol and chloroform and calls this "digitaline cristallisée." This is probably identical with the digitoxin later isolated by SCHMIEDEBERG. This preparation was used widely in France under the name "Digitaline Nativelle."

1871 CARL LIEBERMEISTER (1833–1901), professor of internal medicine in Tübingen, uses digitalis in preference to other substances as it is supposed to be a particularly effective therapy for all hyperpyretic infectious illnesses.

1872 RUDOLF BOEHM (1844–1926), pharmacologist in Dorpat, investigates the effects of digitalis on cardiac muscle in animal experiments. The pulsus bigeminus, otherwise known as the dicrotic pulse, is described as a toxic symptom.

1875 OSWALD SCHMIEDEBERG (1833–1921), pharmacologist in Strasbourg, isolates a fraction from digitalis leaves, which he calls "digitoxin." In an extremely dangerous experiment, his colleague ROBERT KOPPE takes 3.5 mg crystalline substance on 5 consecutive days, and experiences terrible side effects.

1879 EMANUEL AUGUST MERCK (1855–1923) isolates three different substances from scilla.

1885 THOMAS RICHARD FRASER (1841–1920), pharmacologist in Edinburgh, commences attempts to isolate strophanthin.

1888 BLONDEL CATILLON manages to isolate strophanthin, FRANÇOIS ARNAUD (1858–1927) is also able to crystallize ouabain from the plant Acocanthera ouabaio.

1889 WASSILY DEMITSCH, pharmacologist in Dorpat, reports on the uses of convallaria majalis in Russian medicine in the eighteenth and nineteenth centuries (against dropsy, epilepsy and hysterical convulsion).

1890 FRASER introduces strophanthin, isolated from strophanthus hispidus, into his drug collection.

1892 HEINRICH KILIANI (1855–1945) isolates digitalin and further digitalis glycosides from digitalis seeds.

1898 ELIJAH HOUGHTON (1867–1937) introduces the standardization of digitalis preparations in the frog.

1900 ADOLF KUSSMAUL (1822–1902), professor of internal medicine in Heidelberg, recommends a reduction in dose in the long-time therapy with digitalis.

1904 ALBERT FRAENKEL (1864–1938), physician at the Speyershof near Heidelberg, reports on the cumulative effect of digitalis glycosides. MAX CLOETTA (1868–1940), pharmacologist in Zürich, introduces the parenteral application of digalen.

1905 KURT KOTTMANN and FELIX MENDEL (1862–1925) perform the first investigations with intravenous digitalis therapy using digalen. JAMES MACKENZIE (1853–1925), Scottish cardiologist, reports the first observation of atrial fibrillation as a symptom of digitalis overdosage.

1906 ALBERT FRAENKEL introduces the intravenous strophanthin therapy.

1910 ROBERT ANTHONY HATCHER (1868–1944), laboratory of pharmacology, Cornell University, New York, and his colleague J.G. BRODY introduce the biologic standardization of cardiac glycosides in cats (Hatcherdose).

1912 FRIEDRICH KRAFT (1864–1914) introduces gitalin and gitoxin into pharmacotherapy.

1915 CARY EGGLESTON (1884–1966) recommends in the USA high initial doses of digitalis in order to quickly reach saturation.

1916 WALTER STRAUB (1874–1944), pharmacologist in Munich, studies the effects of digitalis on the isolated heart.

1928 ADOLF WINDAUS (1876–1959) produces almost pure digitoxin and begins its chemical analysis.

1930 SYDNEY SMITH (born 1902) isolates digoxin from the leaves of digitalis lanata.

1932 ALBERT FRAENKEL and RUDOLF THAUER (born 1906) publish their fundamental work concerning strophanthin therapy.

1933 ARTHUR STOLL (1887–1971) starts to purify digitalis lanata and purpurea glycosides and crystallize scillaren A.
ERNST EDENS (1876–1944), professor of internal medicine in Düsseldorf, summarizes the clinical experience of more than a decennium of digitalis therapy in his monograph.

Although EDENS as early as 1933 remarked in his monograph, that the amount of scientific investigation concerning the digitalis pharmacotherapy appeared boundless, we have, in the last 50 years, witnessed an immense increase in publications on this theme. Great progress has at the same time been seen in the clarification of the mechanism of action and pharmacokinetics of cardiac glycosides. Here, some information as to more recent history may be relevant: DE GRAFF, professor of therapy in New York, published his experiences on the use and dosage of cardiac glycosides in congestive heart failure (1950) and recommended, as did GOLD et al. (1942), a rapid digitalization with high initial doses. SCHWIEGK and JAHRMÄRKER (1960) did not propagate rapid saturation, but recommended careful digitalization; they at the same time accentuated the need for accompanying therapeutic measures, such as bed rest, diet, diuretics and sedatives. GROSSE-BROCKHOFF, EDENS' successor in Düsseldorf, also recommended an appropriate therapy (see GROSSE-BROCKHOFF and PETERS, this Handbook, Vol. 56/II, Chap. 12; GROSSE-BROCKHOFF and HAUSAMEN, 1975).

A rejection of the guiding principles for maximal digitalization meant the formation of new concepts in the USA and digitalis is not used in certain forms of cardiac insufficiency. This applies principally to acute cardiac insufficiency. Here the use of vasodilators has been recommended, whereas the application of digitalis is, according to most cardiologists, still indicated in chronic insufficiency (see MASON and LEE, Part II, Chap. 10; GROSSE-BROCKHOFF and PETERS, Part II, Chap. 12).

The dosage instructions of AUGSBERGER (1951, 1954) mark an advance in the rational and calculable digitalis therapy, creating the terms "total body dose," "maintenance dose," and "decay rate."

The investigation of the pharmacokinetics of cardiac glycosides was greatly aided by the use of radioactively labeled glycosides. GEILING and MCINTOSH (1948) were able to synthezise ^{14}C-labeled glycosides in digitalis plants, with which OKITA (1957) carried out pharmacokinetic investigations. For the chemical synthesis of ^{3}H-labeled glycosides we are indebted to WILZBACH (1957) and WARTBURG et al. (1965). The introduction of the radioimmunoassay by SMITH et al. (1969) meant a further advance (see Chap. 5).

The most important discovery concerning the mechanism of cardiac glycosides action, in the last 50 years, is the demonstration of their inhibition of membrane electrolyte transport and the Na^+-, K^+-activated membrane ATPase. SCHATZMANN (1953) observed that cardiac glycosides inhibit the active transport of potassium and sodium in erythrocytes, whereas passive diffusion remains unaffected. KAHN and ACHESON (1955), as well as GLYNN (1957) found that only those glycosides which are cardioactive are able to inhibit ion transport in erythrocytes. SKOU in 1957 discovered an ATPase in the microsomal fraction of the crab nerve, which was activated by K^+ and Na^+ and inhibited by cardiac glycosides. DUNHAM and GLYNN (1961), as well as POST et al. (1960) were able to show that a definite parallel exists between the inhibition of the electrolyte transport and inhibition of the Na^+-, K^+-activated ATPase with increasing glycoside concentrations in erythrocytes. Later the same ATPase was found in membrane fractions of other tissues, in particular cardiac muscle (BONTING et al., 1962; REPKE, 1964; DRANSFELD et al., 1966). Further details are given in Chaps. 13–16.

In 1933 EDENS remarked „Die Wissenschaft verläuft in der Form einer Spirale. So ist die Digitalisforschung heute wieder zu ihrem Ausgangspunkt zurückgekehrt, zur Erfahrung am Krankenbett.“ This reference to the importance of clinical pharmacology, despite more recent experiences in the field of experimental pharmacology, is being taken seriously today and has led to the arrangement of several symposia: DIMOND (1957), GREEFF (1968), STORSTEIN (1973), JAHRMÄRKER (1975), BODEM and DENGLER (1978), GREEFF and RIETBROCK (1979).

In conclusion, several review articles concerning the pharmacology and toxicology of cardiac glycosides may be referred to: WEESE (1936), WOLLENBERGER (1949), LOWN and LEVINE (1954), ROTHLIN and BIRCHER (1954), BAUMGARTEN (1963), GREEFF (1963), GLYNN (1964), BING (1965), CHUNG (1969), FISCH and SURAWICZ (1969), LEE and KLAUS (1971), MARKS and WEISSLER (1972), SMITH and HABER (1973), SCHWARTZ et al. (1975). These publications are not only complementary to the chronological history of cardiac glycosides, but should make it possible to inform oneself outside the bounds of this handbook and to relive older controversial opinions.

References

Arnaud, F.: Sur la matière cristallisée active des flèches empoisonnées de çomalis, extraite du bois ouabaio. Compt. Rend. Acad. Sci. Paris *106*, 1011–1014 (1888)

Augsberger, A.: Quantitatives zur Therapie mit Herzglykosiden. I. Mitteilung. Die Variabilität von Glykosidbedarf und -toleranz. Med. Welt *3*, 1471 (1951)

Augsberger, A.: Quantitatives zur Therapie mit Herzglykosiden. II. Mitteilung. Kumulation und Abklingen der Wirkung. Klin. Wochenschr. *32*, 945 (1954)

Bähr, B.: Digitalis Purpurea in ihren physiologischen und therapeutischen Wirkungen. Leipzig 1859

Baumgarten, G.: Die herzwirksamen Glykoside. Herkunft, Chemie und Grundlagen ihrer pharmakologischen und klinischen Wirkung. Edition Leipzig 1963
Bing, R.J.: Cardiac metabolism. Physiol. Rev. *45*, 171 (1965)
Bock, H.: Kreuter-Buch. Straßburg 1546, Fol. 335
Bodem, G., Dengler, H.J. (eds.): Cardiac glycosides. Berlin, Heidelberg, New York: Springer 1978
Boehm, R.: Untersuchungen über die physiologische Wirkung der Digitalis und des Digitalins. Arch. Ges. Physiol. *5*, 153–191 (1872)
Bonting, S.L., Caravaggio, L.L., Hawkins, N.M.: Studies on sodium-potassium activated adenosinetriphosphatase. IV. Correlation with cation transport sensitive to cardiac glycosides. Arch. Biochem. Biophys. *98*, 413–419 (1962)
Buchheim, R.: Lehrbuch der Arzneimittellehre. 2. Aufl. Leipzig: Leopold Voss 1859
Catillon, B.: Etude chimique du strophanthus. J. Pharm. 5. Ser. *17*, 220–222 (1888)
Chung, E.K.: Digitalis intoxication. Amsterdam: Excerpta Medica Foundation 1969
Cloetta, M.: Über Digalen. Münch. Med. Wschr. *51*, 1466–1468 (1904)
Cox, J.M.: A case of insanity cure by digitalis. Med. Comm. *2*, 261 (1790)
Darwin, E.: An account of the successful use of foxglove in some dropsies. Med. Tr. Roy. Coll. Phys. London 255–286 (1785)
Darwin, E.: Von der Wirkung des rothen Fingerhutes in der Wassersucht. Samml. Auserl. Abh. *6*, 2, Leipzig (1781)
De Graff, A.C.: Digitalis and cardiac glycosides in congestive heart failure. Med. Clin. North Am. *34*, 663 (1950)
Demitsch, W.: Literärische Studien über die wichtigsten russischen Volksheilmittel aus dem Pflanzenreiche. Dorpat (1888), Diss. Med., p. 55ff.
Demitsch, W.: Russische Volksheilmittel aus dem Pflanzenreiche. Histor. Studien aus dem Pharmakol. Institut Kais. Univ. Dorpat (ed. R. Kobert), Band 1, Halle (1889), p. 134–266
Dimond, E.G.: Digitalis. Publ. Springfield, Ill.: Charles C. Thomas 1957
Dodoens, R.: Cruydt-boek. p. 270. Leiden 1608
Dransfeld, H., Greeff, K., Berger, H., Cautius, V.: Die verschiedene Empfindlichkeit der Na^+ und K^+-aktivierten ATPase des Herz- und Skelettmuskels gegen K-Strophanthin. Naunyn-Schmiedebergs Arch. Pharmacol. *254*, 225–234 (1966)
Dunham, E.T., Glynn, I.M.: Adenosinetriphosphatase activity and the active movements of alkali metal ions. J. Physiol. (Lond.) *156*, 274 (1961)
Edens, E.: Die Digitalisbehandlung. 2. und 3. Aufl. Berlin: Urban und Schwarzenberg 1934 und 1948
Eggleston, C.: Digitalis dosage. Arch. Int. Med. *16*, 1–32 (1915)
Fagge, C.H., Stevenson, T.: On the application of physiological tests for certain organic poisons, and especially digitaline. Proc. Roy. Soc. London *14*, 270–274 (1865) (auch London 1866)
Ferriar, J.: An essay on the medical properties of the digitalis purpurea, or foxglove. Manchester 1799
Fisch, Ch., Surawicz, B.: Digitalis. Fisch, Ch., Surawicz, B., Knoebel, S.B., Greenspan, K. (eds.). New York, London: Grune & Stratton 1969
Fraenkel, A.: Vergleichende Untersuchungen über die kumulative Wirkung der Digitaliskörper. Arch. Exp. Path. Pharmak. *51*, 84–102 (1904)
Fraenkel, A.: Zur Digitalistherapie. Über intravenöse Strophanthintherapie. Verh. Dtsch. Ges. Inn. Med. 257–265 (1906)
Fraenkel, A., Thauer, R.: Strophanthin-Therapie, zugleich ein Beispiel quantitativer Digitalisanwendung nach pharmakologischen Grundsätzen. Berlin 1932
Fraser, T.R.: The action and uses of digitalis and its substitutes, with special reference to strophanthus. Brit. Med. J. *1885 II*, 904–910
Fraser, T.R.: Strophanthus hispidus; its natural history, chemistry, and pharmacology. Tr. Roy. Soc. Edinb. *35*, 955–1027 (1890)
Fuchs, L.: Historia stirpium, p. 892. Basel 1542
Geiling, E.M.K., McIntosh, B.J.: Biosynthesis of radioactive drugs using carbon-14. Science *108*, 558 (1948)

Gerhardt, C.F.: Aide-mémoire pour analyse chimique. Paris 1852
Glynn, J.M.: The action of cardiac glycosides and sodium and potassium movements in human red cells. J. Physiol. (Lond.) *136*, 148 (1957)
Glynn, J.M.: The action of cardiac glycosides on ion movements. Pharmacol. Rev. *16*, 387 (1964)
Gold, H., Kwit, N.T., Cattell, M., Travell, J.: Studies on purified digitalis glycosides. IV. The single dose method of digitalization. J. Am. Med. Assoc. *119*, 928–932 (1942)
Greeff, K.: Zur Pharmakologie der herzwirksamen Glykoside. Die Wirkung der Glykoside auf den Stoffwechsel des Herzmuskels. In: Klinische Physiologie. Mueller, W.A. (ed.), Bd. I, p. 340–370. Stuttgart: Georg Thieme 1963
Greeff, K.: Probleme der klinischen Prüfung herzwirksamer Glykoside. Greeff, K., Bahrmann, H., Benthe, H.F., Haan, D., Kreuzer, H. (eds.). Darmstadt: Dietrich Steinkopff 1968
Greeff, K., Rietbrock, N.: Digitoxin als Alternative in der Therapie der Herzinsuffizienz. Greeff, K., Rietbrock, N. (eds.). Stuttgart, New York: F.K. Schattauer 1979
Grosse-Brockhoff, F., Hausamen, T.-U.: 200 Jahre Herztherapie mit Digitalis. Dtsch. Med. Wschr. *100*, 1980–1991 (1975)
Hahnemann, S.: Versuche über ein neues Princip zur Auffindung der Heilkräfte der Arzneisubstanzen. J. Pract. Arzneykd. *2*, 465–561 (1796)
Hamilton, W.: Observations on the preparation, utility, and administration of the digitalis purpurea. London 1807
Harch, J. Enchiridion medicum ..., p. 99. Basel 1573
Hatcher, R.A., Brody, J.G.: The biological standardization of drugs. Am. J. Pharm. *32*, 360–372 (1910)
Hippokrates: Oeuvres complètes d'Hippocrate. Ed. E. Littré, *2*, p. 519, Paris 1840
Home, F.: Clinical experiments, history and dissections. Edinburgh 1780
Homolle, A.-E.: Mémoire sur la digitale pourprée. J. Pharm. Chim. 3 Sér. *7*, 57–83 (1845)
Houghton, E.: The pharmacologia assay of the heart tonics. J. Am. Med. Assoc. *31*, 959–961 (1898)
Jackson, H.: In: J. Worth Estes, Hall Jackson and the Purple Foxglove. Medical Practice & Research in Revolutionary America 1760–1820. Hannover, New Hampshire: University Press of New England 1979
Jahn, F.: Auswahl der wirksamsten, einfachen und zusammengesetzten Arzneimittel, oder praktische Materia Medica. 2. Aufl. Bd. 1, S. 375. Erfurt 1801
Jahrmärker, H.: Digitalistherapie. Beiträge zur Pharmakologie und Klinik. Berlin, Heidelberg, New York: Springer 1975
Kahn, J.B., Acheson, G.H.: Effects of cardiac glycosides and other lactones, and of certain other compounds on cation transfer in human erythrocytes. J. Pharmacol. Exp. Ther. *115*, 305 (1955)
Kiliani, H.: Über Digitonin und Digitogenin. Ber. Dtsch. Chem. Ges. *24*, 339–347 (1892)
Kiliani, H.: Zur Kenntnis des Digitonins. Ber. Dtsch. Chem. Ges. *24*, 3951–3954 (1892)
Koppe, R.: Untersuchungen über die pharmakologischen Wirkungen des Digitoxins, Digitalins und Digitaleins. Arch. Exp. Path. Pharmak. *3*, 274–301 (1875)
Kottmann, K.: Klinisches über Digitoxinum Solubile Cloetta (Digalen). Ein Beitrag zur subkutanen und intravenösen Digitalistherapie. Z. Klin. Med. *56*, 128–166 (1905)
Kraft, F.: Die Glykoside der Blätter der Digitalis purpurea. Arch. Pharm. *250*, 118–141 (1912)
Kreysig, F.L.: Die Krankheiten des Herzens... II, 2. Berlin 1816, p. 715ff.
Kussmaul, A.: Über lange fortgesetzte Anwendung kleiner Digitalisgaben. Ther. Gegenw. *2*, 1–49 (1900)
Lee, K.S., Klaus, W.: The subcellular basis for the mechanism of inotropic action of cardiac glycosides. Pharmacol. Rev. *23*, 193 (1971)
Le Royer, A.: Bibliothèque universelle des sciences. Bd. *27*, 1824
Liebermeister, C.: Über die Behandlung des Fiebers. Samml. Klin. Vortr. Vol. *31*, Leipzig (1871) p. 255ff.
Livingstone, D., Livingstone, C.: Narrative of an expedition to the Zambesi. London, p. 465 (1865)

Lown, B., Levine, S.A.: Current concepts in digitalis therapy. N. Engl. J. Med. *250*, 866 (1954)
Mackenzie, J.: New methods of studying affections of the heart. Brit. Med. J. *1905 I*, 519–521
Marks, B.H., Weissler, A.M.: Basic and clinical pharmacology of digitalis. Springfield, Ill.: Charles C. Thomas Publ. 1972
Mendel, F.: Die intravenöse Digitalisbehandlung. Ther. Gegenw. *7*, 398 (1905)
Merck, E.: Pharm. Z. (Frankf.) 286 (1879)
Nativelle, C.-A.: Sur la digitaline cristallisée. J. Pharm. Chim. 4. Ser. *9*, 255–262 (1869)
Okita, G.T.: Selected studies on the metabolic fate of radioactive digitoxin in man. In: Digitalis. Dimond, E.G. (ed.), pp. 57–85. Springfield, Ill.: Charles C. Thomas Publ. 1957
Papyrus, Ebers: Das älteste Buch über Heilkunde. Ed. H. Joachim, Berlin, pp. 9, 53, 75 (1890)
Parkinson, J.: Theatrum botanicum. London (1640) p. 653ff.
Pelikan, E.W.: Sur un nouveau poison du coeur provenant de l'inée ou onage, et employé au Gabon (Afrique occidentale) comme poison des flèches. Compt. Rend. Acad. Sci. Paris *60*, 1209–1211 (1865)
Pharmacopoeia Londinensis. London 1650
Portius, H.J., Repke, K.: Die Wirkung von Herzglykosiden auf verschiedene ATPasen des Herzmuskels in Abhängigkeit von Ionenmilieu. Arch. Exp. Path. Pharmacol. *243*, 335 (1962)
Post, R.L., Merrit, C.R., Kinsolving, C.R., Albright, C.D.: Membrane adenosine triphosphatase as a participant in the active transport of sodium and potassium in the human erythrocyte. J. Biol. Chem. *235*, 1796 (1960)
Purkinje, J.E.: Über ein Flimmern vor den Augen nach dem Gebrauch des rothen Fingerhutes. In: Beobachtungen und Versuche zur Physiologie der Sinne, p. 120. Berlin 1825
Repke, K.: Über den biochemischen Wirkungsmodus von Digitalis. Klin. Wschr. *42*, 157–165 (1964)
Rothlin, E., Bircher, R.: Pharmakodynamische Grundlagen der Therapie mit herzwirksamen Glykosiden. Erg. Inn. Med. u. Kinderheilk. Vol. *5*, p. 457–552. Berlin, Göttingen, Heidelberg: Springer 1954
Salerne, F.: Observation de Botanique. Hist. Acad. Roy. Sci., Paris, Année 1748 (1752) p. 74–75
Salmon, W.: Botanologia: The English herbals: or, history of plants, p. 399. London 1710
Schatzmann, H.J.: Herzglykoside als Hemmstoffe für den aktiven Kalium- und Natriumtransport durch die Erythrocytenmembran. Helv. Physiol. Pharmacol. Acta *1953 II*, 346
Schiemann, C.C.: De digitali purpurea. Göttingen 1786
Schmiedeberg, O.: Untersuchungen über die pharmakologisch wirksamen Bestandtheile der Digitalis purpurea. Arch. Exp. Path. Pharmak. *3*, 16–43 (1875)
Schroff von, C.: Beiträge zur Kenntnis der Meerzwiebel. Wbl. Ges. Aerzte, Wien *20*, 389 (1865)
Schwartz, A., Lindenmayer, G.E., Allen, J.C.: The sodium-potassium adenosine triphosphatase: Pharmacological, physiological and biochemical aspects. Pharmacol. Rev. *27*, 3–134 (1975)
Schwiegk, H., Jahrmärker, H.: Therapie der Herzinsuffizienz. Handb. Inn. Med. Ed. by Bergmann, G., Frey, W. Schwiegk, H., p. 402–652. Berlin, Göttingen, Heidelberg: Springer 1960
Skou, H.Ch.: The influence of some cations on an adenosine triphosphatase from peripheral nerves. Biochim. Biophys. Acta *23*, 394 (1957)
Smith, S.: Digoxin, a new digitalis glycoside. J. Chem. Soc. 508–510 (1930)
Smith, T.W., Haber, E.: Digitalis. N. Engl. J. Med. *289*, 945, 1010, 1063, 1125 (1973)
Smith, T.W., Butler, V.P., Haber, E.: Determination of therapeutic and toxic serum digoxin concentrations by radioimmunoassay. N. Engl. J. Med. *281*, 1212 (1969)
Sobernheim, J.F.: Handbuch der praktischen Arzneimittellehre. Vol. *2*, p. 28, Berlin 1836
Stannius, F.H.: Zwei Reihen physiologischer Versuche. Arch. Anat. Physiol. Wiss. Med. *1852*, 85–100
Stoll, A.: The cardiac glycosides. London 1937

Storstein, O.: Symposium on digitalis. Storstein, O., Nitter-Hauge, S., Storstein, L. (eds.). Oslo: Gyldendal Norsk Forlag 1973
Straub, W.: Digitaliswirkung am isolierten Vorhof des Frosches. Arch. Exp. Path. Pharmak. *79*, 19–29 (1916)
Swieten van, G.: Commentaria in H. Boerhaave aphorismos de Cognoscendis et curandis morbis. Vol. 4, Turin (1744) p. 258ff.
Theophrast: De historia plantarum (enquiry into plants). Loeb Classical library. Hort, A. (ed.), Vol. 2. Cambridge (USA) and London *129* p. 7, 13, 4 (1961)
Traube, L.: Über die Wirkungen der Digitalis, insbesondere über den Einfluß derselben auf die Körpertemperatur. Ann. Charité Krh. 1, 622–691 (1850). Also in: Gesamte Beiträge zur Pathologie und Physiologie. Berlin *1*, 192–234 and *2*, 97–211 (1871)
Vassal, P.G.: Dissertation sur les effects de la digitale pourprée dans l'hydropisie. Paris 1809
Wartburg, A.V., Kalber, F., Rutschmann, J.: Tritium labeled cardiac glycosides: Digoxin (12 Alpha-3H). Biochem. Pharmacol. *14*, 1883 (1965)
Weese, H.: Digitalis. Leipzig 1936
Wilzbach, K.E.: Tritium-labeling by exposure of organic compounds to tritium gas. J. Am. Chem. Soc. *79*, 1013 (1957)
Windaus, A.: Über die Formel der Digitalisglykoside. Nachr. Ges. Wiss. Göttingen Math. Naturw. Kl. 170–174 (1926/27)
Withering, W.: An account of the foxglove, and some of its medical uses: With practical remarks on dropsy, and other diseases. Birmingham 1785
Wollenberger, A.: The energy metabolism of the failing heart and the metabolic action of the cardiac glycosides. Pharmacol. Rev. *1*, 311–352 (1949)

CHAPTER 2

Chemistry and Structure-Activity Relationships of Cardioactive Steroids

T. W. GÜNTERT and H. H. A. LINDE

A. Introduction

The possible relationships between the three-dimensional structure of a cardioactive steroid (CAS) and its therapeutic and toxic effects haven been examined repeatedly. The reason for this sustained interest has its roots in the medical use of CAS with the ever-increasing number of the elderly, the narrow therapeutic spectrum, and the wide differences in patient responses to this class of compounds (see this Handbook, Vol. 56/II, Chap. 12). It is hoped eventually to find a lead for the synthesis of specific substances with better cardiotonic activity and less toxicity.

When testing CAS, questions of the strength of the methods used and the answers they give have to be critically assessed again and again. It is of great importance that methods which use a model (ATPase test), dipole moment (DITTRICH and REPKE, 1976), or which do not measure the desired effect directly should never be used alone to decide on the fate of a new CAS. They always have to be used together with tests directly answering the question of potency. Only in this way it is possible to decide on the usefulness or ineffectiveness of a new substance.

Since one can find in the literature (GUENTERT, 1975) a great number of different testing methods used on different animals, isolated organs, physiologic solutions etc., it seems desirable to introduce standard test methods. Only then can results from different laboratories be compared with confidence.

From what has been said so far it becomes clear that the validity of structure-activity relationships quoted in the next section is necessarily limited, because they have usually been tested under conditions which were not strictly the same.

B. Structure-Activity Relationships

I. Uncertainties in Structure-Activity Relationships

About 15–20 years ago most of the structural elements which are absolutely necessary for cardiotonic activity seemed to be evident, but subsequent results made the earlier dogmas increasingly untenable. A major event in this process was the synthesis of steroidal mono- and bis-guanylhydrazones (I, Fig. 1) (KRONEBERG et al., 1964) which showed pharmacologic properties very similar to those of CAS (DRANSFELD et al., 1964, 1967). Looking at the structure of I in Fig. 1 it is clear, that most of the rules or dogmas (e. g., TAMM, 1963) dealing with structure-activity relations of CAS had to be modified. According to TAMM (1963) the prerequisites

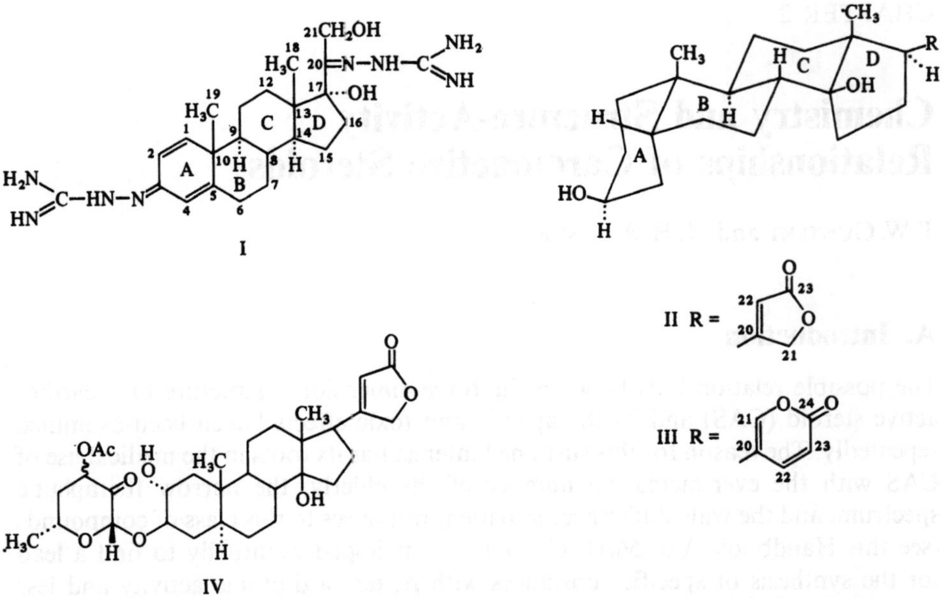

Fig. 1. Structures of *bis*-guanylhydrazone (I), digitoxigenin (II), bufalin (III), and asclepin (IV)

for activity were a steroidal 14β-hydroxyskeleton, in which, unless there are double bonds at C4 or C5 the rings A and B have to be in a *cis*, B and C in a *trans*, and C and D in a *cis* configuration. Furthermore steroids have to carry an unsaturated lactone (γ-lactone or α-pyrone) as a 17β-side-chain and a 3β-oriented oxygen function (OH groups or glycosidic linkage). TAMM (1963) assumed that further hydroxyl groups are of minor importance for a cardiac effect, but he considered the distance between the lactone carbonyl and the oxygen function at C3 to be the critical value for this effect. This distance, defined by the two electronegative poles and rigid steroid skeleton should be closely connected with the properties of the physiologic digitalis receptor.

The sugar component in naturally occuring glycosides is of great importance to the kinetic behavior of the substance in the organism (LUCHI and CONN, 1965). It is generally accepted today that cardenolide monodigitoxosides are inotropically more active than di- and tridigitoxosides (ANGARSKAYA, 1975; BOETTCHER et al., 1975), but the opposite seems to be true for glucosides (TAKIURA et al., 1974). Sugar-free compounds (genins) are also "active". The genins with a five-membered, unsaturated lactone as side-chain (digitalis type) are in contrast to those with a six-membered, double-unsaturated lactone (scilla-bufo type) less active (toxic) than the corresponding glycosides.

Several synthetic modifications in the structure of cardiac glycosides disprove some of these prerequisites for a positive inotropic effect.

II. 3β-OH Group

SAITO et al. (1970) synthesized 3-desoxydigitoxigenin (3-desoxy-II in Fig. 1) and found, on isolated frog heart (Straub's preparation), an activity comparable to that

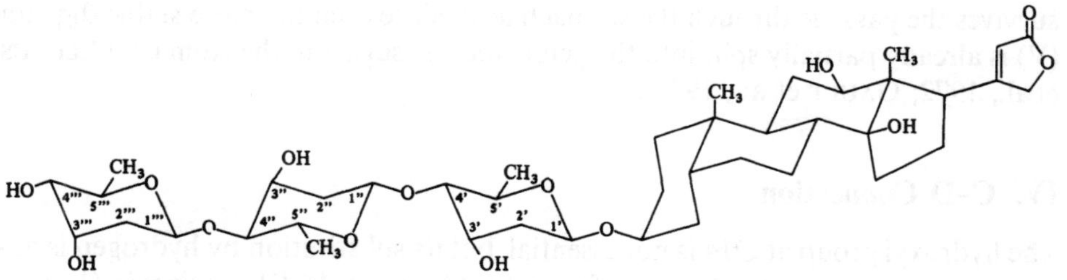

V

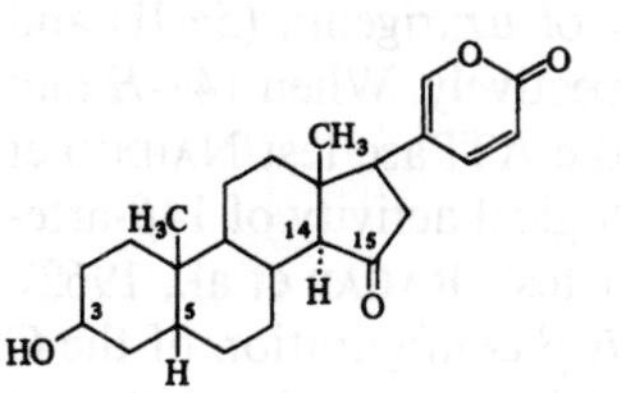

VI

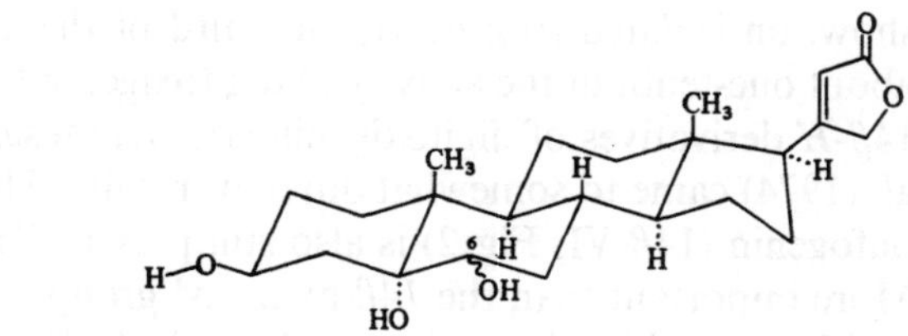

VII

Fig. 2. Structures of digoxin (V), 14α-artebufogenin (VI), and 6α- and 6β-hydroxy derivatives (VII)

of digitoxigenin (II), the structure of which is shown in Fig. 1. ZÜRCHER et al. (1969) showed that in the ATPase test, this derivative loses only 50% activity. WITTY et al. (1975) have confirmed this result. They furthermore demonstrated that hydrogenation of the lactone in 3-desoxydigitoxigenin led to a far greater loss of activity (ATPase test) than that associated with the conversion of digitoxigenin (II) to dihydrodigitoxigenin, a tendency observed in all the 3-desoxycompounds tested. A change in the configuration at C3 of the cardenolide digitoxigenin (II), diminishes the effect in the ATPase test to about $^1/_{30}$ of that of digitoxigenin (ZÜRCHER et al., 1969). The corresponding bufadienolide 3α-bufalin still shows one seventh of the toxicity of bufalin (III, Fig. 1) (Hatcher test, MEYER, 1971).

III. A–B Connection

The A and B rings in natural glycosides generally show a *cis* configuration. A change of configuration at C5 normally yields less active products (OKADA and ANJYO, 1975). According to ISHIKAWA et al. (1974) this loss of activity (frog heart) seems to be less pronounced in the cardenolides than in the bufadienolides. But there are A–B *trans* compounds (e.g., IV, Fig. 1) which have by various in vitro and in vivo experiments been shown to be more potent than digoxin (V, Fig. 2) and to possess a wider safety margin (PATNAIK and DHAWAN, 1978; PATNAIK and KOEHLER, 1978). Our knowledge of the influence of a double bond at C4–C5 or C5–C6 on inotropic and toxic effects is only rudimentary; scillaren A (glucorhamnoside of Δ^4-III) has a strong cardiotonic and cardiotoxic (Hatcher) effect, xysmalogenin (Δ^5-II), however, has weak cardiotoxic activity (Hatcher test only, BAUMGARTEN, 1963). In oral use it is doubtful, that much of the Δ^4-glycosides survives the passage through the stomach acid, since even the more stable digoxin (V) is already partially split into the genin and the sugars in the stomach (ALDOUS et al., 1972; GAULT et al., 1977).

IV. C–D Connection

The hydroxyl group at C 14 is not essential, but its substitution by hydrogen is accompanied by a considerable loss of activity: 14-desoxy-14*βH*-uzarigenin (14-desoxy-14*βH*-5α-II) (OKADA and SAITO, 1968; SHIGEI and MINESHITA, 1968) and 14-desoxy-14*βH*-digitoxigenin (14-desoxy-14*βH*-II) (SHIGEI et al., 1973) continued to show, on isolated frog heart, one-third of the activity of uzarigenin (5α-II) and about one-tenth of the activity of digitoxigenin (II) respectively. When 14α-*H* and 14β-*H* derivatives of digitoxigenin (II) were tested in the ATPase test NAIDOO et al. (1974) came to somewhat different results. The biological activity of 14β-artebufogenin (14β-VI, Fig. 2) is also still present (Hatcher test, RAGAB et al., 1962). More important than the 14β-hydroxyl group is the *cis-β* configuration of the C and D rings. 14-α-digitoxigenin (14α-II) (ATPase test, ZÜRCHER et al., 1969) and 14α-artebufogenin (VI) (Hatcher test, RAGAB et al., 1962) are inactive in contrast to the corresponding β-analogs (II and 14β-VI). A number of 5α, 14α-cardenolides (e.g., VII, Fig. 2) show transient positive activity in the Langendorff test with isolated guinea-pig heart (WAGNER et al., 1976) and synthetic prednisone and prednisolone derivatives (e.g., I) with largely similar effects to those of digitalis, have proved to be biologically active in spite of their 14α-*H* structures (isolated frog heart, isolated guinea-pig atrium, Langendorff heart, guinea-pig, KRONEBERG et al., 1964). It is claimed (FRITSCH et al., 1975), that Δ^{14}-scillarenone (3-keto-$\Delta^{4,14}$-III) which is sterically similar to an A–B and C–D *trans* steroid shows diuretic and inotropic activity.

V. Structure at C14 and C15

Relationships between the structure at C14 and C15 and the biologic activity are still somewhat contradictory. Experiments by SHIGEI et al. (1964, 1968, 1973) and by OKADA and SAITO (1967) showed that the introduction of an oxygen function at C15 in digitoxigenin (II), always resulted in a decrease of pharmacologic activity, when tested on isolated frog heart (Straub), the order of activity being: digitoxigenin (II) (relative activity 1.0) > 15β-hydroxy-II (0.1) > 15-oxo-II (0.02) > 15α-hydroxy-II (inactive). In the series of 14-desoxy-14*βH*-digitoxigenin (14-desoxy-14β-*H*-II), the activity usually decreases also. The 14-desoxy-14*βH*-II itself has a relative activity of only 0.04–0.15, but the 15-oxo derivative still retains 10%–30% of the activity of digitoxigenin (II). This last result, however, is in contrast with the results of HENDERSON and CHEN (1965, Hatcher test), who found 14*βH*-15-oxo-digitoxigenin (14*βH*-15-oxo-II) to be ineffective (see also MEYER, 1971). Disputing the HENDERSON and CHEN result is the finding that the relevant bufadienolide, 15β-artebufogenin (14β-VI), is also active (Hatcher test, MEYER, 1971). Both in the series of digitoxigenin (II) and that of 14-desoxy-14*βH*-digitoxigenin (14-desoxy-14*βH*-II) the 15α-*OH*-compound is ineffective. In the series of 14-desoxy-14β-

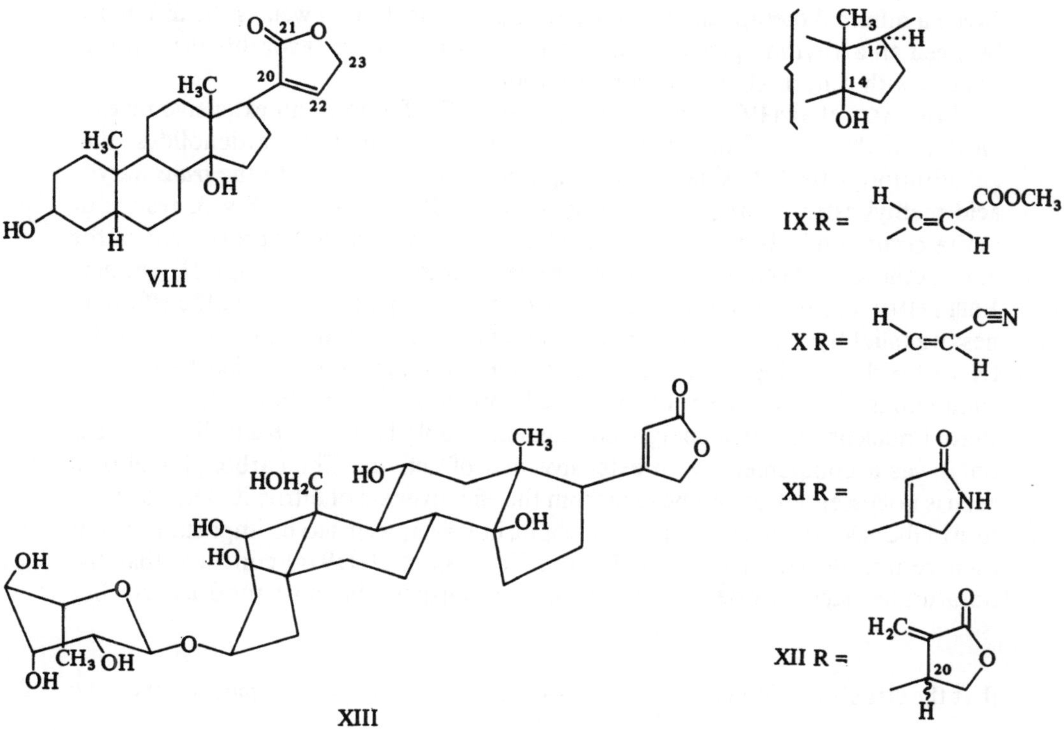

Fig. 3. Structures of "cardenolides": isomeric butenolide (VIII), *trans*-acrylic acid methyl ester (IX), *trans*-acrylonitrile (X), unsubstituted lactam (XI), 20-R and 20-S derivatives (XII), and ouabain (XIII)

chlorodigitoxigenin (14-desoxy-14β-chloro-II) derivatives, however, the 15β-hydroxy and 15-oxo derivatives are both without effect, whereas the 15α-hydroxy-analog still shows 10%–30% of the effect of digitoxigenin (II) (SHIGEI et al., 1973). 14β-15β-epoxides are in general less effective than the respective 14β-*OH* compounds (NAIDOO et al., 1974). Their activity is, however, supposed to be frequently linked to a better therapeutic index (MEYER, personal communication).

VI. Side-Chain

The lactone side-chain is commonly regarded as the most essential functional group of cardiac glycosides. It is undisputed that this C17 side-chain must be β-oriented – a change of configuration yields inactive compounds (REPKE, 1972; SAITO et al., 1970; TAMM, 1963; for a possible exception see CHOAY et al., 1978). DEGHENGHI (1970) has been able to show that isomeric cardenolides (e.g. VIII, Fig. 3), differing from the natural ones only in the position of attachment of the butenolide ring to the steroidal 17β position, are also active and even possess a more favorable therapeutic index than naturally occuring representatives. While the synthetic work is impressive here, pharmacological data are sparse. MENDEZ et al. (1974) noticed in experiments with the same substance, that the difference between a minimal therapeutic and an arrhythmogenic dose as well as the difference between an arrhythmogenic and a lethal dose is here more favorable than in glycosides with a normal lactone configuration.

BOUTAGY et al. (1973) and THOMAS et al. (1974) found in an extensive investigation on the role of the side-chain in the effectiveness of cardenolides, that substitution of the C17 side-chain in digitoxigenin (II), leading to the *trans*-acrylic acid methyl ester IX and the *trans*-acrylonitrile X as shown in Fig. 3, results in active compounds. In the Na^+,K^+-ATPase test as well as in improvement of the contractile force, both showed the same effectiveness as digitoxigenin (II) (see also EBERLEIN et al., 1974). With a guanylhydrazone group as side-chain, the effectiveness was slightly reduced, but was still of comparable size. Going from the 17β-ester IX to the 17α-analog. i.e., alteration of the configuration, also led to inactive compounds. The increase of the distance between the ester carbonyl in IX and the steroid nucleus on introducing a conjugated double bond (all-*trans*-diene system only) was accompanied by an extensive loss of activity. The carbonyl group as such is not essential, as can be seen from the effectiveness of nitrile X. The distribution of the electron density in the side-chain, however, seems to be important. From their results BOUTAGY et al. (1973) and THOMAS et al. (1974) conclude that the unsaturated lactone does not constitute an indispensable structural feature but that the structural element $-\overset{\overset{R''}{|}}{\underset{\delta+}{C}}=CH-\underset{\underset{R'}{|}}{C}=\overset{\delta-}{A}$ in the C17 β-side-chain seems to be a prerequisite for cardiotonic steroids, where A is a heteroatom (e.g., $=O$, $\equiv N$), R′ (and usually R″) being H or an alkoxyl group. But this structural element is lacking in saturated lactones, which also produce – though in higher concentration – a positive inotropic effect in the failing heart of a dog heart–lung preparation (MENDEZ et al., 1974; see also KAISER et al., 1974; BOEHRINGER, 1974; SCHAUMANN et al., 1977). EBERLEIN et al. (1974), who replaced the butenolide ring in 14β-hydroxy C and D *cis*-steroid glycosides and aglycones by open side-chains, obtained similar results. Acrylic acid esters and nitriles showed typical cardioglycosidic effects in vivo (isolated guinea-pig atrium, dog, cat) and in vitro (Na^+, K^+-ATPase inhibition at human erythrocytes, see also THOMAE, 1975; FULLERTON et al., 1976).

Both LINDIG and REPKE (1977) and FULLERTON et al. (1977) prepared C22 methylcardenolides. The latter pointed out, that their 20-R and 20-S derivatives (XII, Fig. 3) showed considerable differences in inhibiting Na^+, K^+-ATPase, but had no inotropic activity in guinea-pig left atria! EBERLEIN et al. (1972) have substituted the vinylic hydrogen at C22 of the lactone with halogen, alkyl, and alkoxyl groups. By introducing fluorine into the butenolide ring of digoxin (V) they achieved an increase of contraction at the isolated guinea-pig atrium, while a strong decrease was noticed after introducing a methoxyl group. They consider – as did BOUTAGY et al. (1973) and THOMAS et al. (1974) – a conjugated four-center Π-system with the negative charge center at its free end as a minimal prerequisite for cardiotonic activity (EBERLEIN et al., 1974). BODEM et al. (1978) reported on the synthesis of 22-fluoro-α-acetyldigoxin (22-fluoro-3‴-*O*-acetyl-V), VOIGTLÄNDER et al. (1976a–c) as well as KAISER et al. (1977) and VOIGTLÄNDER et al. (1978) on the synthesis of C22 alkyl-substituted cardenolides without dealing specifically with the pharmacologic activity of these substances. The same can be said about a number of patents in which the synthesis of C21-substituted (alkyl, bromo, hydroxy, chloro, fluoro, amino, azido, ester) cardenolides is described (DITTRICH et al., 1975; MEGGES et al., 1975a, b; LINDIG and REPKE, 1976a–d) and of the isolation of a highly unsaturated C21 hydroxycardenolide from *Nerium odorum* (YAMAUCHI et al., 1976). Furano derivatives with "cardiotonic activity" have been prepared by DITTRICH et al. (1973) and by ABRAHAM and LEFEBVRE (1976, 1977). The synthesis of unsubstituted lactams (XI, Fig. 3) with no dramatic cardiotonic activity has been reported by GUENTERT et al. (1978) as well as by MEGGES et al. (1978a, b), the synthesis of substituted unsaturated ones by MEGGES et al. (1978c).

C. Influence of Additional Structural Modifications

I. Halogens

Chlorine was introduced into different positions of the steroid nucleus. In doing so, active compounds could in some cases be obtained. None of them, however, surpassed the natural glycosides in their pharmacologic qualities. MEYER (1952) was able to attach a chlorine atom at the 15α position of bufalin (III), but this "resibufogenin hydrochloride" showed no toxicity. STACHE et al. (1974b) substituted the hydrogen at C4 by chlorine in canarigenin (Δ^4-digitoxigenin, Δ^4-II), 14,15β-epoxycanarigenin, and 14-desoxy-14α*H*-canarigenin. When tested for positive inotropic activity on isolated left atria of guinea-pigs, 4-chlorocanarigenin was about as active as canarigenin itself while 14,15β-epoxycanarigenin was devoid of any activity.

II. Branching at C3

Branching at C3 leads to an interesting group of compounds, which was synthesized by ALBRECHT and KUNZ (1975) and by ALBRECHT and NEUGEBAUER (1975, 1976, 1977). Only the inhibition of beef brain and dog heart Na^+, K^+-ATPase was tested by SIEBENEICK and HOFFMANN (1978), hence nothing is known about the positive inotropic action of these substances.

III. N-Analogs

Recently, various *N*-analogs of cardioactive cardenolides have been developed (MEYER, 1971). The compounds concerned are 3α- and 3β-amino-3-desoxydigitoxigenin, 3-amino-3-desoxy derivatives of uzarigenin (5α-II), oleandrigenin (16β-acetyl-II), gitoxigenin (16β-hydroxy-II) and digoxigenin (12β-hydroxy-II) prepared by MEYER's school (SAWLEWICZ et al., 1972; HAUSER et al., 1973). The same class of compounds has been the subject of a patent application (STACHE et al., 1972) and patents (MEYER, 1975, 1976, 1977); its therapeutic applicability is still being evaluated.

Promising pharmacologic results with amino groups in the sugar side-chain were first reported by FOUSSARD-BLANPIN et al. (1973) and confirmed by CALDWELL and NASH (1976, 1978), COOK et al. (1977) and CHOAY et al. (1978). The pharmaceutical industry considered these substances to be useful cardiotonic (and cytostatic) agents as well (STACHE et al., 1974a; PETERSEN, 1977).

We must also mention some 19-nor-10-cyanocardenolides which improve the contraction of the heart (COHNEN, 1977a, b), some nitrates of cardenolide genins, and glycosides (MEGGES et al., 1974) with better inotropic activity than the starting materials.

IV. Structure at C16

In the cardenolide series, gitoxigenin (16β-hydroxy-II) derivatives have attracted increasing interest in recent years. KOVAŘIKOVÁ et al. (1964) concluded from their experiments with gitoxigenin and its 16-epimer that the 16-hydroxyl group, as such, causes an unfavorable effect (electrically stimulated cat papillary muscle). On the other hand REPKE (1972) found for the tridigitoxoside 16α-gitoxigenin (tridigitoxoside of 16α-hydroxy-II), at maximally inotropic concentrations, less arrhythmias than for gitoxin (tridigitoxoside of 16β-hydroxy-II) or ouabain (XIII, Fig. 3) (isolated guinea-pig atrium). Furthermore 16α-gitoxin showed a wider dosage range than gitoxin or ouabain (XIII) between minimal and maximal inotropic effect (HAUSTEIN et al., 1970). A partial separation – at least at the guinea-pig heart – of therapeutic and toxic effects seems to have been achieved with 16α-gitoxin and, according to HAUSTEIN and HAUPTMANN (1974), with 16-acetyl-16α-gitoxin. Both compounds have a stronger effect on the ventricular muscle and a weaker one on the atrium and on the Purkinje system. There have also been reports of the synthesis of 16-methyl and 16-ethyl ethers of gitoxin and 16α-gitoxin and their genins respectively (LINDIG et al., 1975).

V. Other Compounds

Finally we would like to mention some exotic substances, which show different degrees of activity. STACHE et al. (1977) reported on the synthesis of a cardenolide glycoside with an additional steroidal ring (XIV, Fig. 4) ("cardio-propellane") which showed positive inotropic activity in the guinea-pig heart. It is also reported, that a ring A "opened" cardenolide (XV, Fig. 4) had some positive inotropic action (TSURU et al., 1975, 1977). Unfortunately nothing is published on the pharmacology of a cardenolide glycoside, whose C ring is opened, so that the "steroid" is left with only three normal rings (YAMAUCHI and ABE, 1978). GALEL et al. (1974) reported that the sesquiterpene judaicin showed activity similar to compounds such as digoxin (V). Certain steroidal alkaloids are also reported to show positive inotropy (NISHIE et al., 1976).

We would really like to know about the positive inotropic effect of desglucohellebrin derivatives prepared by ISAAC et al. (1976) which show, on oral application in mice, a LD_{50} of 1–2,000 mg/kg when the presumed hemiactonate of β-methyldigoxin (4‴-O-methyl-V) has an LD_{50} of about 11 mg/kg in male mice (TOIDA et al., 1976).

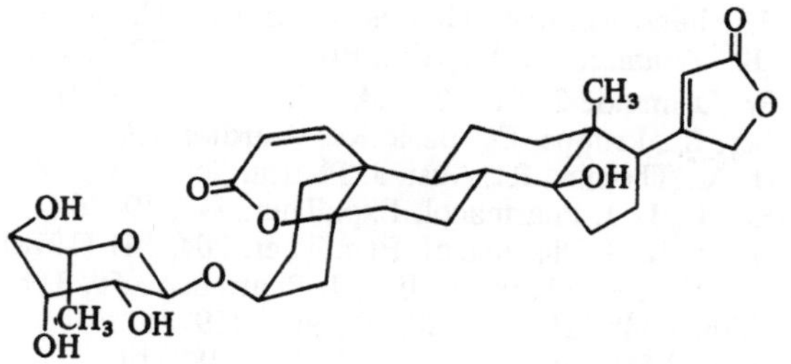

XIV

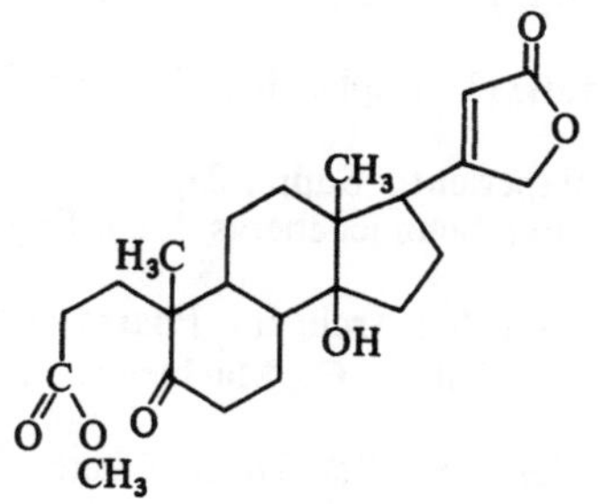

XV

Fig. 4. Structures of "cardio-propellane" (XIV) and a cardenolide after opening of the A ring (XV)

D. Summary

The structure-activity relationships of cardioactive steroids are still obscure, although a number of individual phenomena are known. In elucidating their pharmacologic properties, the array of test methods and test animals and organs used is of great importance. Methods frequently applied today, such as the one determining the toxicity in the cat (Hatcher) or the ATPase test can easily lead to wrong conclusions unless they are used in conjunction with more revealing tests for inotropy.

References

Abraham, N.A., Lefebvre, Y.: US 3'944'541, C.A. *85*, 21'728 p (1976)
Abraham, N.A., Lefebvre, Y.: US 4'011'12, C.A. *87*, 39'738 t (1977)
Albrecht, H.P., Kunz, B.: Liebigs Ann. Chem. *1975*, 2216–2226
Albrecht, H.P., Neugebauer, G.: Ger. Offen. 2'336'445, C.A. *82*, 140'395 a (1975)
Albrecht, H.P., Neugebauer, G.: Ger. Offen. 2'427'977, C.A. *84*, 122'155 j (1976)
Albrecht, H.P., Neugebauer, G.: Ger. Offen. 2'517'293, C.A. *86*, 140'414 p (1977)
Aldous, S., Nation, R., Thomas, R.: Aust. J. Pharm. Sci. N.S. *1*, 61–62 (1972)
Angarskaya, M.A.: Farmakol. Toksikol. *9*, 24–28 (1974); C.A. *82*, 11'027 h (1975)

Baumgarten, J.G.: Die herzwirksamen Glykoside. Leipzig: Thieme 1963
Bodem, G., Wirth, K., Zimmer, A.: Arzneim. Forsch. *28*, 322–325 (1978)
Boehringer, C.H.: Fr. Demande 2'191'902, C.A. *81*, 25'912 j (1974)
Boettcher, H., Fischer, K., Proppe, D.: Basic Res. Cardiol. *70*, 279–291 (1975)
Boutagy, J., Gelbart, A., Thomas, R.: Aust. J. Pharm. Sci. N.S. *2*, 41–46 (1973)
Caldwell, R.W., Nash, C.B.: J. Pharmacol. Exp. Ther. *197*, 19–26 (1976)
Caldwell, R.W., Nash, C.B.: J. Pharmacol. Exp. Ther. *204*, 141 (1978)
Choay, P., Cordboeuf, E., Deroubaix, E.: Eur. J. Pharmacol. *50*, 317–323 (1978)
Cohnen, E.: Ger. Offen. 2'548'525, C.A. *87*, 102'606 r (1977a)
Cohnen, E.: Ger. Offen. 2'558'208, C.A. *87*, 152'524 h (1977b)
Cook, L.S., Caldwell, R.W., Nash, C.B.: Arch. Int. Pharmacodyn. *227*, 220–232 (1977)
Deghenghi, R.: Synthetic cardenolides and related products. In: Chemistry of natural products, Vol. 6. London: Butterworths 1970
Dittrich, F., Megges, R., Portius, H.J., Repke, K.: Ger. (East) 94'616, C.A., *79*, 42'775 t (1973)
Dittrich, F., Megges, R., Portius, H., Repke, K.: Ger. (East) 108' 291, C.A. *83*, 10'737 y (1975)
Dittrich, F., Repke, K.R.H.: Experientia Suppl. *23*, 249–254 (1976)
Dransfeld, H., Greeff, K.: Naunyn-Schmiedebergs Arch. Exp. Path. Pharmacol. *249*, 425–431 (1964)
Dransfeld, H., Galetke, E., Greeff, K.: Arch. Int. Pharmacodyn. *166*, 342–349 (1967)
Eberlein, W., Nickl, J., Heider, J., Dahms, G., Machleidt, H.: Ber. Dtsch. Chem. Ges. *105*, 3686–3694 (1972)
Eberlein, W., Heider, J., Machleidt, H.: Ber. Dtsch. Chem. Ges. *107*, 1275–1284 (1974)
Foussard-Blanpin, O., Hubert, F., Choay, P., Leboeuf, M.: Ann. Pharm. Fr. *31*, 593–600 (1973)
Fritsch, W., Haede, W., Radscheit, K., Stache, U., Inhoffen, H.H., Kreiser, W., Warneke, H.U.: Ger. Offen. 2'361'059, C.A. *83*, 206'484 h (1975)
Fullerton, D.S., Pankaskie, M.C., Ahmed, K., From, A.H.L.: J. Med. Chem. *19*, 1330–1333 (1976)
Fullerton, D.S., Gilman, T.M., Pankaskie, M.C., Ahmed, K., From, A.H.L., Duax, W.L., Rohrer, D.C.: J. Med. Chem. *20*, 841–844 (1977)
Galel, E.E., Kandil, A.M., Abdel-Latif, M., Khedr, T., Rashad, T., Khafagy, S.M.: J. Drug Res. *6*, 63–71 (1974); C.A. *83*, 90944 b (1975)
Gault, M.H., Charles, J.D., Sugden, D.L., Kepkay, D.C.: J. Pharm. Pharmacol. *29*, 27–32 (1977)
Guentert, T.: Dissertation Universität Basel 1975
Guentert, T.W., Linde, H.H.A., Ragab, M.S., Spengel, S.: Helv. *61*, 977–983 (1978)
Hauser, E., Boffo, U., Meister, L., Sawlewicz, L., Linde, H.H.A., Meyer, K.: Helv. *56*, 2782–2795 (1973)
Haustein, K.-O., Markwardt, F., Repke, K.R.H.: Eur. J. Pharmacol. *10*, 1–10 (1970)
Haustein, K.-O., Hauptmann, J.: Pharmacology *11*, 129–138 (1974)
Henderson, F.G., Chen, K.K.: J. Med. Chem. *8*, 577–579 (1965)
Isaac, O., Posselt, K., Uthemann, H.: Ger. Offen. 2'513'370, C.A. *84*, 44'613 r (1976)
Ishikawa, N., Tsuru, H., Shigei, T., Anjyo, T., Okada, M.: Experientia *30*, 1308–1310 (1974)
Kaiser, F., Schaumann, W., Stach, K., Voigtländer, W.: Ger. Offen. 2'233'147, C.A. *80*, 96'230 a (1974)
Kaiser, F., Schaumann, W., Stach, K., Voigtländer, W.: Ger. Offen. 2'550'354, C.A. *87*, 102'608 t (1977)
Kimble, M.A., Elenbaas, R.M.: J. Am. Pharm. Ass. N.S. *14*, 362–375 (1974)
Kovaříková, A., Kolárová, H., Pitra, J.: Experientia *20*, 263–264 (1964)
Kroneberg, G., Meyer, K.H., Schraufstätter, E., Schütz, S., Stoepel, K.: Naturwissenschaften *51*, 192–193 (1964)
Lindig, C., Schmidt, H.J., Repke, K.: Ger. (East) 110'263, C.A. *83*, 179'415 c (1975)
Lindig, C., Repke, K.: Ger. (East) 116'226, C.A. *85*, 21'729 q (1976a)
Lindig, C., Repke, K.: Ger. (East) 116'227, C.A. *85*, 21'730 h (1976b)
Lindig, C., Repke, K.: Ger. (East) 116'613, C.A. *85*, 46'966 c (1976c)

Lindig, C., Repke, K.: Ger. (East) 116′614, C.A. *85*, 46′965 b (1976d)
Lindig, C., Repke, K.: Ger. (East) 119′042, C.A. *86*, 90′148 g (1977)
Luchi, R.J., Conn, H.L.: Prog. Cardiovasc. Dis. *7*, 336–359 (1965)
Megges, R., Franke, R., Streckenbach, B., Kammann, G., Repke, K.: US 3′806′502, C.A. *81*, 25′913 k (1974)
Megges, R., Timm, H., Dittrich, F., Portius, J.H., Repke, K.: Ger. (East) 109′869, C.A. *83*, 179′414 b (1975a)
Megges, R., Timm, H., Thiemann, P., Dittrich, F., Franke, P., Portius, H., Repke, K.: Ger. (East) 109′622, C.A. *83*, 59′166 a (1975b)
Megges, R., Timm, H., Portius, H.J., Franke, P., Hintsche, R., Repke, K.: Ger. (East) 126′092, C.A. *88*, 152′861 m (1978a)
Megges, R., Timm, H., Portius, H.J., Franke, P., Hintsche, R., Repke, K.: Ger. (East) 126′093, C.A. *88*, 170′392 p (1978b)
Megges, R., Timm, H., Portius, H.J., Glusa, E., Repke, K.: Ger. (East) 129′795, C.A. *89*, 147′194 n (1978c)
Mendez, R., Pastelin, G., Kabela, E.: J. Pharmacol. Exp. Ther. *188*, 189–197 (1974)
Meyer, K.: Helv. *35*, 2444–2469 (1952)
Meyer, K.: Planta Medica, Suppl. 4/1971, 1–33
Meyer, K.: Swiss 559′219, C.A. *83*, 28′459 u (1975)
Meyer, K.: Swiss 565′202, C.A. *84*, 59′879 f (1976)
Meyer, K.: Swiss 583′255, C.A. *87*, 6′266 j (1977)
Naidoo, B.K., Witty, T.R., Remers, W.A., Besch, H.R.: J. Pharm. Sci. *63*, 1391–1394 (1974)
Nishie, K., Fitzpatrick, T.J., Swain, A.P., Keyl, A.C.: Chem. Path. Pharmacol., R.C. *15*, 601–607 (1976)
Okada, M., Saito, Y.: Chem. Pharm. Bull. *15*, 352–353 (1967)
Okada, M., Saito, Y.: Chem. Pharm. Bull. *16*, 2223–2227 (1968)
Okada, M., Anjyo, T.: Chem. Pharm. Bull. *23*, 2039–2043 (1975)
Patnaik, G.K., Dhawan, B.N.: Arzneim. Forsch. *28*, 1095–1099 (1978)
Patnaik, G.K., Koehler, E.: Arzneim. Forsch. *28*, 1368–1372 (1978)
Peterson, R.: Ger. Offen. 2′603′046, C.A. *87*, 202′044 z (1977)
Ragab, M.S., Linde, H.H.A., Meyer, K.: Helv. *45*, 1794–1799 (1962)
Repke, K.R.H.: Pharmazie *27*, 693–701 (1972)
Saito, Y., Kanemasa, Y., Okada, M.: Chem. Pharm. Bull. *18*, 629–631 (1970)
Sawlewicz, L., Weiss, E., Linde, H.H.A., Meyer, K.: Helv. *55*, 2452–2460 (1972)
Schaumann, W., Dietmann, K., Bartsch, W., Kaiser, F., Voigtländer, W.: Ger. Offen. 2′546′779, C.A. *87*, 39′772 z (1977)
Shigei, T., Katori, M., Murase, H., Imai, S.: Experientia *20*, 572–573 (1964)
Shigei, T., Mineshita, S.: Experientia *24*, 466–467 (1968)
Shigei, T., Tsuru, H., Saito, Y., Okada, M.: Experientia *29*, 449–450 (1973)
Siebeneick, H.U., Hoffmann, W.: J. Med. Chem. *21*, 1310–1312 (1978)
Stache, U., Haede, W., Fritsch, W., Radscheit, K., Lindner, E.: Ger. Offen. 2′013′032, C.A. *76*, 14′816 b (1972)
Stache, U., Fritsch, W., Haede, W., Lindner, E.: Ger. Offen. 2′254′060, C.A. *81*, 37′778 h (1974a)
Stache, U., Radscheit, K., Fritsch, W., Haede, W.: Liebigs Ann. Chem. *1974b*, 608–620
Stache, U., Fritsch, W., Haede, W., Radscheit, K.: Liebigs Ann. Chem. *1977*, 1461–1474
Takiura, K., Yuki, H., Okamoto, Y., Takai, H., Honda, S.: Chem. Pharm. Bull. *22*, 2263–2269 (1974)
Tamm, C.: Proc. 1st International Pharmacol. Meeting, Volume 3, 11–26, Stockholm Oxford: Pergamon Press 1963
Thomae, K.: Fr. Demande 2′199′987, C.A. *83*, 59′227 w (1975)
Thomas, R., Boutagy, J., Gelbart, A.: J. Pharm. Sci. *63*, 1649–1683 (1974)
Toida, S., Matsuura, S., Hidano, T., Tanihata, T., Ito, R.: Toho Igakkai Zasshi *23*, 198–202 (1976), C.A. *85*, 104′093 d (1976)
Tsuru, H., Ishikawa, N., Shigei, T., Anjyo, T., Okada, M.: Experientia *31*, 955–956 (1975)
Tsuru, H., Ishikawa, N., Shigei, T., Anjyo, T., Okada, M.: Jpn. J. Pharmacol. *27*, 799–805 (1977)

Voigtländer, W., Kaiser, F., Schaumann, W., Stach, K.: Ger. Offen. 2'418'127, C.A. *84*, 59'963 d (1976a)
Voigtländer, W., Kaiser, F., Schaumann, W., Stach, K.: Ger. Offen. 2'433'563, C.A. *84*, 150'934 f (1976b)
Voigtländer, W., Kaiser, F., Schaumann, W., Stach, K.: Ger. Offen. 2'457'219, C.A. *85*, 124'316 g (1976c)
Voigtländer, W., Kaiser, F., Schaumann, W.: Ger. Offen. 2'614'046, C.A. *88*, 23'346 y (1978)
Wagner, F., Lang, S., Kreiser, W.: Ber. Dtsch. Chem. Ges. *109*, 3304–3317 (1976)
Witty, T.R., Remers, W.A., Besch, H.R.: J. Pharm. Sci. *64*, 1248–1250 (1975)
Yamauchi, T., Abe, F., Takahashi, M.: Tetrahedron Letters *1976*, 1115–1116
Yamauchi, T., Abe, F.: Tetrahedron Letters *1978*, 1825–1828
Zürcher, W., Weiss-Berg, E., Tamm, C.: Helv. *52*, 2449–2458 (1969)

Methods for the Determination of Cardiac Glycosides

CHAPTER 3

Chemical and Chromatographic Methods

H. FLASCH and W. DIEMBECK

A. Introduction

The analytical methods for cardiac glycosides can be divided into two groups. In the μg range the classical methods of photometry and chromatography still have an established place in the pharmacopoeias and are very widely employed in control laboratories for quantitative determination of the content and purity of glycoside preparations. For pharmacokinetic investigations in the ng range the available methods require greater expenditure on apparatus. They comprise: the isotope technique, assays, and GC-MS (a gas chromatograph coupled to a mass spectrometer).

Whereas in most instances the classical methods require preliminary purification – usually by chromatography – with the isotope technique or the assays it is possible to measure the glycosides directly in the presence of other substances, e.g., in biologic materials.

Developments in apparatus for the so-called "cold" analytical methods have recently led to considerable improvements in sensitivity. By coupling a chromatographic unit (GC or high performance liquid chromatograph HPLC) to a sensitive detector (MS or fluorescence detector) is is now possible to achieve reliable measurements in the ng range.

B. Spectroscopic Procedures

Direct measurement by ultraviolet (UV) spectroscopy is difficult because the only chromophores carried by cardiac glycosides have technically inconvenient absorption maxima: cardenolides at 217 nm ($\varepsilon_{mol} = 16{,}595$), bufadienolides at 300 nm ($\varepsilon_{mol} = 5{,}250$) where ε_{mol} is the molar extinction coefficient. For quantitative analytical determination they must therefore be converted into colored derivatives. The components of a cardenolide molecule (digoxin) depicted in Fig. 1 can be converted into colored derivatives by the BALJET or KELLER-KILIANI reactions or by treatment with strong acids, and these can be measured by conventional photometers or fluorimeters (for reviews see PFORDTE and FÖRSTER, 1970; NEUWALD, 1950; ROWSON, 1952; ROWSON and DYER, 1952). However, all chemical methods are liable to fail in the presence of large quantities of other substances and furthermore, because of their low sensitivity, they are seldom suitable for the estimation of cardioactive steroids in biologic samples. Only a few methods are specific for a given cardiac glycoside; exact identification requires supplementary physicochemical analyses.

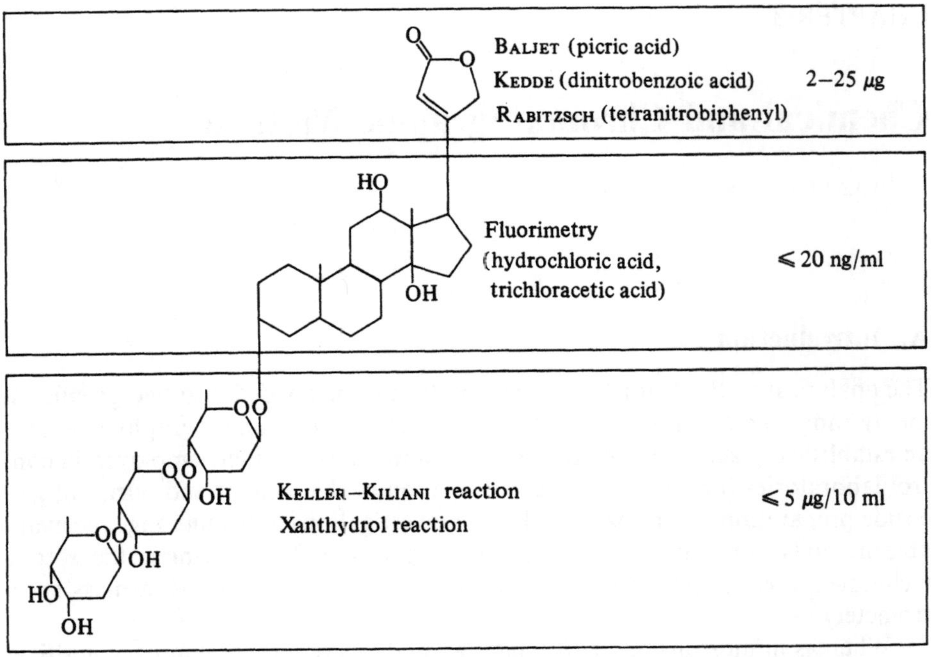

Fig. 1. Chemical procedures for the photometric and fluorimetric determination of cardenolides

I. Alkaline Reagents

The reactions between cardenolides and polynitroaromatic derivatives in alkaline solution – picric acid (BALJET, 1918), 1,3-dinitrobenzene (RAYMOND, 1930), 3,5-dinitrobenzoic acid (KEDDE, 1947) and tetranitrodiphenyl (RABITZSCH, 1967) – are based on the fact that C-C coupling of the unsaturated lactone ring with nitrated aromatic derivatives produces dye complexes which can be measured photometrically. The specificity of the BALJET reaction is low, because many other substances, e.g., ketones, give intense color reactions with picric acid and alkali.

Various reaction mechanisms are suggested in the literature (for review see KOVAR et al., 1977). According to recent studies by BURNS et al. (1977) and KOVAR et al. (1977), splitting off of one proton at C21 produces a carbamine which undergoes nucleophilic linkage to the polynitroaromatic molecule. The resulting dye complexes are known as Meisenheimer compounds of the cyclohexadienate type (Fig. 2).

The molar extinction coefficients of various nitro complexes are listed in Table 1; Fig. 3 shows the typical absorption spectrum of a BALJET reaction. Some methods, however, are disqualified by inadequate sensitivity, too narrow a range, or insufficiently constant color values (BROCKELT, 1963; NEUWALD, 1950; TATTJE, 1958). The reagents which have gained an established place are picric acid and tetranitrodiphenyl. For the latter the lower limit of detection is 5×10^{-10} mol (RABITZSCH and TAMBOR, 1969). This is equivalent to quantitative detection of 0.2–0.3 μg aglycone or 0.3–0.5 μg glycoside. The standard deviation of measurements

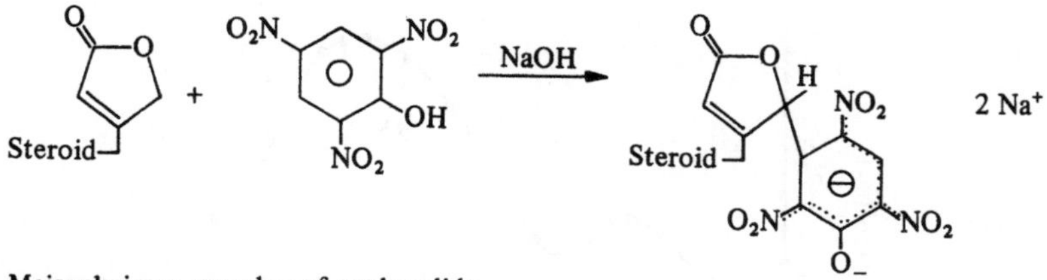

Meisenheimer complex of cardenolide
λ_{max} = 485 nm ϵ_{mol} = 15,000

Fig. 2. Mechanism of the BALJET reaction

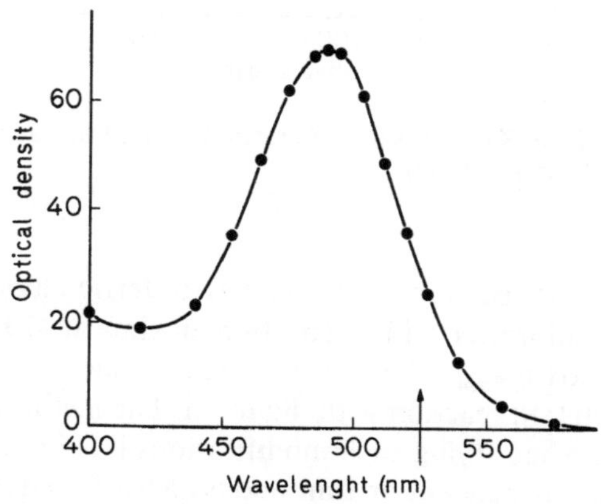

Fig. 3. UV absorption spectrum of the BALJET reaction with digitoxin. (DEMOEN and JANSSEN, 1953)

Table 1. Molar extinction coefficients (ε_{mol}) of various cardenolide ring-specific color reactions (RABITZSCH and TAMBOR, 1969)

Combination	ε_{mol}
Digoxin/3,5-dinitrobenzoic acid	8,300
Digoxin/*m*-dinitrobenzoic acid	28,000
Digoxin/picric acid	16,000
Digoxin/2,4,2',4'-tetranitrodiphenyl	26,600

with a cardenolide content of 1 µmol (assay volume 10 ml) is stated to be $\pm 3\%$. As an example of "genin determination" of digitalis products the work of NEUWALD and DIEKMANN (1951) may be cited.

II. Acidic Reagents

Both the KELLER-KILIANI and the xanthydrol reactions convert 2-desoxy sugars into characteristic colored derivatives. In this way all digitoxose-containing glycosides can be quantitatively determined.

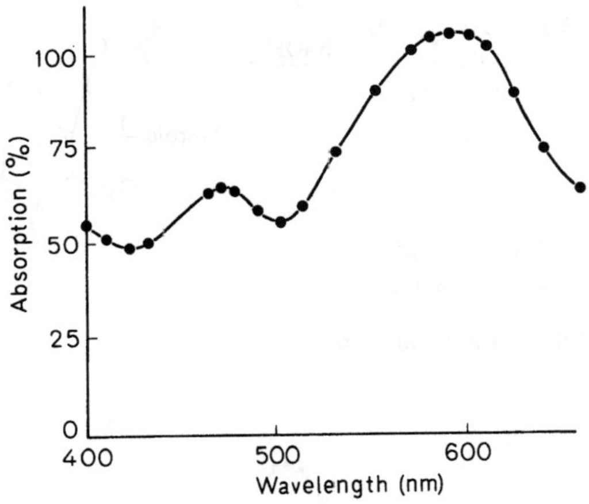

Fig. 4. UV spectrum of the KELLER-KILIANI reaction carried out as in USP XIII with digitoxin. (DEMOEN and JANSSEN, 1953)

The KELLER-KILIANI reaction with acetic acid, ferric chloride, and sulphuric acid produces a blue coloration with absorption maxima at 470 and 590 nm (Fig. 4, for digoxin and digitoxin $\varepsilon_{mol} = 3 \times 27{,}000$). Figure 4 shows the typical UV spectrum of a KELLER-KILIANI reaction with digitoxin. The method has been modified in various ways, but when trying such modifications it is important to remember that color formation is dependent on time and is also affected by moisture content (SOOS, 1948; STOLL and JUCKER, 1955; ROWSON, 1952). Owing to its high specificity for the sugar component of the glycoside, the reaction is scarcely influenced by the nature of the genin: it is the digitoxoses split off under the conditions of the test which are responsible for the color reaction (DEMOEN and JANSSEN, 1953). Digilanidobiose, which is not hydrolyzed, explains the fact that lanatosides and purpurea glycosides display only two-thirds of the color intensity given by digoxin or digitoxin.

The xanthydrol reaction (PESEZ, 1952) with xanthydrol, acetic acid, and hydrochloric acid produces a red coloration with an absorption maximum at 520 nm (for free digitoxose $\varepsilon_{mol} = 15{,}000$). However, the reagent is not very stable, and decomposition products tend to interfere with the color reaction. PÖTTER (1965) therefore suggested the use of the more stable dixanthylurea instead of xanthydrol. The absorption maximum is at the same position, while the sensitivity is somewhat greater (for free digitoxose ε_{mol} = approximately 19,000). Like other acid reagents, this detects only those digitoxoses which are easily hydrolysed under the conditions of the test. The calibration curves reproduced in Fig. 5 demonstrate this relationship.

One important prerequisite for multiple analyses is assured by the fact that the xanthydrol reaction is stable for a period of 75 min. The lower limit of detection is 0.1 µg, and tests in which known amounts (5 µg) of digitoxin were added have shown that the standard deviation of the method is ±1.75% (PFORDTE and FÖRSTER, 1970). When using it for analysing digoxin in tablets, JENSEN (1973) found that it gave more reproducible results than the BALJET method. Over the

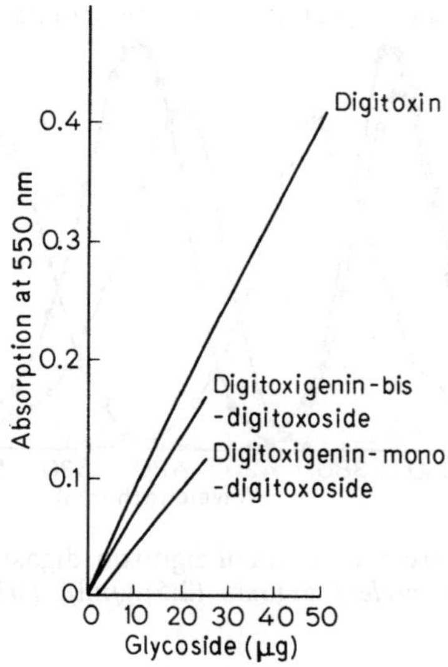

Fig. 5. Dixanthylurea reaction. Reference curves of digitoxin (Dgt-3), digitoxigenin-bis-digitoxoside (Dgt-2) and digitoxigenin-mono-digitoxoside (Dgt-1). (PFORDTE and FÖRSTER, 1970)

range 0–10 μg digoxin per ml solution, the xanthydrol reaction conforms to the Beer-Lambert law.

The most recently introduced reagent in this group is thiobarbituric acid. According to MESNARD and DEVANT (1961) desoxy sugars are oxidized by HIO_4 to dialdehydes, the reaction being accompanied by opening of the ring. With 2-thiobarbituric acid these dialdehydes form colored complexes which can be measured by spectrophotometry at 532 nm. MYRICK (1969) describes a modification of this procedure for the automated determination of digoxin in tablets of 0.25 mg nominal content.

II. Fluorescence Spectroscopy

In absorption photometry with the generally employed double-beam spectrophotometers it is the intensity differences between the test beam and the reference beam that are measured. For technical reasons the signal differences have a less favorable signal-to-noise ratio than the fluorescence signals, from which only the dark current has to be subtracted. This means that fluorescence spectroscopy is 10–100 times more sensitive (RINGHARDTZ, 1980).

The reaction between cardiac glycosides and strong acids gives a lower limit of detection in the ng range. For digoxin determinations an activating wavelength of 340 nm is used and the fluorescence emitted at 420 nm is measured. The underlying reaction mechanism has been known for some time (for review see WELLS et al., 1961): exposure to phosphoric acid, concentrated hydrochloric acid, or trichloracetic acid splits off hydroxyl groups from the steroid nucleus by a dehydration reaction. WINDAUS and SCHWARTE (1925) investigated the structure of the reaction product produced by treating gitoxin in this way and found that it was 14,16-dian-

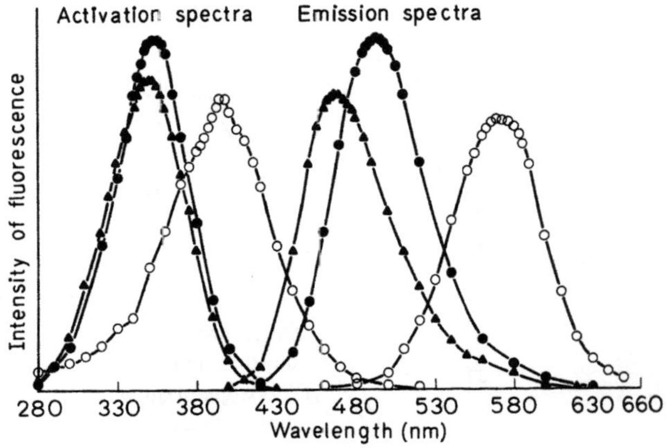

Fig. 6. Activation and fluorescence spectra of digitoxin, digoxin and gitoxin: *(open circles)* digitoxin (10 μg/ml); *(solid circles)* digoxin (2.5 μg/ml); *(triangles)* gitoxin (10 μg/ml). (WELLS et al., 1961)

hydrogitoxigenin. For digoxin and digitoxin the following sequence of reactions has been suggested (JELLIFFE, 1967): hydrolysis of the glycoside to the genin, formation of the 14-anhydrogenin, reaction to the corresponding 14-anhydro-16-hydroxy- or -16-chloro derivative, dehydration or dehalogenization to the 14,16-dianhydrogenin.

In 1950 PESEZ demonstrated that the fluorescence emitted in response to UV irradiation can be used for quantitative determination of cardiac glycosides. Certain modifications of the procedure were described shortly afterwards by SCIARINI and SALTER (1951) and by JENSEN (1952). Working with a modification of JENSEN'S method and using a spectrofluorimeter, WELLS et al. (1961) succeeded in measuring digitoxin independently of digoxin or gitoxin in the same sample. They did this by selecting suitable activation and fluorescence wavelengths, the latter being specific for a given hydroxylation configuration in the genin (Fig. 6).

BRITTEN and NJAU (1975) described a fluorimetric procedure for the specific determination of digoxin in tablets in the presence of digitoxin, gitoxin, and ouabain. After treatment with alcoholic trichloracetic acid, the lower limit of detection is 20 ng/ml. Over the range 0.02–1.0 μg/ml the relationship is linear and the coefficients of variation amount to 0.4%. The fluorescence method can be used in automated analyzers for serial measurements of "content uniformity" and of glycoside release from tablets (SCHWARZE and HITTINGER, 1974 a, b).

IV. Quantitative Determination After Chromatography

The chemical methods so far described are nonspecific (group reagents) and most of them are extremely susceptible to foreign substances. Glycosides accompanied by other substances (e.g., in pharmaceutical preparations) or glycoside mixtures have to be submitted to preliminary chromatographic separation or purification. TSCHESCHE et al. (1953) studied the sensitivity of various procedures. The test with nitroaromatic derivatives is of much the same sensitivity as the KELLER-KILIANI reaction, while the BALJET test is five times and the xanthydrol reaction seven times more sensitive.

The xanthydrol method has the advantage that the reagent gives zero blank readings with filter paper. Direct determination by thin layer chromatography is impracticable, because the carrier material adsorbs 40%–50% of the dye which is formed (PFORDTE and FÖRSTER, (1970). According to TSCHESCHE et al. (1953) the detection limit is around 0.25 µg, and in 0.1 mg lanatoside it is possible to determine lanatoside A and C in amounts as low as 1–2 µg.

PÖTTER and BARISCH (1972) have described a method for thin layer chromatographic determination of digitalis glycosides in tablets and other pharmaceutical preparations with the aid of the dixanthylurea reagent. After extraction with acetic acid or methylene chloride, separation is carried out in solvent mixtures (tetrahydrofuran/chloroform/formamide or methylethylketone/xylene) on formamide-impregnated kieselguhr layers. The thin layer carrier is eluted directly with the reagent. The most advantageous range for analysis lies between 10 and 120 µg/10 ml. The coefficient of variation over ten samples is stated to be 1.3% for pure glycosides, 3%–4% for glycoside mixtures, and 4%–6% for extracts from pharmaceutical preparations.

According to RABITZSCH et al. (1969), for the reaction with tetranitrobiphenyl, preceded by elution of the substrate from paper chromatograms, the assay range for digitoxin is 2.5–75 nmol. Elution of a silica gel carrier with chloroform/ethanol (8:2) gave an 80% yield.

The fluorimetry reagent (HCl/H_2O_2 or trichloracetic acid/chloramine) is suitable for direct use as a spray reagent for photofluorimetric analysis of chromatograph plates (DOELKER et al., 1969; JELLIFFE, 1967). The sensitivity limit is 10–20 ng per application. After elution of the fluorescent spots and measurement in a fluorimeter it proved possible to refine the technique to give an analytical error of only ± 0.003 µg (JELLIFFE, 1967).

Techniques for the photometric assay of scilla drugs and preparations after separation by paper or thin layer chromatography are given by KRAUS et al. (1969) and WICHTL and FUCHS (1962).

V. Quantitative Determination in Biologic Material

The chemical determination of therapeutic amounts of glycosides in biologic samples requires not only an assay sensitivity in the ng range, but also an appropriate method for removing interfering substances. The first step is enrichment by extraction, followed by column or layer chromatography for purification. The recovery during processing must be ascertained by carrying out the procedure after adding a known amount of glycoside. Although these methods for the most part give unsatisfactory results (HEINROTH et al., 1972) and are now no longer in use, a few references will be given.

REPKE (1957) reported the first attempts to devise methods for chemical determination of digitoxin in tissue and excreta. COX and WRIGHT (1959) and COX et al. (1959) carried out quantitative determinations of glycosides in urine and bile samples from experimental animals by means of the xanthydrol reaction or dinitrobenzene. After extraction and paper chromatography it proved possible to determine amounts as low as 10 µg with a recovery of 95% and an accuracy of ±2 µg.

The first clinically applicable technique for chemical measurement of digoxin and digitoxin in urine was presented by JELLIFFE in 1966
Extraction of the 24-h urine sample with chloroform
Washing of the extract with alkali and water
Thin layer chromatography
Location of the glycoside spots
Elution and xanthydrol reaction
Reference curves between 5 and 50 µ

With recovery rates between 67% and 89% (digoxin or digitoxin) for the extraction and between 89% and 78% for the TLC elution, the range of analytical error (±2 standard deviations above and below the mean) is stated to be 18 µg. From our present day standpoint, where for example 100 digoxin radioimmunoassays (RIA) can easily be carried out in one day, it is amusing to note that the chemical procedure is described as "sufficiently rapid to permit one unassisted technican to perform one determination per day" JELIFFE (1966).

SEIPEL et al. (1968) described a method in which thin layer chromatography and fluorimetry were used in combination for measuring digitoxin in blood from digitalized patients. In a 30 ml sample of whole blood it was possible to measure digitoxin accurately in amounts as low as 50 ng.

C. Chromatographic Procedures

The qualitative or quantitative analysis of cardiac glycosides and their breakdown products or metabolites in plants, pharmaceutical preparations, or biologic material usually requires preliminary isolation and further processing. For this purpose chromatographic methods are of great importance. Developments began in the 1950 s with the introduction of paper chromatography (PC), followed by thin layer chromatography (TLC), and liquid chromatography with open columns (LC). In the last few years gas chromatography (GC), often in conjunction with mass spectrometry (MS) has come to the fore, and high performance liquid chromatography (HPLC) in columns under high pressure has gained special prominence.

Chromatographic methods for the identification of drugs, nutrients, and metabolites have been established as standard methods in many laboratories for a considerable time. However, only in the last few years has it been possible to carry out quantitative in situ determinations of chromatographically separated substances on a large scale with adequate sensitivity and accuracy and reasonable economy in apparatus.

An outstanding feature of chromatographic techniques is that both the separation of the desired substances from the matrix of accompanying materials and their quantitative determination can usually be accomplished in a single working step. The choice of a suitable chromatographic method thus depends on the nature of the drug, the matrix surrounding it, the sample volume, the concentrations of the substances, and their detectability and stability. Furthermore, the technical resources and time required must in any event be kept in reasonable proportion to the problem which has to be solved.

Detection is carried out:

Qualitatively:	Visually	With or without spraying with suitable reagents
Quantitatively:	Microanalytically	After removal of the layer
	In situ	By measurements of UV absorption, fluorescence quenching or diminution

I. Paper Chromatography and Thin Layer Chromatography

Since 1950 or thereabouts paper chromatographic techniques have been used for qualitative and quantitative analysis of the glycosides of *Digitalis lanata*, *D. purpurea*, and other digitalis species (KAISER, 1966). Formamide-impregnated paper is generally employed, the choice of mobile phase being governed by the glycoside type (KAISER, 1955).

For the principal glycosides the composition ratios are xylene:methylethylketone = 50:50, or chloroform:tetrahydrofuran:formamide = 47:47:6.

Some 60 glycosides are known to occur in *D. lanata* and *D. purpurea* and if it is desired to identify them the recommended procedure is preliminary fractionation in a silica gel column followed by paper chromatography (KAISER, 1966). For processing chloroform extracts of urine, blood, bile, and viscera the following systems are still of interest. For polar glycosides, chloroform:benzene:butanol = 84:11:5 (Zafforoni type); for less polar glycosides, methylisobutylketone:isopropyl ether = 84:16. (Development times approximately, 3–4 h for 28 cm or approximately 14 h for 45 cm runs; capacity approximately 200–1,000 µg or 500–2,000 µg).

Once the substances have been separated they can be identified either by photometric methods (the reaction between the butenolide ring of the aglycones with nitroaromatics usually gives a deep blue coloration, the detection limit being approximately 5 µg) or fluorimetrically (reaction with trichloracetic acid or antimony trichloride; detection limit approximately 0.5 µg) (FUCHS et al., 1958). A comprehensive review of paper chromatographic investigations of cardiac glycosides and genins together with their characteristic R_f values in various mobile phase systems is given by NOVER et al. (1968).

In consequence of great improvements in silica gel manufacture (small and uniform grain size) and plate production (reproducible layer thicknesses and surfaces), TLC has steadily gained favor in recent years. It is faster and gives sharper separation than PC and allows quantitative in situ measurements even of micro amounts by UV absorption and fluorescence quenching directly on the plate.

In addition to TLC on hydrophilic silica gel plates, there is the technique of reversed phase chromatography (RPC) with hydrophobic separating materials and hydrophilic eluents (COHNEN et al., 1978). From conventional TLC with runs of 100–200 mm and development times of 0.5–1 h the technique known as high performance thin layer chromatography (HPTLC) has been developed for analytical purposes. It gives a considerable improvement in separation performance in relation to the length of run and the time required. It utilizes a more closely fractionated range of grain sizes (5–10 µm as compared with 5–25 µm in ordinary TLC). For a 40–50 mm run linear HPTLC requires approximately 6–9 min while circular or anticircular techniques require 1–4 min for a run of 20–25 mm. These methods are particularly suitable for quality control in pharmaceutical manufacturing because they allow rapid simultaneous analysis of several samples. In these circumstances the concentrations (mg or μg range) are always close to a given value and hence their quantitative evaluation presents no problem and can be performed fully automatically with commercially available scanners.

Preparative TLC is usually carried out on conventional plates with a 200 mm migration distance and a layer thickness of 1–2 mm (as against approximately 0.2 mm for analytical TLC plates).

The advantages of TLC are
Inexpensive separating medium
Single use, i.e., elaborate sample preparation is often unnecessary
Simultaneous processing of numerous samples (e.g., 72 separation paths on 200 × 100 mm HPTLC plates)
Much faster for qualitative analyses than any competing method
Plates can be stored, so that readings can be repeated subsequently
Off-line quantitation, i.e., only one quantitating instrument is required for a considerable number of separating chambers (computer controlled quantitation is also feasible)
Rapid evaluation (scanning speed 1–5 mm/s)
Usable spectral range for quantitation not restricted by the spectral properties of the mobile phase.

The disadvantages of TLC are
Sample application and separation cannot be completely automated
Major dynamics of the measurement range may make several standardizations necessary.

Numerous mobile phases are given in the literature, most of them intended for the qualitative and quantitative determination of glycosides in plant extracts or pharmaceutical preparations (KAISER et al., 1966; STEIDLE, 1961; ZÜLLICH et al., 1975; STORSTEIN, 1976; SZELECZKY, 1979; STAHL, 1961). CARVALHAS and FIGUEIRA (1973) suggest three systems for the separation of digoxin, digitoxin and their principal metabolites on silica gel plates, ethyl acetate : chloroform : acetic acid = 90:5:5, cyclohexane : acetone : acetic acid = 49:49:2, and chloroform : isopropanol : acetone = 80:5:15.

For a 15 cm run approximately 1 h is required.

Better resolution of chemically similar cardenolides can be achieved by two-dimensional development (CLARKE and COBB, 1979), e.g., for the first direction, ethyl acetate : dichlormethane : methanol : water = 59:35:4:2 and for the second direction: dichlormethane : methanol = 90:10.

The measurement of therapeutic blood levels by in situ evaluation of the TLC plate is feasible only when the glycosides (as in the case of digitoxin) are present in concentrations of between 10 and 50 ng/ml. After preliminary extraction with chloroform digitoxin in amounts as small as 1–2 ng can be demonstrated on the plate by fluorescence measurements. However, FABER et al. (1977) state that not more than 20–24 samples can be analyzed in one day.

As digoxin is present in blood levels of approximately 1.3 ± 0.5 ng/ml direct determination by this technique is not feasible and the method could only be used

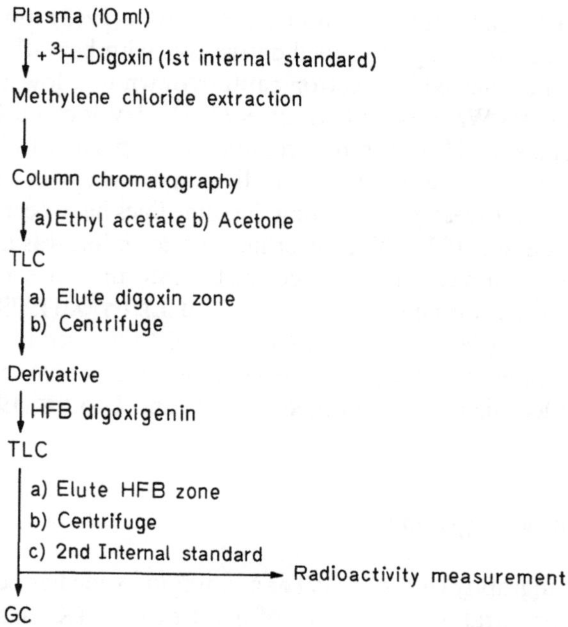

Fig. 7. Sample preparation for GC determination of digoxin. HFB = heptafluorobutyrate. (WATSON and KALMAN, 1971)

after elaborate concentration of correspondingly larger serum volumes. As regards HPTLC for glycosides, not many papers have as yet been published. KARTNIG and KOBOSIL (1977) suggest the following mobile phases for formamide-impregnated silica gel plates, xylene : methylethyketone = 40:60 and chloroform : tetrahydrofuran : formamide = 40:67:3.

For 7 cm runs development takes approximately 10–15 min. After application of 100 nl, the amounts detectable are claimed to be 30 ng by UV adsorption or 3 ng by fluorescence measurements.

II. Gas Chromatography

The principle advantages of this method are
Sensitivity (μg–ng: or, in conjunction with MS, down to pg)
Accuracy
Various detectors can be used
Numerous separation methods have already been worked out.
The disadvantages of GC are
Before separation it must be possible to volatilize the sample without decomposition
Elaborate sample preparation (preliminary purification)
Conversion into a derivative is usually essential; recovery of the sample is not always possible
The apparatus required for mass spectrometry (MS) is very elaborate and expensive.

Experimental work by JELLIFFE and BLANKENHORN (1963) showed that the cardenolides of interest to analysts yielded derivatives with heptafluorobutyrate and that these could be detected by electron capture down to a lower limit of 2 ng/injection. On this basis WATSON and KALMAN (1971) developed a GC assay for digoxin in human plasma. However, they required 10 ml plasma and had to carry out elaborate processing in order to obtain the digoxin derivative in a state suitable for injection (see Fig. 7). Investigations along the same lines have been carried out with digitoxin (KLEHR et al., 1977). The lower limit of detection was around 0.5 ng/ml. Digoxin and digitoxin have also been converted into the corresponding acetates. Using a flame ionization detector, MEILINK and LENSTRA (1979) found a lower limit of approximately 20 ng/injection. Nevertheless, GC used in conjunction with MS has proved of great value in glycoside research. This technique also enables metabolites to be identified and quantified (BODEM and UNRUH, 1978).

III. Liquid Chromatography

Column chromatography with silica gel or aluminium oxide has been employed for many years for preparative separations of plant extracts (KAISER, 1966). Among the solvents employed for elution are water-saturated ethyl acetate, chloroform, or chloroform/ethanol. Other column filling materials have been used including Sephadex G-200 (GRADE and FÖRSTER, 1967), Sephadex LH-20 (GAULT et al., 1976) and DEAE-Sephadex LH-20 (SUDGEN et al., 1976).

For analytical determinations of cardiac glycosides, in particular in pharmaceutical manufacturing, high performance liquid chromatography (HPLC) has made great strides in recent years. As a result of the application of adequately sensitive detection methods, HPLC is now finding increasing use in pharmacologic and pharmacokinetic studies.

The advantages of HPLC are

Much faster than LC at normal pressure

Separation at room temperature, i.e., suitable even for thermolabile substances

Preparation of derivatives is necessary in exceptional cases only

Substances are not decomposed and can be reclaimed and used for other methods (e.g., MS)

Wide range of stationary and mobile phases

Small sample requirements

Outstanding accuracy and reproducibility

Separation can be followed by direct visual inspection of the detector output

Various detection modes (UV, fluorescence, radioactivity, light refraction, electrochemical) can be used simultaneously by arranging the detectors in series

Immediate evaluation is feasible

Complete automation including sample preparation, injection, separation, and evaluation is possible, i.e., 24-h operation is feasible

Simultaneous determination of several constituents.

The disadvantages of HPLC are

Apparatus for automation is elaborate and expensive

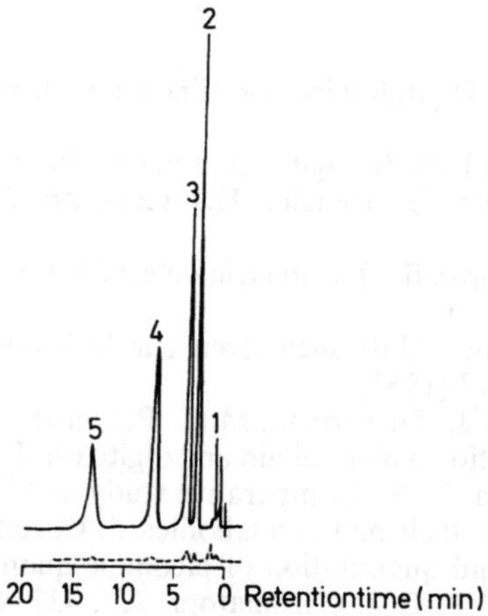

Fig. 8. HPLC separation of digoxin and metabolites with RP-18 column. 25% Acetonitrile/water; 3 ml/min; UV detection wavelength 220 nm; injection 10 µg cardenolide in 100 µl ethanol. Peaks are identified as follows: 1 solvent; 2 digoxigenin; 3 digoxigenin-monodigitoxoside; 4 digoxigenin-bis-digitoxoside; 5 digoxin. (CASTLE, 1975)

Because of the need for repeated use of the same separating column, careful sample separation is necessary when analyzing biologic material and care must be taken to avoid irreversible damage to the column.

Glycoside separations by HPLC are carried out on columns of various kinds including silica gel (NACHTMANN et al., 1976; LINDNER and FREI, 1976), ion exchangers (EVANS, 1974), and reversed phase columns (e.g., CASTLE, 1975; see also Fig. 8).

Because of their almost universal applicability (at present approximately 90% of all separations) and their insensitivity towards polar contaminants, reversed phase columns are gaining wide acceptance in many fields including the HPLC of digitalis glycosides. With RP-8 or RP-18 columns and acetonitrile/water or methanol/water for elution, followed by UV detection at 220 nm, the majority of separation problems can be solved. As the lower limit of detection for cardenolides is around 10 ng/injection, it is even possible to use urine samples, and after appropriate concentration to assay the original substance and any metabolites which may be present. Various attempts have been made to heighten the detection sensitivity. NACHTMANN et al. (1976) chose the 4-nitrobenzoate derivatives and used UV detection. GFELLER et al. (1977) exploited a fact known for TLC, namely that glycosides can be converted into fluorescent substances by treatment with concentrated mineral acids, and applied it to HPLC.

By the process known as post-column derivatization, i.e., treatment of the column eluate with concentrated hydrochloric acid and continuous measurement in a fluorescence detector, it has been found possible to detect amounts as small as 500 pg/20 µl injection. This method also has interesting possibilities for cardenolides (dihydrodigoxin), as UV detection is not practicable for these substances.

References

Baljet, H.: Glykoside mit Digitaliswirkung. Eine neue Identitätsreaktion. Pharm. Weekbl. *55*, 457 (1918)

Bodem, G., Unruh, E. v.: Dihydrodigitoxin, a metabolite of digitoxin in humans. In: Cardiac glycosides. Bodem, G., Dengler, H.J. (eds), pp. 74–84. Berlin, Heidelberg, New York: Springer 1978

Britten, A.Z., Njau, E.: Specific fluorimetric determination of digoxin. Anal. Chim. Acta *76*, 409–415 (1975)

Brockelt, G.: Vergleichende Untersuchungen zur Mikrobestimmung von Cardenoliden. Pharmazie *18*, 673–677 (1963)

Burns, L.B., Stedmann, R.J., Tuckerman, M.H.: Polynitro aromatic compounds in analytical chemistry. I: Reaction with quabain and digitoxin. J. Pharm. Sci. *66*, 753–754 (1977)

Carvalhas, M.L., Figueira, M.A.: Comparative study of TLC techniques for separation of digoxin, digitoxin and their main metabolites. J. Chromatogr. *86*, 254–260 (1973)

Castle, M.C.: Isolation and quantitation of picomole quantities of digoxin, digitoxin and their metabolites by HPLC. J. Chromatogr. *115*, 437–445 (1975)

Clarke, C.J., Cobb, P.H.: Two-dimensional TLC of digitalis cardenolides using a continuous development technique. J. Chromatogr. *168*, 541–549 (1979)

Cohnen, E., Flasch, H., Heinz, N., Hempelmann, F.W.: Verteilungskoeffizienten und R_m-Werte von Cardenoliden. Drug Res. *28*, 2179–2182 (1978)

Cox, E., Wright, S.E.: The hepatic excretion of digitalis glycosides and their genins in the rat. J. Pharmacol. Exp. Ther. *126*, 117–122 (1959)

Cox, E., Roxburgh, G., Wright, S.E.: The metabolism of ouabain in the rat. J. Pharm. Pharmacol. *11*, 535–539 (1959)

Demoen, P., Janssen, P.: Digitoxin. J. Am. Pharm. Ass. *42*, 635–640 (1953)

Doelker, E., Kapetanidis, J., Mirimanoff, A.: Application de la rémission et de la fluorescence au dosage direct sur chromatoplaque des digitoxosides. Pharm. Acta Helv. *44*, 647–651 (1969)

Evans, R.I.: HPLC of digitoxigenin and its glycosides. J. Chromatogr. *88*, 411–412 (1974)

Faber, D.B., De Kok, A., Brinkman, V.A.Th.: TLC method for the determination of digitoxin in human serum. J. Chromatogr. *143*, 95–103 (1977)

Fuchs, L., Wichtel, M., Jachs, H.: Die quantitative papierchromatographische Bestimmung von Digitalisglykosid-Gemischen. Arch. Pharm. *291*, 193–208 (1958)

Gault, M.H., Ahmed, M., Symes, A.L., Vance, J.: Extraction of digoxin and its metabolites from urine and their separation by sephadex LH-20. Column chromatography. Clin. Biochem. *9*, 46–52 (1976)

Gfeller, J.C., Frey, G., Frei, R.W.: Post-column derivatization in HPLC using the air segmentation principle: application to digitalis glycosides. J. Chromatogr. *142*, 271–281 (1977)

Grade, K., Förster, W.: An improved method for the isolation of heart glycosides from tissues by using sephadex G-200. Biochem. Pharmacol. *16*, 1299–1303 (1967)

Heinroth, J., Sziegoleit, W., Schäbitz, J., Förster, W., Krosch, H.: Ergebnisse einer chemischen Bestimmung der renalen Digitoxinausscheidung bei Patienten unter Digitoxinbehandlung. Z. Gesamte Inn. Med. *27*, 523–526 (1972)

Jelliffe, R.W., Blankenhorn, D.H.: Gaschromatography of digitoxigenin and digoxigenin. J. Chromatogr. *12*, 268 (1963)

Jelliffe, R.W.: A chemical determination of urinary digitoxin and digoxin in man. J. Lab. Clin. Med. *67*, 694–708 (1966)

Jelliffe, R.W.: An ultramicro fluorescent spray reagent for detection and quantification of cardiotonic steroids on thin-layer chromatograms. J. Chromatogr. *27*, 172–179 (1967)

Jensen, K.B.: Fluorimetric determination of gitoxigenin. Acta Pharmacol. Toxicol. *8*, 101–109 (1952)

Jensen, K.: Analysis of digoxin in tablets. Arch. Pharm. Chem., Sci. Ed. *1* (3), 447–451 (1973)

Kaiser, F.: Die papierchromatographische Trennung von Herzgiftglykosiden. Chem. Ber. *4*, 556–563 (1955)

Kaiser, F.: Chromatographische Analyse der herzwirksamen Glykoside von Digitalis-Arten. Arch. Pharm. *299*, 263 (1966)
Kartnig, Th., Kobosil, P.: Zur Trennung der Digitalis-Cardenolide mit Hilfe der Hochleistungs-Dünnschichtchromatographie. J. Chromatogr. *138*, 238–242 (1977)
Kedde, D.L.: The chemical investigation of digitalis preparations. Pharm. Weekbl. *82*, 741–757 (1947)
Klehr, H.U., v. Unruh, G., Bodem, G.: Digitoxinstoffwechsel bei niereninsuffizienten Patienten. Verh. Dtsch. Ges. Inn. Med. *83*, 1656–1659 (1977)
Kovar, K.-A., Frankas, G., Seidel, R.: Zum Mechanismus der Reaktionen nach Raymond, Kedde und Baljet. Arch. Pharm. *310*, 40–47 (1977)
Kraus, K., Mutschler, E., Rochelmeyer, H.: Zur Analytik und Wertbestimmung von Extrakten aus Bulbus Scillae. Arzneim. Forsch. *19*, 322–328 (1969)
Lindner, W., Frei, R.W.: Partition HPLC systems for the separation of digitalis glycosides of the cardenolide group. J. Chromatogr. *117*, 81–86 (1976)
Mellink, J.W., Lenstra, J.B.: Gas-liquid chromatography of digitoxigenin and digoxigenin as acetates of their epoxygenic acid methyl esters. J. Chromatogr. *170*, 35–41 (1979)
Mesnard, P., Devant, G.: Application of the reaction between malonic dialdehyde and thiobarbituric acid to heterosides and deoxyoses. Compt. Rend. *253*, 497–499 (1961)
Myrick, J.W.: Automated assay of single tablets of digoxin. J. Pharm. Sci. *58*, 990–993 (1969)
Nachtmann, F., Spitzy, H., Frei, R.W.: Ultraviolett derivatization of digitalis glycosides as 4-nitrobenzoates for liquid chromatographic trace analysis. Anal. Chem. *48*, 1576–1579 (1976)
Neuwald, F.: The photometric assay of digitoxin by the Baljet reaction. J. Am. Pharm. Ass. *39*, 172 (1950)
Neuwald, F.: Chemische Wertbestimmung von Digitalisdrogen und herzwirksamen Glykosiden. Arch. Pharmacol. *283*, 93–102 (1950)
Neuwald, F., Diekmann, A.: Chemische Wertbestimmung von Digitalisdrogen und herzwirksamen Glykosiden. 3. Mitteilung: Vergleichsuntersuchungen mit verschiedenen chemischen Methoden. Arch. Pharm. (Weinheim) *284*, 19–22 (1951)
Nover, L, Baumgarten, G., Luckner, M.: Über die Beziehungen zwischen chemischer Struktur und chromatographischem Verhalten bei Herzglykosiden. I. Mitteilung: Papierchromatographische Untersuchungen an Herzglykosiden und ihren Geninen. J. Chromatogr. *32*, 93–122 (1968); II. Mitteilung: Die Anwendung der R_m-Wert-Theorie bei Herzglykosiden und ihren Geninen. J. Chromatogr. *32*, 123–140 (1968); III. Mitteilung: Der Einfluß der Zuckerkette auf das chromatographische Verhalten der Genine. J. Chromatogr. *32*, 141–161 (1968)
Pesez, M.: Dosage fluoromatrique du gitoxoside. Ann. Pharm. Fr. *8*, 746–750 (1950)
Pesez, M.: Le xanthydrol réactif des désoses. Ann. Pharm. Fr. *10*, 104–108 (1952)
Petit, A., Pesez, M., Bellet, P., Amiarel, G.: Pure digitoxin. Bull. Soc. Chim. Fr. *1950*, 288–291
Pfordte, K., Förster, W.: Quantitative determination of cardenolides and their metabolites in biological material. Z. Med. Lab. Diagn. *11*, 272–282 (1970)
Pötter, H.: Dixanthylharnstoff, ein neues Reagens zur quantitativen Bestimmung von Digitalis-Glykosiden. Pharmazie *20*, 737 (1965)
Pötter, H., Bärisch, H.: Versuche zur dünnschichtchromatographischen Bestimmung der Digitalis-Glykoside. Pharmazie *27*, 315–328 (1972)
Rabitzsch, G., Juengling, S., Tambor, U.: Untersuchungen zur Kolorimetrie und quantitativen Papier- und Dünnschichtchromatographie herzwirksamer Genine und Glykoside zum Digitalis- und Strophanthus-Typ. Wiss. Beitr. Martin-Luther-Univ. Halle-Wittenberg, No. 6, 925–330 (1969)
Rabitzsch, G., Tambor, U., Juengling S.: Colorimetry of digitalis glycosides and its combination with methods of chromatographic analysis. Monatsber. Dtsch. Akad. Wiss. *11*, 116–122 (1969)
Rabitzsch, G., Tambor, U.: Verfahren zur quantitativen Bestimmung der Herzglykoside und Genine vom Cardenolidtyp mit 2.4.2′,4′-Tetranitrodiphenyl. Pharmazie *24*, 262–269 (1969)

Raymond, W.P.: Detection and estimation of ouabain and strophanthin. Analyst *63*, 478–482 (1938)
Repke, K.: Eine Methode zur chemischen Bestimmung von Digitoxin in Geweben und Exkreten. Naturwissenschaften *44*, 619–620 (1957)
Ringhardtz, J.: Was leistet die Fluoreszenz-Spektroskopie in der Biochemie? Labor-Praxis *4*, April, 32–39 (1980)
Rowson, J.M., Dyer, F.J.: Studies in the genin digitalis. Part II. A comparison of the colorimetric and biological methods for the evaluation of digitalis purpurea. J. Pharm. Pharmacol. *4*, 831–843 (1952)
Schwarze, P., Hittinger, H.: Verfahren für die Bestimmung der Auflösungsgeschwindigkeit von in wäßrigem Milieu schwerlöslichen Arzneistoffen. Pharm. Ind. *36*, 732–735 (1974a)
Schwarze, P., Hittinger, H.: Bestimmung der Einzeldosierung und Auflösungsgeschwindigkeit von Digitoxin aus Tabletten und Dragees. Pharm. Ind. *36*, 794–797 (1974b)
Sciarini, L.J., Salter, W.T.: Chemical correlatives of digitalis potency in man, cat, and pigeon. J. Pharmacol. *101*, 167–175 (1951)
Seipel, H., Hueber, E.F., Deutsch, E., Lutz, U., Wichtl, M., Jentsch, K.: Quantitative Digitoxinbestimmung im Blut routinemäßig digitalisierter Patienten. Klin. Wochenschr. *46*, 1257–1260 (1968)
Soos, E.: Chemische Wertbestimmung von Digitalisblättern. Sci. Pharm. *16*, 1, 29 (1948)
Stahl, E.: Schnelltrennung von Digitalis- und Podophyllum Glykosidgemischen. J. Chromatogr. *5*, 458–460 (1961)
Steidle, W.: Quantitative Bestimmung von Scilla-Glykosiden. Planta Med. *9*, 435–441 (1961)
Stoll, A., Jucker, E.: Phytosterin, Steroidsaponine und Herzglykoside. In: Methoden der Pflanzenanalyse, III. (Ed.: Peach und Tracey), Springer 1955, S. 216–271
Storstein, L.: Studies on digitalis, a method for RC separation and determination of digitoxin and cardioactive metabolites in human blood and urine. J. Chromatogr. *117*, 87–96 (1976)
Sudgen, D., Ahmed, M., Gault, M.H.: Fractionation of tritiated digoxin and dihydrodigoxin with DEAE-sephadex LH-20. J. Chromatogr. *121*, 401–404 (1976)
Szeleczky, Z.: Thin-layer chromatography of cardenolides in the presence of boric acid. J. Chromatogr. *178*, 453–458 (1979)
Tattie, D.H.E.: Colorimetric determination of digitalis glycosides with 2-4-dinitrophenyl sulfone. J. Pharm. Pharmacol. *10*, 493–498 (1958)
Tschesche, R., Grimmer, G., Seehofer, F.: Über pflanzliche Pilzgifte, XXIV. Mitteil.: Die quantitative Trennung und Identifizierung von Herzglykosiden aus Digitalis purpurea und lanata durch „echte Verteilungschromatographie an Papier". Chem. Ber. *86*, 1235–1241 (1953)
Watson, E., Kalman, S.M.: Assay of digoxin in plasma by gaschromatography. J. Chromatogr. *56*, 209–218 (1971)
Wells, D., Katzung, B., Meyers, F.H.: Spectrofluorometric analysis of cardiotonic steroids. J. Pharm. Pharmacol. *13*, 389–395 (1961)
Wichtel, M., Fuchs, L.: Eine photometrische Methode zur Gehaltsbestimmung von Scilla-Drogen und Scilla-Präparaten nach papierchromatographischer Trennung der herzwirksamen Glykoside. Arch. Pharm. *295*, 361–373 (1962)
Windaus, A., Schwarte, G.: Über ein in Chloroform unlösliches Glykosid aus Digitalis-Blättern, das Gitoxin. Chem. Ber. *58*, 1515–1519 (1925)
Züllich, G., Braun, W., Lisboa, B.P.: Thin-layer chromatography for the separation of digitoxin, digitoxigenin and related compounds. J. Chromatogr. *103*, 396–401 (1975)

CHAPTER 4

Use of Radioactively Labeled Glycosides

H. FLASCH

A. Introduction

We owe our present knowledge of the pharmacokinetics and biotransformations of cardiac glycosides primarily to the successful use of the isotope technique. As in the case of other drugs, radioisotope labeled cardiac glycosides offer valuable analytical advantages in the study of absorption, distribution, metabolic breakdown, and elimination. They can be used to measure glycoside concentrations in all forms of biological material. The technique is easy, inexpensive, highly sensitive, and extremely precise. As impurities and accompanying substances do not interfere with the radioactivity measurements, there is no need for preliminary purification or concentration of the kind required in most physicochemical methods.

However, another characteristic feature of radioactivity measurement, namely its nonspecificity, may mislead those who adopt an uncritical approach to the results of radioactivity counts. What is measured is not the drug itself but the radioactivity. Though the latter was originally attached to the labeled molecule, at the time when the sample is collected the radioactivity is not necessarily identical to the latter. It may represent the sum total of an assortment of different molecular species which may have been formed by chemical, radiochemical, or biologic decomposition or by metabolic transformation.

At the present day the pharmacokinetic research worker is subject to two conflicting forces: from one side there are demands for profounder knowledge of familiar or newly developed glycosides, much of which can be obtained only by the use of radioactive isotopes, while on the other side the exigencies of radiation protection and environmental health impose restrictions on the use of radioactive materials. This conflict emphasizes the need for a more critical attitude towards the use of radioisotope labeled drugs. The best prospects for the future seem to lie in a rational selection of isotope techniques and their use in conjunction with other analytical methods. There are of course some problems which would remain unsolvable without the use of radioactive tracers, but aside from these there are good economic reasons for the planned and selective use of isotope techniques, especially in animal experiments designed to provide preliminary orientation in the development of new drugs.

B. Prerequisites for the Use of Isotope Techniques

The results of tracer experiments cannot be properly evaluated unless the following conditions have been fulfilled:

1) The radioactive atoms at the labeling sites must be chemically and biologically stable. This applies in particular to tritium-labeled compounds. It is essential to ensure that tritium-hydrogen exchange does not occur, either in the body or during subsequent processing of the material.

2) The labeled molecules must not be recognized as such by the body. The likelihood of biological discrimination between labeled and unlabeled molecules is always to be reckoned with, and becomes more probable as the chemical and physical properties of the two molecular species diverge. For example, tritium has an atomic weight three times that of hydrogen and the possibility of isotopic effects cannot be ignored. However, in the case of cardiac glycosides such effects need not as a rule be feared, because no biologic transformations take place in their well known tritiation positions.

3) The radioactive isotope used to label the drug must not harm the body. There is no likelihood of any adverse effects from the usual doses used in animal experiments. The question of radiation protection in human experiments is discussed in Sect. G.

C. Production of Radioisotope Labeled Glycosides

If 100% of the molecules in a cardiac glycoside are labeled with one tritium atom each, a maximal specific radioactivity of 29 Ci/mmol can be achieved. Such compounds are detectable in amounts down to 1 pg. For ^{125}I-labeled derivatives the detection limit is lower by a factor of ten. Labeled compounds of high specific activity are generally used for radioassays only; for pharmacokinetic studies dilute preparations are adequate. For example, 100 µCi in 0.5 mg ^{3}*H*-digoxin is sufficient for experiments in humans. In this material 0.54% of the digoxin molecules are labeled with a tritium atom.

In view of the varied uses of labeled glycosides, the producer must give due regard to three important aspects: *specific radioactivity*, *site of labeling*, and *practicability*.

I. Biosynthesis

Some years after the discovery of radiocarbon by RUBEN and KAMEN (1940), GEILING et al. (1948), and OKITA et al. (1954) produced biosynthetic ^{14}*C*-digitoxin by cultivating plants of *Digitalis purpurea* in an atmosphere enriched with $^{14}CO_2$. Harvested after a growing period of 4–6 weeks, the plants yielded ^{14}*C*-digitoxin (U)[1] with a specific radioactivity of 0.26–0.54 µCi/mg. This product was the first labeled drug ever used in a pharmacokinetic study (GEILING, 1950).

So far as cardiac glycosides are concerned, this biological labeling technique is now of little more than historical significance. As it is extremely time consuming and as it yields products with low specific activity – usually too low for pharmacokinetic investigations – it is nowadays seldom employed. However, it is still the

1 During the biosynthetic incorporation of carbon into the molecule, the distribution of the isotopes throughout all positions in the molecule is statistically uniform or almost uniform. This fact is denoted by the designation (U)

only method producing ^{14}C-labeled glycosides. For example BRETSCHNEIDER et al. (1962), working with *Digitalis lanata*, used it to produce ^{14}C-lanatoside A(U) and ^{14}C-lanatoside C (U).

II. Wilzbach Labeling

If an organic compound is exposed to the action of tritium gas of high specific radioactivity, hydrogen atoms will be exchanged for tritium atoms. This radiation-induced isotope exchange was developed by WILZBACH (1957). The advantages of the technique are the simplicity and economy with which even the most complicated natural substances can be labeled with high specific radioactivity, e.g., digoxin 100 mCi/mmol (SEGEL, 1961).

One serious disadvantage is the emergence of labeled byproducts, formed under the intense irradiation. In the case of the cardenolides one particularly troublesome side reaction is the hydrogenation of the 20, 22 double bond of the α, β-unsaturated lactone ring, which in contrast to the chemically unaltered glycoside produces almost carrier-free contaminants, i.e., substances with a very high degree of specific labeling. According to RABITZSCH (1969) up to 55% of hydrogenated contaminants can originate in this way, especially if the glycoside is attached to silica gel for the purpose of Wilzbach labeling. In the light of this fact, we must now regard with extreme reserve all pharmacokinetic results based on work with Wilzbach-labeled cardenolides unless they have been checked for hydrogenated contaminants.[2]

These drawbacks of the technique, insofar as they apply to cardiac glycosides, seem to have occasioned an editorial recently published by DI CARLO (1979) under the title "Goodbye Wilzbach!"

III. Catalytic Exchange with Tritium Water

The hydrogen atoms of the unsaturated butenolide ring of the cardenolides are accessible to base-catalyzed isotope exchange. In the labeling method of HABERLAND and MAERTEN (1969, 1971), the glycoside is dissolved in dimethylformamide and treated with tritium water in the presence of triethylamine at a raised temperature. In the course of equilibration reactions a small proportion of the base can split off protons from positions 21 and 22 (Fig. 1). The resulting resonance-stabilized anions can then react with tritons from the tritium water. The structure of the labeled glycosides was elucidated by using deuterated β-acetyldigoxin as an example (HABERLAND and MAERTEN, 1969). In the NMR spectrum of the original substance there are proton signals at C21 (doublet) and C22 (singlet), but in the new substance they do not appear. This means that all three protons of the butenolide ring have been exchanged and that tritiation will produce 21,22-^{3}H-labeled cardenolides. After splitting off the sugars by acid saponification 97% of the radioactivity can be localized in the genin, while radiochromatography in suitable

2 As a result of the differing reactivity of the various C–H bonds, the labeling positions taken up during the process of radiation-induced hydrogen-tritium exchange are nonuniformly distributed throughout the molecule. This fact is denoted by the designation (G) for "generally labeled."

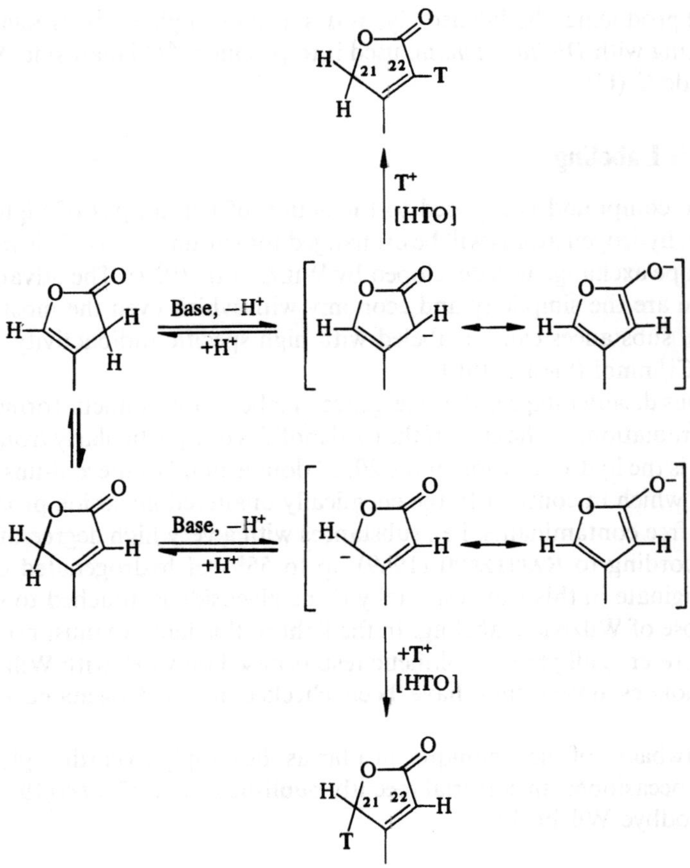

Fig. 1. Catalytic tritium labeling of a cardenolide with HTO. To demonstrate the reaction principle the lactone ring only is shown. (HABERLAND and MAERTEN 1969)

systems (RABITZSCH, 1968 a) has shown that the glycosides are free from hydrogenated impurities. As the lactone ring is one part of the molecule which is essential for the biological action of cardiac glycosides, it was thought necessary to check the latter with 21,22-labeled compounds. Using deuterated glycosides as an example, it was succesfully demonstrated that the pharmacological activity remained intact after isotope exchange (HABERLAND and MAERTEN, 1971).

The main advantages of the catalytic exchange technique are the simplicity of the apparatus and the low cost. It produces remarkably high specific radioactivities of the order of 1 Ci/mmol. Owing to the gentleness of the reaction conditions, even chemically sensitive glycosides can be specifically labeled without significant decomposition and without the formation of hydrogenated impurities.

IV. Reductive Tritiation

WARTBURG et al. (1965) described a synthetic tritiation method which is suitable for digoxin and its derivatives. In a preliminary reaction acetyldigoxin is oxidized

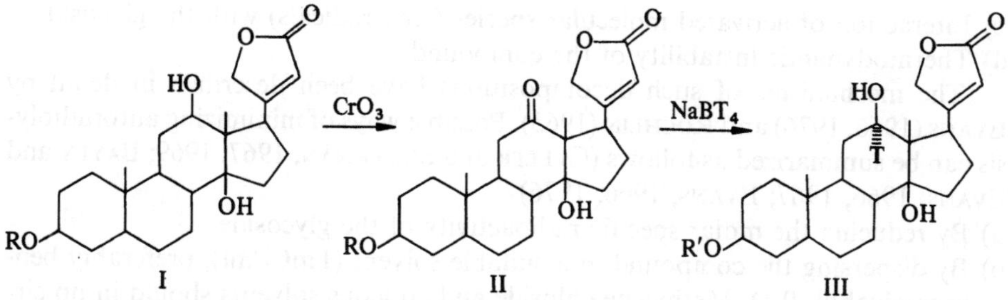

Fig. 2. Partial synthesis of 12α-H-digoxin R = acetyl-(digitoxose)$_3$; R′ = digitoxose$_3$; I = acetyldigoxin; II = 12-dehydroacetyldiogoxin; III = 12α-^3H-digoxin. (WARTBURG et al., 1965)

with chromium trioxide to 12-dehydroacetyldigoxin (Fig. 2). The subsequent reduction of the carbonyl group with NaB^3H_4 takes place stereospecifically and, after the acetyl residue has been split off, leads to the formation of 12α-^3H-digoxin.

This technique also provides labeled compounds with high specific radioactivity at a specific position. Hardly any impurities are formed. However, the reaction is restricted to those cardiac glycosides which either possess a carbonyl group or in which a carbonyl group can be formed by oxidation and subsequently reduced by a stereoselective procedure. A variant of reductive tritiation was described by SCHENK et al. (1978) for the labeling of methylproscillaridin.

V. Partial Synthetic Procedures

For the purpose of selective ^{14}C labeling, partial synthetic procedures can be utilized to build up the lactone ring. Reviews describing such syntheses have been published by THOMAS et al. (1974) and BOUTAGY and THOMAS (1973).

By using partial syntheses of this kind, FLASKAMP and BUDZIKIEWICZ (1977) produced ^2H- and ^{18}O-cardenolides with a label in the lactone ring. Techniques for the synthesis of ^{125}I- or ^{131}I-labeled glycoside derivatives will not be dealt with here; the reader is referred to the references sited in connection with the corresponding radioimmunoassays.

D. Stability of the Radioactive Label

If labeled cardiac glycosides are to be of any value it is essential that the radioactive isotope should remain in its position in the molecule throughout the investigation, i.e., from the time when the product is dispatched until the radioactivity measurements are carried out. The stability of the label must therefore be guaranteed during storage, under the biological conditions of the experiment, and during the processing of the samples.

During storage, cardiac glycosides, like all steroids, are subject to *decomposition by self irradiation*. The causes are

a) Natural radioactive decay

b) Interaction of radioactive emission with glycoside molecules

c) Interaction of activated molecular species (e.g., radicals) with the glycoside
d) Thermodynamic instability of the compound

The mechanisms of such decompositions have been described in detail by EVANS (1966, 1976) and ROCHLIN (1965). Possible ways of minimizing autoradiolysis can be summarized as follows (GELLER and SILBERMAN, 1967, 1969; BAYLY and EVANS, 1966, 1967; EVANS, 1966, 1976)

a) By reducing the molar specific radioactivity of the glycoside
b) By dispersing the compound in a suitable solvent (1 mCi/ml), preferably benzene/alcohol (9:1). Methylene chloride and aqueous solvents should in no circumstances be employed, because they may form free radicals when exposed to radiation
c) By adding a substance which will capture radicals
d) By storing at the lowest possible temperature, without causing the separation by freezing of any solvent present

These recommendations for storage apply to the labeled glycoside between manufacture and experimental use, but are even more important for the storage of biological samples awaiting investigation of metabolites. For example, organic solvent extracts may contain residues which tend to accelerate radiolytic decomposition. During tests of 6,7-^{3}H-estrone, which was stored in the presence of solvent residues, decomposition rates of up to 50% in six weeks were observed (GELLER and SILBERMAN, 1969). It is therefore advisable to analyse biological material for metabolite investigations immediately after it has been collected.

When using tritiated cardiac glycosides the *biological stability of the label* must be checked. In the course of metabolic processes in which enzymes are involved hydrogen exchange can take place even in chemically stable labels. Among the recognized methods of testing for metabolically labile tritium (EVANS, 1966) the simplest is the measurement of "volatile" radioactivity in urine or plasma samples. Isolation of water from biological material should be carried out either by freeze drying or by vacuum distillation at low temperature. Under less gentle experimental conditions such as benzene distillation at boiling point (WERBIN et al., 1959) pH shifts and hence decomposition may occur. For example, ROESCH et al. (1973), who used this technique for studies with 21,22-^{3}H-digitoxin in rats and mice, found up to 50% of HTO in the plasma. On checking the results, the writer found that during the progressive withdrawal of water from plasma or tissue samples the pH rose above 8 and increasing proportions of HTO were distilled off (FLASCH, unpublished results). In keeping with the mechanism of catalytic exchange labeling for 21,22-^{3}H-labeled cardenolides illustrated in Fig. 1, the release of tritium can be explained by a corresponding reversal of the synthesis reaction in an alkaline medium. Under the gentle conditions of freeze drying, glycosides labeled at the lactone ring have been proved to have a biological stability of over 95%. During biological investigations with other tritiated glycosides volatile radioactivity readings of over 5% have not so far been observed.

E. Purity Testing of Labeled Glycosides

The following terms are used to define the purity of radioactive compounds (CATCH, 1968).

Carrier-free. A preparation of a radioisotope to which no carrier has been added, and for which precautions have been taken to minimize contamination with other isotopes.

Radiochemical purity of a radioactive material. The proportion of the total activity that is present in the stated chemical form.

Specific activity. The radioactivity per unit mass of a compound containing a radioactive nuclide.

As labeled compounds are all to some degree unstable, stored preparations must be checked to ensure their radiochemical purity. For most cardiac glycosides a storage time of some three months presents no problems provided suitable conditions are maintained. Even though the manufacturer may have issued a declaration of purity, the user should carry out a quantitative determination before employing the preparation and should ascertain whether its radiochemical purity is adequate for the planned investigation.

Physical methods of purity testing (e.g., melting point, UV, IR, and NMR spectroscopy) provide information only on the chemical purity of the material, and for most tracer experiments this is not critical. They are not sensitive enough to detect traces of impurities which may in some circumstances carry the bulk of the radioactivity, as may be the case in compounds labeled by the Wilzbach method.

Chemical methods – such as isotope dilution analysis and chromatography – are more suitable for the purpose. In "reverse dilution analysis" a small amount (approximately 0.1 mg) of the preparation to be assayed is mixed with at least 1,000 times its weight of the equivalent inactive compound. The dilution is homogenized in a suitable solvent, the solvent is then removed and the specific activity of the preparation is measured s_o. The preparation is then recrystallized several times until a constant specific activity is attained (s_p). The radiochemical purity of the original compound can be calculated from the following formula (EVANS, 1966).

$$P\,[\%] = \frac{s_p}{s_o} \times 100\,.$$

This method is applicable only in cases where no isotopic effect occurs and where there is no tendency to mixed crystal formation with the radioactive impurities.

Chromatographic methods such as paper and thin layer chromatography (TLC) are most frequently used for the analysis of labeled cardiac glycosides. (Examples of suitable chromatography systems are given by the following authors: CARVALHAS and FIGUEIRA, 1973; KAISER 1966; KARTWIG and KOBOSIL, 1977; NOVER et al., 1969; STORSTEIN, 1976; ZELNIK et al., 1964.)

Whatever procedure is adopted, a careful watch must be kept for possible artifacts (SHEPPARD, 1972)

a) Decomposition during application
b) Absorption to support due to inadequate carrier loading
c) Overloading with carrier
d) "Double-peaking" due to poor initial application

After measuring the radioactivity distribution on the developed chromatogram, the radiochemical purity is expressed as a percentage of the principal component of the total activity along the "solvent track." One of the earlier methods for measuring radioactivity distribution is autoradiography. Quantitative measurement by densitometry is feasible, but not very exact – especially for minor impurities. Some amplification of "radiation output" can be achieved by spraying the chromatogram with fluorescent compounds (PRYDZ et al., 1970; RANDERATH, 1970; WILSON and SPEDDING, 1965). Radio-TLC scanning is the technique nowadays most widely employed for the evaluation of chromatograms (for reviews see CATCH, 1968; FIGGE et al. 1970; RABITZSCH, 1968b; SHEPPARD, 1972).

When using ^{14}C-labeled glycosides, the scanner diagram, in differential or integral registration, obtained with the aid of a rate meter coupled to an X–Y recorder, can be used for direct quantitative evaluation. In such circumstances the numerical yields amount to up to 30%. In the evaluation of integral radioactivity diagrams of ^{3}H-labeled compounds (numerical yield up to 3%) there is in general a tendency to overestimate substances with higher R_F-values. For example, as the R_F value increases from 0.2 to 0.7, there is a relative increase of 10% in the numerical yield (FIGGE et al., 1970). To ensure more exact measurement it is therefore advisable to scrape off the radioactive-zones of the thin layer chromatogram, to convert the material into gel form, and measure its radioactivity in a scintillation counter.

"Low levels of impurity" can be estimated by "comparative peak area measurement." An aliquot of the sample to be analyzed is diluted 1 : 100 and chromatographed together with the undiluted sample. The areas beneath the peaks are scanned under identical conditions, and on comparison the dilution will represent 1% of the main peak. If the total of the peak areas produced by the impurities in the undiluted sample is smaller than the "1% peak," the material must have a radiochemical purity of at least 99%. This of course assumes that the TLC system employed has adequate separating power. One highly efficient technique for purity testing and also for metabolite investigations is HPLC (high pressure liquid chromatography) in conjunction with a radioactivity monitor (HODENBERG et al., 1977).

F. Pharmacokinetic Investigations with Labeled Cardiac Glycosides

To establish the pharmacokinetic attributes of a cardiac glycoside the investigator will begin by studying its fundamental parameters in experimental animals: the radioactivity concentration in the blood, in the excretion compartments, and in the tissues. Liquid scintillation counters are prefered for radioactivity measurements because of their precision and because they are largely unaffected by interference. With the aid of such an instrument, the necessary analyses are easily performed and seldom present any technical problems. When following the temporal changes in blood or plasma concentrations the investigator must take special note of the background count, especially at low concentrations, and must carefully consider which readings should or should not be accepted into the body of data. Especially in the terminal segment of a blood concentration curve, any analytical error can falsify the so-called β-slope and hence vitiate the calculation of the biological half-life.

The radioactivity of tissue or feces samples can be measured in the form of $^{3}H_2O$ or $^{14}CO_2$ after combustion in a "sample oxidizer" (DOBBS, 1963; GLEIT,

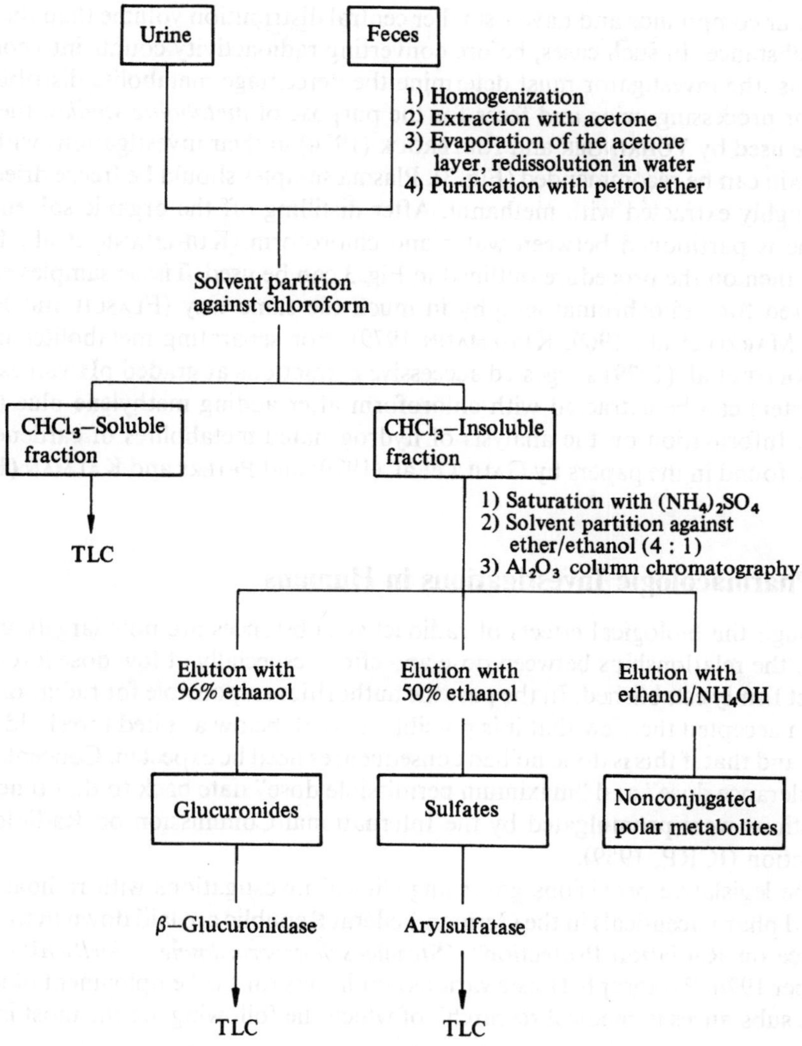

Fig. 3. Extraction procedure for metabolites of cardiac glycosides

1963; KUHLMANN et al. 1979; WEGNER and WINKELMANN, 1970). When dissolving tissue in detergents (MARZO et al., 1969) or acids (MAHIN and LOFBERG, 1966) it is important to allow sufficient time for chemiluminescence and phosphorescence effects to subside.

Feces samples should be homogenized, extracted with acetone, centrifuged, and decolorized with H_2O_2 (RIETBROCK et al., 1975). As in the case of tissue samples, it is advisable to use scintillation "cocktails" with quench additives to reduce chemiluminescence and to establish blank values with equivalent nonradioactive materials.

When metabolites are present, pharmacokinetic analysis of data for total blood radioactivity concentration often leads to erroneous interpretations. One important source of error lies in the fact that most of the metabolites of cardiac glycosides are polar compounds and have a smaller central distribution volume than the original substance. In such cases, before converting radioactivity counts into concentrations, the investigator must determine the percentage metabolite distribution.

For processing urine and feces for the purpose of *metabolite studies*, the procedure used by VÖHRINGER and RIETBROCK (1974) in their investigations with ^{3}H-digitoxin can be recommended (Fig. 3). Plasma samples should be freeze dried and thoroughly extracted with methanol. After distilling off the organic solvent, the residue is partitioned between water and chloroform (KUHLMANN et al., 1974). From then on the procedure outlined in Fig. 3 can be used. Tissue samples can be prepared for radiochromatography in much the same way (FLASCH and HEINZ 1976; MARZO et al., 1969; KUHLMANN 1979). For separating metabolites in bile SCHMOLDT et al. (1979) suggested successive extractions at graded pH values. Sulfate esters can be extracted with chloroform after adding methylene blue (ROY, 1956). Information on the analysis of hydrogenated metabolites of cardenolides will be found in the papers by GAULT et al. (1979) and PETERS and KALMAN (1978).

G. Pharmacologic Investigations in Humans

Although the biological effects of radioactive substances are now largely understood, the relationships between dose and effect, especially at low dose levels, are not yet firmly established. In the past the authorities responsible for radiation protection accepted the view that it is possible to work below a stated threshold of an effect and that if this is done no bad consequences need be expected. Concepts such as "tolerance dose" and "maximum permissible dose" date back to that time. The definitions were promulgated by the International Commission on Radiological Protection (ICRP, 1959).

The legislative provisions governing clinical investigations with radioactively labeled pharmaceuticals in the German Federal Republic are laid down in the "Ordinance on Radiation Protection" *(Strahlenschutzverordnung – StrlSchV)* of 13 October 1976. Paragraph 41 lists various conditions for the "employment of radioactive substances in medical research" of which the following are the most important

a) A committee of experts has to certify that the existing research results on the drug are inadequate and that the use of radioactive labeling is necessary
b) The lowest practicable radioactivity and the radionuclide with the least radiotoxicity must be employed
c) The number of experimental subjects must not be more than absolutely necessary
d) The radiation exposure of any individual subject must not exceed 500 mrem per year
e) The doctor in charge of the investigation must have at least two years' experience in the administration of radioactive substances to human beings
f) The enlistment of volunteers under the age of 50 years is subject to certain restrictions
g) The subjects must receive full information regarding the trial and must give their written consent beforehand

As regards clinical trials with radioisotope labeled pharmaceuticals, the reader is refered to paragraph 83 of the provisions in the "Act for the reorganization of the law governing pharmaceutical products" *(Gesetz zur Neuordnung des Arzneimittelrechts)* dated 24 August 1976. Interpretations of the regulations quoted in this chapter are formulated in a circular from the German Federal Ministry for Home Affairs *(Bundesministerium des Innern)* dated 20 July 1976 (*sRS* II 1–515 260/2).

Radiation exposure can be calculated by the method given by HINE and BROWNELL (1956) or from the ICRP publication (1959, 1978).

$$D_{\beta(\infty)} = 73.8 \times \bar{E}_\beta \times \Sigma C_i \times T_i .$$

$D_{\beta(\infty)}$ is the total absorbed β-particle dose in rads received by the tissue during complete decay or elimination of a radioisotope with an effective half-life of T days. E_β is the mean energy of the β-radiation in MeV, and C_i is the initial concentration of the component (μCi/g). The effective half-life can be calculated from the physical half-life (T_p) and the biological half-life (T_b).

$$T = \frac{T_p \times T_b}{T_p + T_b}.$$

MILLER (1971) gives examples of the calculations for various drugs. For instance, an investigation with 100 μCi ^3H-digoxin carried out in a subject weighing 70 kg will result in radiation exposure (body dose) of approximately 1 mrad. Figures for the radiation exposure of individual organs can be calculated in a similar way with the aid of pharmacokinetic data (WANG and ROBBINS, 1956).

References

Bayly, R.J., Evans, E.A.: Storage and stability of compounds labeled with radioisotopes. Part II. J. Comp. *3*, 349–374 (1967)

Bayly, R.J., Evans, E.A.: Stability and storage of compounds labeled with radioisotopes. J. Comp. *2*, 1–34 (1966)

Boutagy, J., Thomas, R.: Cardenolide Analogues III. Synthesis of C17α- and C17β-(α,β-unsaturated), esters, ketones, nitriles and related derivatives from digixogenin. Aust. J. Pharm. Sci. *NS2*, 9–20 (1973)

Bretschneider, H.J., Doering, P., Eger, W., Haberland, G., Kochsiek, K., Mercher, H., Scheler, F., Schulze, G.: Arterielle Konzentration, arterio-venöse Differenz im Coronarblut und Organverteilung von ^{14}C-markiertem Lanatosid C nach rascher intravenöser Injektion. Naunyn-Schmiedebergs Arch. Exp. Path. Pharmak. *244*, 117–144 (1962)

Carvalhas, M.L., Figueira, M.A.: Comparative study of thin-layer chromatographic techniques for separation of digoxin, digitoxin and their metabolites. J. Chromatogr. *86*, 254–260 (1973)

Catch, J.R.: Purity and analysis of labelled compounds. Review 8. The Radiochemical Centre Amersham, England 1968

Di Carlo, F.J.: Goodbye Wilzbach. Drug Metab. Rev. *9*, 1 (1979)

Dobbs, H.E.: Oxygen flask method for the assay of tritium, carbon-14 and sulfur-35 labeled compounds. Anal. Chem. *35*, 783–786 (1963)

Evans, E.A.: Tritium and its compounds, pp 306–379. London: Butterworths 1966

Evans, E.A.: Self-decomposition of radiochemicals. Review 16. The Radiochemical Centre Amersham, England 1976.

Figge, K., Piater, H., Ossenbrüggen, H.: Radio-Schichtchromatographie schwacher β-Strahler. GIT-Fachz. Lab. *14*, 900–910 (Teil I), 1013–1026 (Teil II) (1970)
Flasch, H., Heinz, N.: Konzentration von Herzglykosiden im Myocard und im Gehirn. Arzneim. Forsch. *26*, 1213–1216 (1976)
Flaskamp, E., Budzikiewicz, H.: Darstellung von spezifisch D- und ^{18}O-markierten Cardenoliden. Z. Naturforsch. *32b*, 187–192 (1977)
Gault, M.H., Sugden, D., Maloney, C., Ahmed, M., Tweeddale, M.: Biotransformation and elimination of digoxin with normal and minimal renal function. Clin. Pharmacol. Ther. *25*, 499–513 (1979)
Geiling, E.M.K., Kelsey, F.E., McIntosh, B.J., Ganz, A.: Biosynthesis of radioactive drugs using carbon-14. Science *108*, 558–559 (1948)
Geiling, E.M.K.: Biosynthesis of radioactive medicinally important drugs with special reference to digitoxin. Trans. Assoc. Am. Physicians *63*, 191–195 (1950)
Geller, L.E., Silberman, N.: The effect of chemical impurities on the stability of labelled steroids. Steroids *9*, 157–161 (1967)
Geller, L.E., Silberman, N.: Some factors involved in the decomposition of labelled steroids on storage. J. Comp. *5*, 66–71 (1969)
Gleit, C.E.: Electronic apparatus for ashing biologic specimens. Am. J. Med. Electronics *2*, 112–114 (1963)
Haberland, G., Maerten, G.: Spezifischer Deuterium- und Tritium-Austausch in Cardenoliden und Cardenolidglykosiden. Naturwissenschaften *56*, 516 (1969)
Haberland, G., Maerten, G.: Verfahren zur Herstellung von durch Deuterium oder Tritium substituierten Cardenoliden und Cardenolidglykosiden. Deutsches Patent (FRG) Offenlegungsschrift 1.959.064 (1971)
Hine, G.J., Brownell, G.L.: Radiation Dosimetry. New York: Academic Press 1956
Hodenberg, A.V., Kleimisch, W., Vollmer, K.-O.: Metabolismus und Pharmakokinetik von Piprozolin bei Ratte, Hund und Mensch. Arzneim. Forsch. *27*, 508–511 (1977)
ICRP Publication 2: Report of Committee II on permissible dose for internal radiation. Publication of the International Commission on Radiological Protection. Oxford: Pergamon Press 1959
ICRP Publication 10 and 10a (Deutsche Ausgabe des Bundesgesundheitsamtes Berlin): Ermittlung der Körperdosis bei beruflich strahlenexponierten Personen nach Inkorporation radioaktiver Stoffe. Stuttgart: G. Fischer 1978
Kaiser, F.: Chromatographische Analyse der herzwirksamen Glykoside von Digitalis-Arten. Arch. Pharm. *299*, 263–274 (1966)
Kartwig, T., Kobosil, P.: Zur Trennung der Digitalis-Cardenolide mit Hilfe der Hochleistungs-Dünnschichtchromatographie. J. Chromatogr. *138*, 238–242 (1977)
Kuhlmann, J., Abshagen, U., Rietbrock, N.: Pharmacokinetics and metabolism of digoxigenin-mono-digitoxoside in man. Eur. J. Clin. Pharmacol. *7*, 87–94 (1974)
Kuhlmann, J., Rietbrock, N., Schnieders, B.: Tissue distribution and elimination of digoxin and methyldigoxin after single and multiple doses in dogs. J. Cardiovasc. Pharmacol. *1*, 219–234 (1979)
Mahin, D.T., Lofberg, R.T.: A simplified method of sample preparation for determination of tritium, carbon-14, or sulfur-35 in blood or tissue by liquid scintillation counting. Anal. Biochem. *16*, 500–509 (1966)
Marzo, A., Sardini, D., Merlo, L., Marchetti, G.: Quantitative determination of tritium-labelled ouabain in organs and biological fluids by the liquid scintillation technique. Biochim. Biol. Sper. *8*, 263–271 (1969)
Miller, J.P.: Health hazards of radioactive pharmaceuticals. In: Radionuclides in pharmacology. Cohen, Y. (ed.), Vol. II, pp. 883–908. Oxford: Pergamon Press 1971
Nover, L., Jüttner, G., Noack, S., Baumgarten, G., Luckner, M.: Über die Beziehung zwischen chemischer Struktur und chromaotgrafischem Verhalten bei Herzglykosiden. V. Mitteilung. Dünnschichtchromatographische Untersuchungen an Herzglykosiden und ihren Geninen. J. Chromatogr. *39*, 419–449 (1969)
Okita, G.T., Kelsey, F.E., Walaszek, E.J., Geiling, E.M.K.: Biosynthesis and isolation of carbon-14 labelled digitoxin. J. Pharmacol. Exp. Ther. *110*, 244–250 (1954)

Peters, U., Kalman, S.M.: Dihydrierte Metaboliten des Digoxins: Klinische Bedeutung und Nachweisverfahren. Z. Kardiol. *67*, 342–345 (1978)
Prydz, S., Melö, T.B., Eriksen, E.L., Koren, J.F.: Tritium detection by β-radioluminescence in standard media for thinlayer chromatography. J. Chromatogr. *47*, 157–175 (1970)
Rabitzsch, G.: Separation of the cardiac glycosides, digitoxin and digoxin from their 20.22-dihydro derivatives by multiple thin-layer chromatography on cellulose films. J. Chromatogr. *35*, 122–125 (1968a)
Rabitzsch, G.: Untersuchungen zur quantitativen Radiodünnschicht-Chromatographie tritiummarkierter Verbindungen. J. Chromatogr. *37*, 476–486 (1968b)
Rabitzsch, G.: Spezifische Tritierung von Herzglykosiden unter Wilzbach-Bedingungen. Naturwissenschaften *56*, 328 (1969)
Randerath, K.: An evaluation of film detection methods for weak β-emitters, particularly tritium. Anal. Biochem. *34*, 188–205 (1970)
Rietbrock, N., Abshagen, U., Bergmann, K.v., Rennekamp, H.: Disposition of β-methyldigoxin in man. Eur. J. Clin. Pharmacol. *9*, 105–114 (1975)
Rochlin, P.: Self-decomposition of carbon-14 labelled organic compounds. Chem. Rev. *65*, 685–696 (1965)
Roesch, A., Koch, K., Schaumann, W.: β-Methyldigoxin V. Protein binding, tissue distribution and extra-cardiac effects in rats and mice. Naunyn Schmiedebergs Arch. Pharmacol. *279*, 211–226 (1973)
Roy, A.B.: The enzymic synthesis of steroid sulphates. Biochem. J. *63*, 294–300 (1956)
Ruben, S., Kamen, M.D.: Radioactive carbon of long half-life. Phys. Rev. *57*, 549 (1940)
Schenk, G., Albrecht, H.P., Lietz, H.: Radioaktive Markierung von Meproscillaridin in der C-19-Methylgruppe mit Tritium. Arzneim. Forsch. *28*, 518–529 (1978)
Schmoldt, A., Benthe, H.F., Haberland, G.: Impaired biliary excretion of digitoxin and its metabolites after treatment with polychlorinated biphenyls. Toxicol. Appl. Pharmacol. *47*, 483–491 (1979)
Segel, K.H.: Die Markierung von Herzglykosiden mit Tritium. Acta Biol. Med. Ger. Suppl. *1*, 202–207 (1961)
Sheppard, G.: The radiochromatography of labelled compounds. Review 14. The Radiochemical Centre Amersham, England 1972
Storstein, L.: Studies on digitalis IV. A method for thin-layer chromatographic separation and determination of digitoxin and cardioactive metabolites in human blood and urine. J. Chromatogr. *117*, 87–96 (1976)
Thomas, R., Boutagy, G., Gelbart, A.: Synthesis and biological activity of semisynthetic digitalis analogs. J. Pharm. Sci. *63*, 1649–1683 (1974)
Vöhringer, H.F., Rietbrock, N.: Metabolism and excretion of digitoxin in man. Clin. Pharmacol. Ther. *16*, 796–806 (1974)
Wang, C.C., Robbins, L.L.: Biological and medical effects of radiation. In: Radiation Dosimetry. Hine, G.J., Brownell, G.L. (eds.), pp. 125–152. New York: Academic Press 1956
Wartburg, A.v., Kalberer, F., Rutschmann, J.: Tritium-labelled cardiac glycosides: Digoxin-[12α-^{3}H]. Biochem. Pharmacol. *14*, 1883–1889 (1965)
Wegner, L.A., Winkelmann, H.: Die Verbrennung ^{14}C- oder ^{3}H-haltiger Proben als Vorstufe zur LS-Messung. Atompraxis *16*, 1–7 (1970)
Werbin, H., Chaikoff, I.L., Imada, M.R.: Rapid sensitive method for determining ^{3}H-water in body fluids by liquid scintillation spectrometry. Proc. Soc. Exp. Biol. Med. *102*, 8–12 (1959)
Wilson, A.T., Spedding, D.J.: Detection of tritium on paper and thin-layer chromatograms. J. Chromatogr. *18*, 76–80 (1965)
Wilzbach, K.E.: Tritium-labeling by exposure of organic compounds to tritium gas. J. Am. Chem. Soc. *79*, 1013 (1957).
Zelnik, R., Ziti, L.M., Guimaraes, C.V.: A chromatographic study of the bufadienolides isolated from the venom of the parotid-glands of Bufo paracnemis Lutz. J. Chromatogr. *15*, 9–14 (1964)

CHAPTER 5

Radioimmunologic Methods

K. STELLNER

A. Radioimmunoassay

I. Basic Principles

Based on the work of YALOW and BERSON (1959) on the detection of insulin, radioimmunoassay (RIA) has now developed to be the best and most versatile method for the detection of trace amounts of biologically active substances (hereinafter termed antigens). To enable the detection of an antigen using RIA, the following conditions must be fulfilled

a) Availability of purified antigen for standardization
b) Availability of antibodies reacting specifically with the antigen to be determined
c) Availability of radioactively labeled antigen with sufficient specific activity
d) Availability of a system allowing separation of the antibody-bound and free antigen.

In the assay itself, a constant amount of antibody (which is selected in order to bind only part of the amount of added antigen) is incubated with a constant amount of tracer and the antigen to be determined. In the course of the competitive reaction, more labeled antigen will be bound by the antibody with decreasing amounts of unlabeled antigen present in the system, and vice versa, according to the law of mass action. After a set time – ideally at equilibrium – the antibody-bound antigen is separated from the free antigen and the radioactivity in one of the fractions measured. In order to determine the unknown concentration of the antigen from the measured count rates, it is necessary to set up a calibration curve. This requires that standard and unknown antigen behave identically in their ability to displace labeled antigen from an antibody–labeled-antigen immunce complex, but identical behaviour of tracer and unknown or standard antigens is not necessary.

It should be stressed that the specificity of the underlying reaction is immunochemical and unrelated to the biologic properties of the antigen.

The detection limit of competitive RIA is determined by the affinity constant of the antibody; this is usually in the range 10^9–10^{10} l/mol. As a general rule of experience, the detection limit is approximately one-tenth of the reciprocal of the affinity constant, consequently concentrations in the picomolar range can be detected. In order to reach this detection limit in practice, the concentration of tracer used must also be of this order of magnitude, i.e., high specific activities are necessary (BERSON et al., 1968; SKELLEY et al., 1973; SÖNKSEN, 1974; PARKER, 1976; CHARD, 1978; BUTLER, 1978a).

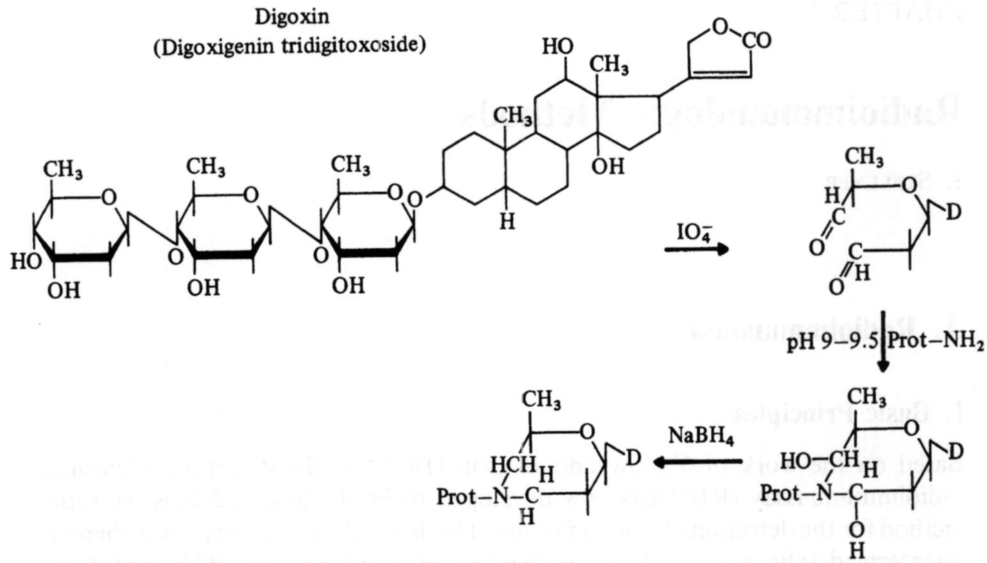

Fig. 1. Proposed scheme of conjugation of digoxin to protein carrier. D = digoxigenin didigitoxoside. (BUTLER and CHEN, 1967)

II. Antibodies

1. Immunogens and Immunization

Cardiac glycosides are relatively small molecules with molecular weights between 500 and 1,000 d. They are too small to be immunogenic by themselves. To obtain antibodies it is hence necessary to couple them as haptens to protein carriers.

BUTLER and CHEN (1967) were the first to succeed in obtaining antibodies against digoxin. The method of ERLANGER and BEISER (1964) was used to prepare the immunogen. Both OH groups at C3 and C4 of the terminal digitoxose of digoxin are first oxidized with sodium metaperiodate, yielding a dialdehyde derivative. Excess periodate is deactivated by addition of ethylene glycol. The dialdehyde is then coupled to albumin under mildly alkaline conditions, whereby free NH_2 groups, principally the amino groups of lysine in the protein, form an unstable Schiff base type addition product with the dialdehyde. Reduction with sodium borohydride leads to a stable end product with a proposed structure shown in Fig. 1. Finally, low molecular weight reaction products and unconjugated digoxin are removed by a series of precipitation and dialysis steps; see Table 1.

2. Characterization

Direct evidence for the presence of digoxin-specific antibodies was obtained in equilibrium dialysis experiments employing tritiated digoxin (BUTLER and CHEN, 1967). The digoxin specificity of the antisera was confirmed by the demonstration that digoxin is a potent inhibitor of the binding of tritiated digoxin by anti-BSA-

Table 1. Preparation of protein–digoxin conjugate (from SMITH et al., 1970, based on preparation of BUTLER and CHEN, 1967)

1) Dissolve 436 mg (0.056 mmol) digoxin in 20 ml absolute ethanol
2) Add dropwise 20 ml of 0.1 *M* sodium periodate and stir for 25 min
3) Add 0.6 ml of 1 *M* ethylene glycol and stir for 5 min
4) Add this reaction mixture dropwise to 560 mg HSA in 20 ml H_2O at pH 9.5 maintain pH by dropwise addition of 2 ml of 5% K_2CO_3
5) After 45 min add 0.3 g sodium borohydride, freshly dissolved in 20 ml H_2O
6) 3 h later lower pH to 6.5 by addition of 7.6 ml of 1 *M* formic acid, 1 h later raise pH to 8.5 by addition of 1.5 ml of 1 *M* NH_4OH
7) Dialyse the reaction mixture against H_2O overnight
8) Lower pH to 4.5 by addition of 2.8 ml of 0.1 *N* HCl, keep 1 h at room temperature and 4 h at 4 °C
9) Centrifuge suspension at 4 °C for 1 h and discard supernatant
10) Dissolve precipitate in 5 ml of 0.15 *M* $NaHCO_3$, dialyse against H_2O for 4 days
11) Lyophilize the solution remaining in the dialysis bag (yield: 461 mg)

digoxin serum, and far more effective than digitoxin or ouabain. In detailed studies SMITH et al. (1970) showed that average intrinsic affinity constants of 1.7×10^{10} l/mol could be reached. The antisera tested reacted principally with the C, D, and lactone rings; compounds with the steroid nucleus of digoxin such as digoxigenin and deslanoside reacted almost identically with the antibody, while only a small change such as the loss of the OH group in the C12 position that occurs in digitoxin reduced the cross reactivity to about one-tenth. Increasing substitution of chemical groups on the steroid nucleus reduced the ability of the compound to bind to the antibody. Thus the corticosteroids and estrogens cross reacted less than did progesterone. SMITH also showed that specificity for antigenic determinants of the steroid nucleus tended to increase with time following immunization. On the other hand there was an apparent decrease in discrimination between digoxin and deslanoside.

Anti-digitoxin antibody (OLIVER et al., 1968) was found to bind digoxin only one-tenth as well as digitoxin, again implying an important role in determining antibody–hapten binding of the structure most distal to the site of coupling to carrier protein. PARK et al. (1973), using a commercially available RIA system report interference with progesterone, cortisone, testosterone, cortisol, and cholesterol to an extent that falsely high digoxin levels might result in the patient receiving massive doses of steroid medication. In an assay very similar to that of SMITH, LARBIG and KOCHSIEK (1972) show identical behavior of digoxin, β-acetyldigoxin, and β-methyldigoxin. Of the metabolites, digoxigenin bis-and monodigitoxoside cross react almost completely, digoxigenin to a very small extent less. The same results were obtained by STOLL et al. (1972), – with a slightly less cross reacting digoxigenin – using a modified commercially available procedure, essentially the same as that reported by SMITH. MARCUS et al. (1975) even report a slightly (but definitely higher) affinity of digoxigenin bis- and monodigitoxoside as compared with digoxin, also in a SMITH-type assay. Dihydrodigoxin was about ten times less reactive than digoxin. In a similarly constructed digitoxin assay, analogous reactivities of the corresponding digitoxin metabolites were found. PHILLIPS (1973) found considerable cross reaction with levels of spironolactone, aldactone, prednisone, prednisolone, and digitoxin which could be encountered in vivo. Altered reaction conditions, especially a "cold" preincubation, influenced the degree of cross reaction. ZEEGERS et al. (1973) also reported unacceptably high digoxin values in a commercial test after spironolactone dosage in vivo. HUFFMAN (1974) has suggested that canrenone, a major active metabolite, is responsible for the interference. SILBER et al. (1979) report that spironolactone metabolites other than canrenone must be involved, stressing that assay specificity should be established by using digoxin-free serum from patients ingesting spironolactone and not by using spironolactone or canrenone-spiked digoxin-free serum.

In the course of the wider distribution of commercial RIA kits, a number of preliminary evaluations have been published confirming (LICHEY et al., 1977) or not confirming the interference by spironolactone and metabolites (RAVEL, 1975; DOERING and BLÜMEL, 1978; MÜLLER et al., 1978). The same holds for dihydrodigoxin, for which negligible (LADER and JOHNSON, 1974) or significant (KRAMER et al., 1976) values have been reported.

Detailed predictions of the specificity of an anti-digoxin antiserum cannot be made. It is therefore essential to conduct extensive cross reactivity studies before using the antiserum. Antisera can be raised against digoxin, which as a rule react 100% with the cardioactive metabolic breakdown products (i.e., digoxigenin bis- and monodigitoxosides), less strongly with digoxigenin, negligibly with the cardioinactive dihydrodigoxin, and with a panel of physiological steroid hormones plus spironolactone and derivatives. Digitoxin will always cross react between 1% and 10%, which is however not a serious limitation; alternatively this cross reaction can be exploited to set up a digitoxin assay using suitable anti-digoxin antisera (SMITH, 1970).

When examined spectrophotometrically in concentrated H_2SO_4, it was estimated that the HSA-digoxin conjugate contained an average of 6.7 digoxin residues per molecule of HSA (SMITH et al., 1970). The original preparation of BUTLER and CHEN (1967) had only an average ratio of 1.4. Most antisera for digoxin RIA, subsequently described by other authors (for example CHAMBERLAIN et al., 1970; HOESCHEN and PROVEDA, 1971; LARBIG and KOCHSIEK, 1971; OLIVER et al., 1971) were obtained in the same way.

Digoxin was also conjugated to polylysine by a modification of the periodate oxidation technique yielding an immunogen with up to 11 digoxin residues per molecule polylysine (EVERED et al., 1970). In our case (BATZ et al., 1977; STELLNER et al., 1975) a BSA-digoxin conjugate, carrying a glutaryl bridge between carrier and intact terminal digitoxose of digoxin, could be successfully used. With the periodate oxidation method, digoxigenin bis-digitoxoside conjugates are strictly speaking obtained.

To raise antibodies against digitoxin its aglycone, digitoxigenin, was allowed to react with succinic anhydride to yield 3-*O*-succinyldigitoxigenin. This derivative was then conjugated to HSA or BSA by the carbodiimide and mixed anhydride methods (OLIVER et al., 1968). The conjugates contained at least 5 or 13 molecules digitoxigenin per molecule of protein respectively (*m*-dinitrobenzene assay).

Ouabain has been conjugated via its glycosidic rhamnose residue to the free NH_2-groups of poly (DL-alanyl)-HSA by the periodate oxidation method (SMITH, 1972; VERSPOHL et al., 1978). Concerning the preparation of immunogens with the less important cardiac glycosides, see the review of BUTLER (1978 a).

To elicit antibodies for digoxin, rabbits were injected with BSA-digoxin, 1 mg/ml in complete Freund's adjuvant, three injections of 0.4 ml in the foot pads over a four week period. A single booster injection was given intramuscularly. Terminal bleedings were taken about 10 weeks later (BUTLER and CHEN, 1967). Essentially the same schedule was followed by SMITH et al. (1970), with the higher substituted conjugate. Another series of rabbits was also immunized for a longer period (up to 97 weeks) during which intramuscular injections were given in intervals of 1–2 weeks.

III. Tracers

1. General Remarks

The original digoxin RIA method (SMITH et al., 1969) employed ^{3}H-digoxin, tritium labeled in the 12α-position with a specific activity of 3.2 Ci/mmol. Immunologically this represented the ideal tracer due to its chemical identity with unlabeled digoxin, however there are several practical disadvantages. Owing to the low specific activity long counting times are required. Hemolysis products and bilirubin cause color quenching (CERCEO and ELLOSO, 1972), which makes quench corrections obligatory, and chemiluminescence may interfere (BUTLER, 1971). ^{3}H as a β-emitter requires expensive scintillation "cocktails," so that the unit cost per assay is higher compared with labels allowing γ-counting. For these reasons virtually only ^{125}I-labeled cardiac glycosides are used today. The concomitant shorter shelf life (2–3 months for specific activities in the range 50–2,300 Ci/mmol) and the necessity for more frequent labeling does not represent a serious restriction.

2. Conjugates for Labeling with ^{125}I

Since it is impossible to iodinate digoxin or digitoxin directly, and furthermore undesirable to block the C and D rings or the unsaturated β-lactone ring, which are essential for the reaction with antibody, a derivative capable of iodination must first be synthesized. Table 2 lists the most important precursors described, which then enable a direct iodination.

It is basically possible to couple onto the C3 position of the genin or onto one or both of the OH groups in the C3 and C4 position at the terminal sugar of digoxin/digitoxin itself. The terminal digitoxose can either retain or lose its ring structure, in which case digoxin/digitoxin or digoxigenin/digitoxigenin bis-digitoxoside derivatives are formed. In most cases tyrosine derivatives have been introduced, occasionally histidine derivatives also, mostly via additional spacer groups using coupling methods known from the literature. The first assay for determination of digitoxin (OLIVER et al., 1968), also the first RIA for any cardiac glycoside, employed 3-*O*-succinyldigitoxigenin tyrosine methyl ester, obtained by coupling of tyrosine methyl ester (TME) to 3-*O*-succinyldigitoxigenin using the mixed anhydride method. Minimum specific activities of 400 mCi/mg and 98% immunoreactivity have been reported. In an analagous fashion tyrosine derivatives of digoxigenin (RUTNER et al., 1971 a; BARBIERI and GANDOLFI, 1977) and digitoxigenin (RUTNER et al., 1971 b) have been produced, and also histamine digoxigenin derivatives (PAINTER, 1978) using the carbodiimide method. Digoxigenin previously oxidised in the 3 position was reacted as the carboxymethyl derivative with TME in the same way (LADER et al., 1972 b). OSLAPAS and HERRIN (1976) reacted tyramine with 3-chloroformyldigoxigenin.

The terminal digitoxose in digoxin has been oxidised – using periodate – to the dialdehyde and subsequently directly (WILKINSON, 1974) or via a lysine bridge (PIASIO and WOISZWILLO, 1975) coupled to TME or directly to histamine (BLAZEY, 1977). Reduction with sodium borohydride led in both cases to stable compounds.

The dialdehyde has also been converted to the Di-*O*-carboxymethyloxime followed by coupling to histamine or TME (TOVEY and FINDLAY, 1976) or after oxidation to a dityramide derivative (TERRANCE, 1976).

Imidazoleacetyldigoxin (POLITO, 1976), a TME derivative after its conversion to the isocyanate (EISENHARDT et al., 1977) and TME-succinyldigoxin (COOMBES, 1976), all retaining the terminal digitoxose, have been described. Also, a tyramine derivative was reacted with CNBr-activated digoxin (WEILER and ZENK, 1979) with a claimed shelf life of up to 6 months. In our hands, a commercial grade hydroxyethyl tyrosinamide derivative with an additional glutaryl spacer has provided excellent results since 1976 (BATZ et al., 1977, 1979).

Table 2. Digoxin/digitoxin derivatives suitable for iodination

Digoxigenin (R)/digitoxigenin (R′) derivatives	References
HO⟨◯⟩$CH_2CHNHCOCH_2CH_2COOR'$ | $COOCH_3$	OLIVER et al. (1968)
HO⟨◯⟩$CH_2CHNHCOCH_2CH_2COOR'$ or R | COOH	RUTNER et al. (1971 a, b); BARBIERI and GANDOLFI (1977)
HO⟨◯⟩$CH_2CHNHCOCH_2ON{=}R$ | $COOCH_3$	LADER et al. (1972 b), WILKINSON (1976)
HO⟨◯⟩$CH_2CHNHCOCH_2CH_2CH_2COOR$ | COOH	BERNSTEIN et al. (1979a)
HO⟨◯⟩CH_2CHCH_2COOR | CH_2COOH	BERNSTEIN et al. (1979b)
HO⟨◯⟩$CH_2CH_2NHCOOR$	OSLAPAS and HERRIN (1976)
(imidazolyl, N / N–H ring)$CH_2CH_2NHCOCH_2CH_2COOR$	PAINTER (1978)

Digoxigenin (R_1) and digitoxigenin-bis-digitoxoside (R_1') derivates	
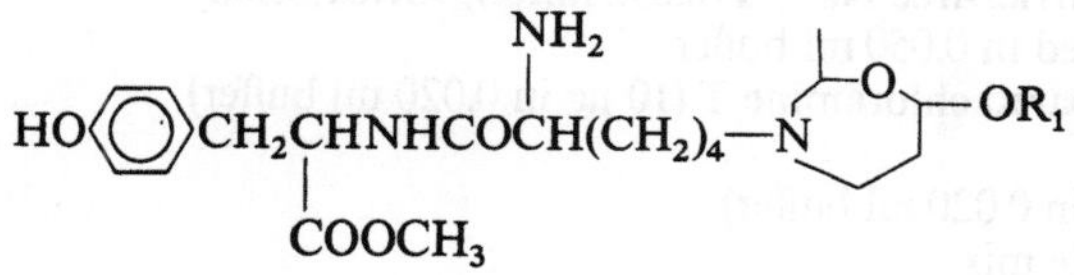	Piasio and Woiszwillo (1975)
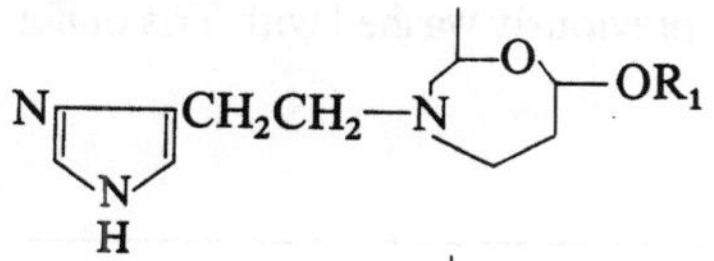	Blazey (1977)
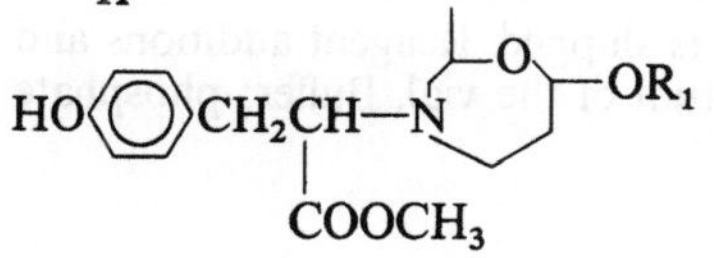	Wilkinson (1974)
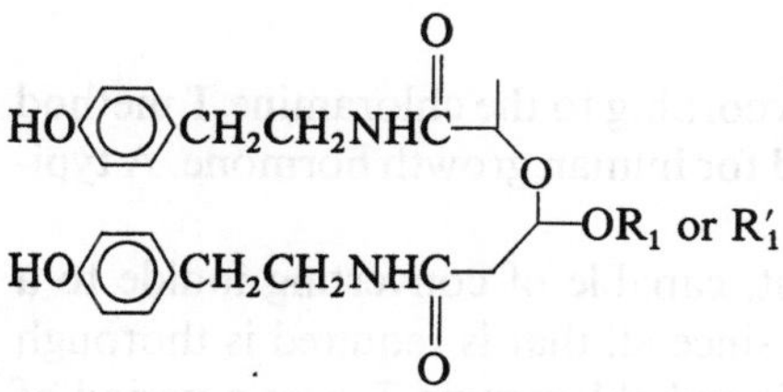	Terrance (1976)
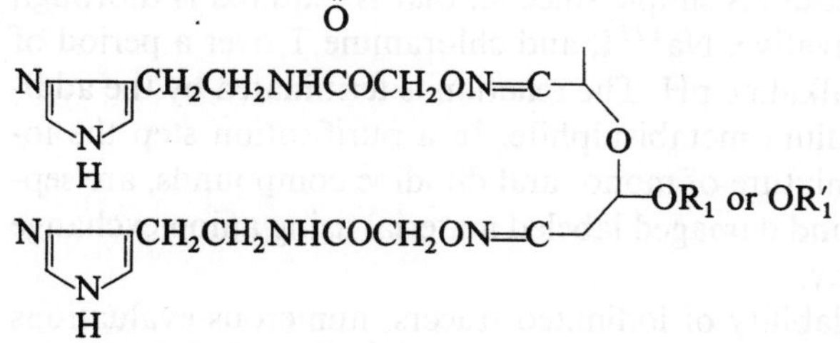	Tovey and Findlay (1976)
Digoxin (R_2) and digitoxin (R_2') derivates	
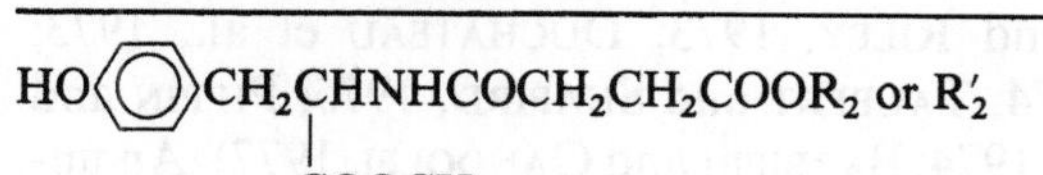	Coombes (1976)
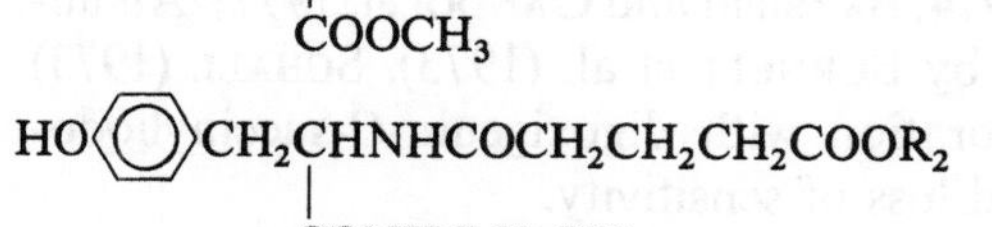	Batz et al. (1977), (1979)
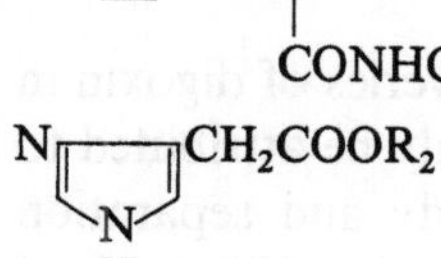	Polito (1976)
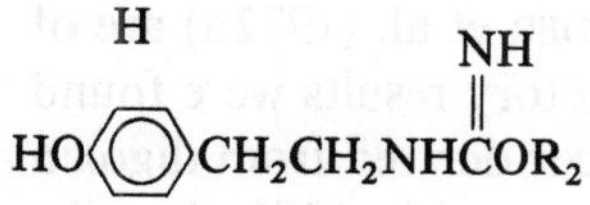	Weiler and Zenk (1979)
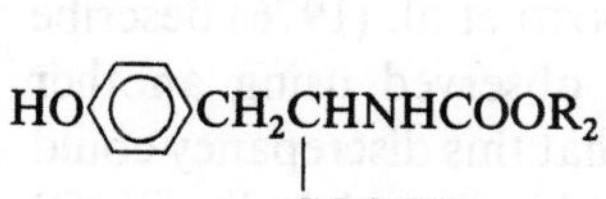	Eisenhardt et al. (1977)

Table 3. Preparation of iodinated digoxin derivative

1) Add 0.030 ml of buffer to 2 mCi of carrier-free $Na^{125}I$ (0.092 nmol), vortex briefly
2) Add 5 µg (5 nmol) derivative, dissolved in 0.050 ml buffer
3) Start reaction by adding freshly prepared chloramine T (10 µg in 0.020 ml buffer)
4) Mix thoroughly for 30 s
5) Add sodium metabisulphite (100 µg in 0.020 ml buffer)
6) Add 0.5 ml buffer/dimethylformamide mix
7) Transfer to 1 × 30 cm column of DEAE-Sephadex A-25, previously washed with Tris buffer pH 7.6. Collect fractions of 1 ml
8) Evaluate tracer
9) Store lyophilized or deep frozen in closed vials

Iodination is done in the sealed vial in which the nuclide is shipped. Reagent additions and transfers are made by injections through the rubber septum of the vial, Buffer: phosphate 0.05 mol/l, pH 7.4)

3. Iodination

The radioiodination is generally performed according to the chloramine T method of HUNTER and GREENWOOD (1962), described for human growth hormone. A typical labeling protocol can be seen in Table 3.

Chloramine T is a potent oxidizing agent, capable of converting iodide to a more reactive form. The procedure is simple since all that is required is thorough mixing of solutions of the derivative, $Na^{125}I$, and chloramine T over a period of approximately 30 s at slightly alkaline pH. The reaction is terminated by the addition of the reducing agent sodium metabisulphite. In a purification step the iodinated derivatives, mostly a mixture of mono- and diiodine compounds, are separated from free iodide, salts, and damaged labeled material using anion exchange – or thin layer chromatography.

Since the commercial availability of iodinated tracers, numerous evaluations have been published, especially in comparison with assays using ^{3}H-digoxin as tracer. The general superiority of ^{125}I-labeled tracers has been emphasized (CERCEO and ELLOSO, 1972; HORGAN and RILEY, 1973; DUCHATEAU et al., 1973; GUTCHO et al., 1973; CHAMBERS, 1974; TAUBERT and SHAPIRD, 1975; PIPPIN and MARCUS, 1976; DREWES and PILEGGI, 1974; BARBIERI and GANDOLFI, 1977). An unfavorable compatibility was described by BURNETT et al. (1973). SCHALL (1973) found a severe immunological deterioration with digoxigenin-*O*-succinyliodotyrisine leading to variable binding and loss of sensitivity.

It should be stressed that results concerning differing recoveries of digoxin in clinical samples using ^{3}H- and ^{125}I-digoxin derivatives can only be attributed to the tracer when all other test conditions (especially antibody and separation method) are identical. In this respect the findings of LADER et al. (1972a) are of interest in the comparison of two ^{125}I tracers. Unsatisfactory results were found with a digoxigenin-based tracer, good results with a tracer derived from digoxin (for structures see WILKINSON, 1974, 1976 in Table 2). SOTO et al. (1976) describe the same phenomenon, although this was no longer observed using another antiserum. KROENING and WEINTRAUB (1976) speculate that this discrepancy could be caused by differing binding of the tracer to thyroxine binding globulin (TBG) and prealbumin. Also, PAINTER and VADER (1979) postulate TBG implication in the previously unexplained interference of radioiodinated haptens, especially pronounced in hypothyroid sera. They could demonstrate that digoxigenin-3-*O*-succinyl iodo-tyrosine, in contrast to the monoiodinated derivative, binds strongly to TBG, the binding being inhibited by the structurally related tetraiodotyrosine (T4). In cases where diiodotyrosine tracers could not be separated out, it is advisable to add the well-known TBG-blocker 1,8-anilinonaphthalenesulfonate or to use iodine-labeled histamine derivatives.

IV. Standards

Owing to the accessibility of pure crystalline cardiac glycosides which are free from potentially cross reactive structural analogues, e.g., digoxin *BP* 68, *USP* XIX, Ph. Eur., and digitoxin *BP* 68, *D*eutsches *A*rzneibuch, 7. Auflage, *USP* XVIII, Ph. Eur., standards can be relatively simply prepared by careful weighing. The substance is dissolved first in ethanol and then further diluted in human serum or plasma.

Owing to the requirement for identity of standard and sample matrix, the use of serum or plasma is essential. The use of aqueous standards would lead to false recovery values in samples. It has been reported that, for example, digoxin dissolved in various, individual human plasmas or sera does not result in identical preparations (ANGGARD et al., 1972; BURNETT et al., 1973). It is therefore necessary to screen plasma or serum for a possible "pseudo-digoxin" content before adding digoxin. Preparations showing lower binding values (pseudo-digoxin concentrations) compared with a previously optimized zero digoxin standard are unsuitable as matrices.

V. Separation Methods

Virtually all generally applicable bound–free separation methods for RIA such as adsorption, chemical or immunologic preciptation of one fraction, or solid phase techniques, have also been described for digoxin/digitoxin. A detailed listing is given in Table 4.

The dextran-coated charcoal method (DCC) – especially in the early phase of the digoxin determination, the double antibody method, ammonium sulfate precipitation, and the solid phase versions have found a wider application, the latter having attained prominence owing to their practicability for the operator.

The first assay for digoxin (SMITH et al., 1969) used DCC in the form described for insulin (HERBERT et al., 1965). DCC would be a good choice in terms of specificity and cost, but some drawbacks have to be considered. One of the basic assumptions for the suitability of a separation method is noninterference in the antigen–antibody reaction. MEADE and KLEIST (1972) and KUNO-SAKAI and SAKAI (1975) were able to demonstrate a definite drop in activity of the bound fraction as a function of the duration of DCC in the reaction mixture. Since digoxin itself is bound to DCC in a matter of seconds, additional digoxin from the digoxin–anti-digoxin complex must have been bound to the DCC (the so-called stripping effect). This effect is dependent on the quality of the antiserum employed (SMITH and HA-

Table 4. Methods of separating bound and free digoxin/digitoxin

Method or material used	References
Dextran-coated charcoal	Smith et al. (1969)
Albumin-coated charcoal	Lader et al. (1972b)
Gelatin-coated charcoal	Brock (1973)
Anion-exchange resin (column)	Brooker et al. (1976)
Anion-exchange resin (strips)	Brown and Lyle (1976)
Sephadex G-25 (column)	Christiansen and Nielsen (1972)
	Ertinghausen et al. (1975)
Gel equilibration (batch)	Greenwood et al. (1974)
Somogyi precipitation ($ZnSO_4 + NaOH$)	Meade and Kleist (1972)
Ammonium sulfate	Drewes and Pileggi (1974)
Sodium sulfite	Dwenger et al. (1976)
Polyethylene glycol	Barrett and Cohen (1972)
	Shaw et al. (1975)
Protein-A-bearing *Staphylococcus aureus* cells	Gauldie et al. (1980)
Double antibody	Oliver et al. (1968)
	Lader and Johnson (1974)
	Berk et al. (1974)
Membrane filtration	Gershman et al. (1972)
Capillary chromatography tubes	Wagner et al. (1979)
Solid phase	
Porous glass beads	Yauerbaum et al. (1973)
	Line et al. (1973)
Agarose	Line et al. (1973)
Sepharose 4B (column)	Boguslaski (1975)
Cellulose	Arndts (1975)
Plastic tubes	Stellner et al. (1975)
	Gutcho et al. (1977)
	Litt (1977)
	Brown et al. (1978)
Capillary tubings	Friedel and Dwenger (1975)
Polymer-coated magnetic Fe_3O_4 particles	Hersh and Yauerbaum (1975)
	Nye et al. (1976)
18-Finned stick as tube insert	Piasio et al. (1979)
Microencapsulated antibodies	Halpern and Bordens (1979)
Polyacrylamide entrapped antibodies	Sjöholm et al. (1976)
Polymerized antibody tablet	Adams (1976)

ber, 1973); the higher the antibody affinity, i.e., the more strongly it binds digoxin, the lower will be the digoxin exchange and the longer the DCC can remain in the system before and after centrifugation. Normally the assay is performed at 4 °C thus further reducing the effect. Before using a new antiserum the phenomenon must be experimentally investigated and if necessary another separation method employed. It is possible that other anomalies caused by unknown plasma factors (Burnett et al., 1973) or proteins in general (Anggard et al., 1972; Brock, 1973) are attributable to the use of DCC (Drewes and Pileggi, 1974).

The first assay for the determination of digitoxin (Oliver et al., 1968) used a second antibody (goat anti-rabbit gamma globulin) to achieve separation. Also donkey anti-rabbit gamma globulin has been used (Lader and Johnson, 1974). In the case of ammonium sulfate a relatively high nonspecific binding in the absence

Table 5. Typical separation procedures

a) Using dextran-coated charcoal
   1) Prepare a suspension containing 2 g activated charcoal, 0.05 g Dextran T-70, 0.28 g K_2HPO_4, 0.055 g $NaH_2PO_4H_2O$, and 1.75 g NaCl in 200 ml distilled water
   2) Provide for continous mixing, with the container in ice during use (magnetic stirrer)
   3) After a 30 min incubation of the primary reaction mixture (0.700 ml PBS, 0.100 ml standard or sample, 0.100 ml of ^{125}I-digoxin solution, and 0.100 ml of an appropriate dilution of antiserum) add 0.500 ml of the dextran-coated charcoal suspension, keep for 5 min at room temperature
   4) Centrifuge at 2,000 *g* for 15 min at 4 °C
   5) Decant immediately the clear supernatant into counting vials
   6) Determine radioactivity

b) Using antibody-coated tubes
   1) Fill polystyrene tubes with 1:8,000 dilution of digoxin antiserum (or gammaglobulin fraction) in phosphate buffer pH 7.4 and incubate for 3 h at room temperature
   2) Aspirate contents, fill tubes with 1% solution of BSA in saline
   3) Aspirate contents and dry tubes overnight at room temperature
   4) Add 0.100 ml of standard or sample and 0.500 ml of ^{125}I-digoxin solution and incubate for 1 h at room temperature
   5) Aspirate contents and discard, wash once with water
   6) Count all tubes in a γ-counter

of antibody, amounting to 20%–30% of the total counts has been reported (Drews and Pileggi, 1974). Sheep anti-digoxin serum has been coupled covalently to porous silica glass of various pore diameters and particle sizes (Yaverbaum et al., 1973). Antisera coupled to glass isothiocyanate retained about 50% of its immunologic activity upon attachment, 7% when coupled to CNBr-activated agarose and only 1.5% when coupled to alkylamino glass (Line et al., 1973).

The simplest test procedure can be achieved by binding the antibody to the test tube in which the reaction is carried out, so that the separation step can be effected simply by aspirating the solution. Digoxin antisera, or the gamma globulin fraction obtained by ammonium sulfate precipitation, have been coupled to polystyrene tubes purely by adsorption (Stellner et al., 1975), or to polypropylene tubes to which a second antibody had previously been fixed (Litt, 1977). The production of anti-digoxin tubes has also been described in the presence of blocking agents such as ouabain, which are assumed to prevent binding of the immunologically active sites to the tube wall, and additionally with denaturing agents such as glycine, which are supposed to prevent aggregation of antibodies during adsorption (Gutcho et al., 1977). A further modification is the covalent coupling using glutaraldehyde of anti-digoxin to BSA which had initially been coupled by adsorption to the lower portion of tubes of 92% polyethylene and 8% acrylic acid (Brown et al., 1978).

The more sophisticated methods of binding antibody to tubes do not show better assay performance or stability compared with tubes produced by mere adsorption of antibody. In view of the complications of preparing large numbers of coated tubes, simple adsorption seems therefore to be the most attractive alternative.

Detailed separation procedures using DCC and antibody-coated tubes respectively, are described in Table 5.

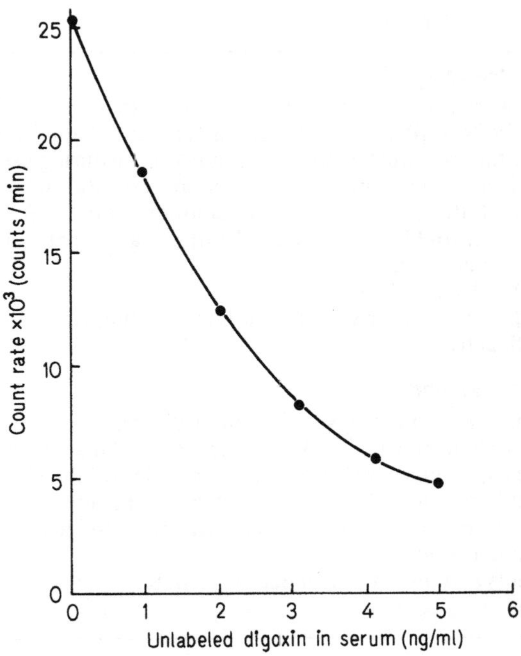

Fig. 2. Typical standard curve for digoxin radioimmunoassay

VI. Assay Performances

The first published RIA for digoxin (SMITH et al., 1969) employed high affinity anti-digoxin antibody, ^3H-digoxin as tracer, and dextran-coated charcoal as separation system. A serum volume of 1 ml was required with a 15 min incubation. After centrifugation of the free digoxin bound to DCC, the antibodies in the supernatant were denatured by heating at 60 °C and centrifuged, followed by measurement of radioactivity in the supernatant. Quench corrections were carried out. The calibration curve ranged from 0.2 to 20 ng/ml digoxin, 0.2 ng/ml being the lower detection limit of the system. Starting from approximately 40% tracer binding in the absence of unlabeled digoxin, 2–3 ng/ml digoxin caused a 50% displacement of tracer. CVs of 3%–4% over the entire range were found. There was complete cross reaction with deslanoside, a compound having an identical steroid aglycon, some cross reaction with digitoxin, and no cross reaction with several steroids. Assay performances in terms of precision, shape of the standard curve, specificity, and accuracy (evaluated in terms of the toxic and nontoxic ranges established), and overall assay time would be very acceptable today.

In contrast modern assays use less serum (0.1 ml or less) and by the almost exclusive use of iodinated tracers the counting process has become much more convenient and less vulnerable (see Sect. a. III). As a consequence of the presently accepted toxic and nontoxic ranges the range of the calibration curve can be restricted to approximately 5 ng/ml digoxin.

A typical standard curve is seen in Fig. 2.

For plotting of the calibration curve and determination of the concentrations of samples, all methods, applicable to RIA can be used (EKINS, 1974; CHARD, 1978). For a detailed assessment of the digoxin/digitoxin parameters see BESCH and WATANABE (1975).

Emergency "STAT" procedures for digoxin with shorter incubation times have been described (GREENWOOD et al., 1974; BESCH and WATANABE, 1975; CHEN et al., 1978).

In various systems digoxin values are occasionally found which do not agree with the clinical picture. These discrepancies, which are not attributable to poor antibody specificity have not yet been completely clarified. Such conflicting data have been reported on the effect of differing albumin concentration in samples on digoxin values (CERCEO and ELLOSO, 1972; BROCK, 1973; HOLTZMAN et al., 1974; HOLTZMAN and SHAFER, 1975; SHAW, 1975; VOSHALL et al., 1975; AHLUWALIA and KUCZALA, 1976; SOTO et al., 1976), also free fatty acids have been suggested as a cause in this respect (SHAW et al., 1977; O'LEARY et al., 1979). See also Sects. III.3 and V. Before routine determinations are initiated with a digoxin RIA, inhibition studies with a panel of normal human sera are essential to show that these do not interfere (BUTLER, 1978 b).

The general capabilities and limitations of digoxin RIA have been summarized (BUTLER, 1972; SMITH, 1976; SHAPIRO et al., 1975; RAVEL, 1976; RAVEL and SCHALL, 1976). In the course of increasing availability of commercial kits, numerous reports have been published with comparisons of results obtained using various kits (e.g., HALL, 1974; KUBASIK et al., 1974 a, b, 1975, 1976; SOTO et al., 1975; MACKINNEY et al., 1975; BELPAIRE et al., 1975; SHAPIRO et al., 1975; VOSHALL et al., 1975; BLEND and FERRY, 1975; MÜLLER et al., 1975, 1976a; RAVEL and ESPINOLA, 1976; KUCZALA and AHLUWALIA, 1976; BOINK et al., 1977; WOOD, 1977; WOOD and WACHTER, 1979; DWENGER et al., 1978; DWENGER andTRAUTSCHOLD, 1978; LARSON et al., 1978; BERGDAHL et al., 1979), in part with considerable differences in terms of precision, accuracy, and specificity. Improvements have been proposed (OJALA et al., 1972; HOFMAN, 1973; CALESNICK and DINAN, 1976; VIJGH, 1975; WITHERSPOON et al., 1978; WOOD, 1977; DWENGER and TRAUTSCHOLD, 1978). Nevertheless there is agreement that with kits employing antisera of high affinity and specificity reliable and reproducible results are obtained (BUTLER, 1978 b). Most of the digoxin determinations worldwide are done with commercially available kits.

Characteristics and criteria for judging the acceptability of methods for measuring digoxin have been made (CARLSON et al., 1975), also for quality control of digoxin RIA (LEUNG et al., 1975; WOODCOCK et al., 1975).

Several organizations offer programs for interlaboratory comparisons in RIA. For example the College of American Pathologists (CAP) included digoxin in its survey program in 1972. Today approximately 1,000 laboratories participate on a regular basis, reporting their results according to methodology. The large number of participants enables kit-specific, statistically valid performance data to be evaluated. After initial discrepancies between the methods (HANSELL, 1976) more uniform results have been reported more and more regularly and the results today are rather satisfactory.

VII. Automation

All manual test methods consist of a series of reagent pipeting steps, possibly with mixing, a variably complicated bound–free separation, the determination of radioactivity in one fraction, and lastly the evaluation of the concentration corresponding to the individual count rate.

It can easily be recognized that in processing larger numbers of samples there is a danger of reduced precision; furthermore, manual procedures are labor intensive and therefore expensive. A partial or total automation is therefore desirable.

Amongst the presently available, more or less completely automated systems are some which also enable the determination of digoxin, together with other parameters.

The Centria system (ERTINGSHAUSEN et al., 1975) consists of three integrated modules: an automated sample–reagent dispenser, an incubator/separator employing centrifugation to initiate and terminate up to 30 radioassay incubations and separations simultaneously, and a γ-counter/computer which can simultaneously count three tubes and converts counts into concentration units. The bound-free separation takes place on prepackaged Sephadex G-25 columns. For clinical evaluation see CASTRO et al. (1976).

The Gammaflow system of BROCKER et al. (1976) employs a column containing an anion-exchange resin adsorbing free digoxin. All reagents pass through the column in a continuous flow system. The self-contained instrument can execute 42 determinations per hour. For clinical evaluation see VALDES et al. (1979).

The Concept 4 automatic radioimmunosassay (JOHNSON et al., 1976; PAINTER and HASLER, 1976) also automates RIA protocols from sample introduction to printing of final results. The system employs 8 × 50 mm antibody-coated polypropylene tubes (see also CARLTON et al., 1978).

GREENWOOD et al. (1977) describe another fully automated RIA for digoxin which uses the continuous flow Autoanalyzer system of Technicon. The antibody is covalently linked to a polymer-coated iron oxide (Enzacryl). The magnetic properties of the solid phase are used to separate free and bound tracer and to facilitate an on-line washing procedure.

It is obvious that automated systems can be cost-effectively utilized only in large clinical laboratories with more than 100 samples per day.

B. Enzyme Immunoassay

I. Introduction

Owing to the utilization of radioactive reagents RIA possesses a number of disadvantages, which have occasioned the search for other marker substances.

Thus, working with RIA requires an official license, a specially designed laboratory is necessary, and waste disposal is complicated and expensive. In particular, expensive isotope counting equipment is needed and the relatively short lifetime of the tracer is restrictive (although this disadvantage is often overestimated). In all these respects improvements can be achieved by substituting an enzyme for the ra-

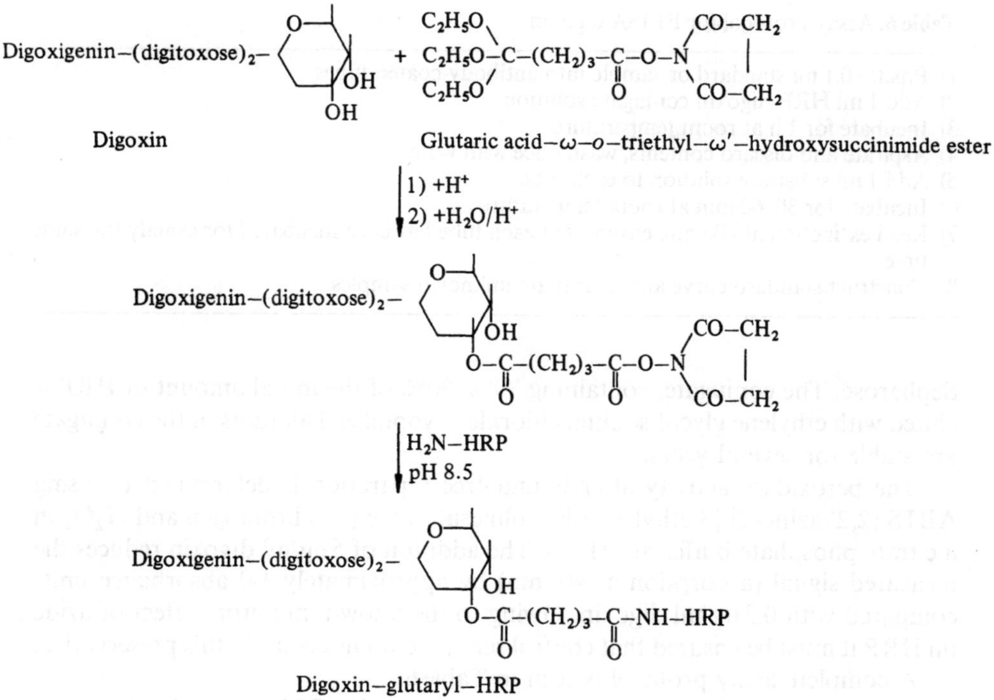

Fig. 3. Procedure for linking digoxin to horseradish peroxidase. (BATZ et al., 1977)

dioactive isotope (SCHUURS and VAN WEEMEN, 1977; WISDOM, 1976; SCHARPE et al., 1976; O'SULLIVAN et al., 1979).

At the present time two enzyme immunoassays have been described in the literature for the determination of digoxin and are commercially available as complete test kits. The first version is completely equivalent in reaction principle to the competitive form of RIA and is usually termed ELISA, or heterogeneous enzyme immunoassay. This is to differentiate ELISA from the other version which does not require a bound–free separation and is hence termed homogeneous enzyme immunoassay.

II. Heterogeneous Enzyme Immunoassay

The assay described here (KLEINHAMMER et al., 1976; STELLNER, 1978) is designed for use with the practical solid-phase tube technique, employing the same polystyrene tubes coated with sheep anti-digoxin antibodies already described for a digoxin RIA (STELLNER et al., 1975). A horseradish peroxidase (HRP)-digoxin conjugate was synthesized using an asymmetrical reactive dicarboxylic ester, whereby an orthoester group reacts with the hydroxyl groups on C3 or C4 of the terminal digitoxose of digoxin. The succinimide ester (a cyanomethyl ester can also be employed) combines with free NH_2 groups of the HRP (BATZ et al., 1977), see Fig. 3.

The immunogen was synthesized in the same fashion using BSA instead of HRP. The HRP conjugate was purified by hydrophobic chromatography on PBA-

Table 6. Assay protocol for ELISA digoxin

1) Pipette 0.1 ml standard or sample into antibody-coated tubes
2) Add 1 ml HRP-digoxin conjugate solution
3) Incubate for 1 h at room temperature
4) Aspirate and discard contents, wash once with water
5) Add 1 ml substrate solution to each tube
6) Incubate for 30–60 min at room temperature
7) Read extinction at 405 nm, ensure that each tube has been incubated for exactly the same time
8) Construct standard curve and quantitate unknown samples

Sepharose. The conjugate, containing 70%–80% of the initial amount of HRP is eluted with ethylene glycol/sodium chloride. Lyophilized aliquots of the conjugate are stable for several years.

The peroxidase activity after bound–free separation is determined by using ABTS (2,2′-azino-di-[3-ethylbenzthiazolinesulfonate (6)] chromogen and H_2O_2 in a citrate phosphate buffer at pH 5.0. The addition of 5 ng/ml digoxin reduces the measured signal (absorption at 405 nm) by approximately 1.0 absorbance units compared with 0.3 ng/ml digoxin. Owing to the known inhibitory effect of azide on HRP it must be ensured that control sera used do not contain this preservative.

A complete assay protocol is seen in Table 6.

Cross reactivity studies in direct comparison with RIA using identical antibody-coated tubes and standards yielded identical values for all relevant compounds with the exception of digitoxin, which showed an approximately two-fold better displacement in the enzyme immunoassay.

Two and a half years production experience has proved the equality of this test with RIA. A series of comparative studies has been published (BORNER and RIETBROCK, 1978; OELLERICH et al., 1978; MUNZ et al., 1979; ONG et al., 1979).

III. Homogeneous Enzyme Immunoassay

In 1975 CHANG et al. announced the development of an EMIT (enzyme multiplied immunoassay technique) for digoxin, which is at present commercially available. Using this technique (RUBENSTEIN et al., 1972) drugs in urine and anticonvulsants in serum can at present be determined. The key substance is a hapten–enzyme conjugate consisting of digoxin coupled to glucose-6-phosphate dehydrogenase (G6PDH) from *Leuconostoc mesenteroides* (EC 1.1.1.49) (ULLMAN and RUBENSTEIN, 1977).

Digoxin is covalently bound near the active site of the enzyme such that a sufficient proportion of the enzyme activity is retained. On binding of anti-digoxin antibodies the enzyme is partially deactivated since the substrates (glucose-6-phosphate and NAD^+) have no (or only restricted) access to the active site. The deactivation is reversed by free digoxin, which competes with the digoxin–enzyme conjugate for antibody binding sites. The activity of the G6PDH is thus directly related to the concentration of digoxin. Since the extent of binding of the antibody to the label is directly reflected in the enzymatic activity, no separation step is required.

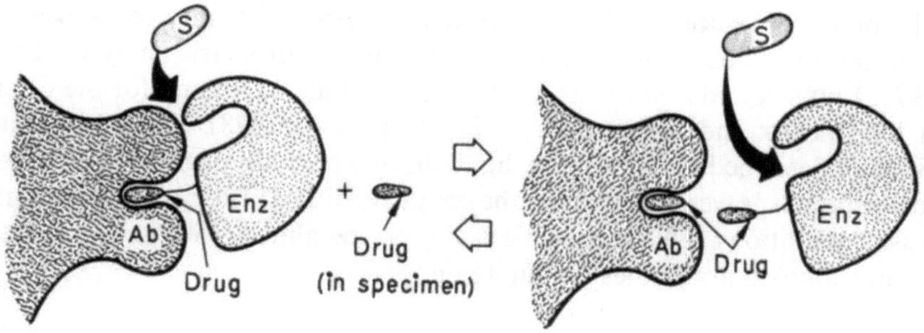

Fig. 4 a, b. Principle of the EMIT homogeneous assay system **a** substrate excluded, enzyme inactive. **b** substrate admitted, enzyme active. (S = substrate, Ab = antibody, Enz = enzyme.)

Table 7. Manual assay protocol for EMIT digoxin

1)	Transfer 0.200 ml of each serum sample or calibrator into test tubes
2)	Dispense 0.050 ml of NaOH solution into each tube
3)	Vortex entire rack of tubes for 10 s and wait 5 min before proceeding
4)	Dispense 1.25 ml of antibody/substrate reagent solution into each tube
5)	Vortex entire rack of tubes for 10 s and wait 15 min before proceeding
6a)	Dispense 0.500 ml of digoxin–enzyme conjugate buffer solution to first assay tube
b)	Vortex tube, purge flow cell and aspirate 0.500–0.900 ml from the tube into the spectrophotometer, wipe aspiration tubing
c)	Immediately place first tube in a temperature block or water bath at 30 °C
d)	Read initial absorbance and record, start timer at instant of spectrophotometer reading
Repeat cycle 6a–d at exactly 30 s intervals for each remaining tube	
7)	Purge flow cell, remove first tube from temperature block, aspirate into spectrophotometer and read final absorbance when timer indicates exactly 30 min
Repeat cycle for each remaining tube at exactly 30 s intervals	
8)	Construct standard curve and quantitate unknown samples

Interference by G6PDH activity in serum is avoided by using the coenzyme NAD^+ which reacts only with the bacterial enzyme.

As is apparent from the assay protocol in Table 7 all samples are pretreated with NaOH in order to denature interfering proteins. Extremely accurate timing and thermostating are necessary since the accuracy of the results depend on measurement of relatively small differences between two fairly large absorbances; a ΔE of 0.085 is allowed between lowest and highest standards.

The residual activity of the digoxin G6PDH–antibody complex is appreciable since the inhibition of the antibody-bound enzyme is far from complete.

Rosenthal et al. (1976) could only obtain agreement with two established RIA after introducing some important modifications. Sun and Spiehler (1976) found acceptable agreement with RIA albeit with a somewhat higher coefficient of variation. Müller et al. (1976 b) found in some cases considerable deviation from RIA values. Further comparative studies with RIA have been published (Drost et al., 1976).

A shorter and simpler version for a "stat" procedure is described by Chang et al. (1976).

The potential of this method lies probably in its suitability for automation. Adaptations have been described for the ABA 100 bichromatic analyzer (LEVY et al., 1977; ERIKSEN and ANDERSEN, 1978; CASTRO et al., 1979) and for the Centrifichem 300 (BRUNK and MALMSTADT, 1977; VOGT et al., 1978). LESNE and KADIMA (1978) have described an EMIT for digitoxin in which all reagents except for the digitoxin standards were taken from the commercial EMIT digoxin kit with modified assay conditions. This is possible owing to the almost 100% cross reactivity of the anti-digoxin antibodies for digitoxin.

References

Adams, R.J.: Copolymers of globulin with specific binding proteins – for use as specific adsorbents for detecting the presence of a serum constituent by radioassay. US patent 3980764 (1976)

Ahluwalia, G.S., Kuczala, Z.J.: Absence of hypoalbuminemia in patients undergoing digoxin therapy, as related to use of digoxin radioimmunoassay kits. Clin. Chem. *22*, 1539–1540 (1976)

Anggard, E.E., Chew, L.F., Kalman, S.M.: A source of error in digoxin radioimmunoassay. N. Engl. J. Med. *287*, 935 (1972)

Arndts, D.: A solid phase radioimmunoassay for digoxin and its acylated derivatives. Naunyn-Schmiedebergs Arch. Pharmacol. *287*, 309–319 (1975)

Barbieri, U., Gandolfi, C.: Digoxin radioimmunoassay: a comparison of the assay performances using tritiated and ^{125}I radioiodinated tracers. Clin. Chim. Acta *77*, 257–267 (1977)

Barrett, M.J., Cohen, P.S.: Radioimmunoassay of serum renin activity and digoxin concentrations, with use of polyethylene glycol to separate free and antibody-bound ligand. Clin. Chem. *18*, 1339–1342 (1972)

Batz, H.-G., Linke, H.-R., Stellner, K., Weimann, G.: Reaktive unsymmetrische, einen Digoxin- bzw. Digitoxinrest enthaltende Dicarbonsäureester, Verfahren zu deren Herstellung sowie ihre Verwendung zur Herstellung von Testreagentien. German patent 2537129 (1977)

Batz, H.-G., Stellner, K., Linke, H.-R., Weimann, G.: Hydroxylalkylamide von Hydroxyphenylalaninen, ein Verfahren zu ihrer Herstellung und deren Verwendung. German patent 2616724 (1979)

Belpaire, F.M., Bogaert, M.G., De Broe, M.E.: Radioimmunoassay of digoxin in renal failure: a comparison of different commercial kits. Clin. Chim. Acta *62*, 255–261 (1975)

Bergdahl, B., Molin, L., Lindwall, L., Dahlström, G., Scherling, I.-L., Bertler, A.: Four kits for plasma digoxin radioimmunoassay compared. Clin. Chem. *25*, 305–308 (1979)

Berk, L.S., Lewis, J.L., Nelson, J.C.: One-hour radioimmunoassay of serum drug concentrations, as exemplified by digoxin and gentamicin. Clin. Chem. *20*, 1159–1164 (1974)

Bernstein, J., Varma, R.K., Vogt, B.R., Weisenborn, F.L.: Steroid derivatives and their use in radioimmunoassay. Brit patent 2002385 (1979 a)

Bernstein, J., Varma, R.K., Vogt, B.R., Weisenborn, F.L.: Radiolabelled digoxigenin derivatives for radioimmunoassay of digoxin in plasma, and their unlabelled precursors. Brit patent 2003480 (1979 b)

Berson, S.A., Yalow, R.S.: General principles of radioimmunoassay. Clin. Chim. Acta *22*, 51–69 (1968)

Besch, H.R., Watanabe, A. M.: Radioimmunoassay of digoxin and digitoxin. Clin. Chem. *21*, 1815–1829 (1975)

Blazey, N.D.: The preparation of ^{125}I-labelled digoxin suitable for use in the digoxin radioimmunoassay. Clin. Chim. Acta *80*, 403–407 (1977)

Blend, M.J., Ferry, J.D.: RIA kit induced errors in serum digoxin measurement: a critical evaluation of two systems. Clin. Chem. *21*, 966 (1975)

Boguslaski, R.C.: A column radioimmunoassay method for the determination of digoxin. Biochem. Med. *14*, 83–92 (1975)

Boink, A.B.T.J., Kruyswijk, H.H., Willebrands, A.F., Maas, A.H.J.: Some factors affecting a commercial kit for radioimmunoassay of digoxin using tritiated digoxin. J. Clin. Chem. Clin. Biochem. *15*, 261–266 (1977)
Borner, K., Rietbrock, N.: Bestimmung von Digoxin im Serum – Vergleich von Radioimmunoassay und heterogenem Enzymimmunoassay. J. Clin. Chem. Clin. Biochem. *16*, 335–342 (1978)
Brock, A.: Use of gelatin-coated charcoal in digoxin immunoassay. Clin. Chem. *19*, 140–141 (1973)
Brooker, G., Terasaki, W.L., Price, M.G.: Gammaflow: a completely automated radioimmunoassay system. Science *194*, 270–276 (1976)
Brown, J.L., Lyle, L.R.: Radioimmunological titration of digoxin in non-extracted blood serum using I-125 hapten in aqueous buffer, antiserum and separation with ion-exchange membrane. US patent 4064227 (1976)
Brown, J.L., Lin, W.H.-T., Woods, J.W.: Länglicher hohler Behälter für immunochemische und enzymatische Methoden. Patent applied for (DT-OS) 2738183 (1978)
Brunk, S.D., Malmstadt, H.V.: Adaptation of the EMIT serum digoxin assay to a mini-disc centrifugal analyzer. Clin. Chem. *23*, 1054–1056 (1977)
Burnett, G.H., Conklin, R.L., Wasson, G.W., MacKinney, A.A.: Variability of standard curves in radioimmunoassay of plasma digoxin. Clin. Chem. *19*, 725–726 (1973)
Butler, V.P., Chen, J.P.: Digoxin-specific antibodies. Proc. Natl. Acad. Sci. U.S.A. *57*, 71–78 (1967)
Butler, V.P.: Digoxin radioimmunoassay. Lancet *1971 I*, 186
Butler, V.P.: Assays of digitalis in the blood. Progr. Cardiovasc. Dis. *14*, 571–600 (1972)
Butler, V.P.: The immunological assay of drugs. Pharmacol. Rev. *29*, 103–184 (1978a)
Butler, V.P.: Evaluation of different methods for determining serum concentrations of cardiac glycosides. In: Cardiac glycosides, part I. Bodem, G., Dengler, H.J. (eds.), pp. 1–21. Berlin, Heidelberg, New York: Springer 1978b
Calesnick, B., Dinan, A.: Improved micro-radioimmunoassay of digoxin in serum, with use of ^{125}I-labeled digoxin. Clin. Chem. *22*, 903–905 (1976)
Carlson, I.H., Westgard, J.O., Joiner, B.L.: Characteristics and criteria for judging acceptability of RIA methods which measure digoxin. Clin. Chem. *21*, 966 (1975)
Carlton, M.T., Witty, T.R., Hasler, M.J., Bjornsen, R.E., Painter, K.H.: The development of a fully automated solid phase radioimmunoassay for digoxin. J. Radioanal. Chem. *43*, 389–397 (1978)
Castro, A., McCreary, S., Wunsch, C., Malkus, H.: Totally automated radioimmunoassay systems for digoxin and thyroxine. Clin. Chem. *22*, 1166 (abstract) (1976)
Castro, A., Seckinger, D., Cid, A., Buschbaum, P., Noto, R.: A simple automatic kinetic enzyme immunoassay for digoxin. Clin. Chim. Acta *99*, 289–295 (1979)
Cerceo, E., Elloso, C.A.: Factors affecting the radioimmunoassay of digoxin. Clin. Chem. *18*, 539–543 (1972)
Chamberlaine, D.A., White, R.J., Howard, M.R., Smith, T.W.: Plasma digoxin concentrations in patients with atrial fibrillation. Brit. Med. J. *1970 III*, 429–432
Chambers, R.E.: Digoxin radioimmunoassay: improved precision with iodinated tracer. Clin. Chim. Acta *57*, 191–193 (1974)
Chang, J.J., Crowl, C.P., Schneider, R.S.: Homogeneous enzyme immunoassay for digoxin. Clin. Chem. *21*, 967 (abstract) (1975)
Chang, J.J., Cobb, M.E., Haden, B.H., McNeil, K.G., Schneider, R.S.: Stat procedure for digoxin based upon homogeneous enzyme immunoassay. Clin. Chem. *22*, 1185 (abstract) (1976)
Chard, T.: An introduction to radioimmunoassay and related techniques. In: Laboratory techniques in biochemistry and molecular biology, Work, T.S., Work, E. (eds.), Vol. 6, part II, pp. 293–534. Amsterdam, New York, Oxford: North-Holland Publ. Comp. 1978
Chen, I.W., Sperling, M., Volle, C., Maxon, H.R.: Emergency procedure for digoxin radioassay. Clin. Chem. *24*, 1564–1567 (1978)
Christiansen, N.J.B., Nielsen, M.D.: Digoxin radioimmunoassay: sephadex separation of free from antibody-bound digoxin. Clin. Chim. Acta *42*, 125–129 (1972)

Coombes, R.F.: Labelled cardiotonic glycosides for use in radioimmunoassay. US patent 396397 (1976)
Doering, W., Blümel, E.: Vereinfachte Serum-Digoxin-Bestimmung mit einem Jod-125-Solid-Phase-Test. Klin. Wochenschr. *56*, 497–502 (1978)
Drewes, P.A., Pileggi, V.J.: Faster and easier radioimmunoassay of digoxin. Clin. Chem. *20*, 343–347 (1974)
Drost, R.H., Plomp, T.A., Teunissen, A.J., Maas, A.H.J., Maes, R.A.A.: A comparative study of the homogeneous enzyme immunoassay (EMIT) and two radioimmunoassays (RIA-S) for digoxin. Clin. Chim. Acta *79*, 557–567 (1977)
Duchateau, A.M.J.A., Van Megen, T., Merkus, F.W.H.M.: Digitoxin radioimmunoassay, I – comparison of three assays for the determination of serum digitoxin levels. Pharm. Weekbl. *108*, 193–200 (1973)
Dwenger, A., Zic, R., Friedel, R., Trautschold, I.: Separation techniques in radioimmunoassays for digoxin and insulin. Z. Anal. Chem. *279*, 108–109 (1976)
Dwenger, A., Friedel, R., Trautschold, I.: Performance of radioimmunoassays for digoxin as evaluated by a group experiment. In: Proceedings, Int. symposium on radioimmunoassay and related procedures in medicine, Berlin 1977, International Atomic Energy Agency, Vienna, Vol. II, pp. 141–148 (1978)
Dwenger, A., Trautschold, I.: Untersuchungen zur Vergleichbarkeit der Ergebnisse unterschiedlicher Methoden zur radioimmunologischen Digoxin-Bestimmung – Ein Beitrag zur Qualitätskontrolle. J. Clin. Chem. Clin. Biochem. *16*, 587–595 (1978)
Eisenhardt, W.A., Hedaya, E., Theodoropulos, S.: Protected hydroxyphethyl isocyanatesfor isotopically labelling compounds for use in assays. US patent 4115539 (1977)
Ekins, R.P.: Basic principles and theory. In: Radioimmunoassay and saturation analysis. Sönksen, P.H. (ed.). Brit. Med. Bull. *30*, 3–11 (1974)
Eriksen, P.B., Andersen, O.: Homogeneous enzyme immunoassay of serum digoxin with use of a bichromatic analyzer. Clin. Chem. *25*, 169–171 (1978)
Erlanger, B.F., Beiser, S.M.: Antibodies specific for ribonucleosides and ribonucleotides and their reaction with DNA. Proc. Natl. Acad. Sci. U.S.A. *52*, 68–74 (1964)
Ertingshausen, G., Shapiro, S.I., Green, G., Zborowski, G.: Adaptation of a T_3-uptake test and of radioimmunoassays for serum digoxin, thyroxine, and triiodothyronine to an automated radioimmunoassay system – "Centria". Clin. Chem. *21*, 1305–1313 (1975)
Evered, D.C., Chapman, C., Hayter, C.J.: Measurement of plasma digoxin concentration by radioimmunoassay. Brit. Med. J. *1970 III*, 427–428
Friedel, R., Dwenger, A.: A simple, fast and sensitive microtechnique for radioimmunoassays. Clin. Chem. *21*, 967 (abstract) (1975)
Gauldie, J., Tang, H.K., Corsini, A., Walker, W.H.C.: Solid-phase radioimmunoassay with protein-A-bearing Staphylococcus aureus cells used to assay a protein (ferritin) and a hapten (digoxin). Clin. Chem. *26*, 37–40 (1980)
Gershman, H., Powers, E., Levine, L., Van Vunakis, H.: Radioimmunoassay of prostaglandins, angiotensin, digoxin, morphine, and adenosine-3′,5′-cyclic-monophosphate with nitrocellulose membranes. Prostaglandins *1*, 407–423 (1972)
Greenwood, H., Howard, M., Landon, J.: A rapid, simple assay for digoxin. J. Clin. Path. *27*, 490–494 (1974)
Greenwood, H., Landon, J., Forrest, G.C.: Radioimmunoassay for digoxin with a fully automated continuous-flow system. Clin. Chem. *23*, 1868–1872 (1977)
Gutcho, S., McCarter, H., Rapun, R.: Radioimmunoassay of digoxin: an intercomparison of results with three methods. Clin. Chem. *19*, 1058–1059 (1973)
Gutcho, S., McCarter, H., Chanod, E.: Immobilising receptor for ligand analysis – by depositing on a plastics support in the presence of blocking and antiaggregation agents. US patent 4105410 (1977)
Hall, C.: Critical review of digoxin reagents and methodologies. Lab. Med. *5*, 42–44 (1974)
Halpern, E.P., Bordens, R.W.: Microencapsulated antibodies in radioimmunoassay – I. Determination of digoxin. Clin. Chem. *25*, 860–862 (1979)
Hansell, J.R.: Three-years' experience in interlaboratory testing of commercial digoxin kits. Am. J. Clin. Path. *66*, (1. Suppl), 234–237 (1976)

Herbert, V., Lau, K.-S., Gottlieb, C.W., Bleicher, S.J.: Coated charcoal immunoassay of insulin. J. Clin. Endocr. Metab. *25*, 1375–1384 (1965)
Hersh, L.S., Yaverbaum, S.: Magnetic solid-phase radioimmunoassay. Clin. Chim. Acta *63*, 69–72 (1975)
Hoeschen, R.J., Proveda, V.: Serum digoxin by radioimmunoassay. Can. Med. Ass. J. *105*, 170–173 (1971)
Hofman, L.F.: More accurate radioassay of digoxin. Clin. Chem. *19*, 1413–1414 (1973)
Holtzman, J.L., Shafer, R.B., Erickson, R.R.: Methodological causes of discrepancies in radioimmunoassay for digoxin in human serum. Clin. Chem. *20*, 1194–1198 (1974)
Holtzman, J.L., Shafer, R.B.: Radioimmunoassay of digoxin: effect of albumin. Clin. Chem. *21*, 636–637 (1975)
Horgan, E.D., Riley, W.J.: Radioimmunoassay of plasma digoxin, with use of iodinated tracer. Clin. Chem. *19*, 187–190 (1973)
Huffman, D.H.: The effect of spironolactone and canrenone on the digoxin radioimmunoassay. Res. Commun. Chem. Path. Pharmacol. *9*, 787–790 (1974)
Hunter, W.M., Greenwood, F.C.: Preparation of iodine-131 labelled human growth hormone of high specific activity. Nature *194*, 495–496 (1962)
Johnson, E.G., Sturgis, B.E., Stonecypher, T.E.: Development of Micromedic systems' "Concept 4 automatic radioassay" system. Clin. Chem. *22*, 1164 (abstract) (1976)
Kleinhammer, G., Lenz, H., Linke, R., Gruber, W.: Enzyme immunoassay for determination of serum digoxin in antibody coated tubes. (Abstract) 2nd European congress on clinical chemistry, Prague 1976
Kramer, W.G., Bathala, M.S., Reuning, R.H.: Specificity of the digoxin radioimmunoassay with respect to dihydrodigoxin. Res. Commun. Chem. Path. Pharmacol. *14*, 83–88 (1976)
Kroening, B.H., Weintraub, M.: Reduced variation of tracer binding in digoxin radioimmunoassay by use of ^{125}I-labeled tyrosin-methyl-ester derivative: relation of thyroxine concentration to binding. Clin. Chem. *22*, 1732–1734
Kubasik, N.P., Schauseil, S., Sine, H.E.: Comparison of commercial kits for radioimmunoassay: I. The radioimmunoassay of serum digoxin using tritium tracer. Clin. Biochem. *7*, 206–211 (1974a)
Kubasik, N.P., Norkus, N.S., Sine, H.E.: Comparison of commercial kits for radioimmunoassay: II. The radioimmunoassay of serum digoxin using iodinated tracer. Clin. Biochem. *7*, 307–312 (1974b)
Kubasik, N.P., Brody, B.B., Barold, S.S.: Problems in measurement of serum digoxin by commercially available radioimmunoassay kits. Am. J. Cardiol. *36*, 975–977 (1975)
Kubasik, N.P., Hall, J.L., Barold, S.S., Volosin, M.T., Sine, H.E.: Evaluation of the sensitivity of commercially available digoxin radioimmunoassay kits. Chest *70*, 217–220 (1976)
Kuczala, Z.J., Ahluwalia, G.S.: Evaluation of two digoxin radioimmunoassay procedures in which ^{125}I-labeled digoxin is used. Clin. Chem. *22*, 193–197 (1976)
Kuno-Sakai, H., Sakai, H.: Effects on radioimmunoassay of digoxin of varying incubation periods for antigen-antibody reaction and varying periods of adsorption by dextran-coated charcoal. Clin. Chem. *21*, 227–229 (1975)
Lader, S., Bye, A., Marsden, P.: The measurement of plasma digoxin concentration: a comparison of two methods. Eur. J. Clin. Pharmacol. *5*, 22–27 (1972a)
Lader, S., Court, G., Johnson, B.F., Hurn, B.A.L.: Radioimmunoassay of digoxin with iodinated tracer. Scand. J. Clin. Lab. Invest. *29*, Suppl. 126:14.10 (abstract) (1972b)
Lader, S., Johnson, B.: Interpretation of results obtained using digoxin radioimmunoassay. Postgrad. Med. J. *50* (Suppl. 6):18–23 (1974)
Larbig, D., Kochsiek, K.: Radioimmunchemische Bestimmung von Digoxin im menschlichen Serum. Klin. Wochenschr. *49*, 1031–1032 (1971)
Larbig, D., Kochsiek, K.: Zur radioimmunologischen Bestimmung von Digoxin und Digoxinderivaten. Dtsch. Med. Wochenschr. *97*, 1310–1312 (1972)
Larson, J.H., Beckala, H.R., Homburger, H.A.: Assay characteristics of four commercial ^{125}I-digoxin radioimmunoassay kits. Am. J. Clin. Path. *69*, 208–209 (1978)

Lesne, M., Kadima, L.: Assay of digitoxin in plasma by an enzyme-immunoassay technique. Clin. Chim. Acta *84*, 45–47 (1978)
Leung, F.Y., Pomeroy, J., Capling, G.: Quality control of testosterone and digoxin radioimmunoassays and cortisol competitive protein binding assay by the reference sample methode. Clin. Biochem. *8*, 118–123 (1975)
Levy, S., Dioso, C., Baron, R.: Automated determination of digoxin by enzymimmunoassay utilizing the micro-automated ABA-100 analyzer. Clin. Chem. *23*, 1169 (abstract) (1977)
Lichey, J., Schröder, R., Rietbrock, N.: The effect of oral spironolactone and intravenous canrenoate-K on the digoxin radioimmunoassay. Int. J. Clin. Pharmacol. *15*, 557–559 (1977)
Line, W.F., Siegel, S.J., Kwong, A., Frank, C., Ernst, R.: Solid-phase radioimmunoassay for digoxin. Clin. Chem. *19*, 1361–1365 (1973)
Litt, G.J.: Solid phase immunoassay of antigens- by reaction of labelled and unlabelled antigen with an antibody complexed to an anti-antibody. US patent 4092408 (1977)
MacKinney, A.A., Burnett, G.H., Conklin, R.L., Wasson, G.W.: Comparison of five radioimmunoassays and enzyme bioassay for measurement of digoxin in blood. Clin. Chem. *21*, 857–859 (1975)
Marcus, F.I., Ryan, J.N., Stafford, M.G.: The reactivity of derivatives of digoxin and digitoxin as measured by the Na-K-ATPase displacement assay and by radioimmunoassay. J. Lab. Clin. Med. *85*, 610–620 (1975)
Meade, R.C., Kleist, T.J.: Improved radioimmunoassay of digoxin and other sterol-like compounds using Somogyi precipitation. J. Lab. Clin. Med. *80*, 748–754 (1972)
Müller, H., Graul, E.H., Müller, L.: Digoxinbestimmung im Serum: Klinische Bedeutung, Ergebnisse und Fehlerdiskussion an Hand vergleichender Untersuchungen mit verschiedenen Radioimmunoassays. Z. Kardiol. *64*, 1123–1139 (1975)
Müller, H., Bräuer, H., Tratz, A.: Radioimmunologische Fehlbestimmung von Digoxin im Schwangerenserum. Ärztl. Lab. *22*, 355–356 (1976a)
Müller, H., Bräuer, H., Reinhardt, M., Förster, G.: Serumdigoxinbestimmung mit EIA und RIA. Ärztl. Lab. *22*, 399–402 (1976b)
Müller, H., Bräuer, H., Resch, B.: Cross reactivity of digitoxin and spironolactone in two radioimmunoassays for serum digoxin. Clin. Chem. *24*, 706–709 (1978)
Munz, E., Kessler, A., Koller, P.U., Busch, E.W.: Erfahrungen mit einem Enzymimmunoassay für Digoxin. Lab. Med. *3*, 71–76 (1979)
Nye, L., Forrest, G.C., Greenwood, H., Gardner, J.S., Jay, R., Roberts, J.R., Landon, J.: Solid-phase, magnetic particle radioimmunoassay. Clin. Chim. Acta *69*, 387–396 (1976)
Oellerich, M., Haindl, H., Haeckel, R.: Die Bestimmung von Digoxin im Serum mit Enzymimmunotests. Internist *19*, 188–190 (1978)
Ojala, K., Karjalainen, J., Reissell, P.: Radioimmunoassay of digoxin. Lancet *1972 I*, 150
O'Leary, T.D., Howe, L.A., Geary, T.D.: Improvement in a radioimmunoassay for digoxin. Clin. Chem. *25*, 332–334 (1979)
Oliver, G.C., Parker, B.M., Brasfield, D.L., Parker, C.W.: The measurement of digitoxin in human serum by radioimmunoassay. J. Clin. Invest. *47*, 1035–1042 (1968)
Oliver, G.C., Parker, B.M., Parker, C.W.: Radioimmunoassay of digoxin, technic and clinical application. Am. J. Med. *51*, 186–192 (1971)
Ong, T.S., Baumann, H., Pohle, W., Schött, D., Hobrecker, M., May, B., Fritze, E.: Digoxin: vergleichende Bestimmung mit Radioimmun- und Enzymimmunassay. Diagnostik *12*, 139–142 (1979)
Oslapas, R., Herrin, T.R.: Radioimmunological determination of digoxin in serum – by incubating with antibodies and labeled digoxin-tyramin analogue. US patent 3981982 (1976)
O'Sullivan, M.J., Bridges, J.W., Marks, V.: Enzyme immunoassay: a review. Ann. Clin. Biochem. *16*, 221–240 (1979)
Painter, K., Hasler, M.J.: Development of chemistries for a fully automated radioimmunoassay instrument, Micromedic systems "concept 4 automatic radioassay". Clin. Chem. *22*, 1164 (abstract) (1976)

Painter, K.: Verfahren zum Messen des Digoxingehalts einer Serumprobe. Dt-OS 2743446 (1978)
Painter, K., Vader, C.R.: Interference of iodine-125 ligands in radioimmunoassay: evidence implicating thyroxine-binding globulin. Clin. Chem. *25*, 797–799
Park, H.M., Chen, I.W., Manitasas, G.T., Lowey, A., Saenger, E.L.: Clinical evaluation of radioimmunoassay of digoxin. J. Nucl. Med. *14*, 531–533 (1973)
Parker, C.W.: Radioimmunoassay of biologically active compounds. Englewood Cliffs, New Jersey: Prentice-Hall Inc. 1976
Phillips, A.P.: The improvement of specificity in radioimmunoassays. Clin. Chim. Acta *44*, 333–340 (1973)
Piasio, R.N., Woiszwillo, J.E.: Radioiodine labelled peptide derivatives of digoxin – for radioimmunoassay, having same antibody affinity as digoxin. US patent 3925355 (1975)
Piasio, R.N., Perry, D.A., Nayak, P.N.: Verfahren und Vorrichtung zur Durchführung klinischer Diagnose-Tests in vitro unter Verwendung eines Festphasen-Bestimmungs-Systems. DT-OS 2824742 (1979)
Pippin, S.L., Marcus, F.I.: Digoxin Immunoassay with use of ^{3}H-Digoxin vs. ^{125}I-tyrosin-methyl-ester of digoxin. Clin. Chem. *22*, 286–287 (1976)
Polito, A.J.: Digoxin carboxylic ester derivatives – pref. labelled with radioiodine for use in radioimmunoassay. US patent 4021535 (1976)
Ravel, R.: Negligible interference by spironolactone and prednisone in digoxin radioimmunoassay. Clin. Chem. *21*, 1801–1803 (1975)
Ravel, R., Schall, R.F.: Pitfalls in iodine-125 digoxin measurement. Ann. Clin. Lab. Sci. *6*, 365–371 (1976)
Ravel, R., Espinola, A.F.: Comparison of ^{125}I digoxin kits. Lab. Med. *7*, 19–24 (1976)
Rosenthal, F., Vargas, M.G., Klass, C.S.: Evaluation of enzymemultiplied immunoassay technique (EMIT) for determination of serum digoxin. Clin. Chem. *22*, 1899–1902 (1976)
Rubenstein, K.E., Schneider, R.S., Ullman, E.F.: "Homogeneous" enzyme immunoassay. A new immunochemical technique. Biochem. Biophys. Res. Commun. *47*, 846–851 (1972)
Rutner, H., Rapun, R., Lewin, N.: Labelled 3-acetyldigoxigeninaminoacids- for digoxin test from digoxigenin-12-acetate-3-hemisuccinate and aminoacids. US patent 3855208 (1971 a)
Rutner, H., Rapun, R., Lewin, N.: Labelled digitoxigenin-3-aminoacids- for digitoxin test from mixed anhydrides of hemisuccinates and pivalylchloride coupled. US patent 3810886 (1971 b)
Schall, R.F.: Digoxin radioimmunoassay: dealing with a loss of sensitivity. Clin. Chem. *19*, 688–689 (1973)
Scharpe, S.L., Cooreman, W.M., Blomme, W.J., Laekeman, G.M.: Quantitative enzyme immunoassay: current status. Clin. Chem. *22*, 733–738 (1976)
Schuurs, A.H.W.M., Van Weemen, B.K.: Enzyme-Immunoassay. Clin. Chim. Acta *81*, 1–40 (1977)
Shapiro, B., Kollmann, G.J., Heine, W.I.: Pitfalls in the application of digoxin determinations. Semin. Nucl. Med. *5*, 205–220 (1975)
Shaw, W.: Radioimmunoassay of digoxin: effect of albumin. Clin. Chem. *21*, 636 (1975)
Shaw, W., Schulman, L., Spierto, F.W.: The carrier effect of gamma globulin when a polethylene glycol separation technique is used in digoxin radioimmunoassay. Clin. Chim. Acta *60*, 385–389 (1975)
Shaw, W., Powell, M.K., Bayse, D.: The influence of serum albumin concentration on fatty acid interference in the radioimmunoassay for serum digoxin. Clin. Chem. *23*, 1124–1125 (abstract) (1977)
Silber, B., Sheiner, L.B., Powers, J.L., Winter, M.E., Sadee, W.: Spironolactone-associated digoxin radioimmunoassay interference. Clin. Chem. *25*, 48–50 (1979)
Sjöholm, I.G.H., Lindmark, N.R., Ekman, B.M.: Immobilisierte biologisch aktive Substanz und deren Verwendung. Dt-OS 2546379 (1976)
Skelley, D.S., Brown, L.P., Besch, P.K.: Radioimmunoassay. Clin. Chem. *19*, 146–186 (1973)

Smith, T.W., Butler, V.P., Haber, E.: Determination of therapeutic and toxic serum digoxin concentrations by radioimmunoassay. N. Engl. J. Med. *281*, 1212–1216 (1969)

Smith, T.W., Butler, V.P., Haber, E.: Characterization of antibodies of high affinity and specificity for the digitalis glycoside digoxin. Biochemistry *9*, 331–337 (1970)

Smith, T.W.: Radioimmunoassay for serum digitoxin concentration: methodology and clinical experience. J. Pharmacol. Exp. Ther. *175*, 352–360 (1970)

Smith, T.W.: Ouabain-specific antibodies: immunochemical properties and reversal of Na^+, K^+-activated ATPase inhibition. J. Clin. Invest. *51*, 1583–1593

Smith, T.W., Haber, E.: Clinical value of the radioimmunoassay of the digitalis glycosides. Pharmacol. Rev. *25*, 219–228 (1973)

Smith, T.W.: Digoxin radioimmunoassay: capabilities and limitations. J. Pharmac. Biopharm. *4*, 197 (1976)

Sönksen, P.H. (ed.): Radioimmunoassay and saturation analysis. Brit. Med. Bull. *30*, 1–103 (1974)

Soto, A.R., Brotherton, M., Castellanos, M.E., Chambliss, K.W.: Comparison of digoxin values obtained using several commercially available tritiated and iodinated digoxin methods. Clin. Chem. *21*, 966 (1975)

Soto, A.R., Brotherton, M., Castellanos, M.E., Chambliss, K.W.: Causes of variability of assay values in radioimmunoassay of digoxin. Clin. Chem. *22*, 1183 (abstract) (1976)

Stellner, K., Glatz, C., Linke, R.: Solid-phase tube radioimmunoassay zur Bestimmung von Digoxin. Z. Klin. Chem. Biochem. *13*, 257 (abstract) (1975)

Stellner, K.: A new simple assay for determining digoxin serum levels. In: Cardiac glycosides. Bodem, G., Dengler, H.J. (eds.), part I, pp. 22–27. Berlin, Heidelberg, New York: Springer 1978

Stoll, R.G., Christensen, M.S., Sakmar, E., Wagner, J.G.: The specificity of the digoxin radioimmunoassay procedure. Res. Commun. Chem. Path. Pharmacol. *4*, 503–510 (1972)

Sun, L., Spiehler, V.: Radioimmunoassay and enzyme immunoassay compared for determination of digoxin. Clin. Chem. *22*, 2029–2031 (1976)

Taubert, K., Shapiro, W.: Serum digoxin levels using an ^{125}I-labelled antigen: validation of method and observations on cardiac patients. Am. Heart J. *89*, 79–86 (1975)

Terrance, G.: Tetra-^{125}iodo-dityramine of digitalis derivative. US patent 3997525 (1976)

Tovey, K.C., Findlay, J.W.A.: Digoxin- und Digitoxinderivate und Verfahren zu ihrer Herstellung. DT-OS 2526984 (1976)

Ullman, E.F., Rubenstein, K.E.: Cardiac glycoside enzyme conjugates. US patent 4039385 (1977)

Valdes, R., Savory, G., Bruns, D., Renoe, B., Savory, J., Wills, M.R.: Performance assessment of the gammaflow ™ automated radioimmunoassay system by assaying for digoxin. Clin. Chem. *25*, 1254–1258 (1979)

Verspohl, E., Strobach, H., Greeff, K.: Eine Methode zur radioimmunologischen Bestimmung von g-Strophanthin. Arzneim. Forsch. *28*, 1694–1698 (1978)

Vijgh, W.J.F. van der: Modification of a commercially available digoxin radioimmunoassay kit for higher precision. Clin. Chem. *21*, 1172–1174 (1975)

Vogt, W., Tausch, A., Jacob, K., Knedel, M.: Die Bestimmung von Digoxin im Serum mit einem schnellen Analyser mit der EMIT-Technik. In: Enzymimmunoassay. Vogt, W. (Hrsg.), S. 27–35. Stuttgart: Thieme 1978

Voshall, D.L., Hunter, L., Grady, H.J.: Effect of albumin on serum digoxin radioimmunoassays. Clin. Chem. *21*, 402–406 (1975)

Wagner, D., Alspector, B., Feingers, J., Pick, A.: Direct radioimmunoassay with capillary chromatography tubes. Clin. Chem. *25*, 1337–1339 (1979)

Weiler, E.W., Zenk, M.H.: Iodinated digoxin derivatives with improved reactivity and stability, for use in radioimmunoassay. Clin. Chem. *25*, 44–47 (1979)

Wilkinson, S.: Reagentien zur Untersuchung von Herzglycosiden. DT-OS 2331922 (1974)

Wilkinson, S.: Radio-iodine-labelled steroid 3-oximes for radioimmunoassay of genins and their 3-glycosides. US patent 3954739 (1976)

Wisdom, G.B.: Enzyme-Immunoassay. Clin. Chem. *22*, 1243–1255 (1976)

Witherspoon, L.R., Shuler, S.E., Garcia, M.M.: Spurious underestimation of results for digoxin radioimmunoassay with a commercial kit. Clin. Chem. *24*, 494–498 (1978)
Wood, W.G.: Test of six commercial ^{125}I-labelled digoxin radioimmunoassay kits. J. Clin. Chem. Clin. Biochem. *15*, 679–685 (1977)
Wood, W.G., Wachter, C.: A critical appraisal of a further three new commercial digoxin radioimmunoassay kits with reference to cross-reacting substances. J. Clin. Chem. Clin. Biochem. *17*, 77–83 (1979)
Woodcock, B.G., Rodgers, M., Gibbs, A.C.: Quality control of digoxin radioimmunoassay. In: Radioimmunoassay in clinical biochemistry. Pasternak, C.A. (ed.), pp. 131–138. London, New York, Rheine: Heyden 1975
Yalow, R.S., Berson, S.A.: Assay of plasma insulin in human subjects by immunological methods. Nature *184*, 1648–1649 (1959)
Yaverbaum, S., Vann, W.P., Findley, W.F., Piasio, R.A.: A rapid and specific porous glass solid-phase radioimmunoassay of digoxin. Fed. Proc. *32*, 962 (abstract) (1973)
Zeegers, J.J.W., Maas, A.H.J., Willebrands, A.F., Kruyswijk, H.H., Jambroes, G.: The radioimmunoassay of digoxin. Clin. Chim. Acta *44*, 109–117 (1973)

CHAPTER 6

ATPase for the Determination of Cardiac Glycosides

URSULA GUNDERT-REMY and ELLEN WEBER

A. Introduction

The existence of an Na^+, K^+-activated ATPase (Mg^{2+}-dependent, Na^+, K^+-activated ATP-phosphohydrolase, Ec 3.6.1.3) was discovered by SKOU in 1957. Soon afterwards he also found that cardiac glycosides are able to inhibit ATPase activity by binding to the enzyme (SKOU, 1960). Extensive work was done by different groups to purify and characterize this enzyme which is specifically associated with active cation transport. Recently the molecular weight of the enzyme, which is composed of two types of polypeptide chain, was found to be 250,000 daltons (HOPKINS et al., 1976). The first chain has a molecular weight of 95,000 daltons and contains both an aspartic acid residue that becomes phosphorylated during the transport reaction as well as a binding site for ouabain (BASTIDE et al., 1973). The second polypeptide has a molecular weight of 50,000 daltons and is characterized as being a glycoprotein (KYTE, 1972). There appear to be one ouabain binding site and either one or two phosphorylation sites per 250,000 daltons molecular weight (WILSON, 1978).

Both reactions – the dephosphorylation of ATP and the binding of ouabain – may be employed as tools to measure the concentration of cardiac glycosides. The first method measures the inhibition of ATP hydrolysis; the second is based on the displacement of radiolabeled ouabain from its binding site by other cardiac glycosides.

B. Preparation of ATPase

Many proposals have been made for the preparation of the enzyme. Various authors used ATPase isolated from heart tissue in order to be able to compare the inotropic action of cardiac glycosides, which is thought to be mediated by ATPase inhibition, and their inhibition of the enzyme activity in vitro (AKERA et al., 1970; BESCH et al., 1970; MATSUI and SCHWARTZ, 1968).

Because of the high activity per milligram of protein ATPase from brain tissue was employed to determine cardiac glycosides in human plasma. Repeated high speed centrifugation and treatment with desoxycholate of brain tissue homogenate led to an enzyme preparation which is well characterized and used for the enzyme activity technique (SKOU and HILBERG, 1969). ATPase preparation for displacement assay are obtained by repeated centrifugation at low speed followed by dialysis and further centrifugation. The activity of the enzyme preparation obtained by the last technique has not been determined, the enzyme characteristics have not been described in detail (BROOKER and JELLIFFE, 1972).

Table 1. Inhibition of Na^+, K^+-stimulated ATPase by different cardiac glycosides with and without extraction procedure (activity in % of the control), and % extraction efficiency (a) by 4 pmol (MARDH, 1973), (b) by 12.5 pmol (GUNDERT-REMY et al., (1974)

Glycoside	With extraction	Without extraction	Extraction efficiency (%)
(a)			
Digoxin	58.5±2.6	59.0±2.5	101
Acetyl digoxin	52.2±1.5	57.1±1.2	91
Digitoxin	55.7±0.8	55.7±1.6	100
Acetyl digitoxin	36.1±1.2	55.7±1.0	65
Lanatoside C	19.5±2.7	47.0±1.0	42
Strophanthin G	0	55.7±1.7	0
(b)			
Digoxin	44.0±5.6	44.6±4.1	98.7
Methyldigoxin	44.8±6.2	44.7±2.9	100
Digoxigenin-mono-digitoxosid	50.2±1.8	53.0±1.7	94.7
Digoxigenin-bis-digitoxosid	46.7±1.9	49.4±1.0	94.5

The activity of the control was determined using a sample without content of cardiac glycoside

C. Extraction Procedure from Biological Fluids

To be able to measure concentrations in biological fluids with both methods an extraction step has to be performed in order to concentrate the cardiac glycoside and to remove other substances which may interfere with the reaction. Polar glycosides can not be extracted by dichloromethane or chloroform (which have usually been used as the organic solvents) because of the low extraction efficiencies. However, the extraction efficiency of nonpolar glycosides is sufficient to give reproducible results (Table 1). Re-extraction into buffered solution is necessary in order to carry out the final ATPase reaction.

D. Determination Based on Measurement of Enzyme Activity

The principle of this method is based on the inhibition of the enzyme activity by cardiac glycosides. Since the hydrolysis of ATP to ADP is the basic reaction, the enzyme activity is measured as moles of inorganic phosphate or ADP released per unit time. The concentration of the cardiac glycoside is proportional to the inhibition of ATP hydrolysis and can be estimated from standards run with the procedure.

I. Measurement of the Hydrolysis of ATP

The rate of hydrolysis of ATP may be estimated by measurement of the inorganic phosphate resulting from the reaction. For this purpose the method based on the spectrophotometric estimation of a phosphomolybdate complex after re-extraction into methanol as described by MARSH (1959) is preferably employed since the results obtained by the Fiske–Subbarrow method (FISKE and SUBBARROW, 1925)

Table 2. Recovery of different cardiac glycosides after extraction according to isotope displacement assay. (MARCUS et al., 1975)

Glycoside	Recovery (%)
Digoxin	88.6 ± 5.2
Digoxigenin-bis-digitoxoside	95.1 ± 9.4
Digoxigenin-mono-digitoxoside	72.3 ± 10.6
Digoxigenin	83.4 ± 9.1
Digitoxin	50.5 ± 3.1
Digitoxigenin-bis-digitoxoside	63.4 ± 7.2
Digitoxigenin-mono-digitoxoside	51.5 ± 8.7
Digitoxigenin	27.9 ± 4.7

are less reproducible. Since ATP concentration decreases during the hydrolysis reaction the substrate concentration varies.

The second possibility is to couple the hydrolysis of ATP (Eq. 1) to a pyruvate kinase system (Eq. 2) and measure the pyruvate formed during the incubation period by the method of FRIEDEMANN and HAUGEN (1943) for example. Other authors assess the pyruvate concentration by using the lactate reaction (Eq. 3) (SCHONER et al., 1967).

$$ATP + H_2 0 \rightarrow ADP + P_i, \quad (1)$$

$$ADP + \text{phosphoenolpyruvate} \rightarrow ATP + \text{pyruvate}, \quad (2)$$

$$\text{pyruvate} + NADH + H^+ \rightarrow \text{lactate} + NADH. \quad (3)$$

P_i represents inorganic phosphate.

The advantage of the latter method is that the concentration of ATP is maintained at a constant level.

II. Inhibition of ATPase by Different Cardiac Glycosides

The activity of the enzyme is impaired by different glycosides and their metabolites when they are incubated with the enzyme without preceding extraction procedure. Table 3 shows that 50% inhibition can be achieved with concentrations between 11.0×10^{-9} and 35.25×10^{-6}. In the *lanata* group the efficiency of inhibition is clearly dependent on the structure of the molecule. Substances with at least one sugar residue are more effective than the related aglycones, a finding in good agreement with the biological effect. During the measurement of plasma levels of cardiac glycosides an extraction step with an organic solvent precedes the ATPase reaction so that polar metabolites cannot influence the result. However, because of the high extraction efficiency the mono- and bis-digitoxosides are able to falsify the results of digoxin and digitoxin measurements in biological samples since they inhibit the enzyme to the same extent as the parent drugs.

III. Precision and Sensitivity of the Assay

BURNETT and CONKLIN (1968) described a procedure for the measurement of digitoxin plasma levels based on the inhibition of ATPase. They were able to estimate

Table 3. Inhibition of ATPase from ox brain by different cardiac glycosides (GUNDERT-REMY et al., 1978)

Glycoside	ED_{50} (nmol/l)[a]
Lanatoside A	11.25
Digitoxin	26.75
Digitoxigenin-bis-digitoxoside	12.25
Digitoxigenin-mono-digitoxoside	14.25
Digitoxigenin	162.0
Lanatoside B	80.0
Gitoxin	46.5
Gitoxigenin	970.0
Lanatoside C	20.75
Digoxin	12.25
Digoxigenin-bis-digitoxoside	13.75
Digoxigenin-mono-digitoxoside	15.05
β-Methyldigoxin	11.0
Digoxigenin	920.0
Dihydrodigoxin	35,250.0
Strophanthoside K	18.75
Cymarin	15.75
Cymarol	24.75
Convallatoxin	11.0
Peruvosid	27.0
Strophanthin K β	33.0
Strophanthidin K	265.0

[a] Concentration at which the enzyme is 50% inhibited

the concentration of digitoxin down to 3.3 ng/ml. The coefficient of variation ranged from 65.4%–4.6% measuring amounts of 10–100 ng which corresponds to 3.3–33.3 ng/ml since 3 ml of plasma were extracted. Three years later the same authors (BURNETT and CONKLIN, 1971) and later on MARDH (1973) reported a modification by which it became possible to measure plasma levels of digoxin. The coefficient of variation given by MARDH (1973) for the range 1–5 µmol/l (corresponding to 1.3–6.4 ng/ml) amounted to 15%–4.2%. In our hands the method in the modification of MARDH (1973) had a similar precision. The lowest detectable level was about 0.2 ng/ml according to MARDH (1973).

IV. Comparison of Results Obtained by ATPase Activity and Radioimmunoassay

Since some known metabolites of digoxin and digitoxin are able to inhibit ATPase and to bind the antibodies respectively, both methods lack specifity. However, with respect to the minor extent to which the metabolites occur in plasma (MARCUS et al., 1966; VÖHRINGER and RIETBROCK, 1974) the results obtained by both assays should be comparable. As can be seen from Fig. 1, in which plasma levels from patients on digoxin maintenance therapy are depicted, this assumption holds true.

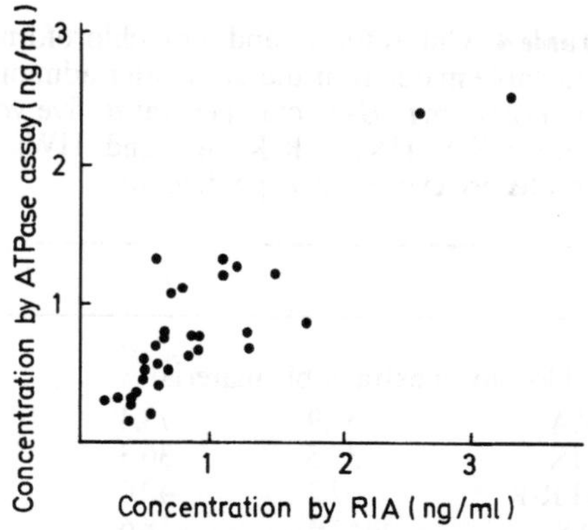

Fig. 1. Correlation of digoxin measurements obtained by radioimmunoassay and ATPase avtivity assay in plasma samples from patients on maintenance therapy. (GUNDERT-REMY et al., 1974)

Results obtained from urine samples by both methods might show differences between the methods since polar and nonpolar metabolites occur in remarkably high concentrations in the urine (MARCUS et al., 1966; VÖHRINGER and RIETBROCK, 1974).

After application of a dose of 0.5 mg tritiated digoxin to each of five human volunteers the total radioactivity was determined in the polar and in the chloroform-extractable fractions of the urine. Measurements using ATPase activity as well as radioimmunoassay were performed for the same samples (GUNDERT-REMY et al., 1978). Table 4 shows that no systematic deviation could be detected between the results obtained from the three different assays. This indicates that both radioimmunologic and enzymatic assays are nonspecific to the same extent since the results are not statistically different from the data obtained by measurement of total radioactivity.

E. Determination Based on Isotope Displacement

BROOKER and JELLIFFE (1972) were able to show that tissue homogenate rich in ATPase binds radiolabeled ouabain. Unlabeled digoxin, ouabain, digitoxin, and other cardiac glycosides displace the radiolabeled ouabain from the binding sites. As in the radioimmunoassay technique, the amount of radiolabeled ouabain bound to the enzyme will decrease when the amounts of unlabeled glycosides added are increased. As in the case of ATPase activity assay an extraction procedure has to be performed before the binding reaction to the enzyme can be carried out. In order to separate bound from unbound ^3H-ouabain the samples have to be centrifuged. The radioactivity in the supernatant representing the amount of unbound ouabain corresponds to the concentration of unlabeled cardiac glycosides in the sample.

I. Binding Affinities of Different Cardiac Glycosides

The affinities of digoxin and digitoxin metabolites relative to their parent compounds were investigated by MARCUS et al. (1975). As Table 5 shows the mono- and

Table 4. Chloroform- and non-chloroform-extractable material in the urine after administration of 0.5 mg ^3H-digoxin peroral to five volunteers CKA, HK, H.R-K, AS and HW). The results are expressed as μg digoxin

	1	2	3
Chloroform extractable material			
KA	51.9	73.4	72.0
HK	32.5	36.3	30.5
H.R-K	37.2	40.0	39.0
AS	51.4	58.6	50.3
HW	34.9	38.5	29.8
$\bar{x} \pm SD$ (mean and standard deviation)	41.6 ± 8.4	49.4 ±14.4	44.5 ±15.7
Non-chloroform-extractable material			
KA	18.7	20.6	17.8
HK	10.1	4.8	10.4
H.R-K	8.1	7.1	10.3
AS	13.9	14.5	11.7
HW	14.4	5.5	11.0
$\bar{x} \pm SD$ (mean and standard deviation)	13.1 ± 3.7	10.5 ± 6.1	12.2 ± 2.8

1)=radiochemical determination, 2)=RIA, 3)=enzymatic assay (GUNDERT-REMY et al., 1978)

bis-digitoxosides of digoxigenin have higher binding affinities to ATPase than digoxin whereas digoxigenin and dihydrodigosin bind to the enzyme with lower affinities compared with digoxin. The bis-digitoxoside of digitoxigenin shows a lower binding affinity than the parent drug, in contrast to the monodigitoxoside. Digitoxigenin reveals the same binding affinity as its parent drug.

The similar extraction efficiency and binding profile of possibly occuring metabolites led to the assumption that, besides the parent compounds, the metabolites influence the estimated level.

II. Precision and Sensitivity of the Assay

In their paper MARCUS et al. (1975) gave values of 0.8 and 4.1 ng/ml as limits of detection of digoxin and digitoxin respectively. The accuracy of the estimation of digoxin was stated to be ±0.2 ng/ml over a concentration range of 0–5 ng/ml, the respective values of digitoxin determination was ±0.7 ng/ml over a concentration

Table 5. Relative binding affinities of digoxin and digitoxin and their metabolites to ATPase as assessed by ATPase displacement assay (MARCUS et al., 1975)

Compound	Assay range (pmol)	Relative binding affinities
Digoxin	0–12.5	1.00
Digoxigenin-bis-digitoxoside	0–11.5	0.71
Digoxigenin-mono-digitoxoside	0– 5.8	0.56
Digoxigenin	0–12.5	2.44
Dihydrodigoxin	0– 9.6	10.00
Digitoxin	0–11.9	1.00
Digitoxigenin-bis-digitoxoside	0– 8.0	1.82
Digitoxigenin-mono-digitoxoside	0–11.9	0.53
Digitoxigenin	0–10.1	1.14

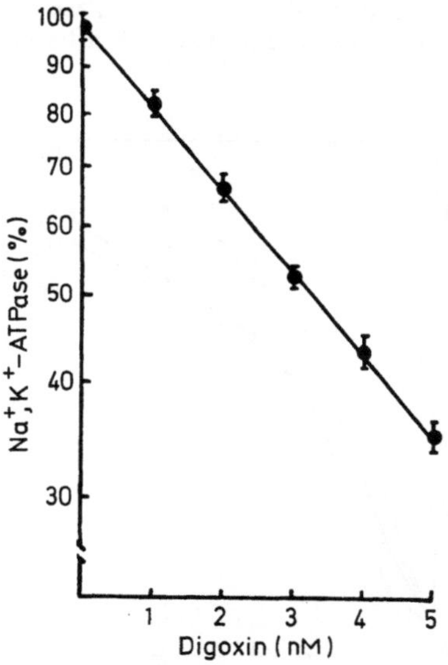

Fig. 2. Standard curve for digoxin obtained by ATPase activity assay. (MARDH, 1973)

range of 0–50 ng/ml. The precision of the determination was found to be ± 0.01–0.6 n/ml for digoxin and ± 0.5–6.0 ng/ml for digitoxin in the range mentioned above.

III. Comparison of Results Obtained by Isotope Displacement and Other Assays

Since data from biological samples measured by isotope displacement assay as well as other techniques are not available, direct comparison is not possible. MARCUS et al. (1975) concluded from their data that the determination of digoxin plasma

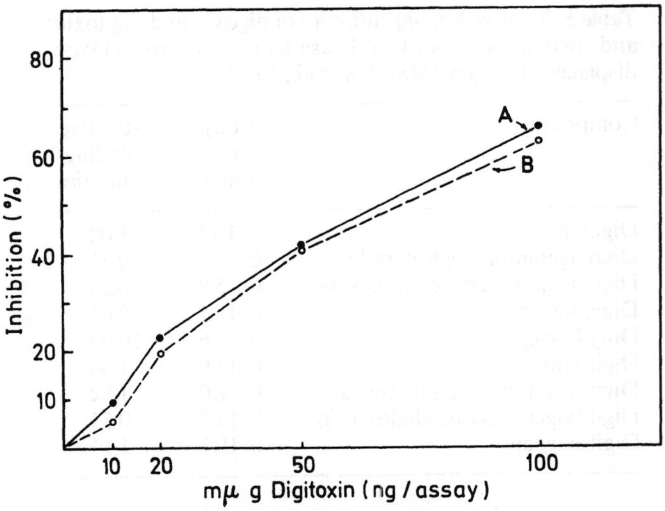

Fig. 3. Standard curves for digitoxin obtained by ATPase activity assay. (BURNETT and CONKLIN, 1968) Curve A without extraction. Curve B after extraction

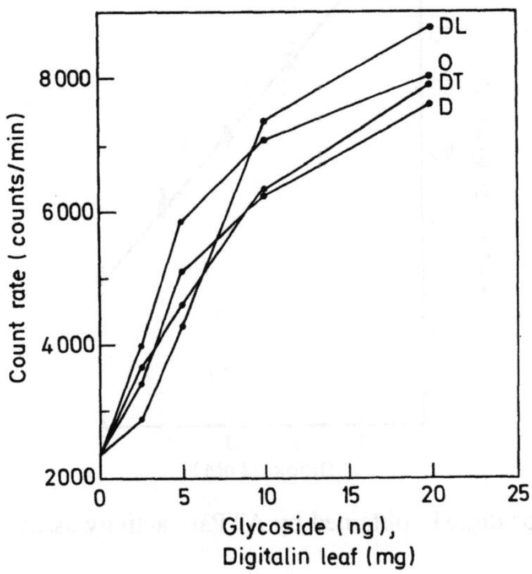

Fig. 4. Standard curves for digoxin (D), digitoxin (DT), ouabain (O), and digitalis leaf (DL) obtained by isotope displacement assay. ^{3}H-labeling efficiency 20%. (BROOKER and JELLIFFE, 1972)

levels should give results comparable to those obtained by radioimmunoassay. They stated, however, that differences might be expected if the ATPase or immunoassay techniques are used to determine the concentration in the urine of patients given digoxin.

Since digitoxin undergoes extensive biotransformation in humans (VÖHRINGER et al., 1974) and striking differences in the displacement of ^3H-ouabain by the metabolites of digitoxin are obvious, MARCUS et al. (1975) assumed that the digitoxin levels in plasma and urine will vary depending on the type of assay used. On the other hand BENTLEY et al. (1970) reported digitoxin plasma levels of patients on maintenance therapy which are comparable to the results obtained by other methods.

F. Commentary

A direct comparison of both methods which employ ATPase in order to measure cardiac glycoside concentrations is not possible since specimens have never been simultaneously determined. The comparison could therefore only be based on data concerning the sensitivity, precision, and specificity. The time required to carry out each type of assay should also be taken into consideration.

Since the ATPase enzyme is stable for not more than 3 months, it is necessary to prepare batches of enzyme from time to time and to characterize them for use in the assay procedure. MARDH (1973) gave a values for the lower level of sensitivity of 0.2 ng/ml digoxin, whereas in our laboratory using the same technique of preparation as MARDH the corresponding value was 0.4 ng/ml, which is in agreement with the result of BURNETT and CONKLIN (1971). The digoxin concentration at which 50% of the enzyme activity is inhibited is reported to be 2.5, 4.5, and 4.6 ng/ml by different authors (BURNETT and CONKLIN, 1971; GUNDERT-REMY et al., 1974; MARDH, 1973). It can be concluded from these data that the assay sensitivity strongly depends on the respective enzyme preparation and that the preparation of the enzyme is a crucial step in the assay. The same is true for the ATPase preparation used in the displacement assay. BELZ et al. (1975) claimed that the discontinuous displacement of ^3H-ouabain and the steep increase of unbound ^3H-ouabain in the interesting range make it difficult to use the displacement assay for routine laboratory determinations.

Both methods are nonspecific since known metabolites can contribute to glycoside levels measured by both techniques. In order to reach effective amounts in the assay several milliliters of plasma have to be extracted which is another disadvantage. The precision of ATPase inhibition as well as displacement assay as reflected by the data cited above can be judged as satisfactory in the therapeutic range though the results from subtherapeutic levels showed a coefficient of variation of more than 10%.

According to BROOKER and JELLIFFE (1972) it takes 60 min and according to MARDH (1973) 90 min to run a sample by displacement and activity assay respectively. From the aspect of time ATPase methods are therefore comparable to the commonly used technique of radioimmunoassay. However, large series of plasma samples can be determined more conveniently by radioimmunoassay because of the extraction procedure necessary for ATPase assay. On the other hand, a major advantage of ATPase assays over radioimmunoassay may be seen in the possibility of estimating levels of different cardiac glycosides with the same instrumentation.

In summary both ATPase assays have the same advantages and disadvantages. In comparison with radioimmunoassay the handling of large amounts of samples in less convenient whereas the possibility of measuring different cardiac glycosides with the same instrumentation is the major advantage of ATPase assay techniques.

Appendix

I. Preparation of ATPase

1. ATPase Activity Assay
(SKOU and HILBERG, 1969)

A sample of 200 g gray matter of ox brain is homogenized with buffer solution (0.24 *M* sucrose, 0.03 *M* histidine, 0.1% Na-desoxycholate, pH 7.2, 2 °C) in a Potter–Elevelijem homogenizator for about 2 min. After centrifugation (10,000 *g*, 20 min, 2 °C) the supernatant is centrifuged again (20,000 *g*, 60 min. 2 °C). The pellet is resuspended in buffer solution (390 ml, 0.25 *M* sucrose, 0.03 *M* histidine, 0.05% Na-desoxycholate, 0.005 *M* EDTA, adjusted to pH 7.2 with 2-amino-2-methyl-1,3-propanediol) and homogenized for 5 min. After centrifugation (10,000 *g*, 20 min, 2 °C) the supernatant is stored overnight at 2 °C. The final enzyme preparation is obtained by fractionated centrifugation in three steps (20,000 *g*, 100,000 *g*, 100,000 *g*, 60 min each, 2 °C) using the above mentioned buffer and may be stored in the final buffer solution (50 ml, 0.25 *M* sucrose, 0.03 *M* histidine, pH 7.2) at −20 °C for 3 months.

2. Isotope Displacement Assay
(BROOKER and JELLIFFE, 1972)

The brain tissue of 3 male guinea-pigs (aged 9–12 months) weighing about 13 g, are homogenized in 20 ml 0.25 *M* sucrose for 1 min at 0 °C. After centrifugation (2,600 *g*, 20 min, 4 °C), the pellet is homogenized with 15 ml 5 m*M* Tris-HCl, pH 7.2, for 1 min at 0 °C. The pellet obtained after a second centrifugation (2,600 *g*, 20 min, 4 °C) is resuspended with 15 ml 5 m*M* Tris-HCl, pH 7.2. This procedure is repeated three times and the resulting pellet is resuspended in 50 ml 50 m*M* Tris-HCl, pH 7.2, and 50 ml of a solution containing 10 m*M* $CaCl_2$ and 5 m*M* Tris-HCl, pH 7.2. After an incubation period of 1 h at 30 °C and centrifugation (2,600 *g*, 20 min, 4 °C) the pellet is taken up with 50 ml 0.6 *M* sucrose, which was diluted to a final concentration of 0.3 *M* sucrose with water. After centrifugation 220 *g*, 5 min, 4 °C) the supernatant is dialyzed against 2 ml 5 m*M* Tris-HCl, pH 8.0 at 4 °C for 12 h. The dialyzed homogenate is concentrated to a volume of approximately 50 ml by centrifugation (2,600 *g*, 20 min, 4 °C), resuspended and quick frozen in 1 ml portions. The enzyme homogenate is stable for 3 months when stored at −20 °C.

II. Extraction Procedure

1. Dichlormethane
(MARDH, 1973)

A volume of 1.25 ml plasma is mixed with 5 ml dichloromethane for 20 s on a vortex mixer. After centrifugation (600 *g*, 5 min) 4 ml of the organic layer are transferred to a test tube and evaporated to dryness (water bath at 43 °C under an air stream). The residue is dissolved in 25 µl benzene. After adding 100 ml water the benzene phase was evaporated in a stream of air.

2. Chloroform
(BROOKER and JELLIFFE, 1972)

A volume of 5 ml plasma is shaken with 30 ml chloroform for 30 s on a vortex mixer. After centrifugation (750 *g*, 2 min) the chloroform layer is decanted into another centrifuge tube and evaporated to dryness at 60 °C (Buchler evapomixer under vacuum). The residue is rinsed down to the bottom with 1 ml of ethyl ether and evaporated to dryness (50 °C, water bath, under an air stream). 130 µl 100 m*M* Tris-HCl (pH 8.0) and 20 µl toluene are then added. After mixing for 20 s and centrifuging (750 *g*, 3 min) the top toluene layer is removed.

III. Assay Procedure

1. ATPase Activity Assay
(MARDH, 1973)

To 100 µl of the final extract 100 µl of a solution containing 0.007 U ATPase, 6 m*M* $MgCl_2$, 6 m*M* orthophosphate (Tris-salt) and Tris-HCl buffer (30 m*M* Tris, pH 7.4) are added. After incubation at 37 °C for 45 min the reaction is started by adding 200 µl of a solution containing 2.8 m*M* ATP, 4.0 m*M* phosphoenolpyruvate, 3 m*M* $MgCl_2$, 240 m*M* NaCl, 40 m*M* HCL, and 30 m*M* Tris-HCl, pH 7.4, and 0.2 U of pyruvate kinase. After 20 min at 37 °C the reaction is stopped by adding 500 µl 2,4-dinitrophenylhydrazine solution. Ten minutes later 500 µl 20 *M* NaCl are added and the reaction mixture diluted with 4 ml 96% ethanol. The absorbance is then measured at 520 min.

2. Isotope Displacement Assay
(BROOKER and JELLIFFE, 1972)

To 50 ml of the buffer solution 20 µl of the enzyme and 20 µl of a mixture of 22.5 m*M* Na_2ATP, 22.5 m*M* $MgCl_2$, 22.5 m*M* EDTA, 450 m*M* NaCl, 0.04 µCi ^3H-ouabain and 100 m*M* Tris (pH 8.0) are added. After incubation for 15 min at 45 °C the bound ouabain is separated from the ouabain not bound to the enzyme by centrifuge (2 min). A sample of 50 µl of the supernatant containing unbound ^3H-ouabain is counted by liquid scintillation.

References

Akera, T., Larsen, F.S., Brody, T.M.: Correlation of cardiac sodium- and potassium-activated adenosine triphosphat activity with ouabain-induced inotropic stimulation. J. Pharmacol. Exp. Ther. *173*, 145 (1970)

Bastide, F., Meissner, G., Fleischner, S., Post, R.L.: Similarity of the active site of phosphorylation of the adenosine triphosphatase for transport of sodium and potassium ions to that for transport of calcium ions in the sarcoplasmic reticulum of muscle. J. Biol. Chem. *248*, 8385 (1973)

Belz, G.G., Pflederer, W.: Studies on a plasma cardiac glycoside assay based upon displacement of ^{3}H-ouabain from Na^+K^+-ATPase. Basic Res. Cardiol. *70*, 142 (1975)

Bentley, J.D., Burnett, G.H., Conklin, R.L., Wasserburger, R.H.: Clinical applications of serum digitoxin levels. Circulation *41*, 67 (1970)

Besch, H.R., Allen, J.C., Glick, G., Schwartz, A.: Correlation between the inotropic action of ouabain and its effect on subcellular enzyme systems from canine myocardium. J. Pharmacol. Exp. Pher. *171*, 1 (1970)

Brooker, G., Jelliffe, R.W.: Serum cardiac glycoside assay based upon displacement of ^{3}H-ouabain from Na-K-ATPase. Circulation *45*, 20 (1972)

Burnett, G.H., Conklin, R.L.: The enzymatic assay of plasma digitoxin levels. J. Lab. Clin. Med. *71*, 1040 (1968)

Burnett, G.H., Conklin, R.L.: The enzymatic assay of plasma digoxin. J. Lab. Clin. Med. *78*, 779 (1971)

Fiske, S., Subbarrow, Y.: A colorimetric method for phosphorus. J. Biol. Chem. *66*, 375 (1925)

Friedemann, T.E., Haugen, G.E.: Pyruvic acid. II. The determination of keto acids in blood and urine. J. Biol. Chem. *147*, 415 (1943)

Gundert-Remy, U., Thorade, B., Karacsonyi, P., Weber, E.: Inhibition of a NaKMg activated ATPase isolated from ox brain by different cardiac glycosides. Naunyn Schmiedebergs Arch. Pharmacol. *284*, R24 (1974)

Gundert-Remy, U., Koch, K., Hristka, V.: Chloroform-extractable and polar metabolites examined with different assays. In: Cardiac glycosides. Bodem, G., Dengler, H.J. (eds.), p. 28. Berlin, Heidelberg, New York: Springer 1978

Hopkins, B.E., Wagner, J.H., Smith, T.W.: Sodium- and potassium-activated adenosine triphosphatase of the nasal salt gland of the duck, purification, characterization and NH_2-terminal amino acid sequence of the phosphorylating polypeptide. J. Biol. Chem. *251*, 4365 (1976)

Kyte, J.: The titration of the cardiac glycoside binding site of the ($Na^+ + K^+$)-denosine-triphosphatase. J. Biol. Chem. *247*, 7642 (1972)

Marcus, F.J., Burkhalter, L., Cuccia, C. Pavlovich, J., Kapadia, G.G.: Administration of tritiated digoxin with and without loading dose. A metabolic study. Circulation *34*, 865 (1966)

Marcus, F.I., Ryan, J.N., Stafford, M.G.: The reactivity of digoxin and derivates of digoxin and digitoxin as measured by the Na-K-ATPase displacement assay and by radioimmunoassay. J. Lab. Clin. Med *85*, 610 (1975)

Mardh, S.: A simple enzymatic assay of cardiac glycosides and its application to analysis of glycoside levels in plasma. Clin. Chim. Acta *44*, 165 (1973)

Marsh, B.B.: The estimation of anorganic phosphate in the presence of adenosine triphosphate. Biochim. Biophys. Acta *32*, 357 (1959)

Matsui, H., Schwartz, A.: Mechanism of cardiac glycoside inhibition of the ($Na^+ + K^+$)-dependent ATPase from cardiac tissue. Biochim. Biophys. Acta *151*, 655 (1968)

Schoner, W., von Ilberg, C., Kramer R., Seubert, W.: On the mechanism of Na^+- and K^+-stimulated hydrolysis of adenosine triphosphate. Eur. J. Biochem. *1*, 334 (1967)

Skou, J.C.: The influence of some cations on an adenosine-triphosphatase from peripheral crab nerves. Biochim. Biophys. Acta *23*, 394 (1957)

Skou, J.C.: Further investigations on a Mg^{++}-Na^+ activated adenosine-tri-phosphatase possibly related to the active, linked transport of Na^+ and K^+ across the nerve membrane. Biochim. Biophys. Acta *42*, 6 (1960)

Skou, J.C., Hilberg C.: The effect of cations, g-strophanthin, and oligomycin on the labeling from (32P)-ATP of the Na^+, K^+-activated enzyme system and the effect of cations and g-strophanthin on the labeling from (32P)-ITP and 32 Pi. Biochim. Biophys. Acta *185*, 198 (1969)

Vöhringer, H.F., Rietbrock, N.: Metabolism and excretion of digitoxin in man. Clin. Pharmacol. Ther. *16*, 796 (1974)

Wilson, D.B.: Cellular transport mechanisms. Ann. Rev. Biochem. *47*, 933 (1978)

CHAPTER 7

Rubidium Uptake in Erythrocytes

G. G. BELZ

A. Introduction and Principle of the Method

Active uptake by „Kälteerythrocyten" (i.e., human erythrocytes stored at 4 °C for 5–8 days) of photometrically measured potassium has been demonstrated to be inhibited by the cardiac glycosides (c.g.), k-strophanthoside, and digitoxin and their aglucons (SCHATZMANN, 1953). The drug concentrations necessary to yield this effect ranged from 0.1 µg/ml to 10 µg/ml. This effect was shown to be due to an inhibition of membrane Na^+-K^+-ATPase activity (POST et al., 1960; DUNHAM and GLYNN, 1961; GLYNN, 1957).

Potassium uptake is inhibited at a range of concentrations of glycosides present in plasma of patients receiving digitalis therapy. Therefore LOWENSTEIN (1965) introduced the principle of inhibition of active cation uptake by erythrocytes to measure plasma concentrations of cardiac glycosides. Using isotope techniques the sensitivity of the methodological principle was much increased, so that „Kälteerythrocyten" could be replaced by untreated erythrocytes which showed significant effects and were much easier to obtain and to standardize. Potassium, however, had to be replaced by rubidium, since available potassium isotopes are impractical for standard laboratory techniques. Since ^{86}Rb has a half-life of 19.7 days and emits β- and γ-rays, it is very appropriate for routine laboratory use. $^{86}Rb^+$ is taken up by human erythrocytes in the same way as potassium (LOVE and BURCH, 1953).

Figure 1 shows the curve of ^{86}Rb uptake of erythrocytes with and without c.g.

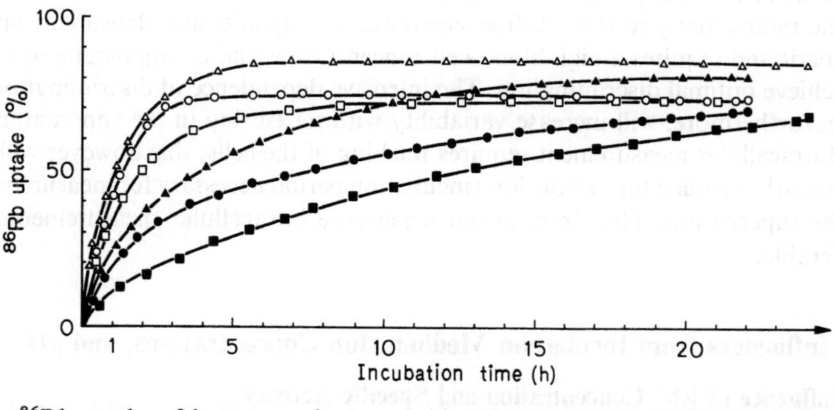

Fig. 1. ^{86}Rb-uptake of human erythrocytes with 100 mg of digitoxin (full symbols) and without cardiac glycoside in the assay system. Age of the erythrocytes: ○ = 24 h; △ = 1 week; □ = 3 weeks (according to BELZ et al., 1972)

The inhibitory effect of c.g. upon active ^{86}Rb uptake by erythrocytes (using the ^{86}Rb erythrocyte assay = ^{86}REA) can be used (1) for measuring c.g. concentrations in various specimes (e.g. plasma, urine) and (2) for testing unknown substances for their bioactivity in respect to ion transport. From this property – with certain reservations – prognoses on the cardioactivity of the substances may be made.

B. Factors Affecting the ^{86}Rb Uptake of Human Erythrocytes

The original method described by LOWENSTEIN (1965) for measuring glycoside plasma levels was imperfect in at least two aspects:

Accuracy and reproducibility were not optimal (BERTLER and REDFORS, 1970, 1973), and a false-positive glycoside-like activity was detected in about 7% of these cases (LOWENSTEIN, 1965). LOWENSTEIN and CORRILL (1966) introduced a dichlormethane extraction procedure to improve the method. Besides this improvement other modifications of various parameters affecting the assay resulted in an accurate and reproducible method.

In the following the influence of some well-documented parameters on the method is shown.

I. Measurement of ^{86}Rb-Activity

^{86}Rb has β- as well as γ-emission. At 1.08 and 0.68 MeV γ-emission accounts for 8.5% and β-emission for 91.5% of the total radioactivity (WILSON, 1966).

In principle measurement of both γ- and β-activity is suitable for the ^{86}REA (BELZ et al., 1972). The higher amount of β-activity would favor using β-counting. Practical experience, however, has shown that using β-counting many problems arose with quench correction etc. and therefore γ-counting using well-type γ-spectrometers has been shown to be most convenient.

Another principal question is whether one should measure radioactivity in the red cells or in the supernatant. Data from GJERDRUM (1972) show that recording of the radioactivity in the cell-free supernatant is significantly dependent on hematocrit and requires a high blood-cell concentration and a long incubation time to achieve optimal discrimination. The intensive dependence of discrimination on time, furthermore, will increase variability with measuring in the supernatant.

Intracellular measurement requires washing of the cells; this however will not take nearly as much time as the long incubation period necessary for measurements in the supernatant. Therefore, in our experience intracellular measurements are preferable.

II. Influences from Incubation Medium, Ion Concentrations, and pH

1. Influence of Rb^+ Concentration and Specific Activity

By varying the specific activity of Rb–^{86}RbCl mixture from 232 to 4.0 µCi/mg Rb it was shown that uptake of Rb was independent of specific activity when it was above 50–60 µCi/mg Rb. With lower specific activities (higher concentrations of stable Rb) the stable Rb supresses the incorporation of ^{86}Rb (GJERDRUM, 1970). Increasing the concentration of "cool" RbCl in the assay leads to a decrease of ^{86}Rb uptake into the red cells. Compared to an assay with glycoside-inhibited cells, the discrimination between the two decreases with increasing Rb^+ concentrations. Within the range between 0 to 5 mg% of RbCl, however, the influence is negligible (BELZ et al., 1972). Potassium in the incubation medium will result in the same effect, as it is known that especially with concentrations >4 mEq/l a slower and non-linear intracellular transport has been found in samples without digitalis (DUNHAM and GLYNN, 1961; HOFFMAN, 1966).

High K^+ or Rb^+ concentrations may further decrease the glycoside-induced inhibition of ATPase (DUNHAM and GLYNN, 1961; HOFFMAN, 1966).

The results indicate that for practical use of the ^{86}REA concentrations of K^+ or Rb^+ should be kept both constant and low (i.e., above 5 mg%).

2. Influence of Sodium Concentration

The uptake of ^{86}Rb in cells increases with decreasing concentration of sodium however, discrimination decreases with low sodium concentrations (GJERDUM, 1970). For practical purposes physiologic 0.9% saline has been shown to yield sufficient results (BELZ et al., 1973).

3. Influence of Calcium and Magnesium Concentration

The presence of these cations alone or together was without influence on ^{86}Rb uptake and glycoside discrimination (GJERDRUM, 1970).

4. Influence of pH During Incubation

The intracellular uptake of ^{86}Rb has a maximum at pH 7.2–7.4. This value was obtained by standard procedures (GJERDRUM, 1970; BELZ, unpublished).

III. Influence of the Erythrocyte Preparation

1. Concentration of Erythrocytes in the Incubation Medium

GJERDRUM (1970) has analyzed the influence of the concentration of red blood cells in the incubation medium in this method. With concentrated blood suspensions a maximum uptake was obtained after 3–4 h. When the incubation period was kept constant (4 h) and varying concentrations of digitoxin were added to the assay, the discrimination increased with the erythrocyte concentration up to about 15 g Hb/100 ml. With higher erythrocyte concentration (21.8 g) a lower discrimination was found. Optimal discrimination is found at about 15 g Hb/100 ml in the incubation tubes.

2. Age of the Erythrocyte Preparation

Figure 1 shows the influence of the age of erythrocyte preparation on the ^{86}REA. The percentage of ^{86}Rb uptake with or without digitoxin in the assay shows in principle a decrease with the age of the cell preparation. The same is true for the discrimination in the groups with and without digitalis. However, the influence of age is only minor and will not play a great role if the parameter is kept constant within certain limits.

3. Source of Erythrocyte Samples

No data are available on the influence of individual differences among the erythrocytes, e.g., blood group, age, or race of the donor. In our experience these seem not to play a significant role.

It should be noted, however, that some diseases may have distinct influences on the active ion transport of erythrocytes. Reticulocytes have a much more intensive metabolism and will eagerly take up the $^{86}Rb^+$. Erythrocyte preparations from patients with significant reticulocytosis (e.g., hemolytic anemia) will therefore have quite different properties and must be excluded from standard use (BELZ and KLEEBERG, unpublished).

With erythrocytes from patients receiving digoxin treatment a significant decrease in the 60-min uptake of $^{86}Rb^+$ was shown compared to the control. The amount of inhibition under this condition is variable in time and seems not to be dependent on digoxin plasma level or cardiac effect (GRAHAME-SMITH and ARONSON, 1978).

To exclude any error resulting from the erythrocytes used in the test, it is best to use only preparations from healthy young people not taking any drugs.

IV. Influence of Incubation Procedures on the ^{86}REA

1. Incubation Temperature

The uptake of ^{86}Rb by the cells reaches a maximum at about 40 °C when glycosides are absent. With digitoxin the temperature is somewhat lower. The discrimination between different concentrations of digitoxin increases with increasing temperature. However, as increased hemolysis takes place at 40 °C and for simplicity in the laboratory, a temperature of 37 °C is sufficient (GJERDRUM, 1970).

2. Time of Incubation of Erythrocytes with $^{86}Rb^+$

Discrimination in assays with and without digitoxin depends on the incubation time, as is shown in Fig. 2. Optimal discrimination is reached after about 3 h (BELZ et al., 1972). For routine use a 2 h period is sufficient.

3. Influence of Preincubation of Erythrocytes with Digitalis

Preincubation of digitoxin with the red cells prior to administration of ^{86}Rb increases glycoside-induced inhibition of cation uptake (BELZ et al., 1972). Preincubation for 120 min seems to result in a nearly asymptotic effect. Results from BERTLER and REDFORS (1970, 1973) correspond well to this observation.

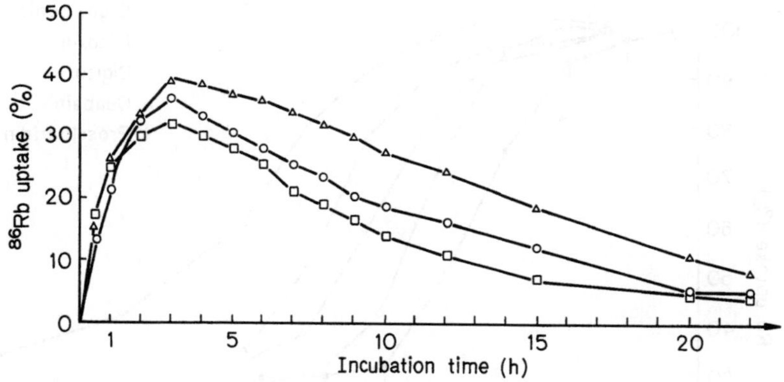

Fig. 2. Influence of incubation time on the discrimination of the assays with and without digitalis (BELZ et al., 1972). For symbols see Fig. 1

To minimize variations for routine use a preincubation period between 60 and 120 min is optimal.

V. Separation of Erythrocytes from Incubation Medium After Incubation

For measuring radioactivity in the supernatant a single centrifugation after incubation is sufficient.

For intracellular measurements it is necessary to eliminate the extracellular ^{86}Rb. It was shown by washing the erythrocytes three times with double their volume that less than 5% of the radioactivity remained after the first washing and almost none after the second washing (BELZ et al., 1972).

According to this, washing the erythrocytes twice with double their volume of saline at 0 °C is sufficient to eliminate the extracellular ^{86}Rb in routine use of the method.

C. Influence of Various Cardiac Glycosides, Genins, and Conjugates on the ^{86}REA

Using a standardized ^{86}REA the inhibitory effects of a series of c.g., genins, and derivates were tested.

Method (BELZ et al., 1973). Laboratory work was done with an Eppendorf Microliter system, including microliter pipettes, except application of erythrocytes which were measured with a 500-µl tuberculin syringe. A Telefunken γ-counter was used (zero effect 50–60 cpm).

Each determination of Rb uptake was done in triplicate. For preparation of the erythrocytes venous blood was taken in 3.8% Na citrate (10:1) from fasting healthy men. After centrifugation plasma and leukocytes were separated from the erythrocytes by suction.

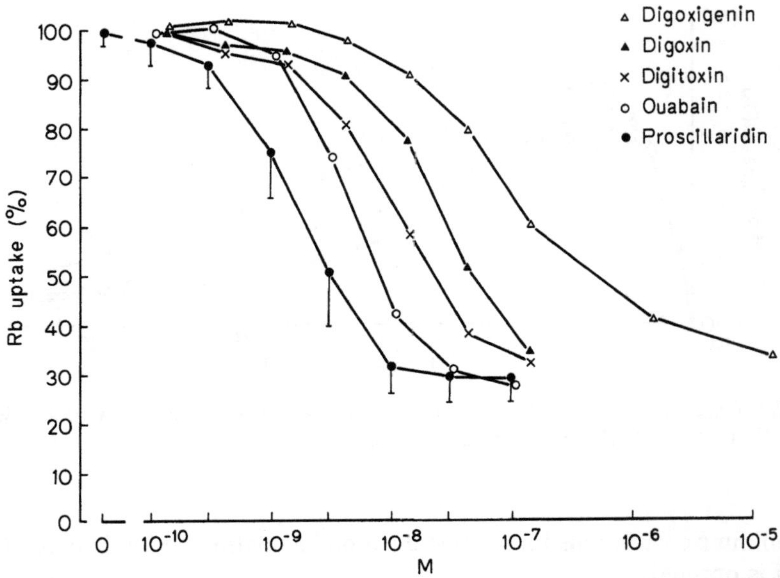

Fig. 3. Inhibition of ^{86}Rb-uptake induced by some c.g. (Belz et al., 1973). For proscillaridin a mean (S.D.) of 13 standard curves is shown

Erythrocytes were washed twice with five times their volume of sterile saline. Erythrocytes packed at 1,200 × *g* (tip) were used on the same day. Each reaction vessel was filled with 1 ml sterile saline containing the glycoside and 400 µl of erythrocytes. The mixture was preincubated for 2 h at 37 °C.

Thereafter, 1 µCi ^{86}Rb activity and 1 mg glucose in 50 µl saline was added and incubated for 2 h at 37 °C. After centrifugation the supernatant was discarded. The remaining erythrocytes were washed twice with 1 ml saline at 0 °C in order to remove the extracellular radioactivity. After the last suctioning, ^{86}Rb-activity incorporated by the erythrocytes was directly counted for 1 min. Impulse rates ranged from 1,500 to 3,500 cpm.

Calculations. The median of three determinations without glycosides was defined to be 100% ^{86}Rb uptake, and the percentage uptake of ^{86}Rb by the remaining samples was calculated.

The glycoside concentration having half-maximal effect is denoted IC_{50}.

The typical dose-response curves for c.g. in the ^{86}REA are shown in Fig. 3. It is evident that in general the inhibition curves are similar and independent of the glycoside. None of the substances tested completely inhibits ^{86}Rb uptake by the erythrocytes. The maximal inhibition depends on the individual erythrocytes and not on the substances use. Generally, the ^{86}Rb-uptake is maximally inhibited by glycoside at about 30% of the glycoside-free control. From the individual incubation curves the half-maximal inhibitory concentrations (IC_{50}) were determined and the results are shown in Table 1. A wide range of potency of the various substances is shown; with the proscillaridin group half-maximal inhibition occurs at the lowest concentrations, i.e., these glycosides show the greatest potency in inhibiting active cation transport.

Table 1. Molar concentrations necessary to achieve half-maximal inhibition of ^{86}Rb uptake by human erythrocytes (IC_{50}). Data from different studies with somewhat different assay techniques are compiled. A, pre- and main incubation 2 h (BELZ et al., 1972); B, 1 h pre- and 3 h main incubation (BELZ and HEINZ, 1977)

Bufadienolides	IC_{50} ($\times 10^{-9}$M)	
	A	B
Proscillaridin (P.)	1.6	./.
P.-3'acetate	1.7	./.
P.-4'acetate	1.8	./.
P.-3'methyl ether	2.0	./.
P.-4'methyl ether	2.2	./.
P.-β-epoxide	2.4	./.
P.-α-epoxide	3.6	./.
3 β-Scillarenin	3.0	./.
3 α-Scillarenin	59.0	./.
Scillarenin 3 β-methyl ether	4.0	./.
Scillarenin 3 α-methyl-ether	55.0	./.
Scillarenin 3 β-methyl-ether 4-β-5 epoxide	13.0	./.
Scillarenin 3 β-methyl-ether 4-α-5 epoxide	140.0	./.
Cardenolides	IC_{50} ($\times 10^{-9}$M)	
	A	B
Ouabain	5.0	./.
Lanatoside C	14.5	./.
Digoxin	24.0	19.5
Digoxin-16'-βD-glucuronide	./.	45.0
Digoxigenin-bis-digitoxoside	15.0	14.0
Digoxigenin-mono-digitoxoside	8.5	9.0
Digoxigenin	140.0	67.0
Digoxigenin-3-β-D-glucuronide	./.	1580.0
Digitoxin	10.0	7.6
Digitoxin-16'-β-D-glucuronide	./.	14.0
Digitoxigenin-bis-digitoxoside	./.	3.2
Digitoxigenin-mono-digitoxoside	./.	4.6
Digitoxigenin-mono-digitoxoside-4'-glucuronide	./.	10.9
Digitoxigenin	./.	29.3
Digitoxigenin-3-β-D-glucuronide	./.	172.0
Digitoxigenin-3-sulfate	./.	183.0
Epidigitoxigenin-3-β-D-sulfate	./.	4800.0
Canarigenin 3-β-methyl ether	110.0	./.
Canarigenin-3-β-methyl ether 4-α-5-epoxide	7500.0	./.

D. Specificity of the Inhibition of ^{86}Rb Uptake

I. Diverse Drugs

The influence of a variety of substances and drugs on ^{86}Rb uptake by human erythrocytes was analyzed (see Table 2). None of the substances tested showed any effect upon the assay within the concentration range tested.

Table 2. Substances which had no influence on the ^{86}REA in the tested range of concentrations (Vollmer et al., 1972)

	Range of concentrations tested	
	From (molar)	To (molar)
Coenzymes		
Ergothioneine	7×10^{-7}	7×10^{-4}
Folic acid	3.5×10^{-7}	3.5×10^{-4}
NAD^{+}	1×10^{-7}	1×10^{-3}
Pyridoxal phosphate	7×10^{-7}	7×10^{-4}
Amino acids and related amines		
Arginine	1×10^{-6}	1×10^{-3}
Creatinine	1×10^{-6}	1×10^{-3}
Cysteine	1×10^{-6}	1×10^{-3}
Histamine	1×10^{-6}	1×10^{-3}
Histidine	1×10^{-6}	1×10^{-3}
Serotonin	1×10^{-6}	1×10^{-3}
Spermidine	1×10^{-7}	1×10^{-4}
Spermine	1×10^{-7}	1×10^{-4}
Tryptamine	1×10^{-6}	1×10^{-3}
Tyramine	1×10^{-6}	1×10^{-3}
Proteins and peptides		
Bradykinin	1×10^{-8}	1×10^{-5}
Clupeine sulfate	1×10^{-8}	1×10^{-5}
Fibrinogen (human)	1×10^{-9}	1×10^{-6}
Glucagon	1×10^{-9}	1×10^{-5}
Glycylglycine	5×10^{-7}	5×10^{-4}
Histone sulfate	2×10^{-7}	2×10^{-4}
Insulin (cattle)	1×10^{-9}	1×10^{-5}
Protamine sulfate	1×10^{-8}	1×10^{-5}
Serum albumin (cattle)	1×10^{-9}	1×10^{-5}
Trypsin inhibitor (soya bean)	1×10^{-9}	1×10^{-5}
Vasopressin	1×10^{-9}	1×10^{-5}
Drugs, primarily of the following groups: *sympathicolytics and sympathicomimetics, vagolytics and vagomimetics, barbiturates, steroid hormones, antiarrythmics, etc.*		
Acetazolamide	1×10^{-6}	1×10^{-3}
Acetylcholine	1×10^{-6}	1×10^{-3}
Atropine sulfate	5×10^{-7}	5×10^{-4}
Curarine	2.5×10^{-7}	2.5×10^{-4}
Epinephrine	5×10^{-9}	2.5×10^{-5}
Ethyl-methyl-butylbarbituric acid	1×10^{-6}	1×10^{-3}
Glycerine	1×10^{-6}	1×10^{-3}
Heparin	1.2×10^{-7}	1.2×10^{-4}
Hydrocortisone acetate	2.5×10^{-7}	2.5×10^{-4}
Hydroxyethyltheophylline	7.5×10^{-7}	7.5×10^{-5}
Isoproterenol	5×10^{-9}	2.5×10^{-5}
Mepyramine	3×10^{-7}	3×10^{-4}
Mescaline sulfate	3×10^{-7}	3×10^{-4}
Methyl-cyclohexenyl-methyl-barbituric acid	1×10^{-7}	1×10^{-3}
Thymocamine	7.5×10^{-7}	7.5×10^{-4}
L-Norepinephrine	1×10^{-6}	1×10^{-3}

Table 2 (continued)

	Range of concentrations tested	
	From (molar)	To (molar)
Oxytocin	1×10^{-9}	1×10^{-5}
Pethidine	1×10^{-6}	1×10^{-3}
Phentolamine	5×10^{-6}	5×10^{-3}
Phenylbutazone	5×10^{-7}	5×10^{-3}
Phenylephrine	5×10^{-9}	2.5×10^{-5}
Physostigmine	5×10^{-7}	5×10^{-4}
Pilocarpine	7.5×10^{-7}	7.5×10^{-4}
Procaine	3×10^{-5}	3×10^{-2}
Procainamide	3×10^{-5}	3×10^{-2}
Pyridamol	5×10^{-6}	0.5×10^{-3}
Verapamil	1×10^{-7}	1×10^{-4}
Others		
1,4-Diaminobutane (Putrescin)	2×10^{-5}	2×10^{-2}
1,6-Diaminohexane	2×10^{-5}	2×10^{-2}
1,4-Diaminopentane (Cadaverin)	2×10^{-5}	2×10^{-2}
Polyvinylsulfate K-salt	1×10^{-6}	1×10^{-3}

The results of GRAHAME-SMITH and EVEREST (1969) indicate that neither drugs such as chlorothiazide, ethacrynic acid, furosemide, methyldopa, guanethidine, bethanidine, and prednisolone nor quinidine or bishydroxycoumarin interfere with the method. It was shown that allocymarin, verapamil, tocopherol (WISSLER et al., 1972) as well as quinidine (DOERING and BELZ, 1981) do not interfere with the glycoside-induced inhibition of ^{86}Rb uptake.

II. Spironolactone

In contrast to early results (GRAHAME-SMITH and EVEREST, 1969), it has been shown that spironolactone and its metabolites (canrenone and canrenoate) inhibit ^{86}Rb uptake by human erythrocytes (BELZ and KLEEBERG, 1973; KLEEBERG and BELZ, 1974).

Figure 4 shows the influence of various derivatives of spironolactone on the ^{86}REA. The highest potency is seen with spironolactone itself. Differences between the Na^+ and K^+ salts of canrenoate are caused by the potassium ions. With canrenone it is possible to increase the half-maximal inhibition of ^{86}Rb uptake induced by proscillaridin (KLEEBERG and BELZ, 1974). Maximum glycoside-induced inhibition of cation transport, however, cannot be changed by the addition of canrenone.

III. Human Plasma

In the first paper on the method, LOWENSTEIN (1965) reported an unspecific glycoside-like inhibitory effect of plasma from patients, most of whom were suffering

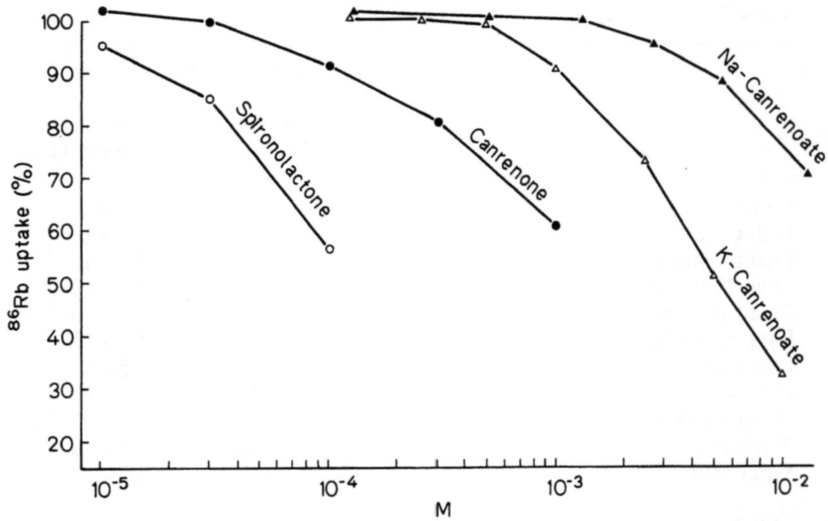

Fig. 4. Influence of spironolactone and some of its derivatives on the ^{86}REA (KLEEBERG and BELZ, 1974)

from metastatic carcinoma, severe fluid and electrolyte disturbances, or heart failure. With dichlormethane extraction these high "false positive" glycoside levels were no longer detectable. One can assume the substances producing this activity to be polar. There are no data available about their nature.

Another problem arises from the glycoside-binding properties of plasma proteins. Increasing plasma concentrations in the incubation system resulted in two different groups of glycosides (BELZ and SCHREITER, 1974). With digitoxin, proscillaridin, meproscillarin, and peruvoside, addition of plasma induced a shift of the curves to the right, i.e., higher glycoside concentrations were needed to produce an equal inhibition of ^{86}Rb uptake; with ouabain, digoxin, β-methyldigoxin, and lanatoside C, plasma did not inhibit the influence of c.g. on the ^{86}Rb uptake. Intensive protein binding of the glycosides of the first group is responsible for this effect.

E. Correlation of Activity of Cardiac Glycosides in ^{86}REA and Cardioactivity

As the relative potency of different c.g. in respect to the inhibition of ^{86}Rb uptake show distinct differences, studies were done to correlate the activity of the c.g. in this system with the cardioactivity of the drugs. For measuring the latter parameter cardiotoxicity (expressed as the minimum lethal dose in cats and guinea pigs) (BELZ et al., 1973) and positive inotropic effects (expressed as the concentration necessary to double the contractile force of guinea pig papillary muscles) (BELZ and HEINZ, 1977) were used.

Figure 5 shows the correlation between IC_{50} and positive inotropism ($C_{+100\%}$) in guinea pig papillary muscles. The potency of glycosides and deriv-

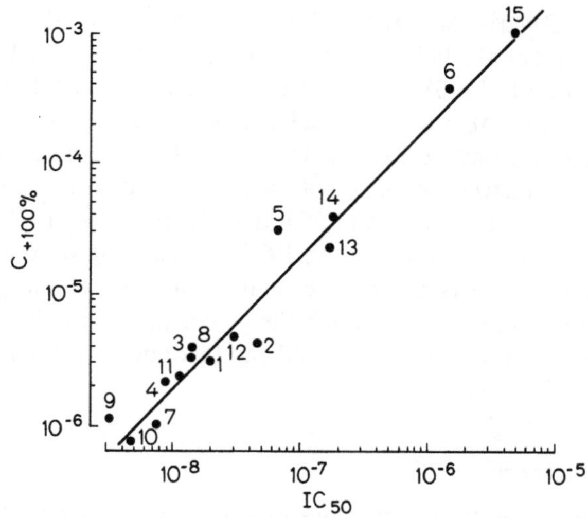

Fig. 5. Molar concentrations of various cardenolides and cardenolide metabolites to achieve half-maximal inhibition of ^{86}Rb-uptake in human erythrocytes (IC_{50} on abscissa) and to achieve doubling of force of concentration in isolated guinea pig papillary muscle ($C_{+100\%}$, on the ordinate) (BELZ and HEINZ, 1977). Substances are identical to group B in Table 1

atives in both systems runs largely parallel. The correlation coefficients were very high (0.9996 and 0.9819) according to lin/linear or log/logarithmic calculation, respectively, the slope for the latter correlation was nearly unity. The ratio of potency of 200:1 between the Rb uptake and positive inotropic action resulted from a different time-course pattern in the two experiments, as well as from species differences.

The correlation analysis finally results in the conclusion that cardioactive glycosides and derivatives are detected (and eventually measured) analogous to their activity. Using the ^{86}REA for concentration measurements in body fluids or tissues one can assume that the detected activity represents cardioactive compounds analogous to their cardioactivity. With glycosides undergoing metabolism this may be a crucial advantage of this assay over competitive methods.

F. Use of ^{86}REA for Measurement in Body Fluids

As an extraction step is part of the method, these can be adapted to nearly any desired measurement. Only the volumes of samples and the volume of extraction medium must be adapted to measure, if possible, in the steep part of the dose-effect curves.

I. Plasma Glycoside Concentrations

In principle the method can be used with any glycoside since it detects a biological activity rather a specific substance. Methodological modifications of the assay have been developed for measurement of plasma concentrations of digitoxin (BELZ and ERBEL, 1979; GJERDRUM, 1970, 1972; RASMUSSEN et al., 1971; SHAPIRO et al., 1970), digoxin (BELZ et al., 1974c; BERTLER and REDFORS, 1970, 1973; BINNION et al., 1969; SHAPIRO et al., 1970); and its derivatives (BELZ, 1974; BELZ et al., 1974b; BELZ and NÜBLING, 1975), proscillaridin (ANDERSSON et al., 1975b; ANDERSSON et al., 1977a, d; BELZ and BRECH, 1974; BELZ et al., 1974c, d), meproscillarin (BECKMANN et al., 1978; BECKMANN et al., 1979; BELZ and BADER, 1974; BELZ and BELZ, 1978; BELZ et al., 1974a; BELZ et al., 1976a, b; RUPP et al., 1978; TWITTENHOFF et al., 1978), and pengitoxin (HAUSTEIN, 1978). Depending on the individual glycoside concentration in plasma and the relative potency of the glycoside in the ^{86}REA, the plasma volume, from which the glycoside has to be extracted, varies. The following techniques of the assay itself can be done in a standardized way, independent of the glycoside.

1. Extraction Procedure

As the assay is disturbed by the presence of plasma (compare Sect. III) extraction of the glycoside was introduced. Following the original method of LOWENSTEIN and CORRILL (1966) dichlormethane is used as extraction medium.

The required volumes of plasma and extraction medium are summarized as follows:

Glycoside	Digitoxin	Digoxin and derivatives	Proscillaridin, meproscillarin
Plasma volume (ml)	0.4	10	10
Dichlormethane volume (ml)	6	15	15

Plasma and dichlormethane are shaken intensively by hand for 30–60 s.

The relatively high volumes needed for digoxin and the squill glycosides are a consequence of the low plasma concentrations of these glycosides. Modifications of the ^{86}REA will also work with 1–2 ml of plasma (BERTLER and REDFORS, 1970). However, the inhibition induced by therapeutic glycoside concentrations in these assays will be in the initial parts of the dose-effect curve.

2. Preparation of the Extract

Aliquots of 10 ml with digoxin and squill glycosides and of 5 ml digitoxin are evaporated to dryness at 37 °C, and the dried extract is transferred quantitatively into an Eppendorf reaction vessel with three lots of 500 µl of dichlormethane.

The extract is further evaporated to dryness at room temperature and normal atmospheric air pressure. With digoxin and squill glycosides 400 µl cyclohexane are added and shaken for 10 min. This procedure allows the elimination of nonpolar substances (e.g., cholesterol) which were extracted from plasma with dichlormethane and which increase hemolysis in the ^{86}REA. The glycosides are not extracted by means of cyclohexane. Thereafter, 1,100 µl 1% methanol-saline are added and shaken for 10 min, the upper cyclohexane phase is removed by suction, and 1,000 µl of the saline phase is pipetted into a fresh reaction vessel.

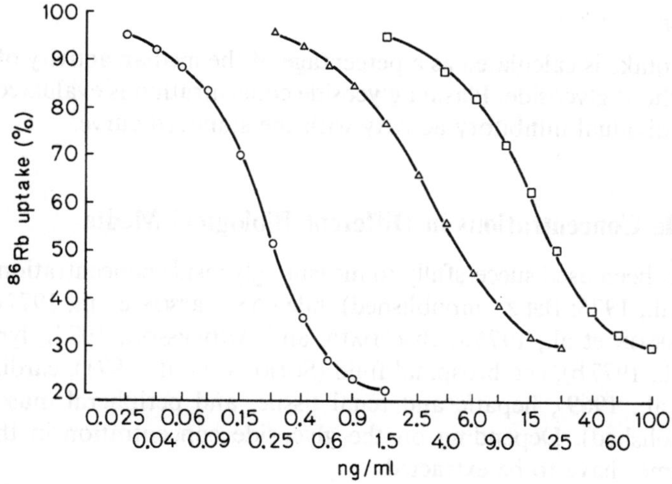

Fig. 6. Standard curves for digitoxin-□-, digoxin-△-, and meproscillarin -○- in the method used for serum concentration measurements

3. Preparation of Erythrocytes

Blood of healthy man is taken on heparin. The erythrocytes are separated from the plasma by centrifugation (1,200 *g*). Plasma and leukocytes are sucked off. Next the cells are washed twice with three times their own volume of sterile saline and packed by centrifugation at 1,200 *g*.

The fresh erythrocytes are used for the assay. In our experience, erythrocytes prepared in this way can be used for about 2 days after which hemolysis increases.

4. Incubation Assay

The packed erythrocytes (400 µl) are immediately pipetted into the reaction vessel containing the glycoside extract in 1% methanol-saline and incubated for 1 h.

Thereafter 0.01 µCi of ^{86}RbCl (initial specific activity 75–100 µCi/mg) are added in a volume of 50 µl in saline. Afterwards the tubes are incubated for 3 h with shaking of the reaction vessels every 30 min. Centrifugation at 9,700 *g* separates the erythrocytes from the supernatant, which is discarded by suction. The remaining erythrocytes are washed twice with 1 ml saline at 0 °C to remove the extracellular ^{86}RbCl. After the last suction the ^{86}Rb gamma activity incorporated by the red cells is counted directly for 10 min (we use a well-type γ-spectrometer "Auto-Logic" Abbott Laboratories).

5. Standard Curves

Standard curves with varying concentrations of the glycoside are run in parallel with each assay. For this purpose the glycoside is directly added to 10 ml (for digoxin and proscillaridin) or 0.4 ml (for digitoxin) portions of pooled human plasma (from healthy men). These values are handled identical with the samples. Some standard curves for digitoxin, digoxin, and meproscillarin are shown in Fig. 6.

6. Calculation

Basic ^{86}Rb uptake is calculated as a percentage of the median activity of three estimations without glycoside. Plasma glycoside concentration is evaluated by comparing the individual inhibitory activity with the standard curve.

II. Glycoside Concentrations in Different Biological Media

The assay has been used successfully to measure glycoside concentrations in urine (BERTLER et al., 1975; BELZ, unpublished), bile (ANDERSSON et al., 1977c), gastric juice (ANDERSSON et al., 1975a; BERGDAHL and ANDERSSON, 1977), lymph (ANDERSSON et al., 1977b), cerebrospinal fluid (SOMOGYI et al., 1971), cardiac muscle (BINNION et al., 1969), hepatic and renal tissue, and peripheral muscle tissues (BELZ, unpublished). Depending on the glycoside concentration in the sample varying volumes have to be extracted.

After extraction the ^{86}REA is performed as usual. Standard curves must be run in parallel using extraction from analogous glycoside-free samples.

An additional modification of the method was used by L. Storstein to evaluate the concentration of glycoside in chromotographic samples to analyze the metabolism of digitoxin (see this Handbook, Vol. 56/II, Chap. 1).

G. Comparison of Plasma Glycoside Measurements Using ^{86}REA and Immunochemical Methods

I. Determination of Plasma Digoxin

BERTLER and REDFORS (1973) analyzed 100 routine digoxin serum samples from clinic patients by using ^{86}REA and RIA. The results obtained with the two showed a rather good correspondence in about three-fourths of the cases. A marked difference was found in one-fourth. In 13 of the cases the radioimmunochemical value exceeded the ^{86}Rb result by more than 0.6 ng/ml. Nine of these patients had simultaneously been treated with spironolactone. These "false high" results with RIA are of course dependent on the specificity of the antibody used for the assay.

It can be concluded from the results that there is good agreement between the results of ^{86}REA and RIA if certain interfering treatments are excluded.

II. Determination of Plasma Digitoxin

Comparative measurement of plasma digitoxin was done using a digoxin ELISA system modified for digitoxin measurements (BELZ and BELZ, 1979) and the ^{86}REA. In Fig. 7 the correlation between the results obtained with the two methods is shown for 23 patients on chronic digitoxin treatment.

Correlation of plasma digitoxin levels obtained by RIA on the one hand and by ^{86}REA on the other from samples obtained from healthy volunteers under 7 days of treatment (BELZ et al., 1978) yielded a correlation with $r = 0.80$ (BELZ and ERBEL, 1979).

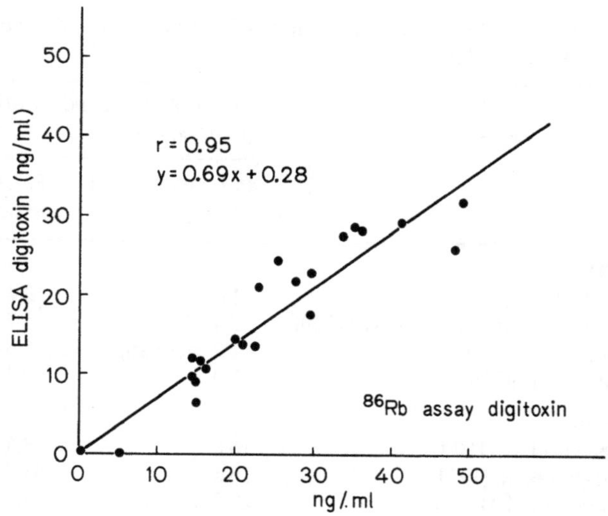

Fig. 7. Correlation between serum digitoxin concentrations measured by ^{86}REA and an ELISA system (values from n = 24 patients under maintenance digitoxin treatment) (BELZ and BELZ, 1979)

It is noteworthy that the slope of the correlation lines is not unity but in both cases about 0.7; i.e., the results obtained with the ^{86}REA are somewhat higher. This may be caused partly by the greater range of the standard curve for the latter assay, which allows exact determinations above 35 ng/ml.

H. Criticism of the Method as Used for Serum and Tissue Glycoside Concentration Determination

I. Extraction

Dichlormethane extraction is a crucial step in the method. It decreases variability and frequency of false-positive reactions of the assay (LOWENSTEIN and CORRILL, 1966). Only those c.g. and their metabolites are extracted which have a sufficient solubility in dichlormethane. High polar compounds, e.g. glucuronides, are not extracted. If these substances are still cardioactive, and there are arguments that it could be so (BELZ and HEINZ, 1977), the method in cases of c.g. with high degree of polar metabolites (e.g., scilla glycosides) would lead to a marked underestimation of the cardioactive fraction in plasma. With digoxin which is only to a minor amount converted to polar metabolites, this plays no role.

II. Plasma Volume

Except for digitoxin great volumes (> 2 ml, up to 10 ml) of plasma are needed. This property excludes the use of the method in cases where the sample material is limited, e.g., in children.

Table 3. Plasma glycoside concentrations from patients with and without signs of toxicity (^{86}REA)

References	Mean plasma concentration (ng/ml)	
	Patients without toxicity	Patients with toxicity
Digoxin		
BERTLER et al. (1975)	1.4	2.4
GRAHAME-SMITH and EVEREST (1969)	2.4	5.7
RITZMANN et al. (1972)	1.2 (median)	5.5 (median)
LADER et al. (1972)	1.0	2.2
Digitoxin		
HILLESTAD et al. (1973)	16.8	28.3
RASMUSSEN et al. (1971)	16.6	48.7
RITZMANN et al. (1972)	20.5 (median)	37.0 (median)

III. Biological Standard

As the method compares inhibition of ^{86}Rb uptake induced by an extract from a sample to a standard inhibition induced by a known amount of c.g. it really does not give a drug concentration in ng/ml, but rather an equivalent biological activity. This property which is also known from other biological and immunological methods should be kept in mind if considering the results obtained with ^{86}REA.

On the other hand the method only detects those parts of the glycoside and its metabolites which are biologically active and those parts analogous to the relative potency of their metabolites.

Therefore, one main advantage of the ^{86}REA is that the results obtained with this method can be used as biological reference values.

IV. Various Cardiac Glycosides

As the method works with any c.g. it can be widely used in the field of research on new c.g. or those c.g. for which no RIA is available.

V. Range of Discrimination

The standard curve covers a dose range of about factor 20. This is in favor of the method, especially when great variations of the levels occur, e.g. in pharmacokinetic studies.

VI. Use of the Method

The performance of the complete analysis for plasma determinations needs about 8 h. In this respect the method is much more time-consuming then RIA methods or ELISA.

VII. Precision and Accuracy

A comparative study using RIA and ^{86}REA for digoxin plasma level determinations (LADER et al., 1972) could not find significant differences between the recoveries at different plasma concentrations, but lower standard deviations were documented for the latter method.

J. Plasma Concentrations of Cardiac Glycosides

Various authors have published glycoside concentrations (^{86}REA) under maintenance therapy with c.g. In Table 3 the values for some c.g. are summarized.

References

Andersson, K.-E., Bertler, A., Redfors, A.: In vitro stability of proscillaridin A. Eur. J. Clin. Pharmacol. *8*, 135–139 (1975a)

Andersson, K.-E., Bertler, A., Redfors, A.: On the pharmacokinetics of proscillaridin A in man. Eur. J. Clin. Pharmacol. *8*, 421–425 (1975b)

Andersson, K.-E., Bergdahl, B., Bertler, A., Redfors, A.: On the absorption of proscillaridin A after single oral doses to normal and achlorhydric subjects. Acta Pharmacol. Toxicol. (Kbh.) *40*, 153–160 (1977a)

Andersson, K.-E., Bergdahl, B., Dencker, H., Wettrell, G.: Activities of proscillaridin A in thoracic duct lymph after single oral doses in man. Acta Pharmacol. Toxicol. (Kbh.) *40*, 280–284 (1977b)

Andersson, K.-E., Bergdahl, B., Wettrell, G.: Biliary excretion and enterohepatic recycling of proscillaridin A after oral administration to man. Eur. J. Clin. Pharmacol. *11*, 273–276 (1977c)

Andersson, K.-E., Bergdahl, B., Dencker, H., Wettrell, G.: Proscillaridin activity in portal and peripheral venous blood after oral administration to man. Eur. J. Clin. Pharmacol. *11*, 277–281 (1977d)

Beckmann, H., Belz, G.G., Quellhorst, E.: Die Eliminationsgeschwindigkeit von Meproscillarin nach wiederholter Applikation bei Patienten mit eingeschränkter Nierenfunktion. Arzneim. Forsch. *28*, 565–567 (1978)

Beckmann H., Belz, G.G., Quellhorst, E.: Wirkung und Plasmakonzentration von Meproscillarin unter Daueranwendung bei Niereninsuffizienz mit gleichzeitig vorliegender Herzinsuffizienz. Med. Klin. *74*, 1761–1766 (1979)

Belz, G.G.: Plasma concentrations of intravenous β-methyl digoxin with and without oral charcoal. Klin. Wochenschr. *52*, 749–750 (1974)

Belz, G.G., Bader, H.: Effect of oral charcoal on plasma levels of intravenous methyl proscillaridin. Klin. Wochenschr. *52*, 1134–1135 (1974)

Belz, G.G., Belz, G.: Untersuchungen zur Pharmakokinetik von Meproscillarin. Arzneim. Forsch. *28*, 535–539 (1978)

Belz, G.G., Belz, G.: Bestimmung der Digitoxin-Serumkonzentration mit einem Solid-Phase ELISA. Med. Klin. *74*, 620–623 (1979)

Belz, G.G., Brech, W.J.: Plasmaspiegel und Kumulationsverhalten von Proscillaridin bei Niereninsuffizienz. Klin. Wochenschr. *52*, 640–644 (1974)

Belz, G.G., Erbel, R.: Na^+-K^+-ATPase Assay und ^{86}Rb-Erythrozyten-Assay zum Nachweis von Digitoxin. In: Digitoxin. Greeff, K., Rietbrock, N. (eds.), pp. 13–20. Stuttgart, New York: Schattauer 1979

Belz, G.G., Heinz, N.: The influence of polar and nonpolar digoxin and digitoxin metabolites on the ^{86}Rb-uptake of human erythrocytes and the contractility of guinea pig papillary muscels. Arzneim. Forsch. *27*, 653–655 (1977)

Belz, G.G., Kleeberg, U.R.: Inhibition of Na^+-K^+-ATPase and ^{86}Rb-uptake by Canrenone. Horm. Metab. Res. *5*, 312 (1973)
Belz, G.G., Nübling, H.: Half life in plasma following repetitive applications of β-acetyl digoxin in man. Klin. Wochenschr. *53*, 543–544 (1975)
Belz, G.G., Schreiter, H.: Influence of plasma on the glycoside induced inhibition of ^{86}Rb-uptake of human erythrocytes. Z. Kardiol. *63*, 475–479 (1974)
Belz, G.G., Vollmer, K.O., Wissler, J.H.: Zur Hemmwirkung von Herzglykosiden auf die ^{86}Rb-Aufnahme der Erythrocyten. I. Methodische Untersuchungen zur Konzentrationsbestimmung von Cymarin und Digitoxin. Eur. J. Clin. Pharmacol. *4*, 92–98 (1972)
Belz, G.G., Stauch, M., Belz, G., Kurbjuweit, H.G., Oberdorf, A.: The effect of various cardenolides and bufadienolides with different cardiac activity on the ^{86}Rb-uptake of human erythrocytes. Naunyn Schmiedebergs Arch. Pharmacol. *280*, 353–362 (1973)
Belz, G.G., Nübling, H., Schmidt-Wiederkehr, P., Franz, H.E.: Plasmakonzentrationen und Elimination von Methylproscillaridin bei Niereninsuffizienz. Klin. Wochenschr. *52*, 1078–1081 (1974a)
Belz, G.G., Rudofsky, G., Belz, G.: Glykosidplasmaspiegel und Elektrokardiogramm während mittelschneller und langsamer Sättigung mit β-Methyldigoxin bei gesunden Probanden. Dtsch. Med. Wochenschr. *99*, 329–332 (1974b)
Belz, G.G., Rudofsky, G., Lossnitzer, K., Wolf, G., Stauch, M.: Plasmaspiegel und Elektrokardiogramm nach intravenöser Applikation von Proscillaridin und Digoxin. Z. Kardiol. *63*, 201–211 (1974c)
Belz, G.G., Stauch, M., Rudofsky, G.: Plasma levels after a single oral dose of proscillaridin. Eur. J. Clin. Pharmacol. *7*, 95–97 (1974d)
Belz, G.G., Nübling, H., Belz, G.: Plasma concentrations during repetitive intravenous and oral methyl proscillaridin application in man. Arzneim. Forsch. *26*, 277–278 (1976a)
Belz, G.G., Schreiter, H., Wolf, G.: Studies on pharmacokinetics and pharmacodynamics of methyl proscillaridin in healthy man. Eur. J. Clin. Pharmacol. *10*, 101–108 (1976b)
Belz, G.G., Erbel, R., Schumann, K., Gilfrich, H.J.: Dose-response relationships and plasma concentrations of digitalis glycosides in man. Eur. J. Clin. Pharmacol. *13*, 103–111 (1978)
Bergdahl, B., Andersson, K.E.: Stability in vitro of methyl proscillaridin. Eur. J. Clin. Pharmacol. *11*, 267–271 (1977)
Bertler, A., Redfors, A.: An improved method of estimating digoxin in human plasma. Clin. Pharmacol. Ther. *11*, 665–673 (1970)
Bertler, A., Redfors, A.: The ^{86}Rb method for digoxin assay. Comparison with radioimmunoassay. In.: Symposium on digitalis. Storstein, O. (ed.), pp. 64–70. Oslo: Gyldendal Norsk Forlag 1973
Bertler, A., Monti, M., Ohlin, P., Redfors, A.: Cardiac arrhythmias, electrolytes and digoxin concentration in plasma and urine in patients treated with digoxin. Acta Med. Scand. *197*, 391–401 (1975)
Binnion, P.F., Morgan, L.M., Stevenson, H.M., Fletcher, E.: Plasma and myocardial digoxin concentrations in patients on oral therapy. Br. Heart J. *31*, 636–640 (1969)
Doering, W., Belz, G.G.: Quinidine-digoxin interaction: Effect of quinidine on ^{86}Rb-uptake of human erythrocytes. Klin. Wochenschr. *59*, 95–96 (1981)
Dunham, E.T., Glynn, I.M.: Adenosintriphosphatase activity and the active movements of alkali metal ions. J. Physiol. London *156*, 274–293 (1961)
Gjerdrum, K.: Determination of digitalis in blood. Acta Med. Scand. *187*, 371–379 (1970)
Gjerdrum, K.: Digitoxin studies. Acta Med. Scand. *191*, 25–34 (1972)
Glynn, I.M.: The action of cardiac glycosides on sodium and potassium movements in human red cells. J. Physiol. London *136*, 148–173 (1957)
Grahame-Smith, D.G., Aronson, J.K.: Assessment of digoxin action by a pharmacodynamic biochemical method. In: Cardiac glycosides. Bodem, G., Dengler, H.J. (eds.), p. 242. Berlin, Heidelberg, New York: Springer 1978
Grahame-Smith, D.G., Everest, M.S.: Measurement of digoxin in plasma and its use in diagnosis of digoxin intoxication. Br. Med. J. *1969*, 286–289
Haustein, K.O.: Measurement of plasma glycoside level following pengitoxin administration. Eur. J. Clin. Pharmacol. *13*, 389–391 (1978)

Hillestad, L., Hansteen, V., Hatle, L. et al: Digitalis intoxication. In: Symposium on Digitalis, Oslo, Norway. Storstein, O. (ed.), pp. 281–286. February 22–23. Oslo: Gyldendal Norsk Forlag 1973

Hoffman, J.F.: The red cell membrane and the transport of sodium and potassium. Am. J. Med. *41*, 660 (1966)

Kleeberg, U.R., Belz, G.G.: Die Hemmung der $Na^{+}K^{+}$-Membran-ATPase und der ^{86}Rb-Aufnahme menschlicher Erythrozyten durch Spironolacton und seine Derivate. Verh. Dtsch. Ges. Inn. Med. *80*, 1521–1523 (1974)

Lader, S., Bye, A., Marsden, P.: The measurement of plasma digoxin concentration: A comparison of two methods. Eur. J. Clin. Pharmacol. *5*, 22–27 (1972)

Love, W.D., Burch, G.E.: A comparison of potassium^{42}, rubidium^{86}, and cesium^{134} as tracers of potassium in the study of cation metabolism of human erythrocytes in vitro. J. Lab. Clin. Med. *41*, 351–362 (1953)

Lowenstein, J.M.: A method for measuring plasma levels of digitalis glycosides. Circulation *31*, 228–233 (1965)

Lowenstein, J.M., Corrill, E.M.: An improved method for measuring plasma and tissue concentrations of digitalis glycosides. J. Lab. Clin. Med. *67*, 1048–1052 (1966)

Post, R.L., Merrit, C.R., Kinsolving, C.R., Albright, C.D.: Membrane adenosine triphosphatase as a participant in the active transport of sodium and potassium in the human erythrocytes. J. Biol. Chem. *235*, 1796–1802 (1960)

Rasmussen, K., Jervell, J., Storstein, O.: Clinical use of a bio-assay of serum digitoxin activity. Eur. J. Clin. Pharmacol. *3*, 236–242 (1971)

Ritzmann, L.W., Bangs, C.C., Coiner, D., Custis, J.M., Walsh, J.R.: Serum glycoside levels by rubidium assay. Arch. Intern. Med. *132*, 823–830 (1973)

Rupp, M., Brass, H., Belz, G.G.: Die Hämoperfusion zur Elimination von Meproscillarin. Arzneim. Forsch. *28*, 567–569 (1978)

Schatzmann, H.J.: Herzglykoside als Hemmstoffe für den aktiven Kalium- und Natriumtransport durch die Erythrozytenmembran. Helv. Physiol. Acta *11*, 346–354 (1953)

Shapiro, W., Narahara, K., Taubert, K.: Relationship of plasma digitoxin and digoxin to cardiac response following intravenous digitalization in man. Circulation *42*, 1065–1072 (1970)

Somogyi, G., Káldor, A., Jankovics, A., Faix, L.: Studies of the digitalis level of the cerebrospinal fluid. The effect of digitalis glycosides on the nervous system. Int. J. Clin. Pharmacol. *4*, 421–423 (1971)

Twittenhoff, W.D., Brittinger, W.D., Deckert, D.W., Belz, G.G., Schubert, I.: Zur Frage der Kumulation von Meproscillarin bei Niereninsuffizienz. Arzneim. Forsch. *28*, 563–565 (1978)

Vollmer, K.O., Wissler, J.H., Belz, G.G.: Zur Hemmwirkung von Herzglykosiden auf die ^{86}Rb-Aufnahme der Erythrocyten. II. Spezifität der Hemmwirkung unter besonderer Berücksichtigung der Cymarin-Reihe. Eur. J. Clin. Pharmacol. *4*, 99–103 (1972)

Wilson, B.J. (ed.): The radiochemical manual, 2nd ed. Amersham: Radiochemical Centre 1966

Wissler, J.H., Belz, G.G., Vollmer, K.O.: Zur Hemmwirkung von Herzglykosiden auf die ^{86}Rb-Aufnahme der Erythrozyten. III. Über die Beeinflußbarkeit der durch Cymarin bewirkten ^{86}Rb-Aufnahmehemmung der Erythrozyten. Eur. J. Clin. Pharmacol. *4*, 104–106 (1972)

Biological Methods for the Evaluation of Cardiac Glycosides

CHAPTER 8

Evaluation of Cardiac Glycosides in the Intact Animal

H. BAHRMANN and K. GREEFF

A. Introduction

For many decades biologic evaluation in experimental animals was the only way of standardizing extracts from digitalis or other plants, in order to obtain guidelines for their therapeutic dosage. Subsequently, as knowledge of the drugs advanced, animal experiments served chiefly to demonstrate and elucidate the actions of the glycosides. More recently, experimental animals have been employed mainly to determine the therapeutic range, intestinal absorption, and duration of action of new glycosides or semisynthetic derivatives. In addition, numerous investigations have been carried out to study interactions with other drugs which potentiate or attentuate their effects or which alter their therapeutic range.

B. Toxicity as a Parameter of Biologic Efficacy

When dealing with chemically pure glycosides of known toxicity it is nowadays unnecessary to carry out toxicity testing in order to check their biologic efficacy or potency. However, when studying glycoside mixtures, new cardiac glycosides, semisynthetic derivatives or digitalis-like compounds, determination of the lethal dose is a necessary part of the biologic evaluation. Admittedly, the toxicity of a drug is not as a rule a good parameter of its therapeutic effect. However, in the light of previous animal experiments it may be assumed that for cardiac glycosides there is usually a constant relation between those doses which have a positive inotropic action and those which cause toxic arrhythmias or cardiac arrest. Toxicity determinations are also widely employed to study interactions with other drugs or to detect other factors which may modify the efficacy of the glycosides (NADEAU and DE CHAMPLAIN, 1973; NEUGEBAUER, 1977; SOMBERG et al., 1979).

I. Determination of the Lethal Dose in Anesthetized Animals by Intravenous Infusion Continued Until Cardiac Arrest

1. Cats

A method for determining the toxicity of cardiac glycosides by intravenous infusion in cats was first described by HATCHER and BRODY (1910) and has been modified on numerous occasions since then. (For reviews see LENDLE, 1933 a, b; WEESE, 1936; BIRCHER et al., 1947; ROTHLIN and BIRCHER, 1954; BAUMGARTEN, 1963; HOLLAND and BRIGGS, 1964; KLAUS, 1966; KATZUNG, 1968; FISCH and SURAWICZ,

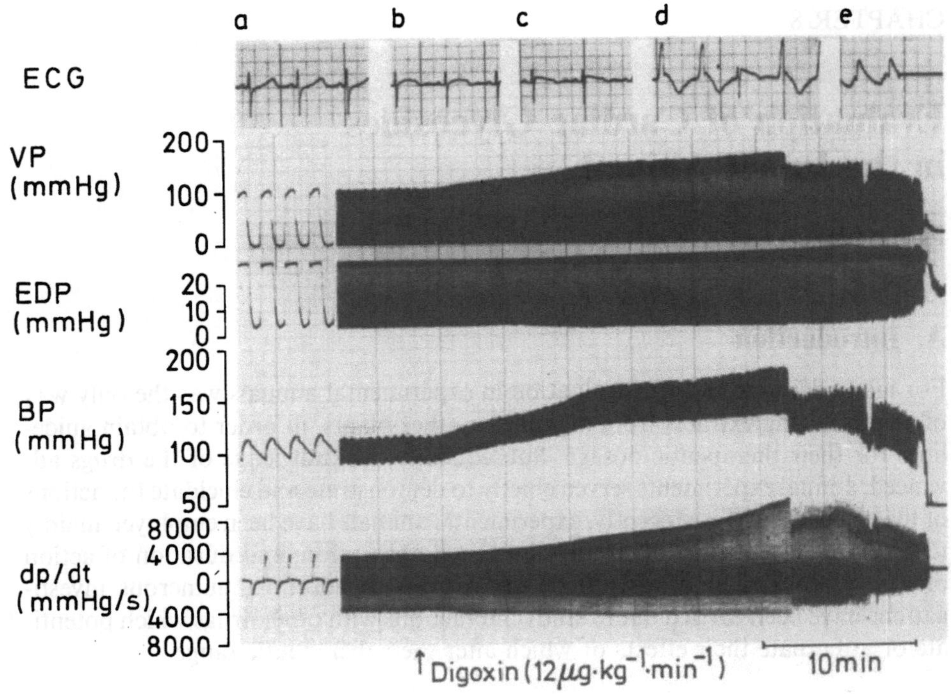

Fig. 1a–e. Effects of digoxin in a cat during an intravenous infusion continued until cardiac arrest. Pentobarbital anesthesia (40 mg/kg), artificial ventilation, alcuronium (0.2 mg/kg), heparin (500 IU), and phenoxybenzamine (1 mg/kg). dP/dt = rate of pressure rise in left ventricle, BP = blood pressure, EDP = end-distolic pressure, and VP = ventricular pressure. The ECG was recorded: **a** before the start of the infusion; **b** at 22 min (264 µg/kg); **c** at 24 min (288 µg/kg); **d** at 27 min (324 µg/kg); **e** at 33 min (396 µg/kg). The arrhythmogenic dose was 295 and the lethal dose 402 µg/kg. Features of note are the early rise in blood pressure, which points to a peripheral vascular effect of digoxin, and the limitation of the positive inotropic effect caused by the onset of arrhythmias

1969; BAHRMANN et al., 1973.) It is still preferred for the standardization of digitalis extracts (see STEWART, this Handbook, Vol. 56 II, Chap. 9) and for determining the efficacy and toxicity of new cardiac glycosides or digitalis-like compounds (GREEFF et al., 1965; SCHAUMANN and WEGERLE, 1971; CHEN and HENDERSON, 1954; HENDERSON, 1969; RASCHACK et al., 1978).

In the same experiment it is possible to determine the onset of toxic arrhythmias during the infusion or, by recording the rate of pressure rise in the left ventricle, to measure the positive inotropic effect. Figure 1 is a recording made in a cat during a digoxin infusion; it demonstrates the changes of the maximum rate of pressure rise (dP/dt_{max}) in the left ventricle, together with the changes in blood pressure, end-diastolic pressure and the ECG. One remarkable feature is the early rise in blood pressure, which is probably caused by an increase in peripheral resistance and, contrary to the surmise of BELLER and SMITH (1975), cannot be accounted for by activation of the adrenergic system, because the cat was given phenoxybenzamine before the infusion. It is also noteworthy that dP/dt_{max} continues to rise until

Table 1. Influence of infusion rate on the lethal dose of various cardiac glycosides. Interpolated lethal dose (μg/kg) at a lethal time of 19 min/kg (A), or 90 min/kg (B). In the case of a rapid acting glycoside such as strophanthoside K, the lethal dose remains practically unaltered between *A* and *B*, while with slower acting glycosides the difference becomes greater and greater. (GRAUWILER et al., 1966)

	Lethal dose (μg/kg)		*A/B*
	A (19 min/kg)	*B* (90 min/kg)	
Digitoxin	399	193	2.1
Digoxin	247	180	1.4
Strophanthoside K	127	113	1.1

arrhythmia begins, indicating that the positive inotropic action of digoxin is limited by the onset of disorders of cardiac rhythm.

Technique. Cats weighing 1.5–2.5 kg are anesthetized with ether or urethane–chloralose and given the glycoside by infusion into the jugular or femoral vein. As respiration may become shallow or even cease toward the end of the experiment, it is advisable to ventilate the animals artificially. The infusion rate is so adjusted that the survival time of the cats is 40–60 min. Cardiac action is monitored by recording the ECG. The endpoint of the titration is taken as the time at which action potentials cease to be detectable over a 10 s period (Hatcher dose/kg body weight).

Careful attention must be given to the infusion rate if reproducible results for the i.v. lethal dose are to be obtained. This question has been discussed in detail by numerous earlier workers (BAUER and FROMHERZ, 1933; ACHELIS and KRONEBERG, 1956; GRAUWILER et al., 1966; VOGEL and BAUMANN, 1969; KOSSWIG and ENGELHARDT, 1967). If the infusion is too rapid, overtitration will result. Only if the infusion time exceeds 40 min can it be assumed that the lethal dose for most glycosides will remain constant. Nevertheless, it must be noted that different glycosides – depending on their pharmacokinetics – have different optimal infusion times for the determination of i.v. lethal dose (KRAUPP et al., 1959). These variations are due to differences in their physicochemical properties, which affect inter alia their distribution volumes, the rate and extent of their binding to serum proteins or cardiac receptors, and their elimination. At very slow infusion rates the lethal dose may rise again, especially if the glycosides are rapidly eliminated (KRAUPP et al., 1959; BACH and REITER, 1964).

GRAUWILER et al. (1966) postulate that the lethal dose will remain constant only if the infusion time is set at not less than 90 min, and accordingly recommend that the titration should always be carried out at two different infusion rates. They also recommend that the lethal dose should be standardized in terms of body weight. Table 1 contains some examples showing how a change in infusion time has profound consequences in the case of the slowly acting glycosides digitoxin and digoxin, but lesser effects in the case of strophanthoside K.

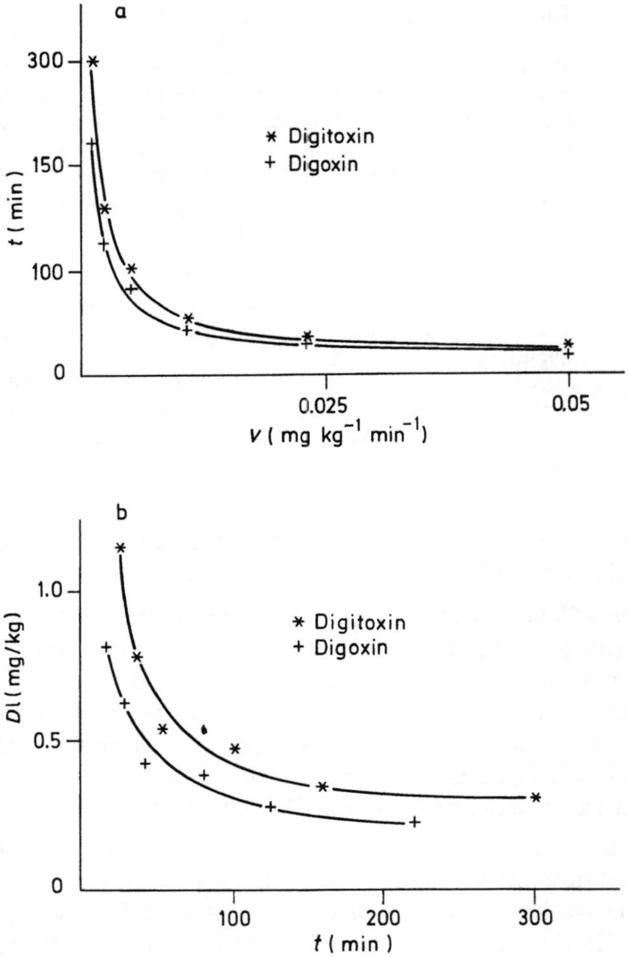

Fig. 2. a Relationship between dosage rate v ($mg\ kg^{-1}\ min^{-1}$) and survival time t (min) during infusion of digitoxin and digoxin at diminishing infusion rate by the method of LENKE and SCHNEIDER (1969, 1970 a, b) in cats. **b** The same values as in **a** plotted in the conventional manner, i.e., lethal (DL) dose against lethal time *(t)*

Hatcher's method was modified by introducing the dosage rate as an independent variable by LENKE and SCHNEIDER (1969, 1970 a, b). By plotting the dosage rate v ($mg\ kg^{-1}\ min^{-1}$) against the survival time t (min) it is possible to construct a curve which gives a time-dependent "lethal dose" D (mg/kg) and a minimal survival time t_0 (min), which provide data more suitable for comparison (Fig. 2). According to HENDERSON (1969) the results obtained in cats are transferable to humans with certain reservations. Triple the lethal dose (Hatcher dose) has proved suitable as the initial clinical dose; five times the Hatcher dose is approximately equivalent to the clinical body pool for maximum effect in humans (GOLD et al., 1941).

For a compilation of i.v. lethal doses in the cat and in other species see Table 5.

2. Guinea Pigs

KNAFFL-LENZ (1926) applied the technique of i.v. infusion, as described by Hatcher in cats, to guinea pigs usually anesthetized with urethane 1.5 g/kg subcutaneous (s.c.) or intraperitoneal (i.p.). This method is still used for standardizing extracts and for determining the toxicity of new glycosides (BRAUN and SIEGFRIED, 1947; ACHELIS and KRONEBERG, 1956; BROWN et al., 1962; GREEFF et al., 1965; GRAUWILER et al., 1966; SCHAUMANN and KAISER, 1967; SCHAUMANN and WEGERLE, 1971; RASCHACK et al., 1978). In guinea pigs, as in cats, the infusion rate is a crucial factor in determining the toxic dose and the constancy of the results (VOGT, 1967; VOGEL et al., 1970). As a rule, guinea pigs are less sensitive towards cardiac glycosides than cats (see Table 5), but interindividual variations in the lethal dose are usually somewhat larger (VOGEL et al., 1970).

3. Dogs

The expense of i.v. toxicity testing in dogs is justifiable only if the i.v. infusion can be used to measure other parameters of glycoside effect, e.g., dP/dt_{max}, cardiac output, coronary and peripheral resistance, or cardiac O_2 consumption (KATZ et al., 1938; WALTON et al., 1950; ROSS et al., 1960 a, b; HORWITZ et al., 1969, 1977; DEAVERS et al., 1979).

In dogs, as in cats, the infusion rate is an important factor in the determination of toxicity (ACHESON et al., 1964). The lethal dose is somewhat higher than in cats; for example, HASKELL et al. (1928), working with ouabain, found a value of 178 μg/kg in dogs as compared with 99μg/kg in cats.

4. Pigs

VOGEL et al. (1970) measured the i.v. toxicity of various glycosides in miniature pigs under pentobarbital anesthesia (15 mg/kg). For this purpose they used various infusion rates and found that with long survival times (approximately 135 min) toxicity was lower than with short survival times (approximately 65 min). In pigs, the lethal doses for short (and long) survival times were as follows: for digitoxin 400 ± 20 μg (540 ± 30 μg), for digoxin 230 ± 10 μg (300 ± 20 μg), and for convallatoxin 50 ± 10 μg (80 ± 10 μg) per kg body weight. For cats the corresponding figures were 620 ± 50 μg (340 ± 20 μg), 450 ± 50 μg (160 ± 10 μg), and 110 ± 40 μg (80 ± 10μg) per kg body weight. The lethal doses for pigs are hence somewhat lower than for cats.

II. Determination of the Lethal Dose in Unanesthetized Animals

When working with unanesthetized animals by the customary methods in current use, increasing individual doses are administered i.v., s.c. or orally and, after an observation period of 3–14 days, the median lethal dose (LD_{50}) is calculated statistically. This dose is designated as a "time-independent lethal dose" and has been ascertained for purposes of comparison in the same species in which the i.v. toxicity has been determined by infusion (WHITE, 1955; KRONEBERG et al., 1962; SCHAUMANN and STOEPEL, 1963; BENTHE and CHENPANICH, 1965). The LD_{50} is lower than

Table 2. Comparison of the LD_{50} (time-independent i.v. lethal dose) with the Hatcher dose (infusion dose) as determined by (*A*) BENTHE and CHENPANICH (1965) and (*B*) SCHAUMANN and WEGERLE (1971)

	Lethal dose (µg/kg)	
	LD_{50}	Hatcher
A Digoxin	180	298
β-Acetyldigoxin	220	413
B Digoxin	180	445
β-Methyldigoxin	210	332

Table 3. LD_{50} values for digitoxin, digoxin, and ouabain. In brackets relative efficacy in relation to the mortality in rats = 1. (RUSSELL and KLAASSEN, 1972, 1973)

Species	Digitoxin	Digoxin	Ouabain
	Lethal dose (LD_{50}) (µg/kg)		
Rat	5,300 (1)	18,000 (1)	13,000 (1)
Rabbit	590 (9.0)	730 (24.7)	210 (61.9)
Dog	380 (13.9)	190 (94.7)	100 (130)

the Hatcher dose determined by i.v. infusion; a few examples are given in Table 2. RUSSELL and KLAASSEN (1972, 1973) determined the LD_{50} by an "up and down" procedure. This method consists of giving a dose of a compound to one animal and noting if it dies within 24 h. If the animal dies, one decreases the dose and repeats the trial on a second animal. If the animal lives, the dose is increased and the trial repeated. This is repeated three times after one death and one survival have been obtained within two successive doses.

Table 3 shows the time-independent lethal doses (LD_{50}) for digitoxin, digoxin, and ouabain obtained by this method in dogs, rabbits, and rats.

1. Frogs

The principle of this method depends on the occurrence of cardiac arrest in systole, brought about by the cumulative effect of successive increasing doses injected into the lymph sac. According to the time-independent method, cardiac death should be reached within 24 h, while other time-dependent procedures call for cardiac death after 1 or 4 h (for review see LENDLE, 1933 a, b; WEESE, 1936; GOLD et al., 1941; CHEN and HENDERSON, 1954). When using this method it must be remembered that the susceptibility of frogs is dependent on the species, the season, and the external temperature (VOGEL and LEHMANN, 1959; HERRMANN et al., 1964).

HENDERSON and CHEN (1965) used frogs to study glycosides which were ineffective in cats under ether anesthesia in order to find out whether they might possibly be active in another species, but they did not find any differences. LIU and BENTLEY (1971) investigated ouabain in frogs and snails. The lethal dose (LD_{100}) for ouabain amounted to 0.6 mg/kg in frogs and 60 mg/kg in snails.

Although frogs are comparatively sensitive in their reactions to digitalis glycosides the method is nowadays little used, chiefly because of the pronounced seasonal fluctuations in the experimental results. Furthermore, the doses obtained in frogs are less readily transferable to humans than those obtained in cats (GOLD, 1946).

2. Pigeons

In pigeons, the toxic dose is determined by giving fractionated, i.e., repeated i.v. injections. Pigeons of assorted strains weighing 300–400 g are lightly anesthetized with ether and then, as laid down in *USP* XVII, given repeated injections into a wing vein at intervals of 5 min; cardiac arrest should ensue after 13–19 injections. This method is seldom used (HENDERSON and CHEN, 1965; SCHAUMANN and WEGERLE, 1971). With this technique, GRAUWILER et al. (1966) found the following lethal doses in pigeons: digitoxin 480 µg, digoxin 337 µg, and strophanthoside K 214 µg/kg (the concentration of the solutions was adjusted so that an average of 16 single doses was necessary before cardiac arrest occurred). The vomiting which regularly occurs in this species was formerly employed as an indicator of glycoside action (HANZLIK, 1929). ROSKE (1969) reported mean vomiting doses of 110 µg/kg for digoxin and 300µg/kg for digitoxin. There is probably no fixed correlation between the emetic and the cardiotoxic effects of different glycosides (for review see WEESE, 1936; BRAUN and LUSKY, 1948). The advantage of this technique of toxicity determination in pigeons lies in the fact that sublethal doses are sufficient to evaluate the glycoside effect, and cardiac death is not required. It is therefore possible, by giving repeated doses, to determine the decay rate. Using this technique, ROSKE (1969) found a daily decay rate of 45% for digoxin and 25% for digitoxin. HAAG and WOODLEY (1934) AND HANZLIK (1929) found good correlation between the i.v. lethal doses in pigeons and cats.

3. Mice and Rats

Mice and rats are of little use for measuring the effects of cardiotoxic glycosides, because it is always uncertain whether death is due to cardiac arrest or to some extracardiac action. The low toxicity of cardiac glycosides in rats and mice, as compared with cats, guinea pigs and dogs, has often been described (FARAH, 1946; GREEFF and KASPERAT, 1961 a, b; ALLEN and SCHWARTZ, 1969; SCOTT et al., 1971; RUSSELL and KLAASSEN, 1973; see Table 3). The low toxicity is due to the fact that the Na^+, K^+-ATPase of the myocardium is less sensitive (REPKE et al., 1965; DRANSFELD et al., 1966) and that the glycosides are more rapidly eliminated than in other species (see this Handbook, Vol. 56 II, Chap. 3).

Because of the low sensitivity of their hearts towards digitalis, rats are well suited for the study of extracardiac effects. After i.v. injection there may be paralyses or convulsions (GREEFF and KASPERAT, 1961 a, b). The paralyses appear to be caused predominantly by an effect on skeletal muscle and may lead to respi-

Table 4. Lethal dose and symptoms of poisoning in mice and rats after intravenous injection. Mean values and scatter by the method of LITCHFIELD and WILCOXON (1949), GREEFF and KASPERAT (1961a)

	Lethal dose (LD_{50}) (μM/kg)	Symptoms[a]
	Mouse	
Strophanthidin K	135.0 (100 –182)	P
α-Strophanthidin K	5.1 (3.0 – 9.0)	P (C)
Strophanthoside K	0.95 (0.77– 11.8)	P (C)
Digitoxigenin	3.5 (3.2 – 4.0)	C (P)
Digitoxigenin-mono-digitoxoside	8.6 (6.0 – 12.5)	P (C)
Digitoxin	14.5 (9.1 – 23.0)	P
Lanatoside A	22.0 (14.6 – 33.2)	P
	Rat	
Strophanthidin K	>100	P
Strophanthoside K	9.1 (5.2 – 15.9)	P
Digitoxigenin	1.7 (1.4 – 2.1)	C
Digitoxin	>100	

[a] P = paralysis; C = convulsions; ( ) = less pronounced

ratory arrest (GREEFF and WESTERMANN, 1955). The convulsions are of central origin and also occur after intracerebral injections, if these are given by the method described by HALEY and MCCORMICK (1957) (GREEFF and KASPERAT, 1961 a, b). There are considerable differences in the signs and symptoms of poisoning produced by different glycosides and genins and also in their lethal doses (Table 4). AFIFI and AMMAR (1974) compared the effects of glycosides after i.v. and intracerebral administration to conscious mice. They found that intracerebral injections of 0.069–0.241 mg/kg produced central convulsions. By i.v. injection the lethal doses were 7.6 mg/kg for digoxin, 6.9 mg/kg for digitoxin, and 2.3 mg/kg for ouabain; they state that the cause of death was cardiac arrest and respiratory paralysis.

III. Factors Which Modify Toxicity

1. Anesthesia

The anesthetics usually employed for i.v. titration experiments are ether, pentobarbital, or urethane–chloralose. STICKNEY (1974), working with cats, compared the effect of anesthesia induced by urethane–α-chloralose with that of sodium pentobarbital on the arrhythmogenic dose and lethal dose of ouabain and found that both doses were elevated by pentobarbital. Corresponding observations have been made with digoxin (KÖHLER and GREEFF, 1972; GREEFF and KÖHLER, 1975; Table 6). From this work it is apparent that the arrhythmogenic doses and lethal doses of glycosides under pentobarbital anesthesia are higher than those under urethane–chloralose anesthesia. One possible reason is the lower level of sympathetic activity under pentobarbital anesthesia. In his experiments STICKNEY (1974) observed a decrease in toxicity after bilateral sinus and vagus nerve section, apart from the effects of the two anesthetics. Halothane anesthesia has been found to cause a further increase in the lethal dose of digoxin. It amounted to 394 ± 15 µg/kg as compared with 265 ± 13 µg/kg obtained under pentobarbital anestesia (45 mg/kg, HEEG and GREEFF, 1968).

Overdoses of barbiturates have occasionally been given with the aim of producing myocardial failure. For example, DIEDEREN and KADATZ (1970) gave cats pentobarbital infusions at a rate of 3 mg kg^{-1} min^{-1} until a 50% depression of dP/dt_{max} had been reached. Under these conditions they found no significant changes in the lethal doses of ouabain, β-acetyldigoxin, and lanatoside C.

In guinea pigs, too, the choice of anesthetic has some influence on the toxicity of cardiac glycosides. For example, the lethal doses of digoxin and β-methyldigoxin under pentobarbital anesthesia are almost twice as high as under urethane anesthesia (SCHAUMANN and KOCH, 1974a; SCHAUMANN and WEGERLE, 1971).

In experiments in dogs, MORROW (1970) studied the influence of anesthesia with peptobarbital, thiopental, and halothane on the arrhythmogenic dose and lethal dose of digoxin. The dogs were given 44 or 57 mg/kg pentobarbital i.v., 50 or 100 mg/kg thiopental i.v., or were ventilated with 1% or 2% halothane. As compared with unanesthetized control animals, the arrhythmogenic doses under pentobarbital or thiopental anesthesia were not significantly different, but under halothane anesthesia they were considerably higher. The lethal dose was unaltered by thiopental or pentobarbital anesthesia, irrespective of the amount of anesthetic given, but it was higher under halothane anesthesia (Table 7).

2. Hypothermia

AKHTAR et al. (1971) investigated the effect of hypothermia at 29 °C on the toxicity of digoxin in dogs under pentobarbital anesthesia. After pretreatment with atropine 0.1 mg/kg the dogs were given digoxin i.v. at regular 15 min intervals until cardiac arrest ensued. The lethal dose of digoxin under hypothermia was 0.502 mg/kg – considerably higher than under normothermia (0.221 mg/kg). In conformity with these results, BEYDA et al. (1961) found that the lethal dose of digoxin by i.v. infusion was 0.489 mg/kg under hypothermia at 26 °C as compared with 0.239 mg/kg at 38 °C. To account for the reduced toxicity, these workers point to the decrease in heart rate from 167 to 80 beats/min and to the diminution in myocardial metabolism. On the other hand, ANGELAKOS and HURWITZ (1961), working with dogs under pentobarbital anesthesia, observed that the lethal consequences of an i.v. ouabain injection (0.07–0.1 mg/kg) could not be averted by hypothermia, though it certainly delayed the onset of toxic arrhythmias.

SATOSKAR and TRIVEDI (1956) found that by lowering the rectal temperature from 37° to 25°–26 °C the i.v. toxicity of lanatoside C in cats could be reduced by approximately 50%. Similar observations were made by SZEKELY and WYNNE (1960), using a digitalis tincture.

3. Hypoxia

Working with conscious dogs, BELLER and SMITH (1975b) found that acute hypoxia caused a slight but significant increase in toxicity amounting to approximate-

Table 5. Lethal doses of digitalis glycosides by intravenous infusion. Mean values ± standard errors

Reference	Species	LD_{100} i.v. (µg/kg)
	Digitoxin	
HATCHER and BRODY (1910)	Cat	300
FROMHERZ and WELSCH (1931)	Cat	410
HOTOVY (1951)	Cat	447 ± 10
REINERT (1952)	Cat	345
ROTHLIN et al. (1953)	Cat	386
ROTHLIN and BIRCHER (1954)	Cat	390
ACHELIS and KRONEBERG (1956)	Cat	620
	Guinea pig	1,200
KRONEBERG (1959)	Cat	480
	Guinea pig	1,200
LINGNER et al. (1963a)	Cat	462 ± 9
FÖRSTER (1963)	Cat	470 ± 150
	Guinea pig	850 ± 350
SCHAUMANN and STOEPEL (1963)	Pigeon	443
GRAUWILER et al. (1966)	Cat	399
	Pigeon	480
SCHAUMANN and KAISER (1967)	Guinea pig	2,300 ± 150
VOGEL et al. (1970)	Cat	620 ± 50
	Pig	540 ± 30
	Guinea pig	1,280 ± 50
SCHAUMANN and WEGERLE (1971)	Guinea pig	1,233 ± 79
KÖHLER et al. (1971)	Cat	587 ± 20
MATSUMURA et al. (1977)	Pigeon	520 ± 30
	Digitoxigenin	
FROMHERZ and WELSCH (1931)	Cat	500
FÖRSTER (1963)	Cat	440 ± 20
	Guinea pig	1,400
	Digoxin	
STOLL and KREIS (1953)	Cat	280
ROTHLIN and BIRCHER (1954)	Cat	280
KRONEBERG (1959)	Guinea pig	550
BEYDA et al. (1961)	Dog	239 ± 10
FÖRSTER (1963)	Cat	255 ± 25
	Guinea pig	790 ± 110
LINGNER et al. (1963)	Cat	267 ± 4
BACH and REITER (1964)	Guinea pig	1,020 ± 30
BENTHE and CHENPANICH (1965)	Cat	298 ± 16
GREEFF et al. (1965)	Cat	314 ± 9
	Guinea pig	1,220 ± 50
GRAUWILER et al. (1966)	Cat	247
	Pigeon	337
SCHAUMANN and KAISER (1967)	Guinea pig	770 ± 50
VOGEL et al. (1970)	Cat	450 ± 50
	Pig	300 ± 20
	Guinea pig	980 ± 40
KÖHLER et al. (1971)	Cat	381 ± 21
SCHAUMANN and WEGERLE (1971)	Cat	445 ± 40
	Guinea pig	667 ± 33

Table 5 (continued)

Reference	Species	LD$_{100}$ i.v. (µg/kg)
FILLMORE and DETWEILER (1973)	Dog	177 ± 8
PATNAIK and DHAWAN (1978)	Cat	289 ± 14
	Guinea pig	868 ± 31
	Lanatoside C	
SCHAUMANN and KAISER (1967)	Guinea pig	980 ± 40
DIEDEREN and KADATZ (1970)	Cat	375 ± 17
SCHAUMANN and WEGERLE (1971)	Guinea pig	639 ± 38
	β-Acetyldigoxin	
CHEN and HENDERSON (1954)	Cat	429 ± 23
BENTHE and CHENPANICH (1965)	Cat	413 ± 42
GREEFF et al. (1965)	Cat	385 ± 14
	Guinea pig	1,380 ± 90
GRAUWILER et al. (1966)	Cat	395
	Pigeon	562
SCHAUMANN and KAISER (1967)	Guinea pig	1,300 ± 20
DIEDEREN and KADATZ (1970)	Cat	386 ± 9
SCHAUMANN and WEGERLE (1971)	Cat	385 ± 28
	Cat	754 ± 49[a]
	β-Methyldigoxin	
SCHAUMANN and WEGERLE (1971)	Cat	332 ± 10
	Cat	455 ± 25[a]
	Guinea pig	847 ± 41
RASCHACK et al. (1978)	Guinea pig	653 ± 43
	Cat	190 ± 35
	Ouabain	
HASKELL (1928)	Dog	178
	Cat	99
FROMHERZ and WELSCH (1931)	Cat	118
FARAH (1946)	Rat	77,110
REINERT (1952)	Cat	100
KRAUPP et al. (1959)	Guinea pig	246
FÖRSTER (1963)	Cat	110 ± 20
	Guinea pig	280 ± 50
KURBJUWEIT (1964)	Guinea pig	290
	Cat	110
DIEDEREN and KADATZ (1970)	Cat	117 ± 7
BELLER et al. (1975)	Dog	780 ± 12
	Strophanthin K	
FÖRSTER (1963)	Cat	130
	Guinea pig	670
GRAUWILER et al. (1966)	Cat	127
	Pigeon	214
KÖHLER et al. (1971)	Cat	110 ± 5

[a] Intraduodenal application

Table 6. Influence of anesthesia on the toxicity of ouabain and digoxin in cats

	Ouabain		Digoxin	
Narcosis	Urethane – chloralose	Pentobarbital	Urethane – chloralose	Pentobarbital
Arrhytmogenic dose (μg/kg)	77.5±3.0	112±18	169±15	268±18
Lethal dose (μg/kg)	98.5±4.5	141±25	258±14	381±21

The ouabain infusion was given under anesthesia with urethane (500 mg/kg) and chloralose (60 mg/kg) i.v. or with pentobarbital (35 mg/kg) i.v. or i.p. (STICKNEY, 1974). The digoxin infusion was given under anesthesia with urethane (356 mg/kg) and chloralose (75 mg/kg) i.p. or with pentobarbital (50 mg/kg) i.p. (KÖHLER and GREEFF, 1972; GREEFF and KÖHLER, 1975)

Table 7. Influence of anesthesia on the toxicity of digoxin given by intravenous infusion to dogs. (MORROW, 1970)

		Digoxin (μg/kg)	
		Arrhythmogenic dose	Lethal dose
Without narcotic		91±14	
Pentobarbital	44 mg/kg	100±15	177±18
	75 mg/kg	89± 6	166±13
Thiopental	50 mg/kg	77± 5	169± 8
	100 mg/kg	70±12	167±19
Halothane	1%	138±12	232±16
	2%	168±18	207±17

ly 10% (a decrease in the arrhythmogenic dose of ouabain from 79.2 to 71.7 μg/kg). The enhanced sensitivity towards digitalis caused by hypoxia seems not to be due to any increase in the myocardial uptake of digitalis or to increased binding to ATPase FRIEDMAN et al., 1972). In dogs under pentobarbital anesthesia, ventilated with 8% oxygen, HARRISON et al. (1968) observed a 19% increase in the toxicity of acetylstrophanthidin given by i.v. infusion, as compared with the results in animals ventilated with ordinary air. The arrhythmogenic dose fell from 74.5 to 55.2 μg/kg.

4. Acidosis

In cats under pentobarbital anesthesia, the induction of respiratory acidosis causes a significant increase in the toxicity of digitoxin, digoxin, and strophanthin K (KÖHLER et al., 1971; KÖHLER and GREEFF, 1972). In the experiments summarized in Table 8 the blood pH of cats under pentobarbital anesthesia was reduced from pH 7.4 to 7.0 by raising the CO_2 content of the respired air to 20%. This resulted in a significant decrease in the arrhythmogenic doses and lethal doses of the glycosides. As this effect was reduced by reserpine pretreatment or by propranolol,

Table 8. Experiments in anesthetized cats. Arrhythmogenic doses and lethal doses of digitoxin, strophanthin K and digoxin in relation to blood pH. Number of animals is indicated by *n*. KÖHLER and GREEFF (1972)

Blood pH	*n*	Arrhythmogenic dose		Lethal dose	
		(μg/kg)	(%)	(μg/kg)	(%)
Digitoxin					
7.40	10	436±38	100	587±20	100
7.25	4	420±31	96	539±28	92
7.15	6	290±18	67	381±20	65
7.00	4	248±14	57	322±11	55
Digitoxin after 2.5 mg/kg reserpine					
7.40	9	572±46	131	848±50	144
7.15	9	529±40	121	709±42	121
Digitoxin after 10 mg/kg practolol					
7.40	5	537±11	123	808±71	138
7.00	5	394±40	90	642±49	109
Strophanthin K					
7.40	7	94± 5	100	110± 5	100
7.00	7	62± 4	66	86± 6	78
Digoxin					
7.40	10	268±18	100	381±21	100
7.00	10	193±24	72	321±11	84

it would appear that the increase in toxicity can be in part explained by activation of the sympathoadrenal system.

In comparative experiments on the isolated guinea pig atrium, the same workers found that the toxicity of digitoxin, digoxin, and strophanthin K was not significantly altered by changes in the pH of the Tyrode solution between 7.0 and 7.8 (KÖHLER and GREEFF, 1972). This supports the surmise that extracardiac factors are involved in the enhancement of toxicity.

Under conditions of metabolic acidosis BLISS et al. (1963) found no change in the sensitivity of dogs towards acetylstrophanthidin.

5. Alkalosis

WARREN et al. (1968) found some prolongation of the duration of toxic arrhythmias in dogs with metabolic alkalosis. In their experiments the alkalosis (pH 7.5) was produced by a NaOH infusion and acetylstrophanthidin (123 μg/min) was then infused intravenously. Galmarini et al. (1973) induced alkalosis in dogs by hyperventilation or by infusion of 1.42 *M* sodium carbonate solution and found that the toxicity of ouabain was enhanced (infusion of 50 μg/ml ouabain continued until cardiac arrest ensued within 60–120 min). The mean lethal dose of ouabain determined by i.v. infusion was 126.5±5.7 mg/kg in control animals and was significantly reduced after hyperventilation (107.2±4.6) or after infusion of sodium carbonate (93.6±6.3). In cats with alkalosis of pH 7.6 produced by $NaHCO_3$ (KÖHLER and GREEFF (1972) found a slight increase in the toxicity of digitoxin.

6. Age

From human pharmacology it has long been known that children in the first few years of life tolerate higher doses of glycosides than adults and require higher doses to produce a therapeutic effect. Similar age-linked differences have also been found in experimental animals. WOLLENBERGER et al. (1953) found that the lethal dose of ouabain given by i.v. infusion was 242±13 µg/kg in 20–25-day-old guinea pigs as compared with 177±15 µg/kg in 5–6.5-year-old animals. LIESE (1981) likewise found that the lethal doses of digitoxin, digoxin, and ouabain in elderly guinea pigs were lower than in young animals.

HALLORAN et al. (1970), working with dogs under pentobarbital anesthesia, investigated the influence of age on the arrhythmogenic dose of acetylstrophanthidin. In young dogs, 17–56 days old, they found a mean arrhythmogenic dose of 169±12 µg/kg for ouabain, as compared with a dose of 64±15 µg/kg for adult dogs. GLANTZ et al. (1976) found that the distribution volume of ouabain in young dogs is larger than in old dogs, a fact which may partially explain the lower sensitivity.

GUARNIERI et al. (1979) compared the positive inotropic effect of acetylstrophanthidin in healthy adult beagles (age 2–3 years) and old beagles (age 12–14 years) under pentobarbital anesthesia and found that the increase in contractility in the young animals was twice as great as in the old ones. This difference persisted even under β-blockade with 2–4 mg/kg practolol i.v. On the other hand, in unanesthetized dogs, arrhythmias occurred in young and old at the same doses and at the same serum concentrations of acetylstrophanthidin (65±12 as compared with 70±15 ng/ml). From their results they conclude that the toxicity of acetylstrophanthidin is not age dependent, whereas the inotropic effect decreases in old age.

SCOTT et al. (1971) found that young rats (age 7 days) were 200–1,000 times more sensitive to digitoxin and proscillaridin than old rats. As BOOR et al. (1976) found that young rats took up less digoxin in the liver and eliminated less in the bile than old rats, this could explain the heightened sensitivity of the younger members of this species toward glycosides.

Working with unanesthetized rabbits, KELLIHER and ROBERTS (1976) found that the arrhythmogenic dose by i.v. infusion was 278±23 µg/kg at an age of up to 3 weeks, as compared with 130±14 µg/kg in mature animals. As the noradrenalin content of the ventricular muscle rises by approximately 300% with advancing age, the authors postulate that this is the main cause of the increase in glycoside sensitivity.

BERMAN and MUSSELMAN (1979) compared the rates of metabolic breakdown of ^3H-digoxin in old and newborn sheep and found no differences.

7. Seasons

The seasonal variations in the sensitivity of frogs towards cardiac glycosides have long been known. According to ROSENKRANZ (1933) and ROTHLIN and SUTER

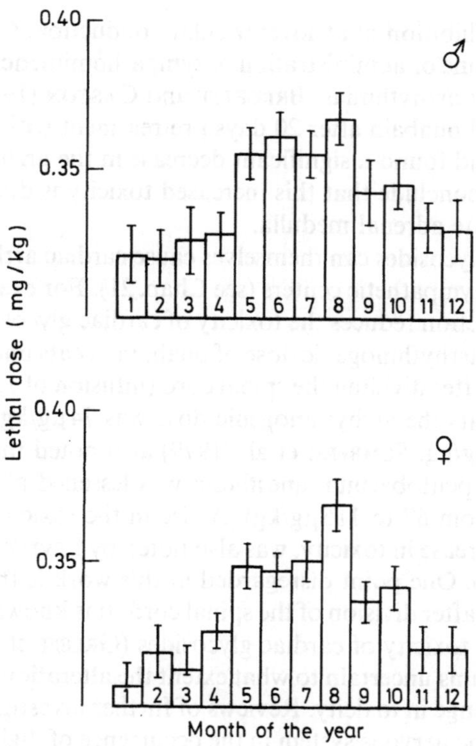

Fig. 3. Monthly variations in the lethal dose of ouabain given by intravenous infusion to guinea pigs. (SELECKY et al., 1970)

(1947) the maximum difference between February and July may be as great as 250%. CHEN et al. (1953) confirmed this seasonal fluctuation in frogs (412 µg/kg in January, 633 µg/kg ouabain in July), but in corresponding monthly investigations in cats they found no seasonal variation. There is also some seasonal fluctuation in pigeons. Using the digitalis standard *USP*, XIV, SACHS et al. (1956), in experiments extending over the course of 4 years, found that the lethal dose in pigeons was higher in summer than in winter.

SELECKY et al. (1976) followed the seasonal differences in the sensitivity of guinea pigs towards strophanthin G over the course of 4 years and found that resistance was higher in summer than in winter (Fig. 3). There was also a positive correlation between toxicity and external temperature and between toxicity and the duration of daylight. There was no evidence of any correlation between toxicity and atmospheric pressure or relative humidity.

8. Autonomic Tone

The interactions of digitalis glycosides with drugs which act upon the autonomic nervous system are described in detail in Chap. 24.

Parasympathomimetic and β-adrenolytic agents diminish the toxicity of cardiac glycosides. The reason is probably that the bradycardiac effect of digitalis glycosides is potentiated by vagal excitation, while the occurrence of conduction disorders is favored by inhibition of atrioventricular conduction. Conversely, an increase in sympathetic tone or administration of sympathomimetic agents facilitates the onset of ventricular arrhythmias. BRISSETTE and GASCON (1978) gave dogs an intravenous infusion of ouabain after 20 days pretreatment with diazepam (0.25–25 mg/kg by mouth) and found a significant decrease in the arrhythmogenic dose and lethal dose. They conclude that this increased toxicity is due to liberation of catecholamines from the adrenal medulla.

However, cardiac glycosides can themselves cause cardiac arrhythmias, probably by stimulating the sympathetic centers (see Chap. 21). For example, it has been observed that spinal section reduces the toxicity of cardiac glycosides. GILLIS et al. (1972) determined the arrhythmogenic dose of ouabain in cats under dial-urethane anesthesia before and after dividing the spinal cord (infusion of 1.6 µg kg^{-1} min^{-1} ouabain). In normal cats the arrhythmogenic dose was 74 µg/kg, but after spinal section it rose to 150 µg/kg. SOMBERG et al. (1979) also noted that the toxicity of ouabain in cats under pentobarbital anesthesia was lessened after spinal section (the lethal dose rose from 67 to 115 µg/kg). A rise in the toxic doses of ouabain, or in other words a decrease in toxicity, was also noted by LEVITT et al. (1973) after spinal cord transection. One point disregarded in this work is that the heart rate decreased significantly after division of the spinal cord. It is known that an increase in heart rate raises the toxicity of cardiac glycosides (GREEFF et al., 1971; LEVITT et al., 1973) and it remains uncertain to what extent the alteration in heart rate was responsible for this change in toxicity. Reviews of further investigations of the role played by the autonomic nervous system in the occurrence of digitalis arrhythmias will be found in the paper by SOMBERG et al. (1979) and in Chap. 21 and 24.

C. Sublethal Parameters of the Efficacy of Cardiac Glycosides

I. The Inotropic Effect

In earlier investigations the inotropic effect of cardiac glycosides was usually analyzed in heart–lung preparations, but nowadays modern techniques have made it possible to determine the glycoside effect quantitatively in situ, e.g., (1) with strain gauges sutured to the epicardial surface of the left ventricle, the thorax being either open or closed (WALTON and BRODIE, 1947; COTTEN and STOPP, 1958; MORAN, 1972; BENTHE et al., 1973); (2) by making intracardiac pressure measurements and determining maximum rate of rise of ventricular pressure (dP/dt_{max}) or some other parameter (HAMACHER, 1960; VERAGUT and KRAYENBUHL, (1965); HEEG and GREEFF, 1968; VATNER et al., 1971 a, b; SCHAUMANN and KOCH, 1974 b; BRAUNWALD et al., 1976; DEAVERS et al., 1979); or (3) by determining cardiac output (HARRISON et al., 1969; HIGGINS et al., 1972; SCHAUMANN and KOCH, 1974 a; BELLER et al., 1975). A few examples of the new methods for demonstrating the inotropic effect are given below.

WALTON and BRODIE (1947) were the first to develop a technique for measuring the contractile force of the heart with the aid of a strain gauge sutured to the right ventricle. Using this method, WALTON et al. (1950) compared the inotropic effects of various glycosides (e.g., digitoxin, digoxin, ouabain, acetylstrophanthidin). After inserting the strain gauge they closed the dog's thorax and were able to demonstrate a positive inotropic digitalis effect 2–5 days later, when the dog was not anesthetized. This fact proved that cardiac glycosides are effective even in the nonfailing heart of an unanesthetized animal.

DAGGETT and WEISFELDT (1965) analyzed the positive inotropic effect of 25 µg/kg acetylstrophanthidin (infused i.v. at a rate of 50 µg/min) in dogs under pentobarbital anesthesia (30 mg/kg) or urethane–chloralose anesthesia (a mixture of 60 and 600 mg/kg). Cardiac contractility (dP/dt_{max}) increased to much the same extent in the nonfailing heart, in the chronically denervated heart and in animals in which nervous control had been suspended by ganglion block. However, when stroke volume and end-diastolic pressure were used as parameters of the glycoside effect, no response was demonstrable in the nonfailing heart, whereas there was a response in the chronically denervated animals and after ganglion block. The authors conclude that the effect of digitalis on stroke volume in healthy animals is masked by a counter-regulatory decrease in cardiac and peripheral vascular sympathetic activity.

The experiments of SIMON et al. (1969 a, b) can be interpreted in the same way. These authors investigated the effects of strophanthin K and ouabaigenin in a group of dogs anesthetized with morphine (2 mg) and chloralose (90 mg/kg) and in another group in which cardiac failure had been induced by administering sodium pentobarbital (250–600 mg in repeated injections until dP/dt_{max} dropped to approximately 60% of the initial value). Left ventricular peak dP/dt rose significantly in both groups in response to the cardiac glycosides, but maximum systolic pressure and mean aortic pressure increased only in the animals pretreated with pentobarbital. BUSSMANN et al. (1969) likewise found that in the nonfailing heart the only change was a rise in contraction velocity (dP/dt_{max}), while in propranolol-pretreated heart (0.5 mg/kg) there was also an improvement in stroke volume and end-diastolic pressure as evidence of a positive inotropic effect. In the authors' opinion pretreatment with propranolol is of value in the testing of digitalis glycosides, first because its β-adrenolytic action prevents counter-regulation and second because its nonspecific action produces myocardial depression. As opposed to the myocardial depression produced by barbiturates, this has the advantage of longer duration (SIEGEL, 1969; KOCH-WESER, 1970).

HORWITZ et al. (1977) studied the effects of ouabain in repeated i.v. doses from 0.01 to 0.04 mg/kg in dogs with permanently implanted gauges, at rest and during exercise on a treadmill. Under resting conditions ouabain produced a 38% rise in dP/dt_{max} and a 16% rise in stroke volume as compared to controls, while end-diastolic volume and maximum pressure in the left ventricle remained unaffected. The rises in heart rate and maximum systolic pressure normally induced by exercise were partially suppressed by pretreatment with ouabain; an effect on dP/dt_{max} was observed only in animals which had been given propranolol (1 mg/kg) in addition to ouabain. The authors conclude that during exercise the effect of ouabain is masked by activation of the sympathetic system.

II. The Arrhythmogenic Effect

In 1959 LOWN and LEVINE described the acetylstrophanthidin tolerance test in dogs. In this test the cardenolide is given to dogs under pentobarbital anesthesia

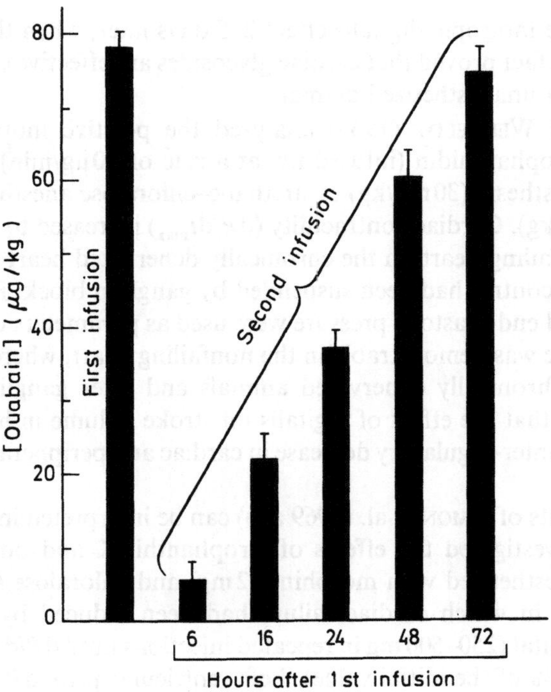

Fig. 4. Measuring the elimination of ouabain by giving repeated intravenous infusions to dogs until the onset of arrhythmia. The second infusion was given to different dogs at the time stated. Number of animals for the second infusion 3–5. Standard errors of the mean are indicated *above* the *solid bars*. (GREEFF et al., 1969)

(30 mg/kg) by slow i.v. infusion until arrhythmia begins. After its onset and persistence for 1 min (nodal or supraventricular tachycardia, ventricular premature beats, or ventricular tachycardia) the infusion is terminated. The toxic manifestations rapidly subside and the infusion can be repeated a few hours later.

Acetylstrophanthidin has the merit as a test substance that its cardiac effect comes on quickly and subsides rapidly. The plasma half-life amounts to 1.2 h in dogs and 2.3 h in humans (SELDEN et al., 1973). Repeated daily infusions given to the same animal yield reproducible arrhythmogenic doses. GREEFF et al. (1969), using this method in dogs, found that the biologic half-life of digitoxin was 10 h and of ouabain 26 h. As will be seen from Fig. 4, the same glycoside can also be used for the second and subsequent infusions. The use of the same glycoside for determining the decay rate may give more reliable results, as it is known that supra-additive or infra-additive effects can arise from administration of two different glycosides (SCHAUMANN, 1963).

LUCCHESI and SHIVAK (1964), using the acetylstrophanthidin tolerance test (infusion rate 80 µg/min), demonstrated that quinidine and procaine both raise the dose of acetylstrophanthidin required to induce arrhythmias. CHAPPLE et al. (1976) carried out the acetylstrophanthidin tolerance test in unanesthetized dogs and found an arrhythmogenic dose of 135 ± 11 µg/kg. However, pentobarbital anesthesia seems to have no significant effect on the arrhythmogenic dose (MORROW, 1970;

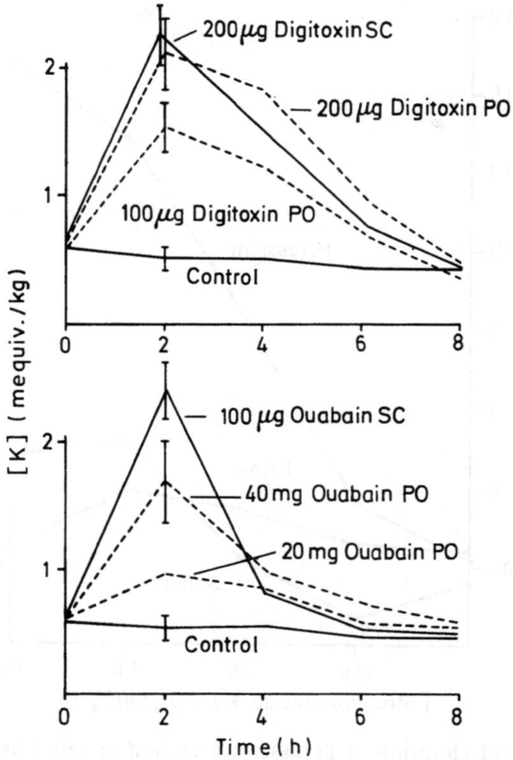

Fig. 5. Renal potassium excretion in rats after administration of digitoxin or ouabain. A method for measuring the intestinal absorption of cardiac glycosides with the aid of an extracardiac parameter. Results from 4 to 6 groups each containing 4 rats, which were kept in diuresis cages for 8 h. Standard errors of the mean are indicated by the *vertical bars*. PO = oral, SC = subcutaneous. (GREEFF, 1958 a, b)

Table 7). Some data for arrhythmogenic doses by intravenous infusion in dogs, cats, and guinea pigs are summarized in Table 8.

III. The Kaliuretic Effect

Glycosides in doses 10% of the lethal dose induce a significant rise in renal potassium excretion in rats (GREEFF, 1958 a, b). For example, after s.c. injection of digitoxin the kaliuretic response reaches its maximum within as short a time as 2 h wears off in the course of the subsequent 4 h (Fig. 5). The amount of potassium excreted depends on the dose of the glycoside given (Fig. 6). By comparing the effects after subcutaneous and oral administration, this extracardiac effect can be used to determine intestinal absorption (see Sect. E).

IV. Subacute Poisoning

FILLMORE and DETWEILER (1973) developed a method for maintaining continuous digitalis intoxication in dogs for several weeks (17–20 days). They gave beagles an

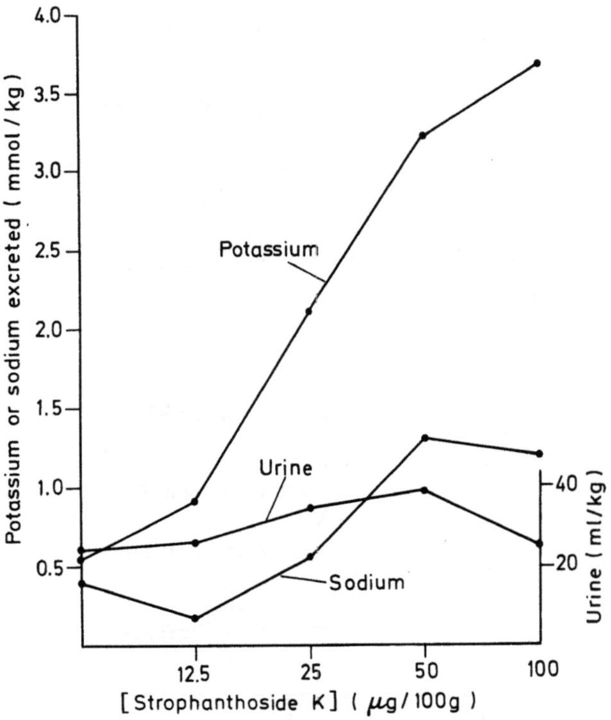

Fig. 6. Dose-dependent elevation of potassium excretion in rats 2 hs after subcutaneous injection of strophantoside K. A dose of 50 μg/100 g strophantoside K produces an approximately 600% increase in renal potassium excretion. (GREEFF, 1956)

initial i.v. loading dose of 0.125–0.150 mg/kg digoxin followed by daily doses of 0.015 mg/kg i.v. The signs of poisoning were ECG changes, reduced food intake, and weight loss. TESKE et al. (1976) induced subacute poisoning in dogs by i.v. injections of digoxin given by the following dosage schedule: initial dose 50 μg/kg, 1 h later a further dose of 50 μg/kg, 3 h later 25 μg/kg, and thereafter daily doses of 15–35 μg/kg. The course of the poisoning was followed by ECG monitoring and biochemical and histologic investigations; large individual variations in the toxic effects were observed. In mild poisoning the serum digoxin concentration was below 2.5 ng/ml, in moderately severe poisoning it was between 2.5 and 6 ng/ml, while concentrations above 6.0 ng/ml caused severe intoxication; there was good correlation between the serum digoxin concentration and the degree of intoxication.

Dogs eliminate digitoxin comparatively rapidly, but by giving repeated intravenous injections of 400 μg/kg at intervals of 2 days, it is possible to produce a cumulative effect which is characterized by increasing bradycardia (GREEFF et al., 1969).

D. Determination of the Therapeutic Range

The therapeutic range of a drug is defined as the ratio between the dose which produces an optimal therapeutic effect and the dose which causes toxic side effects. From practical experience in human beings it has long been known that cardiac

Table 9. Comparison of arrhythmogenic doses and lethal doses in guinea pigs, cats, and dogs by intravenous administration. Mean values ± standard errors

Reference	Species	*A* Arrhythmogenic dose (μg/kg)	*B* Lethal dose (μg/kg)	*A/B* × 100
	Digitoxin			
SCHAUMANN and WEGERLE (1971)	Guinea pig[b]	727 ± 82	1,233 ± 79	58
	Cat[b]	239	359 ± 15	66
GREEFF et al. (1973)	Cat[c]	436 ± 38	587 ± 20	74
	Guinea pig[b]	1,330 ± 120	2,320 ± 117	57
	Digoxin			
BENTHE and CHENPANICH (1965)	Cat	220 ± 48	298 ± 16	74
SMALL et al. (1971)	Guinea pig[a]	549 ± 55	725 ± 41	75
GREEFF et al. (1973)	Cat[c]	294 ± 48	428 ± 48	68
	Guinea pig[b]	590 ± 70	1,200 ± 130	49
GREEFF and KÖHLER (1975)	Guinea pig[e]	169 ± 15	258 ± 14	65
SCHAUMANN and WEGERLE (1971)	Guinea pig[b]	407 ± 20	667 ± 32	61
SCHAUMANN and KOCH (1974a)	Guinea pig[d]	750 ± 60	1,300 ± 70	57
	Ouabain			
FROMMER et al. (1966)	Dog[c]	54 ± 10	97 ± 19	55
BRISSETTE and GASCON (1978)	Dog[c]	53 ± 6	108 ± 9	49
CALDWELL and NASH (1976)	Dog[c]	70	114	61
SMALL et al. (1971)	Cat[a]	80 ± 8	150 ± 9	53
	Guinea pig[a]	157 ± 7	240 ± 10	75
GREEFF et al. (1973)	Cat[c]	87 ± 9	132 ± 11	45
	Guinea pig[b]	233 ± 13	312 ± 12	66
GREEFF and KÖHLER (1975)	Guinea pig[e]	133 ± 11	237 ± 17	55
HASHIMOTO et al. (1973)	Dog[b]	64 ± 4	98 ± 10	65
	Strophanthin K			
FRATZ et al. (1967)	Guinea pig[b]	345 ± 20	566 ± 16	61
	β-Acetyldigoxin			
BENTHE and CHENPANICH (1965)	Cat	300 ± 103	413 ± 42	72
SCHAUMANN and WEGERLE (1971)	Cat[b]	253	385 ± 28	65
	β-Methyldigoxin			
SCHAUMANN and WEGERLE (1971)	Guinea pig[b]	450 ± 20	847 ± 41	53
	Cat[b]	253	332 ± 10	51
SCHAUMANN and KOCH (1974a)	Guinea pig[d]	740 ± 65	1,510 ± 125	49

The following anesthetics were used in the experiments:
[a] Dial-urethane
[b] Urethane
[c] Pentobarbital
[d] Barbital
[e] Urethane – chloralose

glycosides have only a narrow therapeutic range. Experimental pharmacologists have therefore made great efforts to measure the therapeutic range in experimental animals. However, in the sense of the definition of the range this is hardly feasible because the therapeutic effect of cardiac glycosides consists in their effectiveness to correct cardiac failure, while their toxic side effects are predominantly extracardiac. For this reason, in animal experiments it is necessary to substitute other parameters or to employ methods for inducing experimental cardiac failure.

I. Arrhythmogenic Dose and Lethal Dose

Table 9 summarizes various experiments in which measurements of these two doses have been used as an index of the therapeutic range of various cardiac glycosides. The expression $A/B \times 100$ represents the arrhythmogenic dose expressed as a percentage of the lethal dose. This ranges between 49% and 75%. No clear-cut differences are apparent, as regards either species or glycoside. The experiments shown in this table served in part as controls for testing whether the toxicity or the therapeutic range of cardiac glycosides is influenced by other drugs.

II. Inotropic Dose and Lethal Dose

As the maximal positive inotropic dose by i.v. infusion is practically identical with the arrhythmogenic dose (see Fig. 1), submaximal effective doses have also been measured. CADWELL and NASH (1976), working with dogs under pentobarbital anesthesia, compared the dose which increased contractile force by 50% with the arrhythmogenic and lethal doses; for ouabain these three doses amounted to 48, 70, and 114 µg/kg respectively. ACHESON et al. (1964), using a similar technique in dogs, ascertained the dose at which the inotropic effect began, the arrhythmogenic dose and the dose which evoked ventricular fibrillation; for ouabain they obtained the following results: 25, 75, and 135 µg/kg.

Working with guinea pigs, SCHAUMANN and KOCH (1974a) compared the effects of digoxin and β-methyldigoxin on arterial blood pressure, cardiac output (thermodilution method), peripheral resistance, heart rate, and cardiac performance. So as to eliminate counter-regulatory phenomena, both vagus nerves were divided. The glycosides were infused i.v. until cardiac arrest ensued. The therapeutic effect was measured by a rise in arterial blood pressure and, more clearly, by the increase in cardiac performance. Arrhythmias and cardiac arrest were assessed as toxic effects. These workers found no significant difference in therapeutic range between digoxin and β-methyldigoxin.

III. Experimental Cardiac Failure

Spontaneous heart failure may arise in dogs as a result of old age and can be treated with digitalis (for review see KATZUNG, 1968). A surgical method for inducing cardiac failure is to create pulmonary stenosis (DAVIS et al., 1955; DAVIS, 1965; CHANDLER et al., 1967; SPANN et al., 1967, 1972; HAMLIN et al., 1971).

VOGEL and MARKS (1964) induced experimental cardiac failure with edema in guinea pigs by ligating the apical third of the heart. Only some 20% of the animals survived 14 or 28 days. Treatment with 3 μg/kg digoxin raised the survival rate by 50%.

DIEDEREN and KADATZ (1970) induced cardiac failure in cats by continuous infusion of pentobarbital (3 mg kg^{-1} min^{-1}). Using various glycosides, they determined the doses necessary to restore cardiac compensation, and compared them with the arrhythmogenic and lethal doses. SCHORSCHER et al. (1968) used phenylbutazone (50 mg/kg, followed by repeated smaller doses) to damage the myocardium. As inhalation anesthetics suitable for inducing cardiac failure in cats, halothane and chloroform may be used (BENTHE et al., 1973). SCHAUMANN and KOCH (1974a) induced heart failure in guinea pigs by administration of barbital or urethane (SCHAUMANN, 1964).

E. Intestinal Absorption

Measurement of the intestinal absorption of cardiac glycosides is an important prerequisite for their application in human beings. For this reason several methods for the determination of intestinal absorption in experimental animals have been developed, various approaches having been employed (Table 10).

1) By comparing the efficacy of cardiac glycosides, given by the enteral and intravenous or subcutaneous routes. Their efficacy is measured in terms of mortality, toxicity, or some other subtoxic parameter.

2) By giving a known amount e.g., the i.v. LD_{100} of glycoside orally or intraduodenally and measuring the proportion absorbed by subsequent determination of the intravenous "supplementary dose", i.e., the dose which is necessary after the pretreatment to produce effects which are observed after i.v. injection or infusion alone, e.g., cardiac arrest, toxic arrhythmias, or other glycoside effects.

3) By measuring the residue in the intestine at definite times after an oral or intraduodenal dose.

4) By measuring intestinal absorption with the aid of radioimmunologic or radiochemical methods.

I. Comparison of Oral with Intravenous or Subcutaneous Efficacy in Unanesthetized Animals

1. By Determining the Lethal Dose

Cats are given increasing doses of glycosides i.v. and orally. So as to prevent or reduce vomiting, the animals must be given no food for 24 h beforehand. With the aid of conventional statistical methods, the time-independent lethal doses (LD_{50}) by the intravenous and oral routes are determined and from these results the absorption ratio is calculated. The time which elapses until death occurs gives some indication of the rate of absorption (WHITE and GISVOLD, 1952; WHITE, 1955; SCHAUMANN and WEGERLE, 1969). The signs of poisoning (vomiting, defecation, paralyses, tremor, convulsions, etc.) provide information regarding extracardiac side effects.

Table 10. Intestinal absorption of various cardiac glycosides in cats

Method[a]	Reference		LD_{100} µg/kg	Rate of absorption %
		Digitoxin		
1	WHITE (1955)	i.v.	310*	
		p.o.	320*	100
2	GRAUWILER et al. (1966)	i.v.	237	
		i.d.	315	75
2	SCHAUMANN and WEGERLE (1971)	i.v.	359	
		i.d.	478	75
3b	HOROVY (1951)	i.v.	462	100
3b	REINERT (1952)	i.v.	345	100
3b	LINGNER et al. (1963)	i.v.	462	~100
2	VOGEL et al. (1970)	i.v.	340	
		i.d.	380	90
2	ZECHEL and BROCK (1974)	i.v.	429	
		i.d.	508	84
		Digoxin		
3b	LINGNER et al. (1963)	i.v.	267	~100
1	BENTHE and CHENPANICH (1965)	i.v.	180	
		p.o.	800	22
3b	GREEFF et al. (1965)	i.v.	314	53
2	GRAUWILER et al. (1966)	i.v.	178	
		i.d.	441	43
2	VOGEL et al. (1970)	i.v.	160	
		i.d.	500	32
2	SCHAUMANN and WEGERLE (1971)	i.v.	445	
		i.d.	895	50
2	ZECHEL and BROCK (1974)	i.v.	307	
		i.d.	780	39
		β-Acetyldigoxin		
1	BENTHE and CHENPANICH (1965)	i.v.	220	
		p.o.	630	35
3b	GREEFF et al. (1965)	i.v.	385	92
				83
2	GRAUWILER et al. (1966)	i.v.	260	
		i.d.	628	41
2	SCHAUMANN and WEGERLE (1971)	i.v.	385	
		i.d.	754	51
2	ZECHEL and BROCK (1974)	i.v.	330	
		i.d.	504	78

2. By Demonstrating ECG Changes

Instead of the lethal dose it is possible to measure the dose required to evoke toxic arrhythmias. This can be done by giving increasing i.v. or oral doses to unanesthetized cats or dogs and recording the ECG. This method can also be used to determine the rate of onset of action or the rate of absorption (BENTHE and CHENPANICH, 1965).

Table 10 (continued)

Method[a]	Reference		LD$_{100}$ µg/kg	Rate of absorption
		β-Methyldigoxin		
2	SCHAUMANN and WEGERLE (1971)	i.v.	332	
		i.d.	455	73
3b	RASCHACK et al. (1978)	i.v.	190	74
		Ouabain		
2	ZECHEL and BROCK (1974)	i.v.	106	3
		i.d.	4,306	3
		Gitalin		
1	WHITE (1955)	i.v.	1,040	
		p.o.	1,230	85
		Proscillaridin		
2	ZECHEL and BROCK (1974)	i.v.	193	
		i.d.	1,054	22
3a	KURBJUWEIT (1964)	i.v.	284	≅30
3b	RASCHACK et al. (1978)	i.v.	120	24
		Meproscillarin		
3b	RASCHACK et al. (1978)	i.v.	137	85

[a] Results were obtained by the following methods: 1) By comparing LD$_{50}$ (time-independent lethal dose) by the oral (p.o.), intraduodenal (i.d.), or intravenous (i.v.) routes; 2) By comparing the lethal doses by intravenous or intraduodenal infusion; 3a) By giving an oral dose to the conscious animal and subsequently determining the i.v. supplementary dose in the anesthetized animal; 3b) By giving a predetermined intraduodenal dose to the anesthetized animal and subsequently determining the i.v. supplementary dose

[b] LD$_{50}$

3. By Determining the Kaliuretic Effect in Rats

Administration of cardiac glycosides to rats, even in subtoxic doses, is followed by a steep rise in renal potassium excretion (GREEFF, 1956, 1958 a, b). By giving increasing oral and subcutaneous doses it is possible to ascertain the dose which raises renal potassium excretion by a definite amount (see Sect. C). By this method the following values for relative efficacy by the enteral route have been calculated: for digitoxin 110% and for cymarin 30% (GREEFF, 1958 a, b), for digoxin 74% and for β-acetyldigoxin 87% (GREEFF et al., 1965), for β-methyldigoxin 118% and for digoxin 50% (CZERWEK et al., 1971), and for proscillaridin 30% (HAAS, (1967).

4. Tolerance Test with Acetylstrophanthidin After Oral Pretreatment

After receiving an oral dose of glycoside, dogs, either anesthetized or unanesthetized, are given an i.v. infusion of acetylstrophanthidin, a short-acting cardenolide, in a dose of, e.g., 95 µg/min until the onset of toxic arrhythmias. The absorption ratio is calculated from the difference between this and the arrhythmogenic dose in nonpretreated animals (CHAPPLE et al., 1976). This test has also been used for determining the arrhythmogenic dose and for detecting cumulation (see Sect. C).

II. Comparison of the Lethal Dose or Arrhythmogenic Dose by Intraduodenal and Intravenous Infusion in Anesthetized Animals

Laparotomy is carried out in anesthetized cats or dogs, a tube is introduced into the duodenum, and the pylorus is ligated to prevent reflux. The lethal dose or arrhythmogenic dose is determined by intraduodenal infusion. The absorption ratio can be calculated by comparing the results with the lethal or arrhythmogenic doses by i.v. infusion. For this purpose it is necessary to adjust the rate of the i.v. infusion so that cardiac arrest or arrhythmias ensue at approximately the same time as they do during the intraduodenal infusion. Comparisons of doses have been reported by REINERT (1952), GRAUWILER et al. (1966); SCHAUMANN and WEGERLE (1971) (cats), LINDNER and SCHÖNE (1972) (dogs); VOGEL et al. (1970) (guinea pigs, cats, miniature pigs).

III. Oral or Intraduodenal Pretreatment Followed by Determination of the Supplementary Dose in Anesthetized Animals

Unanesthetized cats or dogs are given a predetermined oral dose of the glycoside, this dose being chosen in the light of the intravenous lethal dose. One or several hours later the animals are anesthetized and given an i.v. infusion of the glycoside with the purpose of determining the supplementary dose which will induce cardiac arrest. From the difference between this and the i.v. lethal dose in nonpretreated animals it is possible to calculate the absorption ratio (HOTOVY, 1951; KURBJUWEIT, 1964; LINGNER et al., 1963a; GREEFF et al., 1965; RASCHACK et al., 1978).

Instead of giving pretreatment to conscious animals, an alternative is to anesthetize them at the beginning of the experiment. After laparotomy, the cats or dogs are given a predetermined dose of the glycoside by injection into the duodenum; the pylorus is ligated to prevent reflux. After 1–3 h, or at the time at which maximum absorption is expected, the lethal supplementary dose is determined by intravenous infusion. From the difference between this and the i.v. lethal dose in nonpretreated animals it is possible to calculate intestinal absorption (REINERT, 1952; KRONEBERG et al., 1962; LINDNER et al., 1979). In this experimental technique, the arrhythmogenic dose can be used instead of the lethal dose as a measure of glycoside effect.

IV. Determination of the Residue After Intraduodenal Administration

The glycosides are injected into an intestinal segment which has previously been tied off. At predetermined times thereafter the intestinal segment is emptied and the glycoside content measured by chemical methods. Using this technique in cats, DILLE and WHATMORE (1942) compared intestinal absorption, as measured by determination of the supplementary dose, with the results obtained by analysis of the residual contents. They administered lanatoside C or tincture of digitalis and found certain differences between these values, from which they concluded that a proportion of the glycosides is destroyed in the intestine. LINGNER et al. (1963 b) employed the same method in cats and came to the conclusion that the difference in intestinal absorption as determined by the titration method and by measurement of the residual amount can be explained by assuming that destruction takes place in the liver.

V. Determination of Intestinal Absorption by Radiochemical or Radioimmunologic Methods

After oral or intraduodenal administration of ^{3}H-labeled glycosides it is possible to measure blood levels and urinary or fecal excretion and to compare these results with the corresponding values as found after i.v. injection. These methods correspond to those which are used for measuring intestinal absorption in humans (see this Handbook, Vol. 56 II, Chaps. 6 and 18). In experimental animals it is also possible to measure the residue of ^{3}H-labeled glycosides in the excised intestinal segments or to estimate unlabeled glycosides by radioimmunoassay (GREENBERGER and CALDWELL, 1972).

VI. Determination of Hepatic Extraction

As noted in the foregoing chapter, when a drug is found to be of poor biologic availability it is almost impossible to decide whether the glycoside is poorly absorbed, or wheter it is destroyed in the gastrointestinal tract, inactivated in the intestinal wall or extracted in the liver by a "first-pass effect," i.e., metabolized or excreted in the bile. Hepatic extraction can be measured by injecting the glycoside into the portal vein. IGA and KLAASSEN (1979) compared plasma levels and biliary excretion after injecting ^3H-ouabain into the femoral or portal veins and found that approximately 50% of the glycoside injected into the portal vein was extracted in the liver. However, this high extraction rate probably applies only to rats, as they are known to excrete ouabain mainly in the bile (see Chap. 29).

F. Measurement of Cumulation and Duration of Action

The cumulation of cardiac glycosides and their duration of action are dependent on their elimination rate. Measurements of their duration of action can be performed in anesthetized or unanesthetized animals by methods similar to those used for studying intestinal absorption.

I. Repeated Administration of Sublethal Doses to Unanesthetized Experimental Animals

Cats are given subtoxic doses of glycoside i.v., i.p., or s.c. at intervals of 1–3 days. It is advisable to give 10%–50% of the time-independent i.v. lethal dose. By comparing the total amount of glycoside given up to the time of death with the time-independent i.v. lethal dose, it is possible to calculate the duration of action of the glycoside (ROTHLIN and BIRCHER, 1954; KRONEBERG, 1959; SCHAUMANN and STOEPEL, 1963; BENTHE and CHENPANICH, 1965; SCHAUMANN and WEGERLE, 1971; RASCHACK et al., 1978). When evaluating these experiments it must be remembered that they take several days and that intercurrent illnesses may arise during this time or the animals' sensitivity to glycosides may change as the result of myocardial necrosis or of some extracardiac glycoside effect.

II. Single Administration of a Sublethal Dose to Unanesthetized Animals Followed by Intravenous Titration Under Anesthesia

In this technique the animal is given a single dose of the glycoside by i.v. injection and 1–10 days later the supplementary dose is determined by i.v. infusion given to the same animal under anesthesia. The elimination rate can be calculated from the supplementary dose. According to the experimental procedure advocated by Hatcher, cats are given 33% or 60% of the intravenous lethal dose parenterally. After an interval of 5, 10, or 15 days the supplementary dose is determined with strophantin G. Using this method, Hatcher found a decay rate of 6%/24 h for digitoxin. The technique has been critized by other workers (BAUER and FROMHERZ, 1933; BAUER, 1934 a, b; LENDLE, 1935). The preliminary subtoxic dose of the glycoside should not exceed 50% of the i.v. lethal dose, as larger doses may damage the heart and cause changes in glycoside sensitivity. SCHAUMANN (1963) recommended that the glycoside used for the supplementary dose should be the same as that used for the preliminary dose. Instead of measuring the lethal dose by i.v. infusion, an alternative is to determine the time-independ lethal dose 1, 2, or 3 days after the initial dose (KRONEBERG et al., 1962).

III. Titration at Different Infusion Rates

This method is suitable only for determining the duration of action of those glycosides which have a high elimination rate. As noted in Sect. B, when the infusion time is short (e.g., between 30 and 90 min) the lethal dose by i.v. infusion decreases, the fall being dependent on the rate of action of the relevant glycoside. If the infusion time is then prolonged, the lethal dose may rise again, because the glycosides are eliminated during the infusion. This rise in the lethal dose can be employed in cats or guinea pigs as a guide to the elimination rate. However, when using this experimental procedure it must again be remembered that the toxicity of glycosides may alter during a prolonged experiment (VOGEL and KRÜGER, 1967; GREVEN, 1967).

IV. Determination by Radiochemical or Radioimmunologic Methods

In experimental animals as in humans the excretion rates and half-lives of glycosides can be measured by giving ^{3}H-labeled glycosides and following the radioactivity in the urine and feces, or by giving unlabeled glycosides and estimating them by radioimmunologic methods (DOHERTY and PERKINS, 1966; HARRISON et al., 1966; VAN ZWIETEN, 1967; FORTH et al., 1969; ZIELSKE et al., 1969; GREENBERGER and CALDWELL, 1972).

References

Achelis, J.D., Kroneberg, G.: Die Pharmakologie des Gitalins. I. Mitt. Arzneim. Forsch, *6*, 182–188 (1956)

Acheson, G.H., Kahn, J.B. jr., Lipicky, R.J.: A comparison of dihydro-ouabain, dihydrodigoxin, dihydrodigitoxin, 3-acetyl strophanthidin, erysimin, and ouabain given by continuous infusion into dogs. Naunyn-Schmiedebergs Arch. Exp. Path. Pharmakol. *248*, 247–260 (1964)

Afifi, A.M., Ammar, E.M.: Neurological, respiratory and cardiac effects of Cardiac glycosides administered intracerebrally to conscious mice. Pharmacol. Res Commun. *6*, 417–425 (1974)

Akhtar, M., Chakravarti, R.N., Sarkar, A.K., Wahi, P.L.: Effect of hypothermia on digitalis toxicity: An experimental study. Indian J. Med. Res. *59*, 58–63 (1971)

Allen, J.C., Schwartz, A.: A possible biochemical explanation for the sensitivity of the rat to cardiac glycosides. J. Pharmacol. Exp. Ther. *168*, 42–46 (1969)

Angelakos, E.T., Hurwitz, B.S.: Influence of induced hypothermia on digitalis toxicity. Circ. Res. *9*, 1144–1147 (1961)

Bach, E.J., Reiter, M.: The difference in velocity between the lethal and inotropic action of dihydrodigoxin. Naunyn-Schmiedebergs Arch. Exp. Path. Pharmakol. *248*, 437–449 (1964)

Bahrmann, H., Benthe, H.F., Greeff, K.: Pharmakologische Methoden. In: Greeff, K., Probleme der klinischen Prüfung herzwirksamer Glykoside. 2nd ed. Darmstadt: Steinkopf 1973

Bauer, H., Fromherz, K.: Zur Digitaliswertbestimmung an der Katze. Naunyn-Schmiedebergs Arch. Exp. Path. *172*, 693–698 (1933)

Bauer, H.: Zur Kenntnis der Ursachen der Kumulierungserscheinungen der Digitalisglykoside. II. Mitteilung: Reversible und irreversible Digitaliswirkungen. Arch. Exp. Path. *176*, 65–73 (1934a)

Bauer, H.: Zur Kenntnis der Ursachen der Kumulierungserscheinungen der Digitalisglykoside. III. Mitteilung: Sekundäre Schädigungen des Herzmuskels. Naunyn-Schmiedebergs Arch. Exp. Path. *176*, 74–77 (1934b)

Baumgarten, G.: Die herzwirksamen Glykoside. Leipzig: Edition 1963

Beller, G.A., Giamber, S.R., Saltz, St.B., Smith, Th.W.: Cardiac and respiratory effects of digitalis during chronic hypoxia in intact conscious dogs. Am. J. Physiol. *229*, 270–274 (1975)

Beller, G.A., Smith, Th.W.: Digitalis toxicity during acute hypoxia in intact conscious dogs. J. Pharmacol. Exp. Ther. *193*, 963–968 (1975)

Benthe, H.F., Chenpanich, K.: Vergleich der enteralen Wirksamkeit von Digoxin, Acetyldigoxin und Digitoxin. Arzneim. Forsch. *15*, 486–489 (1965)

Benthe, H.F., Göthert, M., von Klinggräff, G.: Zur negativ inotropen Wirkung von Inhalationsnarkotika und zur Kompensation dieses Effekts durch Herzglykoside. Anaesthesist *22*, 62–68 (1973)

Berman, W. jr., Musselman, J.: The relationship of age on the metabolism and protein binding of digoxin in sheep. J. Pharmacol. Exp. Ther. *208*, 263–266 (1979)

Beyda, E.J., Jung, M., Bellet, S.: Effect of hypothermia on the tolerance of dogs to digitalis. Circ. Res. *9*, 129–135 (1961)

Bircher, R., Rothlin, E., Suter, E.: Glykosidwirkung auf Elektrokardiogramm und Myokard. II. Mitteilung. Untersuchungen am Elektrokardiogramm und am Myokard bei tödlicher Vergiftung von Katzen nach einmaliger subcutaner Verabreichung herzwirksamer Glykoside. Helv. Physiol. Acta *5*, 322–332 (1947)

Bliss, H.A., Fishman, W.E., Smith, P.M.: Effect of alterations of blood pH on digitalis toxicity. J. Lab. Clin. Med. *62*, 53–58 (1963)

Boor, P.J,, Reynolds, E.S., Moslem, M.T.: Neurotoxicity of digitoxin in adults and newborn rats: drug distribution. Arch. Int. Pharmacodyn. Ther. *224*, 4–12 (1976)
Braun, H.A., Siegfried, A.: The assay of digitalis. V. The guinea-pig method. J. Am. Pharm. Assoc. *36*, 363–368 (1947)
Braun, H.A., Lusky, L.M.: A comparative study of the intravenous pigeon and the intravenous cat method in the assay of digitalis. J. Pharmacol. *93*, 81–85 (1948)
Braunwald, E.J., Ross, J., Sonnenblick, E.H.: Methods for assessing cardiac contractility. In: Mechanisms of contraction of the normal and failing heart, 2nd ed., pp. 130–165. Boston: Little, Brown and Company 1976
Brissette, Y., Gascon, A.L.: Increase in cardiac toxicity of ouabain in dogs after repetitive treatment with diazepam. Toxicol. Appl. Pharmacol. *44*, 127–135 (1978)
Brown, B.T., Stafford, A., Wright, S.E.: Chemical structure and pharmacological activity of some derivatives of digitoxigenin and digoxigenin. Brit. J. Pharmacol. *18*, 311–324 (1962)
Bussmann, W.D., Wirz, P., Lüthy, E., Krayenbühl, P.: Die Wirkung von herzwirksamen Glykosiden und Aglykonen am suffizienten Herzen. Dtsch. Med. Wochenschr. *94*, 779–785 (1969)
Caldwell, R.W., Nash, C.B.: Pharmacological studies of a new 4-aminosugar cardiac glycoside (ASI-222). J. Pharmacol. Exp. Ther. *197*, 19–26 (1976)
Chandler, B.M., Sonnenblick, E.H., Spann, J.R. jr., Pool, P.E.: Association of depressed myofibrillar adenosinetriphosphatase and reduced contractility in experimental heart failure. Circ. Res. *21*, 717–725 (1967)
Chapple, D.J., Hughes, R., Johnson, B.F.: The relationship between cardiotoxicity and plasma digoxin concentration in conscious dogs. Br. J. Pharmacol. *57*, 23–27 (1976)
Chen, K.K., Henderson, F.G., Robbins, E.B.: Seasonal variation in response to ouabain. J. Pharmacol. Exp. Ther. *107*, 131–133 (1953)
Chen, K.K., Henderson, F.G.: Pharmacology of sixty-four cardiac glycosides and aglycones. J. Pharmacol. Exp. Ther. *111*, 365–383 (1954)
Cotten, M.de V., Stopp, P.E.: Action of digitalis on the nonfailing heart of the dog. Am. J. Physiol. *192*, 114–120 (1958)
Czerwek, H., Hardebeck, K., Kaiser, F., Schaumann, W.: β-Methyl-Digoxin. II. Extracardiale Wirkungen bei enteraler und parenteraler Gabe. Arzneim. Forsch. (Drug Res.) *21*, 231–234 (1971)
Daggett, W.M., Weisfeldt, M.L.: Influence of the sympathetic nervous system on the response of the normal heart to digitalis. Am. J. Cardiol. *16*, 394–405 (1965)
Davis, J.O., Hyatt, R.E., Howell, D.S.: Right-sided congestive heart failure in dogs produced by controlled progressive constriction of the pulmonary artery. Circ. Res. *3*, 252–258 (1955)
Davis, J.O.: The physiology of congestive heart failure. In: Handbook of Physiology. Hamilton, W.F., Dow, Ph. (eds.), Sec. 2, Circulation III, pp. 2071–2122. (1965)
Deavers, S., Rosborough, J.P., McCrady, J.D.: The effects of oleandrin on cardiac contractility in the normal dog. Arch. Int. Pharmacodyn. *239*, 283–295 (1979)
Diederen, W., Kadatz, R.: Quantitative Untersuchungen der therapeutischen Breite von Herzglykosiden. Ärztl. Forsch. *24*, 149–155 (1970)
Dille, J.M., Whatmore, G.B.: The gastro-intestinal absorption of lanatoside C. J. Pharmacol. Exp. Ther. *75*, 350–355 (1942)
Doherty, J.E., Perkins, W.H.: Tissue concentration and turnover of tritiated digoxin in dogs. Am. J. Cardiol. *17*, 47–52 (1966)
Dransfeld, H., Greeff, K., Berger, H., Cautius, V.: Die verschiedene Empfindlichkeit der Na^+ und K^+-aktivierten ATPase des Herz- und Skelettmuskels gegen K-Strophanthin. Naunyn-Schmiedebergs Arch. Exp. Path. Pharmakol. *254*, 225–234 (1966)
Farah, A.: On the elimination of g-strophanthin by the rat. J. Pharmacol. Exp. Ther. *86*, 248–257 (1946)
Fillmore, G.E., Detweiler, D.K.: Maintenance of subacute digoxin toxicosis in normal beagles. Toxicol. Appl. Pharmacol. *25*, 418–429 (1973)
Fisch, C., Surawicz, B.: Digitalis. New York: Grune & Stratton 1969
Förster, W.: Struktur-Wirkungsbeziehungen bei Cardenoliden und Bufadienoliden. In: Die herzwirksamen Glykoside. Baumgarten, G., pp. 199–242. Leipzig: Edition 1963

Forth, W., Furukawa, E., Rummel, W.: The estimation of the intestinal resorption of cardiac glycosides through measurement of tritium labeled glycoside in portal venous blood and in intestinal lymph in cats. Naunyn-Schmiedebergs Arch. Pharmakol. Exp. Path. *264*, 406–419 (1969)

Fratz, R., Greeff, K., Wagner, J.: Über den Einfluß der β-Rezeptorenblocker des Iproveratrils, Chinidins und Reserpins auf die Wirkung des k-Strophanthins am Meerschweinchenherzen. Naunyn-Schmiedebergs Arch. Pharmakol. Exp. Path. *256*, 196–206 (1967)

Friedman, J.P., Harris, C.N., Goldman, R.H.: Digitalis toxicity in normal and hypoxemic dogs: Correlation with myocardial Na^+, K^+-activated ATPase activity. Clin. Res. *20*, 205 (1972)

Fromherz, K., Welsch, A.: Vergleich der Toxizität verschiedener herzwirksamer Reinsubstanzen und Fraktionen aus Digitalis purpurea für den Frosch und die Katze bei verschiedenen Applikationsbedingungen. Arch. Exp. Path. *161*, 266–305 (1931)

Frommer, P.L., Robinson, B.F., Braunwald, E.: Studies on digitalis. XII. The effect of paired electrical stimulation on digitalis-induced arrhythmias. J. Pharm. Exp. Ther. *151*, 1–6 (1966)

Galmarini, D., Campodonico, J.F., Wenk, R.D.: Effect of alkalosis on ouabain toxicity in the dog. J. Pharmacol. Exp. Ther. *186*, 199–203 (1973)

Gillis, R.A., Raines, A., Sohn, Y.J., Levitt, B., Standaert, F.G.: Neuroexcitatory effects of digitalis and their role on the development of cardiac arrhythmias. J. Pharmacol. Exp. Ther. *183*, 154–168 (1972)

Glantz, S.A., Kernoff, R., Goldman, R.H.: Age-related changes in ouabain pharmacology. Circ. Res. *39*, 407–414 (1976)

Gold, H., Cattell, McK., Kwit, N.T., Kramer, M.: The relative activity of digitalis preparations in the frog, the cat, and man, and its bearing on the problem of bio-assay and socalled deterioration. J. Pharmacol. Exp. Ther. *73*, 212–228 (1941)

Gold, H.: Pharmacologic basis of cardiac therapy. J. Am. Med. Assoc. *132*, 547–554 (1946)

Grauwiler, J., Schalch, W.R., Taeschler, M.: Pharmakologische Eigenschaften von Acetyldigoxin-α. Schweiz. Med. Wochenschr. *41*, 1381–1388 (1966)

Greeff, K., Westermann, E.: Untersuchungen über die muskellähmende Wirkung des Strophanthins. Naunyn-Schmiedebergs Arch. Exp. Path. *226*, 103–113 (1955)

Greeff, K.: The action of digitalis glycosides on electrolyte metabolism. Dtsch. Med. Wochenschr. *81*, 666 (1956) and Germ. Med. Monthly. *1*, 311 (1956)

Greeff, K.: Vergleich der kaliuretischen Wirkung verschiedener Glykoside des Strophanthidins und Digitoxigenins bei parenteraler und enteraler Applikation. Naunyn-Schmiedebergs Arch. Exp. Path. *233*, 468–483 (1958a)

Greeff, K.: Tierexperimentelle Untersuchungen über die Resorption von Strophanthin bei oraler Verabfolgung. Verh. Dtsch. Ges. Kreisl. Forsch. *24*, 310–317 (1958b)

Greeff, K., Kasperat, H.: Vergleich der neurotoxischen Wirkung von Digitalisglykosiden und Geninen bei intracerebraler und intravenöser Injektion an Mäusen, Ratten und Meerschweinchen. Naunyn-Schmiedebergs Arch. Exp. Path. *242*, 76–89 (1961a)

Greeff, K., Kasperat, H.: Konvulsive und paralytische Wirkungen von Digitalis-Glykosiden und Geninen bei intracerebraler und intravenöser Injektion an Mäusen. Arzneim. Forsch. *11*, 908–909 (1961b)

Greeff, K., Schwarzmann, D., Waschulzik, G.: β-Acetyldigoxin und Digoxin. Vergleichende pharmakologische Untersuchungen ihrer Wirksamkeit und enteralen Resorption. Arzneim. Forsch. *15*, 483–486 (1965)

Greeff, K., Greven, G., Osswald, W., Viana, A.P.: Studies on the elimination of digitoxin und ouabain in the dog, using the tolerance test of repeated intravenous infusions. Arch. Int. Pharmacodyn. *179*, 326–335 (1969)

Greeff, K., Pereira, E., Wagner, J.: Die Wirkung des Strophanthins bei Änderung der Schlagfrequenz und der extrazellulärer K^+- und Ca^{++}-Konzentration. Arch. Int. Pharmacodyn. *190*, 219–228 (1971)

Greeff, K., Köhler, E., Fortmüller, H.-W., Schmidt, R.: Wechselwirkung zwischen herzwirksamen Glykosiden und β-Sympathicomimetika. Arzneim. Forsch. *23*, 759–764 (1973)

Greeff, K., Köhler, E.: Tierexperimentelle Untersuchungen über den Einfluß von Triamteren und Amilorid auf Herz, Kreislauf und Toxizität des Digoxins. Arzneim. Forsch. *25*, 1766–1769 (1975)
Greenberger, N.J., Caldwell, J.H.: In: Marks, B.H. (eds.). Intestinal absorption of digitalis glycosides in rats and guinea-pigs. In: Basic and clinical pharmacology of digitalis, pp. 15–47. Springfield: Charles C. Thomas 1972
Greven, G.: Vergleich der Wirksamkeit der Digitalis lanata-Standarddroge mit Digoxin und Cymarin an Meerschweinchen. Arzneim. Forsch. *17*, 1239 (1967)
Guarnieri, Th., Spurgeon, H., Froehlich, J.P., Weisfeldt, M.L., Lakatta, E.G.: Diminished inotropic response but unaltered toxicity to Acetylstrophanthidin in the senescent beagle. Circulation *60*, 1548–1554 (1979)
Haag, H.B., Woodley, J.D.: The use of pigeons in the estimation of digitalis potency. J. Pharmacol. *51*, 360–369 (1934)
Haas, H.: Scillaglykoside. Med. Klin. *62*, 121–126 (1967)
Haley, T.J., McCormick, W.G.: Pharmacological effects produced by intracerebral injection of drugs in the conscious mouse. Brit. J. Pharmacol. *12*, 12–15 (1957)
Halloron, K.H., Schimpff, St.C., Nicolas, J.G., Talner, N.S.: Digitalis tolerance in young puppies. Pediatrics *46*, 730–736 (1970)
Hamacher, J.: Die Steilheit des intracardialen Druckanstiegs in der isometrischen Phase der spontanen Herzaktion als Kriterium einer pharmakologischen Wirkungsanalyse am Warmblüterherzen in situ. Naunyn-Schmiedebergs Arch. Exp. Path. Pharmakol. *238*, 73–74 (1960)
Hamlin, R.L., Dutta, S., Smith, C.R.: Effects of digoxin and digitoxin on ventricular function in normal dogs and dogs with heart failure. Amer. J. Vet. Res. *32*, 1391–1398 (1971)
Hanzlik, P.J.: A new method of estimating the potency of digitalis: Pigeon-Emesis. J. Pharmacol. Exp. Ther. *35*, 363–391 (1929)
Harrison, T.R., Leonard, B.W.: The effect of digitalis on the cardiac output of dogs and its bearing on the action of the drug in heart disease. J. Clin. Invest. *3*, 1–36 (1926)
Harrison, C.E. jr., Brandenburg, R.O., Ongley, P.A., Orvis, A.L., Owen, C.A. jr.: The distribution and excretion of tritiated substances in experimental animals following the administration of digoxin-^{3}H. J. Lab. Clin. Med. *67*, 764–777 (1966)
Harrison, D.C., Robinson, M.D., Kleiger, R.E.: Role of hypoxia in digitalis toxicity. Am. J. Med. *256*, 352–359 (1968)
Harrison, L.A., Blaschke, J., Phillips, R.S., Price, W.E., De V. Cotten, M., Jacobson, E.D.: Effects of ouabain on the splanchnic circulation. J. Pharmacol. Exp. Ther. *169*, 321–327 (1969)
Hashimoto, K., Kimura, T., Kubota, K.: Study of the therapeutic and toxic effects of ouabain by simultaneous observations on the excised and blood-perfused sinoattrial node and papillary muscle preparations and the in situ heart of dogs. J. Pharmacol. Exp. Ther. *186*, 463–471 (1973)
Haskell, C.C., Copenhauer, J.R., Stone, G.E., Yost, O.R.: Laboratory methods: The use of dogs in the assay of digitalis. J. Lab. Clin. Med. *14*, 155–159 (1928)
Hatcher, R.: The persistence of action of the digitalins. Arch. Intern. Med. *10*, 268–296 (1912)
Hatcher, R.A., Brody, J.: The biological standardisation of drugs. Am. J. Pharmacy. *82*, 360 (1910)
Heeg, E., Greeff, K.: Analyse der Digitaliswirkung durch fortlaufende Messung des intraventrikulären Druckes an Katzen. Naunyn-Schmiedebergs Arch. Exp. Path. Pharmakol. *260*, 134–135 (1968)
Henderson, G., Chen, K.K.: Cardiac glycosides and aglycones by synthesis and microbiological conversion. J. Med. Chem. *8*, 577–579 (1965)
Henderson, F.G.: Chemistry and biological activity of the cardiac glycosides. In: Digitalis. Fisch, Ch. (ed.), pp. 3–12. New York: Grune & Stratton 1969
Herrmann, I., Portius, H.J., Repke, K.: Untersuchungen zur Ursache der Digitalisresistenz der Kröte. Naunyn-Schmiedebergs Arch. Exp. Path. Pharmakol. *247*, 1–18 (1964)
Higgins, C.B., Vatner, S.F., Braunwald, E.: Regional hemodynamic effects of a digitalis glycoside in the conscious dog with and without experimental heart failure. Circ. Res. *30*, 406–417 (1972)

Holland, W.C., Briggs, A.H.: Cardioactive agents. Evaluation of drug activities: Pharmacometrics Ed. Laurence, A., Bacharach, A (eds.), pp. 601–612. New York: Academic Press 1964

Horwitz, L.D., Bishop, V.S., Stone, H.L., Stegall, H.F.: Cardiovascular effects of lowoxygen atmosphere in conscious and anesthetized dogs. J. Appl. Physiol. *27*, 370–373 (1969)

Horwitz, L.D., Bishop, V.S., Stone, H.L., Stegall, H.F.: Effect of digitalis on left ventricular function in exercising dogs. Circ. Res. *41*, 370–373 (1977)

Hotovy, R.: Die enterale Resorption von Digitoxin und Gitoxin. Arzneim. Forsch. *1*, 160–164 (1951)

Iga, T., Klaassen, C.D.: Hepatic extraction of nonmetabolizable xenobiotics in rats. J. Pharmacol. Exp. Ther. *211*, 690–697 (1979)

Katz, L.N., Rodbard, S., Friend, M., Rottersman, W.: The effect of digitalis on the anesthetized dog. J. Pharmacol. Exp. Ther. *62*, 1–15 (1938)

Katzung, B.: Evaluation of drugs affecting the contracility and the electrical properties of the heart. In: Selected Pharmacological Testing Methods. Mantegazza, P., Piccinini, F. (eds.). New York: Marcel Dekker 1968

Kelliher, G.J., Roberts, J.: Effect of age on the cardiotoxic action of digitalis. J. Pharmacol. Exp. Ther. *197*, 10–18 (1976)

Klaus, W.: Evaluation of cardiac glycoside-like activity. In: Methods in drug evaluation. Burger, A. (ed.), pp. 193–234. Amsterdam: North-Holland Publ. Comp. 1966

Klupp, H.: Quantitativer Vergleich der positiv inotropen und der toxischen Wirkung einiger Herzglykoside. Naunyn-Schmiedebergs Arch. Exp. Path. *252*, 314–331 (1966)

Knaffl-Lenz, E.: The physiological assay of preparation of digitalis. J. Pharmacol. Exp. Ther. *29*, 407–425 (1926)

Koch-Weser, J.: Beta-receptor blockade and myocardial effects of cardiac glykosides. Circ. Res. *28*, 109–118 (1970)

Köhler, E., Greeff, K.: Der Einfluß des Blut-pH auf die Toxicität herzwirksamer Glykoside. Res. Exp. Med. *159*, 65–74 (1972)

Köhler, E., Greeff, K., Noack, E., Wirth, K.: Tierexperimentelle Untersuchungen über die Änderung der Glykosidempfindlichkeit bei Verschiebung des Blut-pH. Verhandlg. Dtsch. Ges. Inn. Med. *77*, 994–996 (1971)

Kosswig, W., Engelhardt, A.: Biometrischer Beitrag zur Bestimmung der Infusionstoxizität von Digitalis-Glykosiden. Arzneim. Forsch. (Drug Res.) *17*, 1258–1263 (1967)

Kraupp, O., Obenaus, B., Pillat, B., Stumpf, Ch.: Die Abhängigkeit der letalen Dosis von der Infusionsgeschwindigkeit am Beispiel des g-Strophanthin und des Thevetin B. Naunyn-Schmiedebergs Arch. Exp. Path. Pharmakol. *237*, 388–400 (1959)

Kroneberg, G.: Zuckergehalt und Wirksamkeit der Digitoxigenin- und Digoxigenindigitoxoside. Naunyn-Schmiedebergs Arch. Exp. Path. Pharmakol. *237*, 222–228 (1959)

Kroneberg, G., Schaumann, W., Stoepel, K.: Wirkungsdauer und enterale Resorption von Digitoxigenin-monodigitosid. Naunyn-Schmiedebergs Arch. Exp. Path. *243*, 91–98 (1962)

Kurbjuweit, H.G.: Zur Pharmakologie eines neu in die Therapie eingeführten Reinglykosids. Arzneim. Forsch. *14*, 716–720 (1964)

Lendle, L.: Digitalis. Berlin: Springer 1933a

Lendle, L.: Über herzwirksame Glykoside. Pharmakologische Beeinflussung der Strophanthinwirkung am Herzen bei der Titration nach Hatcher. Arch. Exp. Path. *169*, 583–603 (1933b)

Lendle, L.: Über die Eliminationsgeschwindigkeit und Kumulationsneigung von Digitalisglykosiden und Strophanthin. Naunyn-Schmiedebergs Arch. Exp. Path. *180*, 518–538 (1935)

Lenke, D., Schneider, B.: Zur Bestimmung der Cardiotoxizität von Herzglykosiden an der Katze (I. Mitteilung). Arzneim. Forsch. *19*, 687–693 (1969)

Lenke, D., Schneider, B.: Zur Bestimmung der Cardiotoxizität von Herzglykosiden an der Katze (II. Mitteilung). Arzneim. Forsch. *20*, 1199–1206 (1970a)

Lenke, D., Schneider, B.: Zur Bestimmung der Cardiotoxizität von Herzglykosiden an der Katze (III. Mitteilung). Arzneim. Forsch. *20*, 1765–1770 (1970b)

Levitt, B., Cagin, N.A., Somberg, J., Bounous, H., Mittag, Th., Raines, A.: Alteration of the effects and distribution of ouabain by spinal cord transection in the cat. J. Pharmacol. Exp. Ther. *185*, 24–28 (1973)
Liese, R.: Tierexperimentelle Untersuchungen über die Bedeutung des Alters für die Toxizität der Digitalisglykoside. Dissertation Düsseldorf 1980.
Lindner, E., Schöne, H.H.: Änderungen der Wirkungsdauer und Wirkungsstärke von Herzglykosiden durch Abwandlung der Zucker. Arzneim. Forsch. (Drug Res.) *22*, 428–435 (1972)
Lindner, E., v. Reitzenstein, G., Schöne, H.H.: Das 14,15-β-Oxido-Analoge des Proscillaridins (HOE 040). Ein neues Herzglykosid mit geringer Arrhythmie-Wirkung und hoher Resorptionsquote. Arzneim. Forsch. (Drug Res.) *29* (I), 221–226 (1979)
Lingner, K., Irmscher, K., Küssner, W., Hotovy, R., Gillissen, J.: Enterale und parenterale Wirksamkeiten von Derivaten der Herzglykoside. Arzneim. Forsch. *13*, 142–149 (1963a)
Lingner, K., Hotovy, R., Gillissen, J., Küssner, W.: Resorption und Abbau herzwirksamer Glykoside. 3. Mitteilung: Chemische Bestimmung der enteralen Resorption von Cardenolidglykosiden an Katzen. Arzneim. Forsch. *13*, 764–767 (1963b)
Litchfield, J.T., Wilcoxon, F.: A simplified method of evaluating dose-effect experiments. J. Pharmacol. Exp. Ther. *96*, 99–113 (1949)
Liu, A.Y.-C., Bentley, P.J.: A comparison of the toxicity of ouabain in vivo and in vitro in the frog, rana pipiens, and the toad, bufo marinus. Comp. Gen. Pharmacol. *2*, 476–479 (1971)
Lown, B., Levine, S.A.: Current concepts in digitalis therapy. New Engl. J. Med. *250*, 866–874 (1954)
Lucchesi, B.R., Shivak, R.: Effect of quinidine and Procain amide upon acetylstrophantidin cardiotoxicity. J. Pharmacol. Exp. Ther. *143*, 366–373 (1964)
Matsumura, S., Kimoto, S., Uno, O., Minesita, T., Ueda, M.: Some pharmacological studies on the cardiotonic effects of furanosteroidal glycosides. Arch. Int. Pharmacodyn. *225*, 246–256 (1977)
Moran, N.C.: The effect of cardiac glycosides on mechanical properties of heart muscle. In: Basic and clinical pharmacology of digitalis. Marks, B.H., Weissler, A.A. (eds.), pp. 94–117. Springfield: Charles C. Thomas 1972
Morrow, D.H.: Anesthesia and digitalis toxicity. VI. Effect of barbiturates and halothane on digoxin toxicity. Anesth. Analg. *49*, 305–309 (1970)
Nadeau, R., de Champlain, J.: Comparative effects of 6-hydroxy-dopamine and of reserpine on ouabain toxicity in the rat. Life Sci. *13*, 1753–1761 (1973)
Neugebauer, G.: Einfluß hoher i.v. Dosen von D-Penicillamin auf toxische Herzglykosidwirkungen am Meerschweinchen. Arzneim. Forsch. (Drug. Res.) *27* (II), 2073–2074 (1977)
Patnaik, G.K., Dhawan, B.N.: Pharmacological investigations on asclepin – a new cardenolide from asclepius curassavica. Arzneim. Forsch. (Drug Res.) *28* (II), 1095–1099 (1978)
Raschack, M., Haas, H., Neugebauer, G., Sipos, J.: Kardiale Wirkungen, relative enterale Wirksamkeit und Abklingquote des Herzglykosids Meproscillarin. Arzneim. Forsch. (Drug Res.) *28*, 495–502 (1978)
Reinert, H.: Die enterale Resorption des g-Strophanthins, K-Strophanthols und Digitoxins bei der Katze. Naunyn-Schmiedebergs Arch. Exp. Path. Pharmak. *215*, 1–7 (1952)
Repke, K., Est, M., Portius, H.J.: Über die Ursache der Speciesunterschiede in der Digitalisempfindlichkeit. Biochem. Pharmacol. *14*, 1785–1802 (1965)
Rosenkranz, S.: Über die Schwankungen des Wirkungswertes herzwirksamer Glykosiddrogen beim Frosch. Naunyn-Schmiedebergs Arch. Exp. Path. Pharmak. *172*, 18–25 (1933)
Roske, U.: Die pharmakologische Prüfung einiger Helveticosidderivate. Untersuchungen mit Cyclohexanon-, Acetophenon-, Cycloheptanon- und Methylpropylketon-helveticosid. Dissertation Düsseldorf 1969
Ross, J. Jr., Waldhausen, J.A., Braunwald, E.: Studies on digitalis. I. Direct effects on peripheral vascular resistance. J. Clin. Invest. *39*, 930–936 (1960a)
Ross, J. Jr., Braunwald, E., Waldhausen, J.A.: Studies on digitalis. II. Extracardiac effects on venous return and on the capacity of the peripheral vascular bed. J. Clin. Invest. *39*, 937–942 (1960b)

Rothlin, E., Suter, E.: Glykosidwirkung auf Elektrokardiogramm und Myokard. (I. Mitt.). Vergleichende elektrokardiographische Untersuchungen verschiedener herzwirksamer Glykoside an der Katze bei intravenöser Infusion. Helv. Physiol. Acta *5*, 298–321 (1947)
Rothlin, E., Bircher, R., Schalch, W.R.: Zur Pharmakologie des Acetyl-Digitoxin. Schweiz. Med. Wochenschr. *83*, 267–269 (1953)
Rothlin, E., Bircher, R.: Pharmakodynamische Grundlagen der Therapie mit herzwirksamen Glykosiden. Ergebn. Inn. Med. Kinderheilk. *5*, 457–552 (1954)
Russell, J.Q., Klaassen, C.D.: Species variation in the biliary excretion of ouabain. J. Pharmacol. Exp. Ther. *183*, 513–520 (1972)
Russell, J.Q., Klaassen, C.D.: Biliary excretion of cardiac glycosides. J. Pharmacol. Exp. Ther. *186*, 455–462 (1973)
Sachs, R.A., Highstrete, J.D., Pabst, M.L.: Digitalis studies by the U.S.P. XIV method of assay I. Seasonal variation in response of the pigeon to digitalis. J. Am. Pharm. Ass. *45*, 248–250 (1956)
Satoskar, R.S., Trivedi, J.C.: Effect of intravenous cedilanid on cats under hypothermia. Arch. Int. Pharmacodyn. *104*, 417–423 (1956)
Schaumann, W., Stoepel, K.: Intensität und Dauer der Herzwirkung von Formylgitoxigenin-mono-digitoxosid (Lanadoxin) bei enteraler und parenteraler Gabe. Naunyn-Schmiedebergs Arch. Exp. Path. *245*, 518–540 (1963)
Schaumann, W.: Über- und unteradditive Wirkung von Herzglykosiden. Naunyn-Schmiedebergs Arch. Exp. Path. Pharmak. *246*, 152–162 (1963)
Schaumann, W.: Kreislaufwirkungen von g-Strophanthin am normalen und herzinsuffizienten Meerschweinchen. Naunyn-Schmiedebergs Arch. Exp. Path. Pharmak. *247*, 229–242 (1964)
Schaumann, W., Kaiser, F.: Schwierigkeiten bei der biologischen Wertbestimmung von Digitalis-Extrakten. Arzneim. Forsch. (Drug Res.) *17*, 1264–1266 (1967)
Schaumann, W., Wegerle, R.: Verbesserung der Resorption von Helveticosid durch Substitution der Hydroxylwasserstoffe an der Digitoxose. Naunyn-Schmiedebergs Arch. Pharmak. Exp. Path. *262*, 73–86 (1969)
Schaumann, W., Wegerle, R.: β-Methyl-Digoxin. I. Cardiotoxizität bei enteraler und parenteraler Gabe. Arzneim. Forsch. *21*, 225–234 (1971)
Schaumann, W., Koch, K.: β-Methyl-Digoxin. VI. Tissue distribution and therapeutic ratio in guinea pigs in comparison with digoxin. Naunyn-Schmiedebergs Arch. Pharmak. *282*, 9–14 (1974a)
Schaumann, W., Koch, K.: β-Methyl-digoxin. VII. Tissue distribution, positive inotropic, and central action in cats in comparison with other digitalis glycosides. Naunyn-Schmiedebergs Arch. Pharmacol. *286*, 195–210 (1974b)
Schorscher, E., Sommer, S., Wild, A.J.N., Block, A.: Untersuchungen über die Wirkung von Peruvosid auf die Herzfunktionen im Tierversuch. Arzneim. Forsch. *18*, 1582–1590 (1968)
Scott, W.J. Jr., Bellies, R.P., Silverman, H.L.: The comparative acute toxicity of two cardiac glycosides in adult and newborn rats. Toxicol. Appl. Pharmacol. *20*, 599–601 (1971)
Selden, R., Klein, M.D., Smith, T.W.: Plasma concentration and urinary excretion kinetics of acetylstrophanthin. Circulation *47*, 744–751 (1973)
Selecky, F.V., Buran, L., Babulova, A.: Der Einfluß exogener Faktoren auf die Toxizität von Herzglykosiden. Arzneim. Forsch. (Drug Res.) *20*, 1048–1051 (1970)
Siegel, J.H.: The myocardial contractile state and its role in the response to anesthesia and surgery. Anesthesiology *30*, 519–564 (1969)
Simon, H., Fricke, G., Turina, M., Noseda, G., Krayenbühl, H.P., Lüthy, E.: Die Wirkung von Ouabagenin auf die Hämodynamik des Hundeherzens bei geschlossenem Thorax. Arzneim. Forsch. *19*, 693–697 (1969a)
Simon, H.J., Turena, M., Fricke, G., Medici, T., Krayenbühl, H.P., Lüthy, E.: Über die Wirkung von k-Strophanthin auf die Haemodynamik des Hundeherzens bei geschlossenem Thorax. Arzneim. Forsch. *19*, 697–701 (1969b)
Small, A., McErroy, H., Ide, R.S.: Studies of the electrocardiogram and the toxicity of cardiac glycosides in animals exposed to hyperbaric helium. Toxicol. Appl. Pharmacol. *20*, 44–56 (1971)
Somberg, J.C., Bounous, H., Levitt, B.: The antiarrhythmic effects of quinidine and propranolol in the quabain-intoxicated spinally transected cat. Eur. J. Pharmacol. *54*, 161–166 (1979)

Spann, J.F. Jr., Buccino, R.A., Sonnenblick, E.H., Braunwald, E.: Contractile state of cardiac muscle obtained from cats with experimentally produced ventricular hypertrophy and heart failure. Circ. Res. *21*, 341–354 (1967)

Spann, J.E. Jr., Covell, J.W., Eckberg, D.L., Sonnenblick, E.H., Ross, J. Jr., Braunwald, E.: Contractile performance of the hypertrophied and chronically failing cat ventricle. Am. J. Physiol. *223*, 1150–1157 (1972)

Szekely, P., Wynne, N.A.: The effect of digitalis on hypothermic heart. Brit. Heart J. *22*, 647–650 (1960)

Stoll, A., Kreis, W.: Acetyl-digitoxin. Schweiz. Med. Wochenschr. *11*, 266–267 (1953)

Stickney, J.L.: The effect of different types of anesthesia on digitalis toxicity. Am. Heart J. *87*, 734–739 (1974)

Teske, R.H., Bishop, S.P., Righter, H.F., Detweiler, D.K.: Subacute digoxin toxicosis in the beagle dog. Toxicol. Appl. Pharmacol. *35*, 283–301 (1976)

Vatner, S.F., Higgins, C.B., Franklin, D., Braunwald, E.: Effects of a digitalis glycoside on coronary and systemic dynamics in conscious dogs. Circ. Res. *28*, 470–479 (1971 a)

Vatner, S.F., Higgins, C.B., Patrik, T., Franklin, D., Braunwald, E.: Effects of cardiac depression and of anesthesia on the myocardial action of a cardiac glycoside. J. Clin. Invest. *50*, 2585–2595 (1971 b)

Veragut, U.P., Krayenbuhl, H.P.: Estimation and quantification of myocardial contractility in the closed-chest dog. Cardiologia *47*, 96–112 (1965)

Vogel, G., Lehmann, H.-D.: Über den Einfluß der Umgebungstemperatur auf die Toxizität und Geschwindigkeit der Wirkung von Convallatoxin und Digitoxin bei Rana temporaria. Med. Exp. *1*, 373–380 (1959)

Vogel, G., Marks, K.H.: Untersuchungen an der experimentellen Herzinsuffizienz des Meerschweinchens. Naunyn-Schmiedebergs Arch. Exp. Path. Pharmak. *247*, 337 (1964)

Vogel, G., Krüger, S.: Kann der Digitalis lanata-Drogenstandard durch ein Reinglykosid aus Digitalis lanata ersetzt werden? Arzneim. Forsch. *17*, 1237–1239 (1967)

Vogel, G., Baumann, I.: Untersuchungen zur Frage des „wahren" Wirkwertes cardiotoner Steroide – die Bestimmung des molaren Optimaltiters an Katzen. Drug Res. *19*, 657–659 (1969)

Vogel, G., Temme, I., Grundei, J.: Vergleichende Untersuchungen zum unterschiedlichen Verhalten von Meerschweinchen, Katze und Zwergschwein gegenüber Convallatoxin, Digoxin und Digitoxin. Arzneim. Forsch. (Drug Res.) *20*, 229–233 (1970)

Vogt, W.: Probleme der Standardisierung herzwirksamer Glykoside. Arzneim. Forsch. (Drug Res.) *17*, 1237 (1967)

Walton, R.P., Brodie, O.J.: The effect of drugs on the contractile force of a section of the right ventricle under conditions of an intact circulation. J. Pharmacol. Exp. Ther. *90*, 26–41 (1947)

Walton, R.P., Leary, J.S., Jones, H.P.: Comparative increase in ventricular contractile force produced by several cardiac glycosides. J. Pharmacol. *98*, 346–357 (1950)

Warren, M.C., Gianelly, R.E., Cutler, Sh.L., Harrison, D.C., Alto, P.: Digitalis toxicity. II. The effect of metabolic alkalosis. Am. Heart J. *75*, 358–363 (1968)

Weese, H.: Digitalis. Leipzig: G. Thieme 1936

White, W.F., Gisvold, O.: Absorption rate studies of orally administered cardiac glycosides in cats. J. Am. Pharmacol. Assoc. *41*, 42–46 (1952)

White, W.F.: Absorption of orally administered cardiac glycosides in cats. J. Am. Pharm. Ass. *44*, 607–610 (1955)

Wollenberger, A., Jehl, J., Karsh, M.: Influence of age on the sensitivity of the guinea pig and its myocardium to quabain. J. Pharmacol. Exp. Ther. *108*, 52–60 (1953)

Zechel, H.J., Brock, N.: Zur Pharmakologie des Kombinationspräparates Diazep/β-Acetyldigoxin. Arzneim. Forsch. (Drug Res.) *24*, 1905–1914 (1974)

Zielske, F., Voigtländer, W., Schaumann, W.: Resorption, Verteilung und Galle-Ausscheidung einiger ^{3}H-markierter Derivate des Helveticosols. Naunyn-Schmiedebergs Arch. Pharmak. *265*, 49–66 (1969)

Zwieten van, P.A.: Über die Gewebsverteilung, den Metabolismus und die Ausscheidung von Herzglykosiden. Dtsch. Med. Wochenschr. *92*, 1684–1687 (1967)

CHAPTER 9

The Use of the Isolated Papillary Muscle for the Evaluation of Positive Inotropic Effects of Cardioactive Steroids

M. REITER

A. The Inotropic Potency

For the quantitative evaluation of the positive inotropic effect of a cardioactive steroid, the measurement of its influence on the isometric contraction curve is, at present, the most suitable method (for review see REITER, 1972). Although isometric contraction curves can, in principle, be obtained from a complete isolated heart contracting against an intraventricular balloon (Fig. 1 b of Chap. 11) as demonstrated by MAGNUS and SOWTON (1910), the use of papillary muscles isolated from the right ventricle of cats or guinea pigs is, for various practical reasons, more appropriate (CATTELL and GOLD, 1938, 1941).

Since the positive inotropic effect of a cardioactive steroid (i.e., the increase in peak force of contraction, ΔF_c) is the consequence of an acceleration of force development (i.e., of a positive klinotropic effect; see Fig. 1), the quantitative measurement of the drug action can be made in terms of both parameters. For comparison of the potency of different cardioactive steroids it would be sufficient to measure the ΔF_c values. Besides the peak force of contraction (F_c), the resting force of the muscle (F_R) has to be observed since its increase under very high drug concentrations indicates the appearance of contracture (Chap. 11, Sect. A.III); under these conditions, total muscle force (F_M) may be composed of variable heights of F_R and F_c (Fig. 6 of Chap. 11). For a more detailed analysis of the isometric contraction curve, the time to peak force (t_1) and the relaxation time (t_2) must be measured. The quotient F_c/t_1 indicates the mean velocity of force development (S_1), and F_c/t_2 the mean velocity of relaxation (S_2). The maximum velocities of both force development and relaxation may be obtained by electrical differentiation of the contraction curve. However, it is questionable whether much is gained by the determination of the maximal velocity of force development $(dF_c/dt_1)_{max}$ in comparison with the mean velocity (S_1) since under most circumstances S_1 is directly related to $(dF_c/dt_1)_{max}$. Both, the continuous measurement of F_c and the complete analysis of the contraction curve are conveniently carried out by a computerized on-line data acquisition system (KÖHLER, 1974).

The superimposed contraction curves in Fig. 1 were obtained by the addition of cumulatively increasing concentrations of digoxin to the bath medium around a papillary muscle. The respective steady-state values of the increases in force of contraction were reached at each concentration step after approximately 1 h. The curves clearly indicate that 2 µmol/l is the maximally effective concentration of digoxin. By relating the maximum increase in peak force of contraction to the respective increments (ΔF_c) at the various concentrations, a concentration–effect curve

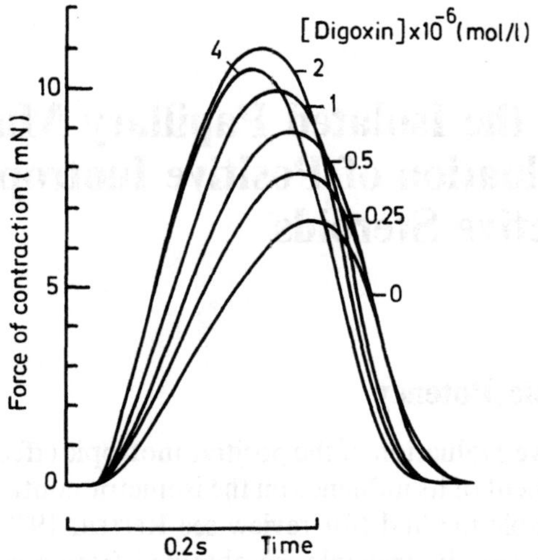

Fig. 1. Superimposed isometric contraction curves of guinea-pig papillary muscle as obtained by cumulatively increased digoxin concentrations

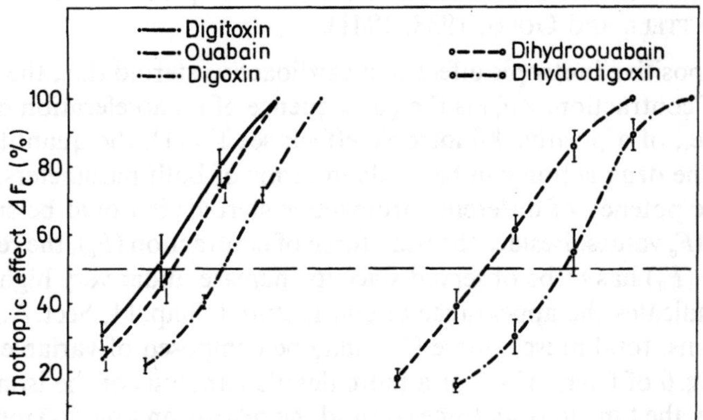

Fig. 2. The positive inotropic effect (ΔF_c) of five different cardioactive steroids as a function of concentration. Guinea-pig papillary muscle contracting under the following conditions: stimulation frequency 1 Hz; stimulation intensity slightly above threshold; resting force (F_R) 4 mN; Krebs-Henseleit solution containing 3.2 mmol/l Ca^{2+} and oxygenated by a gas mixture of 95% O_2 and 5% CO_2; temperature 35 °C. Number of muscles = 6–8. Means ± standard error of the mean. (REITER, 1967)

is obtained (Fig. 2). From interpolation at the 50% level, the half-maximally effective concentration (EC_{50}) can be derived which provides the most accurate measure of the positive inotropic potency of a cardioactive drug. From the concentration–effect curves of the five different substances depicted in Fig. 2, the following EC_{50} values were obtained (REITER, 1967): digitoxin 2.8×10^{-7}; ouabain 3.4×10^{-7}; digoxin 6.2×10^{-7}; dihydroouabain 1.4×10^{-5}; dihydrodigoxin 3.5×10^{-5} (all in mol/l). A statistically valid estimate of the EC_{50} can be obtained by fitting a logistic function to the experimental data (PARKER and WAUD, 1971).

B. Methodologic Considerations

To obtain reliable data for the comparison of the inotropic potency of the various cardioactive steroids it is important to perform the experiments under strictly identical conditions. In order to get a useful concentration–effect curve the muscle should contract under conditions that allow a sufficient increase of F_c over the control value. The following methodologic factors which influence the inotropic effect should be considered.

I. Selection and Preparation of Muscle

In order to secure sufficient oxygen diffusion for an adequate mechanical performance, the cross section of the papillary muscle should not exceed 0.6 mm^2, corresponding to a diameter of 0.87 mm (Koch-Weser, 1963). The diameter has to be even smaller (0.64 mm or less) if maximal oxygen uptake is to be obtained (Cranefield and Greenspan, 1960). This implies that the muscle should not be unduly conical in its structure. In preparing the papillary muscle from the right ventricle of guinea pigs or cats and in mounting it in its holding device care should be taken not to impose any stretch upon it.

II. Incubation Medium

1. Bicarbonate

It was shown by White and Salter (1946) that a papillary muscle contracts optimally for many hours if it is incubated in bicarbonate-buffered solution, whereas its force of contraction declines rapidly in phosphate-buffered solution. In regard to its effect on force of contraction, bicarbonate is also superior to Tris [tris(hydroxymethyl)aminomethane] buffer (Durrett and Adams, 1979). Bicarbonate buffer is not only optimal for constancy of the control force of contraction but also for the inotropic effect of cardioactive steroids (Sanyal and Saunders, 1957). Of all the different solutions used in experimental pharmacology for isolated tissue preparations, Krebs–Henseleit solution contains the highest bicarbonate concentration (25 mmol/l) and is therefore preferable for experiments with isolated heart tissue (Krebs and Henseleit (1932).

2. Potassium

The positive inotropic effect of cardioactive steroids depends in its magnitude on the outside potassium concentration (Chap. 11, Sect. B.II); the inotropic effect decreases with increasing extracellular potassium concentration $[K^+]_o$. This should be considered if potency values are compared which were obtained with different incubation media (e.g., 5.9 mmol/l K^+ in Krebs–Henseleit solution versus 2.7 mmol/l K^+ in Tyrode solution) (Tyrode, 1910).

3. Calcium

One difficulty in obtaining a complete concentration–effect curve of a cardioactive steroid is the appearance of spontaneous contractions. In our experience (REITER, 1967), the incidence of automaticity can be reduced by using a calcium concentration which is somewhat higher (3.2 mmol/l) than in the normal Krebs–Henseleit solution.

III. Temperature

In selecting the temperature of the bath medium it should be considered that by reducing the temperature below physiologic levels, both time to peak force of contraction and peak force itself increase considerably (REITER, 1972). Thereby the F_c value may reach its inherent maximum thus not allowing a sufficient increase of F_c over the control value. At low temperatures (e.g. 27 °C) inotropic effects of cardioactive steroids are either negligible or absent (SAUNDERS and SANYAL, 1958; MEYER and KUKOVETZ, 1962) unless extracellular calcium (REITER and STICKEL, 1970) or frequency of contractions (SAUNDERS and SANYAL, 1958) were reduced in order to lower F_c. Therefore, the temperature should be kept constant somewhere in the physiologic range (35°–37 °C).

IV. Frequency of Contraction

The positive inotropic effect of cardioactive steroids depends on contraction frequency of the heart muscle (Chap. 11, Sect. C.II.1). This must be taken into account if potency values are compared which were obtained at different contraction frequencies. In any case, the selected frequency (1 or 0.5 Hz) must be kept constant throughout the experiment.

V. Stimulation Intensity

This should be kept only slightly above the stimulation threshold in order to avoid additional inotropic effects through the release of stored catecholamines (FURCHGOTT et al., 1959; JEWELL and BLINKS, 1968). It can be achieved better by punctate electrodes than by field stimulation.

VI. Length–Force Relationship and Plasticity

Both force of contraction (F_c) and resting force (F_R) of the heart muscle depend on muscle length (L_M) in a characteristic manner (ABBOTT and MOMMAERTS, 1959; SONNENBLICK, 1962, BRADY, 1964) known as the Frank–Starling relationship (Fig. 3). F_c increases with increasing L_M up to a maximum and declines with a further stretch of the muscle (Fig. 3a). Likewise, the absolute amount of an inotropic effect (ΔF_c) varies with a change in L_M (SONNENBLICK, 1962). The difference in ΔF_c dependence on L_M becomes more prominent at drug concentrations in the upper range of the concentration–effect curve (e.g. 5×10^{-5} mol/l dihydroouabain in Fig. 3a, b) than at lower concentrations. For quantitative measurement of the inotropic

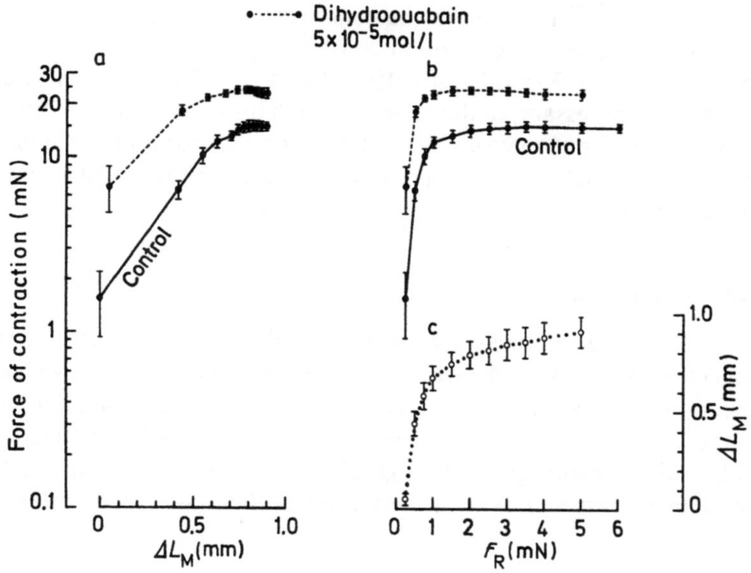

Fig. 3a–c. Length–force relationship of cardiac ventricular muscle under the influence of a cardioactive steroid. Guinea-pig papillary muscles contracting at a frequency of 1 Hz (stimulation intensity slightly above threshold) in oxygenated (95% O_2 + 5% CO_2) Krebs–Henseleit solution containing 3.2 mmol/l Ca^{2+}; temperature 35 °C. Number of muscles = 4. **a** peak force of contraction (F_c) (logarithmic scale) plotted against change in muscle length (ΔL_M) as obtained by stretch of the muscle relaxed to a resting force of 0.25 mN. **b** relation between force of contraction (F_c) and resting force (F_R) as obtained by increasing L_M. Same experiment as in **a**. F_c (logarithmic scale), plotted against F_R. **c** relation between ΔL_M and F_R. From the same experiment in the presence of dihydroouabain. Means ± standard error of the mean

effect and for the purpose of comparison, therefore, the role of L_M in the contractile performance of cardiac muscle has to be considered.

In conducting the experiment of Fig. 3, the control curve was established first. L_M was then reduced until a resting force (F_R) of 4 mN was obtained. Dihydroouabain in a final concentration of 5×10^{-5} mol/l was then added and, after its positive inotropic effect reached the steady state, the muscle was relaxed to $F_R = 0.25$ mN. This corresponded to the original value at $\Delta L_M = o$ for the control and was now located at $\Delta L_M = 0.05$ mm in consequence of a plastic elongation of the muscle during the applied stretch in the control experiment.

The situation is complicated by the fact that, on account of the plasticity of heart muscle (ABBOTT and MOMMAERTS, 1959; REICHEL, 1960; BLINKS and KOCH-WESER, 1963) the values of both F_R and F_c, as obtained at a particular L_M, slowly decline if the muscle is kept under isometric conditions over a longer period of time. This indicates that, owing to the prolongation of the plastic elements, the length of the sarcomere is reduced, and the whole length–force relationship is shifted to the right towards higher values of L_M. In order to compensate for the shift one can make use of the dependence of F_R on L_M (Fig. 3c). By applying an appropriate stretch the length of the muscle is increased until the original value of F_R is obtained, thereby restoring the original sarcomere length which secures the constancy of F_c. Therefore, in order to obtain the concentration effect curve under strictly identical conditions one has to secure a constant F_R over the length of the whole experiment (BLINKS and KOCH-WESER, 1963; REITER, 1967). This may be achieved by readjustment, if necessary, of the applied stretch by hand or, more conveniently, by an automatic device (STAMPFL, 1979) if one provides for a possible increase of F_R as the sign of contracture due to supramaximal concentrations of a cardioactive steroid which, of course, should not be overlooked.

VII. Stray Compliance

A weak point for the exact measurement of force development during contraction is the connection between the muscle end and the isometric lever. Any elasticity of this connection reduces the amount of force which is transmitted to the lever (BLINKS, 1965, REITER, 1967). The most secure way to avoid such elasticity is to use a small chain of metal (preferably gold) (REITER, 1967). This implies that, for precise force measurements, vertical mounting of the muscle is a prerequisite.

VIII. The Two-Chambered Bath

In order to provide the necessary oxygen saturation of the incubation medium at a relatively high temperature (35°–37 °C) it is necessary to aerate the solution vigorously with a mixture of O_2 (95%) and CO_2 (5%). Since a vigorous bubbling of the surface of the medium disturbs the proper tracing of the contraction curve, two-chambered baths have been designed in which the gas is applied to one chamber and the oxygenated solution is circulated by the gas stream to the second adjacent chamber in which the muscle is mounted (GREEN et al., 1953; BLINKS, 1965; REITER, 1965, 1967).

References

Abbott, B.C., Mommaerts, W.F.H.M.: A study of inotropic mechanisms in the papillary muscle preparation. J. Gen. Physiol. *42*, 533–551 (1959)

Blinks, J.R.: Convenient apparatus for recording contractions of isolated heart muscle. J. Appl. Physiol. *20*, 755–757 (1965)

Blinks, J.R., Koch-Weser, J.: Physical factors in the analysis of the actions of drugs on myocardial contractility. Pharmacol. Rev. *15*, 531–599 (1963)

Brady, A.J.: The development of tension in cardiac muscle. In: Pharmacology of cardiac function, Second Int. Pharmacol. Meeting. Krayer, O. (ed.), Vol. 5, pp. 15–23. Oxford: Pergamon Press 1964

Cattell, McKeen, Gold, H.: The influence of digitalis glucosides on the force of contraction of mammalian cardiac muscle. J. Pharmacol. Exp. Ther. *62*, 116–125 (1938)

Cattell, McKeen, Gold, H.: Studies on purified digitalis glucosides III. The relationship between therapeutic and toxic potency. J. Pharmacol. Exp. Ther. *71*, 114–125 (1941)

Cranefield, P.F., Greenspan, K.: The rate of oxygen uptake of quiescent cardiac muscle. J. Gen. Physiol. *44*, 235–249 (1960)

Durrett, L.R., Adams, H.R.: Inotropic responsiveness of atrial myocardium bathed in Tris- or bicarbonate-buffered solutions. Am. J. Physiol. *237*, H318–H325 (1979)

Furchgott, R.F., de Gubareff, T., Grossman, A.: Release of autonomic meditors in cardiac tissue by suprathreshold stimulation. Science *129*, 328–329 (1959)

Green, J.P., Riley, J.A., White, J.M.: A suspension bath for the study of isolated tissues in the presence of surface active agents. Rev. Sci. Inst. *24*, 183–184 (1953)

Jewell, B.R., Blinks, J.R.: Drugs and mechanical properties of heart muscle. Ann. Rev. Pharmacol. *8*, 113–130 (1968)
Koch-Weser, J.: Effect of rate changes on strength and time course of contraction of papillary muscle. Am. J. Physiol. *204*, 451–457 (1963)
Köhler, H.-W.: On line-Datenaufnahme und real-time-Auswertung von isometrischen Kontraktionskurven des Herzmuskels mit Hilfe eines PDP-12-Computers. Arzneim. Forsch. *24*, 815–819 (1974)
Krebs, H., Henseleit, K.: Untersuchungen über die Harnstoffbildung im Tierkörper. Hoppe-Seylers Z. Physiol. Chem. *210*, 33–66 (1932)
Magnus, R., Sowton, S.C.M.: Zur Elementarwirkung der Digitaliskörper. Naunyn Schmiedebergs Arch. Exp. Path. Pharmak. *63*, 255–262 (1910)
Meyer, H.F., Kukovetz, W.R.: Die Wirkung von g-Strophanthin auf den Papillarmuskel der Katze bei Hypothermie, Normaltemperatur und Hyperthermie. Naunyn Schmiedebergs Arch. Exp. Path. Pharmak. *242*, 409–413 (1962)
Parker, R.B., Waud, D.R.: Pharmacological estimation of drug-receptor dissociation constants. Statistical evaluation. I. Agonists. J. Pharmacol. Exp. Ther. *177*, 1–12 (1971)
Reichel, H.: Muskelphysiologie, S. 26–30. Berlin, Heidelberg, New York: Springer 1960
Reiter, M.: Die Untersuchung inotrop wirkender Substanzen. In: Festschrift 75 Jahre Hommel, S. 480–484. Zürich 1965
Reiter, M.: Die Wertbestimmung inotrop wirkender Arzneimittel am isolierten Papillarmuskel. Arzneim. Forsch. *17*, 1249–1253 (1967)
Reiter, M.: Drugs and heart muscle. Ann. Rev. Pharmacol. *12*, 111–124 (1972)
Reiter, M., Stickel, F.J.: Zur Frage einer therapeutischen Steigerung der Kontraktilität bei Abkühlung des Herzens. Klin. Wochenschr. *48*, 935–938 (1970)
Sanyal, P.N., Saunders, P.R.: Action of ouabain upon normal and hypodynamic myocardium. Proc. Soc. Exp. Biol. Med. *95*, 156–157 (1957)
Saunders, P.R., Sanyal, P.N.: Effect of temperature upon the positive inotropic action of ouabain. J. Pharmacol. Exp. Ther. *123*, 161–163 (1958)
Sonnenblick, E.H.: Force-velocity relations in mammalian heart muscle. Am. J. Physiol. *202*, 931–939 (1962)
Stampfl, A.: Nachregelung der mechanischen Vorspannung isolierter Präparate. Elektronik *28*, 82–84 (1979)
Tyrode, M.V.: The mode of action of some purgative salts. Arch. Int. Pharmacodyn. *20*, 205–223 (1910)
White, W.F., Salter, W.T.: The response of hypodynamic myocardium to known concentrations of cardiac glycosides. J. Pharmacol. Exp. Ther. *88*, 1–9 (1946)

CHAPTER 10

Evaluation of Cardiac Glycosides in Isolated Heart Preparations Other than Papillary Muscle

K. GREEFF and D. HAFNER

A. Introduction

Complementary to Chap. 9, which dealt primarily with the evaluation of cardiac glycosides in isolated papillary muscle, further methods often used in the quantitative and qualitative analysis of the action of cardiac glycosides are to be discussed. Details of the methods have been included not only to encourage experimental work, but also to give information to those wishing to evaluate results without previous practical experience. Further details of different methods are described by LENDLE (1935), WEESE (1936), KLAUS (1966), THORP and COBBIN (1967), KATZUNG (1968), and LEVY (1971). In addition to the particular methods, some experimental results have also been included to provide some idea of the usefulness of the procedures and to indicate several possible modifications.

A criticism, which often arises upon evaluating investigations with isolated organs, is that the results are not necessarily transferable to conditions existing within the intact animal. This is correct, when regarding the action of cardiac glycosides not only as organ specific but as a more complex action, which in vivo may be comprised of cardiac and extracardiac components. Investigations on isolated organs enable an evaluation of the cardiac action of glycosides, independent of extracardiac factors and under comparably constant experimental conditions. Temperature, frequency, extracellular ionic concentrations, and the oxygensupply can be kept constant. Preload and afterload can be controlled in isometric contraction experiments.

Table 1. Standard solutions for cold-blooded (RINGER, 1883) and warm-blooded animals (TYRODE, 1910; KREBS and HENSELEIT, 1932)

	Ringer (mmol/l)	Tyrode (mmol/l)	Krebs–Henseleit (mmol/l)
NaCl	111	136.9	118.05
KCl	2.02	2.68	4.69
$CaCl_2$	1.09	1.80	2.52
$MgCl_2$		1.05	
$MgSO_4$			1.19
NaH_2PO_4		0.42	
KH_2PO_4			1.18
$NaHCO_3$	2.40	11.9	24.99
Glucose		5.05	10.1

Suprising effects such as differing potency have been noticed upon accidentally varying the electrolyte composition of the incubation medium. The standard solutions and their composition are summarized in Table 1, for cold-blooded (RINGER, 1883; for review see LOCKWOOD, 1961) or warm-blooded animals (TYRODE, 1910; KREBS and HENSELEIT, 1932). Furthermore, the existence of countless modifications further complicates the comparison of experimental results. Variations in substrate have been recommended (KOCH-WESER, 1971 b; BÜNGER et al., 1975) as well as variations in electrolyte concentrations (LAURENCE and BACHARACH, 1964) and the influence of changing the osmolarity by nonelectrolytes (succrose, mannitol; KOCH-WESER, 1963; BLINKS and KOCH-WESER, 1963) has been investigated. This point will be discussed in detail later.

B. Isolated Atrial Preparations

Atrial preparations are particularly suitable for studies on the isolated heart, as their thinness allows an adequate supply of oxygen and substrates, which in turn ensures optimal function over a period of hours. Furthermore, the technique of preparation is relatively simple and therefore causes little harm. The best preparations are obtained from guinea pigs and young rabbits or cats, whereas in larger species such as dog, pig, or human, atrial tissue can only be used in the form of strips.

I. Methods

The freshly dissected heart should be placed quickly, preferably within 1 min, into a physiologic salt solution (room temperature 20°–22 °C, oxygenated). Both atria are cut away from the ventricle and then separated. The right atrium beats spontaneously and can be used to study effects on frequency; it is however unsuitable for studying the effects of inotropic agents with chronotropic side effects, as it is well known that a change in frequency influences the force of contraction. The left atrium is secured at its base with a thread to a holder containing two electrodes, the opposite end being secured with a thread to the transducer. Atrial strips from left atria (0.5 mm thick, 10 mm long, and 5 mm wide) are fixed in a similar way. Two field electrodes, positioned on either side of the atrium, can be used for excitation. This so-called field stimulation has, however, an additional effect; through simultaneous excitation of myocardial nerve fibers, neurotransmitters, particularly noradrenaline, can be released. The electrical stimulus normally takes the form of a square wave pulse of 1–5 ms duration and a strength 1.5–3 times the excitation threshold. Experimental results on the action of cardiac glycosides çan only be compared when the preload, excitation frequency, and temperature are kept constant as these factors influence the positive inotropic effect. The preload is usually 0.5–1.0 g, frequency of stimulation 1–3 Hz, and temperature 30°–37 °C (see Sects. B.II.1. c, d).

The composition of the incubation medium should also be taken into account when evaluating results. Therefore, when comparing experiments one should take note of the buffer and in particular the K^+, Ca^{2+}, and Mg^{2+} concentrations. $NaHCO_3$ is for instance more suitable as a buffer than phosphate buffer (WHITE and SALTER, 1946) or Tris buffer which has been shown to depress contractility of isolated guinea pig atria (DURRETT and ADAMS, 1979). An increase in the K^+ or Mg^{2+} concentration decreases the potency and toxicity of digitalis glycosides, whereas an increased Ca^{2+} concentration increases the toxicity. A medium of the following composition is usually used for isolated atrial preparations (concentrations in mmol/l): NaCl (120–140), KCl (2–5), $MgSO_4$ (1–2), $CaCl_2$ (1–3), $NaHCO_3$ (10–25), $NaHPO_4$ (0.2–1.0), glucose (5–15). KOCH-WESER (1971 b) recommended the addition of fumarate, pyruvate, L-glutamate (5 mmol/l each), and insulin (5 IU/l). The solution is con-

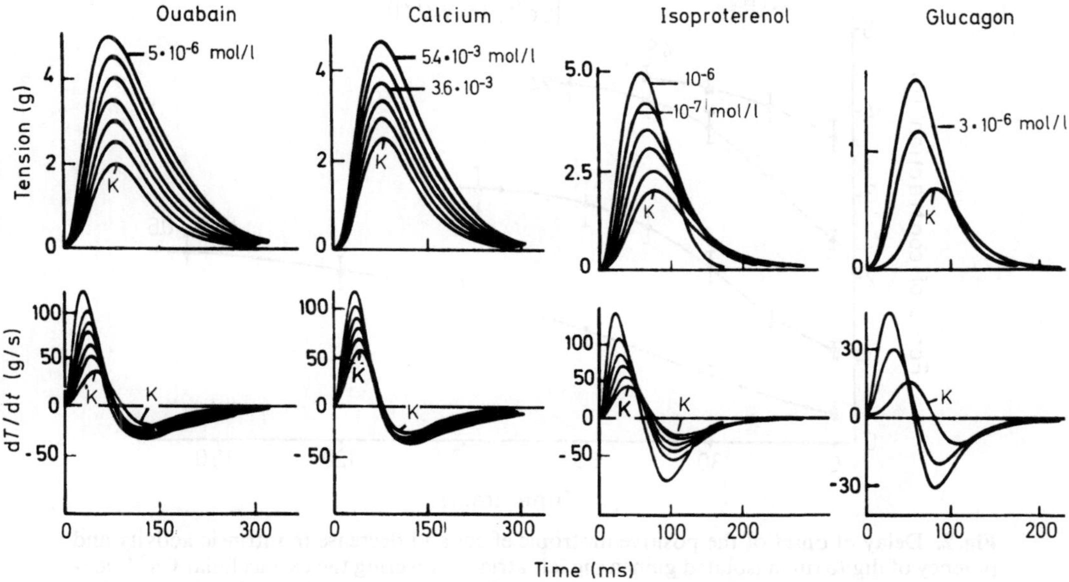

Fig. 1. Superimposed isometric contraction curves and their first derivatives, d*T*/d*t* for guinea pig atria. Glucagon was tested on a guinea pig atrium, pretreated with reserpine (2 × 1 mg/kg within 48 h before the experiment). The preparations were suspended in Krebs–Henseleit solution with 2.4 mmol/l Ca^{2+} at 31 °C and stimulated electrically (1 ms, 1 Hz, intensity 50% above threshold); preload 1.0 g (GREEFF, 1977)

tinuously ventilated with a gas mixture of 5% CO_2 and 95% O_2, resulting in a constant pH of 7.4. Further details as to methods may be found in THORP and COBBIN (1967) and LEVY (1971).

II. Evaluation of Effects

Investigations on isolated atria enable a qualitative differentiation and characterization of cardiac glycosides in comparison with other positive inotropic agents. On the other hand, it is possible to compare quantitatively the effectivity of various glycosides, i.e., with respect to the activity, potency, and also the speed of onset and duration of action.

1. Qualitative Evaluation

A differentiation of the action of cardiac glycosides with respect to other positive inotropic substances, e.g., catecholamines, xanthine derivatives, or glucagon is possible by studying changes in the dynamics of contraction and also by observing the influence of extracellular Ca^{2+} concentration, frequency of stimulation, or temperature on the action of cardiac glycosides. Further information is given when studying the interactions with antagonistically active substances as described later.

a) Dynamics of Contraction

Digitalis glycosides accelerate the velocity of contraction and delay relaxation (Fig. 1). A similar effect may be produced by increasing the Ca^{2+} concentration

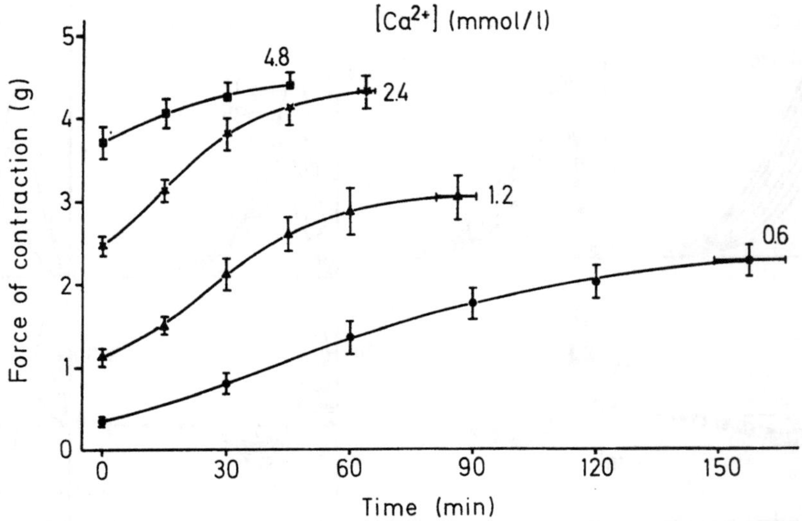

Fig. 2. Delay of onset of the positive inotropic effect and decrease in intrinsic activity and potency of digitoxin in isolated guinea pig left atria by lowering the extracellular Ca^{2+} concentration. In these experiments the maximum effective concentrations of digitoxin, taken from concentration–response curves, were achieved by single injection into the bath fluid. The concentration of digitoxin necessary for the maximal effects increased with decreasing Ca^{2+} concentrations: 1.8×10^{-7} mol/l (Ca^{2+}: 4.8 mmol/l), 2.6×10^{-7} mol/l (Ca^{2+}: 2.4 mmol/l), 3.8×10^{-7} mol/l (Ca^{2+}: 1.2 mmol/l), 5.6×10^{-7} mol/l (Ca^{2+}: 0.6 mmol/l). The experimental conditions the same as described in Fig. 1

of the bathing medium. In contrast, a characteristic of catecholamines mediated by β-adrenoceptors is an acceleration of relaxation; this also applies to glucagon. An increase in force of contraction with a concomitant acceleration of relaxation is also found upon application of dibutyryl cyclic AMP (GREEFF, 1977), an observation which is in agreement with the theory that catecholamines produce their positive inotropic effect via an activation of adenyl cyclase, whereas this enzyme remains unaffected by cardiac glycosides (DÖNGES et al., 1977).

b) Extracellular Calcium Concentration

It is well known that the positive inotropic effect of cardiac glycosides can be characterized by changing the concentration of extracellular calcium (for review see Chaps. 11 and 18; LEE and KLAUS, 1971). An increase in extracellular Ca^{2+} concentration increases the toxicity of cardiac glycosides on isolated atria (GREEFF et al., 1971; WAGNER and SALZER, 1976). The inotropic effects of cardiac glycosides and the rate of onset are reduced with decreasing extracellular Ca^{2+} concentrations as seen in Fig. 2 (see also PARK and VINCENZI, 1976). In this respect cardiac glycosides behave quite differently from catecholamines as their positive inotropic effect is less sensitive to the lowering of the extracellular Ca^{2+} concentration (DUDEK and MANTEL, 1979; DAHMEN and GREEFF, 1981). But also amongst different glycosides the dependence on the extracellular Ca^{2+} concentration may vary. As shown in Fig. 3 dihydroouabain at low Ca^{2+} concentrations (0.6 or 1.2 mmol/l) is much more effective than ouabain. In isolated rabbit atria PARK and VINCENZI (1976)

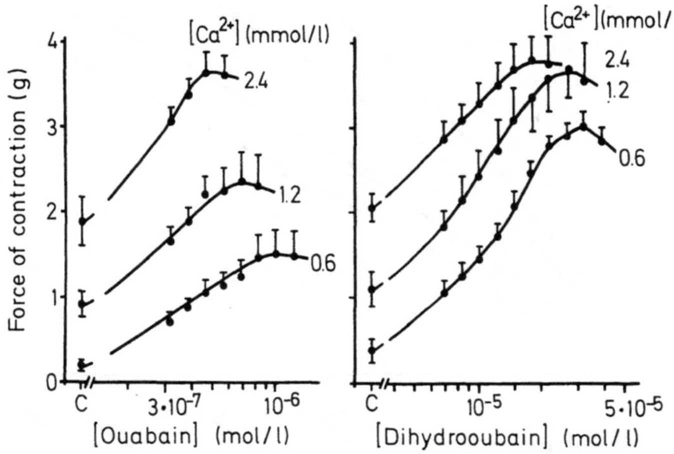

Fig. 3. The effect of ouabain and dihydroouabain on force of contraction in isolated guinea pig atria depending on the extracellular Ca^{2+} concentration. The positive inotropic effect of dihydroouabain is less diminished after lowering the Ca^{2+} concentration than the effect of ouabain. The experimental conditions the same as described in Figs. 1 and 2

compared the effects of glycosides (strophanthoside K and ouabain) and aglycones (acetylstrophanthidin and ouabaigenin) and observed that the rate of onset of aglycone-induced positive inotropism was less dependent on calcium than the rate of onset induced by the glycosides. A specific calcium–glycoside interaction was also observed in isolated papillary muscle by ALKEN et al. (1974) using different cardiac glycosides. Rat atria are comparatively insensitive to cardiac glycosides and the action of ouabain is less affected by changes in extracellular Ca^{2+} concentration (REFSUM and LANDMARK, 1977).

c) Beat Frequency

It has been confirmed in numerous investigations that the positive inotropic action of cardiac glycosides in isolated atria is dependent on frequency (KOCH-WESER and BLINKS, 1962, 1963; TUTTLE and FARAH, 1962; MORAN, 1967). For example, increasing the frequency from 1 to 3 Hz, the positive inotropic action of glycosides is reduced (KOCH-WESER and BLINKS, 1962) and the toxicity increased (GREEFF et al., 1971). These effects are probably correlated to frequency-dependent changes in action potentials, changes in calcium uptake, or calcium exchange. Frequency also affects the time course of development of the positive inotropic action of cardiac glycosides (Fig. 4; KOCH-WESER, 1971 a). Further details are given in Chaps. 11 and 12.

d) Temperature

The sensitivity of isolated atrial preparations to sympathomimetic amines has often been investigated (for review see BROADLEY and DUNCAN, 1977). Their positive inotropic action on guinea pig atria is increased by decreasing the bath temperature from 37° to 25 °C, whereas the optimal temperature for rate responses is 37 °C (DUNCAN and BROADLEY, 1978). In contrast to isoprenaline the authors found that

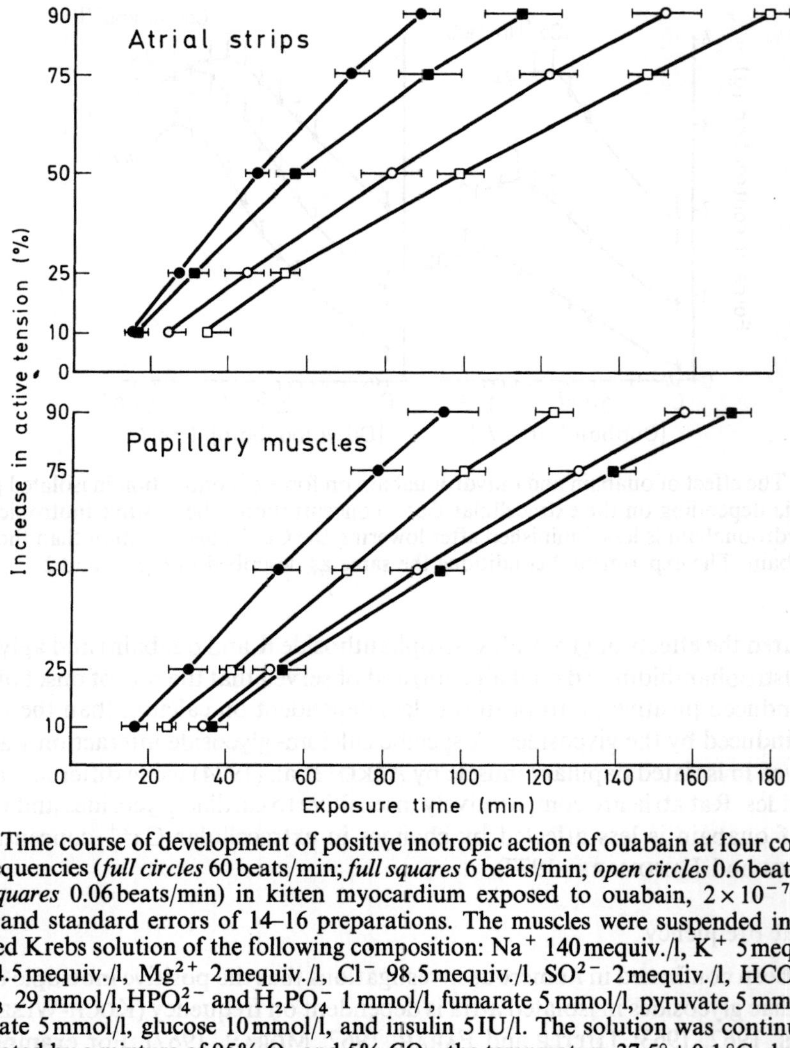

Fig. 4. Time course of development of positive inotropic action of ouabain at four contraction frequencies (*full circles* 60 beats/min; *full squares* 6 beats/min; *open circles* 0.6 beats/min; *open squares* 0.06 beats/min) in kitten myocardium exposed to ouabain, 2×10^{-7} mol/l; means and standard errors of 14–16 preparations. The muscles were suspended in 50 ml modified Krebs solution of the following composition: Na^+ 140 mequiv./l, K^+ 5 mequiv./l, Ca^{2+} 4.5 mequiv./l, Mg^{2+} 2 mequiv./l, Cl^- 98.5 mequiv./l, SO^{2-} 2 mequiv./l, HCO_3^- and H_2CO_3 29 mmol/l, HPO_4^{2-} and $H_2PO_4^-$ 1 mmol/l, fumarate 5 mmol/l, pyruvate 5 mmol/l, L-glutamate 5 mmol/l, glucose 10 mmol/l, and insulin 5 IU/l. The solution was continuously oxygenated by a mixture of 95% O_2 and 5% CO_2, the temperature 37.5° ± 0.1 °C, duration of rectangular pulses 5 ms (KOCH-WESER, 1971 a)

the optimal temperature for the positive inotropic action of ouabain lay between 35° and 40 °C and did not increase upon cooling from 38° to 25 °C (BROADLEY and DUNCAN, 1977). PARK and VINCENZI (1975) also reported that the rate of onset of the action of ouabain in rabbit atria was dependent on the temperature. The onset of action was delayed at a stimulus frequency of 0.1 Hz, when the bath temperature was reduced from 37° to 27 °C; this does not apply to acetylstrophanthidin.

2. Quantitative Evaluation

Investigations on isolated atria also enable a quantitative evaluation of the activity and potency of cardiac glycosides. Figure 5 demonstrates the activity in guinea pig atria of several glycosides, of which digitoxin is the most potent, followed by oua-

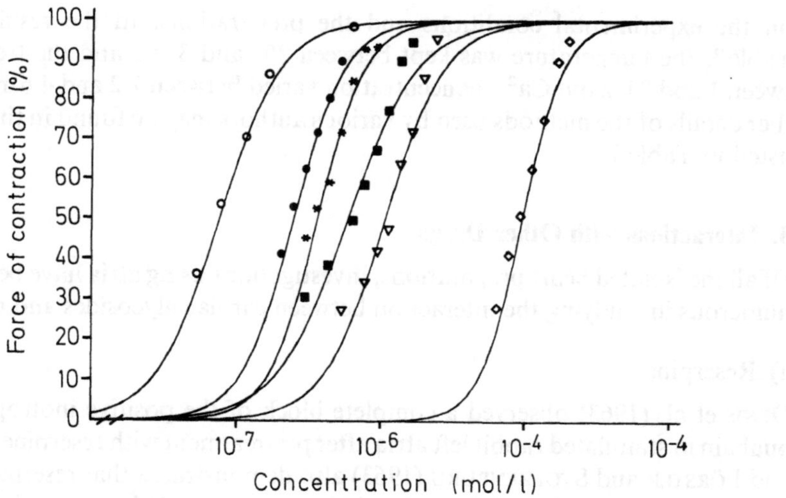

Fig. 5. Concentration–response curves in isolated guinea pig left atria suspended in Krebs–Henseleit solution with 1.2 mmol/l Ca^{2+} as described in Fig. 1. (*Open circles* digitoxin; *full circles* ouabain; *asterisks* digoxin; *squares* strophanthoside K; *diamonds* dihydroouabain.) Data calculated from these concentration–response curves are given in Table 2

Table 2. Comparison of the potency and intrinsic activity of different cardiac glycosides in guinea pig left atria. The corresponding concentration–response curves are shown in Fig. 5. The ED_{50} values were obtained by analysing the concentration–response curves using the logit function described by HAFNER et al. (1977)

	Potency		Intrinsic activity half-maximal effect	
	ED_{50} (mol/l)	Relative	p	(% of control)
Digitoxin	7.6×10^{-8}	1	1.2	92
Ouabain	2.5×10^{-7}	0.31	1.1	83
Digoxin	3.6×10^{-7}	0.21	1.0	90
Cymarin	5.9×10^{-7}	0.13	1.2	92
Strophanthoside K	1.1×10^{-6}	0.07	1.0	82
Dihydroouabain	9.5×10^{-6}	0.008	1.4	107

p = pond

bain, digoxin, cymarin, strophanthoside K, and dihydroouabain (Table 2). Further results of investigations on the actions of several glycosides on atria of various animal species are shown in Table 3. Concentrations of about 1 μmol/l are in general necessary to evoke the positive inotropic effect. PRASAD (1974) found an increase in force of contraction in human atrial preparations at a concentration of 0.01 μmol/l, whereas investigations of ITO (1975) and GREEFF and SCHLIEPER (1967) found a concentration of 1 μmol/l to be necessary. In chicken atria, ouabain is particularly effective in activity and potency (LOCK, 1963). Much higher concentrations are needed to produce a positive inotropic action in rat atria (ILLANES and MARSHALL, 1964; GREEFF and SCHLIEPER, 1967; AKERA et al., 1979). The positive inotropic activity of glycosides varies greatly and seems to be essentially dependent on the experimental conditions and the preparations. In the results given in Table 3, the temperature was kept between 30° and 37 °C and the frequency between 1 and 2 Hz but Ca^{2+} concentration varied between 1.2 and 4.6 mmol/l. Further details of the methods used by various authors may be found in the references listed in Table 3.

3. Interactions with Other Drugs

Of all the isolated heart preparations, investigations using atria have been the most numerous in studying the interaction between cardiac glycosides and other drugs.

a) Reserpine

Denis et al. (1963) observed a complete block of the positive inotropic effect of ouabain in stimulated rabbit left atria after pretreatment with reserpine (0.6 mg/kg) and Förster and Stolzenburg (1963) also demonstrated that reserpine pretreatment weakens the effect of digitoxigenin in spontaneously beating guinea pig atrial preparations. Levy and Richards (1965) observed in electrically driven left atria of rabbits pretreated with intravenous (i.v.) reserpine (1 mg/kg) that the positive inotropic effect of a single concentration (8.3×10^{-7} mol/l) of ouabain was attenuated compared with controls. However, the maximal contractile force developed with higher ouabain concentrations (3.35×10^{-6} mol/l) and was not significantly different from that seen in normal atria. The same authors observed that reserpine pretreatment (5 mg/kg i.v.) did not prevent the toxic effects of ouabain. In spontaneously beating right or stimulated left guinea pig atria, Fratz et al. (1967) found that reserpine pretreatment (1 mg/kg on 2 days) affected neither positive inotropic nor toxic action of strophanthoside K. Govier (1965) reported that the refractory period in rabbit atria was first decreased by ouabain (10^{-6} mol/l), then increased. Atria treated with reserpine or β-adrenergic blocking agents showed an increase in refractory period alone.

b) β-Adrenergic Blocking Agents

Förster and Kalsow (1965) investigated the influence of dichlorisoproterenol (DCI) on the action of various glycosides in spontaneously beating guinea pig atrial preparations. The positive chronotropic action of all glycosides and genins was decreased to the same extent by DCI, whereas DCI had a varying effect on the positive inotropic action: that of ouabain, ouabaigenin, and digitoxigenin was significantly decreased, whereas that of bufalin and digoxigenin was not influenced. In contrast investigations by Levy and Richards (1965) showed that the positive inotropic action of ouabain in isolated rabbit atria was not influenced by propranolol or pronethalol. The investigations of Fratz et al. (1967) also showed that propranolol does not influence the inotropic action of strophanthoside K on guinea pig atria but delays the occurrence of toxic arrhythmia. In atrial strips from kitten or guinea pig atria, the positive inotropic effect of ouabain evaluated from cumulative concentration–effect curves was not altered by 10^{-6} mol/l propranolol (Koch-Weser, 1971 b). Further results on the interaction of reserpine or β-blocking agents observed with other methods are given in this handbook, part II, Chap. 14.

Table 3. The positive inotropic effect of different cardiac glycosides in atria of different animal species

Reference	Species	Glycoside	Stimulation Frequency (Hz)	$[Ca^{2+}]$ (mmol/l)	Temperature (°C)	Effect	Concentration (μmol/l)	Increase of force of contraction Δ (%)
LOCK (1963)	Rabbit	Ouabain	1	2.2	30	e	2.4	~ 25
	Chicken	Ouabain	1	2.2	30	e	1.1	229
ILLANES and MARSHALL (1964)	Ground squirrel	Ouabain	sp	2.2	31	EC_{max}	0.3	42
	Rabbit	Ouabain	sp	2.2	31	EC_{max}	2.0	86
	Rat	Ouabain	sp	2.2	37	EC_{max}	80.0	40
LEVY and RICHARDS (1965)	Rabbit	Ouabain	2	2.5	37.5	e	0.8	82
GREEFF and SCHLIEPER (1967)	Guinea pig	Strophoside K	1	1.8	30	EC_{max}	0.7	~ 40
	Rabbit	Strophoside K	1	1.8	30	EC_{max}	1.2	~ 60
	Human	Strophoside K	1	1.8	30	EC_{max}	0.9	~ 40
	Rat	Strophoside K	1	1.8	30	EC_{max}	18.5	~ 30
PRASAD (1974)	Human	Ouabain	1	4.6	37	e	0.01	~100
ITO (1975)	Human	Ouabain	1	1.8	30	e	1.0	~ 70
CALDWELL and NASH (1976)	Rabbit	Ouabain	1	2.5	37	EC_{50}	0.35	100
BROADLEY and DUNCAN (1977)	Guinea pig	Ouabain	2	1.9	38	EC_{50}	0.52	
COOK et al. (1977)	Rabbit	Digoxin	1	2.5	37	EC_{50}	1.2	~250
MATSUMURA et al. (1977)	Guinea pig	Digitoxin	sp	2.5		EC_{max}	1.3	~ 70
SHIBATA et al. (1978)	Guinea pig	Ouabain	1.6	1.2		EC_{50}	0.34	60
AKERA et al. (1979)	Guinea pig	Ouabain	1.5	2.5	30	e	0.5	~120
	Rat	Ouabain	1.5	2.5	30	e	50.0	~ 55

EC_{50} = half maximal effective concentration
EC_{max} = maximal effective concentration
sp = spontaneously beating
e = effective as single dose

c) Anti-Arrhythmic Agents

Quinidine delays the toxic effect of strophanthoside K whereas the positive inotropic effect on spontaneously beating guinea pig atria is unaffected (FRATZ et al., 1967). The same authors found that the inotropic effect of the glycoside in electrically stimulated atria is also uninfluenced, whereas arrhythmias are delayed, but the concentration necessary for asystolia was not significantly different compared with controls.

According to the observations of SCHÜMANN et al. (1977), the calcium antagonist, verapamil has no effect on the positive inotropic action of digoxin or digitoxin in electrically stimulated guinea pig atria, it reduces toxicity, however. In spontaneously beating rabbit atria, verapamil (0.15 mg/l) was found to reduce the positive inotropic response to ouabain. SHIBATA et al. (1978) studied the influence of verapamil and nifedipine on the positive inotropic effect of ouabain and the polypeptide anthopleurine-A from sea anemone in electrically driven isolated atria from guinea pigs or rabbits. After treatment with the calcium antagonists the median effective dose (ED_{50}) of ouabain dose–response curves were significantly higher and the maximal inotropic response diminished, whereas the positive inotropic effect of anthopleurine-A was not modified. SINGH and VAUGHAN WILLIAMS (1972) also found that verapamil reduces the response to ouabain in isolated rabbit atria.

The potassium-retaining diuretics, triamterene and amilorid lengthen the functional refractory period in the isolated guinea pig atrium and decrease the toxicity of digoxin (GREEFF and KÖHLER, 1975). The anesthetic ketamin, which is known to cause an activation of the cardiovascular system in anesthesia, inhibits the positive inotropic response to ouabain in guinea pig left atria stimulated at a frequency of 1 Hz, whereas the effect of epinephrine in the same preparation was significantly enhanced (ADAMS et al., 1977).

C. Isolated Perfused Heart Preparation

In 1895 LANGENDORFF described a method to keep isolated mammalian heart alive by perfusing the coronary blood vessel. In this preparation cardiac glycosides can be characterized qualitatively and quantitatively and this method can also be used to study interactions with other drugs.

I. Method

The aorta is cannulated in the anesthetized animal or on the dissected heart, at this point it is important not to harm the valves. All the vessels are removed and the heart is placed in a warmed chamber. The perfusion takes place either at constant pressure (BROADLEY, 1970) or constant flow (STICKNEY and BALL, 1979). Pressure or flow are adjusted according to the species; edema occurs when these values are too high (REICHEL, 1976; ARNOLD et al., 1968). Blood or oxygenated and modified Krebs–Henseleit solution at 30°–37 °C are used as a perfusion solution.

BÜNGER et al. (1975) recommended the addition of 2 mmol/l pyruvate in the isolated, isovolumetrically beating guinea pig heart after careful analysis of the hemodynamics, energy balance, and time of survival. In subsequent investigations, BÜNGER et al. (1979) recommended the addition of 40 IU insulin. These authors also suggested chilling the heart in ice-cold saline immediately after opening the thorax in order to arrest the heart for extra protection.

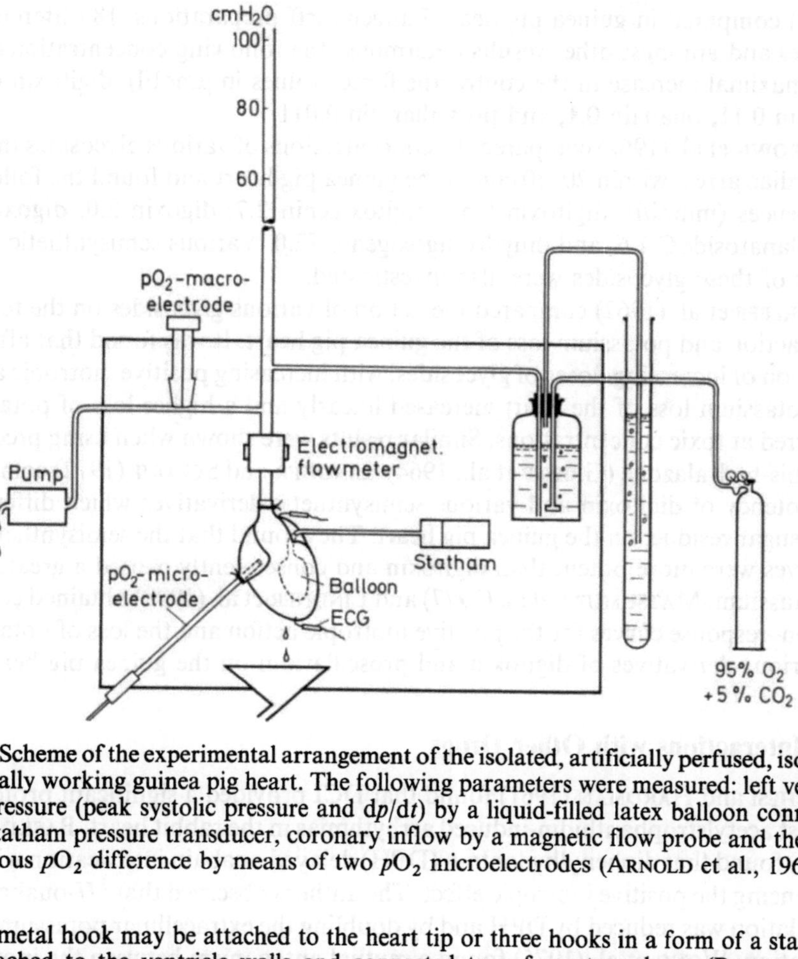

Fig. 6. Scheme of the experimental arrangement of the isolated, artificially perfused, isovolumetrically working guinea pig heart. The following parameters were measured: left ventricular pressure (peak systolic pressure and dp/dt) by a liquid-filled latex balloon connected to a Statham pressure transducer, coronary inflow by a magnetic flow probe and the arteriovenous pO_2 difference by means of two pO_2 microelectrodes (ARNOLD et al., 1968)

A metal hook may be attached to the heart tip or three hooks in a form of a star may be attached to the ventricle walls and connected to a force transducer (BAILEY, 1977, BECKETT, 1970). The coronary flow is measured either by means of a flow meter or by collecting the perfusate.

A liquid-filled latex balloon attached to a pressure transducer can be placed in the left ventricle in order to improve the analysis of the inotropic action under physiologic pressure (WEISBERG et al., 1963; GREEFF et al., 1969; BÜNGER et al., 1975). ARNOLD et al. (1968) measured in addition the arteriovenous oxygen difference and thereby the oxygen consumption (Fig. 6).

In order to estimate the stroke volume, FLYNN et al. (1978) and BÜNGER et al. (1979) perfused the left ventricle through the left atrium at constant pressure and measured the aortic flow with a flow meter against a constant peripheral resistance. Preload and afterload may be controlled in this preparation. The heart rate may be kept constant with two electrodes, which are attached to the tip of the heart and the right atrium and stimulate the heart electrically. The electrocardiogram can be recorded via these electrodes (WONG et al., 1975). Glycosides are applied either in known concentrations to the perfusion fluid, as single injections, or alternatively as an infusion.

II. Quantitative Evaluation

The measurement of concentration–response curves is only possible in a low concentration range, as arrhythmias occur relatively early. HAUSTEIN and HAUPTMANN

(1974) compared in guinea pig heart Langendorff preparations, 18 different glycosides and amongst other results determined the following concentrations for a half-maximal increase in the contractile force (values in µmol/l): digitoxin 0.015, digoxin 0.11, ouabain 0.1, and proscillaridin 0.011.

BROWN et al. (1962) compared the concentrations of various glycosides that led to cardiac arrest within 20–50 min in the guinea pig heart and found the following differences (mmol/l): digitoxin 0.52, digitoxigenin 2.7, digoxin 2.0, digoxigenin 20.0, lanatoside C 1.6, and dihydrodigoxigenin 72.0. Various semisynthetic derivatives of these glycosides were also investigated.

GREEFF et al. (1962) compared the action of various glycosides on the force of contraction and potassium loss of the guinea pig heart. It was found that after the injection of increasing doses of glycosides, with increasing positive inotropic action the potassium loss of the heart increased linearly and a higher loss of potassium occurred at toxic concentrations. Similar results were shown when using prednisolone-bis-hydralazone (GREEFF et al., 1964). LINDNER and SCHÖNE (1972) compared the potency of digitoxin and various semisynthetic derivatives which differed in their sugar residue, on the guinea pig heart. They found that the semisynthetic derivatives were more potent than digitoxin and consequently caused a greater loss of potassium. MATSUMURA et al. (1977) and LINDNER et al. (1979) obtained concentration–response curves for the positive inotropic action and the loss of potassium of various derivatives of digitoxin and proscillaridin on the guinea pig heart.

III. Interactions with Other Drugs

LUCCHESI and HARDMAN (1961) found that DCI provided a significant protection against acetylstrophanthidin-induced arrhythmias in the rabbit heart. BASKIN et al. (1973) found that diphenylhydantoin (DPH) delayed ouabain intoxication without influencing the positive inotropic effect. The authors observed that ^{3}H-ouabain accumulation was reduced by DPH and by doubling the extracellular potassium concentration. WONG et al. (1975) found a mutual antagonism between the inotropic effects of morphine and ouabain.

D. Heart–Lung Preparations

The most important characteristic of the action of digitalis is the increase in cardiac performance, but the quantitative estimation of this effect is limited in isolated organs. The heart preparation of KNOWLTON and STARLING (1912), and PATTERSEN and STARLING (1914) enabled the evaluation of the therapeutic action of digitalis and was used by many authors as early as the 1920s (ANITSCHKOW and TRENDELENBURG, 1928; KRAYER, 1931; for reviews see WEESE, 1936; KATZUNG, 1968; THORP and COBBIN, 1967).

I. Method

The principle of this method is that in animals undergoing artificial respiration the peripheral circulation is replaced by an extracorporal circulation with a controlled resistance. The preload and afterload of the heart with this method can be controlled. The pulmonary circulation stays intact, so that physiologic gas exchange can still take place via the lung. In this way the heart may be kept functioning over a period of 24 h. The heart becomes progressively insufficient over a longer experimental period (REICHEL, 1976). This can be seen as an increase in atrial pressure or a decrease of the stroke volume. In this way, as well as by administering poisons, the model can be used for inducing a defined degree of insufficiency, so that the therapeutic effect of the digitalis glycosides may become more distinct.

The pressure in the atria, in the pulmonary artery, and in the aorta, as well as the frequency and stroke volume can be measured. The contractility characterizing such parameters as left ventricular pressure and its rate of rise may also be used. There have been many investigations as to the action of digitalis as well as its interactions with other drugs on heart–lung preparations, as follows:

1) The comparison of the therapeutic effects and doses for arrhythmias and lethality of various glycosides in the evaluation of the potency and therapeutic range (VICK et al., 1957; GRUHZIT and FARAH, 1953; KLUPP, 1966; BÖTTCHER et al., 1975; BUBNOFF et al., 1955).

2) Demonstration of interactions with electrolytes, anti-arrhythmic agents, and other cardiotropic substances (STANBURY and FARAH, 1950; FAROOQ and RABAH, 1977; FAWAZ, 1955).

3) Characterization of the therapeutic and toxic actions of digitalis on the force of contraction, conduction system, oxygen consumption, coronary resistance, and time of survival after the occurrence of an insufficiency (SARNOFF et al., 1964; KAMMERMEIER et al., 1974; KURBJUWEIT, 1964; SIMAAN et al., 1968).

The following investigations are discussed as examples.

II. Therapeutic Effect and Therapeutic Range

KLUPP (1966) compared the action of a large number of cardiac glycosides on the heart–lung preparation of the guinea pig after inducing an insufficiency with barbiturates. In control experiments the dose of barbiturate (pentobarbital or butallylonal) is determined which, upon infusion into the venous reservoir causes a 50% reduction in cardiac output (insufficiency dose) or damages the heart to such an extent that the stroke volume decreases to zero (lethal dose). The glycosides are administered as soon as barbiturate-induced insufficiency occurs. The action of the glycoside is measured as the increase in the lethal dose of the barbiturate. Concentration–response curves may be obtained by increasing the glycoside concentrations, and used to determine the relative potency. Compared to digitoxigenin (relative potency $r=1$) the following values were calculated: digitoxin $r=12.7$; lanatoside C $r=3.9$; ouabain $r=9.9$; strophanthoside K $r=4.7$; and proscillaridin $r=9.5$ (Fig. 7). GRUHZIT and FARAH (1953) compared therapeutic, arrhythmogenic, and lethal doses of various glycosides on the dog heart–lung preparation; they found considerable quantitative differences in the strength of action, e.g., between ouabain, digitoxin, and digoxin; but no qualitative differences were found when the therapeutic (15%) and the arrhythmogenic dose (60%) were compared with the lethal dose (100%). In similar investigations VICK et al. (1957) found that the dihydro derivatives of ouabain, digitoxin, and digoxin possessed a larger thera-

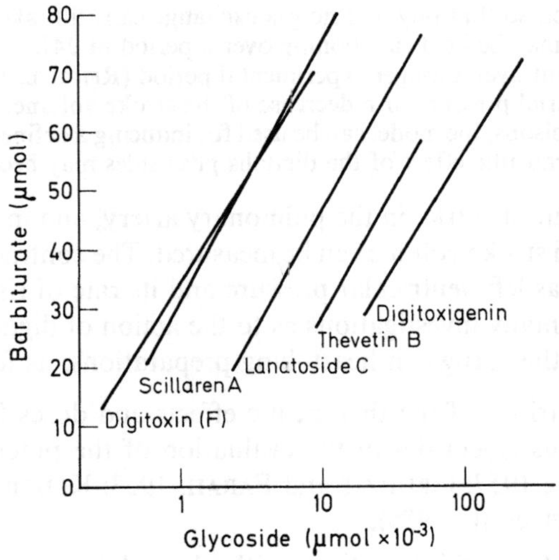

Fig. 7. Comparison of potency of cardiac glycosides in heart–lung preparation of the guinea pig. The regression lines describe the influence of digitalis glycosides on the amount of barbiturate necessary to induce zero stroke volume measured after onset of failure (Klupp, 1966)

peutic range than the nonhydrated glycosides. Böttcher et al. (1975) found that digoxin-mono-digitoxide possessed a larger therapeutic range than digoxin on the cat heart–lung preparation.

III. Interactions

Stanbury and Farah (1950) investigated the action of magnesium and the effect of digoxin on heart–lung preparations of the dog. They found that magnesium in increasing concentrations decreased the heart rate and the systemic output; electrically induced or spontaneous atrial flutter could be depressed. Toxic or lethal digoxin effects were, however, hardly affected. The addition of magnesium after the occurrence of toxic digoxin effects produced a temporary restoration of normal rhythm. Similar results were reported by Farooq and Rabah (1977). Fawaz (1955) demonstrated that procaine amide or quinidine had no effect on the toxicity of ouabain in the dog heart–lung preparation.

IV. Metabolic Characterization of Therapeutic Effects

The heart–lung preparation of the dog has often been used to investigate the influence of digitalis glycosides upon energetics and substrate turnover in the normal and failing heart. From earlier experiments it was supposed that digitalis glycosides improved cardiac efficiency (Gremels, 1933; for review see Weese, 1936). In recent investigations Sarnoff et al. (1964) found an increase in myocardial contractility by acetylstrophanthin without increase in oxygen consumption. The same results were obtained when cardiac failure was induced by reducing coronary blood flow. These experiments were performed by controlling the hemodynamic parameters which determine oxygen consumption: aortic pressure, heart rate, and stroke volume.

In isolated atria KLAUS and KREBS (1968) observed a close relationship between increase of contractile force and oxygen consumption under the influence of cardiac glycosides (for references see Chap. 13).

E. Isolated Frog Heart

The isolated frog heart was first used for the quantitative estimation of digitalis glycosides and the identification of similar substances, e.g., strophanthus glycosides by STRAUB (1910) and SCHMIEDEBERG (1910). A systolic cardiac arrest was found to be characteristic of the toxic effects of digitalis. The advantages of using this preparation are its stability at room temperature, its anatomy (coronary circulation is lacking, there is only a single ventricle), a simple dissection, and a small volume of nutritive solution.

I. Method

The heart is removed and a cannula is inserted into the aorta, which is filled with 1–2 ml Ringer solution (pressure 1–2 cm H_2O). The tip of the heart is attached to a force transducer. The preparation can either beat spontaneously or be stimulated electrically. The sensitivity of the frog heart to applied glycosides lies between 10^{-6} and 10^{-4} g/ml. This is however dependent on the type of frog, the season, and the temperature. For detailed reviews see STRAUB (1924), WEESE (1936), LENDLE (1935), KATZUNG (1968).

II. Application

Many authors have used the frog heart to standardize digitalis glycosides or extracts, characterizing their effect by the time necessary for systolic cardiac arrest and the concentration of glycoside necessary. The heart of the common frog *Rana temporaria* is approximately ten times as sensitive as that of *Rana esculenta* Recently this preparation has been used to compare the potency of various cardenolides, bufadienolides, and their semisynthetic derivatives (ISHIKAWA et al., 1974; ABRAMSON et al., 1976; SHIGEI et al., 1976; TSURU et al., 1975). MATSUMURA et al. (1977) estimated dose–response curves for contractility and frequency, and compared the potency of various glycosides. Prednisolone-bis-guanylhydrazone (10^{-4} g/l) also possesses a characteristic inotropic action with subsequent cardiac arrest (KRONEBERG and STOEPEL, 1964).

F. Embryonic Chicken Heart

PICKERING (1893) was the first to demonstrate an effect of digitalis glycosides on a chicken embryo heart by injection into the yolk sac of an egg. The application of 0.012 mg strophanthin led to cardiac arrest. PAFF (1940) tested the toxicity of digitalis glycosides on isolated chicken heart and used cardiac arrest as a parameter. He estimated the potency of unknown test solutions in the form of concentration–response curves, against a known standard. This method allows the estimation of an extremely small amount of glycosides (5 ng/0.05 ml, WRIGHT, 1960; SHEPHEARD et. al., 1954). The investigations can be carried out within 3–9 min and give reproducible results, if incubation period and temperature are kept constant.

I. Method

The eggs of purebred hens are taken on the day of laying and incubated for 48 h at a constant temperature (37°–39 °C). The heart is then removed and placed in Tyrode solution at room temperature. After washing off any protein, the heart is placed under a microscope. The maturity of the preparation is estimated and unsuitable preparations are discarded. No more than 6 min should elapse between opening the egg and applying the digitalis solution. As a parameter of the glycoside action, the time is estimated up to the occurrence of atrioventricular block or dropped beats. As many as 25 preparations are necessary for the quantitative estimation of the concentration of an unknown glycoside solution using a standard.

II. Application

LEHMAN and PAFF (1942), using statistical methods, were able to specify potency ratios between unknown and known concentrations to great accuracy with this technique. SHEPHEARD et al. (1954) applied this method in pharmacokinetic investigations, and found that digoxin was renally excreted, up to about 10%, in rats within the first 24 h and by simultaneous chromatographic analysis of the urine, they detected a digoxin metabolite. FRIEDMAN and BINE (1949) were the first to use the duck embryo heart for the detection of cardiac glycosides in human blood. Lanatoside C could be detected up to 15 min after application of 1.6 mg i.v. FRIEDMAN et al. (1954) indicated that digitoxin should be extracted from biologic solutions before estimation is carried out, in order to exclude nonspecific effects on the contractile behavior of the embryo heart.

BARRY (1950) and MCCARTY et al. (1960) refined the method of registering the contractile force. The contraction of the heart was transmitted to the movement of a small mirror, which reflected a light source in proportion to the amplitude of contraction, onto a photocell. Using this technique the authors were able to differentiate between hearts with no nervous control (3-day-old) and those, with innervation; this is because physostigmine potentiates the acetylcholine effect only after the fourth day. LELORIER et al. (1975) reported that the 4-day-old heart showed no reaction to tyramine or cocaine, but 7-day-old heart reacted with an increase in frequency and contractility. They supposed that the sympathetic nervous system was developed after the seventh day. Ouabain (7×10^{-7} mol/l) led to a similar increase in contractility after 3 and 7 days, it was therefore concluded that the effect of cardiac glycosides was independent of sympathetic innervation (Fig. 8).

KEYL and DRAGSTEDT (1954) investigated the influence of insulin on the action of cardiac glycosides. Ouabain (0.06–0.16 mg) was applied in the yolk sac of 3-day-old embryos. Cardiac arrest occurred only in the presence of 16 IU insulin, whereas ouabain in the absence of insulin was ineffective. An effect of glycosides in the absence of insulin was noticed after 10 days incubation, it was therefore supposed that endogenous insulin was necessary for the glycoside effect.

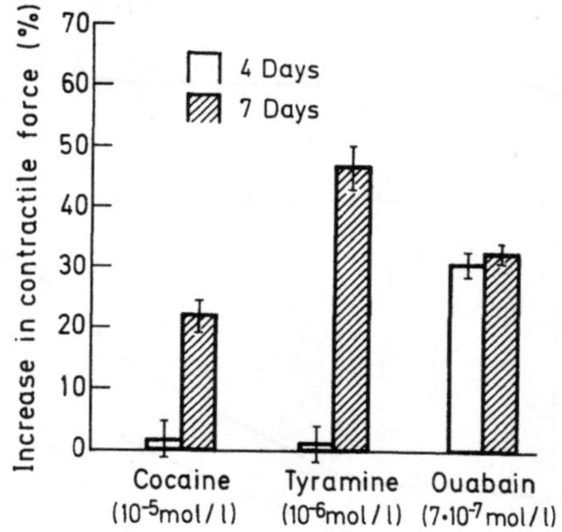

Fig. 8. Inotropic effect of cocaine, tyramine, and ouabain in chicken embryo heart before (4-day-old) and after (7-day-old) sympathetic innervation. Each column represents the mean of eight observations. The vertical bars are standard errors of the mean. The 4-day-old heart did not respond to cocaine or tyramine. The response to ouabain was identical at both stages Chicken embryo heart incubated in Tyrode solution of the following composition (in g/l): NaCl, 7.000; KCl, 0.354; $CaCl_2 \cdot 2H_2O$, 0.350; KH_2PO_4, 0.081; $MgSO_4 \cdot 7H_2O$, 0.147; $NaHCO_3$, 2.100; glucose, 0.900. Under a stereoscopic microscope, heart was carefully freed of other tissue and transferred to a bath containing 50 ml Tyrode solution

G. Isolated Cultures of Heart Cells

The use of isolated, cultured heart cells originates from work of CAVANAUGH (1955), FÄNGE et al. (1956), and HALLE (1959). The heart cells retain contractility and electrical activity after fragments of heart are digested with trypsin and placed in nutritive medium. The cells have certain advantages over the whole heart, for instance they lack extracellular exchange compartments and vegetative innervation.

I. Method

As described by FÄNGE et al. (1956), chicken embryo heart is prepared under sterile conditions with the aid of a microscope. Atria and ventricles are separated and cut into small cubes of volume 0.5 mm^3. The tissue is then washed in a Ca^{2+}-free and Mg^{2+}-free salt solution at 37 °C for 20 min and placed for 15–20 min in a salt solution containing 2% trypsin. A suspension is created by use of a pipette and is then centrifuged for 10 min at 1,000 rpm. The cells are subsequently placed in a nutritive solution, which contains 10% cell-free embryo extract, 20% denaturated horse serum, and 70% Tyrode solution (PUCK et al., 1958). The cell cultures are incubated for 24–48 h at 37 °C before the experiments are commenced. The composition of the trypsin and nutritive solution varies. An alternative is described by RINALDINI (1959).

The contractile behavior can be observed under the microscope as distinct pulsations. Furthermore, investigations have been described, which include electrophysiologic measurements, signals being obtained via a conventional glass microelectrode.

II. Application

HALLE (1967) investigated the influence of digitoxin ($10^{-7}-10^{-3}$ mmol/l) on the frequency of contraction of isolated heart cells and found a concentration-depen-

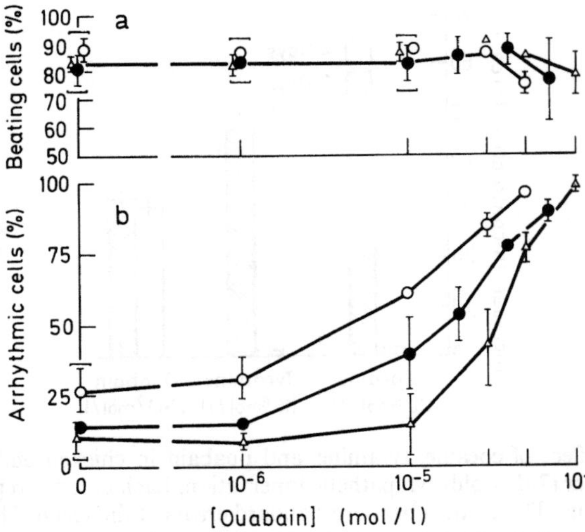

Fig. 9a, b. Influence of ouabain and calcium concentration on **a** beating cells (rhythmic plus arrhythmic) as a percentage of the total single isolated myocardial cells and **b** arrhythmic cells as a percentage of the total beating cells. Cells were incubated with stepwise increasing concentrations of ouabain and the indicated calcium concentrations and observed 20–39 min after each addition of ouabain. Points and bars show means ± standard deviations of values in three experiments. Totals of 100–150 cells were observed.

	Calcium (mmol/l)	Potassium (mmol/l)
(*Open circles*)	2.8	1.0
(*Solid circles*)	1.8	1.0
(*Triangles*)	0.6	1.0

(GOSHIMA, 1977)

dent increase in the pulsation frequency and also arrhythmias at higher concentrations. MERCER and DOWER (1966) compared cell cultures from atrial and ventricular cells. They found that atrial cells beat more strongly but with more arrhythmias. Digoxin (0.2–50 μg/ml) significantly increased the tendency to arrhythmia; the arrhythmia could not be suppressed by the application of quinidine or procaine amide.

LEHMKUHL and SPERELAKIS (1963) in electrophysiologic investigations compared the action potentials of cultured heart cells from embryos with intact heart of the adult chickens; they obtained comparable results. JOSEPHSON and SPERELAKIS (1977) investigated the influence of ouabain on the slow calcium inward current after blocking the fast sodium channels with tetrodotoxin. It was demonstrated that ouabain in concentrations from 10^{-8} to 10^{-4} mol/l did not increase the slow response, but with concentrations of 10^{-6} to 10^{-4} mol/l it was possible to reduce the isoproterenole-induced or caffeine-induced increase in the slow response. These results seem to exclude the possibility that the positive inotropic action of ouabain is due to an increase in the calcium inward current.

GOSHIMA (1976, 1977) investigated the percentage occurrence of arrhythmias in isolated single heart cells and cell clusters. The number of arrhythmias was increased by ouabain (0.25 mmol/l), a reduction of K^+ from 5.4 to 0.4 mmol/l, and an increase of Ca^{2+} from 0.6 to 20.0 mmol/l. The toxic effect of ouabain was enhanced by elevating the Ca^{2+} concentration (Fig. 9). Anti-arrhythmic substances, such as quinidine or procaine amide were able to reduce arrhythmias in both single heart cells and clusters.

A considerable improvement in the registration techniques for isolated embryo heart cells in cell cultures was described by KAUFMANN et al. (1969). A method was developed which enabled the contractile activity of each single cell to be registered objectively. A reverse phase contrast microscope with a built-in microscope photometer was used. The pulsations of the muscle cells, focused through an aperture of variable width, were photometrically converted into electrical signals. It is possible with this equipment to obtain information on the contractile processes as a function of time and in correlation to electrophysiologic measurements. The influence of Ca^{2+} (positive inotropic effect in concentrations up to 4 mmol/l, inhibition of relaxation and contracture 5–10 mmol/l) and also temperature (22 °–37 °C) can be investigated. The addition of digitalis increases the amplitude of cell pulsations (KAUFMANN et al., 1968).

References

Abramson, H.N., Huang, C.L., Wu, T.F., Tobin, T.: Thiocardinolides I: Synthesis and biological actions of 3-β-thiocyanato-14-β-hydroxy-5-β-card-20(22)-enolide. J. Pharm. Sci. *65*, 765–768 (1976)

Adams, H.R., Parker, J.L., Mathew, B.P.: The influence of ketamine on inotropic and chronotropic responsiveness of heart muscle. J. Pharmacol. Exp. Ther. *201*, 171–183 (1977)

Akera, T., Yamamoto, S., Chubb, J., McNish, R., Brody, Th.M.: Biochemical basis for the low sensitivity of the rat heart to digitalis. Naunyn Schmiedebergs Arch. Pharmacol. *308*, 81–88 (1979)

Alken, R.G., Fricke, U., Klaus, W.: Divergent influences of Ca^{2+} on the action of several cardiotonic steroids in isolated heart muscle preparations. Eur. J. Pharmacol. *26*, 331–337 (1974)

Anitschkow, S.W., Trendelenburg, P.: Die Wirkung des Strophanthin auf das suffiziente und auf das insuffiziente Warmblüterherz. Dtsch. Med. Wochenschr. *54*, 1672 (1928)

Arnold, G., Kosche, F., Miessner, E., Neitzert, A., Lochner, W.: The importance of the perfusion pressure in the coronary arteries for the contractility and the oxygen consumption of the heart. Pflügers Arch. *299*, 339–356 (1968)

Bailey, L.E.: Changes in myocardial calcium and E-C-coupling associated with failure and ouabain treatment. Arch. Int. Pharmacodyn. *226*, 118–131 (1977)

Barry, A.: The effect of epinephrine on the myocardium of the embryonic chick. Circulation *1*, 1362–1368 (1950)

Baskin, S.I., Dutta, S., Marks, B.H.: The effects of diphenylhydantoin and potassium on the biological activity of ouabain in the guinea pig heart. Br. J. Pharmacol. *47*, 85–96 (1973)

Beckett, P.R.: The isolated perfused heart preparation: two suggested improvements. J. Pharm. Pharmacol. *22*, 818–822 (1970)

Blinks, J.R., Koch-Weser, J.: Physical factors in the analysis of the actions of drugs on myocardial contractility. Pharmacol. Rev. *15*, 531–599 (1963)

Böttcher, H., Fischer, K., Proppe, D.: Untersuchungen über die Wirkung von Digoxigenin-mono-, bis- und Tridigitoxosid am Herz-Lungen-Präparat der Katze. Basic Res. Cardiol. *70*, 279–291 (1975)

Broadley, K.J.: An analysis of the coronary vascular responses to catecholamines using a modified Langendorff heart preparation. Br. J. Pharmacol. *40*, 617–629 (1970)

Broadley, K.J., Duncan, C.: The contribution of metabolism to the hypothermia-induced supersensitivity of guinea-pig isolated atria; selective supersensitivity for β-adrenoceptor agonists and their positive inotropic responses. Gen. Pharmacol. *8*, 305–310 (1977)
Brown, B.T., Stafford, A., Wright, S.E.: Chemical structure and pharmacological activity of some derivatives of digitoxigenin and digoxigenin. Br. J. Pharmacol. *18*, 311–324 (1962)
Bubnoff, M. von, Krieg, J., Amiri, H.: Über die Wirkung von K-Strophanthin am Herz-Lungen-Präparat der Ratte und des Meerschweinchens. Arch. Exp. Path. Pharmakol. *227*, 111–122 (1955)
Bünger, R., Haddy, F.J., Querengaesser, A., Gerlach, E.: An isolated guinea pig heart preparation with in vivo like features. Pflügers Arch. *353*, 317–326 (1975)
Bünger, R., Sommer, O., Walter, G., Stiegler, H., Gerlach, E.: Function and metabolic features of an isolated perfused guinea-pig heart performing pressure-volume work. Pflügers Arch. *380*, 259–266 (1979)
Caldwell, R.W., Nash, C.B.: Pharmacological studies of a new 4-aminosugar cardiac glycoside (ASI-222). J. Pharmacol. Exp. Ther. *197*, 19–26 (1976)
Cavanaugh, M.W.: Pulsation, migration, and division in dissociated duck embryo heart cells in vitro. J. Exp. Zool. *128*, 573–589 (1955)
Cook, L.S., Caldwell, R.W., Nash, C.B.: Comparison of the cardiac effects of ASI-222 HCL, an aminosugar cardiac glycoside, and digoxin. Arch. Int. Pharmacodyn. *227*, 220–232 (1977)
Dahmen, M., Greeff, K.: Analysis of the positive-inotropic activity of the benzimidazole derivate AR-L 115 BS in isolated guinea pig atria. Arzneim. Forsch. *31*, 611–617 (1981)
Denis, F., Cession-Fossion, A., Dresse, A. (presentee par J. Lecomte): Inhibition de l'action tonicardiaque de l'ouabaine par la reserpine et la guanethidine. Compt. Rend. Soc. Biol. *157*, 206–208 (1963)
Dönges, C., Heitmann, M., Jungbluth, H., Meinertz, Th., Schmelze, B., Scholz, H.: Effectiveness of theophylline to increase cyclic AMP levels and force of contraction in electrically paced guinea-pig auricles. Comparison with isoprenaline, calcium, and ouabain. Naunyn-Schmiedebergs Arch. Pharmacol. *301*, 87–97 (1977)
Dudek, M., Mantel, H.R.: Vergleichende Untersuchungen über den Einfluß der extrazellulären Calciumkonzentration auf die positiv inotrope Wirkung von Digitoxin, Digoxin, G-Strophanthin, Dihydroouabain und Isoproterenol am isolierten Vorhof des Meerschweinchenherzens. Dissertation Düsseldorf, 1979
Duncan, C., Broadley, K.J.: Possible sites of temperature-dependent changes in sensitivity of the positive inotropic and chronotropic responses to sympathomimetic amines by comparison of the temperature optima for a range of agonists. Arch. Int. Pharmacodyn. *231*, 196–211 (1978)
Durrett, L.R., Adams, H.R.: Inotropic responsiveness of atrial myocardium bathed in Tris- or bicarbonate-buffered solutions. Am. J. Physiol. *237*, H318–H325 (1979)
Fänge, R., Persson, H., Thesleff, S.: Electrophysiological and pharmacological observations on trypsin-disintegrated embryonic chick hearts cultured in vitro. Acta Physiol. Scand. *38*, 173–183 (1956)
Farooq, M.G., Rabah, M.: Effect of magnesium chloride on electrical stability of the heart. Am. Heart J. *94*, 600–602 (1977)
Fawaz, G.: Effect of quinidine, procaine amide, and mersalyl on lethal dose of ouabain in isolated dog heart. Proc. Soc. Exp. Biol. Med. *88*, 654–656 (1955)
Flynn, S.B., Gristwood, R.W., Owen, D.A.A.: Characterization of an isolated working guinea-pig heart including effects of histamine and noradrenaline. J. Pharmacol. Meth. *1*, 183–195 (1978)
Förster, W., Kalsow, H.: Über den unterschiedlichen Einfluß von Dichlorisoproterenol auf den positiv inotropen Effekt verschiedener Digitaliskörper am isolierten Vorhof-Aurikelpräparat des Meerschweinchens. Acta Biol. Med. Germ. *15*, 71–78 (1965)
Förster, W., Stolzenburg, U.: Über Struktur-Wirkungsbeziehungen bei Cardenoliden und Bufadienoliden. III. Einfluß einer Reserpin-Vorbehandlung auf die Digitoxigenin- und Digoxigenin-Herzwirkung. Acta Biol. Med. Germ. *11*, 86–92 (1963)

Fratz, R., Greeff, K., Wagner, J.: Über den Einfluß der Beta-Receptorenblocker, des Iproveratrils, Chinidins und Reserpins auf die Wirkung des K-Strophanthins am Meerschweinchenherzen. Naunyn-Schmiedebergs Arch. Pharmakol. Exp. Path. *256*, 196–206 (1967)

Friedman, M., Bine, R.: A study of the rate of disappearance of a digitalis glycoside (lanatoside C). From the blood of man. J. Clin. Invest. *28*, 32–34 (1949)

Friedman, M., Shirley, S.S., Bine, R.: Behaviour and fate of digitoxin in the experimental animal and man. Medicine *33*, 15–41 (1954)

Goshima, K.: Arrhythmic movements of myocardial cells in culture and their improvement with antiarrhythmic drugs. J. Mol. Cell. Cardiol. *8*, 213–238 (1976)

Goshima, K.: Ouabain-induced arrhythmias of single isolated myocardial cells and cell clusters cultured in vitro and their improvement by quinidine. J. Mol. Cell. Cardiol. *9*, 7–23 (1977)

Govier, W.C.: The mechanism of the atrial refractory period change produced by ouabain. J. Pharmacol. Exp. Ther. *148*, 100–105 (1965)

Greeff, K.: Contraction and relaxation of heart muscle as influenced by CAMP, isoproterenol, glucagon, ouabain, and calcium. In: Myocardial failure. Riecker, G., Weber, A., Goodwin, J. (eds.), pp 293–297. Berlin, Heidelberg, New York: Springer 1977

Greeff, K., Heeg, E.: Die anaphylaktische Reaktion perfundierter Herzpräparate, isolierter Herzvorhofpräparate und isolierter Papillarmuskeln. Arch. Int. Pharmacodyn. *149*, 136–152 (1964)

Greeff, K., Meng, K., Schwarzmann, D.: Digitalis-ähnliche Eigenschaften des Prednison- und Prednisolonbisguanylhydrazons. Naunyn Schmiedebergs Arch. Exp. Path. Pharmakol. *249*, 416–424 (1964)

Greeff, K., Köhler, E.: Tierexperimentelle Untersuchungen über den Einfluß von Triamteren und Amilorid auf Herz, Kreislauf und Toxizität des Digoxins. Arzneim. Forsch. *25*, 1766–1769 (1975)

Greeff, K., Schlieper, E.: Artspezifische Wirkungsunterschiede des K-Strophanthins und Prednisolonbisguanylhydrazons: Untersuchungen an isolierten Vorhofpräparaten und Erythrozyten des Menschen, Meerschweinchens, Kaninchens und der Ratte. Arch. Int. Pharmacodyn. *166*, 350–361 (1967)

Greeff, K., Meng, K., Moog, E.: Der Einfluß nichttoxischer und toxischer Konzentrationen herzwirksamer Glykoside auf die Kaliumbilanz isolierter Herzpräparate. Naunyn-Schmiedebergs Arch. Exp. Path. Pharmak. *244*, 270–282 (1962)

Greeff, K., Mellinghoff, P., Schlieper, E.: Vergleich der maximalen Druckanstiegsgeschwindigkeit und des systolischen Spitzendruckes am isoliert durchströmten Meerschweinchenherzen nach Einwirkung von Noradenalin, Histamin, Ca^{2+} und Digitoxigenin. Arch. Int. Pharmacodyn. *179*, 336–342 (1969)

Greeff, K., Pereira, E., Wagner, J.: Die Wirkung des Strophanthins bei Änderung der Schlagfrequenz und der extrazellulären K^+- und Ca^+- Konzentration. Arch. Int. Pharmacodyn. *190*, 219–228 (1971)

Gremels, H.: Zur Physiologie und Pharmakologie der Energetik des Säugetierherzens. Arch. Exp. Path. *169*, 689–723 (1933)

Gruhzit, C.C., Farah, A.E.: Determination of the therapeutic range of gitalin in the heart-lung preparation of the dog. J. Pharmacol. Exp. Ther. *108*, 113–116 (1953)

Hafner, D., Heinen, E., Noack, E.: Mathematical analysis of concentration-response relationships. Arzneim. Forsch. *27*, 1871–1873 (1977)

Halle, W.: Über die Züchtung von Herzmuskelzellen in einem semi-synthetischen Medium und ihr Verhalten gegenüber Digitoxin. Naturwissenschaften *46*, Z 262 (1959)

Halle, W.: Die in vitro kultivierte Zelle als Modell für morphologische und physiologische Untersuchungen. Morph. J. *111*, 3–28 (1967)

Haustein, K.O., Hauptmann, J.: Studies on cardioactive steroids. II. Structure-activity Relationships in the isolated guinea-pig heart. Pharmacology *11*, 129–138 (1974)

Illanes, A., Marshall, J.M.: The effects of ouabain on isolated atria of the ground squirrel; comparison with rat and rabbit atria. Naunyn Schmiedebergs Arch. Exp. Path. Pharmakol. *248*, 15–26 (1964)

Ishikawa, N., Tsuru, H., Shigei, T., Anjyo, T., Okado, M.: Structure-activity relationship of the cardenolides derived from digoxigenin and digitoxigenin, with special reference to the configuration at C-5. Experientia *30*, 1308–1310 (1974)
Ito, S.: A comparative study on the electrical and mechanical properties of human and guinea-pig atrial muscles in vitro. Jpn. Circ. J. *39*, 37–47 (1975)
Josephson, I., Sperelakis, N.: Ouabain blockade of inward slow current in cardiac muscle. J. Mol. Cell. Cardiol. *9*, 409–418 (1977)
Kammermeier, H., Buenger, R., Ziegler, V., Gerlach, E.: Vergleichende Untersuchungen über den Einfluß von Oxyfedrin, Adrenalin und Digitalisglykosiden auf Funktion und energiereiche Phosphatverbindungen des Meerschweinchenherzens. Arzneim. Forsch. *24*, 928–932 (1974)
Katzung, B.: Evaluation of drugs affecting the contractility and the electrical properties of the heart. In: Selected pharmacological testing methods. Burger, A. (ed.), pp. 193–234. New York: Marcel Dekker 1968
Kaufmann, R., Rodenroth, S., Tritthart, H.: Inactivation of β-receptor groups in trypsin-dispersed heart cells. Pflügers Arch. Ges. Physiol. *300*, 57 (1968)
Kaufmann, R., Tritthart, H., Rodenroth, S., Rost, B.: Das mechanische und elektrische Verhalten isolierter embryonaler Herzmuskelzellen in Zellkulturen. Pflügers Arch. *311*, 25–49 (1969)
Keyl, A.C., Dragstedt, C.A.: Influence of insulin on action of cardiac glycosides on the intact embryonic chick heart. J. Pharmacol. Exp. Therap. *112*, 129–132 (1954)
Klaus, W.: Evaluation of cardiac glycoside-like activity. In: Methods in drug evaluation. Mantegazza, P., Piccinini, F. (eds.), pp. 107–119. Amsterdam: North-Holland 1966
Klaus, W., Krebs, R.: Über den Einfluß von Digitoxigenin und Strophanthin auf mechanische Aktivität und Sauerstoffverbrauch isolierter Herzmuskelpräparate. Naunyn Schmiedebergs Arch. Pharmakol. Exp. Path. *261*, 102–117 (1968)
Klupp, H.: Quantitativer Vergleich der positiv inotropen und der toxischen Wirkung einiger Herzglykoside. Naunyn Schmiedebergs Arch. Exp. Path. Pharmakol. *252*, 314–331 (1966)
Knowlton, F.P., Starling, E.H.: The influence of variations in temperature and blood-pressure on the performance of the isolated mammalian heart. J. Physiol. (Lond.) *44*, 206–219 (1912)
Koch-Weser, J.: Influence of osmolarity of perfusate on contractility of mammalian myocardium. Am. J. Physiol. *204*, 957–962 (1963)
Koch-Weser, J.: Myocardial contraction frequency and onset of cardiac glycoside action. Circulation Res. *28*, 34–48 (1971a)
Koch-Weser, J.: β-receptor blockade and myocardial effects of cardiac glycosides. Circ. Res. *28*, 109–118 (1971b)
Koch-Weser, J., Blinks, J.R.: Analysis of the relation of the positive inotropic action of cardiac glycosides to the frequency of contraction of heart muscle. J. Pharmacol. Exp. Ther. *136*, 305–317 (1962)
Koch-Weser, J., Blinks, J.R.: The influence of the interval between beats on myocardial contractility. Pharmacol. Rev. *15*, 601–652 (1963)
Krayer, O.: Versuche am insuffizienten Herzen. Naunyn Schmiedebergs Arch. Pharmakol. *162*, 1 (1931)
Krebs, A., Henseleit, K.: Untersuchungen über die Harnstoffbildung im Tierkörper. Hoppe-Seylers Z. Physiol. Chem. *210*, 33–66 (1932)
Kroneberg, G., Stoepel, K.: Synthetische Verbindungen mit Digitaliswirkung. Naunyn Schmiedebergs Arch. Exp. Path. Pharmakol. *249*, 393–415 (1964)
Kurbjuweit, H.-G.: Zur Pharmakologie eines neu in die Therapie eingeführten Reinglykosids. Arzneim. Forsch. *14*, 716–720 (1964)
Langendorff, O.: Untersuchungen am überlebenden Säugetierherzen. Pflügers Arch. *61*, 291–332 (1895)
Laurence, D.R., Bacharach, A.L.: The compositions of some organ bath solutions. In: Evaluation of drug activities. Vol. 2, pp. 891–898. London: Academic Press 1964
Lee, K.S., Klaus, W.: The subcellular basis for the mechanism of inotropic action of cardiac glycosides. Pharmacol. Rev. *23*, 193–261 (1971)

Lehman, R.A., Paff, G.H.: A practical technique and design for the assay of digitalis on the embryonic chick heart. J. Pharmacol. Exp. Ther. *75*, 207–218 (1942)
Lehmkuhl, D., Sperelakis, N.: Transmembrane potentials of trypsin-dispersed chick-heart cells in vitro. Am. J. Physiol. *205*, 1213–1220 (1963)
Lelorier, J., Minejima, N., Shideman, F.E.: Effect of ouabain on the innervated and the noninnervated embryonic chick heart. Can. J. Physiol. Pharmacol. *53*, 1005–1006 (1975)
Lendle, L.: Digitaliskörper und verwandte herzwirksame Glykoside (Digitaloide). In: Handb. Exp. Pharmakol. I, Ergänzungswerk, pp. 11–241. Berlin: Springer 1935
Levy, J.V.: Isolated atrial preparations. In: Methods in pharmacology. Schwartz, A. (ed.), vol. 1, pp. 77–104. New York: Appleton-Century-Crofts, Educational Division, Meredith Corporation 1971
Levy, J.V., Richards, V.: The influence of reserpine pretreatment on the contractile and metabolic effects produced by ouabain on isolated rabbit left atria. J. Pharmacol Exp. Ther. *147*, 205–211 (1965)
Lindner, E., Schöne, H.H.: Änderungen der Wirkungsdauer und Wirkungsstärke von Herzglykosiden durch Abwandlung der Zucker. Arzneim. Forsch. *22*, 428–435 (1972)
Lindner, E., Reizenstein, G. von, Schoene, H.H.: Das 14,15-beta-oxido-Analoge des Proscillaridins (HOE 040). Arzneim. Forsch. *29*, 221–226 (1979)
Lock, J.A.: Observations on the use of the hen and rabbit isolated auricles for the determination of inotropic potency. Brit. J. Pharmacol. *21*, 393–401 (1963)
Lockwood, A.P.M.: "Ringer" solutions and some notes on the physiological basis of their ionic composition. Comp. Biochem. Physiol. *2*, 241–289 (1961)
Lucchesi, B.R., Hardman, H.F.: Influence of dichloroisoproterenol (DCI) and related compounds upon ouabain and acetylstrophanthidin induced cardiac arrhythmias. J. Pharmacol. Exp. Ther. *132*, 372–381 (1961)
Matsumura, S., Kimoto, S., Uno, O., Minesita, T., Ueda, M.: Some pharmacological studies on the cardiotonic effects of furanosteroidal glycosides. Arch. Int. Pharmacodyn. *225*, 246–256 (1977)
McCarty, L.P., Lee, W.C., Shideman, F.E.: Measurement of the inotropic effects of drugs on the innervated and noninnervated embryonic chicken heart. J. Pharmacol. Exp. Ther. *129*, 315–321 (1960)
Mercer, E.M., Dower, G.E.: Normal and arrhythmic beating in isolated cultered heart cells and the effects of digoxin, quinidine and procainamide. J. Pharmacol. Exp. Ther. *153*, 203–210 (1966)
Moran, N.C.: Contraction dependency of the positive inotropic action of cardiac glycosides. Circulation Res. *21*, 727–740 (1967)
Paff, G.H.: A micro-method for digitalis assay. J. Pharmacol. Exp. Ther. *69*, 311–315 (1940)
Park, M.K., Vincenzi, F.F.: Rate of onset of cardiotonic steroid-induced inotropism: Influence of temperature and beat interval. J. Pharmacol. Exp. Ther. *195*, 140–150 (1975)
Park, M.K., Vincenzi, F.F.: Influence of calcium concentration of the rate of onset of cardiac glycoside and aglycone inotropism. J. Pharmacol. Exp. Ther. *198*, 680–686 (1976)
Pattersen, S.W., Starling, E.H.: On the mechanical factors which determine the output of the ventricles. J. Physiol. (Lond.) *48*, 357–379 (1914)
Pickering, J.W.: Observations of the physiology of the embryonic heart. J. Physiol. (Lond.) *14*, 383–466 (1893)
Prasad, K.: Effect of prednisolone on the electromechanical activity of isolated human atrial muscle. Clin. Pharmacol. Ther. *16*, 1023–1030 (1974)
Puck, T.T., Cieciura, S.J., Robinson, A.: Genetics of somatic mammalian cells. III. Long term cultivation of euploid cells from human and animal subjects. J. Exp. Med. *108*, 945–956 (1958)
Refsum, H., Landmark, K.: A comparison of the effect of ouabain, noradrenaline and nifedipine on the contractile force of the isolated rat atrium at different calcium levels. Acta Pharmacol. Toxicol. *40*, 259–266 (1977)
Reichel, H.: The effect of isolation on myocardial properties. Basic Res. Cardiol. *71*, 1–16 (1976)
Rinaldini, L.M.: An improved method for the isolation and quantitative cultivation of embryonic cells. Exp. Cell. Res. *16*, 477–505 (1959)

Ringer, S.: A third contribution regarding the influence of the inorganic constituents of the blood on the ventricular contraction. J. Physiol. (Lond.) *4*, 222–225 (1883)
Sarnoff, S.J., Gilmore, J.P., Wallace, A.G., Skinner, N.S., Mitchell, J.H., Daggett, W.M.: Effect of acetyl strophanthidin therapy on cardiac dynamics, oxygen consumption and efficiency in the isolated heart with and without hypoxia. Am. J. Med. *37*, 3–13 (1964)
Schmiedeberg, O.: Untersuchung über die Bestimmung des pharmakologischen Wirkungswertes der getrockneten Blätter von Digitalis purpurea. Arch. Exp. Pharmakol. *62*, 305–328 (1910)
Schümann, H.J., Wagner, J., Springer, W.: The influence of calcium antagonists on the maximum inotropic effect and the toxicity of digoxin and digitoxin in the isolated electrically driven guinea-pig atrium. Arzneim. Forsch. (Drug Res.) *27*, 2353–2357 (1977)
Shepheard, E.E., Thorp, R.H., Wright, S.E.: The excretory products of digoxin in the rat. J. Pharmacol. Exp. Ther. *112*, 133–137 (1954)
Shibata, S., Izumi, T., Seriguchi, D.G., Norton, T.R.: Further studies on the positive inotropic effect of the polypeptide anthopleurin-A from a sea anemone. J. Pharmacol. Exp. Ther. *205*, 683–692 (1978)
Shigei, T., Tsuru, H., Ishikawa, N.: Cardiotonic activities of some new type of bufadienolide- and cardenolide-conjugates. Experientia *33*, 258–260 (1976)
Simaan, J., Slim, M., Fawaz, G.: Effect of chronic cardiac sympathectomy on the therapeutic and toxic actions of digitalis in the dog heart-lung preparation. Nauny Schmiedebergs Arch. Pharmakol. Exp. Path. *261*, 212–217 (1968)
Singh, B.N., Vaughan-Williams, E.M.: A fourth class of anti-dysrhythmic action? Effect of verapamil on ouabain toxicity, on atrial and ventricular intracellular potentials, and on other features of cardiac function. Cardiovasc. Res. *6*, 109–119 (1972)
Stanbury, J.B., Farah, A.: Effects of the magnesium ion on the heart and on its response to digoxin. J. Pharmacol. Exp. Ther. *100*, 445–453 (1950)
Stickney, J.L., Ball, T.: Effect of serotonergic antagonists on digitalis arrhythmias in the isolated heart. J. Pharmacol. Exp. Ther. *209*, 411–414 (1979)
Straub, W.: Quantitative Untersuchungen über den Chemismus der Strophanthinwirkung. Biochem. Z. *28*, 392–406 (1910)
Straub, W.: Die Digitalisgruppe. In: Handbuch der experimentellen Pharmakologie, Vol. II, pp. 1355–1452. Berlin: Springer 1924
Thorp, R.H., Cobbin, L.B.: Cardiac stimulant substances. In: Medicinal Chemistry Series. George de Steves (ed.), vol. 7, pp. 74–83. New York: Academic Press 1967
Tsuru, H., Ishikawa, N., Shigei, T., Anjyo, T., Okado, M.: Cardiotonic activities of 3,5-seco-4-nor-cardenolides in *Rana nigromaculata*. Experientia *31*, 955–956 (1975)
Tuttle, R.S., Farah, A.: The effect of ouabain on the frequency-force relation and on post-stimulation potentiation in isolated atrial and ventricular muscle. J. Pharmacol. Exp. Ther. *135*, 142–150 (1962)
Tyrode, M.V.: The mode of action of some purgative salts. Arch. Int. Pharmacodyn. *20*, 205–223 (1910)
Vick, R.L., Kahn, J.B., Acheson, G.H.: Effects of dihydro-ouabain, dihydrodigoxin and dihydrodigitoxin on the heart-lung preparation of the dog. J. Pharmacol. Exp. Ther. *121*, 330–339 (1957)
Wagner, J., Salzer, W.-W.: Calcium dependent toxic effects of digoxin in isolated myocardial preparations. Arch. Int. Pharmacodyn. *223*, 4–14 (1976)
Weese, H.: Digitalis. Stuttgart: Thieme 1936
Weisberg, H., Katz, L.N., Boyd, E.: Influence of coronary flow upon oxygen consumption and cardiac performance. Circ. Res. *13*, 522–528 (1963)
White, W.F., Salter, W.T.: The response of hypodynamic myocardium to know concentrations of cardiac glycosides. J. Pharmacol. Exp. Ther. *88*, 1–9 (1946)
Wong, K.C., Sullivan, S., Wetstone, D.L.: Antagonistic effect of morphine on the positive inotropic response of ouabain on the isolated rabbit heart. Anesth. Analg. Curr. Res. *54*, 787–791 (1975)
Wright, S.E.: The metabolism of cardiac glycosides. Springfield, Ill.: Charles Thomas 1960

Mode of Action of Cardiac Glycosides

CHAPTER 11

The Positive Inotropic Action of Cardiac Glycosides on Cardiac Ventricular Muscle

M. REITER

A. Effects on the Time Course of the Isometric Contraction

I. The Positive Klinotropic* Effect

The positive inotropic effect of cardioactive steroids is the consequence of an acceleration of the force development during contraction, i.e., of a positive klinotropic effect. This was demonstrated probably for the first time by W. STRAUB (1908) when he recorded the intraventricular pressure development during systole by a suitable method. He observed that the velocity of the rise of the intraventricular pressure was increased under the influence of digitalis, and that, accordingly, the isometric contraction phase was shortened (Fig. 1 a). He concluded that the elementary action on the heart of digitalis is to make the ventricle contract faster („Als

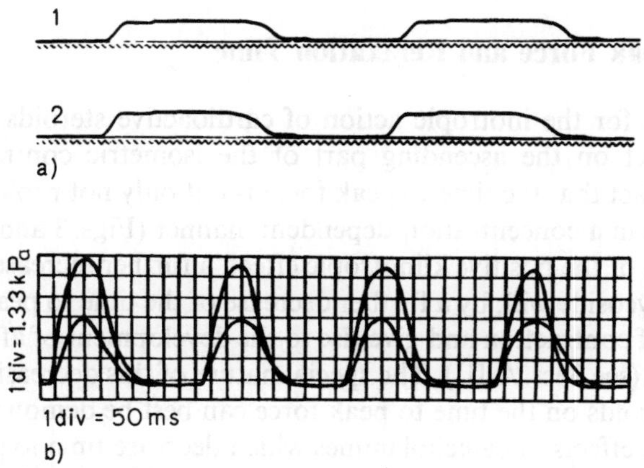

Fig. 1. a Pressure changes in the left ventricle of a cat heart in situ before *1* and after *2* the application of strophanthin. Cannula inserted from outside through the apex of the ventricle; natural resonant frequency of the manometer 31 Hz. (STRAUB, 1908). **b** Isometric pressure curves of an isolated cat heart (Langendorff preparation) contracting against an intraventricular balloon, before (lower trace) and after (upper trace) addition of strophanthin to the perfusion medium. (MAGNUS and SOWTON, 1910)

* The term *inotropic* ("inotrop") was coined by ENGELMANN (1900) to denote influences on the strength of contraction of cardiac muscle. The inotropic effect may be caused either by a change of the duration of the active state or by alterations in contraction velocity leading to changes of the steepness of the contraction curve. Following ENGELMANN (1900), BOHNENKAMP (1922) introduced *klinotropic* ("klinotrop") to denote influences on the steepness of the contraction curve (see also EISMAYER, 1930; REITER, 1972b)

Elementarwirkung der Digitaliskörper auf das Herz dürfte also die anzusehen sein, daß sich der Ventrikel *rascher* zusammenzieht“). These observations were extended two years later by MAGNUS and SOWTON (1910) who recorded complete isometric contraction curves of the left ventricle of the isolated cat heart. Their tracing (Fig. 1 b) shows that, because of the greater speed of the force development under the effect of the glycoside, the height of the contraction curve, representing the peak force of the isometric contraction, was increased.

By STRAUB's observation, the cardioactive steroids are to be classified as inotropic drugs which increase ventricular force by an acceleration of contraction velocity (i.e., by a positive klinotropic effect), in contrast to those interventions which make the heart muscle develop an inotropic response by a prolongation of the duration of the active state during contraction. Examples of the latter possibility of influencing the mechanical properties of the muscle are an increase in the length of the muscle fiber by stretching (Fig. 2 a) or by lowering the temperature (Fig. 2 b). In both cases, the increase in peak force of contraction is the consequence mainly of an increase in the duration of the time to peak force, the change in contraction velocity being of only minor importance. High osmolarity also leads to an increase in force of contraction by a prolongation of contraction time (KOCH-WESER, 1963). Substances which prolong the active state of the heart muscle are fluoride (REITER, 1965), strontium (THOMAS, 1957; REITER, 1964; HARTMANN, 1966; WEYNE, 1966; DE HEMPTINNE et al., 1967; BLINKS et al., 1972), and, of the more commonly known drugs, caffeine and other methylxanthines (BLINKS et al., 1972; KORTH, 1978).

II. Time to Peak Force and Relaxation Time

The importance for the inotropic action of cardioactive steroids of the positive klinotropic effect on the ascending part of the isometric contraction curve is stressed by the fact that the time to peak force is not only not prolonged but is actually shortened in a concentration-dependent manner (Figs. 3 and 4). Eventually, at the maximum of the positive klinotropic effect, a further increase in the concentration of the glycoside will, by a further decrease of the time to peak force, reduce the peak force of contraction and give rise to the development of aftercontractions and contracture (see Sect. A.III). The special nature of this decreasing effect of the cardioactive steroids on the time to peak force can best be demonstrated by comparison with the effects of catecholamines which decrease time to peak force only slightly and in a concentration-independent manner (Figs. 3 and 4). Accordingly, the upper limit of the positive inotropic concentration–effect curve of the catecholamine is not determined by an interference of the time to peak force with the ascending limb of the contraction curve (Fig. 3). Instead, a marked enhancing action of the catecholamine on the relaxation phase of the contraction curve progressively inhibits at higher concentrations the development of peak force, thereby, in comparison with the cardioactive steroid, actually reducing the inotropic efficacy (i.e., the maximum height of the concentration–effect curve). Because of the considerable acceleration of the relaxation process, the relaxation time (Fig. 4) and, therefore, also total contraction time are shortened under the action of catecholamines, as has been repeatedly reported (KROP, 1944; ENGSTFELD et al., 1961; BRADY, 1964;

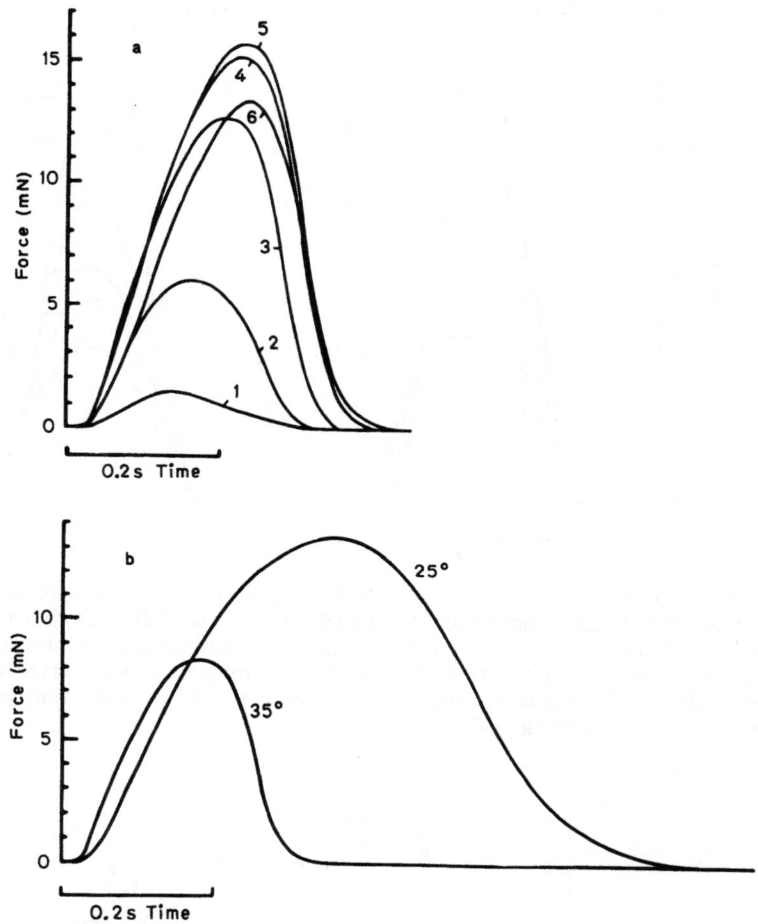

Fig. 2. a Effect of increase in muscle length on the isometric contraction curve of isolated guinea-pig papillary muscle. Superimposed contraction curves of one muscle. Numbers indicate increasing length produced by stretching, leading to a corresponding increase in resting force of: 0.5, 1, 2, 4, 8, 15 mN Temperature 35 °C; contraction frequency 1 Hz. (REITER, 1968). **b** Influence of temperature. Isometric contraction curves of one guinea-pig papillary muscle at 35 °C, and after cooling to 25 °C. Resting force 4 mN; contraction frequency 1 Hz. (REITER and STICKEL, 1970)

KOCH-WESER et al., 1964; REITER and SCHÖBER, 1965a; SONNENBLICK, 1967; GREEFF, 1976).

Cardioactive steroids do not shorten the relaxation time. At high concentrations, they might even prolong it slightly (Fig. 4) which indicates a small relative retardation in relaxation velocity. This could intensify the positive klinotropic and inotropic effects by an interference with the contraction height in just the opposite direction, as that seen with the catecholamines. However, such influences on relaxation time by cardioactive steroids are usually not so great as to lead to an increase in total contraction time, expressed as the sum of time to peak force and relaxation time ($t_1 + t_2$). In this respect, the action of cardioactive steroids resembles that of an increase either in calcium concentration (REITER, 1970a; GREEFF, 1976) or in

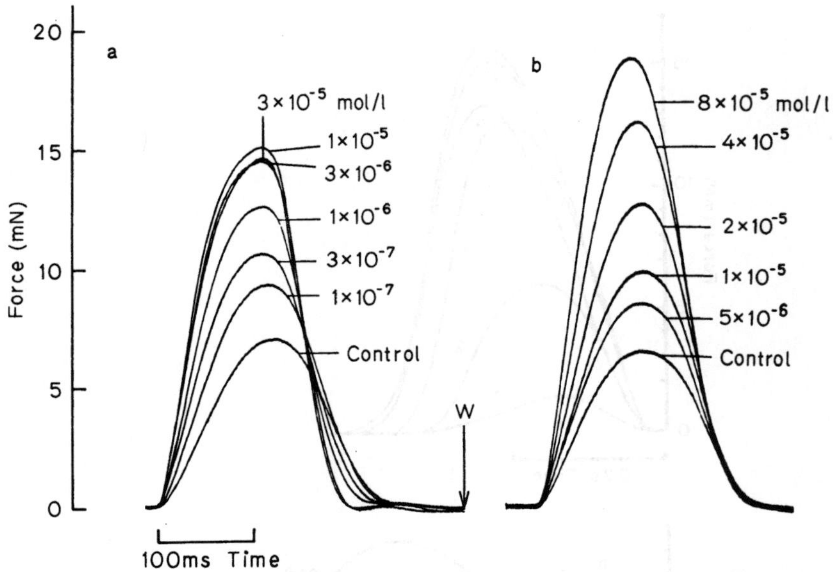

Fig. 3a, b. Isometric contraction curve of guinea-pig papillary muscle as influenced by a cardioactive steroid (dihydroouabain) compared with noradrenaline. Both sets of superimposed contraction curves obtained from the same muscle (diameter 0.77 mm, length at 4 mN resting force 3.0 mm); temperature 35 °C; contraction frequency 1 Hz. Curves represent steady-state effects of either noradrenaline **a** or, after washout (W) and a time interval of 2 h, dihydroouabain **b**. (REITER, 1972a)

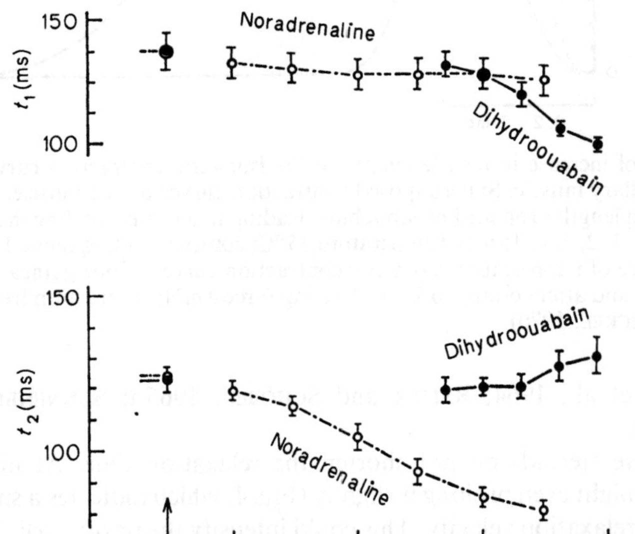

Fig. 4. The influence of dihydroouabain on the time peak force (t_1) and relaxation time (t_2) of the isometric contraction curve of a guinea-pig papillary muscle, compared with the effect of noradrenaline. Mean values (± standard error of the mean) of 12 muscles. Experimental conditions as in Fig. 3. (REITER, 1972a)

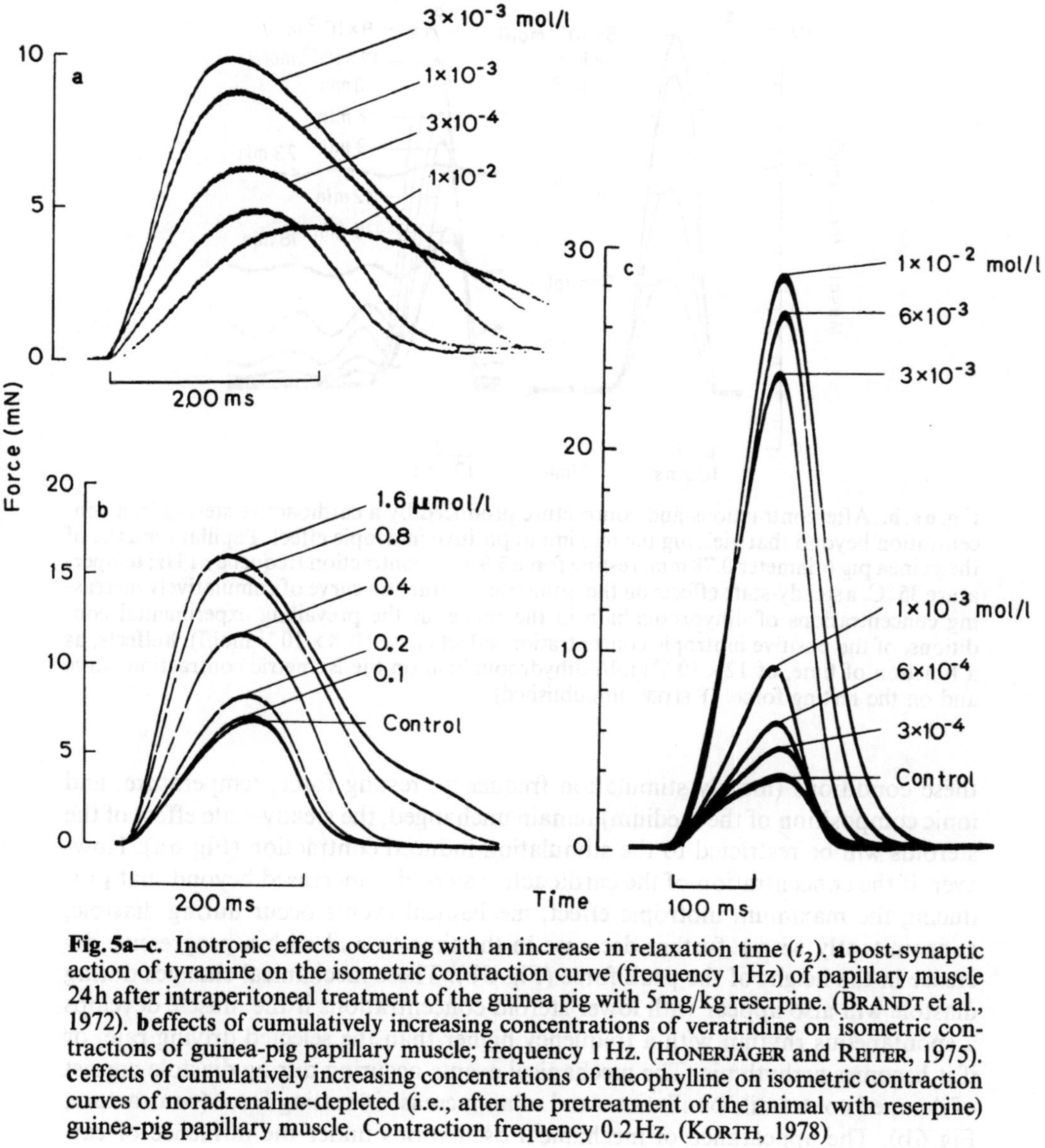

Fig. 5a–c. Inotropic effects occuring with an increase in relaxation time (t_2). **a** post-synaptic action of tyramine on the isometric contraction curve (frequency 1 Hz) of papillary muscle 24 h after intraperitoneal treatment of the guinea pig with 5 mg/kg reserpine. (BRANDT et al., 1972). **b** effects of cumulatively increasing concentrations of veratridine on isometric contractions of guinea-pig papillary muscle; frequency 1 Hz. (HONERJÄGER and REITER, 1975). **c** effects of cumulatively increasing concentrations of theophylline on isometric contraction curves of noradrenaline-depleted (i.e., after the pretreatment of the animal with reserpine) guinea-pig papillary muscle. Contraction frequency 0.2 Hz. (KORTH, 1978)

contraction frequency (REITER, 1966) but is clearly different from the effects on relaxation and total contraction time of the post-synaptically acting tyramine, of veratridine, or of theophylline (Figs. 5a–c).

III. Aftercontractions and Contracture

The diastolic or resting force of the cardiac ventricular muscle is not influenced by those concentrations of cardioactive steroids which are, at constant experimental conditions, in the range of the positive inotropic concentration–effect curve. If

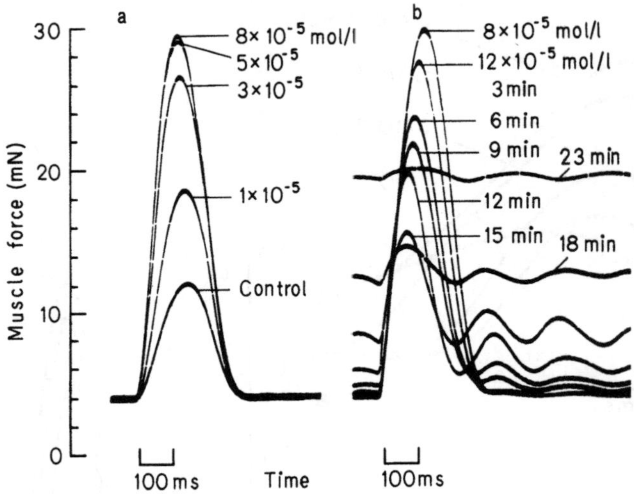

Fig. 6a, b. Aftercontractions and contracture produced by a cardioactive steroid in a concentration beyond that yielding the maximum positive inotropic effect. Papillary muscle of the guinea pig (diameter 0.78 mm; resting force 3.9 mN; contraction frequency 1 Hz; temperature 35 °C. **a** steady-state effects on the isometric contraction curve of cumulatively increasing concentrations of dihydroouabain in the range, at the prevailing experimental conditions, of the positive inotropic concentration–effect curve ($1–8 \times 10^{-5}$ mol/l). **b** effects, as a function of time, of 12×10^{-5} mol/l dihydroouabain on the isometric contraction curve and on the resting force. (REITER, unpublished)

these conditions (mainly stimulation frequency, resting force, temperature, and ionic composition of the medium) remain unchanged, the steady-state effect of the steroids will be restricted to the stimulation-induced contraction (Fig. 6a). However, if the concentration of the cardioactive steroid is increased beyond that producing the maximum inotropic effect, mechanical events occur during diastole, concomitantly with a further decrease in the time to and with a progressive decrease in the height of the peak force (Fig. 6b). These mechanical changes during diastole will also appear with lower steroid concentrations if the muscle develops a spontaneous rhythm with a frequency higher than the selected driving rate, or if it becomes arrhythmic. The mechanical events occuring during diastole consist of damped contractile oscillations and of an increase in resting force (contracture, Fig. 6b). The appearance of mechanical oscillations under the influence of cardioactive steroids was first described by REITER (1961, 1962, 1963a, b). He named them aftercontractions *(Nachkontraktionen)* because they follow a normally triggered contraction and are not associated with action potentials of their own (Fig. 7). Therefore, they can be distinguished clearly from extrasystoles due to automatic depolarizations, as well as from other diastolic contractile phenomena like a delayed or incomplete relaxation from a preceding contraction or a partial contracture (HOFFMAN et al., 1968). These observations were confirmed by various authors using cardiac tissue from different species (KAUFMANN et al., 1963; KATZUNG et al., 1969; FERRIER, 1976, 1977a; WEINGART et al., 1977; KASS et al., 1978a).

Although the development of aftercontractions is a regular event under the influence of high concentrations of cardioactive steroids, the latter are not the only

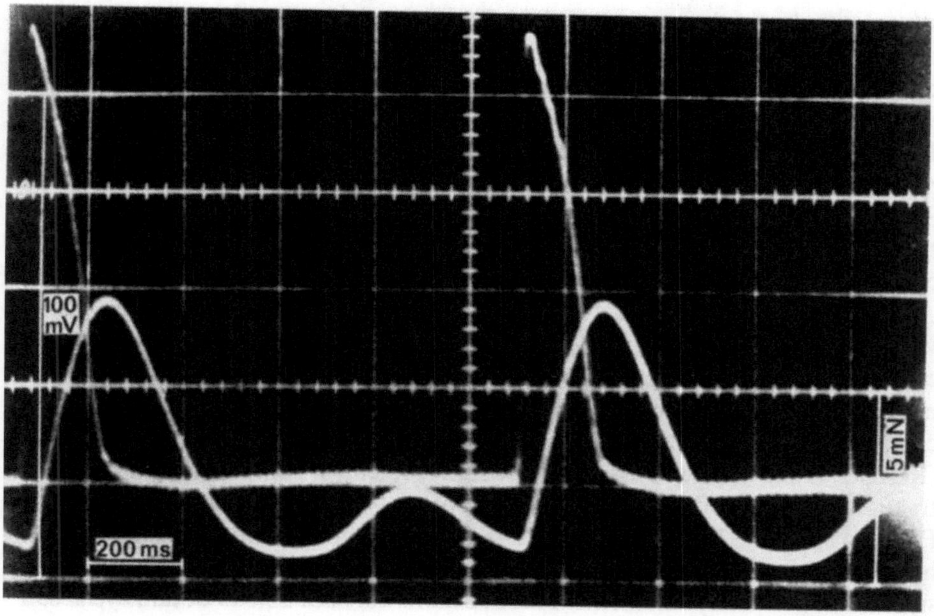

Fig. 7. Aftercontraction without action potential induced by a cardioactive steroid. Simultaneous recordings of transmembrane action potential and force of contraction of guinea-pig papillary muscle. Temperature 25 °C; stimulation frequency 1 Hz:; [Ca^{2+}] 7.2 mmol/l; [dihydroouabain] 2×10^{-5} mol/l. (REITER, 1962)

prerequisite for their appearance. Aftercontractions also occur under the influence of other drugs or inotropic interventions, provided the degree of the inotropic effect is great enough and the drug does not induce a distinct retardation of the relaxation phase of the contraction. Rhythmic variations of tension following regular contractions of a strip of turtle ventricle after treatment with large excess of calcium ions was observed by BOZLER (1943 b), and others reported on similar effects of high calcium concentrations in mammalian cardiac muscle (REITER, 1962; KAUFMANN et al., 1963; KATZUNG, 1964; BRAVENÝ et al., 1966; POSNER and BERMAN, 1969; RYO, 1971; GLITSCH and POTT, 1975). Other changes of the ionic composition of the incubation medium leading to positive inotropic effects likewise produce aftercontractions, namely a reduction of sodium concentration (REITER, 1963 c; KAUFMANN et al., 1963; MASCHER, 1971; SEIBEL and REITER, 1973; GLITSCH and POTT, 1975), the omission of potassium (REITER et al., 1971; EISNER and LEDERER, 1979 a, b), or the addition of lanthanum (RAVENS, 1975). Aftercontractions develop after periods of high frequency stimulation (REITER, 1966; KEDEM et al., 1972), after paired stimulation (FEIGL, 1967 a, b, 1968; KEDEM et al., 1972) or after tetanic contraction (in rat papillary muscle, HENDERSON et al., 1971). Paired-pulse stimulation induced larger aftercontractions in papillary muscles from spontaneously hypertensive than from normotensive rats (HELLER, 1977, 1978, 1979). Period and oscillation height are increased at low temperatures (REITER, 1962; KAUFMANN et al., 1963; BRAVENÝ et al., 1966; WUSSLING and SZYMANSKI, 1975, 1977). In fact the traces presented by LANGENDORFF as long ago as 1895 of the contractions of a cat heart perfused with cold blood (10 °C) demonstrate very clearly the appearance of aftercontractions. At constant temperatures the height and wavelength of the aftercontractions depend on the frequency and, to a certain degree, on the number of preceding contractions (REITER, 1962; KAUFMANN et al., 1963; KATZUNG, 1964; BRAVENÝ et al., 1966; FEIGL, 1967b; FERRIER, 1976, 1977a). Of the nonsteroidal positively inotropic acting drugs, the following have been found to produce aftercontractions: catecholamines (REITER and SCHÖBER, 1965b; RYO, 1971; NATHAN and BEELER, 1975), caffeine in Na-free solution (VERDONCK et al., 1972), dibutyryl cyclic AMP (KAUFMANN, 1977), and the cation ionophore monensin (SUTKO et al., 1977).

The fact that aftercontractions appear concomitantly with, or at least on the verge of, contracture (Fig. 6b), points to a disproportion between the amount of calcium ions to be handled by the cell during the contraction cycle and the capacity of its Ca^{2+} sequestration system as a necessary basis for their development. They can be envisaged as the result of an oscillatory Ca^{2+} release from overloaded stores (REITER, 1962, 1964). Cyclic contractions somehow comparable to aftercontractions have been observed in skinned fibers of cardiac muscle by FABIATO and FABIATO (1972, 1975) which indicates that the function of the sarcolemmal membrane is not necessary for a cyclic release of Ca from cellular stores yielding a transient activation of myofilaments. These authors were not able to obtain cyclic contractions of skinned fibers of frog cardiac muscle which may be related to the paucity of sarcotubules of frog ventricular muscle (STALEY and BENSON, 1968). It points to the similarity between the mechanisms underlying both mechanical phenomena, the cyclic contractions of skinned fibers and the aftercontractions of intact tissues, that KATZUNG et al. (1969) and LENZEN (1973) were not able to obtain aftercontractions in frog myocardium under conditions which were optimal for oscillations in mammalian and turtle myocardium.

Some uncertainty with respect to the primary cellular event responsible for the appearance of aftercontractions came from the observation of oscillatory alterations in membrane potential which sometimes accompany the mechanical oscillations. These alterations in membrane potential (named "transient depolarizations" by FERRIER et al., 1973)[1] usually did not exceed a few mV (KAUFMANN et al., 1963; JENSEN and KATZUNG, 1968; REITER et al., 1971; RYO, 1971; NATHAN and BEELER, 1975; FERRIER, 1976), thereby not reaching the threshold potential for activation in normal myocardial tissue (BEELER and REUTER, 1970). In evaluating the appearance of small potential changes during diastole, it is important to exclude a distortion of the potential records due to the mechanical movement of the tissue as pointed out by MASCHER (1971). Most authors maintain that aftercontractions are not necessarily accompanied by transient depolarizations, i.e., that oscillatory ionic membrane currents responsible for the appearance of the latter cannot be the primary cause of the mechanical oscillations (REITER, 1962; MASCHER, 1971; RYO, 1971; VERDONCK et al., 1972; FERRIER, 1976; WUSSLING and SZYMANSKI, 1977; KASS et al., 1978a). A survey of the pertinent papers reveals that

1 In a later paper, FERRIER (1977b) proposed the use of the term "oscillatory afterpotential" which was already applied by BOZLER (1943a) to diastolic oscillations in membrane potential induced in cardiac tissues by exposure to high concentrations of calcium and low temperature. CRANEFIELD and ARONSON (1974) have applied this name in a general sense to include digitalis-induced oscillations.

atrial muscle and Purkinje fibers are more susceptible to the development of transient depolarizations than ventricular muscle (KAUFMANN et al., 1963; JENSEN and KATZUNG, 1968; FERRIER, 1976; LEDERER and TSIEN, 1976). The resting force (tension) of the tissue is obviously a major determinant for the formation of transient depolarizations in the presence of cardioactive steroids; an increase in resting force was found to promote their appearance in canine ventricular muscle and to increase their amplitude in Purkinje tissue (FERRIER, 1976). There is one experimental condition which leads, in guinea-pig papillary muscle, almost unavoidably to the appearance of transient depolarizations together with aftercontractions – the incubation of the muscle in potassium-free solution for about 25 min (REITER et al., 1971; EISNER and LEDERER, 1979a, b).

A more direct approach to solving the problem of a possible causal relation between transient depolarizations and aftercontractions is the measurement of the respective transient inward currents (LEDERER and TSIEN, 1976) by voltage clamp experiments. The special aim of these studies was to find out whether these currents are carried by Ca ions as suggested by FERRIER and MOE (1973). These experiments, carried out on calf Purkinje fibers (WEINGART et al., 1977; KASS et al., 1978a, b), disclosed that the inversion potential of the transient inward current averaged −5 mV and, therefore, was not compatible with a Ca-specific pathway. The results are rather consistent with an increase in membrane permeability to Na or perhaps K. The authors conclude that an oscillatory release of Ca from an intracellular store is the primary event underlying both the aftercontraction and the conductance change which generates the transient inward current.

Overloading of intracellular Ca stores will then, by means of an oscillatory Ca release, also be responsible for the development of arrhythmia under the effect of cardioactive steroids in high concentrations. A progressively increasing rate of slow diastolic depolarization in Purkinje fibers as a toxic effect of ouabain has already been known for some time (DUDEL and TRAUTWEIN, 1958; VASSALLE et al., 1962). A more thorough investigation of this phenomenon showed that it, in fact, represents oscillatory transient depolarizations which eventually, as already suggested by BOLZER (1943a), can be induced to reach threshold after a driven response (FERRIER et al., 1973, FERRIER and MOE, 1973; FERRIER, 1977b), especially if the preparation is stretched (FERRIER, 1976).

B. Dependence on Extracellular Ion Concentrations

I. Sodium Dependence

If an increase in intracellular sodium concentration due to an inhibition of the sarcolemmal sodium pump is causally related to the positive inotropic effect of cardioactive steroids (SCHATZMANN, 1953; WILBRANDT, 1958; REPKE, 1964), it is to be expected that experimental conditions opposing an increase in $[Na^+]_i$ will reduce the magnitude of the inotropic effect, as other conditions facilitating cellular sodium uptake should potentiate it. A simple procedure to reduce the intracellular sodium activity is to lower the sodium concentration of the extracellular medium (ELLIS, 1977).

In evaluating experiments with low sodium concentrations one has to consider that the reduction of $[Na^+]_o$ by itself increases force of contraction. This might give rise to an erroneous assessment in two ways: first, the positive inotropic effect due to the reduction of $[Na^+]_o$ might be sufficiently great to increase the peak force of contraction to a value near its ceiling, thereby obscuring any possible additional inotropic effect; and second, the measurement of the inotropic glycoside effect in relative terms (i.e., the percentage of increase with respect to the force of contraction before the addition of the drug) would be bound to yield unreliable data. For example, if the absolute increase in force of contraction (i.e., the positive inotropic effect) remains the same as before the reduction of $[Na^+]_o$, its relative value in regard to an increased basal force of contraction must decrease, and, by analogy, it has to increase if the basal force changes in the opposite direction. This is why some papers dealing with the role of extracellular sodium on the inotropic effect of cardioactive steroids are difficult to interpret since the respective authors report on their observations only in relative terms (FARAH and WITT, 1963; CAPRIO and FARAH, 1967; AKERA et al., 1977), neglecting the changes in basal force of contraction. In the experiments with guinea-pig papillary muscles depicted in Fig. 8 (REITER, 1963 b), pains were taken to circumvent these difficulties. This was accomplished by measuring the inotropic effect of the glycoside with either 140 or 70 mmol/l sodium, at different extracellular calcium concentrations thereby covering the whole inotropic calcium concentration–effect curve at the prevailing sodium concentration. It is evident that the inotropic action of dihydroouabain consists of a parallel shift of the calcium concentration–effect curve. The same shift was described for the action of ouabain on the frog heart by SALTER et al. (1949). This implies that the absolute amount of increase in force of contraction by any given drug concentration remains the same over the greater (the steep) part of the calcium concentration–effect curve regardless of the respective control value (Fig. 8 c), a fact which would be obscured by evaluating the relative changes. And, further, it must be deduced that the inotropic action of the cardioactive steroids (i.e., the absolute amount of increase in force of contraction) is almost independent of the outside calcium concentration, with the only exceptions of the two ranges where the different sigmoid concentration–effect curves converge, i.e., at very low and at very high calcium concentrations. This means that there is no inotropic action of the steroids without calcium and also that the inotropic response is reduced if, at a high calcium concentration, the muscle is already near the maximum of its contraction capability (see Fig. 8a, b, and, in Fig. 8c, the reduced inotropic effect, expressed as the change in force of contraction ΔF_c under the influence of 4×10^{-5} mol/l DHO at 4.8 mmol/l Ca as compared with the value at 2.4 mmol/l Ca). While the experiments shown in Fig. 8 demonstrate that the positive inotropic effect of dihydroouabain is essentially independent of the extracellular calcium concentration (as also found to be the case in regard to the development of contracture in guinea-pig atria by GREEFF et al., 1971) they also prove that it depends to a considerable degree on the sodium concentration of the incubation medium. The shift of the calcium concentration–effect curve to the left by the cardioactive steroid (i.e., the positive inotropic effect) is significantly reduced if the outside sodium concentration is only 70 mmol/l instead of 140 mmol/l. In contrast, the positive inotropic effects of substances which do not inhibit the sarcolemmal sodium pump, like

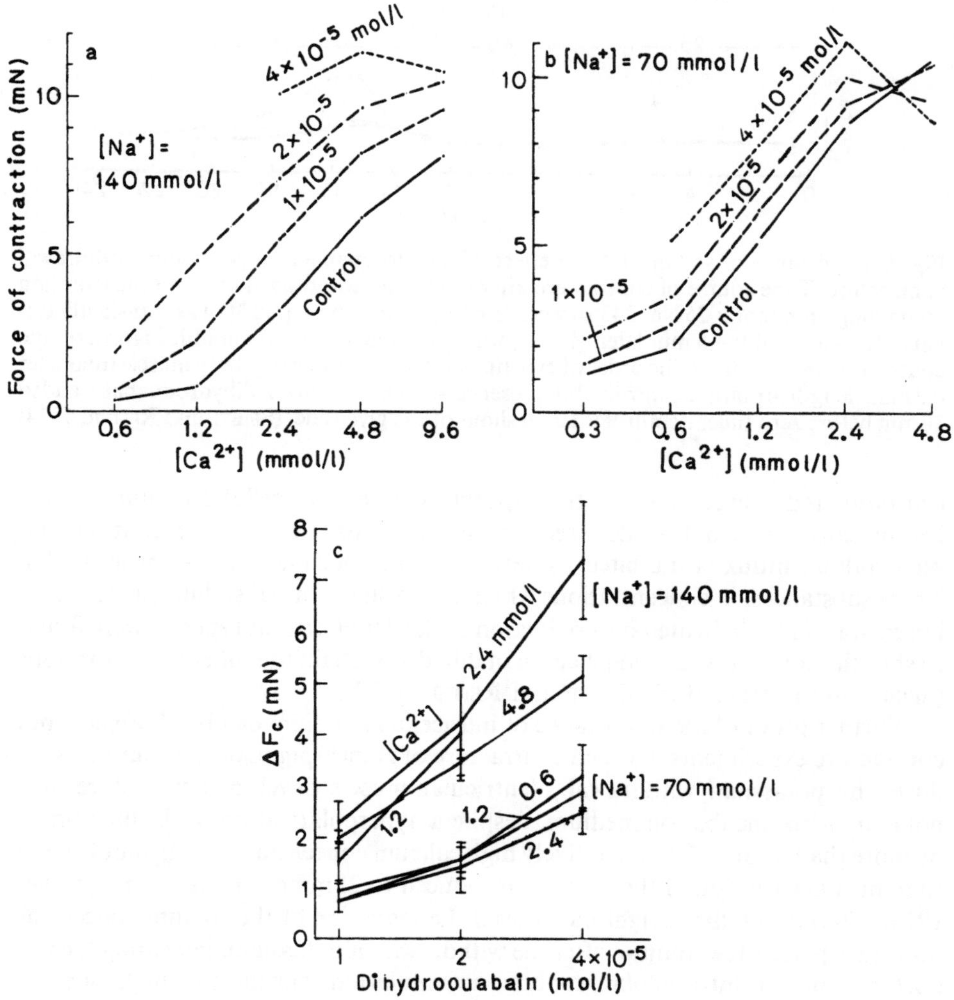

Fig. 8a–c. Positive inotropic effect of dihydroouabain as a function of sodium concentration at different calcium concentrations. Guinea-pig papillary muscle contracting at a frequency of 1 Hz at a temperature of 35 °C.
a calcium concentration–effect curves at 140 mmol/l sodium.
b calcium concentration–effect curves at 70 mmol/l sodium (osmotic adjustment by sucrose.) Mean values from six muscles.
c Inotropic effect of dihydroouabain as a function of sodium and calcium concentrations. ΔF_c = absolute value of increase in force of contraction. Prevailing sodium and calcium concentrations are indicated on the graphs. (REITER, 1963 b)

adrenaline, histamine, or theophylline, were found to be independent of the extracellular sodium concentration (REITER and SCHÖBER, 1965 a; BERNAUER et al., 1971; SCHOLZ and DE YAZIKOF, 1971).

Although experiments like those shown in Fig. 8 clearly demonstrate the sodium dependence of the inotropic effect of cardioactive steroids, they do not prove beyond doubt that the intracellular, and not the extracellular, sodium con-

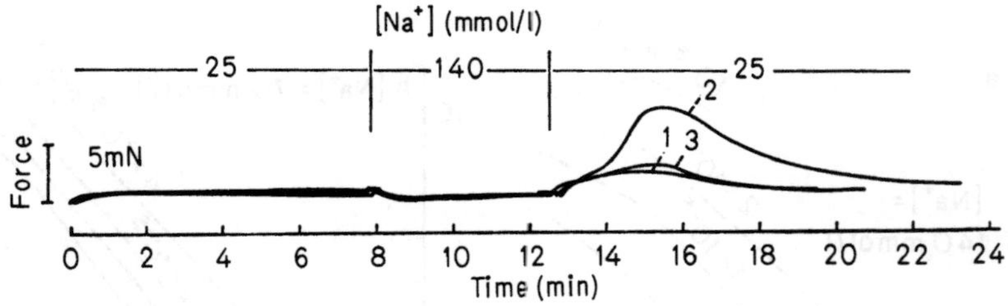

Fig. 9. Importance of sodium for the effect of dihydroouabain on potassium-withdrawal contracture. Time course of contracture after potassium withdrawal at zero time (solution containing, in mmol/l: 25 Na; 243 sucrose; 1.6 Ca). After 8 min, $[Na^+]_o$ was transiently elevated to 140 mmol/l; 4.5 min later, $[Na^+]_o$ was again reduced to 25 mmol/l. Figures on the contracture traces refer to the order of experiments with the same papillary muscle (diameter 0.3 mm, length 4 mm): *1* control; *2* in presence of 3×10^{-5} mol/l dihydroouabain added 10 min before zero time; *3* control after washout of the glycoside. (SEIBEL and REITER, 1974)

centration is decisive. However, the importance of the intracellular sodium concentration can convincingly be deduced from experiments in which the excitation-coupled sodium influx is inhibited by tetrodotoxin, lidocaine, or (+)-propranolol. These substances, in concentrations which diminish the initial sodium current during excitation (as indicated by a reduction of the depolarization speed), were found to shift the concentration–effect curve of dihydroouabain to higher concentrations (SEIBEL and REITER, 1975; EBNER and REITER, 1979).

Further proof of the importance of intracellular sodium can be obtained from contracture experiments in which contracture of guinea-pig papillary muscles is induced by potassium withdrawal. Ventricular muscle develops contracture in a potassium-free incubation medium, despite a hyperpolarization of the membrane by more than 30 mV, if, at a relatively high calcium concentration (6.5 mmol/l), the sodium concentration of the medium is reduced to 25 mmol/l (SEIBEL and REITER, 1973). The contracture height increases if the reduction of the sodium concentration takes place a few minutes after the withdrawal of potassium, indicating the importance of an intracellular sodium accumulation during the high sodium (140 mmol/l) period in consequence of a sodium pump inhibition owing to the lack of extracellular potassium (SEIBEL and REITER, 1973). This potassium-withdrawal contracture is markedly enhanced by dihydroouabain in the same concentrations (10^{-5}–5×10^{-5} mmol/l) which exert positive inotropic effects on regular contractions of papillary muscles at a frequency of 1 Hz (SEIBEL and REITER, 1974; see also Fig. 6). In Fig. 9, an experiment is shown in which the potassium-withdrawal contracture is comparatively weak owing to a relatively low calcium concentration (1.6 mmol/l); the contracture is slightly increased when, after an incubation period with high sodium concentration, sodium is again reduced to 25 mmol/l (traces 1 and 3). The addition of 3×10^{-5} mol/l dihydroouabain 10 min before zero time (trace 2) only produces an increase in contracture force after the period of incubation in high sodium concentration. That is to say, the increase becomes possible after the muscle can gain sodium intracellularly in consequence of the combined effects of its high electrochemical gradient and an inhibition of the sodium pump.

Further evidence of the importance of intracellular sodium for the positive inotropic effect of cardioactive steroids is its potentiation by drugs which increase the

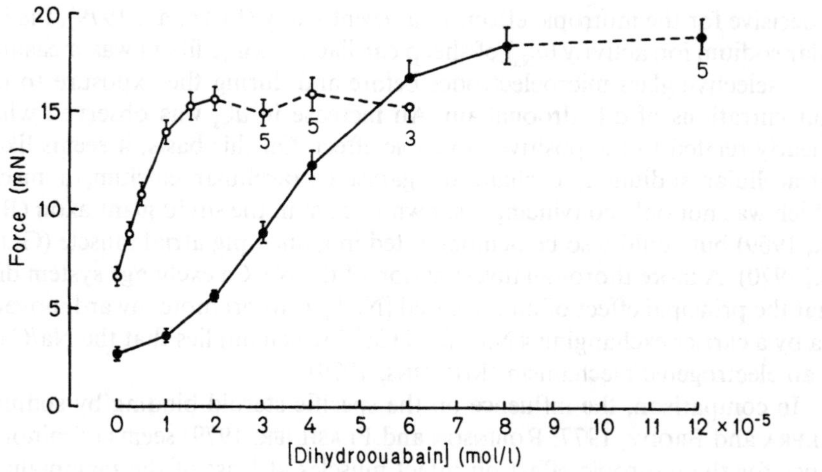

Fig. 10. Potentiation of the positive inotropic effect of dihydroouabain by veratridine. Concentration–effect curves of dihydroouabain in the absence *(solid circles)* or presence *(open circles)* of 0.8 μmol/l veratridine. Cumulative concentration–effect curves obtained from six muscles serving as controls (diameter 0.72 ± 0.03 mm) and from six muscles exposed to veratridine (diameter 0.73 ± 0.02 mm). Mean values, vertical bars indicate standard errors. Time of pre-equilibration in veratridine-containing solution: 3 h (2 h at 1 Hz, 1 h at 0.25 Hz). Control muscles kept for an equal time in veratridine-free Krebs–Henseleit solution under otherwise equal conditions. Figures indicate number of preparations evaluated at the respective dihydroouabain concentration (if < 6); preparations not evaluated developed spontaneous contractions. Contraction frequency 0.25 Hz. (HONERJÄGER and REITER, 1975)

sodium permeability (P_{Na}) of the sarcolemma. Such drugs are the ceveratrum alkaloids like veratridine and germitrine, which increase P_{Na} of excitable membranes by a specific interaction with the fast Na channels (ULBRICHT, 1969). The tetrodotoxin-sensitive prolongation of the cardiac action potential by the alkaloid mixture veratrine (HORACKOVA and VASSORT, 1973, 1974) or the pure alkaloid veratridine (HONERJÄGER and REITER, 1975) as well as the appearance of afterdepolarizations by germitrine (HONERJÄGER and REITER, 1977) indicate that the drug-induced increase in P_{Na} occurs during each excitation. The potentiation of the inotropic effect of dihydroouabain by veratridine is shown in Fig. 10 where two cumulative concentration–effect curves of the cardioactive steroid are depicted one of which was obtained from muscles exposed to 0.8 μmol/l veratridine (HONERJÄGER and REITER, 1975). As Fig. 10 shows, the increase in force of contraction caused by 10^{-5} mol/l dihydroouabain was small (in consequence of the relatively low contraction frequency of 0.25 Hz, see Sect. C.II.1) in the control preparations ($\Delta F_c =$ 0.95 mN). The same concentration of the glycoside caused a 7.7-fold greater effect in the veratridine-treated preparations under otherwise equal conditions ($\Delta F_c =$ 7.33 mN). A similar potentiation was observed with the ceveratrum alkaloid germitrine (HONERJÄGER and REITER, 1977).

The strict parallelism between the magnitude of the positive inotropic effect of a cardioactive steroid and the conditions facilitating sodium entry into the cell, either during excitation or during rest, proves that it is the result of the inhibition of the sodium pump, namely the increase in cellular sodium concentration, which is decisive for the inotropic effect. In a recent study (LEE et al., 1979), the intracellular sodium ion activity (a^i_{Na}) of sheep cardiac Purkinje fibers was measured with Na^+-selective glass microelectrodes before and during the exposure to different concentrations of dihydroouabain. An increase in a^i_{Na} was observed which was linearly related to the positive inotropic effect. On this basis, it seems likely that intracellular sodium is exchanged against extracellular calcium, a mechanism which was not only convincingly shown to exist in the squid giant axon (BAKER et al., 1969) but could also be demonstrated in guinea-pig atrial muscle (GLITSCH et al., 1970). A more thorough investigation of the Na/Ca exchange system disclosed that the principal effect of an increased $[Na^+]_i$ is to promote inward movement of Ca by a carrier exchanging $4\,Na^+$ for $1\,Ca^{2+}$ which implies that the Na/Ca carrier is an electrogenic mechanism (MULLINS, 1979).

In comparison, the influence on the specific steroid binding by sodium (e.g., AKERA and BRODY, 1977; ROBINSON and FLASHNER, 1979) seems of minor importance for the inotropic effect on intact muscles at least of the mammalian heart where the feasible range of a change in sodium concentration is limited. Although the binding of ouabain as well as its inhibitory action on the sodium pump is slowed by a complete replacement of sodium (squid axon), the final steady-state values are obviously not influenced (BAKER and WILLIS, 1972). It was reported by TALBOT (1968) that the rate of development of the inotropic effect of ouabain on frog heart muscle was reduced at low sodium concentrations, an effect which was not observed with guinea-pig ventricular muscle (REITER, 1963 b). Whether this difference indicates particular binding problems for cardioactive steroids in frog heart muscle as compared with mammalian heart muscle or has other causes (possibly temperature) is an open question. In this connection it is noteworthy that the dihydro derivative of ouabain which acts much faster in the mammalian heart than ouabain (REITER, 1967) is extremely slow in acting in the frog heart (LENZEN, 1973).

Further evidence for the main importance of extracellular sodium as the source of its inotropically relevant inctracellular build-up is the dependence on sodium of other inotropic effects which are not being produced by cardioactive steroids. These are the inotropic effects of a reduction of the outside potassium concentration (REITER et al., 1971) and of fluoroacetate (KORTH et al., 1978). In both cases, considerable circumstantial evidence exists for the assumption of an inhibition of the sodium pump as the underlying mechanism. Naturally, steroid binding cannot possibly be involved in the sodium dependence of these effects.

II. The Influence of Potassium

Although the beneficial influence of an increase in the serum potassium level on the toxic effects of cardioactive steroids is well established, the reports on the influence of extracellular potassium on their inotropic action are conflicting (for a review, see LEE and KLAUS, 1971). This might partly be due to the fact that (as pointed out by LEE and KLAUS) changes in $[K^+]_o$ themselves modify myocardial contractile force. Concentrations higher than 10 mmol/l and below 2 mmol/l usually increase the force of contraction. This leaves a relatively small range of potassium concentrations (2–10 mmol/l) in which the basal force of contraction remains more or less constant and which, therefore, lends itself to an investigation of an interference from potassium with the positive inotropic effect of cardioactive steroids. The concentration of potassium beyond which further lowering will lead to an increase in force of contraction depends to some degree on the frequency of contraction: at 1 Hz, guinea-pig ventricular contractile force is already somewhat elevated by 2.0 or 2.4 mmol/l K^+ as compared with 5 mmol/l (Fig. 11; REITER et al., 1966; COHN et al., 1967). This is not the case at 0.5 Hz (KORTH et al., 1978, see their Fig. 14); at 0.25 Hz, force of contraction starts to increase at a potassium concentration below 1.8 mmol/l reaching an almost maximal value at 1.0 mmol/l (HONERJÄGER and REITER, 1975, see their Fig. 15). Since under these conditions myocardial force of contraction is approaching its inherent ceiling, it is evident that any further inotropic effect by a cardioactive steroid is bound to be limited. This was convincingly documented by GREEFF et al. (1971) who showed that, at 1.3 mmol/l K^+, the inotropic effect of strophanthin K was much less than at 5.4 mmol/l, while the absolute values of contractile force obtained under the influence of the steroid were not significantly different. The same consideration applies to the results obtained by COHN et al. (1967) and by PRINDLE et al. (1971). These authors observed that, at 2.0 mmol/l K^+ and 1.5 mmol/l K^+ (cat), respectively, the inotropic effects of ouabain or digoxin, each used only in a single concentration, were equal or even less than at higher potassium concentrations. Therefore, these findings cannot be used as convincing arguments against a possible influence of extracellular potassium on the inotropic effect of cardioactive steroids.

It was reported by REITER et al. (1966) that the accelerating effect of 2×10^{-5} mol/l dihydroouabain on contraction velocity (i.e., the positive klinotropic effect) of guinea-pig papillary muscles was progressively increased with a decrease of $[K^+]_o$ from 9.6 to 4.8 and 2.4 mmol/l (Fig. 11a). The increase in force of contraction between 4.8 and 2.4 mmol/l K^+ was less than that observed during the reduction of $[K^+]_o$ from 9.6 to 4.8 mmol/l. This was the consequence of a diminution of the time to peak force at 2.4 mmol/l K^+ which limited the influence of the acceleration of the contraction velocity on the development of peak force.

The reduction of the time to peak force was the result of a pronounced shortening of the duration of the transmembrane action potential under the influence of the steroid, especially at the 30% repolarization level (Fig. 11 a). In contrast, the variation of $[K^+]_o$ between 2.4 and 9.6 mmol/l did not influence the positive inotropic effect of adrenaline, as shown in Fig. 11 b. Similar results to those depicted in Fig. 11 a were reported by COHN et al. (1967) for the inotropic effect of ouabain on guinea-pig ventricular muscle in the range 5–10 mmol/l K^+. Likewise, AKERA et al. (1979) found a clear-cut negative correlation between the steady-state positive inotropic effect of digitoxigenin on guinea-pig atrial muscle and $[K^+]_o$ between 3.5 and 9.5 mmol/l. With the slowly acting glycoside digoxin, they observed a retardation of the development of the inotropic effect with increasing $[K^+]_o$, the difference in the magnitude of the inotropic effect being significant between 5.8 and 9.5 mmol/l K^+.

The rate of development of the positive inotropic effect of digoxin was slowed much more by potassium (1.5–7.5 mmol/l) in the experiments of PRINDLE et al. (1971) with cat papillary muscle. They worked at a lower contraction frequency (0.2 Hz). This indicates that the influence of $[K^+]_o$ on the inotropic effect of cardioactive steroids increases with decreasing frequencies. After a 45 min incubation

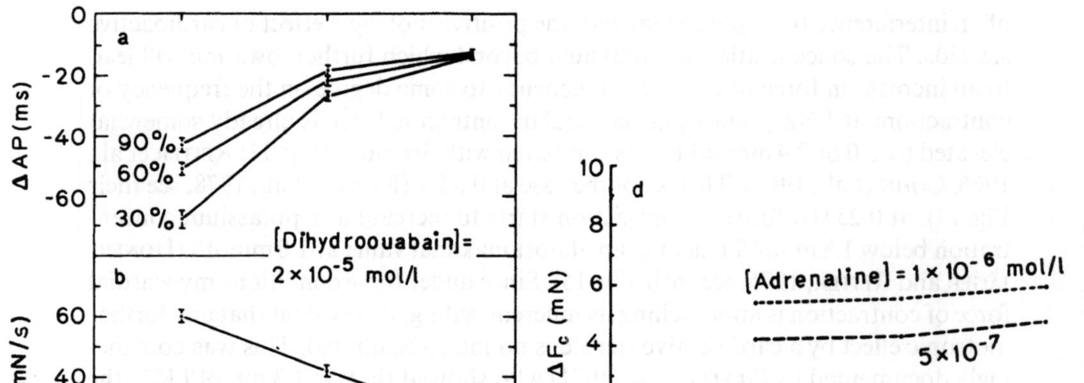

Fig. 11a–e. Influence of $[K^+]_o$ on inotropic effects of dihydroouabain and adrenaline on guinea-pig papillary muscle. **a** shortening of the APduration at different stages of repolarization (30%, 60%, and 90%.) **b**, the positive klinotropic effect (ΔS_1 change in contraction velocity.) **c** absolute values of force of contraction before and after addition of steroid. Mean values from six muscles ± standard error. (REITER et al., 1966). **d** positive inotropic effect. Mean values from six muscles ± standard error. (REITER, 1970b). **e** absolute values of force of contraction before and after addition of adrenaline (concentrations as indicated); For **a–e**: Krebs–Henseleit solution, 140 mmol/l Na^+, 3.2 mmol/l Ca^{2+}, and K^+ as indicated; stimulation frequency 1 Hz; temperature 35 °C

time, the increases in force of contraction produced by 5×10^{-7} mol/l digoxin amounted to 0.7, 1.3, and 2.2 g/mm² at $[K^+]_o$ of 7.5, 4.5, and 1.5 mmol/l, respectively.

That the steady-state increases of force of contraction did not significantly differ between their experiments at 1.5 and 4.5 mmol/l K^+ was probably due to the ceiling situation at 1.5 mmol/l K^+. However, the lack of significance of the reduction of the final inotropic effect at 7.5 mmol/l K^+ most likely had methodological reasons; the steady-state value of force of contraction was reached, at this $[K^+]_o$, only after an incubation period of 150 min. The curve relating the increment of con-

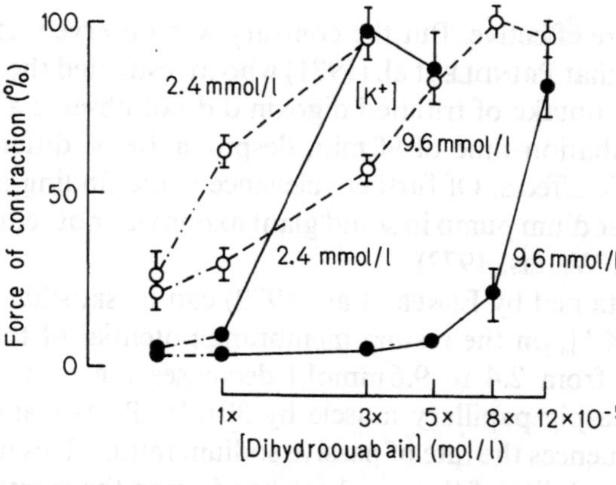

Fig. 12. Antagonism of K^+ against the positive inotropic effect of dihydroouabain as a function of stimulation frequency. Data from four groups each of eight of muscles. Force of contraction is plotted as a percentage of maximum force produced by the glycoside at 1 Hz, 5.9 mmol/l K^+. Concentration of dihydroouabain is plotted on a logarithmic scale. *Open circles* and *broken lines* represent steady-state values at stimulation frequency 0.5 Hz; *solid circles* and *solid lines* the respective values at 0.0033 Hz. The data are shown as arithmetic means ± standard errors of the mean. Standard errors omitted when not exceeding the size of the symbols. Concentrations, in mmol/l 3.2 Ca^{2+}, 140 Na^+, 35 °C, resting force 4 mN. (EBNER, BACHMAIER and REITER, 1979)

tractile force to time shows a sudden increase in slope after 75 min, indicating a change in the experimental working condition of the muscle which was probably due to the influence of its plasticity during the long period of stretch (REITER, 1972b). Since, at the beginning, the resting force was set at the peak of the length–force curve, the muscles very likely contracted, during the late part of the experiment, at a lower point of the length–force curve where the possible increment of contractile force due to an inotropic effect may be greater (see Chap. 9, Sect. B.VI).

The dependence on contraction frequency of the influence of $[K^+]_o$ on the inotropic effect of dihydroouabain was studied by EBNER et al. (1979). Their results are summarized in Fig. 12 where concentration–effect curves are depicted which were obtained at two different contraction frequencies, 0.5 and 0.0033 Hz, at both potassium concentrations: 2.4 and 9.6 mmol/l. Evidently, the inhibitory influence of potassium was drastically enhanced after reduction of the stimulation frequency to 0.0033 Hz where the muscles contracted only once every 5 min. At this low frequency, if $[K^+]_o$ was 2.4 mmol/l, 3×10^{-5} mol/l dihydroouabain produced a maximum inotropic effect whereas it was without effect at 9.6 mmol/l K^+. This shows that the mechanism by which potassium exerts its inhibitory effect is not predominantly connected with the processes of excitation and contraction but with the resting period between contractions.

These results do not support the assumption that the inhibition of the inotropic effect of cardioactive steroids is the result of an interference of potassium with their binding to specific sites of sarcolemmal Na^+, K^+-ATPase, as observed in in vitro enzyme studies (see SCHWARTZ et al., 1975). Since, at the higher stimulation frequency, more sodium has to be pumped, a greater number of enzyme molecules will be operating and therefore be susceptible to an inhibition of drug binding by extracellular potassium, hence the inhibitory effect of this ion on the inotropic effect should be more effective. But the contrary was observed. It is of importance in this connection that PRINDLE et al. (1971) who investigated the influence of $[K^+]_o$ on the myocardial uptake of tritiated digoxin did not observe a significant difference after an incubation time of 90 min, despite a 100% difference between the positive klinotropic effects. Of further relevance is the finding that the inhibition by ouabain of the sodium pump in squid giant axons was not reduced by increasing $[K^+]_o$ (BAKER and WILLIS, 1972).

The results obtained by EBNER et al. (1979) can be satisfactorily explained by the influence of $[K^+]_o$ on the resting membrane potential of the cardiac cell. An increase in $[K^+]_o$ from 2.4 to 9.6 mmol/l decreases the transmembrane resting potential of guinea-pig papillary muscle by 30 mV (REITER and STICKEL, 1968). Such a change influences the rate of passive sodium influx. This influx is dependent upon the Na permeability of the membrane and upon the membrane potential as part of the driving force for Na entry (DEITMER and ELLIS, 1978). In fact, a linear relationship was observed by ELLIS (1977) between the internal Na activity of sheep heart Purkinje fibers and $[K^+]_o$ between 1 and 30 mmol/l, the internal Na activity increasing with decreasing $[K^+]_o$. Therefore, the sodium load of the cardiac cell increases as its resting membrane potential is shifted to more negative values with the lowering of $[K^+]_o$. A partial inhibition of the sodium pump by a cardioactive steroid should then enhance the internal Na activity and facilitate Ca uptake through the Na/Ca exchange system, thereby promoting the inotropic effect. The inhibitory influence of high potassium ought to become less at higher stimulation frequencies in consequence of a stimulation-dependent increase in Na influx.

This hypothesis, according to which the inhibitory effect of $[K^+]_o$ on the inotropic action of cardioactive steroids is not due to an interference with drug binding but is the result of its influence on the potential-dependent driving force for Na entry, allows some predictions which can be examined experimentally. The hypothesis implies that $[K^+]_o$ influences not only the inotropic effect of cardioactive steroids but also the inotropic effects of other substances with a mechanism of action in which an increase in intracellular sodium plays a decisive role. Such substances may directly or indirectly inhibit the sodium pump by mechanisms different from that of cardioactive steroids, like *p*-chloromercuriphenylsulfonic acid (PCMBS) or fluoroacetate; or they may act by an increase of Na influx during stimulation, like the ceveratrum alkaloids. Indeed, the experimental results confirm this prediction. It was reported by HALBACH and SCHÖNSTEINER (1979) that the rate of development of the inotropic effect of PCMBS was considerably decreased as $[K^+]_o$ was increased between 2.4 and 9.6 mmol/l. KORTH et al. (1978) found that the magnitude of the inotropic effect of fluoroacetate was drastically reduced by an increase of $[K^+]_o$ from 2 to 9.6 mmol/l. And of particular significance is the observation (Fig. 13) of an inverse relationship between $[K^+]_o$ (2–6 mmol/l) and the inotropic effect of the ceveratrum alkaloid veratridine (see Sect. B.I; HONERJÄGER and REITER, 1975).

C. Frequency Dependence

A survey of the pertinent reports shows that the role of contraction frequency in regard to the inotropic action of cardioactive steroids is controversial (LEE and

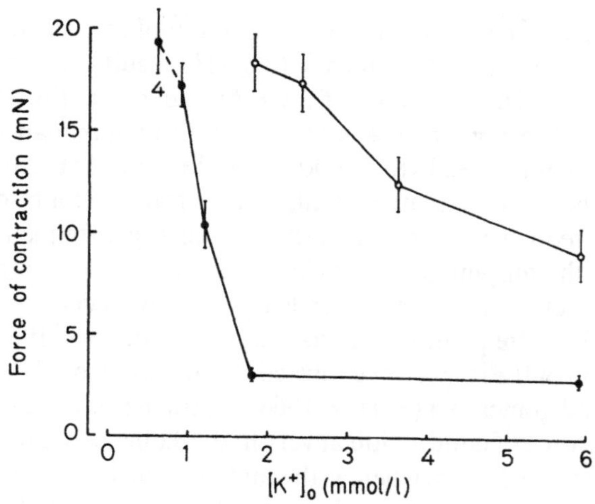

Fig. 13. Effect of reducing potassium concentration, $[K^+]_o$, on force of contraction of guinea-pig papillary muscle in the absence *(solid circles)* or presence *(open circles)* of 0.8 μmol/l veratridine. Mean values of six muscles serving as controls ± standard errors (diameter 0.71 ± 0.05 mm) and of six muscles exposed to veratridine (diameter 0.73 ± 0.04 mm). Pre-equilibration time in veratridine-containing solution 3 h (2 h at contraction frequency 1 Hz, 1 h at 0.25 Hz). Control muscles kept for an equal period in veratridine-free Krebs–Henseleit solution under otherwise equal conditions. Figures below vertical bars indicate number of preparations evaluated at the respective $[K^+]_o$; preparations developing spontaneous contractions not evaluated. Contraction frequency 0.25 Hz. (HONERJÄGER and REITER, 1975)

KLAUS, 1971). However, a thorough inspection of the various papers reveals that the discrepancy between the different opinions can be easily traced to the differences in the experimental conditions of the particular investigations. The main difference concerns the selected tissue: some authors worked with ventricular, others with atrial muscle. Another, likewise important, point is the fact that some authors studied the action of only a single concentration of the drug which is especially misleading if the selected concentration is apt to produce contracture at a moderately high frequency (e.g., 1 Hz). As to the selected cardiac tissue, one has to keep in mind that a distinct difference exists between atrial and ventricular muscle in regard to the regular frequency–force relationship.

I. General Aspects of Drug Effects on the Frequency–Force Relationship of Cardiac Tissue

1. Ventricular Muscle

It was long ago shown by BOWDITCH (1871) that the frequency–force relationship of frog ventricular muscle is a relatively simple one. He noticed that the steady-state force of contraction was highest at a frequency of 0.2–0.25 Hz (which must be regarded as a "high" frequency for the frog heart at room temperature). The force of contraction declined continually with the reduction of frequency, the lowest force value was obtained at a contraction interval of 5 min and was not influenced by a further prolongation of the rest period. Thus, the frequency–force relationship of frog ventricular muscle was monophasic up to a frequency of 0.25 Hz. A further increase of frequency to 0.5 Hz resulted again in a decrease in force of contraction. The same type of force–frequency relationship was found to prevail in cardiac ventricular muscle of almost all mammalian species (with the possible exception of the rat, BENFORADO, 1958), i.e., the contraction elicited under rested-state conditions is extremely small, and the force of contraction is increased with increasing frequencies. Only the range of inotropic frequencies is somewhat higher than for the amphibian ventricle. Like the cardiac ventricle of the frog, mammalian ventricular muscle has a tendency to show a decline of its force of contraction at very high frequencies (KOCH-WESER and BLINKS, 1962; REITER, 1966).

The inspection of the respective contraction curves of frog (NIEDERGERKE, 1956; LENZEN, 1973) and guinea-pig (REITER, 1966) ventricular muscle shows that the decline of the peak force of contraction at very high frequencies is due only to a shortening of the time to peak force whereas the rate of force development (the steepness of the ascending slope of the isometric contraction curve) may be even further increased; the shortening of the time to peak force in turn is probably the consequence of the frequency-dependent shortening of the duration of the transmembrane action potential in both the frog (NIEDERGERKE, 1956) and the mammalian (REITER and STICKEL, 1968) heart. This means that, by the shortening of the time to peak force, the positive inotropic effect of an increase in contraction frequency simply becomes masked at very high frequencies (KOCH-WESER and BLINKS, 1963). It does not seem to be very useful or instructive to elaborate in detail this ultimate range of the frequency–force relationship nor the influence of any drug upon it if the positive inotropic frequency effect, i.e., the increase in the steady-state force of contraction with an increase in contraction frequency, is to be studied. In accordance with these considerations, the frequency–force relationship of cardiac ventricular muscle is to be regarded as predominantly monophasic.

2. Atrial Muscle

In contrast to the cardiac ventricle, mammalian atrium has a biphasic frequency–force relationship (KRUTA, 1937; BRAVENÝ and KRUTA, 1958; BLINKS and KOCH-WESER, 1961; for a review see KOCH-WESER and BLINKS, 1963). Whereas the positive inotropic frequency effect is almost identical with that of the ventricle in the same frequency range, the force of contraction of atrial muscle increases again if the frequency is decreased beyond that with the lowest steady-state force of contraction. This results in rested-state contractions (which cannot be changed by a further prolongation of the stimulation interval, usually 5 min, see BLINKS and KOCH-WESER, 1961) with a strength almost as high as those of the maximum inotropic frequency effect. Accordingly, the difference between the contractile behaviour of the two tissues exists only in the low frequency range (between 0.0033 and about 0.33 Hz).

3. The Atrium-Specific, Rested-State Contraction

In order to assess probable drug effects upon the frequency–force relationship of atrial muscle, one has to consider the qualitative difference between its two peaks, that of the rested-state condition and the other in the high frequency range. Whereas for the latter constant contractions (or depolarizations) at a high frequency are a prerequisite, a long period of rest is necessary to build up the former which is abolished completely by only a few beats (1–3) at a higher frequency (KRUTA, 1937; BRAVENÝ and KRUTA, 1958; BLINKS and KOCH-WESER, 1961). It is obvious, therefore, that different mechanisms must exist for the development of the two peaks of the atrial frequency–force relationship. This was already pointed out by BRAVENÝ and KRUTA (1958) who interpreted the relationship by a competition of two factors, one of which (A) enlarges the contraction as the preceding interval increases, whereas the other (B), on the contrary, stimulates the contraction while shortening the interval.

BLINKS and KOCH-WESER (1961) considered the high rested-state contraction of atrial muscle as the basis of their analysis and described the frequency–force relationship as the result of two opposing effects of activation: a negative inotropic effect of activation (NIEA), and a positive inotropic effect of activation (PIEA), see also KOCH-WESER and BLINKS (1963). In this system, the high rested-state contraction of atrial muscle (which exists neither in amphibian nor in mammalian ventricle) is treated only insofar as it is abolished by a powerful NIEA, whereas BRAVENÝ and KRUTA'S (1958) factor A deals with the mechanism by which the muscle acquires the necessary ability, during a long rest period, to produce a high rested-state contraction. As to the nature of this mechanism a possible key is offered by the observation of VIERLING and REITER (1975) that the monophasic frequency–force relationship of guinea-pig ventricular muscle can easily be transformed into a biphasic atrium-like one with high rested-state contractions, simply by the withdrawal of magnesium from the incubation medium.

The strength of these rested-state contractions of ventricular muscle in magnesium-free solution depends on the extracellular calcium concentration (as is also the case with atrial muscle in Mg-containing solution; ZIEHM, 1972). A reasonable explanation is that, in ventricular muscle, a relatively low magnesium concentration is important for the extrusion of cellular calcium during rest. Such a Mg-dependent mechanism of calcium extrusion has been demonstrated in erythrocytes by SCHATZMANN (1970).

The operation of such a mechanism in cardiac ventricular cells would explain the absence, in normal Mg-containing solution, of any noteworthy contractile activity after long rest periods, even despite high extracellular Ca concentrations. On the other hand, the inhibition by lack of magnesium of such an extrusion mechanism could explain the development of contractile activity, provided the rest period is long enough to allow calcium to accumulate in storage structures from which it can be liberated by stimulation. Such a mechanism would also explain why the strength of the rested-state contraction vanishes with one to three successive beats at 1 s interval. Accordingly, the difference between the two types of cardiac tissue (atrium and ventricle) with regard to the development of rested-state activity would be due to differences in efficiency of their respective Mg-dependent Ca extrusion mechanisms. In regard to drug effects on the rested-state activity of cardiac atria, it must be assumed that their underlying mechanisms are operative during the rest period and not during activity.

II. The Influence of Contraction Frequency on Steady-State Inotropic Effects

1. Ventricular Muscle

In studying the action of digitoxin on the isolated frog heart at different contraction frequencies, HAJDU and SZENT-GYÖRGYI (1952) observed that the "staircase" was abolished when the inotropic action of the steroid was fully developed, i.e., the force of contraction did not increase in a staircase-like manner if the ventricle was stimulated after a long rest period (BOWDITCH, 1871). Instead, the height of the first, rested-state, contraction was the same as that of all the following ones, regardless of their frequency. Similar observations were made by KOCH-WESER and BLINKS (1962) and KOCH-WESER (1971) who studied the inotropic effect of acetylstrophanthidin and of ouabain on the cat papillary muscle over a wide range of contraction frequencies. However, HAJDU and SZENT-GYÖRGYI (1952) as well as KOCH-WESER and BLINKS (1962) and KOCH-WESER (1971) used only a single, relatively high, concentration of the respective steroid which, in the case of ouabain (2×10^{-7} mol/l), led finally to contracture (KOCH-WESER, 1971). With these concentrations, the observed inotropic effects (i.e., the increase in force of contraction measured in absolute or relative terms) were higher with rested-state contractions (stimulation intervals of 5 min and longer) than at a contraction frequency of 1 Hz.

The situation becomes different if lower effective concentrations of a cardioactive steroid are being used. Working with rabbit ventricular muscle, TUTTLE and FARAH (1962) found that the effect of cardioactive glycosides on the frequency–force relationship is dependent upon the drug concentration. This is illustrated in Fig. 14 where the relation is shown between the concentration of dihydroouabain and its positive inotropic effect on guinea-pig papillary muscles as a function of stimulation frequency (EBNER and REITER, 1977). At 1 Hz stimulation frequency, the maximum of the steroid concentration–response curve was obtained, on the average, with 5×10^{-5} mol/l dihydroouabain. With decreasing stimulation frequencies, the concentration–effect curves were shifted to the right; increasingly higher steroid concentrations were required to compensate for the decrease in the control contraction force. At a concentration of 5×10^{-5} mol/l which was maximally effective at 1 Hz, dihydroouabain barely produced any positive inotropic effect when the muscles contracted only once every 5 min (stimulation frequency 0.0033 Hz). At this low frequency, an increase of the steroid concentration to 8×10^{-5} mol/l resulted in a contraction force only slightly higher than the control volue at 1 Hz. In accordance with earlier results obtained in the same laboratory (REITER and STROBL, 1972; STROBL, 1973), the series of experiments depicted in Fig. 14 demonstrates that the extent of the positive inotropic effect of a cardioactive steroid in low concentrations on the guinea-pig ventricular myocardium is strongly dependent on the contraction frequency.

In an attempt to analyze the changes in the shape of the isometric contraction curves of the experiments of Fig. 14 under the various stimulation frequencies and steroid concentrations, the values of peak force were compared with the respective contraction parameters (mean slope of the ascending contraction curve and mean slope of relaxation) and with the time parameters (time to peak force and relaxation time). The relation of the various parameters to peak force was found to be

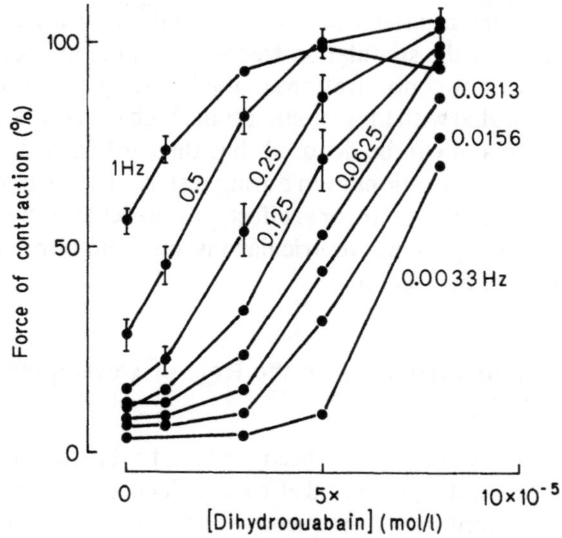

Fig. 14. Dependence of the positive inotropic effect of dihydroouabain on stimulation frequency. Peak force of contraction plotted as percentage of maximal glycoside effect at stimulation frequency 1 Hz. The concentration–effect curve at 1 Hz represents the combined control for three muscle groups. First group 1, 0.5, and 0.125 Hz (eight muscles), second group 1, 0.0625, and 0.0156 Hz (eight muscles), third group 1 and 0.0033 Hz (five muscles). Standard errors of the mean indicated for some points. Resting force 4 mN, $[Ca^{2+}]_o$ 3.2 mol/l, 35 °C. (EBNER and REITER, 1977)

the same, irrespective of whether the changes were obtained by variation of stimulation frequency or by addition of dihydroouabain (EBNER and REITER, 1977). Thus, the isometric contraction curve was altered identically by variation of stimulation frequency or by application of the cardioactive steroid. A similar observation was made by EDMAN and NILSSON (1965) who found that the time course of the active state of rabbit papillary muscle was influenced by both ouabain and increased frequency in a very similar way.

2. Atrial Muscle

The steady-state inotropic action of cardioactive steroids on cardiac atrial muscle differs from that on ventricular muscle insofar as already small concentrations produce almost equal amounts of force increment over the whole frequency range. Thereby the biphasic frequency–force relationship of the atrium (see Sect. C.I.2) is shifted upwards (KOCH-WESER and BLINKS, 1962). With higher concentrations, the inotropic effect is greatest in a medium frequency range (3–30 s stimulation interval), the absolute amount of force increment being reduced at the two peaks of the control frequency–force relationship (KOCH-WESER and BLINKS, 1962; VINCENZI, 1967; KOCH-WESER, 1971). In the ultimate range of the frequency–force relationship (2–6 Hz), the inotropic effect becomes increasingly less and may even be absent in both atrial and ventricular muscle (KOCH-WESER and BLINKS, 1962; GREEFF et al., 1971), in consequence of an increasing shortening of the time to peak force (see Sect. C.I.1).

Accordingly, the main difference between atrial and ventricular muscle in regard to the influence of the stimulation frequency on the inotropic steroid action concerns the effect of low concentrations in the low frequency range (below about 0.33 Hz) and is connected with the atrium-specific high rested-state contraction (see Sect. C.I.3). From this, it can be deduced that the cardioactive steroids act upon cardiac cells in low concentrations even during rest, the low degree of this influence being inhibited to produce an inotropic effect, in ventricular muscle, through a cellular mechanism which prevents ventricular muscle from developing high rested-state contractions (see Sect. C.I.3).

III. The Influence of Frequency on the Rate of Development of the Inotropic Effect

A considerable amount of effort has been devoted to the question of how the rate of development of an inotropic steroid effect is influenced by the activity of the cardiac muscle, i.e., its contraction frequency. The results of the numerous investigations on heart tissue from a great variety of species are fairly uniform, the only major difference of opinion deriving from the limitation in the stimulation frequencies used by some authors working with atrial muscle.

1. Ventricular Muscle

The first paper dealing with the relation between the rate of development of steroid action and contraction frequency was that of WEIZSÄCKER (1913). He found that the time required to obtain, in the frog ventricle, the endpoint of toxicity of strophanthin, i.e., contracture, was shortened if the driving frequency was increased. Similar results were obtained with ventricular muscle of the dog (VASSALLE et al., 1962) and the cat (KOCH-WESER, 1971).

Other authors whose aim was to study the rate of development not of toxicity but of the positive inotropic effect found also a definite dependence on the contraction frequency with ventricles of the frog (WILBRANDT et al., 1953), the guinea-pig (SANYAL and SAUNDERS, 1958; REITER and STROBL, 1972; STROBL, 1973) and the cat (KOCH-WESER, 1971). Of particular interest in regard to a later controversy is the paper of WILBRANDT et al. (1953). These authors related the magnitudes of the positive inotropic response to the total number of contractions during a constant period of exposure (5 min) to strophanthoside. However, a variation of the number of contractions during a constant period of time means a variation in frequency.

2. Atrial Muscle

In the atrium, as in the ventricle, the appearance of contracture due to high concentrations of cardioactive steroids is hastened by an increase in stimulation frequency (HOLLAND and SEKUL, 1961; GREEFF et al., 1971 KOCH-WESER, 1971).

Likewise, the positive inotropic effect develops faster if the stimulation frequency is increased between 0.25 and 3 Hz (KRUTA et al., 1960; MORAN, 1963, 1967; LOCK, 1965; VINCENZI, 1967; BYRNE and DRESEL, 1969; KOCH-WESER, 1971). The influence on the onset of the steroid action is less if the stimulation frequency is varied in the lower range between 0.33 and 0.0033 Hz (VINCENZI, 1967; KOCH-WESER, 1971) where the atrium develops its characteristic high rested-state contraction (see Sect. C.I.3).

Plots of the development of the inotropic effect against the number of beats after the addition and in presence of the steroid by MORAN (1963, 1967) showed that about the same total number of beats was required to reach 50% of the respective maximal inotropic effects. MORAN used left atria of rabbits at a temperature of 30 °C and varied the stimulation frequency between 0.25 and 2 Hz. These results could be confirmed neither by VINCENZI (1967), working with guinea-pig atria, nor by KOCH-WESER (1971) with cat atria. The latter two authors performed their experiments at higher temperatures (35° or 37.5 °C) and investigated a much broader range of frequencies (down to 0.0033 and 0.001 Hz, respectively). In their experiments, the number of beats necessary to reach the respective half maximum inotropic effect increased with increasing frequency. Since MORAN, in his later paper (1967), observed a similar increase of the contraction number with increasing frequency in one set of experiments performed at 37 °C, it is likely that his results obtained at 30 °C were impaired by the influence of the lower temperature on the time parameters of the contraction curve, especially at the highest frequencies.

IV. On the Mechanism of the Frequency Dependence of Inotropic Steroid Action

In regard to the mechanism by which the inotropic effect of cardioactive steroids is influenced by alterations of stimulation frequency one should consider how an increase in stimulation frequency itself produces an inotropic effect. The latter is explained by some authors (BEELER and REUTER, 1970; REUTER, 1973; KAUFMANN et al., 1974) as being merely the consequence, with increasing stimulation frequency, of an increase of the time integral of the transmembrane calcium inward current during the action potential. This ignores the possible role of an increase in sodium load by more frequent depolarizing sodium currents leading subsequently to an increased calcium uptake through Na/Ca exchange (see Sect. B.I).

That the latter mechanism is actually important for the inotropic action of an increase in stimulation frequency must be deduced from the fact that, in guinea-pig papillary muscle, the inotropic frequency effect was found to depend on the extracellular sodium concentration (REITER, 1966), as was the inotropic action of cardioactive steroids (see Sect. B.I). The magnitude of the inotropic frequency effect (expressed as the frequency-dependent shifts of the calcium concentration–effect curve to lower $[Ca^{2+}]_o$) was strongly reduced after diminution of $[Na^+]_o$ by 50%. Such a dependence on $[Na^+]_o$ does not exist with the inotropic effects of catecholamines and other substances (REITER, 1972b) which are generally believed to act by an increase in Ca current during the plateau of the action potential.

The frequency dependence of the inotropic steroid effect can be explained best by the assumption of a stimulation-dependent enhancement of an increase in intracellular sodium concentration resulting from an inhibition of the cellular sodium pump. A slight inhibition, as produced by low concentrations of a cardioactive steroid, should cause only a small increase in $[Na^+]_i$ if the interval between beats is long (i.e., contraction frequency is low). The simultaneous Na/Ca

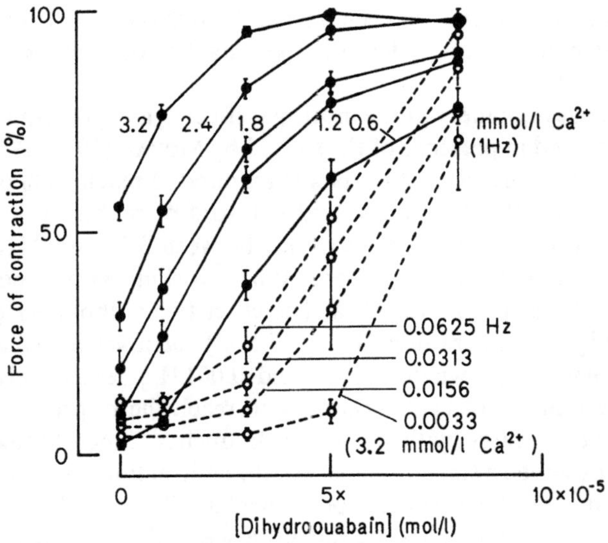

Fig. 15. Dependence of the positive inotropic effect of dihydroouabain on $[Ca^{2+}]_o$ compared with its dependence on stimulation frequency. Peak force of contraction plotted as percentage of maximal glycoside effect at 1 Hz, $[Ca^{2+}]_o$ 3.2 mol/l. *Solid circles, solid lines:* dependence on $[Ca^{2+}]_o$. Concentration–effect curve at 1 Hz, $[Ca^{2+}]_o$ 3.2 mol/l represents the combined control for two muscle groups. First group $[Ca^{2+}]_o$ changed from 3.2 to 0.6 and then to 2.4 mol/l (eight muscles); second group from 3.2 to 1.2 and 1.8 mol/l (eight muscles). Stimulation frequency 1 Hz. *Open circles, broken lines:* dependence on stimulation frequency. $[Ca^{2+}]_o$ 3.2 mol/l, values from Fig. 14. Arithmetic means ± standard errors. Resting force 4 mN, 35 °C. (EBNER and REITER, 1977)

exchange-mediated Ca uptake would be too small to have an inotropic importance in ventricular muscle whose Ca extrusion mechanism prevents the development of any significant rested-state contractions. However, it might be large enough to produce an inotropic effect, even at a very low frequency, in the atrium where contractile capacity accumulates during rest (see Sect. C.I.3). Despite a but slight inhibition of the sodium pump owing to a small glycoside concentration, a larger increase of $[Na^+]_i$ would be expected at higher stimulation frequencies because of the higher rate of sodium influx, thereby causing an increased Ca uptake and, consequently, a positive inotropic effect. On the other hand, an increase in $[Na^+]_i$ would be achieved even in the quiescent ventricle if only the sodium pump were sufficiently inhibited by a very high steroid concentration. According to these considerations, the time lag in sodium extrusion is likely to be the common cause of the inotropic effects of both an increase in contraction frequency and cardioactive steroids (EBNER and REITER, 1977).

The role of intracellular sodium as mediator of the inotropic glycoside effect is corroborated by the comparison of the dependence of the inotropic effect on contraction frequency and on extracellular calcium concentration (Fig. 15). The reduction in force of contraction caused by lowering $[Ca^{2+}]_o$ or by lowering the contraction frequency at unchanged calcium concentration is assumed to be the result of calcium depletion from stores from which it is liberated during excitation. The difference between the two experimental conditions concerns the sodium load, which is affected by stimulation frequency but not by a change in $[Ca^{2+}]_o$. If the access of calcium to the stores were directly affected by the glycoside, the concentration–effect curves should be identical whether a certain control level of force of contraction is accomplished by reduction either in $[Ca^{2+}]_o$ or in frequency. The much higher effectiveness of the glycoside at 1 Hz stimulation frequency, despite a reduction in calcium concentration, therefore, suggests that a constantly high sodium load is necessary for the inotropic effect.

References

Akera, T., Brody, T.M.: The role of Na^+, K^+-ATPase in the inotropic action of digitalis. Pharmacol. Rev. *29*, 187–220 (1977)

Akera, T., Olgaard, M.K., Temma, K., Brody, T.M.: Development of the positive inotropic action of ouabain: effects of transmembrane sodium movement. J. Pharmacol. Exp. Ther. *203*, 675–684 (1977)

Akera, T., Wiest, S.A., Brody, T.M.: Differential effect of potassium on the action of digoxin and digoxigenin in guinea-pig heart. Eur. J. Pharmacol. *57*, 343–351 (1979)

Baker, P.F., Blaustein, M.P., Hodgkin, A.L., Steinhardt, R.A.: The influence of calcium on sodium efflux in squid axons. J. Physiol. (Lond.) *200*, 431–458 (1969)

Baker, P.F., Willis, J.S.: Inhibition of the sodium pump in squid giant axons by cardiac glycosides: dependence on extracellular ions and metabolism. J. Physiol. (Lond.) *224*, 463–475 (1972)

Beeler, G.W., jr., Reuter, H.: The relation between membrane potential, membrane currents, and activation of contraction in ventricular myocardial fibres. J. Physiol. (Lond.) *207*, 211–229 (1970)

Benforado, J.M.: Frequency-dependent pharmacological and physiological effects on the rat ventricle strip. J. Pharmacol. Exp. Ther. *122*, 86–100 (1958)

Bernauer, W., Dörfler, G., Gross-Hardt, M.: Untersuchungen über die Abhängigkeit der Herzwirkungen des Histamins von der Ca- und Na-Konzentration in der Extrazellulärflüssigkeit. Arch. Int. Pharmacodyn. *189*, 72–89 (1971)

Blinks, J.R., Koch-Weser, J.: Analysis of the effects of changes in rate and rhythm upon myocardial contractility. J. Pharmacol. Exp. Ther. *134*, 373–389 (1961)

Blinks, R., Olson, C.B., Jewell, B.R., Bravený, P.: Influence of caffeine and other methylxanthines on mechanical properties of isolated mammalian heart muscle. Circ. Res. *30*, 367–392 (1972)

Bohnenkamp, H.: Über die Wirkungsweise der Herznerven. Pflügers Arch. Ges. Physiol. *196*, 275–310 (1922)

Bowditch, H.P.: Über die Eigenthümlichkeiten der Reizbarkeit, welche die Muskelfasern des Herzens zeigen. Ber. d. sächs. Ges. d. Wiss. Leipzig, Math. Phys. Cl. *23*, 652–689 (1871)

Bozler, E.: The initiation of impulses in cardiac muscle. Am. J. Physiol. *138*, 273–282 (1943a)

Bozler, E.: Tonus changes in cardiac muscle and their significance for the initiation of impulses. Am. J. Physiol. *139*, 477–480 (1943b)

Brady, A.J.: The development of tension in cardiac muscle. In: Pharmacology of Cardiac Function, 2nd Internat. Pharmacol. Meeting Prague 1963. Krayer, O. (ed). Vol. 5, pp. 15–23. Oxford: Pergamon Press 1964

Brandt, W., Reiter, M., Seibel, K.: A postsynaptic effect of tyramine in ventricular muscle. Naunyn-Schmiedebergs Arch. Pharmacol. *273*, 283–293 (1972)

Bravený, P., Kruta, V.: Dissociation de deux facteurs: restitution et potentiation dans l'action de l'intervalle sur l'amplitude de la contraction du myocarde. Arch. Int. Physiol. Biochim. *66*, 633–652 (1958)

Bravený, P., Šumbera, J., Kruta, V.: After-contractions and restitution of contractility in the isolated guinea-pig auricles. Arch. Int. Physiol. Biochim. *74*, 169–178 (1966)

Byrne, J.E., Dresel, P.E.: The effect of temperature and calcium concentration on the action of ouabain in quiescent rabbit atria. J. Pharmacol. Exp. Ther. *166*, 354–363 (1969)

Caprio, A., Farah, A.: The effect of the ionic milieu on the response of rabbit cardiac muscle to ouabain. J. Pharmacol. Exp. Ther. *155*, 403–414 (1967)
Cohn, K., Pirages, S., Harrison, D.C.: The effects of potassium on the positive inotropic action of ouabain. Am. Heart J. *73*, 516–524 (1967)
Cranefield, P.F., Aronson, R.S.: Initation of sustained rhythmic activity by single propagated action potentials in canine cardiac Purkinje fibers exposed to sodium-free solution or to ouabain. Circ. Res. *34*, 477–481 (1974)
Deitmer, J.W., Ellis, D.: The intracellular sodium activity of cardiac Purkinje fibres during inhibition and re-activation of the Na-K pump. J. Physiol. (Lond.) *284*, 241–259 (1978)
Dudel, J., Trautwein, W.: Elektrophysiologische Messungen zur Strophanthinwirkung am Herzmuskel. Naunyn-Schmiedebergs Arch. Exp. Path. Pharmak. *232*, 393–407 (1958)
Ebner, F., Reiter, M.: The dependence on contraction frequency of the positive inotropic effect of dihydro-ouabain. Naunyn-Schmiedebergs Arch. Pharmacol. *300*, 1–9 (1977)
Ebner, F., Bachmaier, A., Reiter, M.: The influence of K^+ on the positive inotropic effect of dihydro-ouabain in relation to stimulation frequency and $[Ca^{2+}]$. Naunyn-Schmiedebergs Arch. Pharmacol. *308*, R 35 (1979)
Ebner, F., Reiter, M.: The influence of (+)-propranolol on the inotropic effect of dihydro-ouabain in relation to stimulation frequency. Naunyn-Schmiedebergs Arch. Pharmacol. *307*, 105–113 (1979)
Edman, K.A.P., Nielsson, E.: The dynamics of the inotropic changes produced by ouabain and increased contraction rate. Acta Physiol. Scand. *63*, 507–508 (1965)
Eismayer, G.: Über Tonus und Kinetik und deren Einfluß auf die Größe und Dynamik des Herzens im muskelkräftigen und muskelschwachen Zustand. Ergebn. Physiol. *30*, 126–152 (1930)
Eisner, D.A., Lederer, W.J.: Inotropic and arrhythmogenic effects of potassium-depleted solutions on mammalian cardiac muscle. J. Physiol. (Lond.) *294*, 255–277 (1979 a)
Eisner, D.A., Lederer, W.J.: The role of the sodium pump in the effects of potassium-depleted solutions on mammalian cardiac muscle. J. Physiol. (Lond.) *294*, 279–301 (1979 b)
Ellis, D.: The effects of external cations and ouabain on the intracellular sodium activity of sheep heart Purkinje fibres. J. Physiol. (Lond.) *273*, 211–240 (1977)
Engelmann, Th. W.: Über die Wirkungen der Nerven auf das Herz. Arch. Anat. Physiol. (Physiol. Abt.) (Leipzig) Jahrg. 1900, S. 315–361
Engstfeld, G., Antoni, H., Fleckenstein, A.: Die Restitution der Erregungsfortleitung und Kontraktionskraft des K^+-gelähmten Frosch- und Säugetiermyocards durch Adrenalin. Pflügers Arch. Ges. Physiol. *273*, 145–163 (1961)
Fabiato, A., Fabiato, F.: Excitation-contraction coupling of isolated cardiac fibers with disrupted or closed sarcolemmas. Calcium-dependent cyclic and tonic contractions. Circ. Res. *31*, 293–307 (1972)
Fabiato, A., Fabiato, F.: Contractions induced by a calciumtriggered release of calcium from the sarcoplasmic reticulum of single skinned cardiac cells. J. Physiol. (Lond.) *249*, 469–495 (1975)
Farah, A., Witt, P.N.: Cardiac glycosides and calcium. In: Proc. 1 st Int. Pharmac. Meeting Stockholm, 1961. Wilbrandt, W., Lindgren, P. (eds.), Vol. 3, pp. 137–171. Oxford: Pergamon Press 1963
Feigl, E.O.: Myocardial diastolic compliance. In: Factors Influencing Myocardial Contractility, Tanz, R.D., Kavaler, F., Roberts, J. (eds.), pp. 86–91. New York: Academic Press Inc. 1967 (a)
Feigl, E.O.: Effects of stimulation frequency on myocardial extensibility. Circ. Res. *20*, 447–458 (1967 b)
Feigl, E.O.: Effects of paired stimulation on cardiac muscle during diastole. In: Paired Pulse Stimulation of the Heart. Cranefield, P.F., Hoffman, B.F. (eds.), pp 155–163. New York, The Rockefeller University Press 1968
Ferrier, G.R.: The effects of tension on acetylstrophanthidin-induced transient depolarizations and aftercontractions in canine myocardial and Purkinje tissues. Circ. Res. *38*, 156–162 (1976)

Ferrier, G.R.: Relationship between acetylstrophanthidin-induced aftercontractions and the strength of contraction of canine ventricular myocardium. Circ. Res. *41*, 622–629 (1977a)
Ferrier, G.R.: Digitalis arrhythmias: role of oscillatory afterpotentials. Prog. Cardiovasc. Dis. *19*, 459–474 (1977b)
Ferrier, G.R., Moe, G.K.: Effect of calcium on acetylstrophantidin-induced transient depolarizations in canine Purkinje tissue. Circ. Res. *33*, 508–515 (1973)
Ferrier, G.R., Saunders, H.J., Mendez, C.: A cellular mechanism for the generation of ventricular arrhythmias by acetylstrophanthidin. Circ. Res. *32*, 600–609 (1973)
Glitsch, H.G., Pott, L.: Spontaneous tension oscillations in guinea-pig atrial trabeculae. Pflügers Arch. Ges. Physiol. *358*, 11–25 (1975)
Glitsch, H.G., Reuter, H., Scholz, H.: The effect of the internal sodium concentration on calcium fluxes in isolated guinea-pig auricles. J. Physiol. (Lond.) *209*, 25–43 (1970)
Greeff, K.: Einfluß von Pharmaka auf die Kontraktilität des Herzens. Verh. Dtsch. Ges. Kreisl. Forsch. *42*, 80–92 (1976)
Greeff, K., Pereira, E., Wagner, J.: Die Wirkung des Strophanthins bei Änderung der Schlagfrequenz und der extrazellulären K^+- und Ca^{++}-Konzentration. Arch. Int. Pharmacodyn. *190*, 219–228 (1971)
Hajdu, S., Szent-Györgyi, A.: Action of digitalis glucosides on isolated frog heart. Am. J. Physiol. *168*, 171–175 (1952)
Halbach, S., Schönsteiner, G.: Two different effects of p-chloromercuriphenylsulfonic acid (PCMBS) on contractility of guinea-pig myocardium. Naunyn-Schmiedebergs Arch. Pharmacol. *308*, R 34 (1979)
Hartmann, D.: Strontium als elektromechanisches Kopplungsglied am Herzmuskel des Meerschweinchens. Inaugural-Dissertation, Ludwig-Maximilian-Universität München, Medizinische Fakultät 1966
Heller, L.J.: Mechanical properties of cardiac muscle from spontaneously hypertensive rats: accentuated aftercontractions (39698). Proc. Soc. Exp. Biol. (N.Y.) *154*, 479–482 (1977)
Heller, L.J.: Cardiac muscle mechanics from doca- and aging spontaneously hypertensive rats. Am. J. Physiol. *235*, H 82–H 86 (1978)
Heller, L.J.: Augmented aftercontractions in papillary muscles from rats with cardiac hypertrophy. Am. J. Physiol. *237*, H 649–H 654 (1979)
De Hemptinne, A., Weyne, J., Leusen, I.: Dynamic parameters of myocardial contractility under influence of calcium and strontium. Arch. Int. Physiol. Biochim. *75*, 96–108 (1967)
Henderson, A.H., Forman, R., Brutsaert, D.L., Sonnenblick, E.H.: Tetanic contraction in mamalian cardiac muscle. Cardiovasc. Res. *5*, 96–100 (1971)
Hoffman, B.F., Bassett, A.L., Bartelstone, H.J.: Some mechanical properties of isolated mammalian cardiac muscle. Circ. Res. *23*, 291–312 (1968)
Holland, W.C., Sekul, A.: Influence of K^+ and Ca^{++} on the effect of ouabain on Ca^{45} entry and contracture in rabbit atria. J. Pharmacol. Exp. Ther. *133*, 288–294 (1961)
Honerjäger, P., Reiter, M.: The relation between the effects of veratridine on action potential and contraction in mammalian ventricular myocardium. Naunyn-Schmiedebergs Arch. Pharmacol. *289*, 1–28 (1975)
Honerjäger, P., Reiter, M.: Sarcolemmal sodium permeability and contractile force of guinea-pig papillary muscle: effects of germitrine. Circ. Res. *40*, 90–98 (1977)
Horackova, M., Vassort, G.: Ionic mechanism of inotropic effect of veratrine on frog heart. Pflügers Arch. Ges. Physiol. *341*, 281–284 (1973)
Horackova, M., Vassort, G.: Excitation-contraction coupling in frog heart. Pflügers Arch. Ges. Physiol. *352*, 291–302 (1974)
Jensen, R.A., Katzung, B.G.: Simultaneously recorded oscillations in membrane potential and isometric contractile force from cardiac muscle. Nature *217*, 961–963 (1968)
Kass, R.S., Lederer, W.J., Tsien, R.W., Weingart, R.: Role of calcium ions in transient inward currents and aftercontractions induced by strophanthidin in cardiac Purkinje fibres. J. Physiol. (Lond.) *281*, 187–208 (1978a)
Kass, R.S., Tsien, R.W., Weingart, R.: Ionic basis of transient inward current induced by strophanthidin in cardiac Purkinje fibres. J. Physiol. (Lond.) *281*, 209–226 (1978b)

Katzung, B.G.: Diastolic oscillation in muscle tension and length. J. Cell. Comp. Physiol. *64*, 103–114 (1964)
Katzung, B.G., Strosberg, A.M., Jensen, R.A.: Relationship of normal and oscillatory epinephrine. Responses to excitation-contraction coupling processes in myocardium. 4th Internat. Congr. on Pharmacol., July 14 to 18, 1969 in Basel, Abstracts p. 268
Kaufmann, R., Bayer, R., Fürniss, T., Krause, H., Tritthart, H.: Calcium-movement controlling cardiac contractility. II. Analog computation of cardiac excitation-contraction coupling on the basis of calcium kinetics in a multi-compartment model. J. Mol. Cell. Cardiol. *6*, 543–559 (1974)
Kaufmann, R., Fleckenstein, A., Antoni, H., Wolf, H.: Ursachen und Auslösungsbedingungen von Myokard-Kontraktionen ohne reguläres Aktionspotential. Pflügers Arch. Ges. Physiol. *278*, 435–446 (1963)
Kaufmann, A.J.: Relaxation of heart muscle by catecholamines and by dibutyryl cyclic adenosine 3′,5′-monophosphate. Naunyn-Schmiedebergs Arch. Pharmacol. *296*, 205–215 (1977)
Kedem, J., Yarom, R., Mahler, Y., Rogel, S.: Contraction without depolarization in the in vivo heart; kinetic and electronmicroscopic studies. Cardiovasc. Res. *6*, 353–359 (1972)
Koch-Weser, J.: Influence of osmolarity of perfusate on contractility of mammalian myocardium. Am. J. Physiol. *204*, 957–962 (1963)
Koch-Weser, J.: Myocardial contraction frequency and onset of cardiac glycoside action. Circ. Res. *28*, 34–48 (1971)
Koch-Weser, J., Berlin, C.M., Jr., Blinks, J.R.: Effects of acetylstrophanthidin, levarterenol and carbachol on the interval-strength relationship of heart muscle. In: Pharmacology of Cardiac Function, 2nd Internat. Pharmacol. Meeting Prague 1963, Vol. 5, pp. 63–72. Krayer, O. (ed.). Oxford: Pergamon Press 1964
Koch-Weser, J., Blinks, J.R.: Analysis of the relation of the positive inotropic action of cardiac glycosides to the frequency of contraction of heart muscle. J. Pharmacol. Exp. Ther. *136*, 305–317 (1962)
Koch-Weser, J., Blinks, J.R.: The influence of the interval between beats on myocardial contractility. Pharmacol. Rev. *15*, 601–652 (1963)
Korth, M.: Effects of several phosphodiesterase-inhibitors on guinea-pig myocardium. Naunyn-Schmiedebergs Arch. Pharmacol. *302*, 77–86 (1978)
Korth, M., Weger, N., Reiter, M.: The positive inotropic action of sodium fluoroacetate on guinea-pig ventricular myocardium. Naunyn-Schmiedebergs Arch. Pharmacol. *303*, 7–14 (1978)
Krop, St.: The influence of "heart stimulants" on the contraction of isolated mammalian cardiac muscle. J. Pharmacol. Exp. Ther. *82*, 48–62 (1944)
Kruta, V.: Sur l'activité rythmique du muscle cardiaque. I. Variations de la réponse mécanique en fonction du rythme. Arch. Int. Physiol. *45*, 332–357 (1937)
Kruta, V., Bravený, P., Husáková, B.: Dependence of the inotropic action of strophanthin on frequency and the postextrasystolic potentiation. Arch. Kreisl. Forsch. *33*, 63–72 (1960)
Langendorff, O.: Untersuchungen am überlebenden Säugetierherzen. Arch. Ges. Physiol. *61*, 291–332 (1895)
Lederer, W.J., Tsien, R.W.: Transient inward current underlying arrhythmogenic effects of cardiotonic steroids in Purkinje fibres. J. Physiol. (Lond.) *263*, 73–100 (1976)
Lee, C.O., Kang, D.H., Sokol, J.H., Lee, K.S.: Relation between intracellular Na ion activity and positive inotropic action of sheep cardiac Purkinje fibers. Biophys. J. *25*, 198a (1979)
Lee, K.S., Klaus, W.: The subcellular basis for the mechanism of inotropic action of cardiac glycosides. Pharmacol. Rev. *23*, 193–261 (1971)
Lenzen, G.: Versuche zur Aufklärung des Wirkungsmechanismus und der Wirkungsbedingungen kontraktilitätsfördernder Maßnahmen am Froschherzen. Inaugural-Dissertation, Technische Universität München, Medizinische Fakultät 1973
Lock, J.A.: The effect of temperature and frequency on the contractile force and inotropic response to ouabain, calcium and some other drugs of the hen isolated driven auricle. Brit. J. Pharmacol. *25*, 557–565 (1965)

Magnus, R., Sowton, S.C.M.: Zur Elementarwirkung der Digitaliskörper. Naunyn-Schmiedebergs Arch. Exp. Path. Pharmak. *63*, 255–262 (1910)

Mascher, D.: Electrical and mechanical events in depolarized cardiac muscle fibers during low sodium perfusion. Pflügers Arch. Ges. Physiol. *323*, 284–296 (1971)

Moran, N.C.: Contraction-dependency of the myocardial binding and positive inotropic action of cardiac glycosides. In: New Aspects of Cardiac Glycosides. Wilbrandt, W., Lindgren, P. (eds.), Vol. 3, pp. 251–257. Oxford: Pergamon Press 1963

Moran, N.C.: Contraction dependency of the positive inotropic action of cardiac glycosides. Circ. Res. *21*, 727–740 (1967)

Mullins, L.J.: The generation of electric currents in cardiac fibers by Na/Ca exchange. Am. J. Physiol. *236*, C 103–110 (1979)

Nathan, D., Beeler, G.W. Jr.: Electrophysiologic correlates of the inotropic effects of isoproterenol in canine myocardium. J. Mol. Cell. Cardiol. *7*, 1–15 (1975)

Niedergerke, R.: The "staircase" phenomenon and the action of calcium on the heart. J. Physiol. (Lond.) *134*, 569–583 (1956)

Posner, Ch.J., Berman, D.A.: Mathematical analysis of oscillatory and non-oscillatory recovery of contractility after a rested-state contraction and its modification by calcium. Circ. Res. *25*, 725–733 (1969)

Prindle, K.H., Skelton, C.L., Epstein, S.E., Marcus, F.I.: Influence of extracellular potassium concentration on myocardial uptake and inotropic effect of tritiated digoxin. Circ. Res. *28*, 337–345 (1971)

Ravens, U.: The effects of lanthanum on electrical and mechanical events in mammalian cardiac muscle. Naunyn-Schmiedebergs Arch. Pharmacol. *288*, 133–146 (1975)

Reiter, M.: The influence of dihydro-ouabain on the contraction of the heart muscle in dependence on temperature and calcium-concentration. Biochem. Pharmacol. *8*, 37–38 (1961)

Reiter, M.: Die Entstehung von „Nachkontraktionen" im Herzmuskel unter Einwirkung von Calcium und von Digitalisglykosiden in Abhängigkeit von der Reizfrequenz. Naunyn-Schmiedebergs Arch. Exp. Path. Pharmak. *242*, 497–507 (1962)

Reiter, M.: "After-contractions" under the action of cardiac-glycosides and calcium. 1 st Internat. Pharmacol. Meeting Stockholm 1961. Wilbrandt, W. (ed.), Vol. 3, pp. 265–270. Oxford: Pergamon Press 1963 a

Reiter, M.: Die Beziehung von Calcium und Natrium zur inotropen Glykosidwirkung. Naunyn-Schmiedebergs Arch. Exp. Path. Pharmak. *245*, 487–499 (1963 b)

Reiter, M.: Die isometrische Kontraktion des Meerschweinchen-Papillarmuskels in Abhängigkeit von der Calciumkonzentration und der Temperatur. Naunyn-Schmiedebergs Arch. Exp. Path. Pharmak. *245*, 551–561 (1963 c)

Reiter, M.: Electrolytes and myocardial contractility. In: Pharmacology of Cardiac Function, 2 nd Internat. Pharmacol. Meeting Prague 1963. Krayer O. (ed.), Vol. 5, pp. 25–42. Oxford: Pergamon Press 1964

Reiter, M.: The effect of various anions on the contractility of the guinea-pig papillary muscle. Experientia *21*, 87–89 (1965)

Reiter, M.: Der Einfluß der Natriumionen auf die Beziehung zwischen Frequenz und Kraft der Kontraktion des isolierten Meerschweinchenmyokards. Naunyn Schmiedebergs Arch. Pharmak. Exp. Path. *254*, 261–286 (1966)

Reiter, M.: Die Wertbestimmung inotrop wirkender Arzneimittel am isolierten Papillarmuskel. Arzneim. Forsch. *17*, 1249–1253 (1967)

Reiter, M.: Pharmakologie der Digitaliswirkung. Verh. dtsch. Ges. Kreisl.-Forsch. „Herzdilatation und Insuffizienz" 34. Tagung, 87–99 (1968)

Reiter, M.: Cardioactive steroids with special reference to calcium. In: Calcium and Cellular Function. Cuthbert, A.W. (ed.), pp. 270–279. London: Macmillan Comp. 1970 a

Reiter, M.: Pharmakologische Grundlagen einer therapeutischen Beeinflussung der Contractilität des Herzens beim Kreislaufversagen. In: Anaesthesiologie und Wiederbelebung. Effert, R., Wiemers, A. (eds.), B. 48, „Intensivtherapie bei Kreislaufversagen", S. 19–27. Berlin, Heidelberg, New York: Springer-Verlag 1970 b

Reiter, M.: Differences in the inotropic cardiac effects of noradrenaline and dihydro-ouabain. Naunyn-Schmiedebergs Arch. Pharmacol. *275*, 243–250 (1972 a)

Reiter, M.: Drugs and heart muscle. Ann. Rev. Pharmacol. *12*, 111–124 (1972b)
Reiter, M., Schöber, H.G.: Die positiv inotrope Adrenalinwirkung auf den Meerschweinchen-Papillarmuskel bei Variation der äußeren Calcium- und Natriumkonzentration. Naunyn-Schmiedebergs Arch. Exp. Path. Pharmak. *250*, 9–20 (1965a)
Reiter, M., Schöber, H.G.: Oscillatorische Kontraktionsphänomene des Herzmuskels unter der Einwirkung von Adrenalin. Naunyn-Schmiedebergs Arch. Exp. Path. Pharmak. *250*, 21–34 (1965b)
Reiter, M., Seibel, K., Stickel, F.J.: Sodium dependence of the inotropic effect of a reduction in extracellular potassium concentration. Naunyn-Schmiedebergs Arch. Pharmak. *268*, 361–378 (1971)
Reiter, M., Stickel, F.J.: Der Einfluß der Kontraktionsfrequenz auf das Aktionspotential des Meerschweinchen-Papillarmuskels. Naunyn-Schmiedebergs Arch. Pharmak. Exp. Path. *260*, 342–365 (1968)
Reiter, M., Stickel, F.J.: Zur Frage einer therapeutischen Steigerung der Kontraktilität bei Abkühlung des Herzens. Klin. Wochenschr. *48*, 935–938 (1970)
Reiter, M., Stickel, F.J., Weber, S.: The influence of the extracellular potassium concentration on the glycoside effects upon contractile force and action potential duration of the guinea-pig papillary muscle. Experientia *22*, 665–666 (1966)
Reiter, M., Strobl, F.: Contraction frequency and concentration in the positive inotropic effect of a cardioactive steroid on ventricular muscle. J. Pharmacol. (Paris) *3*, 22 (1972)
Repke, K.: Über den biochemischen Wirkungsmodus von Digitalis. Klin. Wochenschr. *42*, 157–165 (1964)
Reuter, H.: Divalent cations as charge carriers in excitable membranes. Prog. Biophys. Mol. Biol. *26*, 1–43 (1973)
Robinson, J.D., Flashner, M.S.: The ($Na^+ + K^+$)-activated ATPase enzymatic and transport properties. Biochim. Biophys. Acta *549*, 145–176 (1979)
Ryo, U.Y.: Studies on "after contraction" in cat papillary muscle. In: Research in physiology. Kao, F., Koizumi, K., Vassalle, M. (eds.), pp. 259–273. Bologna: Aulo Gaggi Publ. 1971
Salter, W.T., Sciarini, L.J., Gemmel, J.: Inotropic synergism of cardiac glucoside with calcium acting on the frog's heart in artificial media. J. Pharmacol. Exp. Ther. *96*, 372–379 (1949)
Sanyal, P.N., Saunders, P.R.: Relationship between cardiac rate and the positive inotropic action of ouabain. J. Pharmacol. Exp. Ther. *122*, 499–503 (1958)
Schatzmann, H.J.: Herzglykoside als Hemmstoffe für den aktiven Kalium- und Natriumtransport durch die Erythrocytenmembran. Helv. Physiol. Acta *11*, 346–354 (1953)
Schatzmann, H.J.: Transmembrane calcium movements in resealed human red cells. In: Calcium and Cellular Function. Cuthbert, A.W. (ed.), pp. 85–95. London: Macmillan Comp. 1970
Scholz, H., de Yazikof, E.: The influence of the extracellular Ca, K and Na concentration on the positive inotropic action of theophylline in isolated guinea pig atria. Naunyn-Schmiedebergs Arch. Pharmak. *270*, R 128 (1971)
Schwartz, A., Lindenmayer, G.E., Allen, J.C.: The sodium-potassium adenosine triphosphatase: pharmacological, physiological and biochemical aspects. Pharmacol. Rev. *27*, 3–134 (1975)
Seibel, K., Reiter, M.: Biphasic contractures of guinea-pig cardiac ventricular muscle. Naunyn-Schmiedebergs Arch. Pharmacol. *280*, 295–314 (1973)
Seibel, K., Reiter, M.: Differentiation of inotropic mechanisms by experiments on cardiac contracture. Naunyn-Schmiedebergs Arch. Pharmacol. *286*, 65–82 (1974)
Seibel, K., Reiter, M.: Importance of excitation-induced sodium influx for the positive inotropic effect of dihydro-ouabain. Naunyn-Schmiedebergs Arch. Pharmacol. *287*, R 24 (1975)
Sonnenblick, E.H.: Active state in heart muscle. Its delayed onset and modification by inotropic agents. J. Gen. Physiol. *50*, 661–676 (1967)
Staley, N.A., Benson, E.S.: Ultrastructure of frog ventricular muscle and its relationship to mechanisms of excitation-contraction coupling. J. Cell. Biol. *38*, 99–114 (1968)

Straub, W.: Die Elementarwirkung der Digitalis-Körper. S.-B. phys.- med. Ges. Würzb., Jahrg. 1907, 85–93 (1908)
Strobl, F.E.: Untersuchung über die Abhängigkeit der positiv inotropen Glykosidwirkung von der Kontraktionsfrequenz des Herzmuskels. Inaugural-Dissertation, Technische Universität München, Medizinische Fakultät 1973
Sutko, J.L., Besch, H.R., Jr., Bailey, J.C., Zimmerman, G., Watanabe, A.M.: Direct effects of the monovalent cation ionophores monensin and nigericin on myocardium. J. Pharmacol. Exp. Ther. *203*, 685–700 (1977)
Talbot, M.S.: Sodium dependence of the rate of onset of the ouabain-induced positive inotropic effect on cardiac muscle. Nature *219*, 1053–1054 (1968)
Thomas, L.J., Jr.: An antagonism in the action of calcium and strontium ions on the frog's heart. J. Cell. Comp. Physiol. *50*, 249–264 (1957)
Tuttle, R.S., Farah, A.: The effect of ouabain on the frequency-force relation and on poststimulation potentiation in isolated atrial and ventricular muscle. J. Pharmacol. Exp. Ther. *135*, 142–150 (1962)
Ulbricht, W.: The effect of veratridine on excitable membranes of nerve and muscle. Ergebn. Physiol. *61*, 18–71 (1969)
Vassalle, M., Karis, J., Hoffman, B.F.: Toxic effects of ouabain on Purkinje fibers and ventricular muscle fibers. Am. J. Physiol. *203*, 433–439 (1962)
Verdonck, F., Busselen, P., Carmeliet, E.: Ca-action potentials and contractions of heart muscle in Na-free solutions. Influence of caffeine. Arch. Int. Physiol. Biochim. *80*, 167–169 (1972)
Vierling, W., Reiter, M.: Frequency-force relationship in guinea-pig ventricular myocardium as influenced by magnesium. Naunyn-Schmiedebergs Arch. Pharmacol. *289*, 111–125 (1975)
Vincenzi, F.F.: Influence of myocardial activity on the rate of onset of ouabain action. J. Pharmacol. Exp. Ther. *155*, 279–287 (1967)
Weingart, R., Kass, R.S., Tsien, R.W.: Roles of calcium and sodium ions in the transient inward current induced by strophanthidin in cardiac Purkinje fibers. Biophys. J. *17*, 3a (1977)
Weizsäcker, V.: Über die Abhängigkeit der Strophanthinwirkung von der Intensität der Herztätigkeit. Naunyn-Schmiedebergs Arch. Exp. Path. Pharmak. *72*, 282–294 (1913)
Weyne, J.: Effects of strontium ions on heart muscle. I. Influences on contractility. Arch. Int. Physiol. Biochim. *74*, 449–460 (1966)
Wilbrandt, W., Brawand, K., Witt, P.N.: Die quantitative Abhängigkeit der Strophanthosidwirkung auf das Froschherz von der Tätigkeit des Herzens und von der Glykosidkonzentration. Naunyn-Schmiedebergs Arch. Exp. Path. Pharmak. *219*, 397–407 (1953)
Wilbrandt, W.: Zur Frage der Beziehungen zwischen Digitalis- und Kalziumwirkungen. Wien. Med. Wochenschr. *108*, 809–814 (1958)
Wussling, M., Szymanski, G.: Einige kritische Bemerkungen zur Interpretation von Treppen- und Potentiationsphänomenen am Kaninchenherzmuskel. Acta Biol. Med. Germ. *34*, 1159–1166 (1975)
Wussling, M., Szymanski, G.: Diastolische Oscillationen am Kaninchen-Papillarmuskel. Nova Acta Leopoldina *46*, 635–646 (1977)
Ziehm, E.: Die Wirkungen von Änderungen der Reizfrequenz und der Ca^{++}-Ionenkonzentration in der Außenlösung auf die isometrische Kontraktion und das Aktionspotential am Meerschweinchenvorhof. Inaugural-Dissertation, Technische Universität München, Medizinische Fakultät 1972

CHAPTER 12

Influence of Cardiac Glycosides on Electrophysiologic Processes

R. WEINGART

A. Introduction

An understanding of the effects of cardiac glycosides has been impaired mainly by the limited knowledge of the structures and functions upon which they act. With the introduction of new approaches to investigation, such as the voltage clamp technique or ion-selective microelectrodes, considerable progress has been made over the last decade. It is the purpose of this chapter to review the advances made in this field. The scope of this article has had to be confined to certain aspects of the overall problem. Emphasis will be placed on the mechanical and electrical actions that cardiac glycosides exert on isolated but otherwise "normal" ventricular tissue. The reader is also referred to other review articles (VASSALLE, 1977; BEELER, 1977; FERRIER, 1977a; TSIEN et al., 1978a, b).

Clinically, cardiac glycosides are prescribed for two classes of heart conditions: congestive heart failure and rhythm disturbances mainly of supraventricular origin. In the first, the beneficial action manifests itself as an improvement of the contractile strength, and in the second case in a slowing down of the impulse propagation. Congestive heart failure refers to the working myocardium. Relevant studies focusing on the positive inotropic effect will be discussed in Sect. C. As for rhythm disturbances, little information is available which might provide a clue to the possible underlying mechanism or mechanisms. They will therefore not be covered in this chapter.

The therapeutic value of cardiac glycosides may be limited by the occurrence of severe side effects, e.g., rhythm and conduction disturbances. It is therefore important to understand the basic mechanisms responsible for these detrimental actions. Since such disturbances are usually first observed in the ventricle (ROSEN et al., 1973a), related studies are mainly on Purkinje fibers and ventricular muscle. The relevant work will be described, in Sect. D.

Research performed on healthy isolated preparations is bound to have an impact on the clinical setting. Caution is, however, required in applying information from isolated preparations to the complex behavior of the intact heart. Moreover, the various types of highly specialized cardiac tissue show different patterns of response to cardiac glycosides.

B. Electrophysiologic Background

There are several possible mechanisms by which cardiac glycosides are considered to exert their beneficial as well as detrimental effects on the heart. In this context, it is useful to outline briefly some of the current concepts of cardiac electrophysiology (KATZ, 1977).

I. Ionic Current Components Underlying the Cardiac Action Potential

The heart is composed of divers, highly specialized types of tissue. Action potentials from the various regions differ in their shape, duration, and voltage range. The ionic basis of the action potential of the ventricular working myocardium and of the specialized conducting tissue (Purkinje fiber) are discussed here. These have also been reviewed extensively by BEELER and REUTER (1977) and MCALLISTER et al. (1975)

In short, the transmembrane potential is maintained by ionic gradients and voltage- and time-dependent variations in ionic conductances across a selectively permeable cell membrane. The imbalance resulting from these net ionic movements is compensated by active transport mechanisms that require energy (e.g., Na^+, K^+-pump). It has been suggested that digitalis may produce its effects by producing a change in active ion transport and/or by modifying the passive ionic permeability of the cell membrane.

1. Working Myocardium

The upstroke of the action potential (Fig. 1 a) is generated by a strong and transient sodium inward current (i_{Na}), which is mediated through channels that can be selectively blocked by tetrodotoxin (TTX). This initial depolarization activates another type of channel whose permeability is preferential for Ca^{2+}. The resulting current flow has been referred to as slow inward current (i_{si}), because its time course is considerably slower than that of the initial i_{Na}. Pharmacologically, slow calcium channels can also be distinguished from fast sodium channels, in that they are blocked by Mn^{2+} and substances such as verapamil and D600, but not by TTX.

During normal activity i_{si} plays a large part in generating the action potential plateau. To a first approximation, an elevation in the plateau height indicates an increase in i_{si}. An independent index of i_{si} is provided by experiments where i_{Na} is blocked by TTX or inactivated by depolarization induced either by current flow or high extracellular potassium concentration $[K^+]_o$. Under these conditions regenerative calcium-dependent action potentials can be evoked (for references see CRANEFIELD, 1975). The rate of rise and the amplitude of such slow responses depend on the magnitude of the slow inward current and can be used as a measure of i_{si} under various experimental conditions. The voltage clamp technique provides a more direct way of studying i_{si}. This method controls the membrane potential by holding or "clamping" it at values predetermined by the experimenter. In order to keep the membrane potential at a given level, ionic current is required across the cell membrane. The necessary amount of current is provided by a feedback circuit and can be measured. Using this technique, current systems such as i_{si} can be identified and analyzed on the basis of their time- and voltage-dependent properties.

The termination of the plateau phase of the cardiac action potential could result from either the activation of a time- and voltage-dependent outward current, i_{x1}, carried by potassium ions, or the inactivation of i_{si}. It is generally thought that the decline of i_{si} is the dominant factor.

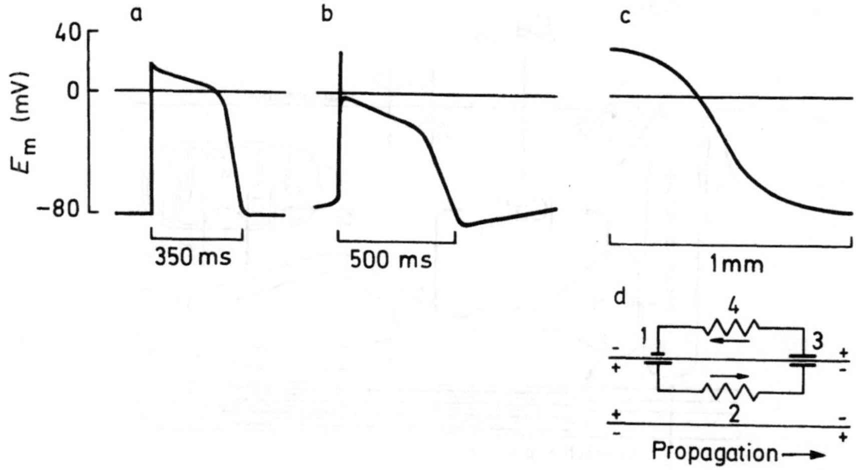

Fig. 1 a–d. Action potential configuration and impulse propagation. **a** Schematic action potential of ventricular muscle, consisting of rapid upstroke, plateau, and repolarization; **b** schematic Purkinje fiber action potential, showing rapid upstroke, early repolarization, secondary depolarization giving rise to plateau, repolarization, and pacemaker depolarization. **c** momentary picture of propagating wavefront of an action potential, revealing the transmembrane voltage distribution as a function of distance. **d** Diagram showing the elements of the corresponding local circuit current. *1* excitatory Na^+ channels, *2* longitudinal intracellular resistance, including nexuses and myoplasm, *3* membrane capacitance, and *4* extracellular resistance. The local circuit extends over several cell lengths as the mean cell length is about 100 µm

2. Specialized Conducting System

A characteristic feature of the specialized conducting tissue is its natural or latent pacemaking ability. In this point it differs from the active myocardium, which shows no tendency towards spontaneous activity under normal conditions.

Purkinje fibers and His bundle preparations (Fig. 1 b) show a rapid upstroke of the action potential, which also depends on the entry of Na^+ via i_{Na}. The initial rapid repolarization results from a decrease of i_{Na} in combination with a calcium-activated transient outward current, i_{qr}, whose ionic nature remains unsettled (SIEGELBAUM and TSIEN, 1977). The initiation of the slow inward current, i_{si}, leaves behind a notch as the membrane again slightly depolarizes. The plateau phase itself is determined by a delicate balance between relatively small currents. The termination of the plateau of pacemaker cells is brought about by an inactivation of i_{si}, combined with initiation of an outward potassium current, i_{x1}. At the end of the repolarization, the membrane potential slowly depolarizes (pacemaker potential) as a consequence of a slow decay of the potassium pacemaker current, i_{K2}, thus unmasking a background inward current which might be carried by Na^+.

II. Excitation–Contraction Coupling

The initiating step of the excitation–contraction coupling cycle is a membrane depolarization which activates the slow inward current (Fig. 2). Since i_{si} is largely carried by Ca^{2+}, it is crucial to the activation of contraction. The amount of calcium

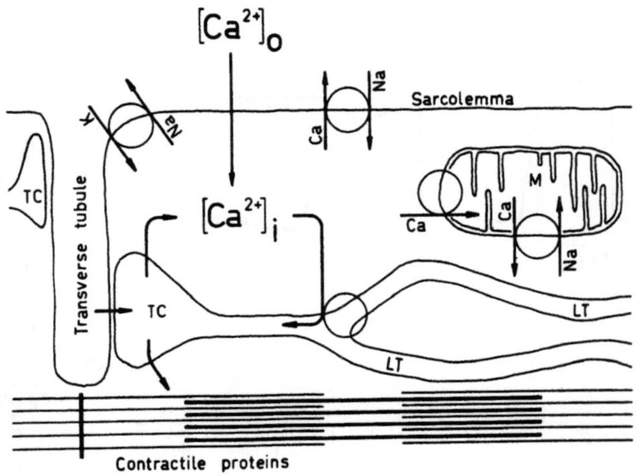

Fig. 2. Diagram of the elements involved in excitation–contraction coupling. The sarcoplasmic reticulum consists of longitudinal tubules (LT) and terminal cisternae (TC). Terminal cisternae and mitochondria (M) form the intracellular Ca^{2+} storages sites. *Circles* represent carrier mechanisms. For further details, see text

that enters the cells during an action potential, however, is about one order of magnitude too small to explain the associated contraction. This discrepancy is generally explained in terms of calcium release from intracellular pools, e.g., the sarcoplasmic reticulum, SR. Two mechanisms have been proposed. One possibility is that depolarization of the sarcolemma induces an increase in the calcium permeability of nearby subsarcolemmal terminal cisternae. The other hypothesis suggests that the calcium release is triggered by the small amount of calcium entering the cells via i_{si}. Such a calcium-induced calcium release has been described for skinned cardiac muscle preparations (FABIATO and FABIATO, 1975, 1977; ENDO and KITIZAWA, 1978). The two proposed mechanisms are not mutually exclusive.

As a result of calcium entry from the extracellular space and release from the SR, the free myoplasmic calcium concentration is elevated. The level of the free calcium determines the amount of tension developed. The free myoplasmic calcium concentration under resting condition is approximately 2×10^{-7} *M*, whereas the maximal activation of the contractile proteins occurs at approximately 10^{-5} *M* calcium. Tension development is a consequence of a complex sequence of events, permitting an interaction between actin and myosin filaments. The key steps of calcium regulation involve two proteins, tropomyosin and troponin, both located on the actin filament. Troponin is calcium sensitive, whereas tropomyosin is not. Calcium binding to troponin is said to cause removal of the tropomyosin inhibition. This allows then myosin cross bridges to attach and detach from available actin sites. The cross bridge cycles produce mechanical activity at the expense of ATP splitting.

Associated with the globular heads of the myosin molecule are the myosin light chains. They are believed to be the target for a Ca^{2+}-dependent phosphorylation via the enzyme myosin light-chain kinase. This mechanism has been suggested as a possible means of modifying contractility (PERRY, 1975).

Relaxation is initiated by reuptake and binding of the calcium by the intracellular storage sites, presumably the longitudinal tubules of the SR. The uptake process involves a Ca^{2+}-ATPase, which transports about two Ca^{2+} ions per molecule of ATP split. As a consequence, the free myoplasmic calcium level is lowered, which in turn reduces calcium binding to troponin and leads to deactivation of the actin sites.

Finally, to close the excitation–contraction coupling cycle, the amount of calcium gained during an action potential must be extruded from the cells in order to maintain balance. It has been proposed that the underlying mechanism is a Na/Ca exchange process (REUTER and SEITZ, 1968), which might be electrogenic (HORACKOVA and VASSORT, 1979; MULLINS, 1979). The sequence of processes involved in the excitation–contraction coupling provides the framework for various possibilities of modulating contractile activity. For more detailed information about excitation–contraction coupling, see CHAPMAN (1979).

III. Electrical Coupling and Impulse Spread

In the heart, individual cardiac cells do not exist in isolation, but form a multicellular continuum. Electron microscopic studies have revealed a specialized structure, the nexus, that presumably represents the morphological basis for functional cell-to-cell communication (MCNUTT and WEINSTEIN, 1973). Electrical coupling seems to be the most likely role for nexuses. Current spread from one cell to another is important for the local circuits, which are the basis for impulse propagation during a heart beat.

Local circuits (Fig. 1 d) can be regarded as loops composed of four elements: (1) ionic channels, allowing an inward excitatory current flow; (2) longitudinal intracellular resistance, providing the path for the current flow along the myoplasm and across the nexuses; (3) membrane capacitance through which current can escape across the surface membrane; (4) longitudinal extracellular resistance, responsible for the return flow of the current.

The conduction velocity is determined by all four elements. Of most physiologic interest are steps 1 and 2, since it is known that they undergo changes under certain conditions, and thereby considerably affect impulse spread. In cardiac preparations, removal of the main ion utilized to carry the inward excitatory current slows the impulse propagation and eventually abolishes the impulse. Conduction may also be impaired by raising the longitudinal intracellular resistance. Such an effect may be obtained by elevating the intracellular calcium concentration, $[Ca^{2+}]_i$ (DE MELLO, 1977). Recently, however, it has been proposed that pH_i instead of $[Ca^{2+}]_i$ might be the factor which controls the degree of cell-to-cell coupling (TURIN and WARNER, 1977).

C. The Positive Inotropic Effect

The aim of this section is to present some of the most attractive hypotheses concerning the positive inotropic effect. Modern concepts in electrophysiology will provide the framework for this discussion.

A transient increase in $[Ca^{2+}]_i$ is generally thought to be the event responsible for activating the contractile machinery (EBASHI, 1976). Modification of contractil-

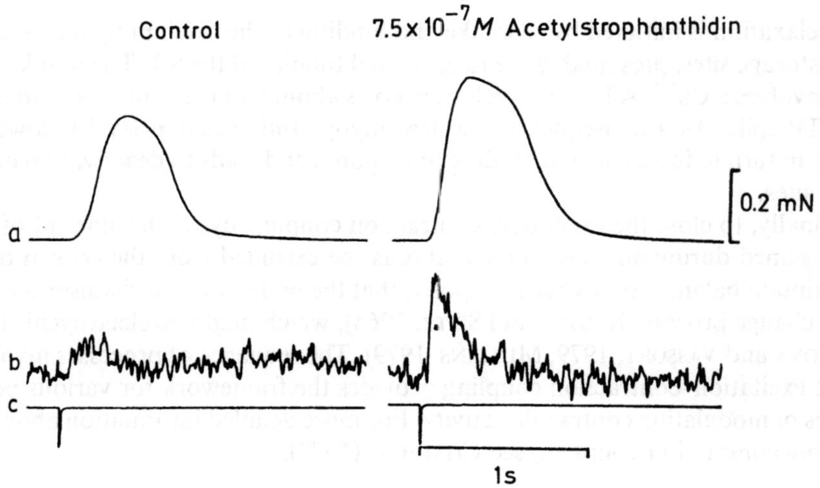

Fig. 3a–c. Influence of acetylstrophanthidin (7.5×10^{-7} *M*) on the twitch tension and the aequorin light signal. **a** isometric twitch tension; **b** aequorin signal (intensity of emitted light; 256 sweeps averaged); **c** stimulus artifact. Exposure to the drug increased both the twitch and the light signal *(right-hand side)* as compared to the controls *(left-hand side)*. Frog atrial trabecula. (ALLEN and BLINKS, 1978)

ity can occur either by a change in this Ca^{2+} transient or by an alteration of the Ca^{2+} sensitivity of the contractile proteins, or by a combination of both. In fact, both possibilities have been envisaged in order to explain the positive inotropic effect of cardiac glycosides.

Supporting evidence for an elevated $[Ca^{2+}]_i$ as the cause for the digitalis inotropism was originally provided by chemical measurements (for references see LANGER, 1977). These earlier observations were recently confirmed convincingly by ALLEN and BLINKS (1978) who performed studies with the Ca^{2+}-sensitive photoprotein aequorin as an indicator for the intracellular Ca^{2+} transient. The positive inotropic effect induced by exposure of frog atrial trabeculae to 7.5×10^{-7} *M* acetylstrophanthidin was accompanied by an increased amplitude and an accelerated rising phase of the aequorin light signal (Fig. 3).

While there is general agreement with this concept, controversy focuses on the source of the extra Ca^{2+} responsible for the tension increment. In principle, the Ca^{2+} could be brought about via transport across the sarcolemma, or via release from intracellular stores and/or binding sites. This section will deal with the possible involvement of surface membrane mechanisms for the inotropic effect (see Chaps. 17 and 18).

An alternative mechanism through which digitalis might exert its inotropic action is via a modification of the properties of the myofilaments. Such a possibility has been suggested by experiments performed with fractionated preparations from cardiac muscle (for references, see LEE and KLAUS, 1971). However, more recent work performed on mechanically or chemically skinned cardiac cells demonstrated that the Ca^{2+} sensitivity of the contractile proteins is not affected by moderate doses of cardiac glycosides (FABIATO and FABIATO, 1973; TRUBE and TRAUTWEIN, 1976; but see also Chap. 19, this volume).

I. Modification of the Na^+, K^+-Pump

Although it is widely agreed that cardiac glycosides inhibit the sarcolemmal Na^+, K^+-ATPase, there have been sporadic reports about stimulation of this enzyme under certain experimental conditions (see also Chap. 14). The leading hypothesis involves a link between the inhibition of the Na^+, K^+-pump and the Na/Ca counter transport system operating in heart muscle (REUTER and SEITZ, 1968). Accordingly, cardiotonic steroids reduce the Na^+ extrusion and allow accumulation of these ions intracellularly (BAKER et al., 1969; LANGER and SERENA, 1970). The decreased transmembrane Na^+ gradient in turn causes a decrease in Ca^{2+} efflux via Na/Ca exchange. This leads to a secondary elevation of the cellular Ca^{2+} level, thus bringing about the enhanced contractility. A modification of this classical "Na-lag" hypothesis predicts that the extra Ca^{2+} is provided by intracellular Ca^{2+} pools, the mitochondria (DRANSFELD et al., 1967; CARAFOLI and CROMPTON, 1978).

Whereas stimulation of the Na^+, K^+-pump by digitalis is expected to steepen the transmembrane gradients for Na^+ and K^+, so far no explicit model has been proposed for the inotropic effect. One possibility is that i_{si} is elevated via the larger K^+ gradient, very similar to the influence of low extracellular potassium $[K^+]_o$ (HOFFMAN et al., 1956; GOTO et al., 1977).

Another hypothesis is that the inotropic action does not bear any relation to a change in the gradients for Na^+ and K^+ at all. It has been suggested that the drug–enzyme interaction activates a labile Ca^{2+} pool located within the sarcolemma (BESCH and SCHWARTZ, 1970; LÜLLMANN and PETERS, 1979; see also Chap. 17, this volume).

1. Correlation Between Pump Inhibition and Positive Inotropy

A substantial body of enzymatic data has accumulated which supports the "Na-lag" concept. The evidence relies on a close correlation between the Na^+, K^+-ATPase activity, the binding of labeled cardiac glycosides to membrane fractions rich in Na^+, K^+-ATPase, and the in vivo inotropic response (for references, see SCHWARTZ et al., 1975; SCHWARTZ, 1976; BRODY and AKERA, 1977; AKERA and BRODY, 1978).

Chemical evidence supporting a correlation between Na^+, K^+-pump inhibition and the positive inotropic effect emerges from studies which have explored the transmembrane ionic exchange and myocardial contractility (see also Chap. 18). A close relationship between Na/K transport and inotropism was suggested by MÜLLER's (1965) experiments, involving simultaneous measurements of the cellular $^{42}K^+$ content, the electrical activity and the twitch from isolated ventricular muscle exposed to 10^{-8} *M* ouabain. He found a temporal parallelism between the loss of intracellular potassium $[K^+]_i$ and the increase of contractile strength which developed over many hours. FRY et al. (1978) have recently confirmed these findings for rabbit interventricular septum and human papillary muscle. Also, HOUGEN and SMITH (1978) confirmed these findings by measuring active $^{86}Rb^+$ uptake of muscle biopsies obtained from open chest dogs, whose ouabain plasma level was held subtoxic. Tracer studies by LANGER and SERENA (1970) indicated that therapeutic or toxic doses of cardiac glycoside produce alterations of the ionic exchange rates which seem closely associated with an inotropic response. Perfusion of rabbit

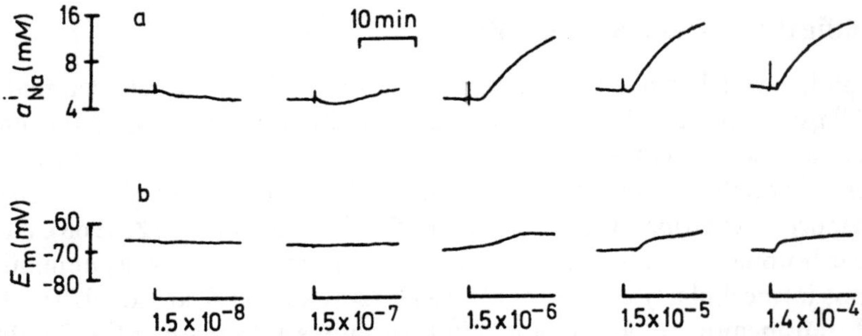

Fig. 4a, b. The effect of strophanthidin on the intracellular Na^+ activity, a^i_{Na}. The traces show the changes in a_{Na} on a logarithmic scale **a**, and the membrane resting potential, E_m **b**, produced by various drug doses. The recovery periods of the individual runs are not shown. Sheep Purkinje fiber; $[K^+]_o = 6\,mM$. (DEITMER and ELLIS, 1978b)

interventricular septae with $1–5 \times 10^{-6}\,M$ acetylstrophanthidin evoked an increment in tissue Ca^{2+}, at net loss of $[K^+]_i$, and a net gain of $[Na^+]_i$.

Direct evidence concerning possible changes in transmembrane ionic gradients comes from studies using ion-selective microelectrodes. DEITMER and ELLIS (1978a) investigated the dose–response relationship for various cardioactive steroids on the intracellular Na^+ activity, a^i_{Na}, of quiescent Purkinje fibers. Figure 4 illustrates the effects of strophanthidin. To produce a rise in a^i_{Na} a drug dose between 10^{-8} and $10^{-7}\,M$ was required, and the half-maximal effect on the rate of a^i_{Na} change was reached with $10^{-6}\,M$. Interestingly, at low drug doses a steady or transient decrease in a^i_{Na} occurred. Such biphasic effects were also observed for acetylstrophanthidin and ouabain (ELLIS, 1977), but not for dihydroouabain (see Sect. C.I.2).

In the experiment shown in Fig. 4, DEITMER and ELLIS (1978b) did not wait long enough for a^i_{Na} to reach a new steady-state value. However, they performed such studies using a supramaximal strophanthidin dose of $10^{-5}\,M$. Surprisingly enough, they found that a^i_{Na} reached a plateau of about 18 mM within 2–4 h, leaving a considerable electrochemical gradient for Na^+ to enter the cells. Assuming an a^i_{Na} coefficient of 0.75 (ELLIS, 1977), this corresponds to a $[Na^+]_i$ of 24 mM. Since the level at which a^i_{Na} eventually settles was found to depend on $[Ca^{2+}]_o$ (DEITMER and ELLIS, 1978a), the authors suggested that a^i_{Na} reaches only a relatively low level upon complete inhibition of the Na^+, K^+-pump mainly because of an exchange of internal Na^+ for external Ca^{2+} via Na/Ca counter transport.

Recently, LEE et al. (1979) have investigated a^i_{Na} and the twitch tension in Purkinje fibers under the influence of dihydroouabain. Over a dose range of 10^{-7}–$5 \times 10^{-6}\,M$ they found a linear relationship between the changes of the two parameters. This strongly suggests that the inotropic action induced by digitalis is related to an increase in a^i_{Na}.

Some investigators have also used K^+-selective microelectrodes. KUNZE (1977) has measured the level of K^+ in the extracellular subendocardial space of stimulated rabbit atria and found it increased during the exposure to ouabain, implying an inhibition of the Na^+, K^+-pump. Similar observations were made by KLINE and MORAD (1976) using acetylstrophanthidin in frog ventricle. MIURA and ROSEN (1978) reported that in Purkinje fibers the intracellular K^+ activity, a^i_K, was decreased from 130 to 112 mM within 30 min of exposure to 2×10^{-7} M ouabain. The concomitantly observed decrease in membrane potential from −84 to −79 mV, however, was too large to be caused by the observed reduction in a_K^i. The authors suggested that the excess depolarization was due to an increase of $[K^+]$ in the extracellular clefts (BAUMGARTEN and ISENBERG, 1977). Experiments performed by WALKER (1978) seem to confirm this view. He reported that the discrepancy between a_K^i and the membrane potential was more pronounced with ouabain doses larger than 10^{-7} M. However, an alternative explanation for at least part of this discrepancy may be an inhibition of an electrogenic Na^+, K^+-pump (this is discussed in the following paragraphs and in Sect. D.II.2.c).

Circumstantial evidence supporting the "Na-lag" hypothesis emerges from voltage clamp experiments. ISENBERG and TRAUTWEIN (1974) reported that in Purkinje fibers treatment with dihydroouabain or ouabain leads to a membrane depolarization and a decreased time- and voltage-independent net outward current (see also COHEN et al., 1976; WEINGART et al., 1978). This is exactly what one would expect from an inhibition of an electrogenic Na^+, K^+-pump (GLITSCH, 1979). While COHEN et al. (1976) feel that the contribution of such an electrogenic pump current is not significant in Purkinje fibers, ISENBERG and TRAUTWEIN (1974) suggest that this mechanism may be responsible for the well-known early prolongation of the action potential duration induced by digitalis (DUDEL and TRAUTWEIN, 1958; REITER, 1968). According to MCDONALD et al. (1975), however, it seems that the electrogenic pump current is much less important in ventricular muscle.

The positive findings are balanced by a number of experimental observations that are inconsistent with an inhibitory role of Na^+, K^+-ATPase in the positive inotropic action of digitalis. OKITA and co-workers (for references see OKITA, 1977) described a marked dissociation emerging during the washout of digitalis, the Na^+, K^+-ATPase inhibition persisting long after the complete loss of the inotropic response. AKERA and BRODY (1978), however, point out that so far there has been no confirmation of this data although several attempts have been made to reproduce the work.

Another line of evidence against pump inhibition focuses on the putative elevation of $[Na^+]_i$, which according to the Na-lag hypothesis is a prerequisite for an increase in $[Ca^{2+}]_i$. There is general agreement that $[Na^+]_i$ is increased and $[K^+]_i$ is decreased at advanced stages of cardiac glycoside action, when toxic effects have occurred (see Sect. D). Some disagreement, however, remains as to whether $[Na^+]_i$ is also elevated under therapeutic conditions, where low drug doses just about produce inotropic actions. GADSBY et al. (1971) claim that in frog ventricle the inotropic action occurs too rapidly to be accounted for by a change in $[Na^+]_i$. Some investigators did not observe an increase in $[Na^+]_i$ at all (for references see LEE and KLAUS, 1971). As illustrated in Fig. 4, DEITMER and ELLIS (1978 b), when using ion-selective microelectrodes, observed a *decrease* in a_{Na}^i rather than an *increase* at very low doses of cardiac glycosides (see Sect. C.I.2).

2. Correlation Between Pump Stimulation and Positive Inotropy

Recently, GHYSEL-BURTON and GODFRAIND (1975) have reported that stimulation of the Na^+, K^+-pump by low doses of ouabain is associated with a positive inotropic effect. The evidence is that in guinea pig atria 10^{-9}–10^{-8} M ouabain enhances the $^{42}K^+$ uptake (GODFRAIND and LESNE, 1972), increases $[K^+]_i$ and decreases $[Na^+]_i$ (GODFRAIND and GHYSEL-BURTON, 1977; see also LOH, 1975). Difficulties for this scheme, however, arise from the observation that an unsaturated lactone ring is a structural requirement for pump stimulation but not for the inotropy (GHYSEL-BURTON and GODFRAIND, 1976).

PETERS et al. (1974) have also found a stimulation of the Na^+, K^+-ATPase activity by low doses of ouabain. However, they point out that the time course of the ATPase stimulation is much faster than the development of tension increase. Another indication of pump stimulation comes from the work by COHEN et al. (1976). They studied the effects of ouabain on the transmembrane K^+ distribution of Purkinje fibers, revealing changes in the reversal potential, E_{rev}, of the pacemaker current, whose direction depends on the drug dose and on $[K^+]_o$. With low doses (5×10^{-8} M) and high $[K^+]_o$ (5.4–8.0 mM), E_{rev} grew more negative, whereas with larger doses (5×10^{-7} M) and low $[K^+]_o$ (2.7–4.0 mM), E_{rev} became more positive. The increased E_{rev} is interpreted as a depletion of K^+ in a restricted extracellular space, brought about by Na^+, K^+-pump stimulation, and the reduced E_{rev} as an accumulation of K^+ in this space, induced by pump stimulation. On kinetic grounds, COHEN et al. (1976) excluded the possibility that the changes in E_{rev} mainly reflect alterations in $[K^+]_i$. Along the same lines, they inferred that in the steady state the changes in electrogenic pump current must be rather small (see also ISENBERG and TRAUTWEIN, 1974).

In some experiments, COHEN et al. (1976) observed transient displacement of E_{rev}. Exposure to 10^{-7} M ouabain and 5.4 mM $[K^+]_o$ produced an initial "low dose" effect (E_{rev} more negative), which eventually gave way to a "high dose" effect (E_{rev} less negative). It is tempting to speculate that this phenomenon could account for the biphasic action of digitalis on the action potential duration and the membrane resistance (DUDEL and TRAUTWEIN, 1958; KASSEBAUM, 1963; REITER, 1968). Depletion of K^+ in the clefts could lead to increased membrane resistance and prolonged action potentials, while accumulation would do the opposite (see also Sects. C.I.1 and D.II). An alternative explanation for the data by COHEN et al. (1976), however, emerges from recent studies by DI FRANCESCO and OHBA (1978). They have shown that alterations in E_{rev} may be associated with changes in the degree of activation of i_{K2}.

The experiments performed on Purkinje fibers by COHEN et al. (1976) have been complemented by inotropic studies (BLOOD and NOBLE, 1977, 1978; BLOOD, 1975). Under conditions indicative of pump stimulation (10^{-8} M ouabain and 4.0 mM $[K^+]_o$), they found small sustained inotropic effects which displayed a rapid onset and decay. When the pump was inhibited with a large ouabain dose (10^{-6} M), a similar small dose response was seen for a few minutes, followed by a much larger inotropic response, which developed slowly and was irreversible. The authors speculate that two different mechanisms might be operating; a high dose action which seems to be related to pump inhibition, and a low dose action which might be related to pump stimulation. BLOOD and NOBLE (1978), however, pointed out that there are quantitative problems when comparing the E_{rev} data with the low dose inotropism. At low drug doses, the increase in K^+ gradient continued to develop for much longer than the few minutes required for the inotropic action to reach its peak. Moreover, very often the time course of E_{rev} was transient while the inotropic action was sustained. BLOOD and NOBLE (1978), therefore, suggested that the low dose inotropism must be due to some action of digitalis other than its effect on the Na^+ and K^+ gradients. However, they have no evidence for an alternative mechanism.

II. Modification of the Slow Inward Current

There is general agreement that Ca^{2+} ions largely contribute to the slow inward current, i_{si}, which flows during the plateau phase of the cardiac action potential (REUTER, 1973; REUTER and SCHOLZ, 1977). Consequently, it has been suggested that digitalis might exert its positive inotropic action by increasing i_{si} (KATZ, 1972; FOZZARD, 1973; see also BAILEY and SURES, 1971). An enhancement of i_{si} would promote a more complete activation of the Ca^{2+}-sensitive regulatory proteins and thus increase contractility.

This simple scheme has evoked a number of studies. According to the experimental approach adopted they can be classified into two groups. In the first one, Ca^{2+}-dependent action potentials or slow responses of partially depolarized preparations were used to assay i_{si} (see Sect. B.I.1). SCHOLZ (1970, 1975) and THYRUM (1974) exposed such preparations from mammalian ventricle or atrium to a moderately acting inotropic dose of digitoxigenin or ouabain (7×10^{-7} *M*). Under these conditions, they found no change or perhaps a small decrease of the apparent i_{si}, as judged from the rate of rise of the slow responses. JOSEPHSON and SPERELAKIS (1977) reported a decrease or blockade of isoproterenol-induced slow responses in embryonic chick heart after exposure to 10^{-8}–10^{-4} *M* ouabain, doses which were inotropic.

In the second group, i_{si} was directly measured by means of voltage clamp methods. VASSORT (1974), using ouabain on frog atrial strips, and MORAD and GREENSPAN (1973) and GREENSPAN and MORAD (1975), working with frog ventricular preparations, found no change or a slight decrease in i_{si} although the twitch tension was increased. This finding was confirmed for mammalian ventricular muscle by MCDONALD et al. (1975). In this study, a large dose of fast-acting cardiac glycosides was used (1.7×10^{-5} *M* dihydroouabain, or 5×10^{-7} *M* ouabain) to circumvent instability problems inherent to the sucrose gap method. Despite this precaution, within 2–4 min of drug application, there was no increase in i_{si} even though the tension–voltage relationship became steeper. After 10 min a decrease in i_{si} was found which was accompanied by a larger tension increase superimposed on a contracture. MCDONALD et al. (1975) inferred that the inotropic effect was due to an increase in the availability of activator Ca^{2+} rather than to an increase in i_{si}.

In contrast to these negative findings, an abstract published by DRAMANE et al. (1971) did report an increase in i_{si} by ouabain in frog atrium. Unpublished experiments by BROWN, GILES, and NOBLE (see WEINGART et al., 1978), carried out on the same tissue, seem to confirm this observation. WEINGART et al. (1978) recently reported that in sheep Purkinje fibers, cardiotonic steroids can increase i_{si} and alter its repriming kinetics under certain experimental conditions. Figure 5 illustrates an experiment, 3 min after exposure to 10^{-6} *M* strophanthidin showing a two-fold increase in twitch tension over control and an overall downward shift

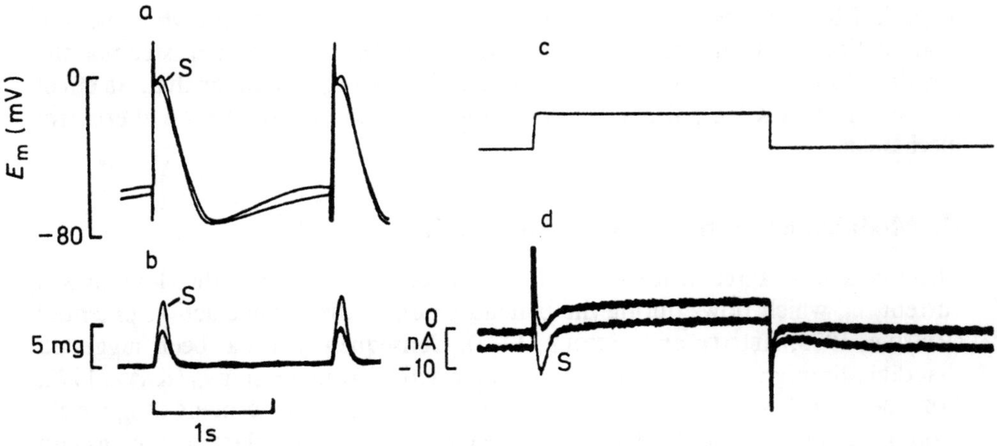

Fig. 5a–d. Comparison of the effects of strophanthidin (10^{-6} M, 3 min exposure) on membrane potential, isometric tension, and membrane current. Action potentials **a** and twitches **b** were elicited by external stimuli. Both traces show the last two responses in a train of fourteen beats. Strophanthidin (S) produced a tension increase, elevated the action potential plateau and enhanced the diastolic depolarization. The voltage clamp was imposed at the point where trace **a** ends. The membrane potential was held at − 38 mV for 2 s before the depolarizing step **c**. The depolarizing clamp pulse to −21 mV was associated with the current records **d**. Strophanthidin induced an increase of the early inward current peak, i_{si}, superimposed on an overall downward shift in membrane current. Calf Purkinje fiber. (WEINGART et al., 1978)

in net membrane current. This shift in time-independent current was assumed to reflect an inhibition of the electrogenic Na^+, K^+-pump (ISENBERG and TRAUTWEIN, 1974; see also COHEN et al., 1976). Superimposed on this net current change there was an increase in the early inward current peak (Fig. 5d), which indicates an augmentation of the time- and voltage-dependent i_{si}. Figure 5a shows an enhancement of the secondary depolarization during the plateau phase of an action potential and thus provides another indication of an increased i_{si}.

WEINGART et al. (1978) also observed that the recovery of i_{si} and the twitch force were strongly influenced by digitalis. This effect is illustrated in Fig. 6. Application of 10^{-6} M strophanthidin increased i_{si} and the twitch force at all intervals. Moreover, the time course of restitution of both parameters depended now on the interpulse interval in a nonmonotonic fashion. Both curves reached a peak at 0.55 s, fell to a valley at 1.3 s, then climbed to a steady level at longer intervals. This kinetic similarity reinforces the correlation between the positive inotropic effect and enhanced i_{si}. The oscillatory repriming of i_{si} reflects a mechanism by which the inotropic action might depend on the basic stimulation frequency (KOCH-WESER and BLINKS, 1963).

Elevated i_{si} and altered repriming were not only found during the early stages of treatment with relatively large drug doses. WEINGART et al. (1978) observed qualitatively similar effects with $2–5 \times 10^{-8}$ M strophanthidin. Because under these conditions no toxic manifestations such as transient inward current and aftercontraction were detectable (see Sect. D.II), they concluded that the actions on the magnitude and the repriming kinetics of i_{si} are genuinely therapeutic and separable

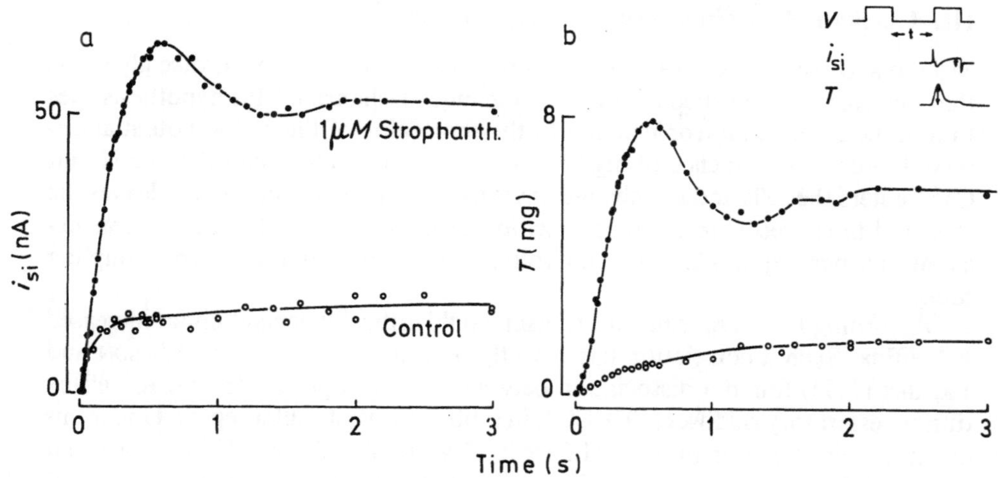

Fig. 6a, b. Time course of recovery of the slow inward current, i_{si}, and the twitch tension, T. A 500 ms depolarizing clamp pulse from −40 mV to −10 mV was followed after a variable interval by an identical test pulse *(inset)*. The maximal amplitudes of i_{si} and T associated with the test pulse were then plotted against the interval. In the control run *(open circles)* the recovery time course of i_{si} **a** and T **b** was exponential. In the test run *(solide circles)*, obtained 15 min after addition of 10^{-6} M strophanthidin, both curves showed distinct oscillations, in addition to increased steady-state values. Calf Purkinje fiber. (WEINGART, KASS and TSIEN, unpublished)

from the toxic domain. This point was further substantiated by time course studies using relatively large drug doses (WEINGART et al., 1978). These revealed a dissociation between the progressive changes in i_{si} and the twitch tension. Up to about 5 min exposure to 10^{-6} M strophanthidin, i_{si} and the twitch tension grew in parallel. Later on i_{si} declined gradually while the twitch tension increased further. This temporal separation suggests that one or more additional mechanisms must come into play during the late inotropic action, such as the "Na-lag" mechanism.

Using a similar protocol to that of WEINGART et al. (1978), MARBAN and TSIEN (1979) recently found reversible increases in i_{si} and twitch tension in ferret papillary muscle, induced by 10^{-6} M ouabain. The effects of the drug could be mimicked by veratridine, a compound known to increase the Na^+ entry into the cells. The authors suggest that an elevated a^i_{Na} might cause the modification in i_{si}, either directly, or indirectly via a secondary increase in $[Ca^{2+}]_i$. ISENBERG (1977b) has performed Ca^{2+} injection experiments with Purkinje fibers which might help to distinguish between the two possibilities. His finding was that i_{si} is enhanced upon Ca^{2+} injection, suggesting that the intracellular Ca^{2+} level sets the Ca^{2+} conductance of the surface membrane.

At present, it is not clear why some investigators have seen an increase in i_{si} while others did not. Explanations such as differences in drug, species, or tissue types seem not very likely. A possible cause of the conflicting reports might be differences in experimental design. More experimental work is needed to answer this question.

III. Calcium/Potassium Exchange Hypothesis

MORAD and GREENSPAN (1973) have proposed a site of action of cardiac glycoside that involves a carrier-mediated Ca/K exchange mechanism. This hypothesis tries to link the positive inotropic action to the shortening of the action potential observed under the influence of digitalis (see MÜLLER, 1965). The rationale is that Ca^{2+} enters the cells and thereby increases the contraction, whereas K^+ leaves the cells and thus speeds up the repolarization process. In order to operate, this exchange carrier requires a time- and voltage-dependence and a variable coupling ratio.

According to this scheme, inotropism would critically depend on an increased K^+ efflux. Such a correlation has recently been questioned. POOLE-WILSON and LANGER (1975) found a dissociation between the inotropic action and K^+ efflux during respiratory acidoses (30% CO_2) in rabbit interventricular septae. Under this condition, acetylstrophanthidin (1.25×10^{-6} *M*) inhibited Na^+, K^+-ATPase and enhanced the twitch tension, but did not elevate the K^+ efflux. Moreover, it would be difficult to explain the biphasic action of digitalis on the duration of the action potential by such a carrier mechanism.

D. Toxic Effects

At drug doses only slightly above the therapeutic level, cardiac glycosides can induce a wide range of electrical and mechanical effects, leading to rhythm or conduction disturbances, or even cardiac arrest. Heart preparations in vitro demonstrate in essence three groups of phenomena, which might explain the toxic manifestations observed in vivo under clinical conditions: altered automaticity, impaired impulse propagation, and development of contractures on which reduced twitch responses are superimposed. Until a few years ago, the interpretation of these effects favored the following sequence of events:

$$D \rightarrow \begin{array}{c}\text{Inhibition}\\ \text{of } Na^+,\, K^+\text{-pump}\end{array} \rightarrow \begin{array}{c}\uparrow[Na^+]_i\\ [K^+]_i\end{array} \rightarrow \begin{array}{c}\text{Gradual membrane}\\ \text{depolarization}\end{array} \rightarrow \begin{array}{l}\text{Spontaneous activity,}\\ \text{conduction disturb-}\\ \text{ances, impaired con-}\\ \text{tractility}\end{array}$$

The initiating step, the blockade of the Na^+, K^+-pump by digitalis, D, seems undisputed. Biochemical studies show that large doses produce an inhibition of the Na^+, K^+-pump (see Chap. 14). The scheme then assumes that cardiac glycosides exert their effects by slowly depolarizing the resting membrane potential, due to a reduction of the K^+ equilibrium potential, E_K. This view is not so much incorrect as incomplete. Recent work has made it obvious that the ultimate consequences of Na^+, K^+-pump inhibition are more complex.

Using the terms circus movement and automaticity in their broad sense, most arrhythmias can be ascribed to one of these two mechanisms or to a combination of the two. Therefore, a discussion of the electrical aspect of digitalis toxicity has been divided into two major sections, circus movements and automatic rhythms. A third section deals with the mechanical aspect of digitalis toxicity.

I. Circus Movement

The essential conditions for the development of circus movements are unidirectional block, short refractory period, and extremely slow impulse propagation. According to SINGER et al. (1967), these phenomena may be explained by localized depolarization in the His–Purkinje system. The depolarization-dependent effect on impulse conduction is controlled by the sequence and magnitude of the ionic conductances of the sarcolemma. This is dealt with first. However, impulse spread is not only determined by the ionic conductances, but also by the ease with which the local current flows from one cell to another. In other words, the passive cable properties represent yet another variable which might be involved in the impaired impulse propagation under the influence of cardiac glycosides. This aspect is discussed in the second part.

1. Ionic Conductances

The rapid rising phase of an action potential is generated by i_{Na} (see Sect. B.I.). Both, i_{Na} and the maximal rate of rise of the action potential, dV/dt_{max}, depend on the membrane potential at the onset of depolarization (WEIDMANN, 1955). Since the rate at which the cells depolarize is a major determinant of the speed of impulse propagation, the conduction velocity, θ, is expected to depend on the resting membrane potential.

Evidence along these lines is provided by experiments performed on open chest dogs where use was made of surface electrodes. After an initial slight improvement, both atrial and ventricular conduction were considerably reduced and eventually blocked by toxic doses of digitaloids (MOE and MÉNDEZ, 1951; MÉNDEZ and MÉNDEZ, 1957). In the ventricles, the above effects were shown to occur primarily in the specialized conducting system. Atrioventricular conduction was also decreased under similar conditions (MÉNDEZ and PISANTY, 1949; MOE and MÉNDEZ, 1951). This tissue seemed to be most affected and ventricular muscle least (see also SWAIN and WEIDNER, 1957). Later on, WATANABE and DREIFUS (1966) confirmed the depressed impulse propagation across the atrioventricular node on the basis of intracellular recordings.

A large number of studies carried out on isolated preparations have demonstrated that high doses of cardiac glycosides cause a fall in the membrane resting potential and dV/dt_{max} (see DUDEL and TRAUTWEIN, 1958; KASSEBAUM, 1963; RUIZ-CERETTI et al., 1977). However, only few investigations have considered the possible effects on θ. BAILEY et al. (1973) and ROSEN et al. (1973a) have shown that in isolated Purkinje fibers the fall in membrane resting potential and dV/dt_{max} are associated with a depression of impulse conduction. The latter group claims that the onset of impaired conduction in the isolated tissue correlates well with the onset of ventricular arrhythmias inferred from the electrocardiogram of an intact heart. PEON et al. (1978) recently reported that during an early stage of digitalis poisoning, Purkinje fibers showed an increase of θ accompanied by oscillatory afterpotentials (see Sect. D.II.1). They agree with MOE and MÉNDEZ (1951) that more advanced toxicity reduces θ and eventually leads to conduction block (see also SAUNDERS et al., 1973).

In severely depolarized preparations, under conditions that almost entirely inactivate the fast Na^+ channels, Ca^{2+}-dependent action potentials can be observed (see Sect. B.I). In Purkinje fibers, such slow responses still conduct in an all-or-nothing manner, although their θ is very much reduced (CRANEFIELD et al., 1972). One could imagine that this mechanism plays a role in heavy digitalis poisoning, however, no experimental information is available.

The observed reduction of θ compatible with the idea that digitalis inhibits the Na^+, K^+-pump and thus eventually decreases the resting membrane potential. The initial increase in θ could either be due to less difference between membrane resting and threshold potential, or to changes in the passive electrical properties of the cardiac cell membrane (see Sect. D.I.2).

2. Passive Electrical Properties

The requirement of local circuit currents for impulse conduction (see Sect. B.III) presupposes that digitalis influences the passive electrical properties of cardiac tissue.

An attempt in this direction was made by WEINGART (1977), who assessed the contribution of various parameters affecting θ. After superfusion of ventricular muscle preparations with 2×10^{-6} *M* ouabain for 90 min, θ was found to be decreased by 40%. At the same time, the intracellular longitudinal resistance, r_i, reflecting electrical cell-to-cell coupling, was increased by 150%. About 60% of the decrease of θ can be accounted for by the observed increase in r_i. The remaining decrease of θ can be explained by the changes to dV/dt_{max}, the amplitude of the action potential and the membrane impedance seen by the propagating impulse. Figure 7 illustrates the effect of r_i on θ.

The action of ouabain on r_i revealed a biphasic response. An initial small drop in r_i was followed by a larger delayed increase in r_i, both effects being dose dependent. The delayed increase in r_i was associated with an increase of diastolic tension (see Sect. D.III), suggesting that the $[Ca^{2+}]_i$ might be involved in the control of the nexal conductance. Since the contracture slightly preceded the decoupling, it was concluded that the threshold $[Ca^{2+}]_i$ for electrical decoupling between cells must be somewhat larger than the threshold level for tension activation (see Sect. B.II). The delayed increase in r_i may be explained as an inhibition of the Na^+, K^+-pump, which leads to an elevation of $[Ca^{2+}]_i$ secondary to $[Na^+]_i$ accumulation. Two possible explanations are offered for the early drop in r_i: either a stimulation of the Na^+, K^+-pump (see Sect. C.I), or a complex relationship between r_i and $[Ca^{2+}]_i$. The transient decrease in r_i would be consistent with the early improvement of θ observed by some investigators (see Sect. D.I.1).

II. Alterations in Automaticity

Arrhythmias caused by alterations in automaticity can be due to changes in the rate of normal or latent pacemaking tissue or induced automaticity in normally quiescent regions of the heart. The basic question therefore is: what areas of the heart in situ might possibly be affected first by digitalis and thus give rise to arrhythmias?

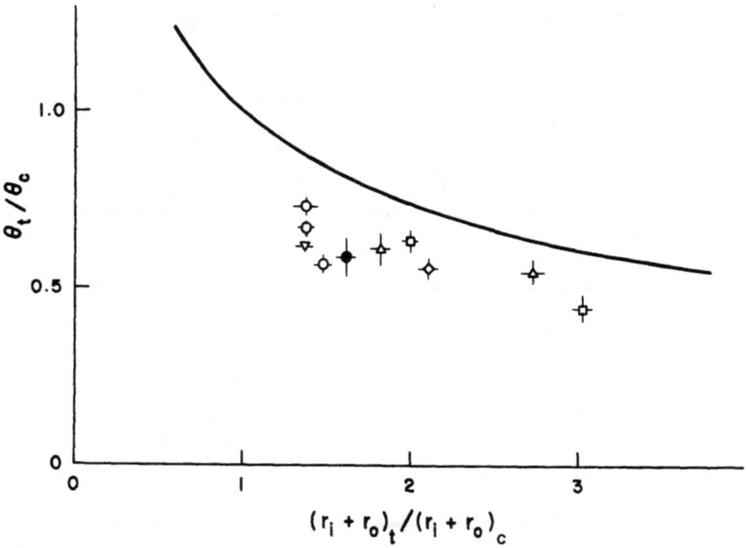

Fig. 7. Influence of the inside longitudinal resistance, r_i, on the conduction velocity, θ. θ and (r_i+r_o), the sum of the inside and outside longitudinal resistance, were measured before and after exposure for 90 min to 2×10^{-6} *M* ouabain. The drug-induced test values, θ_t and $(r_i+r_o)_t$, were normalized with respect to their own control values, θ_c and $(r_i+r_o)_c$, and then plotted against each other. Each symbol represents determinations from a different preparation. The smooth curve illustrates the relationship $\theta \propto (r_i+r_o)^{-1/2}$. The experimental points all lie below that curve, suggesting that the decrease in θ cannot be explained solely on the basis of the observed decrease in (r_i+r_o). Calf ventricular muscle. (WEINGART, 1977)

In isolated tissue, it has been known for many years that cardiac glycosides promote increased automaticity in the specialized conducting system of the ventricle, but not in the working myocardium (see VASSALLE et al., 1962). ROSEN et al. (1973a) were able to make a direct comparison between the toxic effects in vivo and in vitro. Isolated dog Purkinje fibers were superfused with the blood of a donor dog that had been gradually poisoned with ouabain. As a first sign of toxicity, ventricular arrhythmias were observed on an electrocardiogram of the whole heart. The appearance of these disturbances correlated well with the development of increased automaticity of the isolated Purkinje preparations. Using a Purkinje strand attached to a segment of papillary muscle, FERRIER et al. (1973) observed that digitalis enhanced the pacemaker activity of the specialized conducting tissue while the muscle still displayed no spontaneous activity.

These and other results strengthen the idea that digitalis-induced arrhythmias may originate in the specialized conducting system of the ventricle. It is for this reason that Purkinje fibers have been used to study the arrhythmogenic action of cardiac glycosides.

1. Transient Depolarization

Since the 1970s several groups have reported essentially similar observations on the arrhythmogenic activity of cardiotonic steroids (LOWN et al., 1967; DAVIS, 1973; FERRIER et al., 1973; HOGAN et al., 1973; ROSEN et al., 1973b). Figure 8 illus-

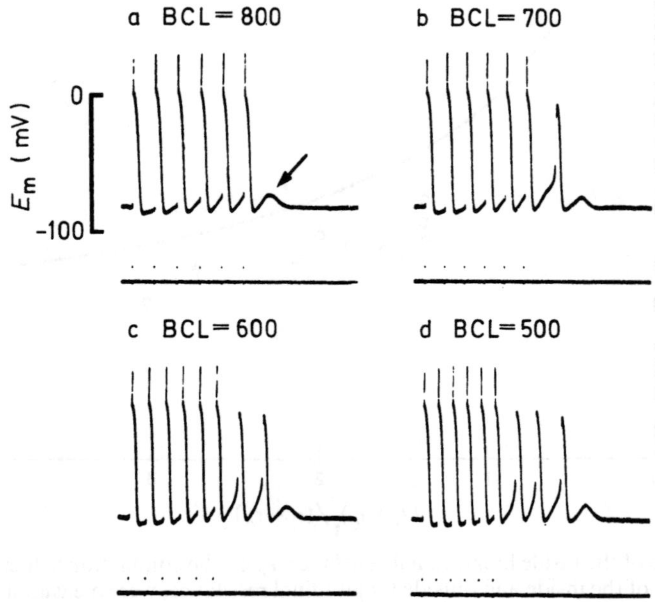

Fig. 8 a–d. Transient depolarizations and extrasystoles. The top traces show records of the transmembrane potential, and the bottom traces show the pattern of external stimuli as applied following a long period of rest. Stimulus trains consisted of six shocks applied with variable basic cycle length, BCL. The arrow in **a** indicates a transient depolarization, TD. Reducing the BCL from 800 to 500 ms **a–d** enhanced the TD to the point of reaching the excitatory threshold for an extrasystolic impulse. Dog Purkinje fiber; 2×10^{-7} *M* acetylstrophanthidin. (FERRIER et al., 1973)

trates the altered electrical activity as recorded from an isolated Purkinje fiber by FERRIER et al. (1973). Exposure to 2×10^{-7} *M* acetylstrophanthidin enhanced pacemaker activity. The diastolic depolarization steepens with every other action potential (Fig. 8 a), a behavior that is not usually observed in pacemaking tissue. A depolarizing afterpotential was found to be the basic event underlying this enhanced pacemaker activity. Following the terminology used by FERRIER et al. (1973), it will be referred to as *transient depolarization*, or TD. Figure 8 a shows such a TD *(arrow)* following a train of six evoked action potentials. Even though this TD falls short of the excitatory threshold, it can influence a propagating impulse via inactivation of i_{Na} (SAUNDERS et al., 1973; see also Sect. D.I.1). Increasing the stimulation frequency (Fig. 8 b) enhances the pacemaker depolarization even further. Concomitantly the TD is increased to the point where it reaches the threshold and thus induces an ectopic beat. Further increases in train frequency (Fig. 8 c, d) give rise to multiple repetitive responses which often last for several minutes (not shown). Such a mechanism, therefore, would not only account for single ectopic beats but also for more complex arrhythmias in vivo (CRANEFIELD, 1977; FERRIER, 1977 a).

The TD depends strongly on Ca^{2+}. Its size varies directly with $[Ca^{2+}]_o$ (FERRIER and MOE, (1973), and it can be abolished by inhibitors of i_{si} such as Mn^{2+} (FERRIER and MOE, 1973), Mg^{2+} (FERRIER and SAUNDERS, 1972) or verapamil

(ROSEN et al., 1974). Recently, FERRIER (1976) has shown that TDs are accompanied by aftercontractions which show a similar dependence on previous activity as the TD. All these findings emphasize the importance that Ca^{2+} plays in the genesis of the arrhythmogenic activity induced by digitalis compounds.

There is evidence that TDs and aftercontractions are more general phenomena, not only restricted to Purkinje fibers. Under appropriate conditions, such signals have also been observed in atrial (KAUFMANN et al., 1963; HASHIMOTO and MOE, 1973; HORDOF et al., 1978) and ventricular muscle (REITER, 1962, 1963; KAUFMANN et al., 1963; FERRIER, 1976, 1977b; TSIEN et al., 1979).

2. Ionic Basis of the Altered Automaticity

In order to understand fully the basis of the enhanced diastolic depolarization seen after the application of cardiac glycosides, it is important to ascertain the action of these drugs on the membrane currents. Recent voltage clamp experiments suggest different mechanisms involving either modifications of already existing current systems, or induction of novel systems.

a) Modification of the Pacemaker Current

In Purkinje fibers under normal conditions, the pacemaker depolarization is caused by the slow decay of a K^+ outward current, i_{K_2}, against a steady background inward leak of Na^+ (see Sect. B.I.2). It is tempting, therefore, to propose that digitalis arrhythmias may be related to modifications in i_{K_2}.

Earlier voltage clamp experiments provide some evidence for an altered pacemaker current. ARONSON et al. (1973) reported that ouabain (2×10^{-7} *M*) reduces the magnitude of i_{K_2} without altering the voltage dependence of its activation (see also GELLES and ARONSON, 1973). They offer two reasons for the decreased i_{K_2}, both being secondary consequences of an inhibition of the Na^+, K^+-pump. First, a decrease in $[K^+]_i$ leads to a reduced driving force for i_{K_2}, and second, accumulation of intracellular Na^+ blocks the i_{K_2} channels and thus reduces their current carrying capacity (GELLES et al., 1975; ARONSON and GELLES, 1977). VASSALLE and MUSSO (1974, 1976) also argue that a decrease in i_{K_2} per se underlies the early arrhythmogenic effects of cardiotonic steroids.

COHEN et al. (1976) found little or no effect of ouabain on the kinetics of i_{K_2}. Like ARONSON and co-workers, they observed changes in the amplitude of i_{K_2}, which they attribute solely to alterations in the reversal potential, E_{rev}, for this current. According to these authors, the changes in E_{rev} mainly reflect an alteration of $[K^+]$ in a restricted extracellular space, caused by an interaction of ouabain with the Na^+, K^+-pump (see Sect. C.I.2).

LEDERER (1976) and LEDERER and TSIEN (1976) could not find significant alterations of i_{K_2} that would have explained the strophanthidin-dependent steepening of the diastolic depolarization. They demonstrated that the time-dependent net membrane current, after drug administration, can be dissected into a more or less unaffected pacemaker current and a novel ionic current, which seems to be the basis for the arrhythmogenic activity (see Sect. D.II.2.d). The authors, however, could not exclude a reduction in i_{K_2} later during the development of the toxicity.

b) Involvement of Plateau Currents

MÜLLER (1963) has proposed that not only the positive inotropy of digitalis, but also its arrhythmogenic activity might be caused by the loss of cellular K^+ via Na^+, K^+-pump inhibition. A gradual loss of K^+ combined with its accumulation in the extracellular clefts will shift the K^+ equilibrium potential, E_K, to less negative values, and thus reduce the membrane potential, E_m. In this way, E_m might eventually reach a range where low voltage oscillations take place (HAUSWIRTH et al., 1969). It is thought that this phenomenon is based on the interaction between the switching off of the plateau current, i_x, and the switching on of the slow inward current, i_{si} (ARONSON and CRANEFIELD, 1974; KASS and TSIEN, 1975).

VASSALLE and MUSSO (1974, 1976) suggest that this mechanism comes gradually into play as the digitalis poisoning progresses and the frequency of spontaneous discharge increases. They noticed that some of their records resembled the repetitive activity seen in partially depolarized Purkinje fibers. GELLES et al. (1975) studied the effect of E_m on digitalis-induced arrhythmias and proposed the existence of two distinct mechanisms. The first, operating at values more negative than -70 mV, is based on a modification of i_{K_2} (see Sect. D.II.2.a) and gives rise to damped oscillatory afterpotentials. The second, functioning at E_m values less negative than -65 mV, invokes the plateau currents and is responsible for the undamped oscillatory activity.

LEDERER and TSIEN (1976) have presented experimental evidence which suggest that in Purkinje fibers cardiac glycosides may promote low voltage oscillations not only by producing partial membrane depolarizations but also by evoking a transient inward current (see Sect. D.II.2.d). No information is available as to how digitalis changes the plateau currents per se under these conditions. These findings suggest that low voltage oscillations may play a role at later stages of cardiac glycoside poisoning. However, they are not usually observed in isolated Purkinje fibers during the early exposure to the drug, nor do they correspond to the kind of rhythm disturbances seen first in patients with digitalis poisoning.

c) Electrogenic Sodium Transport Hypothesis

An alternative possibility to explain the arrhythmogenic drug action is based on the assumption that the Na^+, K^+-pump acts as an electrogenic transport system. If such a system exists (GLITSCH, 1979), then blocking of the pump with cardiac glycosides would be expected to reduce the outward pump current. A reduction in outward current is equivalent to increasing a depolarizing inward current. Since application of small amounts of depolarizing current is known to favor automaticity (TRAUTWEIN and KASSEBAUM, 1961), blocking of an electrogenic pump should have the same effect.

Voltage clamp experiments by ISENBERG and TRAUTWEIN (1974) have revealed a displacement of the steady-state membrane current in the inward direction at all potentials. This shift showed up within 1 min of exposure to dihydroouabain (10^{-5} *M*) or ouabain (10^{-6} *M*). They inferred that electrogenic Na^+ transport must contribute considerably to the membrane potential in Purkinje fibers. COHEN et al. (1976), however, argue that two different factors may be responsible for displacement of the current–voltage relationship: a reduction of the electrogenic

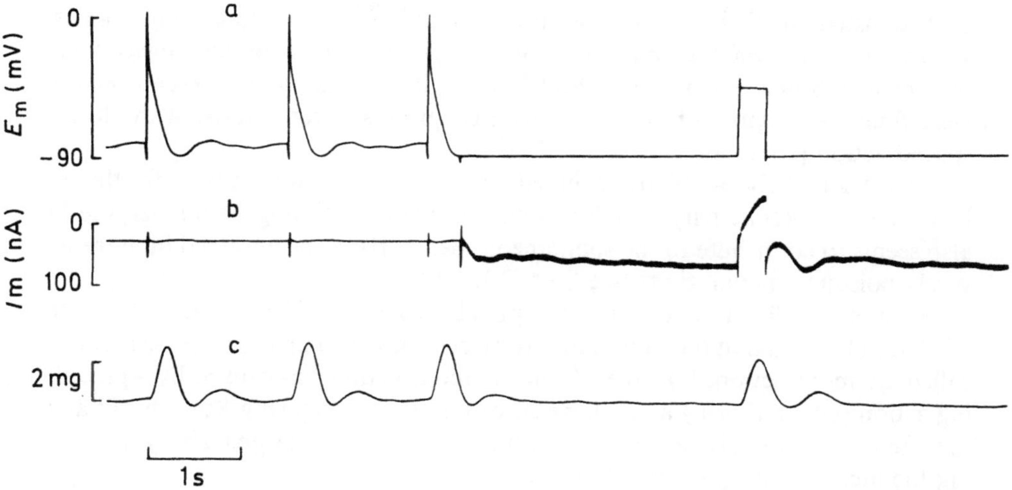

Fig. 9 a–c. Toxic effects on membrane current and contractile activity. **a** membrane potential; **b** membrane current; **c** force. Trace **a** displays the last three action potentials from a train of eleven responses, evoked by external shocks. The action potentials were followed by transient depolarizations, TDs. Trace **c** shows twitches accompanying the action potentials and aftercontractions in association with the TDs. Following the last action potential, the membrane was clamped at the maximum diastolic potential (−88 mV). This revealed a transient inward current, TI **b,** as well as an aftercontraction **c.**. The right-hand side of trace **a** shows a depolarizing clamp pulse from −88 to −44 mV for 300 ms. The depolarizing step induced an inward current surge, i_{si} **b** and a twitch **c,** and the repolarizing step evoked a TI **b** and an aftercontraction **c.** Calf Purkinje fiber; 10^{-6} *M* strophanthidin. (KASS et al., 1978 a)

pump current *and* a less negative reversal potential for the pacemaker current, i_{K_2} (see Sect. C.I.2). The relative significance of the two factors may be difficult to determine.

In general, attempts to explain digitalis-induced arrhythmias on the basis of modified conventional current systems have failed. Each of the hypotheses described above can explain at least one aspect of the observed electrical activity. However, none of them has succeeded in fully explaining the observed increase in diastolic depolarization, especially its unusual dependence on the preceding activity.

d) Transient Inward Current Hypothesis

FERRIER and MOE (1973) have suggested that digitalis-like compounds might induce a current flow, probably carried by Ca^{2+}, which does not participate in normal pacemaker activity. To test the idea, LEDERER and TSIEN (1976) investigated the action of 10^{-6} *M* strophanthidin on the membrane currents of Purkinje fibers under voltage clamp conditions (see also LEDERER, 1976; HIRAOKA, 1977). It was found that the TD is generated by an inward surge of current that follows the repolarization of an action potential or the break of a strong depolarizing voltage clamp pulse (Fig. 9 a, b). This current signal was referred to as *transient inward current* (TI). The TI could be dissociated from the pacemaker current, i_{K_2}, by means of its unusual kinetic properties and its sensitivity towards pharmacologic agents, such as Mn^{2+} or D600. Like the TD, the TI depends strongly on the previous history of membrane potential (see Sect. D.II.1). Its voltage- and time-dependence were found appropriate to explain why TDs grow so dramatically with closely spaced action potentials.

LEDERER and TSIEN (1976) concluded that the TI represents the basis for the enhanced pacemaker activity seen during early digitalis poisoning. Moreover, the TI also seems to contribute to the spontaneous depolarizations observed in more severely poisoned preparations (see Sect. D.II.2.b).

KASS et al. (1978a) investigated the possible role of Ca^{2+} in the genesis of the TI. They observed that the TI was closely associated with a phasic increase in force, called aftercontraction. Like the TI, the aftercontraction was evoked by a preceding action potential or by a repolarizing voltage clamp step (Fig. 9c). The TI and the aftercontraction displayed similar waveforms, the current signal always preceding the mechanical signal by 50–100 ms.

Kinetic and pharmacologic studies revealed that the two transients must be correlated. The size and the time course of both TI and aftercontraction, were always changed in the same direction, whatever the experimental intervention was. Their amplitudes varied directly with the degree of digitalis poisoning, $[Ca^{2+}]_o$, the size and the duration of the depolarizing clamp step, and they were both blocked by Mn^{2+} and D600.

The close relationship between the TI and the aftercontraction can be explained in different ways:

Transient increase in Ca^{2+} permeability → Phasic Ca^{2+} influx (TI) → Mechanical activation (aftercontraction) (1)

Phasic Ca^{2+} release from intracellular store ↗ Transient inward current carried by some unspecified ions; ↘ Mechanical activation (aftercontraction) (2)

The TI might reflect a phasic influx of Ca^{2+} (FERRIER and MOE, 1973); the aftercontraction therefore would be a direct consequence of increased myoplasmic Ca^{2+}. Alternatively, TI and aftercontraction might both be secondary to a primary event such as a phasic release of Ca^{2+} from the SR (REITER, 1962). In this case, the TI could be carried by ions other than Ca^{2+}.

Figure 10 illustrates an experiment that helps distinguish between the two hypotheses. It shows that an aftercontraction can be recorded even when the TI is suppressed by varying the electrical driving force, thus ruling out hypothesis (1). In the range of return potentials between −20 mV and +20 mV, the current signal (Fig. 10b) and force signal (Fig. 10c) were influenced in rather different ways. As the level was displaced towards positive potentials, the aftercontraction grew somewhat larger and was followed by increasing amounts of tonic force. The associated current records showed a different kind of voltage dependence. The transient current was inward at −17 mV, less inward at −8 mV, and nearly flat at −1 mV. An outward transient was recorded at +9 mV and it became larger at +17 mV.

From the voltage-dependent inversion of the TI, a reversal potential, E_{rev}, for this current of about −5 mV was determined. This finding argues against the Ca^{2+}

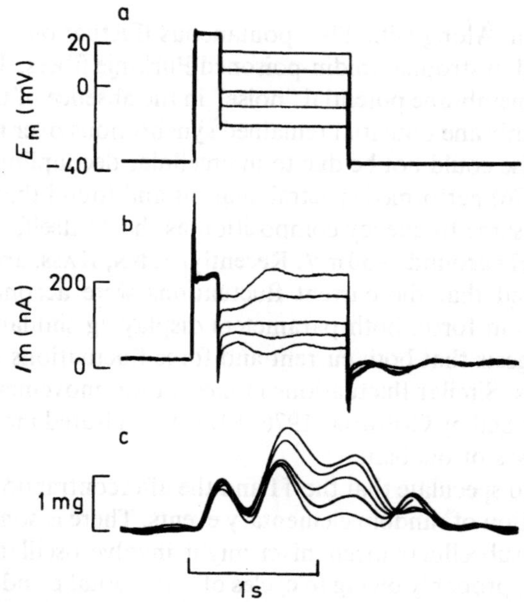

Fig. 10 a–c. Voltage-dependent reversal of the transient inward current. **a** superimposed voltage traces from five trials during a single run. The potential was stepped from a holding potential of −37 mV to +25 mV for 200 ms in order to induce a transient inward current and an aftercontraction. The membrane was then repolarized to −17, −8, −1, +9, and +17 mV; **b** associated records of membrane current (same vertical order as voltage traces); **c** associated contractile force (same vertical order as voltage traces). Calf Purkinje fiber; 10^{-6} *M* strophanthidin. (KASS et al., 1978 a)

influx hypothesis. It is consistent with the idea that $[Ca^{2+}]_i$ regulates the membrane permeability for some unspecified ions as suggested by hypothesis (2).

Obviously, the value of E_{rev} for the TI does not correspond to the Nernst potential for any of the major ions. Therefore, one must consider the possibility of a rather nonselective pathway for the TI. KASS et al. (1978 b) explored the ionic composition of the bathing fluid on E_{rev}. Replacement of external Na^+ with choline or Tris, or substitution of NaCl by sucrose, displaced E_{rev} by −35 mV maximally. On the other hand, E_{rev} was not affected by replacing extracellular Cl^- with methylsulfate. $[Ca^{2+}]_o$ and $[K^+]_o$ could only be varied over a relatively narrow range in order to prevent unacceptable variations in the degree of toxicity. No significant effects on E_{rev} were observed upon varying $[Ca^{2+}]_o$ from 2.7 to 16.2 m*M* and $[K^+]_o$ from 1 to 8 m*M*.

The key question as to the ionic pathway carrying the TI remains open. KASS et al. (1978 b) propose two alternative possibilities. It is conceivable that the TI is carried by leak or background channels whose conductance is modified by $[Ca^{2+}]_i$. Another explanation is that the TI reflects the action of an electrogenic Na/Ca exchange (HORACKOVA and VASSORT, 1979; MULLINS, 1979).

3. Current and Tension Fluctuations

LEDERER and TSIEN (1976) and LEDERER (1976) have reported another interesting aspect of digitalis poisoning which might throw some light on a fundamental subcellular mechanism. Along with TIs, spontaneous fluctuations in membrane current were observed in strophanthidin-poisoned Purkinje fibers. These fluctuations could give rise to membrane potential "noise" in the absence of the voltage clamp. Since the transmembrane potential remained synchronous over many cell lengths, the membrane noise could not be due to intercellular decoupling (see Sect. D.I.2).

KASS et al. (1976) performed spectral analysis and found that the current fluctuations have the same frequency composition as the TI itself, i.e., 0.5–1 Hz near −70 mV and 1–2 Hz around −30 mV. Recently, TSIEN, KASS, and WEINGART (unpublished) observed that the current fluctuations were accompanied by spontaneous variations in force, both parameters displaying similar periodic characteristics. This suggests that both current and force fluctuations reflect a common underlying process. Similar fluctuations in mechanical movement and membrane voltage were described by GOSHIMA (1976, 1977) for cultured mouse heart cells exposed to large doses of ouabain.

It is tempting to speculate that the TI and the aftercontraction represent a synchronized summation of random elementary events. There is some evidence which suggests that the subcellular mechanism might involve oscillatory variations in myoplasmic Ca^{2+}, probably owing to cycles of Ca^{2+} uptake and release by the SR (for references see TSIEN et al., 1979).

III. Influence on the Resting Tension

The actions that cardiac glycosides exert on the contractile activity of heart tissue are manifold. Some aspects related to this topic have already been discussed in this chapter. This section serves as an overview with special emphasis on the toxic manifestations.

Figure 11b, illustrates the various contractile phenomena as recorded from a ventricular muscle preparation exposed to $2 \times 10^{-6}\,M$ ouabain. WEINGART et al. (1978) suggested that after the initial increase in i_{si} another mechanism is responsible for a further increase in twitch tension which might rely on accumulation of $[Ca^{2+}]_i$ via the "Na-lag" mechanism (see Sect. C.II). An explanation along these lines has originally been proposed by LANGER and SERENA (1970) (see Sect. C.I). A manifestation of this effect might be the steepening of the tension–voltage relationship observed by GREENSPAN and MORAD (1975) for frog ventricular muscle and by MCDONALD et al. (1975) for cat papillary muscle. Figure 11 also shows that the positive inotropic action was of a transient nature. This might be explained by the progressive shortening of the action potential (DUDEL and TRAUTWEIN, 1958; KASSEBAUM, 1963; REITER, 1963) owing to an increase of the K^+ conductance via elevated $[Ca^{2+}]_i$. Such a mechanism has been proposed by ISENBERG (1977a, c, d) for Purkinje fibers. So far, however, this mechanism remains speculative for ventricular muscle (BASSINGTHWAIGHTE et al., 1976).

The positive inotropic effect seemed to give way to an increase in resting tension which was barely reversible. LANGER and SERENA (1970) reported that this effect was correlated with a considerable net uptake of Na^+ and Ca^{2+} and a net loss of K^+. GREENSPAN and MORAD (1975) on the other hand, found that under such conditions the voltage–tension relation was shifted in the negative direction. It is conceiveable, therefore, that digitalis-induced contractures might be caused by several

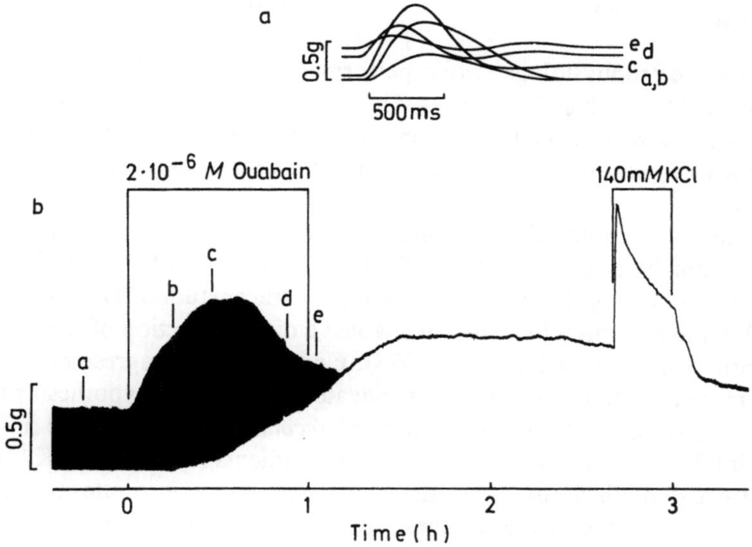

Fig. 11a, b. Effects on the contractile activity. **a** continuous recording of the resting and twitch tension. The horizontal bars indicate the duration of the exposure to the cardiac glycoside (60 min) and the K^+ contracture solution (20 min); **b** selective tension tracings, graphically superimposed: *a* control, *b* inotropic action, before the appearance of any toxic sign, *c* peak inotropic effect, associated with toxic signs, i.e., elevated resting tension and aftercontraction, *d, e* reduced twitch amplitude followed by an aftercontraction, superimposed on a further increased resting tension. Calf ventricular muscle; 2×10^{-6} *M* ouabain. (WEINGART, unpublished)

factors, such as intracellular Ca^{2+} overload via impaired Na/Ca exchange, depolarization-induced Ca^{2+} release from the SR (ENDO, 1977) and shift of the mechanical threshold.

Figure 11a shows tracings of individual twitch records from the run illustrated in Fig. 11b. The observations were that both the rate of tension development and the twitch amplitude are increased initially and decreased later on, whereas the time to peak tension is shortened throughout (see DUDEL and TRAUTWEIN, 1958; KOCH-WESER and BLINKS, 1962). As the resting tension starts to increase, small aftercontractions become apparent (see Sect. D.II.2.d). Both phenomena reflect cellular Ca^{2+} overload. Similar observations have been made by REITER (1963).

The right-hand side of Fig. 11b shows a KCl contracture which was enormously enhanced in the presence of ouabain. GREENSPAN and MORAD (1975) reported that digitalis not only enhanced such contractures in frog ventricle, but also slowed down its relaxation. BREMNER et al. (1977) confirmed these findings using mammalian ventricular muscle. In the presence of 10^{-6} *M* ouabain, they found that the Na^+-free contractures depended on $[Ca^{2+}]_o$, but not on internal Na^+ accumulation. Both groups hypothesize that, besides inhibiting the Na^+, K^+-pump, digitalis might exert an action on the intracellular Ca^{2+} sequestration system. NAYLER et al. (1978) however, could not confirm this idea. They found a normal Ca^{2+} accumulating activity of the SR at a time when Ca^{2+} and the resting tension were elevated. In contrast, they linked the elevated $[Ca^{2+}]_i$ to a concomitantly observed depletion of high energy phosphates.

E. Conclusions

The relevant electrophysiologic work performed during the 1960s was largely dominated by the "Na-lag" hypothesis. This concept proved to be very successful in that it evoked a vast number of studies. However, over the last few years, an increasing body of data has accumulated, which calls for a revision of the classical hypothesis.

It is generally accepted that the inotropic action is caused by an increase in $[Ca^{2+}]_i$, and not by a modification of the Ca^{2+} sensitivity of the myofilaments. Earlier work supporting this idea came from Ca^{2+} tracer studies. The introduction of the bioluminescent protein aequorin enabled a direct detection of the enhanced Ca^{2+} transient (ALLEN and BLINKS, 1978). Considerable disagreement remains about the mechanism by which $[Ca^{2+}]_i$ is elevated. The Na-lag hypothesis predicts that the available cellular Ca^{2+} is augmented *secondary* to an increase in $[Na^+]_i$ via pump inhibition. There is evidence that the gradients for Na^+ and K^+ are reduced by toxic digitalis doses. No doubt, this mechanism contributes to the positive inotropic effect observed under these conditions. However, some experiments suggest that the Na-lag mechanism may not operate in the therapeutic dose range. In this domain, a steepening of both the Na^+ and the K^+ gradient has been reported, consistent with the idea of a stimulation of the Na^+, K^+-pump. These results are in agreement with the observation of an increased Na^+, K^+-ATPase activity. Obviously, such findings raise serious difficulties for any hypothesis which depends on a primary elevation of $[Na^+]_i$.

An attempt to correlate pump stimulation with the low dose inotropy has failed (BLOOD and NOBLE, 1978). It turned out that the inotropic response develops with a time course which differs considerably from those displayed by the Na^+, K^+-ATPase activity (PETERS et al., 1974) and the K^+ gradient (COHEN et al., 1976). Furthermore, it is not easy to grasp the concept of a Na^+, K^+-pump whose transporting capability undergoes directional changes, depending on the concentration of an interfering drug. Does it reflect the existence of two different binding sites (PETERS et al., 1974; FRICKE and KLAUS, 1977)? It is tempting to speculate that some of the actions of digitalis are based on such a system. For instance, the initial changes of both the gradients for Na^+ and K^+ and the inside longitudinal resistance may be explained by a pump stimulation, whereas the late changes point to a pump inhibition. However, it is interesting to mention that BLOOD (1978) has worked out a model which shows that stimulatory sites are not essential in order to explain the observed data. According to his model, it is possible to induce low dose stimulation and high dose inhibition by assuming inhibitory sites only.

Voltage clamp experiments performed on Purkinje fibers as well as ventricular muscle suggest that the low dose inotropy may find its explanation in the enhancement of the slow Ca^{2+} inward current, i_{si} (WEINGART et al., 1978; MARBAN and TSIEN, 1979). At present, however, it is not yet clear how this effect is brought about. An attractive hypothesis is to assume that a small elevation of $[Ca^{2+}]_i$ may induce an increase of the conductance of the i_{si} channels. The primary cause of the elevated $[Ca^{2+}]_i$, however, would be the inhibition of the Na^+, K^+-pump.

Recent electrophysiologic studies suggest that digitalis may interfere with normal rhythmicity through several mechanisms. At the membrane level, net ionic shifts, progressive depolarization, or oscillatory release of Ca^{2+} from intracellular stores may be involved. The important point is that inhibition of the Na^+, K^+-pump is the primary event. Disturbances in impulse propagation, favoring phenomena like circus movements, may be caused either by inactivation of the excitatory inward current, i_{Na} via membrane depolarization, or by increasing cell-to-cell resistance via elevation of $[Ca^{2+}]_i$. In both cases, a slowing of the impulse conduction results. While the first mechanism seems straightforward, the importance of the second has so far been underestimated.

An increase in $[Ca^{2+}]_i$ seems to be a more general requirement for the development of toxic manifestations. It plays a role not only in impaired impulse spread and increased resting tension, but also in the development of the transient inward current, the ionic mechanism underlying the transient depolarization.

Kass et al. (1978a) propose the following sequence of events of the genesis of the transient current:

$$\begin{array}{c}\text{Digitalis}\\ \text{poisoning}\end{array} \xrightarrow{(1)} \uparrow[Na^+]_i \xrightarrow{(2)} \uparrow[Ca^{2+}]_i \xrightarrow{(3)} \begin{array}{c}\text{Oscillatory}\\ Ca^{2+}\text{ release}\\ \text{from SR}\end{array} \begin{array}{l}\nearrow \text{Increased leak conductance, TI}\\ \searrow \text{Aftercontraction.}\end{array}$$

In this scheme, step (1) is caused by the inhibition of the Na^+, K^+-pump, which takes place during digitalis poisoning. Step (2) may be mediated by Na/Ca exchange (Reuter and Seitz, 1968) or by Na^+-dependent release of Ca^{2+} from mitochondria (Carafoli et al., 1974; Carafoli and Crompton, 1978; Gervais et al., 1978). Step (3) occurs after the cells have reached a state of Ca^{2+} overload. It involves a Ca^{2+}-dependent release of Ca^{2+} from the sarcoplasmic reticulum of the type described by Fabiato and Fabiato (1975) and Endo and Kitizawa (1978).

If this scheme is correct, transient phenomena should be evoked by any procedure which leads to an elevation of $[Ca^{2+}]_i$. This is indeed the case. Similar transients have been reported in the absence of cardiac glycosides, simply by increasing $[Ca^{2+}]_o$ (Ferrier and Moe, 1973; Lederer, 1976; Kass et al., 1978a), by reducing $[Na^+]_o$ (Lederer, 1976, Kass et al., 1978b), or by withdrawal of $[K^+]_o$ (Eisner and Lederer, 1979a, b).

Digitalis poisoning promotes spontaneous fluctuations in membrane current and mechanical force, whose frequency components are the same as the one of the transient inward current itself. Tsien et al. (1978a, 1979) suggested the existence of a common underlying mechanism, presumably an oscillatory release of Ca^{2+} from intracellular stores of the type described for skinned preparations. The observation that local anesthetics produce, among other effects, a progressive inhibition of both the transient inward current and the aftercontraction is consistent with this idea (Tsien et al., 1978a).

In conclusion, despite all efforts, neither the inotropy and the arrhythmogenic action of cardiac glycosides are yet fully understood. Recent studies have emphasized that the story is not as simple as one would have thought some years ago. It seems likely that there is more than one mechanism to account for the effects of digitalis. However, the future will show whether or not this view is correct.

Acknowledgements. I am grateful to Professor S. Weidmann and P. Müller and to Dr. C.H. Fry for reading and criticizing the manuscript and for helpful discussion, and Miss Ch. Lehmann for efficient secretarial assistance. This work was supported by the Swiss National Science Foundation (Grant 3.071-0.76).

References

Akera, T., Brody, T.M.: The role of Na^+, K^+-ATPase in the inotropic action of digitalis. Pharmacol. Rev. *29*, 187–220 (1978)

Allen, D.G., Blinks, J.R.: Calcium transients in aequorin-injected frog cardiac muscle. Nature (Lond.) *273*, 509–513 (1978)

Aronson, R.S., Gelles, J.M., Hoffman, B.F.: Effect of ouabain on the current underlying spontaneous diastolic depolarization in cardiac Purkinje fibers. Nature New Biol. *245*, 118–120 (1973)

Aronson, R.S., Cranefield, P.F.: The effect of resting potential on electrical activity of canine cardiac Purkinje fibers exposed to Na-free solution or to ouabain. Pflügers Arch. *347*, 101–116 (1974)

Aronson, R.S., Gelles, J.M.: The effect of ouabain, dinitrophenol, and lithium on the pacemaker current in sheep cardiac Purkinje fibers. Circ. Res. *40*, 517–524 (1977)

Bailey, J.C., Anderson, G.J., Fisch, C.: Digitalis-induced longitudinal dissociation in canine cardiac Purkinje fibers. Am. J. Cardiol. *32*, 202–208 (1973)

Bailey, L.E., Sures, H.A.: The effect of ouabain on the washout and uptake of calcium in the isolated cat heart. J. Pharmacol. Exp. Ther. *178*, 259–270 (1971)

Baker, P.F., Blaustein, M.P., Hodgkin, A.L., Steinhardt, R.A.: The influence of calcium on sodium efflux in squid axons. J. Physiol. (Lond.) *200*, 431–458 (1969)

Bassingthwaighte, J.B., Fry, C.H., McGuigan, J.A.S.: Relationship between internal calcium and outward current in mammalian ventricular muscle; a mechanism for the control of the action potential duration? J. Physiol. (Lond.) *262*, 15–37 (1976)

Baumgarten, C.M., Isenberg, G.: Depletion and accumulation of potassium in the extracellular clefts of cardiac Purkinje fibers during voltage clamp hyperpolarization and depolarization. Pflügers Arch. *368*, 19–31 (1977)

Beeler, G.W.: Ionic currents in cardiac muscle: a framework for glycoside action. Fed. Proc. *36*, 2209–2213 (1977)

Beeler, G.W., Reuter, H.: Reconstruction of the action potential of ventricular myocardial fibres. J. Physiol. (Lond.) *268*, 177–210 (1977)

Besch, H.R., Allen, J.C., Glick, G., Schwartz, A.: Correlation between the inotropic action of ouabain and its effects on subcellular enzyme systems from canine myocardium. J. Pharmacol. Exp. Ther. *171*, 1–12 (1970)

Besch, H.R., Schwartz, A.: On a mechanism of action of digitalis. J. Mol. Cell. Cardiol. *1*, 195–199 (1970)

Blood, B.E.: The influence of low doses of ouabain and potassium ions on sheep Purkinje fibre contractility. J. Physiol. (Lond.) *251*, 69–70P (1975)

Blood, B.E., Noble, D.: Glycoside induced inotropism of the heart – more than one mechanism? J. Physiol. (Lond.) *266*, 76–78P (1977)

Blood, B.E.: Glycoside induced stimulation of membrane Na-K ATPase – fact or artifact? In: Biophysical aspects of cardiac muscle. Morad, M. (ed.), pp. 379–389. New York: Academic Press 1978

Blood, B.E., Noble, D.: Two mechanisms for the inotropic action of ouabain on sheep cardiac Purkinje fiber contractility. In: Biophysical aspects of cardiac muscle. Morad, M. (ed.), pp. 369–378. New York: Academic Press 1978

Bremner, F., Fry, C.H., McGuigan, J.A.S.: Action of ouabain on sodium-free contractures in mammalian ventricular muscle. J. Physiol. (Lond.) *268*, 30–31P (1977)

Brody, T.M., Akera, T.: Relations among Na^+, K^+-ATPase activity, sodium pump activity transmembrane sodium movement, and cardiac contractility. Fed. Proc. *36*, 2219–2224 (1977)

Carafoli, E., Tiozzo, R., Lugli, G., Crovetti, F., Kratzing, C.: The release of calcium from heart mitochondria by sodium. J. Mol. Cell. Cardiol. *6*, 361–371 (1974)

Carafoli, E., Crompton, M.: The regulation of intracellular calcium by mitochondria. In: Calcium transport and cell function. Ann. N.Y. Acad. Sci. *307*, 269–284 (1978)

Chapman, R.A.: Excitation-contraction coupling in cardiac muscle. Prog. Biophys. Mol. Biol. *35*, 1–52 (1979)

Cohen, I., Daut, J., Noble, D.: An analysis of the actions of low concentrations of ouabain on membrane currents in Purkinje fibres. J. Physiol. (Lond.) *260*, 75–103 (1976)

Cranefield, P.F., Wit, A.L., Hoffman, B.F.: Conduction of the cardiac impulse. III. Characteristics of very slow conduction. J. Gen. Physiol. *59*, 227–246 (1972)

Cranefield, P.F.: The conduction of the cardiac impulse. Mount Kisco, New York: Futura Publ. Comp. 1975

Cranefield, P.F.: Action potentials, afterpotentials, and arrhythmias. Circ. Res. *41*, 415–423 (1977)

Davis, L.D.: Effect of changes in cyclic length on diastolic depolarization produced by ouabain in canine Purkinje fibers. Circ. Res. *32*, 206–214 (1973)

Deitmer, J.W., Ellis, D.: Chances in the intracellular sodium activity of sheep heart Purkinje fibres produced by calcium and other divalent cations. J. Physiol. (Lond.) *277*, 437–453 (1978a)

Deitmer, J.W., Ellis, D.: The intracellular sodium activity of cardiac Purkinje fibres during inhibition and re-activation of the Na-K pump. J. Physiol. (Lond.) *284*, 241–259 (1978b)

De Mello, W.C.: Intercellular communication in heart muscle. In: Intercellular communication. De Mello, W.C. (ed.). New York: Plenum Press 1977

Di Francesco, D., Ohba, M.: Dependence of the apparent reversal potential for the pacemaker current i_{K_2} on its degree of activation in cardiac Purkinje fibres. J. Physiol. (Lond.) *280*, 73–74P (1978)

Dramane, K., Driot, P., Garnier, D.: Action de la ouabaïne sur les courants transmembranaires du myocarde sino-auriculaire de grenouille. J. Physiol. (Paris) *63*, 43A (1971)

Dransfeld, H., Greeff, K., Hess, D., Schorn, A.: Die Abhängigkeit der Ca^{++}-Aufnahme isolierter Mitochondrien des Herzmuskels von der Na^{+}- und K^{+}-Konzentration als mögliche Ursache der inotropen Digitaliswirkung. Experientia Basel *23*, 375–377 (1967)

Dudel, J., Trautwein, W.: Elektrophysiologische Messungen zur Strophanthinwirkung am Herzmuskel. Arch. Exp. Path. Pharmakol. *232*, 393–407 (1958)

Ebashi, S.: Excitation-contraction coupling. Ann. Rev. Physiol. *38*, 293–313 (1976)

Eisner, D.A., Lederer, W.J.: Inotropic and arrhythmogenic effects of potassium depleted solutions on mammalian cardiac muscle. J. Physiol. (Lond.) *294*, 255–277 (1979a)

Eisner, D.A., Lederer, W.J.: The role of the sodium pump in the effects of potassium depleted solutions on mammalian cardiac muscle. J. Physiol. (Lond.) *294*, 279–301 (1979b)

Ellis, D.: The effects of external cations and ouabain on the intracellular sodium activity of sheep heart Purkinje fibres. J. Physiol. (Lond.) *273*, 211–240 (1977)

Endo, M.: Calcium release from the sarcoplasmic reticulum. Physiol. Rev. *57*, 71–108 (1977)

Endo, M., Kitizawa, T.: E-C coupling studies on skinned cardiac fibers. In: Biophysical aspects of cardiac muscle. Morad, M. (ed.), pp. 307–327. New York: Academic Press 1978

Fabiato, A., Fabiato, F.: Activation of skinned cardiac cells. Subcellular effects of cardioactive drugs. Eur. J. Cardiol. *1*, 145–155 (1973)

Fabiato, A., Fabiato, F.: Contractions induced by a calcium-triggered release of calcium from the sarcoplasmic reticulum of single skinned cardiac cells. J. Physiol. (Lond.) *249*, 469–495 (1975)

Fabiato, A., Fabiato, F.: Calcium release from the sarcoplasmic reticulum. Circ. Res. *40*, 119–129 (1977)

Ferrier, G.R., Saunders, J.H.: The effects of magnesium on transient depolarizations induced by acetylstrophanthidin. 5th Intern. Congr. Pharmacol. (1972)

Ferrier, G.R., Moe, G.K.: Effect of calcium on acetylstrophanthidin-induced transient depolarizations in canine Purkinje tissue. Circ. Res. *33*, 508–515 (1973)

Ferrier, G.R., Saunders, J.H., Méndez, C.: A cellular mechanism for the generation of ventricular arrhythmias by acetylstrophanthidin. Circ. Res. *32*, 600–609 (1973)

Ferrier, G.R.: The effects of tension on acetylstrophanthidin-induced transient depolarizations and aftercontractions in canine myocardial and Purkinje tissue. Circ. Res. *38*, 156–162 (1976)
Ferrier, G.R.: Digitalis arrhythmias: Role of oscillatory afterpotentials. Progr. Cardiovasc. Dis. *19*, 459–474 (1977a)
Ferrier, G.R.: Relationship between acetylstrophanthidin-induced aftercontractions and the strength of contraction of canine ventricular myocardium. Circ. Res. *41*, 622–629 (1977b)
Fozzard, H.A.: Excitation-contraction coupling and digitalis. Circulation *47*, 5–7 (1973)
Fricke, U., Klaus, W.: Evidence for two different Na^+-dependent [^{3}H]-ouabain binding sites of a Na^+-K^+-ATPase of guinea-pig hearts. Brit. J. Pharmacol. *61*, 423–428 (1977)
Fry, C.H., Galindez, E., Poole-Wilson, P.A.: Potassium exchange and the positive inotropic effect of ouabain in rabbit and human myocardium. J. Physiol. (Lond.) *280*, 72–73P (1978)
Gadsby, D.C., Niedergerke, R., Page, S.: Do intracellular concentrations of potassium or sodium regulate the strength of the heart beat? Nature (Lond.) *232*, 651–653 (1971)
Gelles, J.M., Aronson, R.S.: Effects of ouabain on the pacemaker current in sheep cardiac Purkinje fibers. Fed. Proc. *32*, 709 (1973)
Gelles, J.M., Aronson, R.S., Hoffman, B.F.: Effect of transmembrane potential on the manifestations of ouabain toxicity in sheep cardiac Purkinje fibers. Cardiovasc. Res. *9*, 600–606 (1975)
Gervais, A., Busch, U., Wood, J.M., Schwartz, A.: Effect of a toxic in vivo dose of ouabain on guinea pig heart mitochondria. J. Mol. Cell. Cardiol. *10*, 1003–1015 (1978)
Ghysel-Burton, J., Godfraind, T.: Stimulation and inhibition by ouabain of the sodium pump in guinea-pig atria. Brit. J. Pharmacol. *55*, 249P (1975)
Ghysel-Burton, J., Godfraind, T.: Importance of the lactone ring for the action of therapeutic doses of ouabain in guinea-pig atria. J. Physiol. (Lond.) 76–78P (1976)
Glitsch, H.G.: Characteristics of active Na transport in intact cardiac cells. Am. J. Physiol. *236*, H189–199 (1979)
Godfraind, T., Lesne, M.: The uptake of cardiac glycosides in relation to their actions in isolated cardiac muscle. Brit. J. Pharmacol. *46*, 488–497 (1972)
Godfraind, T., Ghysel-Burton, J.: Binding sites related to ouabain-induced stimulation or inhibition of the sodium pump. Nature (Lond.) *265*, 165–166 (1977)
Goshima, K.: Arrhythmic movements of myocardial cells in culture and their improvement with antiarrhythmic drugs. J. Mol. Cell. Cardiol. *8*, 217–238 (1976)
Goshima, K.: Ouabain-induced arrhythmias of single isolated myocardial cells and cell clusters cultured in vitro and their improvement by quinidine. J. Mol. Cell. Cardiol. *9*, 7–23 (1977)
Goto, M., Tsuda, Y., Yatani, A.: Two mechanisms for positive inotropism of low-K Ringer solution in bullfrog atrium. Nature (Lond.) *268*, 755–757 (1977)
Greenspan, A.M., Morad, M.: Electromechanical studies on the inotropic effects of acetylstrophanthidin in ventricular muscle. J. Physiol. (Lond.) *253*, 357–384 (1975)
Hashimoto, K., Moe, G.K.: Transient depolarizations induced by acetylstrophanthidin in specialized tissue of dog atrium and ventricle. Circ. Res. *32*, 618–624 (1973)
Hauswirth, O., Noble, D., Tsien, R.W.: The mechanism of oscillatory activity at low membrane potentials in cardiac Purkinje fibres. J. Physiol. (Lond.) *200*, 255–265 (1969)
Hiraoka, M.: Membrane current changes induced by acetylstrophanthidin in cardiac Purkinje fibers. Jpn. Heart J. *18*, 851–859 (1977)
Hoffman, B.F., Bindler, E., Suckling, E.E.: Postextrasystolic potentiation of contraction in cardiac muscle. Am. J. Physiol. *185*, 95–102 (1956)
Hogan, P.M., Wittenberg, S.M., Klocke, F.J.: Relationship of stimulation frequency to automaticity in the canine Purkinje fiber during ouabain administration. Circ. Res. *32*, 377–384 (1973)
Horackova, M., Vassort, G.: Sodium-calcium exchange in regulation of cardiac contractility. Evidence for an electrogenic, voltage-dependent mechanism. J. Gen. Physiol. *73*, 403–424 (1979)
Hordof, A.J., Spotnitz, A., Mary-Rabine, L., Edie, R.N., Rosen, M.R.: The cellular electrophysiologic effects of digitalis on human atrial fibers. Circulation *57*, 223–229 (1978)

Hougen, T.J., Smith, T.W.: Inhibition of myocardial monovalent cation active transport by subtoxic doses of ouabain in the dog. Circ. Res. *42*, 856–863 (1978)

Isenberg, G., Trautwein, W.: The effect of dihydro-ouabain and lithium-ions on the outward current in cardiac Purkinje fibers. Pflügers Arch. *350*, 41–54 (1974)

Isenberg, G.: Cardiac Purkinje fibers. Resting, action, and pacemaker potential under the influence of $[Ca^{2+}]_i$ as modified by intracellular injection techniques. Pflügers Arch. *371*, 51–59 (1977a)

Isenberg, G.: Cardiac Purkinje fibers. The slow inward current component under the influence of modified $[Ca^{2+}]_i$. Pflügers Arch. *371*, 61–69 (1977b)

Isenberg, G.: Cardiac Purkinje fibers. $[Ca^{2+}]_i$ controls steady state potassium conductance. Pflügers Arch. *371*, 71–76 (1977c)

Isenberg, G.: Cardiac Purkinje fibers. $[Ca^{2+}]_i$ controls the potassium permeability via the conductance components g_{K_2} and g_{K_2}. Pflügers Arch. *371*, 77–85 (1977d)

Josephson, I., Sperelakis, N.: Ouabain blockade of inward slow current in cardiac muscle. J. Mol. Cell. Cardiol. *9*, 409–418 (1977)

Kass, R.S., Tsien, R.W.: Multiple effects of calcium antagonists on plateau currents in cardiac Purkinje fibers. J. Gen. Physiol. *66*, 169–192 (1975)

Kass, R.S., Lederer, W.J., Tsien, R.W.: Current fluctuations in strophanthidin-treated cardiac Purkinje fibers. Biophys. J. *16*, 25a (1976)

Kass, R.S., Lederer, W.J., Tsien, R.W., Weingart, R.: Role of calcium ions in transient inward currents and aftercontractions induced by strophanthidin in cardiac Purkinje fibres. J. Physiol. (Lond.) *281*, 187–208 (1978a)

Kass, R.S., Tsien, R.W., Weingart, R.: Ionic basis of transient inward current induced by strophanthidin in cardiac Purkinje fibers. J. Physiol. (Lond.) *281*, 209–226 (1978b)

Kassebaum, D.: Electrophysiological effect of strophanthin in the heart. J. Pharmacol. Exp. Ther. *140*, 329–338 (1963)

Katz, A.M.: Increased Ca^{++} entry during the plateau of the action potential: a possible mechanism of cardiac glycoside action. J. Mol. Cell. Cardiol. *4*, 87–89 (1972)

Katz, A.M.: Physiology of the heart. New York: Raven Press 1977

Kaufmann, R., Fleckenstein, A., Antoni, H.: Ursachen und Auslösungsbedingungen von Myokard-Kontraktionen ohne reguläres Aktionspotential. Pflügers Arch. *278*, 435–446 (1963)

Kline, R., Morad, M.: Potassium efflux and accumulation in heart muscle. Biophys. J. *16*, 367–372 (1976)

Koch-Weser, J., Blinks, J.R.: Analysis of the relation of the positive inotropic action of cardiac glycosides to the frequency of contraction of heart muscle. J. Pharmacol. Exp. Ther. *136*, 305–317 (1962)

Koch-Weser, J., Blinks, J.R.: The influence of the interval between beats on myocardial contractility. Pharmacol. Rev. *15*, 601–652 (1963)

Kunze, D.L.: Rate-dependent changes in extracellular potassium in the rabbit atrium. Circ. Res. *41*, 122–127 (1977)

Langer, G.A., Serena, S.D.: Effects of strophanthidin upon contraction and ionic exchange in rabbit ventricular myocardium: Relation to control of active state. J. Mol. Cell. Cardiol. *1*, 65–90 (1970)

Langer, G.A.: Relationship between myocardial contractility and the effects of digitalis on ionic exchange. Fed. Proc. *36*, 2231–2234 (1977)

Lederer, W.J.: The ionic basis of arrhythmogenic effects of cardiotonic steroids in cardiac Purkinje fibers. Ph. D. Thesis, Yale University, New Haven, Conn., 1976

Lederer, W.J., Tsien, R.W.: Transient inward current underlying arrhythmogenic effects of cardiotonic steroids in Purkinje fibers. J. Physiol. (Lond.) *263*, 73–100 (1976)

Lee, C.O., Kang, D.H., Sokol. J.H., Lee, K.S.: Relation between intracellular Na ion activity and positive inotropic action of sheep cardiac Purkinje fibers. Biophys. J. *25*, 198a (1979)

Lee, K.S., Klaus, W.: The subcellular basis for the mechanism of inotropic action of cardiac glycosides. Pharmacol. Rev. *23*, 193–261 (1971)

Loh, C.K.: Effects of cardiac glycosides on myocardial K^+ in amphibian atrium. Fed. Proc. *34*, 434 (1975)

Lown, B., Cannon, R.L., Rossi, M.A.: Electrical stimulation and digitalis drugs: Repetitive response in diastole. Proc. Soc. Exp. Biol. Med. *126*, 698–701 (1967)
Lüllmann, H., Peters, T.: Action of cardiac glycosides. In: Progress in Pharmacology. Grobecker, H., Kahl, G.F., Klaus, W., Schmitt, H., van Zwieten, P.A. (eds.), Vol. 2, No. 2, pp. 1–57. Stuttgart: Gustav Fischer 1979
Marban, E., Tsien, R.W.: Ouabain increases the slow inward calcium current in ventricular muscle. J. Physiol. (Lond.) *292*, 72–73P (1979)
McAllister, R.E., Noble, D., Tsien, R.W.: Reconstruction of the electrical activity of cardiac Purkinje fibres. J. Physiol. (Lond.) *251*, 1–59 (1975)
McDonald, T.F., Nawrath, H., Trautwein, W.: Membrane currents and tension in cat ventricular muscle treated with cardiac glycosides. Circ. Res. *37*, 674–682 (1975)
McNutt, N.S., Weinstein, R.S.: Membrane ultrastructure at mammalian intercellular junctions. Progr. Biophys. Mol. Biol. *26*, 45–101 (1973)
Méndez, C., Méndez, R.: The action of cardiac glycosides on the excitability and conduction velocity of the mammalian atrium. J. Pharmacol. Exp. Ther. *121*, 402–413 (1957)
Méndez, R., Pisanty, J.: Comparative effect of digitoxin and ouabain on auriculoventricular conduction in the dog heart. Fed. Proc. *8*, 320 (1949)
Miura, D.S., Rosen, M.R.: The effects of ouabain on the transmembrane potentials and intracellular potassium activity of canine Purkinje fibers. Circ. Res. *42*, 333–338 (1978)
Moe, G.K., Méndez, R.: The action of several cardiac glycosides on conduction velocity and ventricular excitability in the dog heart. Circulation *4*, 729–734 (1951)
Morad, M., Greenspan, A.M.: Excitation-contraction coupling as a possible site for the action of digitalis on heart muscle. In: Cardiac arrhythmias. Dreifus, L.S., Likoff, W. (eds.), pp. 479–503. New York: Grune & Stratton 1973
Müller, P.: Kalium und Digitalistoxizität. Cardiologia (Basel) *42*, 176–188 (1963)
Müller, P.: Ouabain effects on cardiac contraction, action potential, and cellular potassium. Circ. Res. *17*, 46–56 (1965)
Mullins, L.J., The generation of electric currents in cardiac fibres by Na/Ca exchange. Am. J. Physiol. *236*, C103–110 (1979)
Nayler, W.G., Poole-Wilson, P.A., Williams, A.: Ouabain-induced contractures in mammalian heart muscle. J. Physiol. (Lond.) *275*, 75P (1978)
Okita, G.T.: Dissociation of Na^+, K^+-ATPase inhibition from digitalis inotropy. Fed. Proc. *36*, 2225–2230 (1977)
Peon, J., Ferrier, G.R., Moe, G.K.: The relationship of excitability to conduction velocity in canine Purkinje tissue. Circ. Res. *43*, 125–135 (1978)
Perry, S.V.: The contractile and regulatory proteins of the myocardium. In: Contraction and relaxation in the myocardium. Nayler, W.G. (ed.), pp. 29–77. New York: Academic Press 1975
Peters, T., Raben, R.-H., Wassermann, O.: Evidence for a dissociation between positive inotropic effect and inhibition of the Na^+-K^+-ATPase by ouabain, casaine and their alkylating derivates. Eur. J. Pharmacol. *26*, 166–174 (1974)
Poole-Wilson, P.A., Langer, G.A.: Glycoside inotropy in the absence of an increase in potassium efflux in the rabbit heart. Circ. Res. *37*, 390–395 (1975)
Reiter, M.: Die Entstehung von „Nachkontraktionen“ im Herzmuskel unter Einwirkung von Calcium und von Digitalisglykosiden in Abhängigkeit von der Reizfrequenz. Naunyn-Schmiedebergs Arch. Exp. Path. Pharmak. *242*, 497–507 (1962)
Reiter, M.: After-contractions under the action of cardiac glycosides and calcium. In: Proceedings of the First International Pharmacological Meeting. Wilbrandt, W., Lindgren, P. (eds.), Vol. 3, pp. 265–270. Oxford: Pergamon Press 1963
Reiter, M.: Versuche über die pharmakologische Beeinflussung der elektromechanischen Koppelung. In: Herzinsuffizienz. Reindell, H., Keul, J., Doll, E. (eds.), pp. 102–108. Stuttgart: Thieme Verlag 1968
Reuter, H., Seitz, N.: Dependence of calcium efflux from cardiac muscle on temperature and external ion composition. J. Physiol. (Lond.) *195*, 451–470 (1968)
Reuter, H.: Divalent cations as charge carriers in excitable membranes. In.: Progress in biophysics and molecular biology. Butler, J.A.V., Noble, D. (eds.), Vol. 26, pp. 1–43. Oxford: Pergamon Press 1973

Reuter, H., Scholz, H.: A study of the ion selectivity and the kinetic properties of the calcium dependent slow inward current in mammalian cardiac muscle. J. Physiol. (Lond.) *264*, 17–47 (1977)

Rosen, M.R., Gelband, H., Hoffman, B.F.: Correlation between effects of ouabain on the canine electrocardiogram and transmembrane potentials of isolated Purkinje fibers. Circulation *47*, 65–72 (1973a)

Rosen, M.R., Gelband, H., Merker, C., Hoffman, B.F.: Mechanisms of digitalis toxicity. Effects of ouabain on phase four of canine Purkinje fiber transmembrane potentials. Circulation *47*, 681–689 (1973b)

Rosen, M.R., Ilvento, J.P., Gelband, H., Merker, C.: Effects of verapamil on electrophysiologic properties of canine cardiac Purkinje fibers. J. Pharmacol. Exp. Ther. *189*, 414–422 (1974)

Ruiz-Ceretti, E., Samson, J.P., Reisin, I., Schanne, O.F.: Ionic and electrical effects of ouabain on isolated rabbit hearts. J. Mol. Cell. Cardiol. *9*, 51–61 (1977)

Saunders, J.H., Ferrier, G.R., Moe, G.K.: Conduction block associated with transient depolarizations induced by acetylstrophanthidin in isolated canine Purkinje fibers. Circ. Res. *32*, 610–617 (1973)

Scholz, H.: Ca-abhängige Membranpotentialänderungen und Kontraktion an Herzen unter dem Einfluß von Adrenalin und Digitoxigenin. Naunyn-Schmiedebergs Arch. Pharmakol. *266*, 445–446 (1970)

Scholz, H.: Calcium-dependent depolarizations and contractions in ventricular myocardial fibers: effects of adrenaline and cardiac glycosides. In: Recent advances in studies on cardiac structure and metabolism. Fleckenstein, A., Dhalla, N.S. (eds.), Vol. 5, pp. 51–57. New York: University Park Press 1975

Schwartz, A., Lindenmayer, G.L., Allen, J.C.: The sodium-potassium adenosine triphosphatase: pharmacological, physiological, and biochemical aspects. Pharmacol. Rev. *27*, 3–134 (1975)

Schwartz, A.: Is the cell membrane Na^+,K^+-ATPase enzyme system the pharmacological receptor for digitalis? Circ. Res. *39*, 2–7 (1976)

Siegelbaum, S., Tsien, R.W.: Role of intracellular calcium in the transient outward current of calf Purkinje fibres. Nature (Lond.) *269*, 611–613 (1977)

Singer, D.H., Lazzara, R., Hoffman, B.F.: Interrelationship between automaticity and conduction in Purkinje fibers. Circ. Res. *21*, 537–558 (1967)

Swain, H.H., Weidner, C.L.: A study of substances which alter intraventricular conduction in the isolated dog heart. J. Pharmacol. Exp. Ther. *120*, 137–146 (1957)

Thyrum, P.T.: Inotropic stimuli and systolic transmembrane calcium flow in depolarized guinea-pig atria. J. Pharmacol. Exp. Ther. *188*, 166–179 (1974)

Trautwein, W., Kassebaum, D.G.: On the mechanism of spontaneous impulse generation in the pacemaker of the heart. J. Gen. Physiol. *45*, 317–330 (1961)

Trube, G., Trautwein, W.: Lack of inotropic effect of dihydroouabain in skinned cardiac muscle fibers. Pflügers Arch. *365*, R2 (1976)

Tsien, R.W., Weingart, R., Kass, R.S.: Digitalis: Inotropic and arrhythmogenic effects on membrane currents in cardiac Purkinje fibers. In: Biophysical aspects of cardiac muscle. Morad, M. (ed.), pp. 345–368. New York: Academic Press 1978a

Tsien, R.W., Kass, R.S., Weingart, R.: Calcium ions and membrane current changes induced by digitalis in cardiac Purkinje fibers. Ann. N.Y. Acad. Sci. *307*, 483–490 (1978b)

Tsien, R.W., Kass, R.S., Weingart, R.: Cellular and subcellular mechanisms of cardiac pacemaker oscillations. J. Exp. Biol. *81*, 205–215 (1979)

Turin, L., Warner, A.: Carbon dioxide reversibly abolishes ionic communication between cells of early amphibian embryo. Nature (Lond.) *270*, 56–57 (1977)

Vassalle, M., Karis, J., Hoffman, B.F.: Toxic effects of ouabain on Purkinje fibers and ventricular muscle fibers. Am. J. Physiol. *203*, 433–439 (1962)

Vassalle, M., Musso, E.: Cardiac Purkinje fibers, automaticity and digitalis. Fed. Proc. *33*, 432 (1974)

Vassalle, M., Musso, E.: On the mechanisms underlying digitalis toxicity in cardiac Purkinje fibers. In: Recent advances in studies on cardiac structure and metabolism. Roy, P.-E., Dhalla, N.S. (eds.), Vol. 9, pp. 355–376. Baltimore: University Park Press 1976

Vassalle, M.: Electrophysiological basis of digitalis arrhythmias. J. Cardiovasc. Pulmon. Technol. *4*, 21–33 and 37 (1976); *5*, 25–31 and 34 (1977)
Vassort, G.: Influence of sodium ions on the regulation of frog myocardial contractility. Pflügers Arch. *339*, 225–240 (1973)
Walker, J.L.: Intracellular K^+ activity in sheep cardiac Purkinje fibers. In: Frontiers of biological energetics: From electrons to tissues, Vol. 2. Philadelphia: University of Pennsylvania 1978
Watanabe, Y., Dreifus, L.S.: Electrophysiologic effects of digitalis on A-V transmission. Am. J. Physiol. *211*, 1461–1466 (1966)
Weidmann, S.: The effect of the cardiac membrane potential on the rapid availability of the sodium-carrying system. J. Physiol. (Lond.) *127*, 213–224 (1955)
Weingart, R.: The actions of ouabain on intercellular coupling and conduction velocity in mammalian ventricular muscle. J. Physiol. (Lond.) *264*, 341–365 (1977)
Weingart, R., Kass, R.S., Tsien, R.W.: Is digitalis inotropy associated with enhanced slow inward calcium current? Nature (Lond.) *273*, 389–392 (1978)

CHAPTER 13

Influence of Cardiac Glycosides on Myocardial Energy Metabolism

K. GÜTTLER and W. KLAUS

A. Introduction

According to our present knowledge, myocardial energy metabolism is not involved directly in the mode of action of cardiac glycosides. However, it is well known clinically that the therapeutic efficiency of these drugs is to a certain degree dependent on the metabolic state of the heart; they are most effective in hearts with no defects in energy metabolism such as congestive heart failure due to excessive hemodynamic pressure loads (e.g., hypertension, valvular lesions, arteriosclerotic diseases) whereas they are rather ineffective in cardiac failure due to anemia, hypoxia, thyrotoxicosis, and severe thiamine deficiency (beriberi) in which energy supply is impaired (MOE and FARAH, 1975) and experimentally it has been shown that heart failure caused by poisons that reduce energy-rich phosphate pools (cyanide, dinitrophenol, etc.) is not relieved by digitalis (GRUHZIT and FARAH, 1955).

I. Myocardial Energy Metabolism

The sequence of metabolic events in heart muscle – as far as they might be relevant to cardiac glycoside action – will be briefly discussed on the basis of a scheme (Fig. 1) originally developed by OLSON and PIATNEK (1959) and slightly modified later on by DANFORTH et al. (1960) and OLSON et al. (1971/1972).

1. Energy Liberation

Almost all the energy for cardiac function comes from oxidizing substrates – mainly fatty acids, glucose, lactate, and to a lesser extent amino acids and ketones (BING, 1954, 1961) – by aerobic metabolism in the Krebs tricarboxylic acid cycle (OLSON and BARNHORST, 1974; WILLIAMSON, 1979). The contribution of anaerobic glycolysis (in the presence of 0.3 mmol/l cyanide) to the myocardial energy balance amounts to about 13% (MÜLLER-RUCHHOLTZ, 1973). This process depends highly on the myocardial oxygen and substrate supply and involves the liberation of nascent hydrogen and its oxidation to water via electron transport in the respiratory chain of the mitochondria.

2. Energy Conservation

The energy liberated from this oxidation process is conserved by coupling to the synthesis of ATP (from ADP and inorganic phosphate) and its interaction with

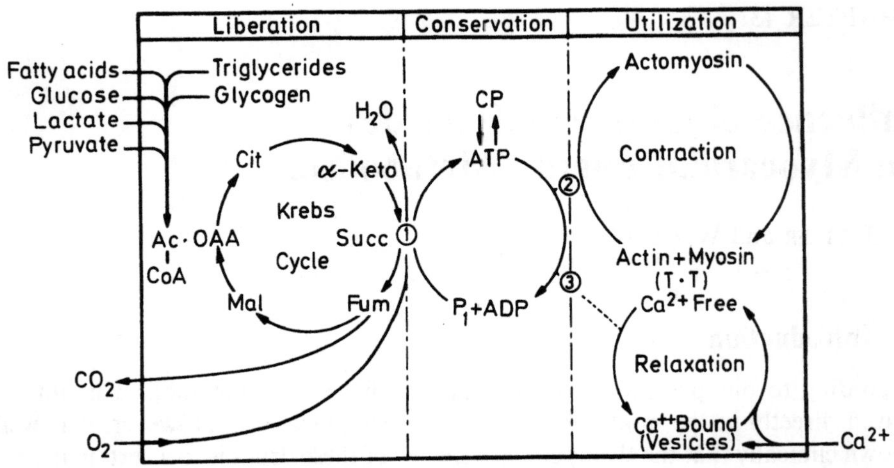

Fig. 1. Schematic representation of myocardial metabolism. Energy coupling points which may be influenced by cardiac glycosides: *1* oxidative phosphorylation, *2* actomyosin ATPase, *3* ATP-dependent calcium pump of the sarcoplasmatic reticulum. Abbreviations: Cit = citrate; α-Keto = α-ketoglutarate; Succ = succinate; Fum = fumarate; Mal = malate; Ac–CoA = acetyl CoA; OAA = oxaloacetate; CP = creatine phosphate; ATP = adenosine triphosphate; ADP = adenosine diphosphate; Pi = inorganic phosphate. (After OLSON et al., 1971/1972)

creatine to form creatine phosphate. This latter compound acts as a high energy phosphate reservoir for resynthesis of ATP (BRAUNWALD et al., 1976).

3. Energy Utilization

This phase deals mainly with the transformation of the ATP-stored energy into mechanical work of the heart (and to a minor extent into another kind of cell work or heat). ATP utilization by cardiac muscle is initiated by electrical depolarization of the myocardial surface membrane and the attached T tubules initiating a slow inward current of calcium from extracellular space which in turn releases intracellularly stored calcium from the sarcoplasmic reticulum (FORD and PODOLSKY, 1970; REUTER, 1973; ENDO, 1977). This free calcium binds to troponin, the specific regulatory protein in the troponin–tropomyosin–actin complex, resulting in conformational changes with bridge formation between the specific binding sites of the myosin and actin filaments. The consequences of this formation of actomyosin are enhanced ATPase activity, translocation of myofilaments, and shortening of the myofibrils. Thus, the chemical energy released by hydrolysis of ATP is finally converted to mechanical work of the heart and part of it is lost as heat (DAVIES, 1963). Relaxation is induced by reversion of the excitation-induced changes in intracellular calcium distribution including (a) ATP-dependent storage of calcium in the sarcoplasmic reticulum and in the mitochondria and (b) facilitated diffusion and active transport of calcium out of the myocardial cell (HASSELBACH and MAKINOSE, 1961; EBASHI and LIPMANN, 1962; CONSTANTIN et al., 1965; EBASHI and ENDO, 1968; JENDEN and FAIRHURST, 1969; LEE and SHIN, 1969; SCHATZMANN and VIN-

CENZI, 1969; UEBA et al., 1971; LANGER, 1973; CARAFOLI, 1974; EBASHI, 1976; CARONI and CARAFOLI, 1980).

II. General Biochemical Aspects of Myocardial Metabolism

The most prominent feature of myocardial metabolism is the predominant oxidation of free fatty acids (FFA) as compared with other substrates such as glucose, lactate, pyruvate, acetate, ketones (ref. see SHIPP et al., 1961; BING, 1965; OPIE, 1968, 1969 a, b; NEELY et al., 1972). However, there are still some uncertainties concerning the relationships between different substrates and the regulation of the preferential oxidation route.

In general, the uptake and turnover of substrates – with due allowance for glucose (OPIE et al., 1962) – is mainly dependent on their arterial concentrations (BERNSMEIER and RUDOLPH, 1961; SCOTT et al., 1962; ÇOWLEY et al., 1969; NEELY et al., 1972; SPITZER, 1974) and to a minor extent on the metabolic condition (THORN et al., 1968) and the physiologic activity (KEDEM et al., 1975) of the heart. But when FFA and carbohydrates are available together the heart will preferentially utilize FFA (BING, 1965; OPIE, 1968, 1969 a, b; NEELY et al., 1972; NEELY and MORGAN, 1974; SEVERSON, 1979). Furthermore, it is well established that besides FFA most other common substrates will all be oxidized before exogenous glucose (SHIPP et al., 1964; WILLIAMSON, 1965; WILLEBRANDS and VAN DER VEEN, 1967; NEELY et al., 1972). The rate-limiting step in glucose utilization appears to be membrane transport, thus, this substrate may contribute to a greater extent to the cardiac energetics if insulin is present (BERNSMEIER and RUDOLPH, 1961; NEELY et al., 1976).

III. Cardiac Function as a Determinant of Myocardial Metabolism

Two components of myocardial energetics are to be distinguished: resting (function-independent) metabolism and active (function-dependent) metabolism. The former is determined by the energetic needs for structural integrity, and the electrophysiologic properties of the cells (BRETSCHNEIDER, 1971), whereas the latter is dependent upon a variety of functional parameters such as:

1) Ventricular pressure work (FEINBERG et al., 1962; BRITMAN and LEVINE, 1964; McDONALD, 1966; NEELY et al., 1967 a; BRAUNWALD, 1971; GIBBS, 1978).
2) Heart rate (WHALEN, 1961; VAN DER VEEN and WILLEBRANDS, 1967; FEINBERG et al., 1968; BRAUNWALD, 1971; GIBBS, 1978; WEBER and JANICKI, 1978, 1979).
3) End-diastolic wall tension (WHALEN, 1961; MONROE and FRENCH, 1961; ROLLETT et al., 1965; COVELL et al., 1966; NEELY et al., 1967 a; FEINBERG et al., 1968; WEBER and JANICKI, 1978, 1979).
4) Contractile state, as reflected by d*p*/d*t* (ROSS et al., 1965; SONNENBLICK et al., 1965; CLANCY et al., 1967; KÜHN and BRACHFELD, 1969; BRAUNWALD, 1971; GIBBS, 1978; WEBER and JANICKI, 1978, 1979).
5) Coronary flow (GREGG et al., 1957; KAHLER et al., 1963; BACANER et al., 1965; OPIE, 1965; MARCHETTI et al., 1966; NEELY et al., 1967 a; FEINBERG et al., 1968; ARNOLD et al., 1968, 1970; BUGGE-ASPERHEIM, 1972; MORGENSTERN, 1973; GIBBS, 1978).

These relationships indicate that drugs such as cardiac glycosides which modify the functional behavior of the heart may – besides their direct effects – also modulate the myocardial metabolism by these indirect mechanisms.

B. Influence of Cardiac Glycosides on Myocardial Energy Liberation

On the basis of numerous observations in experimental and clinical studies a primary influence of cardiac glycosides on myocardial energy liberation has been postulated (ref. see WOLLENBERGER, 1949; KLAUS, 1964). This phase of cardiac metabolism is usually investigated by measuring either the oxygen consumption or the substrate utilization of the heart or by measuring both parameters.

I. Myocardial Oxygen Consumption

The most convenient method of analyzing drug actions on myocardial energy production is to study their influence on oxygen consumption. The data obtained on nonbeating heart muscle preparations are somewhat conflicting with regard to cardiac glycoside action on the resting metabolism. Depending on the drug concentration used, both stimulation ("low range") and inhibition ("high range") of the oxygen consumption have been reported (SALOMON and RIESSER, 1935; GENUIT and HAARMANN, 1940; WOLLENBERGER, 1947, 1949; FINKELSTEIN and BODANSKY, 1948; HERRMANN, 1950; NOWY and HELMREICH, 1950; ROTHLIN and SCHOELLY, 1950; DOULL et al., 1951; BURDETTE, 1952; LEE, 1953, 1958; HERRMANN et al., 1954; SZEKERES et al., 1959; CREVASSE and WHEAT, 1962).

The magnitude of these effects was related to the age of the animals (WOLLENBERGER et al., 1953; LEONARD and HAJDU, 1959), to the oxygen consumption prior to the drug application (FINKELSTEIN and BODANSKY, 1948; LANGEMANN et al., 1953) which in turn is a function of the species and body size (KREBS, 1950; MÜLLER, 1962; BRETSCHNEIDER, 1964) and especially to the substrates available; the increase in oxygen consumption produced by cardiac glycosides is most pronounced in the presence of glucose (WOLLENBERGER, 1947; FINKELSTEIN and BODANSKY, 1948; HERRMANN, 1950; LANGEMANN et al., 1953; CREVASSE and WHEAT, 1962), less if lactate or oxaloacetic acid is used (HERRMANN, 1950; HERRMANN et al., 1954) and absent if the substrates are pyruvate, acetate, malate, or succinate (HERRMANN, 1950; HERRMANN et al., 1954).

Most of these effects were derived from experiments on heart muscle slices or papillary muscle, but they could not be obtained if the structural integrity of the tissue was lost, as in heart muscle homogenates or in isolated subcellular components (SALOMON and RIESSER, 1935; GENUIT and HAARMANN, 1940; WOLLENBERGER, 1947, 1949, 1967; DOULL et al., 1951; REITER and BARRON, 1952; LANGEMANN et al., 1953; ROTHLIN and BIRCHER, 1954; STAUB, 1957; LEE, 1960).

This behavior might reflect the fact that these metabolic effects of cardiac glycosides are mediated by the well-known influence of these drugs on the intra- and extracellular electrolyte distribution (see Chap. 18) which is absent in homogenized tissue. This assumption is further supported by the observation that the cardiac glycoside-induced increase in oxygen consumption is enhanced by the presence of more calcium (or less phosphate) in the incubation medium (FINKELSTEIN and BODANSKY, 1948; HERRMANN, 1950; DUOLL et al., 1951; LANGEMANN et al., 1953; CREVASSE and WHEAT, 1962; WOOD et al., 1972). Thus, it may tentatively be concluded that the stimulating effect of "low" (approximately 10^{-6} mol/l) cardiac glycoside concentrations (which actually are in the toxic range if used on beating preparations) on the myocardial oxygen consumption may be mediated by an increase in cellular calcium content (secondary to the inhibition of Na^+, K^+-ATPase) whereas the inhibitory effect of "high" (10^{-5}–10^{-3} mol/l) cardiac glycoside concentrations is most probably due to progressive cellular damage caused by calcium overload.

Furthermore, in more recent studies on functionally standardized intact heart muscle preparations of different species, no effects of cardiac glycosides on basal oxygen consumption could be observed in the conventional concentration range of cardiac glycosides and of electrolytes in the incubation medium (COLEMAN, 1967; KLAUS and KREBS, 1968; GÜTTLER, 1974). Therefore, it may be assumed that cardiac glycosides do not appreciably influence the myocardial resting oxygen consumption if extreme experimental or artificial conditions are avoided.

The influence of cardiac glycosides on the oxygen consumption of beating heart muscle preparations has been extensively studied (ref. see KLAUS, 1964; KREBS, 1970; LEE and KLAUS, 1971) but these studies have revealed controversial results which may be due to the greatly differing experimental conditions (ROTHLIN and TAESCHLER, 1956; GREEFF, 1963; KLAUS, 1964; KLAUS and KREBS, 1968, 1969).

On the basis of earlier observations on heart–lung preparations or other failing-heart models a specific influence of cardiac glycosides on myocardial metabolism of failing hearts has been postulated (e.g. SARNOFF et al., 1964) since under these conditions cardiac glycosides increase the work output to a greater extent than the oxygen consumption (which is sometimes even reduced) (GREMELS, 1933, 1937; PETERS and VISSCHER, 1936; GOLLWITZER-MEIER and KRÜGER, 1937; BING et al., 1950; PAGE et al., 1951; HAFKENSCHIEL and CERLETTI, 1954; ROTHLIN et al., 1955; BLAIN et al., 1956; GÖKSEL et al., 1963; SARNOFF et al., 1964; KOYAMA et al., 1966; JACOBSON, 1968; SIMAAN et al., 1971).

Simultaneous measurements of the myocardial contractility and oxygen consumption demonstrated an impaired mechanical efficiency in failing hearts which may be normalized or even improved by cardiac glycosides (PETERS and VISSCHER, 1936; BING, 1965; KLAUS and KREBS, 1968, 1969) whereas this parameter was not influenced in nonfailing preparations (BING et al., 1950; SELZER and MALMBORG, 1962; YANKOPOULOS et al., 1968; RABERGER and KRAUPP, 1969).

Studies on nonfailing hearts both in vivo and in vitro did not reveal any specific disturbance of the myocardial metabolism (KLAUS, 1964; BING, 1965; CHIDSEY, 1967; MORKIN and LA RAIA, 1974) and indicated identical inotropic effects of cardiac glycosides (BRAUNWALD et al., 1961; MASON and BRAUNWALD, 1963; WALLACE et al., 1963; KOCH-WESER et al., 1964; WEISSLER et al., 1964; SONNENBLICK et al., 1966; HOOD et al., 1968; KLAUS and KREBS, 1968).

Thus, the controversial observations on the mechanical efficiency are probably due to the difference in the hemodynamic state between the failing and nonfailing heart. More recent observations using improved techniques (polarographic methods and simultaneous determinations of functional and metabolic parameters

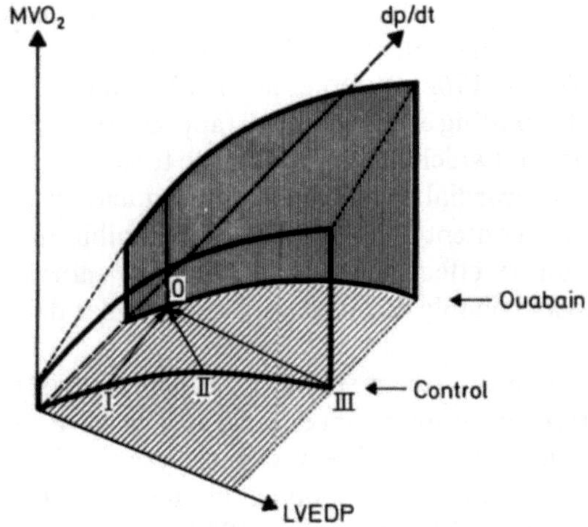

Fig. 2. Schematic representation of the relationship between the left ventricular end-diastolic pressure (LVEDP), contractility (d*p*/d*t*), and oxygen consumption (MVO_2) at different states of heart failure (increased LVEDP : I–III) and under the influence of ouabain on constant working guinea pig heart. (For explanation see text). (After KLAUS and KREBS, 1969)

under standardized conditions) showed clearly that myocardial oxygen consumption followed strictly the improved contractile force under the influence of cardiac glycosides, provided that secondary influences due to variations of hemodynamics were excluded (SMITH and POST, 1950; REITER, 1953; SMITH et al., 1954; NAYLER, 1960; COVELL et al., 1966; KREBS and KLAUS, 1966; CLANCY et al., 1967; COLEMAN, 1967; GOUSIOS et al., 1967; KLAUS and KREBS, 1968, 1969; SIESS, 1968; BOERTH et al., 1978).

The complex relationship between myocardial oxygen consumption and some functional parameters is schematically represented in Fig. 2 to demonstrate the variable influence of cardiac glycosides under different hemodynamic conditions (at constant heart rate and constant afterload). In the absence of ouabain an increase in left ventricular end-diastolic pressure (LVEDP), induced by variation of the preload, results in a slight improvement of contractility and a corresponding enhancement of the oxygen consumption. In the presence of ouabain the same relationship is given at a higher level of contractility and oxygen consumption if no secondary functional influences exist.

In the failing heart at high LVEDP (e.g., III in Fig. 2) and increased myocardial oxygen consumption (LEVINE and WAGMAN, 1962; MESSER and NEILL, 1962; STRAUER et al., 1972) digitalization will increase the contractility resulting in a reduction of the LVEDP (III→O). This shift in the diagram is accompanied by a lower oxygen consumption than in the control state though the work output is enhanced, indicating an improved mechanical efficiency. This action will be the less pronounced the lower the original LVEDP (e.g., II, I), that means, that in the nonfailing heart with only negligible influence of cardiac glycosides on LVEDP the resulting oxygen consumption may be increased owing to the dominant effect of cardiac glycosides on contractility. In vivo this dependence of cardiac glycosides on functional state is even more pronounced because of the additional lowering of the

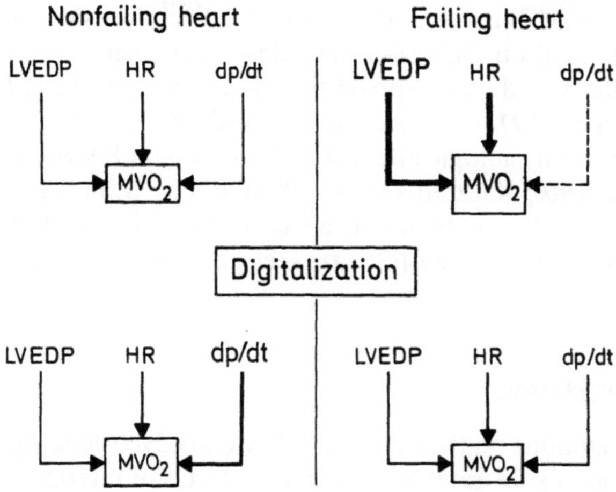

Fig. 3. Influence of digitalization on the myocardial oxygen consumption (MVO_2) in terms of its dependence on the basic oxygen consumption and the different hemodynamic states of nonfailing and failing heart. LVEDP = left ventricular end-diastolic pressure; HR = heart rate; dp/dt = contractility

heart rate (GLEASON and BRAUNWALD, 1962; WALLACE et al., 1963; HEEG, 1966; SONNENBLICK et al., 1966; FEINBERG et al., 1968).

This interpretation is supported by several observations (SIEGEL and SONNENBLICK, 1963; OPIE, 1965; SCHAPER et al., 1965; FEINBERG et al., 1968; SONNENBLICK et al., 1968; KASSEBAUM and GRISWOLD, 1970; GROSS et al., 1977). For two different functional states – failing, and nonfailing heart – the resulting changes in myocardial oxygen consumption after digitalization are illustrated in Fig. 3. In the nonfailing heart the oxygen demand of the positive inotropic effect overrides the moderate reduction due to the minor changes in heart rate and LVEDP, finally resulting in an increase of myocardial oxygen consumption. On the other hand, in the failing heart – characterized by augmented LVEDP and heart rate – the increased myocardial oxygen consumption is reduced to a normal level through lowering of the LVEDP and heart rate secondary to the inotropic effect of cardiac glycosides.

This dependence of metabolic cardiac glycoside action on functional state would explain the improvement of mechanical efficiency in the failing heart whereas in the nonfailing heart a relative increase in oxygen consumption may result. This interpretation is supported by several observations in humans (BING et al., 1950; COVELL et al., 1966; MAROKO et al., 1970; DE MOTS et al., 1976, 1978; GROSS et al., 1977).

A rather controversial discussion was initiated by postulating on this basis a promoting effect of cardiac glycosides on the precipitations of angina pectoris if cardiac glycosides are administered to coronary heart disease patients without manifest signs of cardiac failure. Though this concept is supported by some clinical observations (HOCHREIN et al., 1975; KLEIN et al., 1977; LEHMANN and HOCHREIN, 1978; LEHMANN et al., 1977, 1978, 1979), some other authors could not establish this effect in humans (BRUCE and ZOBL, 1969; KINI et al., 1972; RISLER et al., 1975; KÖTTER et al., 1977; ERBEL, 1977, 1978; TAUCHERT et al., 1977; OCHS et al., 1979).

The assumption of a deleterious effect of cardiac glycosides under these conditions is mainly based on the well-known dose-dependent S–T depression in the ECG which is interpreted as a hypoxic sign (HOCHREIN et al., 1975; KLEIN et al., 1977; LEHMANN et al., 1978) whereas other investigators rather attribute it to digitalis-specific electrophysiologic alterations (KORTH and SPANG, 1937; SCHWIEGK and JAHRMÄRKER, 1960; KJEKSHUS et al., 1972). Recently, in isolated heart preparations a clear dissociation between the effects of hypoxia and digitalis on the S–T segment has been demonstrated (GÜTTLER et al., 1980) not supporting a causal relationship.

II. Substrate Oxidation

Numerous investigations on the influence of cardiac glycosides on myocardial uptake and oxidation of substrates and on cellular levels of intermediate metabolites have failed to reveal any specific effect of cardiac glycosides (ref. see LEE and KLAUS, 1971). However, it has to be considered that most of the earlier studies were based on the measurement of steady-state concentrations only which do not permit the analysis of dynamic changes in the turnover rates of substrates or metabolites. More precise information can be derived by simultaneous measurement of substrate extraction and metabolite release or by following the oxidation rate using radioactively labeled substances. The interpretation of cardiac glycoside effects on several of these parameters is impeded by the complex nature of metabolic events following cardiac glycoside application: (a) the change in functional state may itself modify myocardial metabolism and thereby overlay any specific effect of cardiac glycosides, (b) additional extracardiac effects of cardiac glycosides on several components of substrate utilization and mobilization may interfere with their direct cardiac actions.

1. Carbohydrate Metabolism

The positive inotropic effect of cardiac glycosides seems to depend, among other things, on the presence of oxidizable substrates, especially glucose (BAILEY and DRESEL, 1971). This is in accordance with the observation of an improved inotropic action of digoxin in guinea pig hearts (KIM et al., 1970; KIM and DRESEL, 1970) and of an increased rate of onset of digitalis action in rabbit atria (BAILEY and DRESEL, 1969) in the presence of insulin. These authors, however, relate the observed modifications in cardiac glycoside action to an insulin-induced additional uptake of digoxin into the myocardium rather than to an improved energy metabolism. All factors, including cardiac glycosides, which increase heart work are known to accelerate myocardial glucose extraction and oxidation (KIEN et al., 1960; KREISBERG and WILLIAMSON, 1964; WILLIAMSON, 1964; DAVIS, 1965; ARESE et al., 1967, 1969; NEELY et al., 1967b; CRASS et al., 1969; HOESCHEN, 1971; BENMOUYAL, 1972; BIHLER and SAWH, 1973, 1975a, b). This effect is suggested to be secondary to an enhanced contractile activity since the latter has been reported both to facilitate glucose transport (MORGAN et al., 1965; HOLLOSZY and NARAHARA, 1967; NEELY et al., 1967b) and to stimulate the glycolysis by a rapid activation of phosphofructokinase (NEELY et al., 1976). However, corresponding observations in nonbeating

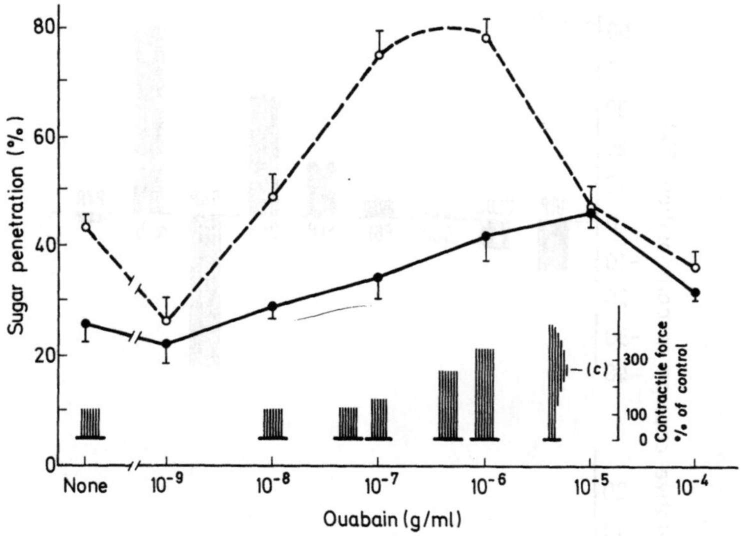

Fig. 4. Effect of ouabain on myocardial uptake of 3-methylglucose and on contractile force in resting *(solid circles)* and beating *(open circles)* guinea pig left atria. C = contracture. (BIHLER and SAWH, 1979)

heart muscle preparations (e.g., left atria, heart slices) indicate an additional function-independent pathway in digitalis-induced glucose uptake and oxidation (WOLLENBERGER, 1951a, b, 1953; CREVASSE and SHIPP, 1962; YUSUF and GANS, 1966a, b; BIHLER and SAWH, 1975a, b) which can even be demonstrated in skeletal muscle (CLAUSEN, 1965; HO and JEANRENAUD, 1967; BIHLER, 1968; BIHLER and SAWH, 1973).

A comparison of the cardiac glycoside influence on the function-dependent and function-independent rate of sugar uptake is presented in Fig. 4. The most plausible explanation for this insulin-like cardiac glycoside effect is based on the well-established assumption that membrane transport is the rate-limiting step of glucose utilization in muscle, including heart muscle (ref. see PARK et al., 1968; NEELY et al., 1972) which is controlled by several regulatory factors adjusting the cellular uptake of glucose to the functional activity and metabolic pattern (ELBRINK and BIHLER, 1975).

An increase in the rate of glucose utilization, as observed in the presence of cardiac glycosides (KIEN et al., 1960; KIEN and SHERROD, 1960), must therefore involve an activation of the glucose transport system which may be mediated by an elevation of the intracellular sodium concentration secondary to the well known inhibition of Na^+, K^+-ATPase by high concentrations of cardiac glycosides (SCHWARTZ et al., 1972, 1975). Consistent with this interpretation is the opposite influence of very low cardiac glycoside concentrations which are known to stimulate the sodium pump and simultaneously depress myocardial glucose transport (BIHLER and SAWH, 1975a, b, 1979).

This analysis of cardiac glycoside effect on sugar transport in cardiac muscle revealed two components: one related to changes in internal sodium level (present in both resting and beating heart muscle preparations and also in other types of

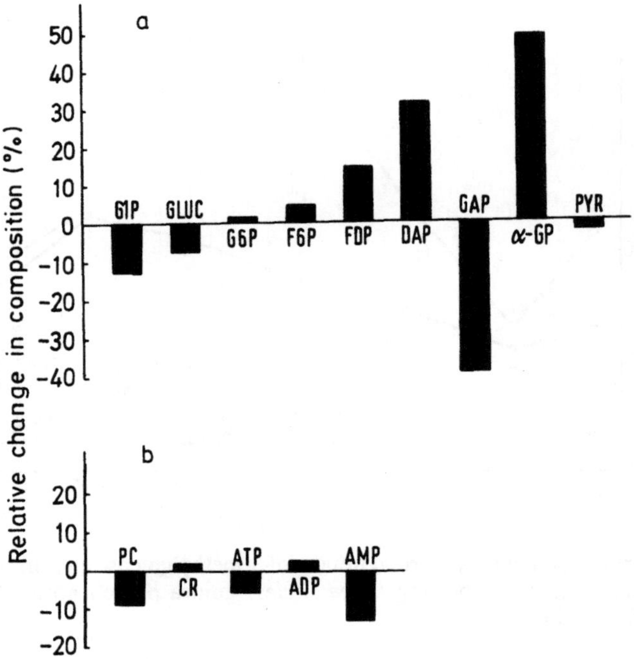

Fig. 5 a, b. Relative changes of the substrates and metabolites which discriminate ouabain-treated from control frog heart; **a** glycolytic pathway; **b** energy-rich compounds. G1P = glucose-1-P; GLUC = glucose; G6P = glucose-6-P; F6P = fructose-6-P; FDP = fructose diphosphate; DAP = dihydroxyacetone-P; GAP = glyceraldehyde-P; α-GP = α-glycero-P; PYR = pyruvate; PC = P-creatine; CR = creatine; ATP = adenosine triphosphate; ADP = adenosine diphosphate; AMP = adenosine monophosphate. (After ARESE et al., 1967)

muscles) and the other related to the contractile function; both may be influenced by cardiac glycosides in a different way, dependent on the concentrations used (BIHLER and SAWH, 1979).

A more detailed analysis of the metabolic steps following the increased glucose uptake in frog heart (ARESE et al., 1967, 1969) disclosed a specific change in metabolite pattern of ouabain-treated preparations (Fig. 5). A cross-over point was noticed within the glucokinase – phosphoglucomutase steps consistent with the increased rate of glucose entry into the cells. The change in glycolytic pattern indicates a relative impairment between phosphofructokinase and pyruvatekinase either due to an activation of phosphofructokinase or to an inhibition in the phosphoglyceromutase – enolase – pyruvatekinase sequence. The increased ATP breakdown occuring during positive inotropic activity may indeed activate phosphofructokinase (WILLIAMSON, 1966; SWYNGHEDAUW and CORSIN, 1969), but an argument against phosphofructokinase stimulation is the simultaneously observed decrease in AMP, a powerful effector of this kinase (LOWRY and PASSONNEAU, 1966). More likely a relative inhibition of the pyruvatekinase may prevail as a consequence of the ionic shifts caused by the subtoxic concentrations of ouabain used: both the loss of the activator potassium and the accumulation of the inhibitor calcium might be responsible for a decrease in this enzyme activity (LEHNINGER, 1977) causing the reported alterations in glycolytic profile.

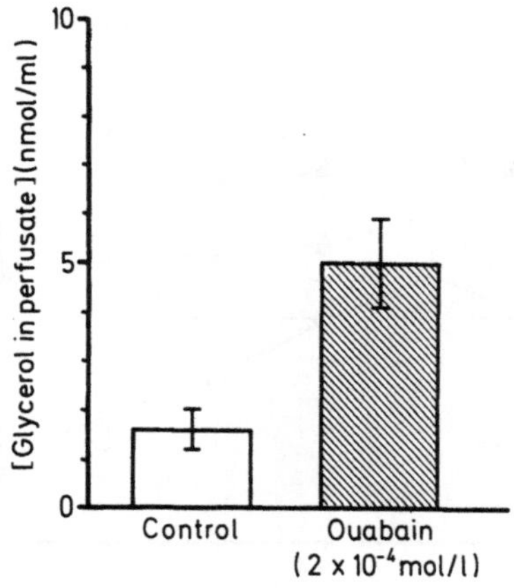

Fig. 6. Gycerol release in isolated perfused rat heart under control conditions and ouabain treatment. (After JESMOK et al., 1977)

The redox ratios α-glycero-P : dihydroxyacetone-P, lactate : pyruvate and malate : oxaloacetate were not appreciably affected by ouabain. The decrease of phosphocreatine as well as the slight increase of creatine are presumably due to functional activation whereas the rather constant values for ATP and ADP may be explained by the buffering effect of phosphocreatine.

However, it should be kept in mind that all these observations on metabolic ouabain effects cannot be generalized owing to the functional and biochemical peculiarities of frog heart (OPIE, 1971/1972).

2. Fatty Acid Metabolism

The reported observations of cardiac glycoside effects on fatty acid metabolism are conflicting but the models and conditions used to test this influence are too different to yield consist results. The analysis of serum levels of free fatty acids and glycerol in humans revealed both lipolytic (SIEDEK et al., 1968a, b) and antilipolytic effects (OGILVIE and KLASSEN, 1972; SCHWARTZ et al., 1972). By measurement of the arteriovenous concentration difference no significant changes in the myocardial free fatty acid extraction were observed at different α-acetyldigoxin doses in dogs (KRAUPP et al., 1968), whereas this parameter was influenced inconsistently by lanatoside C in humans (BLAIN et al., 1956).

The discrepancies in these observations might be caused by extracardiac effects of cardiac glycosides on lipid metabolism in different tissues superimposed on the direct cardiac effects. Thus, several attempts to elucidate the cardiac glycoside action on isolated heart preparations have been performed; both in rabbit heart (GOUSIOS et al., 1967) and in rat heart (JESMOK et al., 1975, 1977) positive inotropic concentrations of ouabain stimulated free fatty acid oxidation (as indicated by an increased glycerol release; Fig. 6) but did not influence the extraction of free fatty acids (GOUSIOS et al., 1967). The lack of effect of ouabain on palmitate metabolism in rat heart (HOESCHEN, 1969, 1971) may be related to the high initial rates of glycerol release obtained in this particular experimental set-up (no stabilization period

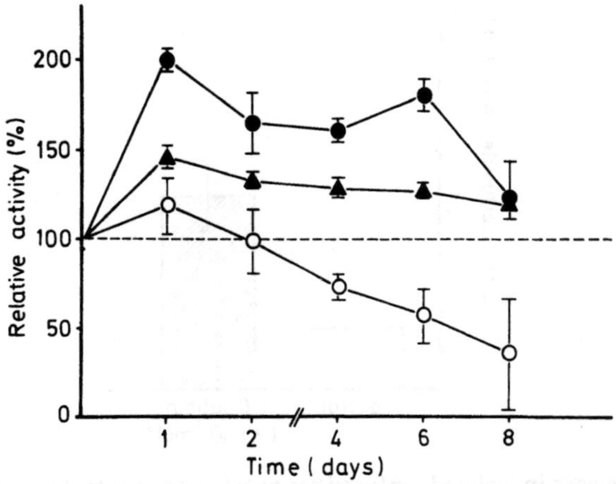

Fig. 7. The relative changes in enzyme activities in rat left ventricular muscle during 8 days digoxin treatment (1 mg/kg subcutaneously daily) expressed as percentages of day-0-value. Succinate dehydrogenase *(solid circles)*; citrate synthase *(triangles)*; phosphofructokinase *(open circles)*. (After PELKONEN, 1978)

after preparation, recirculating system) obscuring any significant drug-induced increase in glycerol release. Contrary to the repeatedly observed inhibition by cardiac glycosides of the stimulated lipolysis in fat cells (KUJALOVA and MOSINGER, 1966; HO et al., 1966, 1967; HO and JEANRENAUD, 1967; SIEDEK et al., 1968 a, b; YU et al., 1969, 1971; YU and TRIESTER, 1972; FAIN et al., 1973) which seems to be mediated by an interference with the adenylate cyclase, there is no experimental evidence for a direct action of cardiac glycosides on biochemical processes governing the lipolysis in heart muscle; both cyclic AMP content (LEE and KLAUS, 1971; MAYER, 1974) and protein kinase activity ratio (JESMOK et al., 1977) were found to be uninfluenced, even by high drug concentrations.

In view of the fact that increased cardiac work per se induced an enhanced uptake, mobilization, and oxidation of exogenous and endogenous fatty acids (CRASS et al., 1971; CRASS and SHIPP, 1972) it seems to be more likely that the increased free fatty acid oxidation might be a consequence of the positive inotropic cardiac glycoside action rather than a direct metabolic effect.

3. Enzyme Activities

All attempts to relate the metabolic effects of cardiac glycosides to a specific influence on enzymes of myocardial energy metabolism have been disappointing as far as isolated purified glycolytic and mitochondrial enzymes or heart cell homogenates have been studied (WOLLENBERGER, 1947, 1949; HELMREICH, 1950; SAUNDERS et al., 1950; SEGRE, 1950; DOULL et al., 1951; REITER and BARRON, 1952).

However, in preparations with intact cellular structures several changes in enzyme activities were reported which presumably are secondary either to the modified metabolic demand following the increased functional activity or to alterations in the intracellular milieu. Most obviously, the few observations of cardiac glycoside-stimulated phosphorylase A activity in the myocardium (WENZEL and

HECHT, 1969; ABIKO and ICHIHARA, 1978) were obtained only in experiments with high doses of cardiac glycosides which are known to release endogenous catecholamines (TANZ and MARCUS, 1966; CHATTERJEE and ROY, 1969; CHERKES and FRANTSUZOVA, 1970; ROY and CHATTERJEE, 1970) and, therefore, they cannot be regarded as cardiac glycoside-specific reactions. This is in accordance with other investigations on the lack of influence of ouabain on the myocardial level of cyclic AMP (MAYER and MORAN, 1960; WATANABE and BESCH, 1974; DÖNGES et al., 1977).

During prolonged digoxin treatment of rats a change in enzyme activity pattern has been reported (PELKONEN, 1978) (Fig. 7) which, in part, is consistent with previously observed cardiac glycoside effects on substrate metabolism (see Sect. B.II.1). A continuous fall in the activity level of the phosphofructokinase, the main regulatory enzyme of glycolysis, indicates that this pathway is not dominant in digoxin-treated animals. On the other hand, the rise in citratesynthetase activity may be interpreted as a shift towards a more aerobic metabolic state. In accordance with this assumption, a significant increase in the level of succinatedehydrogenase activity has been found which seems to be indicative of an increase in the amount of mitochondria (ARCASOY and SMUCKLER, 1969). The linking mechanism might be the stimulated turnover of energy-rich compounds since nucleotides are known to play a role in the activation of succinatedehydrogenase in mitochondria (GUTMAN et al., 1971).

C. Influence of Cardiac Glycosides on Myocardial Energy Conservation

Since the inotropic activity of cardiac glycosides has long been assumed to be related to improved myocardial energetics (ref. see WOLLENBERGER, 1949; ROTHLIN and TAESCHLER, 1956) a great deal of work has been concerned in the past with the action of cardiac glycosides on myocardial energy conservation.

I. Myocardial Content of Energy-Rich Compounds

The analysis of the energy-rich phosphate content of the heart as influenced by cardiac glycosides revealed conflicting results: reduction (HARVEY, 1955; REBAR et al., 1957; FRANTSUZOVA, 1974; DÖRING and HAUF, 1977), increase (WEICKER, 1935; GREINER, 1952; STAUB, 1959; FRANTSUZOVA, 1974), and no change (WOLLENBERGER, 1949, 1951 c; BOGATZKI and STRAUB, 1956; FURCHGOTT and LEE, 1961; PELOSI et al., 1969) have been reported. A conclusive comparison of these data is impeded by the different experimental conditions and methods used. On the basis of standardized investigations on dog heart–lung preparations WOLLENBERGER (1949, 1951 c) came to the conclusion that the myocardial content of energy-rich phosphates is most probably not influenced by therapeutic doses of cardiac glycosides. This, in principle, has been confirmed later on by experiments on guinea pig atria (FURCHGOTT and DEGUBAREFF, 1958) and cat papillary muscle (LEE et al., 1960 b, 1961). This conclusion seems to be reasonable so far as the function-dependent increase in the ATP breakdown can be compensated for by a corresponding increase in the turnover rate of the energy-rich compounds (DÖRING and HAUF, 1977).

Toxic concentrations of cardiac glycosides, however, were uniformly found to reduce the myocardial content of ATP and phosphocreatine with a concomitant increase in the content of ADP, AMP, and inorganic phosphate (WOLLENBERGER, 1949, 1951 c; GREINER, 1952; BOGATZKI and STRAUB, 1956; FURCHGOTT and DE-GUBAREFF, 1958; WILLIAMS and NAYLER, 1977).

II. Oxidative Phosphorylation

An impairment of the oxidative phosphorylation as suggested on the basis of these observations could not be demonstrated in isolated mitochondria, even at extremely high concentrations of cardiac glycosides (WOLLENBERGER, 1949, 1951 c; ROTHLIN and SCHOELLY, 1950; LANGEMANN et al., 1953; GRISOLIA, 1955 a, b; LEE and MCELROY, 1955; PLAUT et al., 1957; LEE et al., 1960 a; GERVAIS et al., 1978). The slight increase of oxygen consumption in the absence of phosphate acceptors observed in mitochondria isolated from cardiac glycoside-intoxicated heart (LEE et al., 1960 a; SCHWARTZ, 1975; GERVAIS et al., 1978) was interpreted as an impaired oxidative phosphorylation in accordance with the above observations on intact heart muscle. The discrepancy between the results obtained on isolated mitochondria and preparations with cellular integrity may indicate the indirect nature of this toxic cardiac glycoside effect (CHANCE, 1965; LEHNINGER et al., 1967; KREBS, 1970; SCHWARTZ, 1975; GERVAIS et al., 1978) which most probably is due to the intracellular accumulation of calcium under these conditions (see Chap. 18). This interpretation is supported by the lack of a stimulatory influence of ouabain on tissue respiration in the absence of external calcium ions (WOOD and SCHWARTZ, 1970; WOOD et al., 1972).

D. Influence of Cardiac Glycosides on Myocardial Energy Utilization

In the foregoing outline of cardiac glycoside actions on various distinct parts of the myocardial energy metabolism their effects on energy utilization have already been referred to briefly. Only a few studies have been concerned with this phase of energy metabolism, the conversion of chemical energy into mechanical work.

I. Utilization of Energy-Rich Compounds

Direct measurement of the parameters involved did not reveal any influence of digoxin on the resting energy utilization of cat papillary muscle but disclosed an increase in contracting preparations, probably related to the increase in myocardial contractility (SKELTON et al., 1970; SNOW, 1979). Similar results on ATP hydrolysis were obtained in frog heart (DAL PRA et al., 1972) and guinea pig heart (DAL PRA and SEGRE, 1980). In a study, which allowed comparative analysis of heart work and energy turnover no significant influence of cardiac glycosides on mechanical efficiency could be demonstrated (GIBBS and GIBSON, 1969).

On the other hand, none of these effects on energy utilization could be established on isolated enzymes, subcellular structures, or in muscle preparations in which the cellular integrity was lost (CHEN and GEILING, 1947; KUSCHINSKY et al., 1952; REITER and BARRON, 1952; LANGEMANN, 1953; WASER and VOLKART, 1954; BOGATZKI and STRAUB, 1956; READ and KELSEY, 1957; ANGKAPINDU et al., 1959; LEE et al., 1960a). The specificity of repeatedly reported slight increase (GUERRA et al., 1946; HEGGLIN et al., 1949; SEGRE, 1949; EDMAN, 1950, 1951; MÜNCHINGER, 1953) and inhibition (KIMURA and DUBOIS, 1947; EDMAN, 1950, 1951; HELMREICH and SIMON, 1952; PROCTOR and MUELLER, 1957; BOGATZKI, 1961) of ATPase activities in such systems remains to be elucidated since, in most cases extremely high concentrations of cardiac glycosides have been used (far above the functionally active range) and similar effects could even be induced by steroids without cardiotonic activity (MOR, 1953, 1955).

This difference in cardiac glycoside effects between intact heart muscle preparations and isolated biochemical systems strongly suggests that the alteration in energy utilization by cardiac glycosides is mediated through some other effects of cardiac glycosides which require the presence of cell membrane or which are directly related to contractile function. Since it is well established that cardiac glycosides primarily affect the myocardial electrolyte metabolism (see Chap. 18) it is reasonable to assume that the metabolic events of cardiac glycosides are (analogous to the functional effects) secondary to the induced electrolyte changes, mainly to the modified calcium distribution.

II. Myocardial Heat Production

The total myocardial energy output comprises work and heat production both of which are related to the major ATPase activities underlying myocardial contraction (GIBBS, 1978; GIBBS and CHAPMAN, 1979). Therefore, it is not surprising that myocardial heat production is closely related to mechanical work (MCDONALD, 1971; THEISOHN et al., 1972, 1977; COULSON and RUSY, 1973; RUSY and COULSON, 1973; COULSON, 1976; GIBBS and CHAPMAN, 1979) but additional components (e.g., maintenance heat, activation heat) have to be considered. Cardiac muscle is characterized by a high resting heat rate, the magnitude of which varies between species and depends on the metabolic substrates (GIBBS and CHAPMAN, 1979). Very few studies have analyzed the influence of cardiac glycosides on these distinct energy parameters (GIBBS and GIBSON, 1969; KLAUS et al., 1972; GÜTTLER and THEISOHN, 1973; DELAUNOIS et al., 1974; GÜTTLER, 1974), but they came to identical conclusions. Whereas the resting heat production remained uninfluenced by cardiac glycosides, the function-dependent heat was increased in proportion to the inotropic effect and was always closely related to the myocardial oxygen consumption (Fig. 8). There is no evidence that cardiac glycosides may alter the efficiency of either the chemomechanical energy transformation or other energy-consuming cellular processes (SARNOFF et al., 1964; COVELL et al., 1966; GIBBS and GIBSON, 1969).

E. Additional Metabolic Effects

In addition to the direct and indirect influences of cardiac glycosides on myocardial energy metabolism a few supplementary effects on other cellular metabolic events will briefly be discussed.

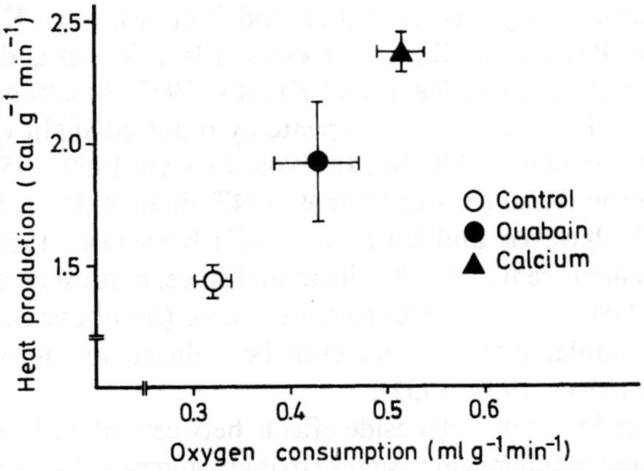

Fig. 8. Relation between oxygen consumption and heat production of beating rabbit heart in different inotropic states: control = 0.9 mmol/l Ca^{2+}; ouabain = 0.4 µmol/l; Ca^{2+} = 3.6 mmol/l. (GÜTTLER, 1974)

Phospholipids have repeatedly been suggested to be involved in the myocardial effects of cardiac glycosides (KUTSCHERA-AICHBERGER, 1929; HAJDU et al., 1957; MARINETTI et al., 1961; BONINSEGNA and DOMINICI, 1972; FRITZ, 1975). Both increase (KUTSCHERA-AICHBERGER, 1929; MARINETTI et al., 1961; NELSON and CORNATZER, 1964; TSAO and CORNATZER, 1967; FRITZ, 1975) and reduction (KUTSCHERA-AICHBERGER, 1929; NICHOLLS et al., 1962; VERNICK et al., 1973) in myocardial phospholipid content and turnover have been reported. These discrepancies may be due, at least in part, to methodologic and species differences as well as to the complex and variable situations of in vivo experiments. In more recent studies on slices of rabbit atria under standardized conditions ouabain was found to decrease the turnover of particular phospholipids (VERNICK et al., 1973) which might be related to an inhibition of the membrane transport function since the phosphatidic acid cycle is suggested to be an integral part of the sodium pump (HOKIN and HOKIN, 1959, 1961).

Myocardial protein metabolism has occasionally been studied as the possible target of cardiac glycoside action (TURTO et al., 1971; TURTO and LINDY, 1973; TURTO, 1977). Pretreatment of rats with digitoxin inhibited protein synthesis (TURTO et al., 1971; TURTO and LINDY, 1973; TURTO, 1977) and probably thereby inhibited development of experimental cardiac hypertrophy (WILLIAMS and BRAUNWALD, 1965; TURTO et al., 1971; TURTO and LINDY, 1973; TURTO, 1977). Furthermore, slight changes in RNA synthesis (POSNER and FANBURG, 1968; TURTO and LINDY, 1973) and in the activity of prolyl hydroxylase (TURTO, 1977) have been demonstrated. Using isolated cardiac ribosomes, however, no influence of cardiac glycosides could be detected (TURTO, 1977). This again may point to the necessity of intact cellular structures for the manifestation of cardiac glycoside action, presumably because of a primary influence of these drugs on transport processes which have only secondary consequences for different metabolic mechanisms (SCHWARTZ et al., 1972).

In accordance with this interpretation is the reported inhibition by high ouabain concentrations of amino acid transport and incorporation into rabbit atrial muscle (KYPSON and HAIT, 1971) and into several kinds of extracardiac tissue (PAINE and HEINZ, 1960; GÜNTHER and WINKELMANN, 1961; GONDA and QUASTEL, 1962; KOSTYO and SCHMIDT, 1963). It seems very likely that the inhibitory effect of cardiac glycosides on protein synthesis is a consequence of the inhibition of transmembrane transport of amino acids. A relationship between Na^{+}, K^{+}-ATPase and amino acid transport has been suggested previously (WINTER and CHRISTENSEN, 1964; YUNIS and ARIMURA, 1965). In this context it is interesting to note that ouabain pretreatment may even modify the myocardial ultrastructure (Golgi membranes) (LINDOWER and MARKS, 1977) similar to ionophore-induced changes in the Golgi complex of smooth muscle (GARFIELD et al., 1974). It therefore seems reasonable to assume that these effects are secondary to the ionic changes induced by ouabain treatment.

F. Conclusions

The several modifications by cardiac glycosides of myocardial energy liberation, conservation, and utilization are closely related to the induced functional changes. With regard to oxygen consumption, substrate oxidation, metabolite turnover, and energy-rich phosphate content there is no specific influence of cardiac glycosides detectable on purified enzymes, subcellular components, cardiac homogenates, and other noncellular systems. In heart preparations with intact cellular integrity, however, the metabolic state is changed according to the drug-dependent alterations in cardiac contractility. This relationship is most probably due to the cardiac glycoside-induced ionic shifts in cardiac muscle which are responsible both for the activation of the contractile system and of some metabolic pathways. Thus, increased myocardial oxygen consumption and heat production, stimulated glucose and fatty acid extraction and oxidation, modified cellular enzyme activity pattern, and energy-rich compound turnover correspond fairly well to the positive inotropic action. In addition, few metabolic processes (such as glucose uptake) are also influenced in a function-independent manner, obviously by the underlying ionic changes. But there is no evidence that cardiac glycosides, as far as the therapeutic concentration range is concerned, may alter the efficiency of either the chemomechanical energy transformation or other energy-consuming cellular processes. In the failing heart, however, a relative reduction in energy demand may result as a consequence of the cardiac glycoside-induced functional and hemodynamic changes.

References

Abiko, Y., Ichihara, K.: Effect of ouabain on myocardial metabolic and contractile responses to coronary ligation. Eur. J. Pharmacol. *47*, 87–94 (1978)

Angkapindu, A., Stafford, A.W., Thorp, R.H.: The influence of cardiac glycosides on the deamination of adenosine and the adenine nucleotides. Arch. Int. Pharmacodyn. *119*, 194–204 (1959)

Arcasoy, M.M., Smuckler, E.A.: Acute effect of digoxin intoxication on rat hepatic and cardiac cells. Lab. Invest. *20*, 190–201 (1969)

Arese, P., Bosia, A., Rossini, L.: Short-term effects of ouabain on energy-rich, glycolytic, and citric-acid-cycle intermediates in frog heart. Biochem. Biophys. Res. Commun. *27*, 138–142 (1967)

Arese, P., Bosia, A., Rossini, L.: Short-term effects of calcium, potassium, and of ouabain on metabolite levels in the frog heart in vivo. Eur. J. Biochem. *11*, 80–88 (1969)
Arnold, G., Kosche, F., Miessner, E., Neitzert, A., Lochner, W.: The importance of the perfusion pressure in the coronary arteries for the contractility and the oxygen consumption of the heart. Pflügers Arch. *229*, 339–356 (1968)
Arnold, G., Morgenstern, C., Lochner, W.: The autoregulation of the heart work by the coronary perfusion pressure. Pflügers Arch. *321*, 34–55 (1970)
Bacaner, M.B., Lioy, F., Visscher, M.B.: Induced change in heart metabolism as a primary determinant of heart performance. Am. J. Physiol. *209*, 519–531 (1965)
Bailey, L.E., Dresel, P.E.: Increase in the rate of onset of the inotropic response to ouabain due to insulin. Life Sci. *8*, 347–351 (1969)
Bailey, L.E., Dresel, P.E.: Role of the sugar transport system in the positive inotropic response to digitalis. J. Pharmacol. Exp. Ther. *176*, 538–544 (1971)
Benmouyal, E.: Effects of ions, ouabain, epinephrine, and tetrodotoxin on glucose metabolism and on contractility in perfused guinea pig heart. Can. J. Physiol. Pharmacol. *50*, 584–593 (1972)
Bernsmeier, A., Rudolph, W.: Myokardstoffwechsel. Verh. Dtsch. Ges. Kreislaufforsch. *27*, 59–76 (1961)
Bihler, I.: The action of cardiotonic steroids on sugar transport in muscle, in vitro. Biochim. Biophys. Acta (Amst.) *163*, 401–410 (1968)
Bihler, I., Sawh, P.C.: The effect of oxidative metabolic substrates on the membrane transport of sugars and on the action of ouabain in skeletal muscle in vitro. Can. J. Physiol. Pharmacol. *51*, 371–377 (1973)
Bihler, I., Sawh, P.C.: Sugar transport in the perifused left atrium: effect of contraction, frequency, digitalis, and ionic alterations. J. Mol. Cell. Cardiol. *7*, 345–355 (1975a)
Bihler, I., Sawh, P.C.: A comparison of the effects of ouabain on sugar and sodium transport and on contractile activity of isolated atria. Proc. West. Pharmacol. Soc. *18*, 106–109 (1975b)
Bihler, I., Sawh, P.C.: Regulation of sugar transport in the myocardium: effects of ouabain in resting and beating isolated guinea pig left atria. J. Mol. Cell. Cardiol. *11*, 407–414 (1979)
Bing, R.J.: Metabolism of the heart. Harvey Lect. *50*, 27–70 (1954)
Bing, R.J.: Über den Stoffwechsel des intakten Herzens. Verh. Dtsch. Ges. Kreislaufforsch. *27*, 145–166 (1961)
Bing, R.J.: Cardiac metabolism. Physiol. Rev. *45*, 171–213 (1965)
Bing, R.J., Maraist, F.M., Dammann, J.F., Draper, A., Heimbecker, R., Daley, R., Gerard, R., Calazel, P.: Effect of strophanthus on coronary blood flow and cardiac oxygen consumption of normal and failing human hearts. Circulation *2*, 513–516 (1950)
Blain, J.M., Eddlemann, E.E., Siegel, A., Bing, R.J.: Studies on myocardial metabolism. V. The effects of lanatoside C on the metabolism of the human heart. J. Clin. Invest. *35*, 314–321 (1956)
Boerth, R.C., Hammermeister, K.E., Warbasse, J.R.: Comparative influence of ouabain, norepinephrine, and heart rate on myocardial oxygen consumption and inotropic state in dogs. Am. Heart J. *96*, 355–362 (1978)
Bogatzki, M.: Adenylpyrophosphatase in the myocardium under the influence of various pharmacological agents. Z. Ges. Exp. Med. *134*, 454–460 (1961)
Bogatzki, M., Straub, H.: ATP-System und übriger Phosphatstoffwechsel im normalen und akut-insuffizienten Herzmuskel unter der Behandlung mit Herzglykosiden. Z. Ges. Exp. Med. *127*, 425–449 (1956)
Boninsegna, A., Dominici, A.: Inotropic effect of cardiac phospholipids on the contraction of rabbit heart perfused with a calcium deficient medium. Biochem. Pharmacol. *21*, 601–605 (1972)
Braunwald, E.: Control of myocardial oxygen consumption. Physiologic and clinical considerations. Am. J. Cardiol. *27*, 416–432 (1971)
Braunwald, E., Bloodwell, R.D., Goldberg, L.T.: Studies on digitalis. IV. Observations in man on the effect of digitalis preparations on the contractility of the nonfailing heart and on total vascular resistance. J. Clin. Invest. *40*, 52–59 (1961)

Braunwald, E., Ross, J. Jr., Sonnenblick, E.H.: Myocardial energetics. In: Mechanisms of contraction of the normal and failing heart, 2nd. ed., pp. 166–199. Boston: Little, Brown and Company 1976
Bretschneider, H.J.: Überlebenszeit und Wiederbelebungszeit des Herzens bei Normo- und Hypothermie. Verh. Dtsch. Ges. Kreislaufforsch. *30*, 11–34 (1964)
Bretschneider, H.J.: Die hämodynamischen Determinanten des Sauerstoffbedarfes des Herzmuskels. Arzneim. Forsch. *21*, 1515–1517 (1971)
Britman, N.A., Levine, H.J.: Contractile element work: a major determinant of myocardial oxygen consumption. J. Clin. Invest. *43*, 1397–1408 (1964)
Bruce, T.A., Zobl, E.: Myocardial oxidative metabolism in normal subjects during ouabain administration. J. Lab. Clin. Med. *74*, 855 (1969)
Bugge-Asperheim, B.: Effects of increased aortic blood pressure on myocardial performance and metabolism during non-adrenergic inotropic stimulation of the heart. Scand. J. Clin. Lab. Invest. *30*, 137–143 (1972)
Burdette, W.J.: Increase in oxygen consumption of human cardiac muscle incubated with lanatosideC. J. Lab. Clin. Med. *40*, 867–871 (1952)
Carafoli, E.: Mitochondria in the contraction and relaxation of heart. In: Myocardial biology. Dhalla, N.S. (ed.), pp. 393–406. Baltimore: University Park Press 1974
Caroni, P., Carafoli, E.: An ATP-dependent Ca^{++}-pumping system in dog heart sarcolemma. Nature (Lond.) *283*, 765–767 (1980)
Chance, B.: The energy-linked reaction of calcium with mitochondria. J. Biol. Chem. *240*, 2729–2748 (1965)
Chatterjee, M.L., Roy, A.R.: Changes in the nor-adrenaline content of rabbit heart after treatment with ouabain. Bull. Calcutta Sch. Trop. Med. *17*, 50–51 (1969)
Chen, G., Geiling, E.M.K.: Some aspects of biochemistry and pharmacology of heart. Schweiz. Med. Wochenschr. *77*, 25–28 (1947)
Cherkes, A.I., Frantsuzova, S.B.: The influence of cardiac glycosides on the catecholamine content in the myocardium and adrenal glands in rats. Biull. Eksp. Biol. Med. *70*, 49–52 (1970)
Chidsey, C.A.: Potential cellular defects responsible for myocardial failure. In: Factors influencing myocardial contractility. Tanz, R.D., Kavaler, F., Roberts, J. (eds.), pp. 37–42. New York: Academic Press 1967
Clancy, R.L., Graham, T.P. Jr., Powell, W.J. Jr., Gilmore, J.P.: Inotropic augmentation of myocardial oxygen consumption. Am. J. Physiol. *212*, 1055–1061 (1967)
Clausen, T.: The relationship between the transport of glucose and cations across the cell membrane in isolated tissues. I. Stimulation of glycogen deposition and inhibition of lactic acid production in diaphragm induced by ouabain. Biochim. Biophys. Acta (Amst.) *109*, 164–171 (1965)
Coleman, H.N.: Role of acetylstrophanthidin in augmenting myocardial oxygen consumption. Circ. Res. *21*, 487–495 (1967)
Constantin, L.L., Franzini-Armstrong, C., Podolsky, R.J.: Localization of calcium-accumulating structures in striated muscle fibers. Science *147*, 158–160 (1965)
Coulson, R.L.: Energetics of isovolumic contractions of the isolated rabbit heart. J. Physiol. (Lond.) *260*, 45–53 (1976)
Coulson, R.L., Rusy, B.F.: A system for assessing mechanical performance, heat production and oxygen utilization of isolated perfused whole hearts. Cardiovasc. Res. *7*, 859–869 (1973)
Covell, J.W., Braunwald, E., Ross, J. Jr., Sonnenblick, E.H.: Studies on digitalis. XVI. Effects on myocardial oxygen consumption. J. Clin. Invest. *45*, 1535–1542 (1966)
Cowley, A.W., Scott, J.C., Spitzer, J.J.: Myocardial FFA metabolism during coronary infusion of norepinephrine in conscious dogs. Am. J. Physiol. *217*, 511–517 (1969)
Crass, M.F. III, McCaskill, E.S., Shipp, J.C.: Effect of pressure development on glucose and palmitate metabolism in perfused heart. Am. J. Physiol. *216*, 1569–1576 (1969)
Crass, M.F. III, McCaskill, E.S., Shipp, J.C., Murthy, V.K.: Metabolism of endogenous lipids in cardiac muscle: effect of pressure development. Am. J. Physiol. *220*, 428–435 (1971)

Crass, M.F. III, Shipp, J.C.: Metabolism of exogenous and endogenous fatty acids in heart muscle. In: Myocardiology. Bajusz, E., Rona, G. (eds.), pp. 115–126. München, Berlin, Wien: Urban & Schwarzenberg 1972
Crevasse, L., Shipp, J.C.: The interactions of calcium and lanatoside C on carbohydrate metabolism in myocardial tissue slices. Mem. IV. Congr. Mund. Cardiol., Mexico City, 1962 vd *5*, 208–215 (1962)
Crevasse, L., Wheat, M.W.: Role of calcium and lanatoside C in increasing oxygen consumption in human myocardial tissue slices. Circ. Res. *11*, 721–726 (1962)
Dal Pra, P., Periti, P., Rossini, L.: Ouabain effect on the levels of nucleotide phosphates in frog heart perfusate. Pharmacol. Res. Commun. *4*, 287–304 (1972)
Dal Pra, P., Segre, G.: Effect of ouabain on the kinetics of ATP hydrolysis in the guinea pig heart. Pharmacol. Res. Commun. *12*, 73–84 (1980)
Danforth, W.H., Ballard, F.B., Kako, K., Choudhury, J.D., Bing, R.J.: Metabolism in heart failure. Circulation *21*, 112–123 (1960)
Davies, R.E.: A molecular theory of muscle contraction: calcium-dependent contractions with hydrogen bond formation plus ATP-dependent extensions of part of the myosin-actin cross-bridges. Nature (Lond.) *199*, 1068–1074 (1963)
Davis, E.J., The effects of insulin and adrenaline on the metabolism of glucose and lactate by perfused guinea pig hearts. Can. J. Biochem. *41*, 1001–1010 (1965)
Delaunois, A.L., Bernard, P.J., Dujardin, J.P.: Influence of cardiotonic and cardioplegic agents on the heat production of the heart. Arch. Int. Physiol. Biochim. *82*, 229–238 (1974)
DeMots, H., Rahimtoola, S.H., Kremkau, E.L., Bennett, W., Mahler, D.: Effects of ouabain on myocardial oxygen supply and demand in patients with chronic coronary artery disease. J. Clin. Invest. *58*, 312–319 (1976)
DeMots, H., Rahimtoola, S.H., McAnulty, J.H., Porter, G.A.: Effects of ouabain on coronary and systemic vascular resistance and myocardial oxygen consumption in patients with heart failure. Am. J. Cardiol. *41*, 88–93 (1978)
Dönges, C., Heitmann, M., Jungbluth, H., Meinertz, T., Schmelzle, B., Scholz, H.: Effectiveness of theophylline to increase cyclic AMP levels and force of contraction in electrically paced guinea-pig auricles. Comparison with isoprenaline, calcium, and ouabain. Naunyn-Schmiedebergs Arch. Pharmacol. *301*, 87–97 (1977)
Döring, H.J., Hauf, G.: Kontraktilität sowie ATP- und Kreatinphosphat-Konzentrationen des Meerschweinchenmyokards bei Normoxie und verschiedenen Hypoxiegraden. Herz/Kreisl. *9*, 926–933 (1977)
Doull, J., Herrmann, R.G., Geiling, E.M.K., Dubois, K.P.: Effects of Bufagin on the respiration of cardiac muscle and other tissues. Arch. Int. Pharmacodyn. *86*, 487–496 (1951)
Ebashi, S.: Excitation – contraction coupling. Ann. Rev. Physiol. *38*, 293–313 (1976)
Ebashi, S., Endo, M.: Calcium ion and muscle contraction. Prog. Biophys. Mol. Biol. *18*, 123–183 (1968)
Ebashi, S., Lipmann, F.: Adenosine triphosphate-linked concentration of calcium ions in a particulate fraction of rabbit muscle. J. Cell. Biol. *14*, 389–400 (1962)
Edman, K.A.P.: Action of ouabain on heart actomyosin. Acta Physiol. Scand. *21*, 230–237 (1950)
Edman, K.A.P.: Wirkung des Ouabains auf das Aktomyosin der Herzmuskulatur. Experientia (Basel) *7*, 71–72 (1951)
Elbrink, J., Bihler, I.: Membrane transport: its relation to cellular and metabolic rates. Science *188*, 1177–1184 (1975)
Endo, M.: Calcium release from the sarcoplasmic reticulum. Physiol. Rev. *57*, 71–108 (1977)
Erbel, R.: Belastungs-EKG unter Digitalistherapie. Dtsch. Med. Wochenschr. *102*, 1892 (1977)
Erbel, R.: Belastungs-EKG unter Therapie von herzwirksamen Glykosiden. Dtsch. Med. Wochenschr. *103*, 632 (1978)
Fain, J.N., Jacobs, M.D., Clement-Cormier, Y.C.: Interrelationship of cyclic AMP, lipolysis, and respiration in brown fat cells. Am. J. Physiol. *224*, 346–351 (1973)

Feinberg, H., Boyd, E., Tanzini, G.: Mechanical performance and oxygen utilization of the isovolumic rabbit heart. Am. J. Physiol. *215*, 132–139 (1968)
Feinberg, H., Katz, L.N., Boyd, E.: Determinants of coronary flow and myocardial oxygen consumption. Am. J. Physiol. *202*, 45–52 (1962)
Finkelstein, M., Bodansky, O.: The effect of cardiac glycosides on the respiration of cardiac muscle. J. Pharmacol. Exp. Ther. *94*, 274–287 (1948)
Ford, L.E., Podolsky, R.J.: Regenerative calcium release within muscle cells. Science *167*, 58–59 (1970)
Frantsuzova, S.B.: Action of cardiac glycosides on the concentrations of adenyl nucleotides and nicotinamide-adenine dinucleotides in the rat myocardium. Biull. Eksp. Biol. Med. *77*, 527–529 (1974)
Fritz, P.J.: Cardiac glycoside effects on rat and guinea pig phospholipids. Res. Commun. Chem. Path. Pharmacol. *12*, 741–749 (1975)
Furchgott, R.F., DeGubareff, T.: High-energy phosphate content of cardiac muscle under various experimental conditions which alter contractile strength. J. Pharmacol. Exp. Ther. *124*, 203–218 (1958)
Furchgott, R.F., Lee, K.S.: High energy phosphate and the force of contraction of cardiac muscle. Circ. Res. *21*, 416–432 (1961)
Garfield, R.E., Somlyo, A.V., Chacko, S., Somlyo, A.P.: Golgi-specific ultrastructural effects of the ionophore X537A in muscle. Fed. Proc. *33*, 1459 (1974)
Genuit, H., Haarmann, W.: Zur Frage der Spezifität der Digitalisstoffwechselwirkung an Herz- und Skeletmuskulatur von Ratte und Meerschweinchen. Arch. Exp. Pathol. Pharmakol. *196*, 481–492 (1940)
Gervais, A., Busch, U., Wood, J.M., Schwartz, A.: Effect of a toxic in vivo dose of ouabain on guinea-pig heart mitochondria. J. Mol. Cell. Cardiol. *10*, 1003–1015 (1978)
Gibbs, C.L.: Cardiac energetics. Physiol. Rev. *58*, 174–254 (1978)
Gibbs, C.L., Chapman, J.B.: Cardiac heat production. Ann. Rev. Physiol. *41*, 507–519 (1979)
Gibbs, C.L., Gibson, W.R.: Effect of ouabain on the energy output of rabbit cardiac muscle. Circ. Res. *24*, 951–967 (1969)
Gleason, W.L., Braunwald, E.: Studies on the first derivative of the ventricular pressure pulse in man. J. Clin. Invest. *41*, 80–91 (1962)
Göksel, F., Katz, L.N., Feinberg, H.: Effect of ouabain on coronary flow, performance of the heart and its oxidative metabolism. Am. J. Physiol. *204*, 21–27 (1963)
Gollwitzer-Meier, K., Krüger, E.: Herzenergetik und Strophanthinwirkung bei verschiedenen Formen der experimentellen Herzinsuffizienz. Pflügers Arch. *238*, 251–278 (1937)
Gonda, O., Quastel, J.H.: Effects of ouabain on cerebral metabolism and transport mechanisms in vitro. Biochem. J. *84*, 394–406 (1962)
Gousios, A.G., Felts, J.M., Havel, R.J.: Effects of ouabain on force of contraction, oxygen consumption, and metabolism of free fatty acids in the perfused rabbit heart. Circ. Res. *21*, 445–448 (1967)
Greeff, K.: Zur Pharmakologie der herzwirksamen Glykoside. Klin. Physiol. *1*, 340–370 (1963)
Gregg, D.E., Rayford, C.R., Khouri, E.M., Kattus, A.A., McKeever, W.P.: Effect of alteration of coronary perfusion pressure on oxygen uptake of left myocardium. Circulation *16*, 888 (1957)
Greiner, T.: Relationship of force of contraction to high-energy phosphate in heart muscle. J. Pharmacol. Exp. Ther. *105*, 178–195 (1952)
Gremels, H.: Zur Physiologie und Pharmakologie der Energetik des Säugetierherzens. Arch. Exp. Pathol. Pharmakol. *169*, 689–724 (1933)
Gremels, H.: Über den Einfluß von Digitalisglykosiden auf die energetischen Vorgänge am Säugetierherzen. Arch. Exp. Pathol. Pharmakol. *186*, 625–660 (1937)
Grisolia, S.: The potentiating effect of digitoxin and quinidine on dinitrophenol uncoupling of oxidative phosphorylation. Biochim. Biophys. Acta (Amst.) *18*, 437–438 (1955a)
Grisolia, S.: Digitoxin and uncoupling of oxidative phosphorylation. Circulation *12*, 716 (1955b)

Gross, G.J., Warltier, D.C., Hardman, H.F., Somani, P.: The effect of ouabain on nutritional circulation and regional myocardial blood flow. Am. Heart J. *93*, 487–495 (1977)
Gruhzit, C.C., Farah, A.E.: A comparison of the positive inotropic effects of ouabain and epinephrine in heart failure induced in the dog heart-lung preparation by sodium pentobarbital, dinitrophenol, sodium cyanide, and sodium azide. J. Pharmacol. Exp. Ther. *114*, 334–342 (1955)
Guerra, F., Eberstadt, P.L., Veerkamp, A.: Acción de la uabaina sobre la liberación enzimática de fósforo en el sistema miosina trifosfato de adenosina. Arch. Inst. Cardiol. Méx. *16*, 449–455 (1946)
Günther, T., Winkelmann, W.: Einfluß von Steroidhormonen auf den aktiven Transport von Glykokoll in Vogelerythrocyten. Naunyn-Schmiedebergs Arch. Pharmacol. *242*, 277–283 (1961)
Güttler, K.: Untersuchungen zum myokardialen Stoff- und Energiewechsel. Wärmeproduktion und Sauerstoffverbrauch schlagender und insbesondere K^+-stillgestellter Herzen unter dem Einfluß von Ca^{++}, Adrenalin und Strophanthin. Dissertation, Pharmakologisches Institut, Hannover (1974)
Güttler, K., Klaus, W., Landgraf, W.: Influence of ouabain and hypoxia on the ECG of the isolated guinea pig heart. Naunyn-Schmiedebergs Arch. Pharmacol. *311* (suppl.), R 41 (1980)
Güttler, K., Theisohn, M.: Oxygen consumption and heat production of K^+-arrested hearts under the influence of Ca^{++}, ouabain and adrenaline. Naunyn-Schmiedebergs Arch. Pharmacol. *277* (suppl.), R 24 (1973)
Gutman, M., Kearney, E.B., Singer, T.P.: Multiple control mechanisms for succinate dehydrogenase in mitochondria. Biochem. Biophys. Res. Commun. *44*, 526–532 (1971)
Hafkenschiel, J.H., Cerletti, A.: Effects of acetyldigitoxin, a new cardiac glycoside, on work performance and O_2-consumption of the canine heart–lung preparation. J. Pharmacol. Exp. Ther. *110*, 23 (1954)
Hajdu, S., Weiss, H., Titus, E.: The isolation of a cardiac active principle from mammalian tissue. J. Pharmacol. Exp. Ther. *120*, 99–113 (1957)
Harvey, S.C.: Radiophosphorus metabolism of guinea pig hearts and actions of digitoxin and pentobarbital. Am. J. Physiol. *183*, 559–564 (1955)
Hasselbach, W., Makinose, M.: Die Calciumpumpe der „Erschlaffungsgrana" des Muskels und ihre Abhängigkeit von der ATP-Spaltung. Biochem. Z. *333*, 518–528 (1961)
Heeg, E.: Untersuchungen über den Einfluß herzwirksamer Pharmaka auf den Druckablauf in der linken Herzkammer der Katze. Habil.-Schrift. Düsseldorf (1966)
Hegglin, R., Grauer, H., Münchinger, R.: Die Beeinflussung der Adenosintriphosphatase-aktivität des Herzmuskels durch verschiedene Substanzen. Experientia (Basel) *5*, 127 (1949)
Helmreich, E.: Der Einfluß von Strophanthin auf das Zytochromo-oxydase-Zytochrom-C-System. Biochem. Z. *321*, 144–151 (1950)
Helmreich, E., Simon, K.: Zur Frage der Wirkung von Digitalisstoffen auf das Aktomyosin und die Adenosintriphosphatase des Herzmuskels. Z. Naturforsch. *7b*, 341–350 (1952)
Herrmann, R.G.: Studies on the effect of ouabain on cardiac muscle respiration. Arch. Int. Pharmacodyn. *81*, 235–241 (1950)
Herrmann, R.G., Flamboe, G.E., Chen, K.K.: The effect of nine cardiac steroids and epinephrine on the respiration of heart muscle slices. J. Pharmacol. Exp. Ther. *112*, 23–28 (1954)
Ho, R.J., Jeanrenaud, B., Renold, A.E.: Ouabain-sensitive fatty acid release from isolated fat cells. Experientia (Basel) *22*, 86–87 (1966)
Ho, R.J., Jeanrenaud, B.: Insulin-like action of ouabain. I. Effect on carbohydrate metabolism. Biochim. Biophys. Acta (Amst.) *144*, 61–73 (1967)
Ho, R.J., Jeanrenaud, B., Posternak, T.H., Renold, A.E.: Insulin-like action of ouabain. II. Primary antilipolytic effect through inhibition of adenyl cyclase. Biochim. Biophys. Acta (Amst.) *144*, 74–82 (1967)
Hochrein, H., Lehmann, H.-U., Helwing, H.-P.: EKG-Veränderungen bei Koronarinsuffizienz und unter dem Einfluß von Digitalis. Klinikarzt *4*, 403–407 (1975)
Hoeschen, R.J.: The effect of ouabain on substrate metabolism in the isolated perfused rat heart. Clin. Res. *17*, 636 (1969)

Hoeschen, R.J.: The effect of ouabain on substrate metabolism in the isolated perfused rat heart. Can. J. Physiol. Pharmacol. *49*, 412–419 (1971)

Hokin, L.E., Hokin, M.R.: Evidence for phosphatidic acid as the sodium carrier. Nature (Lond.) *184*, 1068–1069 (1959)

Hokin, L.E., Hokin, M.R.: Diglyceride kinase and phosphatidic acid phosphatase in erythrocyte membranes. Nature (Lond.) *189*, 836–837 (1961)

Holloszy, J.O., Narahara, H.T.: Enhanced permeability to sugar associated with muscle contraction. J. Gen. Physiol. *50*, 551–562 (1967)

Hood, W.B. Jr., Letac, B., Roberge, G., Lown, B.: Direct digitalization of the myocardium. Hemodynamic effects. Am. J. Cardiol. *22*, 667–671 (1968)

Jacobson, A.L.: Effect of ouabain on the ATPase of cardiac myosin B at high ionic strength. Circ. Res. *22*, 625–632 (1968)

Jenden, D.J., Fairhurst, A.S.: The pharmacology of ryanodine. Pharmacol. Rev. *21*, 1–26 (1969)

Jesmok, G.J., Calvert, D.N., Lech, J.J.: The effect of inotropic agents on glycerol release and protein kinase activity ratios in the isolated perfused rat heart. J. Pharmacol. Exp. Ther. *200*, 187–194 (1977)

Jesmok, G.J., Lech, J.J., Calvert, D.N.: The effects of epinephrine, glucagon, and ouabain on glycerol release in the isolated perfused rat heart. Pharmacologist *17*, 218 (1975)

Kahler, R.L., Braunwald, E., Kelminson, L.L., Kedes, L., Chidsey, C.A., Segal, S.: Effect of alterations of coronary blood flow on the oxygen consumption of the non-working heart. Circ. Res. *13*, 501–509 (1963)

Kassebaum, D.G., Griswold, H.E.: Digitalis in nonfailing cardiac diseases. Prog. Cardiovasc. Dis. *12*, 484–492 (1970)

Kedem, J., Levinger, I.M., Baum, M., Rogel, S.: Heart rate and myocardial substrate preference during normal and hypoxic perfusion of the heart in vivo. Arch. Int. Physiol. Biochim. *83*, 53–62 (1975)

Kien, G.A., Gomoll, A.W., Sherrod, T.R.: Action of digoxin and insulin on transport of glucose through myocardial cell membrane. Proc. Soc. Exp. Biol. Med. *103*, 682–685 (1960)

Kien, G.A., Sherrod, T.R.: The effect of digoxin on the intermediary metabolism of the heart as measured by glucose-C^{14} utilization in the intact dog. Circ. Res. *8*, 188–198 (1960)

Kim, N.D., Bailey, L.E., Dresel, P.E.: The effect of insulin on the subcellular distribution and the inotropic effect of ^{3}H-digoxin in the guinea pig heart. Life Sci. *9*, 1135–1139 (1970)

Kim, N.D., Dresel, P.E.: The effect of insulin on the subcellular distribution and the inotropic effect of ^{3}H-digoxin in the guinea pig heart. Fed. Proc. *29*, 352 (1970)

Kimura, T.E., Dubois, K.P.: Inhibition of enzymatic hydrolysis of ATP by certain cardiac drugs. Science *106*, 370–371 (1947)

Kini, P.M., Willems, J.L., Batchlor, C., Pipberger, H.V.: ST-T changes induced by digitalis and ventricular hypertrophy: differentiation by quantitative analysis. J. Electrocardiol. *5*, 101–110 (1972)

Kjekshus, J.K., Maroko, P.R., Sobel, B.E.: Distribution of myocardial injury and its relation to epicardial ST-segment changes after coronary artery occlusion in the dog. Cardiovasc. Res. *6*, 490–499 (1972)

Klaus, W.: Neuere Aspekte über den Wirkungsmechanismus der Herzglykoside. Z. Grundlagenforsch. *2*, 43–117 (1964)

Klaus, W., Güttler, K., Theisohn, M., Theisohn-Schwedhelm, I.: The relationship between mechanical function and heat production in isolated hearts under the influence of ouabain and epinephrine. 5th. Int. Congress on Pharmacology, San Francisco, Calif. USA, Abstr. *755*, p. 126, 1972

Klaus, W., Krebs, R.: Über den Einfluß von Digitoxigenin und Strophanthin auf mechanische Aktivität und Sauerstoffverbrauch isolierter Herzmuskelpräparate. Naunyn-Schmiedebergs Arch. Pharmacol. *261*, 102–117 (1968)

Klaus, W., Krebs, R.: Über die Abhängigkeit der Strophanthinwirkung auf den myokardialen Sauerstoffverbrauch vom Funktionszustand des Herzens. Naunyn-Schmiedebergs Arch. Pharmacol. *264*, 337–353 (1969)

Klein, W.W., Brandt, D., Pavek, P.: Effekt von K-Strophanthin auf Hämodynamik und Stoffwechsel des koronarkranken Herzens in Ruhe und während frequenzinduzierter Angina pectoris. Verh. Dtsch. Ges. Inn. Med. *83*, 1646–1650 (1977)

Koch-Weser, J., Berlin, C.M. Jr., Blinks, J.R.: Effects of acetylstrophanthidin, levarterenol, and carbachol on the interval-strength relationship of heart. In: Proceedings of 2nd International Pharmacological Meeting, Prague, 1963, Vol. 5, pp. 63–72. Oxford: Pergamon Press 1964

Kötter, V., Schüren, K.-P., Schröder, R.: Der Einfluß von Digoxin auf Coronardurchblutung und myocardialen O_2-Verbrauch bei Patienten mit Angina pectoris. Verh. Dtsch. Ges. Inn. Med. *83*, 1644–1647 (1977)

Korth, C., Spang, K.: Die Wirkung des Digitoxins auf Elektrokardiogramm und Herzmuskel der Katze. Naunyn-Schmiedebergs Arch. Pharmacol. *184*, 349–364 (1937)

Kostyo, J.L., Schmidt, J.E.: Inhibitory effects of cardiac glycosides and adrenal steroids on amino acid transport. Am. J. Physiol. *204*, 1031–1038 (1963)

Koyama, T., Brecht, K., Koyama, Y.: Wirkungen von Strophanthin und Ca^{++} auf den O_2-Verbrauch und die Mechanik (Zuckung und Tonus) von Papillarmuskeln des Meerschweinchens. Z. Kreisl. Forsch. *55*, 838–845 (1966)

Kraupp, O., Raberger, G., Chirikdjian, J.J.: The effects of alpha-acetyldigoxin on cardiac dynamics and myocardial glucose and free fatty acid extraction in nonfailing dogs. Pharmacology *1*, 345–357 (1968)

Krebs, H.A.: Body size and tissue respiration. Biochim. Biophys. Acta (Amst.) *4*, 249–269 (1950)

Krebs, R.: Über die unterschiedliche Wirkung von Calcium und Noradrenalin auf den Sauerstoffverbrauch isolierter Meerschweinchenvorhöfe. Pflügers Arch. *315*, 110–124 (1970)

Krebs, R., Klaus, W.: Über die Beziehung zwischen Sauerstoffverbrauch und Kontraktionskraft isolierter Meerschweinchenvorhöfe unter dem Einfluß von Digitoxin. Naunyn-Schmiedebergs Arch. Pharmacol. *255*, 30 (1966)

Kreisberg, R.A., Williamson, J.R.: Metabolic effects of ouabain in the perfused rat heart. Am. J. Physiol. *207*, 347–351 (1964)

Kühn, P., Brachfeld, N.: Zur Wertigkeit des Tension-Time-Index und der maximalen linksventrikulären Druckanstiegsgeschwindigkeit (dp/dt) in der Korrelation zum myokardialen Sauerstoffverbrauch. Z. Kreisl. Forsch. *58*, 244–251 (1969)

Kujalova, V., Mosinger, B.: Cold- and ouabain-sensitive uptake of glucose stimulated by insulin in incubated rat adipose tissue. Biochim. Biophys. Acta (Amst.) *127*, 255–257 (1966)

Kuschinsky, G., Lange, G., Turba, F.: Über die Wirkung von Digitoxin auf die Freilegung von Actomyosin und seinen Komponenten aus den Strukturen des Muskels. Naunyn-Schmiedebergs Arch. Pharmacol. *215*, 259–269 (1952)

Kutschera-Aichberger, H.: Über Herzschwäche. Wien. Arch. Inn. Med. *18*, 209–358 (1929)

Kypson, J., Hait, G.: The effects of ouabain, calcium, and potassium on the transport and incorporation of some amino acids into rabbit atrial tissue in vitro. J. Pharmacol. Exp. Ther. *177*, 398–408 (1971)

Langemann, H.: Über den Einfluß von herzaktiven Glukosiden auf die Adenosintriphosphatase aus Herzmuskelzellfraktionen. Helv. Physiol. Pharmacol. Acta *11*, C 20–21 (1953)

Langemann, H., Brody, T.M., Bain, J.A.: In vitro effects of ouabain on slices and mitochondrial preparations from heart and brain. J. Pharmacol. Exp. Ther. *108*, 274–280 (1953)

Langer, G.A.: Heart: excitation-contraction coupling. Ann. Rev. Physiol. *35*, 55–86 (1973)

Lee, K.S.: A new technique for the simultaneous recording of oxygen consumption and contraction of muscle: The effect of ouabain on cat papillary muscle. J. Pharmacol. Exp. Ther. *109*, 304–312 (1953)

Lee, K.S.: Sensitive method for simultaneous recording contractility and respiration of heart muscle. Fed. Proc. *17*, 387 (1958)

Lee, K.S.: The relationship of the oxygen consumption to the contraction of the cat papillary muscle. J. Physiol. (Lond.) *151*, 186–201 (1960)

Lee, K.S., Klaus, W.: The subcellular basis for the mechanism of inotropic action of cardiac glycosides. Pharmacol. Rev. *23*, 193–261 (1971)
Lee, K.S., McElroy, W.D.: Effect of ouabain on mitochondria. Fed. Proc. *14*, 362 (1955)
Lee, K.S., Schwartz, A., Burstein, R.: An effect of cardiac glycosides on oxidative phosphorylation by heart mitochondria. J. Pharmacol. Exp. Ther. *129*, 123–127 (1960a)
Lee, K.S., Shin, B.C.: Studies on the active transport of Ca^{++} in human red cells. J. Gen. Physiol. *54*, 713–729 (1969)
Lee, K.S., Yu, D.H., Burstein, R.: The effect of ouabain on the oxygen consumption, the high energy phosphates and the contractility of the cat papillary muscle. J. Pharmacol. Exp. Ther. *129*, 115–122 (1960b)
Lee, K.S., Yu, D.H., Lee, D.I., Burstein, R.: The influence of potassium and calcium on the effect of ouabain on cat papillary muscle. J. Pharmacol. Exp. Ther. *132*, 139–148 (1961)
Lehmann, H.-U., Hochrein, H.: Digitalishaftzeichen im Elektrokardiogramm. Dtsch. Med. Wochenschr. *103*, 315–316 (1978)
Lehmann, H.-U., Janitzki, S., Hochrein, H.: Zur Deutung von Digitalishaftzeichen im Elektrokardiogramm. Dtsch. Med. Wochenschr. *102*, 1335–1341 (1977)
Lehmann, H.-U., Witt, E., Hochrein, H.: Zunahme von Angina pectoris und ST-Strecken-Senkung im EKG durch Digitalis. Z. Kardiol. *67*, 57–66 (1978)
Lehmann, H.-U., Witt, E., Hochrein, H.: Wirkung von Nitroglycerin auf digitalisinduzierte ST-Streckensenkungen bei koronarkranken Patienten. Dtsch. Med. Wochenschr. *104*, 501–506 (1979)
Lehninger, A.L.: Biochemie, pp. 341–361. 2nd. ed. Weinheim, New York: Verlag Chemie 1977
Lehninger, A.L., Carafoli, E., Rossi, C.S.: Energy-linked ion movements in mitochondrial systems. Adv. Enzymol. *29*, 259–320 (1967)
Leonard, C., Hajdu, S.: The effects of potassium on the inotropic action of cardiac glycosides. Clin. Res. *7*, 19–20 (1959)
Levine, H.J., Wagman, R.J.: Energetics of the human heart. Am. J. Cardiol. *9*, 372–383 (1962)
Lindower, J.O., Marks, B.H.: Ultrastructural changes in guinea-pig myocardium after acute ouabain treatment. J. Pharmacol. Exp. Ther. *202*, 76–88 (1977)
Lowry, O.H., Passonneau, J.V.: Kinetic evidence for multiple binding sites on phosphofructokinase. J. Biol. Chem. *241*, 2268–2279 (1966)
Marchetti, G.V., Aguggini, G., Merlo, L., Noseda, V., Santi, A.: Coronary blood flow, oxygen consumption of the myocardium and cardiac work in the sheep. Pflügers Arch. *290*, 80–88 (1966)
Marinetti, G.V., Temple, K., Stotz, E.: The in vivo effect of digitoxin on rat heart phosphatides. J. Lipid Res. *2*, 188–190 (1961)
Maroko, P.R., Braunwald, E., Covell, J.W.: The effect of digitalis on the severity of myocardial ischemic injury following experimental coronary occlusion. Pharmacologist *12*, 212 (1970)
Mason, D.T., Braunwald, E.: Studies on digitalis. IX. Effects of ouabain on the non-failing human heart. J. Clin. Invest. *42*, 1105–1111 (1963)
Mayer, S.E.: The effect of catecholamines on cardiac metabolism. Circ. Res. *34* and *35* (suppl. III), 129–135 (1974)
Mayer, S.E., Moran, N.C.: Relation between pharmacologic augmentation of cardiac contractile force and the activation of myocardial phosphorylase. J. Pharmacol. Exp. Ther. *129*, 271–281 (1960)
McDonald, R.H. Jr.: Developed tension: a major determinant of myocardial oxygen consumption. Am. J. Physiol. *210*, 351–356 (1966)
McDonald, R.H. Jr.: Myocardial heat production: its relationship to tension development. Am. J. Physiol. *220*, 894–900 (1971)
Messer, J.V., Neill, W.A.: Oxygen supply of the human heart. Am. J. Cardiol. *9*, 384–394 (1962)
Moe, G.K., Farah, A.E.: Digitalis and allied cardiac glycosides. In: The pharmacological basis of therapeutics. Goodman, L.S., Gilman, A. (eds.), pp. 653–682. 5th. ed. New York: Macmillan Publ. 1975

Monroe, R.G., French, G.N.: Left ventricular pressure-volume relationships and myocardial oxygen consumption in the isolated heart. Circ. Res. *9*, 362–374 (1961)
Mor, M.A.: ATPase activity of myofibrils and granules isolated from muscle. Experientia (Basel) *9*, 342–343 (1953)
Mor, M.A.: The influence of progesterone on adenosintriphosphatase. Experientia (Basel) *11*, 33–35 (1955)
Morgan, H.E., Neely, J.R., Brineaux, J.P., Park, C.R.: Regulation of glucose transport. In: Control of energy metabolism. Chance, B., Estabrook, R.W., Williamson, J.R. (eds.), pp 347–355. New York: Academic Press 1965
Morgenstern, C., Holjes, U., Arnold, G., Lochner, W.: The influence of coronary pressure and coronary flow on intracoronary blood volume and geometry of the left ventricle. Pflügers Arch. *340*, 101–111 (1973)
Morkin, E., LaRaia, P.J.: Biochemical studies on the regulation of myocardial contractility. New Engl. J. Med. *290*, 445–451 (1974)
Müller, E.R.: Über die aerobe und anaerobe Stoffwechselkapazität des isolierten Warmblüterherzens. Pflügers Arch. *276*, 42–55 (1962)
Müller-Ruchholtz, E.R.: Über die Leistungsfähigkeit des anoxydativen Herzstoffwechsels. Basic Res. Cardiol. *68*, 480–508 (1973)
Münchinger, R.: Untersuchungen über die Aktivität der Adenosintriphosphatase im Herzmuskel, als Beitrag zur Pathogenese der sog. energetisch-dynamischen Herzinsuffizienz. Cardiologia (Basel) *22*, 145–168 (1953)
Nayler, W.G.: Effect of strophanthin G on oxidative metabolism in cardiac muscle. Nature (Lond.) *188*, 70–71 (1960)
Neely, J.R., Liebermeister, H., Battersby, E.J., Morgan, H.E.: Effect of pressure development on oxygen consumption by isolated heart. Am. J. Physiol. *212*, 804–814 (1967a)
Neely, J.R., Liebermeister, H., Morgan, H.E.: Effect of pressure development on membrane transport of glucose in isolated rat heart. Am. J. Physiol. *212*, 815–822 (1967b)
Neely, J.R., Morgan, H.E.: Relationship between carbohydrate and lipid metabolism and the energy balance of heart muscle. Ann. Rev. Physiol. *36*, 413–459 (1974)
Neely, J.R., Rovetto, M.J., Oram, J.F.: Myocardial utilization of carbohydrate and lipids. Progr. Cardiovasc. Dis. *15*, 289–329 (1972)
Neely, J.R., Whitmer, K.M., Mochizuki, S.: Effects of mechanical activity and hormones on myocardial glucose and fatty acid utilization. Circ. Res. *38* (suppl. I), 22–30 (1976)
Nelson, D.R., Cornatzer, W.E.: Effect of digoxin, aldosterone, and dietary sodium chloride on the incorporation of inorganic P^{32} into liver and kidney nuclear and mitochondrial phospholipids. Proc. Soc. Exp. Biol. Med. *116*, 237–242 (1964)
Nicholls, D., Kanfer, J., Titus, E.: The effect of ouabain on the incorporation of inorganic P-32 into phospholipid. J. Biol. Chem. *237*, 1043–1049 (1962)
Nowy, H., Helmreich, E.: Der Einfluß von Strophanthin auf die Gewebeatmung verschiedener Abschnitte des Rattenherzens. Klin. Wochenschr. *28*, 686–687 (1950)
Ochs, H.R., Otten, H., Bodem, G.: Digoxin-induzierte Veränderungen des Belastungs-EKG in Relation zur Digoxinplasmakonzentration. Klin. Wochenschr. *57*, 161–168 (1979)
Ogilvie, R.I., Klassen, G.A.: Metabolic effect and uptake of [^{3}H] digoxin in the forearm of man. Clin. Sci. *42*, 567–577 (1972)
Olson, R.E., Barnhorst, D.A.: The control of energy production and utilization in cardiac muscle. In: Recent. Advanc. Stud. cardiac. Struct. Metab. (Vol. 3) Myocardial metabolism. Dhalla, N.S. (ed.), pp. 11–30. München, Berlin, Wien: Urban & Schwarzenberg 1974
Olson, R.E., Dhalla, N.S., Sun, C.N.: Changes in energy stores in the hypoxic heart. Cardiology *56*, 114–124 (1971/1972)
Olson, R.E., Piatnek, D.A.: Conservation of energy in cardiac muscle. Ann. N.Y. Acad. Sci. *72*, 466–478 (1959)
Opie, L.H.: Coronary flow rate and perfusion pressure as determinants of mechanical function and oxidative metabolism of isolated perfused rat heart. J. Physiol. (Lond.) *180*, 529–541 (1965)
Opie, L.H.: Metabolism of the heart in health and disease. I. Am. Heart J. *76*, 685–698 (1968)

Opie, L.H.: Metabolism of the heart in health and disease. II. Am. Heart J. *77*, 100–122 (1969a)

Opie, L.H.: Metabolism of the heart in health and disease. III. Am. Heart J. *77*, 383–410 (1969b)

Opie, L.H.: Substrate utilization and glycolysis in the heart. Cardiology *56*, 2–21 (1971/1972)

Opie, L.H., Shipp, J.C., Evans, J.R., Leboeuf, B.: Metabolism of glucose-U-C^{14} in perfused rat heart. Am. J. Physiol. *203*, 839–843 (1962)

Page, R.G., Foltz, E.L., Sheldon, W.F., Wendel, H.: Effect of ouabain on coronary circulation and other circulatory functions in intact anesthetized dogs. J. Pharmacol. Exp. Ther. *101*, 112–118 (1951)

Paine, C.M., Heinz, E.: The structural specificity of the glycine transport system of Ehrlich carcinoma cells. J. Biol. Chem. *235*, 1080–1085 (1960)

Park, C.R., Crofford, O.B., Kono, T.: Mediated (non-active) transport of glucose in mammalian cells and its regulation. J. Gen. Physiol. *52*, 296s–318s (1968)

Pelkonen, K.H.O.: Enzyme activities of myocardial energy metabolism during prolonged digoxin treatment in rats. Acta Pharmacol. Toxicol. *42*, 1–6 (1978)

Pelosi, G., Conti, F., Agliati, G.: The effect of hypoxia on cardiac tissue (I) High energy phosphates content and oxygen uptake. Eur. J. Pharmacol. *8*, 19–24 (1969)

Peters, H., Visscher, M.B.: Energy metabolism of the heart in failure and influence of drugs upon it. Am. Heart J. *11*, 273–291 (1936)

Plaut, K.A., Gertler, M.M., Plaut, G.W.E.: Efficiency of oxidative phosphorylation of heart muscle mitochondria from digitalized guinea pigs. Circ. Res. *5*, 226–229 (1957)

Posner, B.I., Fanburg, B.L.: Ribonucleic acid synthesis in experimental cardiac hypertrophy in rats. II. Aspects of regulation. Circ. Res. *23*, 137–145 (1968)

Proctor, C.D., Mueller, G.S.: Digoxin effect on heart ATPase. Fed. Proc. *16*, 328 (1957)

Raberger, G., Kraupp, O.: Herzdynamik und Herzstoffwechsel nach i.v. Verabreichung von alpha-Acetyldigoxin am Hund. Arch. Pharmakol. Exp. Pathol. *263*, 246–247 (1969)

Read, W.D., Kelsey, F.E.: Effect of digoxin on myokinase activity. Science *125*, 120–121 (1957)

Rebar, J. Jr., Rebar, B.T., Omachi, A.: Influence of digitoxin on labile and inorganic phosphates, lactate, glycogen, potassium, and sodium in dog ventricle. Circ. Res. *5*, 504–509 (1957)

Reiter, M.: Wirkung von Strophanthin auf Kontraktionskraft und Sauerstoffverbrauch des Herzstreifens der Ratte. Arch. Exp. Pathol. Pharmakol. *219*, 315–332 (1953)

Reiter, M., Barron, E.S.G.: Über die direkte Fermentwirkung von Herzglykosiden. Arch. Exp. Pathol. Pharmakol. *214*, 341–348 (1952)

Reuter, H.: Divalent cations as charge carriers in excitable membranes. Prog. Biophys. Molec. Biol. *26*, 1–43 (1973)

Risler, T., Grabensee, B., Grosse-Brockhoff, F.: EKG-Veränderungen und Digoxin-Serumkonzentration bei Digitalisintoxikationen. Dtsch. Med. Wochenschr. *100*, 821–825 (1975)

Rollett, E.L., Yurchak, P.M., Hood, W.B., Gorlin, R.: Pressure-volume correlates of left ventricular oxygen consumption in the hypervolemic dog. Circ. Res. *17*, 499–518 (1965)

Ross, J. Jr., Sonnenblick, E.H., Kaiser, G.A., Frommer, P.L., Braunwald, E.: Electroaugmentation of left ventricular performance and oxygen consumption by repetitive application of paired electrical stimuli. Circ. Res. *16*, 332–342 (1965)

Rothlin, E., Bircher, R.: Pharmakodynamische Grundlagen der Therapie mit herzwirksamen Glykosiden. Ergebn. Inn. Med. Kinderheilk. *5*, 457–552 (1954)

Rothlin, E., Schoelly, D.: Einfluß herzwirksamer Glykoside auf die Atmung des Herzens gemessen mit der Warburg-Apparatur. Helv. Physiol. Acta *8*, C69–70 (1950)

Rothlin, E., Taeschler, M., Cerletti, A.: Action of dinitrophenol and lanatoside C on the canine heart-lung preparation. Circ. Res. *3*, 32–38 (1955)

Rothlin, E., Taeschler, M.: Zur Wirkung der herzwirksamen Glykoside auf den Myokardstoffwechsel. Fortschr. Kardiol. *1*, 189–239 (1956)

Roy, A.R., Chatterjee, M.L.: Effect of ouabain on the catecholamine content of heart and adrenal gland of rabbits. Life Sci. *9*, 395–401 (1970)

Rusy, B.F., Coulson, R.D.: Energy consumption in the isolated rabbit heart. Anesthesiology *39*, 428–434 (1973)
Salomon, K., Riesser, O.: Zur Frage des Einflusses von Digitoxin und Strophanthin auf oxydative Vorgänge in Versuchen am Modell sowie am atmenden überlebenden Herzmuskelgewebe. Arch. Exp. Pathol. Pharmakol. *177*, 450–462 (1935)
Sarnoff, S.J., Gilmore, J.P., Wallace, A.G., Skinner, N.S. Jr., Mitchel, J.H., Daggett, W.M.: Effect of acetylstrophanthidin therapy on cardiac dynamics, oxygen consumption and efficiency in the isolated heart with and without hypoxia. Am. J. Med. *37*, 3–13 (1964)
Saunders, P.R., Webb, J.L., Thienes, C.H.: Metabolism of the heart in relation to drug action. V. Action of various drugs and other substances on some dehydrogenase systems of the heart. Arch. Int. Pharmacodyn. *81*, 485–492 (1950)
Schaper, W.K.A., Lewi, R., Jageneau, A.H.M.: The determinants of the rate of change of left ventricular pressure (dp/dt). Arch. Kreisl. Forsch. *46*, 27–41 (1965)
Schatzmann, H.J., Vincenzi, F.F.: Calcium movements across the membrane of human red cells. J. Physiol. (Lond.) *201*, 369–395 (1969)
Schwartz, A.: Effect of a toxic dose of ouabain on guinea pig heart mitochondria. Fed. Proc. *34*, 793 (1975)
Schwartz, A., Lindenmayer, G.E., Allen, J.C.: The Na^+, K^+-ATPase membrane transport system: importance in cellular function. In: Current topics in membranes and transport. Bronner, F., Kleinzeller, A. (eds.), Vol. 3, pp. 1–82. New York: Academic Press 1972
Schwartz, A., Lindenmayer, G.E., Allen, J.C.: The sodium-potassium-adenosinetriphosphatase: pharmacological, physiological, and biochemical aspects. Pharmacol. Rev. *27*, 3–134 (1975)
Schwiegk, H., Jahrmärker, H.: Therapie der Herzinsuffizienz. Glykosidwirkungen auf das EKG. In: Handbuch der inneren Medizin, Band IX/1: Herz und Kreislauf. v. Bergmann, G., Frey, W., Schwiegk, H. (eds), pp. 500–504, 4. th. ed. Berlin, Göttingen, Heidelberg: Springer 1960
Scott, J.C., Finkelstein, L.J., Spitzer, J.J.: Myocardial removal of free fatty acids under normal and pathological conditions. Am. J. Physiol. *203*, 482–486 (1962)
Segre, G.: Ricerche sul meccanismo d'azione de digitalici; effetti metabolici della uabaina. Arch. Int. Pharmacodyn. *80*, 336–346 (1949)
Segre, G.: Sull'azione dei digitalici su alcuni enzimi. Boll. Sco. Ital. Biol. Sper. *26*, 939 (1950)
Selzer, A., Malmborg, R.O.: Hemodynamic effects of digoxin in latent cardiac failure. Circulation *25*, 695–702 (1962)
Severson, D.L.: Regulation of lipid metabolism in adipose tissue and heart. Can. J. Physiol. Pharmacol. *57*, 924–937 (1979)
Shipp, J.C., Opie, L.H., Challoner, D.R.: Fatty acid and glucose metabolism in the perfused heart. Nature (Lond.) *189*, 1018–1019 (1961)
Shipp, J.C., Thomas, J.M., Crevasse, L.: Oxidation of carbon-14-labelled endogenous lipids by isolated perfused rat heart. Science *143*, 371–373 (1964)
Siedek, H., Hammerl, H., Kränzl, C., Pichler, O., Studlar, M.: Untersuchungen über die Stoffwechseleffekte eines parenteral verabreichten Herzglykosides. Wien. Klin. Wochenschr. *80*, 585–586 (1968 a)
Siedek, H., Hammerl, H., Kränzl, C., Pichler, O., Studlar, M.: Zum Einfluß der Herzglykoside auf den Fettstoffwechsel. Verh. Dtsch. Ges. Kreislaufforsch. *34*, 198–201 (1968 b)
Siegel, J.H., Sonnenblick, E.H.: Isometric time-tension relationship as an index of myocardial contractility. Circ. Res. *12*, 597–610 (1963)
Siess, M. Die Bedeutung des Substrates für den Sauerstoffverbrauch isolierter Meerschweinchenvorhöfe nach Leistungssteigerung durch g-Strophanthin oder Adrenalin. Arch. Exp. Pathol. Pharmakol. *261*, 212–213 (1968)
Simaan, J., Fawaz, G., Jarawan, S.: The effect of ouabain-induced contractility on myocardial oxygen consumption. Naunyn-Schmiedebergs Arch. Pharmacol. *271*, 249–261 (1971)
Skelton, C.L., Seagren, S.C., Pool, P.E.: Influence of digoxin on myocardial energy utilization. Circulation *42* (suppl. III), 111 (1970)
Smith, J.A., Glassman, M., Lind, A.H., Post, M., Sohn, H., Warren, S.: Effects of ouabain on beat and oxygen consumption of embryonic chick hearts. Proc. Soc. Exp. Biol. Med. *86*, 747–749 (1954)

Smith, J.A., Post, M.: Effect of k-Strophanthoside on oxygen consumption of embryonic chick hearts as measured in cartesian driver. Am. J. Physiol. *163*, 751 (1950)
Snow, T.R.: The role of Ca^{++} in determining the relation between metabolism and mechanical performance in rabbit papillary muscle. Fed. Proc. *38*, 1048 (1979)
Sonnenblick, E.H., Ross, J. Jr., Covell, J.W., Kaiser, G.A., Braunwald, E.: Velocity of contraction as a determinant of myocardial oxygen consumption. Am. J. Physiol. *209*, 919–927 (1965)
Sonnenblick, E.H., Ross, J. Jr., Braunwald, E.: Oxygen consumption of the heart. Newer concepts of its multifactoral determination. Am. J. Cardiol. *22*, 328–336 (1968)
Sonnenblick, E.H., Williams, J.F. Jr., Glick, G., Mason, D.T., Braunwald, E.: Studies on digitalis. XV. Effects of cardiac glycosides on myocardial force-velocity relations in the nonfailing human heart. Circulation *34*, 532–539 (1966)
Spitzer, J.J.: Effect of lactate infusion on canine myocardial free fatty acid metabolism in vivo. Am. J. Physiol. *226*, 213–217 (1974)
Staub, H.: Zum Wirkungsmechanismus der Herzglykoside. Dtsch. Med. Wochenschr. *82*, 5–13 (1957)
Staub, H.: Pharmacology of cardiac glycoside. Am. J. Cardiol. *3*, 776–793 (1959)
Strauer, B.E., Tauchert, M., Kochsiek, K.: Coronary flow, coronary vascular reserve and oxygen consumption in patients with and without chronic cardiac failure. In: Aktuelle Digitalisprobleme. Schröder, R., Greeff, K. (eds.), pp. 250–256. München, Berlin, Wien: Urban & Schwarzenberg 1972
Swynghedauw, B., Corsin, A.: Les effects d'une perfusion hyper ou hypocalcique sur la glycolyse du coeur isolé de lapin. Arch. Int. Physiol. Biochim. *77*, 181–205 (1969)
Szekeres, L., Lenard, G., Török, T.: Effect of strophanthine on myocardial metabolism of normal and hypoxic rats. Arch. Int. Pharmacodyn. *119*, 102–109 (1959)
Tanz, R.D., Marcus, S.M.: Influence of endogenous cardiac catecholamine depletion on the force and rate of isolated heart preparations and their response to ouabain. J. Pharmacol. Exp. Ther. *151*, 38–45 (1966)
Tauchert, M., Behrenbeck, D.W., Hötzel, J., Jansen, W., Niehues, B., Hilger, H.H.: Der Einfluß von Digoxin auf den myokardialen Sauerstoffverbrauch bei koronarer Herzkrankheit. Verh. Dtsch. Ges. Inn. Med. *83*, 1641–1644 (1977)
Theisohn, M., Friedrich, M., Justus, P., Güttler, K., Klaus, W.: Heat production and oxygen consumption of the isolated rabbit heart: their relation to mechanical function. Basic Res. Cardiol. *72*, 19–33 (1977)
Theisohn, M., Güttler, K., Klaus, W., Theisohn-Schwedhelm, I.: An improved method for simultaneous measurement of heat production and some parameters of mechanical function of isolated hearts. Naunyn-Schmiedebergs Arch. Pharmacol. *274* (suppl.), R 116 (1972)
Thorn, W., Gercken, G., Hürter, P.: Function, substrate supply, and metabolite content of rabbit heart perfused in situ. Am. J. Physiol. *214*, 139–145 (1968)
Tsao, S.-S., Cornatzer, W.E.: Effect of digitonin and digitoxin on the phospholipid metabolism of mammalian tissue culture cells. Biochem. Pharmacol. *16*, 249–262 (1967)
Turto, H.: Experimental cardiac hypertrophy and the synthesis of poly (A) containing RNA and of myocardial proteins in the rat: The effect of digitoxin treatment. Acta Physiol. Scand. *101*, 144–154 (1977)
Turto, H., Lindy, S., Uitto, J.: Effect of digitoxin on myocardial nucleic acids and total proteins in experimental cardiac hypertrophy in rats. Scand. J. Clin. Lab. Invest. *27* (suppl.), 20 (1971)
Turto, H., Lindy, S.: Digitoxin treatment and experimental cardiac hypertrophy in the rat. Cardiovasc. Res. *7*, 482–489 (1973)
Ueba, Y., Ito, Y., Chidsey, C.A. III.: Intracellular calcium and myocardial contractility. I. Influence of extracellular calcium. Am. J. Physiol. *220*, 1553–1557 (1971)
Van der Veen, K.J., Willebrands, A.F.: Effect of frequency and Ca^{++} concentration on oxygen consumption of the isolated rat heart. Am. J. Physiol. *212*, 1536–1540 (1967)
Vernick, R., Sonnenblick, E.H., Lesch, M.: The inhibition by ouabain of phospholipid turnover in the myocardium of the rabbit. J. Mol. Cell. Cardiol. *5*, 553–557 (1973)
Wallace, A.G., Skinner, N.S. Jr., Mitchel, J.H.: Hemodynamic determinants of the maximal rate of rise of left ventricular pressure. Am. J. Physiol. *205*, 30–36 (1963)

Waser, P.G., Volkart, O.: Wirkung von Herzglykosiden auf Actomyosin; Viskositätsmessungen. Helv. Physiol. Pharmacol. Acta *12*, 12–22 (1954)
Watanabe, A.M., Besch, H.R.: Cyclic adenosine monophosphate modulation of slow calcium influx channels in guinea pig hearts. Circ. Res. *35*, 316–324 (1974)
Weber, K.T., Janicki, J.S.: Interdependence of cardiac function, coronary flow and oxygen extraction. Am. J. Physiol. *236*, H 784–H 793 (1978)
Weber, K.T., Janicki, J.S.: The metabolic demand and oxygen supply of the heart: physiologic and clinical considerations. Am. J. Cardiol. *44*, 722–729 (1979)
Weicker, B.: Stoffwechselwirkung des Strophanthins am Warmblütlerherzen. Arch. Exp. Pathol. Pharmakol. *178*, 524–533 (1935)
Weissler, A.M., Gamel, W.G., Goode, H.E., Cohen, S., Schoenfeld, C.D.: The effects of digitalis on ventricular ejection in normal human subjects. Circulation *29*, 721–729 (1964)
Wenzel, J., Hecht, A.: Enzymhistochemische und morphologische Befunde am Meerschweinchenherzmuskel nach Anwendung verschiedener kardiotoxischer Substanzen. Exp. Path. (Jena) *3*, 305–320 (1969)
Whalen, W.J.: The relation of work and oxygen consumption on isolated strips of cat and rat myocardium. J. Physiol. (Lond.) *57*, 1–17 (1961)
Willebrands, A.F., van der Veen, K.J.: Influence of substrate on oxygen consumption of isolated perfused rat heart. Am. J. Physiol. *212*, 1529–1535 (1967)
Williams, A., Nayler, W.G.: High energy phosphate and total calcium concentrations during ouabain induced myocardial contractures. J. Mol. Cell. Cardiol. *9* (suppl.), 64 (1977)
Williams, J.F. Jr., Braunwald, E.: Studies on digitalis. XI. Effects of digitoxin on the development of cardiac hypertrophy in the rat subjected to aortic constriction. Am. J. Cardiol. *16*, 534–539 (1965)
Williamson, J.R.: Metabolic effects of epinephrine in the isolated, perfused rat heart. I. Dissociation of the glycogenolytic from the metabolic stimulatory effect. J. Biol. Chem. *239*, 2721–2729 (1964)
Williamson, J.R.: Glycolytic control mechanisms. I. Inhibition of glycolysis by acetate and pyruvate in the isolated perfused rat heart. J. Biol. Chem. *240*, 2308–2321 (1965)
Williamson, J.R.: Metabolic effects of epinephrine in the perfused rat heart. II. Control steps of glucose and glycogen metabolism. Mol. Pharmacol. *2*, 206–220 (1966)
Williamson, J.R.: Mitochondrial function in the heart. Ann. Rev. Physiol. *41*, 485–506 (1979)
Winter, C.G., Christensen, H.N.: Migration of amino acids across the membrane of the human erythrocyte. J. Biol. Chem. *239*, 872–878 (1964)
Wollenberger, A.: Metabolic action of the cardiac glycosides. I. Influence on respiration of heart muscle and cortex. J. Pharmacol. Exp. Ther. *91*, 39–51 (1947)
Wollenberger, A.: The energy metabolism of the failing heart and the metabolic action of the cardiac glycosides. Pharmacol. Rev. *1*, 311–352 (1949)
Wollenberger, A.: Effect of ouabain on the utilization of ^{14}C-labelled glucose and pyruvate by cardiac muscle. J. Pharmacol. Exp. Ther. *101*, 38–39 (1951 a)
Wollenberger, A.: Utilization of ^{14}C-labelled glucose by cardiac muscle treated with a cardiac glycoside. Science *113*, 64–65 (1951 b)
Wollenberger, A.: Metabolic action of the cardiac glycosides. II. Effects of ouabain and digoxin on the energy-rich phosphate content of the heart. J. Pharmacol. Exp. Ther. *103*, 123–125 (1951 c)
Wollenberger, A.: Metabolic action of cardiac glycosides. III. Influence of ouabain on the utilization of ^{14}C-labelled glucose, lactate, and pyruvate by dog heart slices. Arch. Exp. Pathol. Pharmakol. *219*, 408–419 (1953)
Wollenberger, A.: Introduction of agents influencing myocardial contraction: cardiac glycosides. In: Factors influencing myocardial contractility. Tanz, R.D., Kavaler, F., Roberts, R. (eds.), pp. 506–518. New York: Academic Press 1967
Wollenberger, A., Jehl, J., Karsh, M.L.: Influence of age on the sensitivity of the guinea pig and its myocardium to ouabain. J. Pharmacol. Exp. Ther. *108*, 52–60 (1953)
Wood, J.M., Crow, C.A., Schwartz, A.: Ouabain-induced respiratory changes in guinea pig heart tissue: requirement for calcium. Acta Biol. Med. Germ. *28*, 901–917 (1972)
Wood, J.M., Schwartz, A.: Ouabain-induced respiratory changes in guinea-pig heart tissue: relation of Ca^{++} in the medium. Pharmacologist *12*, 266 (1970)

Yankopoulos, N.A., Kawai, C., Federici, E.E.: The hemodynamic effects of ouabain upon the diseased left ventricle. Am. Heart J. *76*, 466–480 (1968)
Yu, D.H., Triester, S., Gluckman, M.I.: The mechanism of antilipolytic activity of cardioactive drugs in rats. Pharmacologist *11*, 238 (1969)
Yu, D.H., Triester, S., Gluckman, M.I.: Mechanism of antilipolytic effects of cardiac glycosides. J. Pharmacol. Exp. Ther. *177*, 284–290 (1971)
Yu, D.H., Triester, S.: Further studies of the possible mechanism of action of cardiac glycosides on lipolysis. Fed. Proc. *31*, 525 (1972)
Yunis, A.A., Arimura, G.K.: Amino acid transport in blood cells. II. Patterns of transport of some amino acids in mammalian reticulocytes and mature red blood cells. J. Lab. Clin. Med. *66*, 177–186 (1965)
Yusuf, S.M., Gans, J.H.: In vitro metabolism of atrial and ventricular myocardium: effect of ouabain. Arch. Int. Pharmacodyn. *160*, 188–195 (1966a)
Yusuf, S.M., Gans, J.H.: In vitro metabolism of atrial and ventricular myocardium from reserpine treated dogs: effect of ouabain. Arch. Int. Pharmacodyn. *160*, 196–203 (1966b)

CHAPTER 14

Effects of Cardiac Glycosides on Na^+, K^+-ATPase

T. AKERA

A. Introduction

The inhibitory action of cardiac glycosides on active Na^+ and K^+ transport across the cell membrane had been demonstrated in the early 1950s (SCHATZMANN, 1953; see HAJDU and LEONARD, 1959). Thus, with the discovery of Na^+, K^+-ATPase (SKOU, 1957; HESS and POPE, 1957), the effect of these glycosides on the enzyme activity was examined and a specific inhibition was observed (POST et al., 1960; SKOU, 1960; see Fig. 1). The interaction of cardiac glycosides with the enzyme, and its possible relationship to the pharmacologic or toxic actions of these agents have been extensively studied during the last two decades (see the following review articles: GLYNN, 1964; REPKE, 1964, 1965; SKOU, 1965; LANGER, 1971, 1972a; LEE and KLAUS, 1971; SMITH and HABER, 1973; SCHWARTZ et al., 1975; AKERA and BRODY, 1976, 1977; AKERA, 1977; BRODY and AKERA, 1977; LANGER, 1977; OKITA, 1977; WALLICK et al., 1977).

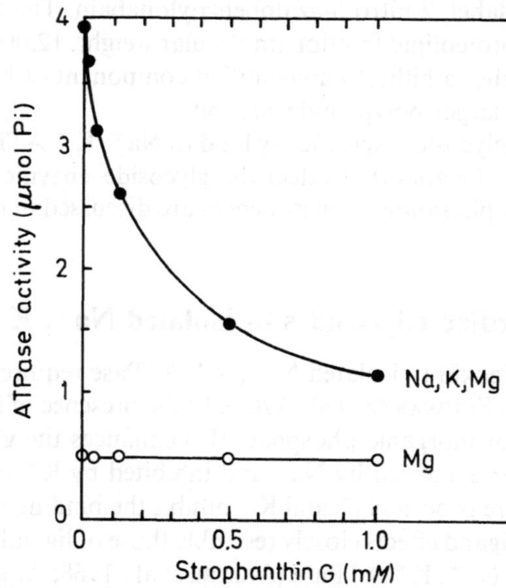

Fig. 1. Effects of strophanthin G (ouabain) on Na^+, K^+-ATPase, and Mg^{2+}-ATPase activity. ATPase enzyme activity assayed at 30 °C in the presence of 6 m*M* $MgCl_2$, 100 m*M* NaCl, 20 m*M* KCl, and 3 m*M* ATP (Na^+, K^+-ATPase) or 3 m*M* $MgCl_2$ and 3 m*M* ATP (Mg^{2+}-ATPase). Ouabain inhibits Na^+, K^+-ATPase without affecting Mg^{2+}-ATPase. Relatively high concentrations of ouabain are required because enzyme is obtained from seashore crab peripheral nerves. (SKOU, 1960)

In hemolyzed and resealed erythrocyte ghosts, ouabain inhibits active ion transport (HOFFMAN, 1966) and Na^+, K^+-ATPase activity (PERRONE and BLOSTEIN, 1973) only when it is in contact with the external surface with respect to the original orientation. ^3H-ouabain also binds to the erythrocyte membrane from the outside (PERRONE and BLOSTEIN, 1973). A purified Na^+, K^+-ATPase obtained from the shark rectal gland exhibits similar characteristics when it is incorporated into phospholipid vesicles (HILDEN et al., 1974). This sidedness of the glycoside-binding site on Na^+, K^+-ATPase contributed to some confusion with regard to the possible presence of a "ouabain-insensitive" Na^+, K^+-ATPase. When tissues are homogenized, inside-out vesicles may be formed. Ouabain then fails to inhibit Na^+, K^+-ATPase totally because of a lack of access to the inhibitory site in the inside-out vesicles. Na^+, K^+-ATPase in such preparations can, however, be made totally "ouabain-sensitive" by appropriate detergent treatment, confirming that the enzyme is essentially ouabain-sensitive (BESCH et al., 1976).

Highly purified enzyme preparations consisting mainly of two major polypeptides are sensitive to ouabain-induced inhibition (KYTE, 1972; HOKIN et al., 1973; LANE et al., 1973; JØRGENSEN, 1974). Photoaffinity labeling of the ouabain-binding site on Na^+, K^+-ATPase by the ethyl diazomalonyl derivative of cymarin and subsequent separation of enzyme subunits revealed that the binding site is on the larger polypeptide subunit (RUOHO and KYTE, 1974) containing phosphorylation sites (KYTE, 1971). Chemical reduction of oxidized ^3H-ouabain previously bound to the enzyme also results in a covalent labeling of the larger catalytic subunit polypeptide (HEGYVARY, 1975). More recently, however, FORBUSH et al. (1978) have reported that the larger polypeptide fraction accounts for only one-half of the covalently bound photoaffinity label, 2-nitro-5-azidobenzoylouabain. The other 50% of the label was found in a proteolipid fraction (molecular weight, 12,000 daltons), which seems to represent either a hitherto unidentified component of Na^+, K^+-ATPase or a fragment of the larger polypeptide subunit.

Thus, the cardiac glycosides specifically bind to Na^+, K^+-ATPase units and inhibit enzyme activity. Factors that affect the glycoside–enzyme interaction, and their biochemical and physiologic consequences are discussed in the following sections.

B. Binding of Cardiac Glycosides to Isolated Na^+, K^+-ATPase

The binding of ^3H-digoxin to isolated Na^+, K^+-ATPase requires Mg^{2+} (MATSUI and SCHWARTZ, 1968; SCHWARTZ et al., 1968). In the presence of Mg^{2+}, adenosine triphosphate (ATP) or inorganic phosphate (Pi) enhances the glycoside binding. The binding is further enhanced by Na^+ and inhibited by K^+ in the presence of Mg^{2+} and ATP, whereas both Na^+ and K^+ inhibit the binding in the presence of Mg^{2+} and Pi. These ligand effects closely resemble those of ligands on the ouabain-induced inhibition of Na^+, K^+-ATPase (ALBERS et al., 1968; AKERA, 1971; AKERA and BRODY, 1971; SKOU et al., 1971; LISHKO et al., 1972). MATSUI and SCHWARTZ (1968), ALBERS et al. (1968), and SEN et al. (1969) concluded that the glycoside inhibits Na^+, K^+-ATPase by *preferentially* binding to a phosphorylated form of the enzyme which is increased by Na^+ and decreased by K^+ in the presence of Mg^{2+} and ATP. BARNETT (1970) reported that the rate constant for the binding of ouabain to the phosphorylated enzyme is independent of the monovalent cation concentrations.

In kinetic studies in the presence of Mg^{2+} and ATP, the initial rate of ^3H-ouabain binding to Na^+, K^+-ATPase follows pseudo-first-order kinetics with respect to ouabain concentrations (BARNETT, 1970; SCHÖNFELD et al., 1972; LINDENMAYER and SCHWARTZ, 1973; WALLICK and SCHWARTZ, 1974), and Na^+ and K^+ apparently compete for a common site to alter the rate of ouabain binding (LINDENMAYER and SCHWARTZ, 1973). Dissociation constants for Na^+ (stimulation) and K^+ (inhibition) for this common site are 13.7 and 0.2 m*M*, respectively. There is an additional high affinity Na^+ site which has a dissociation constant for Na^+ of approximately 0.6 m*M* (INAGAKI et al., 1974). In the presence of a high concentration (10 μ*M*) of ouabain, and after a long incubation period (60 min), the concentration of bound ^3H-ouabain is similar (i.e., the binding sites are saturated by ouabain) for the following combinations of ligands: (Mg^{2+}, Pi), (Na^+, Mg^{2+}, ATP), (Mg^{2+}, ATP), or (Mg^{2+} alone).

In the presence of Mg^{2+} and Na^+, various nucleotide analogs, such as ITP, GTP, CTP, UTP, and ADP, support ^3H-ouabain binding to isolated Na^+, K^+-ATPase (MATSUI and SCHWARTZ, 1968; HOFFMAN, 1969). Since the ability of these nucleotide analogs to form phosphoenzyme is variable, several investigators proposed that the binding of *any* nucleotide to this enzyme rather than the formation of phosphoenzyme is sufficient to support ^3H-ouabain binding (ALBERS et al., 1968; SCHWARTZ et al., 1968; HOFFMAN, 1969; SIEGEL and JOSEPHSON, 1972). In these studies, however, steady-state binding was observed in the presence of relatively high concentrations of ^3H-ouabain. Thus, the observed binding represents the concentration of the binding sites and does not really reflect the rate of ouabain binding. A comparison of the initial velocity of ^3H-ouabain-binding reaction with the concentration of phosphoenzyme formed with ATP, CTP, ITP, or ADP indicates that phosphorylation rather than nucleotide binding is a determinant of the ^3H-ouabain binding (TOBIN et al., 1972a). This concept is supported by the finding that analogs of ATP which have a methylene – or imido – bridge between the β- and γ-phosphate fail to support significant ^3H-ouabain binding (SCHÖNFELD et al., 1972; TOBIN et al., 1973b, 1974a, b). These nonhydrolyzable nucleotides bind to the enzyme but do not yield a significant quantity of the phosphoenzyme. Nonhydrolyzable ATP analogs (β, γ-methylene ATP and ATP-*O*-CH_3) lower the affinity of the enzyme for ouabain in the presence of Mg^{2+}, presumably decreasing the rate of ouabain binding (ERDMANN and SCHONER, 1973b). Additionally, ADP inhibits ouabain binding to rat brain enzyme in the presence of Mg^{2+} and Pi (HANSEN et al., 1971) or Na^+, Mg^{2+} and ATP (TOBIN et al., 1972a). Thus, nucleotide binding per se is insufficient to promote ouabain binding.

During a cycle of enzymatic reaction, Na^+, K^+-ATPase may assume several intermediate forms of the phosphoenzyme, each form having a different reactivity with ouabain. POST et al. (1969) suggested that ouabain combines preferentially with a form of phosphoenzyme, E_2-P, which is highly sensitive to K^+-induced hydrolysis (Fig. 2), based upon the observation that ouabain fails to affect phosphoenzyme concentration in the presence of *N*-ethylmaleimide, an agent which inhibits the conversion of E_1-P to the E_2-P form. The binding of ouabain to Na^+, K^+-ATPase in the presence of Mg^{2+} is promoted by other substrates such as

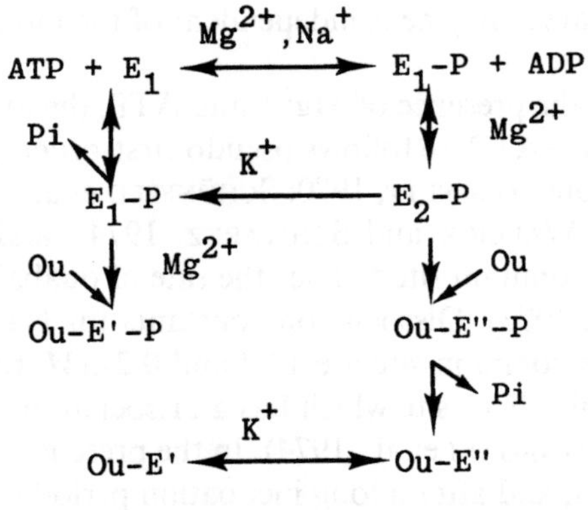

Fig. 2. Tentative reaction sequence for Na^+, K^+-ATPase and ouabain binding. Original proposal by POST et al. (1969) modified to incorporate later findings. Intermediate reaction steps not shown. E_1 and E_2 indicate forms of the enzyme in normal reaction sequence. E′ and E″ indicate forms of the enzyme when combined with and inhibited by ouabain. Two E′ forms of the enzyme shown in the left lower part of the scheme may not be identical. K^+ appears to cause a conformational change in the phosphorylated enzyme, making the phosphate bond more susceptible to hydrolysis, rather than directly causing dephosphorylation

p-nitrophenyl phosphate and carbamyl phosphate (SIEGEL and JOSEPHSON, 1972). Since these organic phosphates form K^+-sensitive phosphoenzyme, the hypothesis that the latter preferentially binds ouabain is further supported.

N-Ethylmaleimide and ethacrynic acid, however, do not inhibit ouabain binding in proportion to their effect on the concentration of the E_2-P form, as judged from the rate of ATP hydrolysis (SIEGEL et al., 1969; ERDMANN and SCHONER, 1973c; TOBIN et al., 1974b; HEGYVARY, 1976). Moreover, Ca^{2+} causes the formation of a K^+-insensitive phosphoenzyme in the presence of Mg^{2+} and ATP, but supports ouabain binding (SCHÖN et al., 1972; TOBIN et al., 1973a). Thus, a phosphoenzyme, apparently different from the K^+-sensitive E_2-P form (Fig. 2), is also capable of combining with ouabain.

It may be the conformation of the enzyme protein associated with a phosphorylation rather than the phosphorylation per se, which favors glycoside binding. Either Mg^{2+} or Mn^{2+} alone or the combination of Mg^{2+} and arsenate also supports ^3H-digoxin binding (SCHWARTZ et al., 1968). Since the enzyme is obviously not "phosphorylated" under these conditions, INAGAKI et al. (1974) suggest that phosphorylation is not an absolute requirement for glycoside binding, and that a particular conformation of the enzyme binds the glycoside when it is combined with Na^+ or K^+, or free from these ligands. In support of this proposition, thermodynamic analysis of the rate of ouabain binding to Na^+, K^+-ATPase in the presence of (Na^+, Mg^{2+}, ATP), (Mg^{2+}, Pi), or (Mg^{2+}, ATP) revealed that the activation energy for ouabain binding is identical under these three conditions (WALLICK and SCHWARTZ, 1974)

SCHÖN et al. (1970) reported that the rate and extent of ouabain binding is identical in the presence of (Na^+, Mg^{2+}, ATP) and (Mg^{2+}, Pi); the complexes formed under these conditions having the same characteristics. Further studies, discussed later in this section indicated that differences may be demonstrated under appropriate conditions. SCHÖN et al. (1970), however, suggested that the dephosphorylated enzyme in a conformation with a high K^+ affinity is the binding form, since phosphoenzymes formed from either ATP or Pi are unlikely to be identical and K^+ reduces the ouabain binding in both conditions. K^+ reduces both the association and dissociation rate constants for the ouabain–enzyme interaction (AKERA and BRODY, 1971; CHOI and AKERA, 1977). It is hence tempting to speculate that the ouabain-binding sites on the enzyme are open and easily accessible in the K^+-sensitive form, and are less accessible after the interaction of the enzyme with K^+. Na^+ and K^+ have been shown to induce conformational changes of the enzyme independent of phosphorylation and dephosphorylation (YODA and HOKIN, 1972; JØRGENSEN, 1975); this may play an important role in ouabain binding *and* release.

There is, however, no a priori reason to believe that only one "form" of enzyme is capable of binding ouabain. HANSEN (1976) proposed that two (or more) populations of enzymes with different affinities bind ouabain in the presence of Mg^{2+} and Pi, and that K^+ makes the enzyme population homogeneous, since ouabain binding to ox brain Na^+, K^+-ATPase exhibits a nonlinear Scatchard plot, and K^+ converts it to a straight line. Contrary reports, however, point to only one type of ouabain-binding site (ERDMANN and SCHONER, 1973a, 1974a; ERDMANN and HASSE, 1975). A nonlinear Scatchard plot was observed only after a long incubation period, probably owing to denaturation of the enzyme (ERDMANN et al., 1976b). The K^+-induced (presumably lower energy) form of the enzyme is possibly more resistant to denaturation during a long incubation at 37 °C.

Nevertheless, different forms of the glycoside–enzyme complex seem to exist. At a high concentration of K^+ or Na^+, most ouabain–enzyme complexes have similar dissociation rates regardless of the condition of binding (LANE et al., 1973). When the complexes are centrifuged and resuspended in a medium with a low ionic strength, however, the dissociation rates are remarkably dissimilar, depending upon the ligands present during the binding reaction (ALLEN et al., 1971a; AKERA et al., 1974b; YODA and YODA, 1974a, b). The complex formed in the presence of Mg^{2+} and ATP is moderately stable and K^+-insensitive (AKERA et al., 1976b). The addition of Na^+ to the mixture increases the ouabain binding. The resulting complex is unstable when the binding ligands are removed, and is highly K^+-sensitive. The addition of K^+ to the mixture containing Na^+, Mg^{2+}, and ATP abolishes the Na^+-stimulated fraction of the ouabain binding, and results in a moderately stable complex which is K^+-insensitive. This characteristic, however, may be peculiar to the rat brain enzyme, since the dog heart enzyme forms a relatively unstable complex which can be stabilized by K^+ under the same ligand conditions. The substitution of p-nitrophenyl phosphate for ATP yields similar results (AKERA et al., 1974b; HANSEN, 1978). The ouabain–enzyme complex formed in the presence of Mg^{2+} and Pi is stable and K^+-insensitive (ALLEN et al., 1971a; AKERA et al., 1974b; YODA and YODA, 1974b).

Which one of the complexes is then formed during the exposure of cardiac tissue to the glycoside? The complex formed in beating hearts is distinctly different from that formed in vitro in the presence of Mg^{2+} and Pi or Mg^{2+} and ATP (ALLEN et al., 1975; AKERA et al., 1976b). (Na^+, Mg^{2+}, ATP)-supported binding is more likely to be analogous to the binding in vivo than (Na^+, Mg^{2+}, ATP, K^+)-supported binding (AKERA et al., 1976b), although K^+ is apparently available to the enzyme. This view is consistent with the finding that hyperkalemia decreases and hypokalemia increases the rate of digoxin binding to cardiac Na^+, K^+-ATPase, as estimated from the development of inhibition of enzyme activity in anesthetized dogs (GOLDMANN et al., 1973, 1975; HALL et al., 1977).

In this case, it is quite likely that intracellular Na^+ stimulates and extracellular K^+ inhibits the binding. Although Na^+, K^+-ATPase activity cannot be directly estimated in intact cardiac muscle cells, sodium pump activity may be assayed from the rate of ouabain-sensitive ^{86}Rb uptake. In atrial preparations of guinea-pig heart, this activity is higher when the frequency of membrane depolarization is higher, or in the presence of a sodium-specific ionophore, monensin, (YAMAMOTO et al., 1979), indicating that intracellular Na^+ indeed enhances the rate of interaction between ouabain and Na^+, K^+-ATPase. BODEMANN and HOFFMAN (1976a, b), however, reported that intracellular Na^+ inhibits, rather than stimulates, ^3H-ouabain binding in the presence of either Pi or ATP in reconstituted human erythrocyte ghosts. The failure of Na^+, both extracellular and intracellular, to stimulate (Mg^{2+}, ATP)-supported ouabain binding directly is unexpected, because Na^+ markedly stimulates ouabain binding to isolated Na^+, K^+-ATPase in the presence of Mg^{2+} and ATP. No explanation for this anomalous finding is available.

Extracellular Rb^+, a more potent analog of K^+, inhibits the glycoside-induced inhibition of the sodium pump (BEAUGE and ADRAGNA, 1971). The various results discussed here generally support the thesis that the binding of the glycoside to a phosphoenzyme taking a particular conformation observed in vitro in the presence of Na^+, Mg^{2+}, and ATP is an appropriate model for the glycoside–enzyme interactions. Since the sidedness of ligands and glycoside binding sites are not maintained in the isolated enzyme studies, any extrapolation of results from these studies to the drug–enzyme interaction in cardiac muscle cells requires caution.

C. Kinetics and Stoichiometry of Glycoside Binding

Few kinetic studies have been performed under experimental conditions in which the sidedness of ligands and glycoside binding sites are maintained, probably because of the difficulty in distinguishing between specific and nonspecific binding of the glycoside to Na^+, K^+-ATPase. HOFFMAN (1969) reported that the percentage inhibition of active K^+ flux is related to the number of molecules of ouabain bound to human erythrocytes. In squid giant axons, BAKER and WILLIS (1972) found that the saturable component of bound ^3H-ouabain parallels the inhibition of Na^+ efflux estimated from the flux of preloaded $^{22}Na^+$. The inhibition developed slowly when the extracellular Na^+ concentration was low. Elevated extracellular Na^+ enhances and Rb^+ antagonizes the glycoside-induced inhibition of the sodium pump (BEAUGE and ADRAGNA, 1971). In contrast, SACHS (1974) reported that the rate of ouabain binding to K^+-free human erythrocytes is unaffected by the external Na^+ concentration. When K^+ was present in the incubation medium, Na^+ enhanced ^3H-ouabain binding by apparently competing with the inhibitory action of K^+ (BODEMANN and HOFFMAN, 1976a).

Extensive kinetic studies have been performed with isolated enzyme preparations. Although asymmetric ligand environments and the sidedness of the glycoside-binding site are not maintained, these studies have yielded useful information on the nature and the mechanisms of interaction between cardiac glycosides and Na^+, K^+-ATPase.

The kinetics of the ouabain–enzyme interaction based on the equilibrium concentration of bound drug is fraught with problems. The binding is a slow process when the concentration of ouabain is below saturation levels (BAKER and WILLIS, 1969, 1970; ALLEN and SCHWARTZ, 1970; ALLEN et al., 1970b; AKERA, 1971; AKERA and BRODY, 1971), hence the binding equilibrium may be underestimated. A long incubation period causes partial inactivation of the enzyme (ERDMANN et al., 1976b), besides decreasing the concentration of ATP. Owing to ATP hydrolysis by Mg^{2+}-ATPase activity, many kinetic analyses are on ouabain binding data obtained after the exhaustion of ATP. Even when ATP is not completely depleted, ADP produced by the hydrolysis of ATP reduces the rate of ouabain binding (TOBIN et al., 1972a). Additionally, the concentration of ouabain may be substantially lowered during incubation owing to the very high affinity of Na^+, K^+-ATPase for ouabain (AKERA, 1971). All these factors contribute to an underestimation of ouabain brinding.

Scatchard analyses of the interaction of ouabain with Na^+, K^+-ATPase in the presence of Mg^{2+} and Pi demonstrated a single type of ouabain-binding site in enzyme preparations from beef brain, kidney and heart, dog heart, and guinea-pig kidney (ERDMANN and SCHONER, 1973a). Homogenized human erythrocytes too have only one type of ouabain binding site in the presence of Mg^{2+} and Pi (ERDMANN and HASSE, 1975). Similar experiments revealed that different cardiac glycosides bind to one type of site (ERDMANN and SCHONER, 1974a). The presence of a single affinity, noninteracting binding site has been confirmed by the labeled-ligand dilution technique (AKERA and CHENG, 1977).

In the presence of Na^+, Mg^{2+}, and ATP, the binding of ouabain to Na^+, K^+-ATPase also adheres to first-order kinetics over a wide range of glycoside and monovalent cation concentrations and over the range 0–37 °C (LINDENMAYER and SCHWARTZ, 1973; WALLICK and SCHWARTZ, 1974). KYTE (1972) reported that *n*-(4′-amino-*n*-butyl)-3-aminoacetylstrophanthidin, a potent inhibitor of Na^+, K^+-ATPase, has one binding site for each molecule of the polypeptide chain. A highly purified enzyme preparation from rabbit kidney binds one mole of ouabain per 282,000 g protein and one mole of phosphate per 132,000 g protein (JØRGENSEN, 1974), suggesting at least two large polypeptide subunits for one ouabain-binding site.

Considerable discrepancy exists regarding the ratio of observed ouabain-binding and phosphorylation sites in Na^+, K^+-ATPase preparations. ERDMANN and SCHONER (1973a) reported that such a ratio varies from one for beef enzymes (brain, kidney, and heart) to four for guinea-pig kidney enzyme. Other reports range from 0.5 for rabbit kidney enzyme (JØRGENSEN, 1974), to unity in ox brain (HANSEN et al., 1971), guinea-pig kidney (HEGYVARY and POST, 1971), rabbit kidney (KYTE, 1972), pig heart (SCHÖN et al., 1972), human heart (ERDMANN et al., 1976a) to three in kidney enzymes, presumably from dogs (KYTE, 1971). Such differences are unlikely to be related to the source or purity of the enzyme. Phosphorylation sites in addition to ouabain binding are probably underestimated, because "maximal phosphorylation" is generally estimated using a single concentration (typically, 0.5 m*M*) of labeled ATP. To some extent, the reported differences may be due to the differences in assay procedures.

Only one phosphorylation and one ouabain-binding site are generally observed for each functional unit of the enzyme which contains at least two large polypeptide subunits, probably because only half the sites are normally active at a given moment (REPKE and SCHÖN, 1973). TANIGUCHI and IIDA (1972) reported that ox brain enzyme preparations contain two apparently different ouabain-binding sites. The binding of ouabain to the high affinity site is sufficient to cause enzyme inhibition. Low affinity sites, present in equal amount as the high affinity sites in the same preparation, may be occluded sites capable of binding ouabain only after long incubation (2 h) in acetate buffer and at low pH (6.1). Hypotheses have been advanced to explain the why binding of one molecule of ouabain can inactivate two enzyme subunits (e.g., REPKE and SCHÖN, 1973). Alternatively, the two larger subunit polypeptides may not be identical, although both apparently can be phosphorylated (JØRGENSEN, 1974).

D. Effects of K^+ on Glycoside Binding

Earlier work implied that ouabain and K^+ compete for the same binding site, since K^+ reduces the ouabain-induced inhibition of Na^+, K^+-ATPase (DUNHAM and GLYNN, 1961; see AKERA and BRODY, 1971) and the binding sites for both K^+ and glycosides are located on the external surface of the cell membrane. According to MATSUI and SCHWARTZ (1966), ouabain-induced inhibition is not competitive with respect to K^+, suggesting different binding sites for ouabain and K^+. Ouabain binding sites on Na^+, K^+-ATPase are remarkably variable in their affinity for the glycoside (REPKE et al., 1965) and in stability of the ouabain–enzyme bond (TOBIN and BRODY, 1972; TOBIN et al., 1972b), depending upon the source of enzyme. In contrast, activation of Na^+, K^+-ATPase by K^+ in the presence of Na^+, Mg^{2+}, and ATP is quite similar in various enzyme preparations having clear species-dependent differences in their affinity for ouabain (AKERA et al., 1969). Thus, binding sites for ouabain and K^+ apparently do not share common characteristics. Furthermore, K^+, but not Na^+, is capable of reversibly stabilizing the ouabain–enzyme complex (AKERA and BRODY, 1971; ALLEN et al., 1971a; AKERA et al., 1974b); K^+ binds to the enzyme when ouabain-binding sites are already occupied. A highly purified Na^+, K^+-ATPase preparation obtained from dog kidney has two moles of K^+-binding sites to one of ouabain-binding site (MATSUI et al., 1977). There is interaction between the two K^+-binding sites, implying they are on the same functional unit of the enzyme. Ouabain and K^+ apparently compete for the same form of the phosphoenzyme, rather than at the same site (MATSUI and SCHWARTZ, 1966). When K^+ combines with the enzyme first, the affinity of the enzyme for ouabain decreases markedly, whereas with the converse, the resulting ouabain–phosphoenzyme complex becomes resistant to K^+-induced dephosphorylation (POST et al., 1969).

Dissociation of ouabain from Na^+, K^+-ATPase is a slow process (SCHÖN et al., 1970; AKERA and BRODY, 1971; ALLEN et al., 1971a) which is further delayed by K^+. This effect of K^+ is usually not apparent in the presence of binding ligands when the binding reaction is terminated (LANE et al., 1973), presumably because the unstable complex constantly formed during the binding reaction is lost almost immediately. A presteady-state kinetic analysis (CHOI and AKERA, 1977) demonstrates that "effective" dissociation rate constants are higher than those estimated after the termination of the binding reaction and are markedly decreased by K^+.

Since K^+ decreases both the forward (MATSUI and SCHWARTZ, 1968; SCHWARTZ et al., 1968; ALLEN and SCHWARTZ, 1970; AKERA, and BRODY, 1971) and backward velocities of the ^3H-ouabain binding reaction, it may or may not affect the equilibrium of ouabain binding, or the affinity of Na^+, K^+-ATPase for glycosides. ALLEN and SCHWARTZ (1970) showed that K^+ delays the rate of ouabain binding to calf heart Na^+, K^+-ATPase without altering the ultimate concentration of the bound ouabain. A similar result (LINDENMAYER and SCHWARTZ, 1970) has been reported for the ouabain binding observed in the presence of Mg^{2+} and Pi. These studies, with relatively high, saturating concentrations of ouabain indicate the inability of K^+ to alter the ouabain-binding capacity of Na^+, K^+-ATPase, but fail to ascertain whether K^+ affects the affinity of Na^+, K^+-ATPase for ouabain. According to SCHÖNFELD et al. (1972), K^+ fails to alter the dissociation rate of the ouabain–enzyme complex formed in the presence of Na^+, Mg^{2+}, and ATP, though opposite results exist (CHOI and AKERA, 1977). Since the rate of binding is apparently reduced by K^+ (SCHÖN et al., 1970), the equilibrium concentration of bound ouabain is likely to be reduced by K^+. In the presence of Na^+, Mg^{2+}, and ATP, an accurate estimation of the effect of K^+ on equilibrium of ^3H-ouabain binding is difficult since K^+ increases both the time to reach equilibrium and the rate of ATP hydrolysis. A K^+-induced decrease in ouabain binding to various enzyme preparations was observed in the presence of Mg^{2+} and Pi (ERDMANN et al., 1975); a condition in which K^+ reduces the rate of binding without affecting the rate of release (ALLEN et al., 1971a).

The apparent affinity of Na^+, K^+-ATPase for digitoxin and strophanthidin K is decreased by an elevation of K^+ concentration from 5 to 16 m*M* in the assay medium (FRICKE and KLAUS, 1971a). The effect of K^+ to increase the dissociation constant of the ouabain–enzyme interaction is also demonstrated directly from ^3H-ouabain binding supported by either ATP or Pi (ERDMANN and SCHONER, 1973b; SKOU, 1973). This ability of K^+ seems to depend on the cardiac glycoside used (FRICKE and KLAUS, 1971a). These results require cautious interpretation. In the assay for ATPase activity to assess the enzyme–glycoside affinity, the extent of the inhibition at a given concentration of glycoside, and therefore the estimated affinity, is dependent on the order of ligand addition (AKERA, 1971). In the presence of Na^+, the glycoside binding requires ATP (TOBIN and SEN, 1970). Thus, in the presence of Na^+, K^+, and Mg^{2+} only, the glycoside does not bind well to the enzyme during the preincubation period. On addition of ATP to the incubation medium the enzyme reaction starts at its uninhibited rate, the reaction velocity decreasing subsequent to the binding (LINDENMAYER and SCHWARTZ, 1973). The enzyme inhibition may not reach equilibrium during a relatively short incubation period. As such, the observed effect is the average of a slowly developing enzyme inhibition, hence the common method for ATPase assay inherently causing an underestimation of the glycoside-induced enzyme inhibition (AKERA, 1971). The underestimation is greater with higher concentrations of K^+. This problem may be eliminated by preincubating the enzyme with the glycoside in the presence of Na^+, Mg^{2+}, and

ATP, i.e., under conditions which favor glycoside binding, and starting the enzyme reaction by adding K^+ to the incubation mixture (AKERA, 1971). When enzyme activity is assayed by measuring the hydrolysis of an organic phosphate compound other than ATP, the underestimation of the inhibitory effect of glycosides is even greater. The rate of glycoside binding supported by phosphatase-type substrates, such as p-nitrophenyl phosphate, carbamyl phosphate, acetyl phosphate, umbelliferone phosphate, or 3-*0*-methylfluorescein phosphate, is markedly slower when the concentration of the substrate is low, which is typical of K^+-stimulated phosphatase assays. This problem may also be avoided by longer preincubation in the absence of K^+ (HUANG and ASKARI, 1976).

The association and dissociation rate constants for the glycoside–enzyme interaction may be estimated from the time course of the glycoside binding during a relatively short incubation period. In such a study with ^3H-ouabain, the binding of ouabain to Na^+, K^+-ATPase in the presence of Na^+, Mg^{2+}, and ATP can be expressed as:

$$E + Ou \underset{k_{-1}}{\overset{k_1}{\rightleftharpoons}} E \cdot Ou, \tag{1}$$

where E is enzyme, Ou is ouabain and E·Ou is the ouabain–enzyme complex (LINDENMAYER and SCHWARTZ, 1973). It follows then that changes in concentrations of bound ouabain with respect to time is (CHOI and AKERA, 1977)

$$\frac{d\,[E \cdot Ou]}{dt} = k_1([E]-[E \cdot Ou])\,([Ou]-[E \cdot Ou]) - k_{-1}[E \cdot Ou]. \tag{2}$$

Fitting this equation to the binding data (Fig. 3) with a polynomial approximation curve-fitting technique, a set of best estimates for k_1, k_{-1}, and the binding site concentration are obtained (Table 1). The results indicate that K^+ fails to alter the binding site concentration, but reduces both the apparent association rate constant (k_1) and the overall dissociation rate constant (k_{-1}). Since the K^+-induced change in k_1 value (approximately 73%) is greater than the change in k_{-1} value (approximately 27%), the apparent dissociation constant ($K_D = k_{-1}/k_1$) is increased, or the affinity of enzyme for ouabain is decreased, in the presence of K^+. Consistent with this conclusion, the addition of K^+ to the incubation medium when the ^3H-ouabain-binding reaction is near equilibrium causes an immediate reduction of the glycoside binding, apparently shifting toward a new, lower equilibrium (BAKER and WILLIS, 1970; HANSEN, 1971; CHOI and AKERA, 1977; AKERA et al., 1978).

Whether K^+ reduces the affinity or the equilibrium interaction depends on the relative extent of K^+-induced changes in the association and dissociation rate constants. In the presence of Na^+, Mg^{2+}, and ATP, the addition of K^+ to the incubation medium reduces the rate of the glycoside binding by decreasing the concentration of K^+-sensitive phosphoenzyme; the binding velocity is thus reduced by K^+ independent of the glycoside. In contrast, K^+ seems to decrease the dissociation rate constant by increasing the effectiveness of a lipid barrier on the glycoside-binding site (AKERA et al., 1974b, 1978, 1979b, c), hence this latter effect of K^+ is dependent on the lipid solubility of the compound. For example, K^+ affects the dissociation rate constant of highly lipid-soluble cardenolides only minimally,

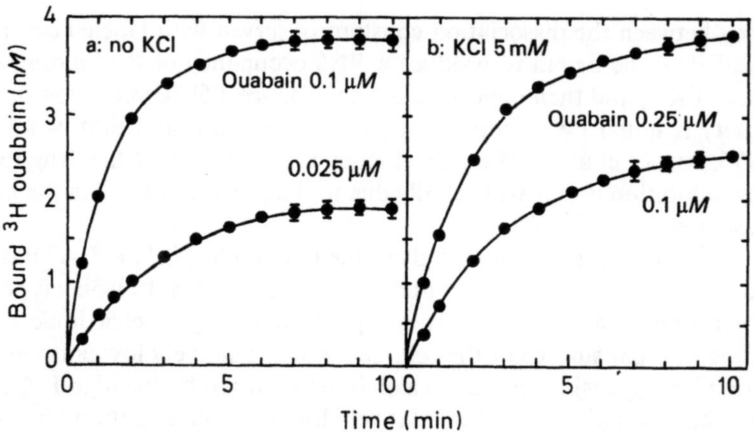

Fig. 3a, b. Time course of ^3H-ouabain binding to Na^+, K^+-ATPase in the absence **a** and presence **b** of KCl. Rat brain Na^+, K^+-ATPase preparations incubated with the ^3H-ouabain at 37 °C in the presence of 100 m*M* NaCl, 5 m*M* $MgCl_2$, and 5 m*M* Tris-ATP with or without KCl. Difference in binding observed in the presence and absence of ATP (ATP-dependent ouabain binding) is shown. Vertical lines indicate standard errors of five experiments. (CHOI and AKERA, 1977)

Table 1. Effect of K^+ on the kinetic parameters for the interaction between ouabain and Na^+, K^+-ATPase

Ligands	Apparent k_1	Overall k_{-1}	K_D[a]	Binding site concentrations[b]
	(n*M* min)$^{-1}$	(min^{-1})	(n*M*)	(n*M*)
Na^+, Mg^{2+}, ATP	0.00486 ± 0.00001	0.254 ± 0.002	52.3	6.00 ± 0.12
K^+, Na^+, Mg^{2+}, ATP	0.00132 ± 0.00001	0.186 ± 0.003	141	5.88 ± 0.10

[a] Apparent dissociation constant; $K_D = k_{-1}/k_1$

[b] Protein concentration = 0.02 mg/ml, hence ouabain-binding sites on enzyme are 300 ± 6 and 294 ± 5 pmol/mg protein in the absence and presence of 5 m*M* KCl.

Kinetic constants were estimated using a polynomial approximation curve-fitting technique from data shown in Fig. 3. K^+ reduces the apparent association rate constant (k_1) for ouabain to a greater extent than the overall dissociation rate constant (k_{-1}), and thus increases the dissociation constant (K_D), or decreases the affinity of the enzyme for ouabain. (CHOI and AKERA, 1977)

since such compounds can readily penetrate the lipid barrier. Therefore, the equilibrium concentration of the bound lipid-soluble drug would be reduced by K^+. Such is the case with aglycones. K^+, however, should effectively reduce both the association and dissociation rate constants for less lipid-soluble compounds, resulting in a relatively minor change in the equilibrium of drug binding. Such is the case with ouabain.

In this context, caution is imperative in any comparison of the inhibitory potency of glycosides on Na^+, K^+-ATPase in vitro estimated in the presence of "optimal" (usually 15 m*M* or higher) concentration of K^+, with that producing positive inotropic effect in isolated heart preparations (FRICKE and KLAUS, 1971 a). A discrepancy between the dissociation constant observed with labeled compounds (concentration of the ligand to produce a 50% occupancy of the binding sites in the absence of K^+) and their concentration to produce a 50% inhibition of the enzyme activity is noted by several investigators (ERDMANN and SCHONER, 1973a; YODA, 1976; AKERA et al., 1979b). The higher concentration of the drug required for enzyme inhibition is at least partially due to the presence of K^+ in the medium for ATPase assay.

Since Na^+ too affects the rate of glycoside binding to Na^+, K^+-ATPase, care should be exercised in selecting Na^+ and K^+ concentrations. In isolated or in situ heart preparations, Na^+, K^+-ATPase is exposed to an asymmetric ionic environment. At the internal surface of the cell membrane, relatively low concentrations of Na^+ stimulate glycoside binding, which is antagonized by the high K^+ concentration. At the external surface, the relatively low K^+ concentration also inhibits glycoside binding. This action of K^+ is partially antagonized by the high extracellular Na^+ concentration. Since this condition cannot be mimicked in studies with isolated enzyme preparations, a quantitative comparison between the concentration of a particular glycoside which causes enzyme inhibition and that which produces a positive inotropic effect appears impossible.

E. Glycoside Binding Sites and Release of Bound Glycosides

The K^+-induced stabilization of the ouabain–enzyme complex formed in the presence of Na^+, Mg^{2+}, and ATP is reversible; however, attempts to convert the stable and K^+-insensitive complex prepared in the presence of Mg^{2+} and Pi to an unstable and K^+-sensitive form were unsuccessful (AKERA et al., 1974b). The addition of Na^+ and ATP to the incubation medium in which the dissociation reaction is monitored has been reported to increase the rate of dissociation of the complex formed in the presence of Mg^{2+} and Pi (YODA, 1974). Na^+ alone appears to be effective (YODA and YODA, 1974b), though ALLEN et al. (1971a) have reported the contrary. Species and tissue-dependent differences cannot be ignored. The ouabain–enzyme complex formed in the presence of Mg^{2+} and Pi has been shown to be stabilized by Na^+, K^+, or Mg^{2+} with enzyme preparations from cattle, rabbit, or guinea-pig kidney but not from beef, rat, rabbit, or guinea-pig brain cortex or rabbit heart (SCHUURMANS STEKHOVEN et al., 1976). Na^+, however, failed to enhance the dissociation of bound ouabain from these enzyme preparations. Ca^{2+} increases the dissociation rate of the complex which is formed in the presence of Na^+, Mg^{2+}, and ATP, when the dissociation reaction is monitored in the presence of K^+ (ERDMANN and SCHONER, 1974b). Since Ca^{2+} is capable of substituting for Na^+ in certain partial reactions of Na^+, K^+-ATPase (SCHÖN et al., 1972), this ligand perhaps alters the characteristics of the glycoside–enzyme complex similarly to Na^+. Alternatively, Ca^{2+} may be competing with K^+. Overall, it is still unclear if the ouabain–enzyme complexes formed in the presence of Na^+, Mg^{2+}, and ATP and those formed in the presence of Mg^{2+} and Pi are interconvertible.

No direct experimental proof presently exists to indicate that (Na^+, Mg^{2+}, ATP)-supported ouabain binding is exactly at the same site as occurs with (Mg^{2+}, Pi). It is the binding of ouabain by either pathway resulting in the inhibition of

Na^+, K^+-ATPase activity (SKOU et al., 1971) that has led to the general assumption that the binding sites on the enzyme are identical. Photoaffinity labeling studies followed by amino acid sequencing at the respective binding sites of covalently bound glycoside derivatives may clarify this point.

The K^+-induced stabilization of the ouabain–enzyme complex formed in the presence of Na^+, Mg^{2+}, and ATP is significantly weakened by deoxycholic acid (AKERA et al., 1974b), and an excess of it further decreases the stability of the subsequently formed complex (CHOI and AKERA, 1978). The conjugates of ouabain or digoxin with a large protein fail to inhibit isolated cardiac sarcolemmal Na^+, K^+-ATPase, active ^{86}Rb uptake by erythrocytes (estimates of sodium pump activity) or solubilized Na^+, K^+-ATPase obtained from erythrocytes (SMITH et al., 1972). When digoxin or ouabain is coupled to the protein via a long, flexible polyamide chain, however, the compound inhibits solubilized (detergent-treated) Na^+, K^+-ATPase. The "chained" glycosides failed to inhibit sarcolemmal Na^+, K^+-ATPase or active ^{86}Rb uptake by erythrocytes, although the glycoside binding site on Na^+, K^+-ATPase is accessible from the external surface of the cell membrane. When digoxin is attached to Sepharose in a similar manner, the compound binds to and inhibits solubilized Na^+, K^+-ATPase obtained from rat brain (OKARMA et al., 1972). The degree of inhibition is related to the length of the hydrocarbon bridge between digoxin and Sepharose. Thus, the glycoside binding site on Na^+, K^+-ATPase is perhaps located in a cleft on the external surface of the cell membrane.

Such a location may account for the relatively stable binding of cardiac glycosides and the high activation energy needed for the binding reaction. SCHÖN et al. (1970) estimated the activation energy for ouabain binding in the presence of either (Na^+, Mg^{2+}, ATP) or (Mg^{2+}, Pi) to be 25 kcal/mol for Na^+, K^+-ATPase obtained from pig heart. Since the affinity of Na^+, K^+-ATPase for cardiac glycosides and related compounds is dependent on the source of the enzyme and also on the chemical structure of the compound, thermodynamic values, too, are dependent on these factors (ERDMANN and SCHONER, 1973a; CLARK et al., 1975). Several investigators report somewhat different values with different compounds and enzyme preparations. Of particular interest is the reduction of the unsaturated lactone of eight cardiotonic steroids or glycosides resulting in a decrease in the free energy of activation of approximately 2–4 kcal/mol, i.e., approximately equal to the energy of one hydrogen bond (CLARK et al., 1975). The possible location of the binding site in a deep cleft of the membrane is consistent with the observation that the ouabain–enzyme complex is more stable than the enzyme itself, and that conditions sufficient for the rapid dissociation of the ouabain over and above the inherently slow rate of release results in an inactivation of the Na^+, K^+-ATPase activity (HARRIS et al., 1973; AKERA et al., 1974a)

The formation of an unstable intermediate complex during the interaction of ouabain with isolated Na^+, K^+-ATPase in the presence of Mg^{2+} and Pi has been reported (YODA, 1973; YODA et al., 1973). Presteady-state kinetic analysis of the ouabain-binding reaction in the presence of Na^+, Mg^{2+}, and ATP with or without K^+ also supports the hypothesis that an unstable intermediate complex is formed preceding the formation of more stable complexes (CHOI and AKERA, 1977). LINDENMAYER and SCHWARTZ (1973), however, observed that a plot of the pseudo-first-order rate constant versus the ouabain concentration is rectilinear over a wide range of ouabain concentrations, and contended that there seems to be no rapidly formed and easily dissociable complex preceding the formation or dissociation of a stable complex. Whether such an unstable intermediate complex is formed and has pharmacologic importance is as yet not confirmed.

Pharmacologically active cardiac glycosides form stable complexes with Na^+, K^+-ATPase (ALBERS et al., 1968; AKERA et al., 1970; YODA and HOKIN, 1970; ERDMANN and SCHONER, 1973a, b, 1974a; AKERA et al., 1974b; WALLICK et al., 1974; YODA and YODA, 1974a), the inhibition being virtually irreversible at low temperatures. (There are several exceptions depending upon the source of enzyme; see ALLEN and SCHWARTZ, 1969; AKERA et al., 1973; SCHWARTZ et al., 1974). At 23 °C and above, the inhibition is slowly reversible (SEN et al., 1969; HEGYVARY and POST, 1971), the release of the glycoside from Na^+, K^+-ATPase being accompanied by the reactivation of the enzyme (AKERA and BRODY, 1971). These characteristics of the glycoside–enzyme interaction were sometimes misinterpreted owing to lack of comprehension of the role of ligands in glycoside binding. FRICKE and KLAUS (1969) preincubated strophanthidin, strophanthidin-3-bromoacetate, or digitoxin with the enzyme without ATP, Na^+, and Mg^{2+}, i.e., under conditions unfavorable for significant glycoside binding. An aliquot was assayed for Na^+, K^+-ATPase activity by adding to above ligands and K^+, resulting in glycoside binding and enzyme inhibition. When another aliquot of the initial incubation was centrifuged, resuspended, and assayed for activity, the results incorrectly suggested a rapidly reversible nature of the inhibition, when actually the glycoside did not bind to the enzyme. Subsequent work of FRICKE and KLAUS (1971c) demonstrated that very high concentrations of these agents (0.1 mM and higher) when incubated with Na^+, K^+-ATPase isolated from guinea-pig hearts results in a reversible inhibition in the presence of Na^+, Mg^{2+}, and ATP. This reversal, however, required repeated resuspension and centrifugation.

The ouabain–enzyme complex formed in vitro with guinea-pig heart enzyme at 37 °C in the presence of Na^+, Mg^{2+}, and ATP dissociates with a half-life of approximately 4 h at 0 °C and several minutes at 27 °C (AKERA et al., 1973). OKITA et al. (1973), however, reported that the inhibition of Na^+, K^+-ATPase by the glycosides is virtually irreversible during the perfusion of isolated rabbit heart; the half-life of recovery being greater than 6 h at 30 °C. Since the ouabain–enzyme complex formed in vitro in the presence of Na^+, Mg^{2+}, and ATP is reversible, OKITA (1977) has suggested that this complex, formed in vitro, and that formed in beating hearts have different stabilities.

In contrast, KU et al. (1974) and SCHWARTZ et al. (1974) observed that the binding of ouabain to Na^+, K^+-ATPase or the inhibition of Na^+, K^+-ATPase or sodium pump activity by the glycoside under similar conditions is readily reversible in guinea-pig or rabbit heart. While OKITA et al. (1973) perfused their Langendorff preparations under constant pressure, the above authors used a constant flow perfusion. During the former, the isolated heart preparation may become hypoxic when exposed to ouabain which reduces the flow of perfusate by causing vasoconstriction, simultaneously increasing the oxygen demand by enhancing the force of contraction. In yet another study, ouabain caused an irreversible cellular Na^+ accumulation in isolated atrial preparations of guinea-pig heart when the frequency of stimulation was high (BENTFELD et al., 1977). Oxygen supply here is relatively limited by diffusion, and therefore the tissue may become hypoxic when stimulated at a high frequency and when ouabain is given to increase the force of contraction. The irreversible Na^+ accumulation has been attributed to a Ca^{2+}-induced inhibition of the sodium pump, secondary to the glycoside-induced increase in intracellular Na^+. Ca^{2+} has been shown to be an inhibitor of isolated Na^+, K^+-ATPase (SKOU, 1957). The concentration of Ca^{2+} which significantly inhibits isolated Na^+, K^+-ATPase is in the range 0.1–1 mM and dependent on Na^+ concentration (REPKE and PORTIUS, 1963). Thus, Ca^{2+} entering the cell in exchange with Na^+ might inhibit the sodium pump if a sufficient concentration is achieved within the cell. The exact cause of the irreversible sodium pump inhibition is presently unknown, though these results indicate that normally the cardiac glycoside reversibly binds to and inhibits Na^+, K^+-ATPase in the beating heart. Under relative hypoxia of cardiac muscle, irreversible inhibition may develop. Irreversible inhibition of cardiac Na^+, K^+-ATPase by hypoxia (BALASUBRAMANIAN et al., 1973), substrate lack (DHALLA et al., 1974), or ischemia (BELLER et al., 1976; WILLERSON et al., 1977) have been reported, although contrasting results also exist with ischemia (SCHWARTZ et al., 1973) or hypoxia (NAGATOMO et al., 1978).

Despite the slow release of bound ouabain from Na^+, K^+-ATPase, other digitalis derivatives may dissociate from the enzyme more rapidly. The release of bound ouabain may also be rapid depending upon the source of the enzyme as discussed later in Section F.

F. Factors that Influence the Interaction of Cardiac Glycosides with Na^+, K^+-ATPase

I. Chemical Structure of the Glycosides

The chemical structure greatly influences how the individual cardiotonic steroid or glycoside interacts with the isolated Na^+, K^+-ATPase (REPKE, 1963; MATSUI and SCHWARTZ, 1968; WILSON et al., 1970; ERDMANN and SCHONER, 1974a; HAUSTEIN and HAUPTMANN, 1974; FLASCH and HEINZ, 1978). Potent inhibitors of Na^+, K^+-ATPase have: (1) an unsaturated lactone ring at C 14, (2) a sugar component attached to the C 3 position (ERDMANN and SCHONER, 1974a), (3) a steroid nucleus with a *cis* configuration of the C–D junction, and (4) a C 14 β-hydroxyl group (NAIDOO et al., 1974). These structural requirements are remarkably similar to those for potent inotropic glycosides. In fact, the relative potency of various compounds for inhibition of isolated Na^+, K^+-ATPase generally parallels that producing positive inotropic effects in isolated heart preparations (REPKE and PORTIUS, 1963; HAUSTEIN, 1974; HAUSTEIN and HAUPTMANN, 1974; BEARD et al., 1975; CALDWELL and NASH, 1977; FLASCH and HEINZ, 1978). Conflicting data are also reported (PETERS et al., 1974; THOMAS et al., 1974). PETERS et al. (1974), however, used different cardiac tissues with known species-dependent differences in the glycoside sensitivity for the enzyme assay and inotropic studies (ALLEN et al., 1971b; AKERA et al., 1973). Results by THOMAS et al. (1974) may be interpreted to *indicate* a relationship between Na^+, K^+-ATPase inhibition and positive inotropic effects, considering the problems of comparing potencies of various compounds to inhibit Na^+, K^+-ATPase in vitro with inotropic potency. It is interesting that the potencies of digitaloids to inhibit Na^+, K^+-ATPase correlate better with those for positive inotropic effects than for cardiac arrhythmias (HAUSTEIN and HAUPTMANN, 1974; BEARD et al., 1975), although it is generally accepted that the direct arrhythmogenic action of digitaloids on cardiac muscle involves Na^+, K^+-ATPase inhibition (see LEE and KLAUS, 1971; OKITA, 1977; ZAVECZ and DUTTA, 1977).

The potency or affinity of the compound is determined primarily by the stability of the complex formed with Na^+, K^+-ATPase. Highly potent compounds, such as hellebrigenin (RUOHO et al., 1968) and *N*-(4′-amino-*n*-butyl)-3-aminoacetylstrophanthidin (KYTE, 1972), produce a relatively stable inhibition of Na^+, K^+-ATPase, i.e., form more stable complexes with the enzyme. Saturation of the lactone ring (dihydroouabain) or its modification (Actodigin) reduces the potency of compounds, and increases the rate of recovery of sodium pump inhibition (ERLIJ and ELIZALDE, 1974) or isolated Na^+, K^+-ATPase inhibition (ROSS and PESSAH, 1975).

MALUR and REPKE (1970) examined the possibility that differences in the affinities of several digitaloids are due to a difference in the attraction between the carbonyl oxygen of the steroid lactone and either NH or OH groups of the enzyme. Infrared spectroscopy, however, failed to detect any relationship between the hydrogen acceptor capacity of the compound and its affinity. They suggested that the difference is probably due to a steric conformation of the compounds such as angles and bridge distances between different parts of the molecule, either favorable or detrimental to the interaction with receptor sites on the enzyme.

The lactone ring itself has a substantial influence on the affinity. Extensive studies on the relationship between the modification of the lactone ring and the inhibitory effect on Na^+, K^+-ATPase have been performed (BOUTAGY et al., 1973). A slight alteration of the lactone ring markedly reduces the potency of the glycoside (ROSS and PESSAH, 1975). Saturation of the lactone ring produces a similar degree of reduction in affinities of several glycosides and aglycones (CLARK et al., 1975) indicating that the enzyme has a binding site which recognizes the unsaturated lactone ring.

In the presence of Mg^{2+} and Pi, the dissociation rate constant is dependent on the sugar moiety (YODA, 1973), whereas the association rate constant is primarily affected by the steroid structure (YODA et al., 1973). For a given steroid, the association rate constants are greatest with monosaccharide derivatives, and decrease as the number of sugar molecules attached to it increases. YODA et al. (1973) suggest that steric hindrance prevents the fast binding of the bis- or tri-glycosides. The first sugar binds to the enzyme and contributes to a tight complex in the presence of (Na^+, Mg^{2+}, ATP) or (Mg^{2+}, Pi) (YODA et al., 1975). Binding sites for the second and third sugars are different from that for the first sugar, and form less tight bonds. Owing to a reduction of the association rate constant, the second and third sugars do not increase the affinity of the compound. WILSON et al. (1970) suggested that the monoglycosides fit the receptor site best since they have the highest affinity. Digoxin monoglycoside is several times more potent an inhibitor than is digoxigenin in Na^+, K^+-ATPase obtained from human heart or guinea-pig heart or brain (DE POVER and GODFRAIND, 1976). The addition of the second, third, and fourth sugar decreases the potency. These results are consistent with the threshold concentration of digitoxigenin mono-, bis- and tri-digitoxosides needed to produce positive inotropic effects in cat heart–lung preparations (BÖTTCHER et al., 1975).

Despite the observation that Na^+, K^+-ATPase has binding sites complementary to the sugar or saccharide moiety of glycosides (YODA, 1973; WALLICK et al., 1974; YODA and YODA, 1974a, 1975, 1977), rhamnose, or digitoxose alone do not apparently bind to these sites with a high affinity, since they fail to inhibit labeled glycoside binding or enzyme activity. It is tempting to speculate that the steroid portion of glycosides binds to the enzyme first, and induces a conformational change of the enzyme, thus activating sugar-binding sites (YODA et al., 1973).

II. Source of the Enzyme

Species-dependent differences in sensitivity of cardiac muscle to pharmacologic and toxic actions of digitalis are well known. Such differences are largely accounted for by the differences in Na^+, K^+-ATPase. The sensitivity of cardiac Na^+, K^+-ATPase has been shown to parallel the reported sensitivity of the animal to lethal or inotropic actions of glycosides (REPKE et al., 1965). The relationship between the concentration of ouabain sufficient to inhibit cardiac Na^+, K^+-ATPase in vitro and that needed to produce mild toxic effects in anesthetized animals has been demonstrated in dog, pig, sheep, guinea-pig, and rat (AKERA et al., 1969). Na^+, K^+-ATPase preparations obtained from rat heart, a species which has an extremely low sensitivity for the glycosides (Fig. 4c), require higher concentrations of ouabain for enzyme inhibition than do enzyme preparations obtained from guinea-pig heart (Fig. 4a). For thallous ion, which does not show a species-dependent difference in enzyme inhibition (Fig. 4b), concentrations required to produce inotropic effects are also similar (Fig. 4d).

The complex formed between ouabain and rat heart Na^+, K^+-ATPase is unstable at 0 °C, whereas a similar complex formed with cardiac Na^+, K^+-ATPase of digitalis-sensitive species such as beef and dog is stable at this temperature (ALLEN and SCHWARTZ, 1969). The complex formed with guinea-pig heart enzyme is intermediate (AKERA et al., 1973). Association rates between ouabain and Na^+, K^+-ATPase obtained from various tissues in the presence of Mg^{2+} and Pi at 37 °C are not different, but dissociation rate constants are highly variable dependent upon the source of the enzyme (TOBIN and BRODY, 1972; TOBIN et al., 1972b). Since the affinity may be expressed as a ratio between association and dissociation rate constants, these differences in dissociation rate constants apparently contribute to the species-dependent differences in affinity of cardiac Na^+, K^+-ATPase for the glycoside.

The stability of the glycoside–enzyme complex reflects the rate of loss of positive inotropic effects in isolated heart preparations. In a highly ouabain-sensitive species, such as dog or cat, the loss of the positive inotropic effect after the termination of ouabain perfusion in Langendorff preparations is slow, corresponding to the slow release of bound ouabain from isolated Na^+, K^+-ATPase. In guinea-pig and rabbit, which are less sensitive to ouabain, both the loss of the inotropic effect and the release of bound ouabain are faster. In either case, the half-life for

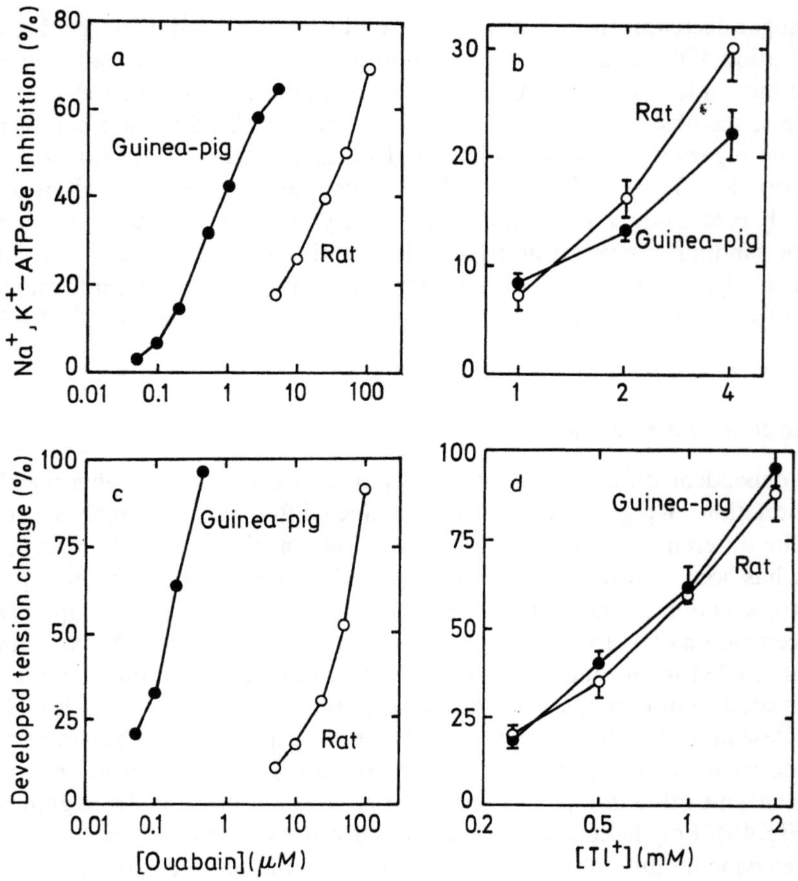

Fig. 4a–d. Na^+, K^+-ATPase inhibition and inotropic action of ouabain and Tl^+. Comparison of guinea-pig and rat heart. Na^+, K^+-ATPase activity assayed in vitro in the presence of 20 m*M* NaCl, 5.8 m*M* KCl, 5 m*M* $MgCl_2$, and 5 m*M* Tris-ATP (pH 7.5). Mg^{2+}-ATPase activity observed in the absence of added NaCl and KCl subtracted. Percentage inhibition of the enzyme activity by either ouabain **a** or Tl^+ **b** calculated against control activity observed in the absence of inhibitor. Left atrial preparations incubated at 30 °C in Krebs–Henseleit solution saturated with 95% O_2 5% Co_2 gas mixture (pH 7.4) and electrically stimulated at 1 Hz. Changes in force of contraction observed 30 min after addition of ouabain **c** or 20 min after the addition of Tl **d**. Control values for force of contraction: 1.57 ± 0.07 (rat) and 0.74 ± 0.09 g (guinea-pig). Each point represents mean of five experiments. Species difference in ouabain sensitivity observed for enzyme inhibition and inotropic effect; no difference observed with Tl^+. (KU et al., 1976)

the loss of inotropic effect is comparable with the half-life for release of bound ouabain from the enzyme in vitro (AKERA et al., 1973).

The low affinity of the rat heart Na^+, K^+-ATPase for digitalis may provide a clue to the structure and geometry of the glycoside-binding sites on the enzyme. A hypothesis that rat heart enzyme does not have binding sites complementary to the sugar moiety of the glycoside, and therefore cannot form a stable complex has been tested (WALLICK et al., 1974). Concentrations of ouabagenin, an aglycone without the sugar moiety, necessary to inhibit Na^+, K^+-ATPase isolated from beef brain and rat hearts are, however, markedly different. Higher concentrations of ouabagenin than of ouabain are required to inhibit rat heart enzyme, indicating that this enzyme has binding sites for the sugar moiety (rhamnose) of ouabain. Rat heart enzymes also require markedly higher concentrations of Actodigin than do guinea-pig heart enzymes for comparable inhibition (AKERA et al., 1979d). Since Actodigin has a modified lactone ring resulting in a reduced affinity for Na^+, K^+-ATPase (i.e., a modified lactone ring does not presumably form a stable bond with the enzyme), the difference between rat and guinea-pig heart enzymes cannot be due to the lack of binding sites for the lactone moiety of the glycoside in rat heart enzymes. Compared with guinea-pig heart enzymes, rat heart Na^+, K^+-ATPase also requires higher concentrations of cassaine, a compound which has digitalis-like actions but has an entirely different chemical structure. Thus, the low affinity of rat heart Na^+, K^+-ATPase for the glycoside cannot be explained by the lack of a particular binding site which recognizes either the steroid nucleus, lactone ring, or the sugar. In rat heart enzymes, glycosides such as ouabain or digoxin, behave like the more lipid-soluble aglycones in response to the effect of K^+ to shift the equilibrium interaction, indicating a low effectiveness of the lipid barrier (AKERA et al., 1979d), which may at least partially account for the failure of the enzyme to form a stable complex with the glycoside.

Na^+, K^+-ATPase of human erythrocyte is more sensitive to strophanthin K than that in the guinea-pig erythrocyte, which in turn has a higher sensitivity than the enzyme from the rat erythrocyte (GROBECKER et al., 1963). The Na^+, K^+-ATPase obtained from human heart has also been shown to be highly sensitive to digitalis (REPKE, 1963; ALLEN and SCHWARTZ, 1972; DE POVER and GODFRAIND, 1976; ERDMANN et al., 1976a).

The structure of Na^+, K^+-ATPase at glycoside-binding sites is different dependent upon the source of the enzyme. Another active site on the enzyme at the external surface of the cell membrane, namely the K^+-binding site, however, does not seem to show a similar magnitude of species-dependent difference (as discussed earlier in Sect. D). No species-dependent differences in the affinity of enzymes for ATP, Mg^{2+}, or Na^+ have been reported. In fact, the intracellular aspects of Na^+, K^+-ATPase must be similar regardless of the source of enzyme, since antibodies prepared against a purified enzyme preparation obtained from pig or dog kidney inhibit the active cation transport or Na^+, K^+-ATPase activity in various tissues and erythrocytes, provided that the antibodies have access to the inner surface of the cell membrane (JØRGENSEN et al., 1973; SMITH and WAGNER, 1975).

Genetic make-up seems to determine the ouabain sensitivity of Na^+, K^+-ATPase. Mutation of cultured pheochromocytoma or Chinese hamster cells from ouabain-sensitive to ouabain-resistant type has been shown to be frequently associated with the alteration in ouabain sensitivity of the Na^+, K^+-ATPase (LELIEVRE et al., 1976; CHANG et al., 1978), although an alteration in the enzyme concentration accounts for ouabain resistance in a few instances (CHANG et al., 1978). Tissue-dependent differences in digitalis sensitivity, however, indicate that other factors are also involved.

ERDMANN et al. (1975) reported that the affinity of Na^+, K^+-ATPase for ouabain is almost identical in all organs in a given species. While this is true in digitalis-sensitive species, such as dog and cat, and perhaps in humans, less sensitive species do exhibit tissue-dependent differences. Thus, rat brain Na^+, K^+-ATPase is highly sensitive to digitalis whereas its heart or kidney enzymes require high concentrations of ouabain for comparable inhibition (AKERA et al., 1972; TOBIN and BRODY, 1972). Comparison of digitalis sensitivity may be complicated because effects of various cations on the stability of the ouabain–enzyme complex are dependent on the tissue from which the enzyme is obtained (SCHUURMANS STEKHOVEN et al., 1976). Moreover, detergent treatment may alter the ouabain sensitivity (affinity) of enzymes dependent upon the tissue from which the enzyme is obtained (CHOI and AKERA, 1978).

Tissue-dependent differences in digitalis sensitivity of Na^+, K^+-ATPase may explain why glycosides primarily affect cardiac muscle despite the ubiquitous presence of this enzyme (DRANSFELD et al., 1966). The lower affinity of Na^+, K^+-ATPase for ouabain in skeletal muscle than in cardiac muscle in rat, guinea-pig, and dog has been observed and offered as an explanation for the failure of the glycoside to increase the force of contraction in skeletal muscle (DRANSFELD et al., 1966; ASH and SCHWARTZ, 1970; CLAUSEN and HANSEN, 1974; ERDMANN et al., 1976b). PITTS et al. (1977), however, argue that the difference in digitalis sensitivity of canine cardiac and skeletal muscle Na^+, K^+-ATPase is relatively minor and cannot account for the lack of the positive inotropic effect of ouabain in the latter muscle. The affinity of Na^+, K^+-ATPase for ouabain in intact cells is difficult to assess from in vitro studies, since it is influenced by intracellular Na^+, which may not be identical in the two types of muscle, and is further affected by the functional state (e.g., frequency of membrane depolarization) of the muscle. Thus, the earlier finding that Na^+, K^+-ATPase in skeletal muscle is not inhibited by ouabain when a significant inhibition of cardiac Na^+, K^+-ATPase and inotropic effect are observed in anesthetized dog may be more relevant to the question (AKERA et al., 1970). Since attempts to produce positive inotropic effects in isolated skeletal muscle, in contrast to isolated rat heart, have largely been unsuccessful even with high concentrations of the glycosides, the low digitalis sensitivity of Na^+, K^+-ATPase alone cannot explain the failure of ouabain to increase the force of contraction in skeletal muscle. PITTS et al. (1977) suggested that a qualitative difference (failure of ouabain to induce a specific conformation of Na^+, K^+-ATPase), rather than a quantitative difference, is responsible for the ineffectiveness of ouabain in skeletal muscle. Other differences such as the source of Ca^{2+} for the activation of contractile protein may also underlie the difference in response of the cardiac and skeletal muscle to glycosides.

Tissue-dependent differences in sensitivity of Na^+, K^+-ATPase to cardiac glycosides are most remarkable in the rat, in which peripheral tissues, including cardiac muscle, have a markedly low affinity for ouabain compared with the enzyme in the brain tissue (AKERA et al., 1972; LIN and AKERA, 1978). In this species, significant inhibition of brain Na^+, K^+-ATPase by the systemic administration of digitoxin has been reported with no lethal cardiotoxicity (GUBITZ et al., 1977). After intravenous infusion of ouabain or digoxin in the dog or cat, however, brain Na^+, K^+-ATPase was not significantly inhibited while the cardiac enzyme was markedly inhibited and cardiac toxicity apparent (AKERA et al., 1970; WEAVER et al., 1977); chronic administration of digoxin in the dog produced similar results (AKERA et al., 1977a).

In rats, the K_D values of Na^+, K^+-ATPase for ouabain (concentration to produce a 50% occupancy of the binding sites) are approximately 30 nM (brain), 290 nM (skeletal muscle), 3,600 nM (kidney), and 12,300 nM (liver) in the presence of Mg^{2+} and Pi (LIN and AKERA, 1978). Studies on the specific uptake of low concentrations of ^3H-norepinephrine, which is dependent on Na^+, K^+-ATPase, indicate that the enzyme of sympathetic nerve terminals probably has a low ouabain sensitivity, similar to that of the rat myocardium in contrast to the brain enzyme (STICKNEY, 1976); a direct estimation of ouabain sensitivity is, however, not available.

III. Pathological Conditions, Temperature, and Membrane Lipids

Several pathological conditions have been shown to affect the Na^+, K^+-ATPase system. For example, the hyperthyroid state has been shown to increase the Na^+, K^+-ATPase of various tissues in rats (EDELMAN, 1975; ASANO et al., 1976; LO et al., 1976) and guinea-pigs (CURFMAN et al., 1977). HEGYVARY (1977) reported that the ouabain sensitivity of cardiac Na^+, K^+-ATPase is decreased in hypothyroid rats but unchanged in thyroid hormone-treated animals. SHARMA and BANERJEE (1978) reported that the triiodothyronine treatment of thyroidectomized rats causes an increase in affinity of cardiac and skeletal muscle Na^+, K^+-ATPase for ouabain, whereas the number of binding sites is unchanged. LO et al. (1976) and LO and EDELMAN (1976), however, reported that triiodothyronine treatment of rats increases the number of Na^+, K^+-ATPase units. This conclusion is based on the finding that the treatment increases the maximal velocity of Na^+, K^+-ATPase, concentrations of phosphorylated intermediate, and the amount of ^3H-ouabain binding simultaneously, and also increases the incorporation of labeled methionine into the polypeptide that represents the large subunit of Na^+, K^+-ATPase. Kinetic analysis indicates that triiodothyronine treatment increases the concentration of (Mg^{2+}, Pi)-supported ouabain-binding sites in particulate fractions obtained from liver, kidney, and skeletal muscle of rats without affecting the affinity of those sites for ouabain (LIN and AKERA, 1978). The Na^+, K^+-ATPase of brain tissue is unchanged by triiodothyronine treatment. Whether triiodothyronine-induced changes in Na^+, K^+-ATPase affect the response of this enzyme to cardiac glycosides in intact cells is unknown.

Hypoadrenalism, in metyrapone-treated guinea pigs or adrenalectomized rats, causes a significant decrease in Na^+, K^+-ATPase of heart and kidney without changes in brain enzyme (BORSCH-GALETKE et al., 1972). It, however, does not change the affinity of the remaining Na^+, K^+-ATPase for strophanthin K as estimated from the inhibition of enzyme activity. The concentration of Na^+, K^+-ATPase in skeletal muscle of genetically obese (Ob/Ob) mice is lower than that of lean littermates (LIN et al., 1978); the affinity of the enzyme for ouabain is similar in the two groups of mice. Chronic potassium deficiency in the diet causes an increase in Na^+, K^+-ATPase of guinea-pig heart muscle (ERDMANN et al., 1971; BLUSCHKE et al., 1976). This treatment does not alter the affinity of the isolated Na^+, K^+-ATPase for ouabain (ERDMANN et al., 1971). BLUSCHKE et al. (1976) and BONN and GREEFF (1978) reported that chronic treatment of guinea-pigs with digitoxin causes an increase in cardiac Na^+, K^+-ATPase without alterations in kidney or brain enzyme. Similar treatment of dogs with digoxin, however, failed to increase cardiac Na^+, K^+-ATPase significantly (KU et al., 1977). The affinity of cardiac Na^+, K^+-ATPase for the glycoside is unchanged either in digitoxin-treated guinea-pigs or in digoxin-treated dogs. Again, the influences of these differences in Na^+, K^+-ATPase on the pharmacologic actions of cardiac glycoside in these tissues are unknown.

Perfusion of isolated rat heart (Langendorff preparations) with an oxygen-free solution for 20 min has been shown to cause a reduction in myocardial Na^+, K^+-ATPase (BALASUBRAMANIAN et al., 1973). The ouabain sensitivity of the enzyme isolated from such hearts is unchanged, when estimated in vitro. Experimental ischemia induced by snaring a branch of coronary artery in anesthetized dogs causes a decrease in Na^+, K^+-ATPase activity with a concomitant decrease in ^3H-ouabain binding in microsomal fractions obtained from the ischemic area (BELLER et al., 1976). The reduction in enzyme activity or ^3H-ouabain binding is not apparent after 1 h occlusion, but is as much as 40% after 6 h occlusion. It is unknown, however, whether hypoxia or ischemia alter the sensitivity of the Na^+, K^+-ATPase for ouabain in situ under the above conditions. Hypoxia and ischemia have been shown to cause a shift in ionic balance with a tendency for a decrease in cellular K^+ and an increase in Na^+ (NAYLER et al., 1971; SINGH et al., 1971; SYBERS et al., 1971). Since intracellular Na^+ should facilitate glycoside binding to Na^+, K^+-ATPase, the effective affinity of the enzyme for ouabain in situ in hypoxic or ischemic hearts may be increased. Combined with a decrease in Na^+, K^+-ATPase activity, these changes may enhance digitalis toxicity in coronary insufficiency.

The interaction between cardiac glycosides and isolated Na^+, K^+-ATPase is affected by the temperature and the pH of the incubation medium. The rate of release of bound ouabain from Na^+, K^+-ATPase observed in vitro is faster at a lower pH (AKERA et al., 1974a). When the ouabain-binding reaction is monitored in the presence of Na^+, Mg^{2+}, and ATP, the concentration of binding sites or the association rate constant is independent of the pH, whereas the dissociation rate constant is higher (the release is faster) at a lower pH (YODA and YODA, 1978). These results predict that the affinity of Na^+, K^+-ATPase for ouabain should be lower at lower pH. ERDMANN and SCHONER (1973b), however, reported that changes in pH below 7.4 do not affect the affinity of Na^+, K^+-ATPase for ouabain, and the affinity is decreased at a higher pH (above 7.4) when the ouabain-binding reaction is observed in the presence of Mg^{2+} and Pi. Changes in pH value between 7.0 and 7.8 fail to alter the toxicity of digitoxin, digoxin, or strophanthin K in isolated atrial preparations of guinea-pig heart (KÖHLER and GREEFF, 1972).

At low temperatures, the inhibitory effect of ouabain is decreased (AHMED and JUDAH, 1965), probably because of a lower rate of glycoside binding (ALLEN and SCHWARTZ, 1970). The effect of Na^+ to stimulate glycoside binding or the effect of K^+ to stabilize the glycoside–enzyme complex is decreased or abolished at low temperatures (SWANSON, 1966; AKERA et al., 1974b). The inhibitory action of ouabain on Na^+, K^+-ATPase isolated from rabbit kidney cortex is highly temperature dependent with a thermal transition at 25 °C (CHARNOCK et al., 1975, 1977). Detergent treatment or a mild lipolysis with phospholipase A eliminates the thermal transition. These results indicate that lipids are involved in the glycoside–enzyme interaction. A low ouabain-binding rate at low temperatures is consistent with the

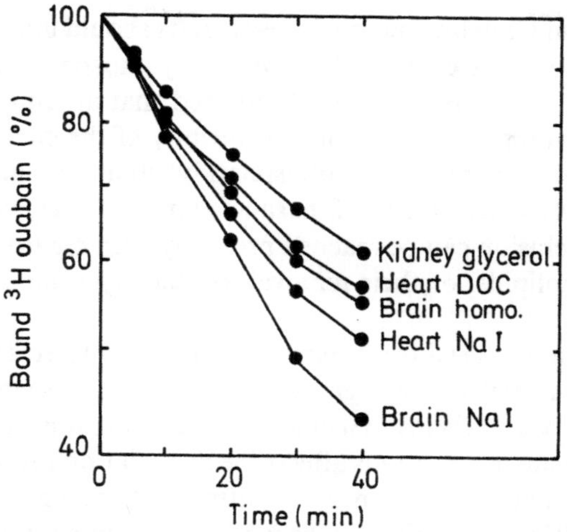

Fig. 5. Effects of purification procedures on dissociation of ^3H-ouabain from Na^+, K^+-ATPase. Na^+, K^+-ATPase preparations from frozen dog tissue by successive treatments with deoxycholate (DOC), NaI, and glycerol. Enzyme preparations at various stages of purification incubated with 0.1 µ*M* ^3H-ouabain in the presence of 100 m*M* NaCl, 5 m*M* $MgCl_2$, and 5 m*M* Tris-ATP (pH 7.5) at 37 °C. After 10-min incubation (time zero), unlabeled ouabain (final concentration, 0.1 m*M*) added to terminate the binding of ^3H-ouabain, and subsequent release of bound ^3H-ouabain monitored. Nonspecific binding observed in the absence of ATP was subtracted. Treatment of heart enzyme with NaI or brain enzyme with deoxycholate and NaI significantly decrease stability of subsequently formed ouabain–enzyme complex. (CHOI and AKERS, 1978)

observation that the development of the positive inotropic action of the cardiac glycosides is slower at lower temperatures (PARK and VINCENZI, 1975). The degree of the positive inotropic effect of the cardiac glycosides is also smaller at lower temperatures (TUTTLE et al., 1962). This phenomenon has been attributed by REPKE (1964) to a lower Na^+, K^+-ATPase activity at the lower temperature.

The involvement of membrane lipids in the interaction between Na^+, K^+-ATPase and the cardiac glycoside is somewhat controversial. TANIGUCHI and IIDA (1971, 1973) reported that phospholipase A treatment of ox brain enzyme decreases ouabain binding in proportion to the decrease in Na^+, K^+-ATPase activity. The initial rate of ouabain binding in phospholipase A treated enzyme is unchanged for either (Mg^{2+}, ATP)- or (Na^+, Mg^{2+}, ATP)-supported binding, but is lower in the presence of (Mg^{2+}, ATP, K^+) or (Mg^{2+}, Pi), with a concomitant change in the affinity of the enzyme for ouabain. TANIGUCHI and IIDA (1973) and GOODMAN and WHEELER (1977) suggest that phosphatidyl serine partially restores the lowered ouabain binding in phospholipid-depleted enzyme preparations. LANE et al. (1977), reported that a 70% removal of the lipid phosphate of lamb kidney enzyme by phospholipase A treatment causes a 70% reduction in ATPase activity and a significant reduction in the rate of ouabain binding observed in the presence of either (Mg^{2+}, Pi), (Na^+, Mg^{2+}, ATP), or (Na^+, Mg^{2+}, ATP, K^+). The number of ouabain-binding sites is unaffected in these studies. CHARNOCK et al. (1977) reported that phospholipase A treatment which causes an 86% decrease in the enzyme activity produces a 51% decrease in the rate of ouabain binding and a 27%

decrease in equilibrium ouabain binding. These results should indicate that the stability of the ouabain–enzyme complex is *increased* by phospholipase A treatment. In contrast, DE PONT and BONTING (1977) reported that an almost complete removal of phosphatidyl serine fails to alter the affinity of the enzyme for ouabain when this phospholipid is converted to phosphatidyl ethanolamine by phosphatidyl serine decarboxylase treatment of hexane extracted enzyme preparations. Thus, no definite conclusion can be reached presently with regard to the influence of membrane phospholipids, or their removal, on the glycoside–enzyme interaction.

The method commonly used for enzyme isolation also affects the characteristics of ouabain binding, such as the degree of Pi-induced enhancement of the binding in the presence of Mg^{2+} (WHITTAM et al., 1976). More importantly, the detergent treatment of enzyme preparations affects the dissociation rate constant for the ouabain–enzyme interaction and hence the affinity of the enzyme for ouabain (CHOI and AKERA, 1978). This effect is dependent on the source of the enzyme (Fig. 5).

These various factors provide an opportunity to modify the glycoside–Na^+, K^+-ATPase interaction, and to examine whether the modification is reflected in the inotropic action of the glycoside. They also make it difficult to compare precisely the affinity of isolated enzymes for cardiac glycosides with the affinity in situ.

G. Consequences of Glycoside Binding in Cardiac Muscle

I. Enzyme and Sodium Pump Activities

The binding of ouabain to Na^+, K^+-ATPase, supported either by the combination of (Na^+, Mg^{2+}, Pi), (Mg^{2+}, Pi, K^+), (Na^+, Mg^{2+}, ATP), or (Mg^{2+}, acetylphosphate), stabilizes the phosphoenzyme by making it resistant to K^+-induced dephosphorylation (CHARNOCK et al., 1963; WHITTAM et al., 1964; LINDENMAYER et al., 1968; SEN et al., 1969; SKOU and HILBERG, 1969; DUDDING and WINTER, 1971). Additionally, ouabain binding inhibits the binding of ATP to the enzyme, i.e., the first step of the forward reaction sequence (CHARNOCK and POTTER, 1969; POST et al., 1969; SKOU and HILBERG, 1969), and hence results in inhibition of Na^+, K^+-ATPase activity. The ouabain binding supported by (Na^+, Mg^{2+}, ATP) or (Mg^{2+}, Pi) results in a stoichiometric inhibition of enzyme (ALLEN et al., 1970b, 1971c; WILSON et al., 1970; HANSEN et al., 1971; HEGYVARY and POST, 1971; ERDMANN and SCHONER, 1974a; WALLICK and SCHWARTZ, 1974; ERDMANN et al., 1976a), and active Na and K transport (HOFFMAN, 1969; GARDNER and KIINO, 1973). Since the fractional occupancy of ouabain-binding sites coincides with the fractional inhibition of Na^+, K^+-ATPase in relatively crude as well as in highly purified enzyme preparations, Na^+, K^+-ATPase seems to be the only ouabain-binding site under these ligand conditions. In line with these observations, solubilization of membrane components obtained from dog heart fails to disclose any high affinity binding site separable from Na^+, K^+-ATPase (SMITH et al., 1974). The capacity of various enzyme preparations to bind the cardiac glycosides strictly parallels the enzyme activity for various enzyme preparations of different purity (SCHWARTZ et al., 1968; BAKER and WILLIS, 1969; HANSEN, 1971; JØRGENSEN and

SKOU, 1971). Moreover, ligands which affect the glycoside binding are either the substrate or product of the Na^+, K^+-ATPase or sodium pump reaction. Thus, the specific ouabain binding observed under the above ligand conditions seems to be the binding of ouabain to the steroid-binding site on Na^+, K^+-ATPase, and the binding results in an inactivation of those units of the enzyme to which ouabain is bound.

It should be noted, however, that other membrane systems are also capable of discriminating between Na^+ and K^+. For example, saturable binding of an opiate antagonist, ^3H-naloxone, is stimulated by Na^+ and inhibited by K^+ (PERT and SNYDER, 1974). There are also lower affinity ouabain-binding sites in tissue homogenates which may be observed in the presence of Na^+ and Mg^{2+} and in the absence of ATP (AKERA et al., 1979a). The binding of ^3H-ouabain to these sites is displaced by unlabeled ouabain. The binding results in a complex which is also K^+-sensitive, but fails to cause Na^+, K^+-ATPase inhibition. Thus, this binding does not seem to involve Na^+, K^+-ATPase. Although their low affinity tentatively rules out the possibility that these sites represent receptors for pharmacological actions of the glycoside, the possibility that there are other high affinity glycoside-binding sites separate from Na^+, K^+-ATPase cannot be ruled out. Those sites, if present, may require hitherto unknown ligands for binding.

Is there any digitalis derivative which binds to the glycoside-binding site on Na^+, K^+-ATPase but does not cause enzyme inhibition? Such a compound displaces active glycosides from their binding sites, and hence should fit into a category of digitalis antagonists, offering advantages in elucidating the mechanisms of actions of cardiac glycosides and also reversing digitalis-induced toxicity. Several steroid derivatives, such as spironolactone or canrenoate, have been claimed to reverse digitalis toxicity. These agents, however, do not antagonize the inhibitory action of cardiac glycosides on isolated Na^+, K^+-ATPase (BASKIN et al., 1973; FRICKE, 1978).

KYTE (1974) reported that an antibody prepared against the larger polypeptide subunit of Na^+, K^+-ATPase binds to the enzyme from the intracellular side but does not inhibit enzyme activity. This antibody, however, binds to sites clearly different from the glycoside-binding sites which are accessible from the extracellular side. MCCANS et al. (1974) reported that when an antibody fraction prepared against a highly purified Na^+, K^+-ATPase of dog kidney is incubated with the excess ouabain–enzyme complex (ouabain-binding sites are masked), approximately one-half of the antibody protein does not interact with the complex and is recovered in the supernatant. This fraction which is claimed to contain antibody against ouabain-binding sites on the enzyme has little inhibitory effect upon the catalytic activity of Na^+, K^+-ATPase when incubated with the enzyme under turnover conditions, but significantly inhibits ^3H-ouabain binding in the presence of Mg^{2+} and Pi. Thus, this fraction of the antibody seemed to be a digitalis antagonist. In a later study by the same group, however, a similar antibody inhibited ouabain binding and ATP hydrolysis to the same extent (MICHAEL et al., 1977). Variable effects of the antibody on ouabain binding are noted dependent upon the ligand conditions. Probably the reactivity of the enzyme with antibodies is influenced by the conformational state of the enzyme, either exposing or occluding the recognition site.

There is a possibility that the highly lipid-soluble digitalis derivatives may be incorrectly assessed as antagonists. When the effects of these compounds on ^3H-ouabain binding are estimated in the presence of (Na^+, Mg^{2+}, ATP) or (Mg^{2+}, Pi), they bind to the enzyme with a high affinity and inhibit ^3H-ouabain binding. When their effects on Na^+, K^+-ATPase activity are estimated, however, they may not bind to the enzyme, because of the presence of K^+ which shifts the equilibrium binding of these lipid-soluble compounds more than an order of magnitide (Akera et al., 1978, 1979b). Thus, these cardenolides may be incorrectly regarded as having a high affinity for Na^+, K^+-ATPase, and yet not inhibiting enzyme activity.

Several investigators reported a stimulation of Na^+, K^+-ATPase or sodium pump activity by the cardiac glycosides under various experimental conditions (see Lee and Klaus, 1971; and Chap. 16). Generally, the stimulation is observed transiently only with low concentrations of the glycoside in relatively unaltered enzyme preparations. In highly purified enzyme preparations, no concentration of ouabain tested stimulated Na^+, K^+-ATPase (Michael et al., 1978). As discussed above, there is a one-to-one relationship between ouabain binding to Na^+, K^+-ATPase observed in the presence of (Na^+, Mg^{2+}, ATP) or (Mg^{2+}, Pi) and enzyme inhibition. These results suggest that the direct interaction (binding) of ouabain to the enzyme protein causes enzyme inhibition. The reported stimulation of Na^+, K^+-ATPase by cardiac glycosides is observed at concentrations of the glycosides two or three orders of magnitude lower than those required to produce a 50% inhibition of the enzyme activity (Lee and Yu, 1963; Godfraind, 1973). Since the positive inotropic action of cardiac glycosides is long lasting, and the inotropic concentrations of the glycosides produce approximately 50% inhibition, rather than stimulation of cardiac Na^+, K^+-ATPase, the physiological significance of the glycoside-induced enzyme stimulation is not clear.

Do pharmacological doses of cardiac glycosides cause a significant occupancy and inhibition of cardiac Na^+, K^+-ATPase? An intravenous infusion of ouabain to the anesthetized dog causes an approximately 50% inhibition of Na^+, K^+-ATPase in the cardiac muscle before the onset of overt toxicity such as complete block of the atrioventricular node or arrhythmias (Akera et al., 1969). The degree of cardiac Na^+, K^+-ATPase inhibition by ouabain or digoxin during an intravenous infusion is related to the positive inotropic effect (Akera et al., 1970; Besch et al., 1970; Goldman et al., 1973, 1975; Hall et al., 1977). In either case, a marked inotropic effect observed before the onset of toxicity is associated with a moderate (approximately 40%–60%) inhibition of cardiac Na^+, K^+-ATPase. K^+ delays the binding of cardiac glycosides to isolated Na^+, K^+-ATPase in the presence of Na^+, Mg^{2+}, and ATP. Similarly, K^+ delays the binding and onset of the inotropic action of digoxin in isolated heart preparations (Prindle et al., 1971). In hyperkalemic dogs, a larger dose of digoxin is required to produce a positive inotropic effect and Na^+, K^+-ATPase inhibition, indicating the similarity of glycoside binding in vivo and in vitro, and also the similarity of the glycoside binding to Na^+, K^+-ATPase and to the inotropic receptor (Goldman et al., 1975).

Several investigators, however, reported that inotropic concentrations of cardiac glycosides or their derivatives fail to inhibit cardiac Na^+, K^+-ATPase (Roth-Schechter et al., 1970; Fujino et al., 1971; Murthy et al., 1974; Rhee et al., 1976). The results of Roth-Schechter et al. (1970) are interpreted by them as due to the rupture of the glycoside–enzyme bond during the purification procedure used to isolate Na^+, K^+-ATPase. MURTHY et al. (1974) reported that ouabain or digoxin increased the force of acetylcholine-induced contractions in isolated rabbit myometrium pretreated with estrogen. Ouabain treatment of these tissues, however, failed to alter Na^+ or K^+ concentrations or suppress the recovery of tissue Na^+ or K^+ concentration which is altered by an exposure to a low K^+ solution. In the sodium pump study, however, the muscle preparation was not stimulated. Since the binding of the glycoside to Na^+, K^+-ATPase is dependent on membrane excitation (YAMAMOTO et al., 1979), it would be anticipated that ouabain should fail to inhibit Na^+, K^+-ATPase in the sodium pump study, but does inhibit the enzyme under the conditions of the inotropic study. RHEE et al. (1976) reported that ouabain infusion in moderately inotropic doses fails to cause a *statistically* significant inhibition of cardiac Na^+, K^+-ATPase in anesthetized dogs. These findings, however, appear to result from inadequate experimental design and technical limitations which make it difficult to detect a relatively small change with large biologic variability (see AKERA and BRODY, 1977). When experiments are performed using a similar dosage schedule to that of RHEE et al. (1976), ouabain-sensitive ^{86}Rb uptake by biopsied ventricular muscle of anesthetized dogs is significantly inhibited indicating that the sodium pump activity is inhibited at the time when the rate of change of left ventricular pressure dP/dt, an index of myocardial contractility, is increased by the glycoside but before the first sign of arrhythmias (HOUGEN and SMITH, 1978). FUJINO et al. (1971) reported that ouabain perfusion of isolated cat heart (Langendorff preparations) fails to cause significant Na^+, K^+-ATPase inhibition, but SCHWARTZ et al. (1974) demonstrated a marked enzyme inhibition under similar experimental conditions. Overall, several studies strongly indicate that the enzyme is inhibited 40%–60% when a marked positive inotropic effect is produced by cardiac glycosides, but before the onset of overt toxicity.

Several investigators have reported that the cardiac glycosides irreversibly inhibit Na^+, K^+-ATPase under certain conditions, but reversibly increase the force of contraction. For example, strophanthidin-3-bromoacetate, an alkylating derivative of strophanthidin, inhibited isolated Na^+, K^+-ATPase irreversibly (FRICKE and KLAUS, 1969, 1971c), but 70% of its positive inotropic effect was reversible in isolated papillary muscle preparations of guinea-pig heart (FRICKE and KLAUS, 1971b). Inotropic concentrations of strophanthidin-3-bromoacetate, however, fail to produce irreversible inhibition of isolated Na^+, K^+-ATPase (TOBIN et al., 1973c). Irreversible enzyme inhibition required substantially higher concentrations of this agent.

OKITA et al. (1973) reported that a perfusion of isolated rabbit heart with strophanthidin-3-bromoacetate, strophanthidin, or ouabain causes a readily reversible inotropic effect. Na^+, K^+-ATPase, however, was inhibited, after 6h of a drug-free perfusion of the preparation. This is in contrast to the observation by SCHWARTZ et al. (1974) who were unable to show enzyme inhibition, even at the peak of the inotropic effect of ouabain, because of the glycoside–enzyme bond rupture during enzyme isolation. In the rabbit, the ouabain–enzyme complex is relatively unstable. OKITA (1977), however, suggests that the detergent treatment used by SCHWARTZ et al. (1974) during the enzyme purification process is responsible for the rupture of the glycoside–enzyme bond. On the other hand, SCHWARTZ et al. (1974) demonstrated that enzyme inhibition and recovery accompanies the development of the inotropic effect of ouabain and its decay in cat heart, in which Na^+, K^+-ATPase forms a stable complex with ouabain. Moreover, substantial occupancy of Na^+, K^+-ATPase by ouabain or digitoxin and its reversal associated with the development and subsequent washout of the positive inotropic effect are observed when fractional occupancy is estimated from the initial velocity of ATP-dependent ^3H-ouabain binding to tissue homogenates without any detergent treatment (KU et al., 1974). Sodium pump activity as estimated from ouabain-sensitive ^{86}Rb uptake by ventricular slices also recovers after the termination of digitoxin perfusion with a time course similar to the loss of inotropic effect. Thus, it seems reasonable to conclude that the inhibition of Na^+, K^+-ATPase and sodium pump resulting from the binding of cardiac glycosides to the enzyme, is reversible either in isolated enzyme preparations or in beating hearts. Perhaps, the irreversible inhibition of Na^+, K^+-ATPase observed by OKITA et al. (1973) is due to an indirect effect of the cardiac glycosides, such as hypoxia; the anticipated increase in the force of contraction resulting from Na^+, K^+-ATPase inhibition is probably masked as a result of hypoxic damage.

Conditions that affect the interaction of the glycosides with Na^+, K^+-ATPase also affect their positive inotropic effect in a parallel fashion. The rate of development of the positive inotropic action of digoxin in isolated heart preparations or in anesthetized dogs is delayed by higher K^+ concentrations in the medium or plasma (PRINDLE et al., 1971; GOLDMAN et al., 1975), similar to the effect of K^+ on the rate of ouabain binding to isolated enzyme. K^+ slightly reduces the maximal inotropic effect obtained with a moderate concentration of ouabain or digoxin, and markedly reduces the degree of positive inotropic effect of digoxigenin (AKERA et al., 1979b, c).

In isolated Na^+, K^+-ATPase, K^+ decreases the binding at equilibrium only slightly for relatively water-soluble glycosides, but markedly decreases equilibrium binding of highly lipid-soluble compounds (AKERA et al., 1978). Intracellular Na^+ enhances the rate of ouabain binding to the sodium pump in intact cardiac muscle (YAMAMOTO et al., 1979). Similarly, membrane depolarizations, or other factors which would increase the transmembrane Na^+ influx, enhance the development of the positive inotropic action of ouabain (AKERA et al., 1977b). As such, the binding of ouabain to Na^+, K^+-ATPase seems to be the determinant of the rate of development of the positive inotropic effect and the extent of the glycoside binding to Na^+, K^+-ATPase is related to the magnitude of the effect.

II. Reserve Capacity of the Sodium Pump and Sodium Transients

Inotropic concentrations of cardiac glycosides do not cause a marked alteration in myocardial Na^+ or K^+ concentrations (see LEE and KLAUS, 1971), although they substantially inhibit the sodium pump. Even when glycoside-induced changes in intracellular Na^+ or K^+ concentrations are reported, the results may be secondary to the glycoside-induced hypoxia. Nevertheless, several carefully executed, recent experiments indicate a small increase in cellular Na^+ concentration (LANGER and SERENA, 1970; BRENNAN et al., 1972; SCHOLZ, 1972; HEINEN and NOACK, 1972). There is, however, a disparity between the observed change in tissue Na^+ or K^+

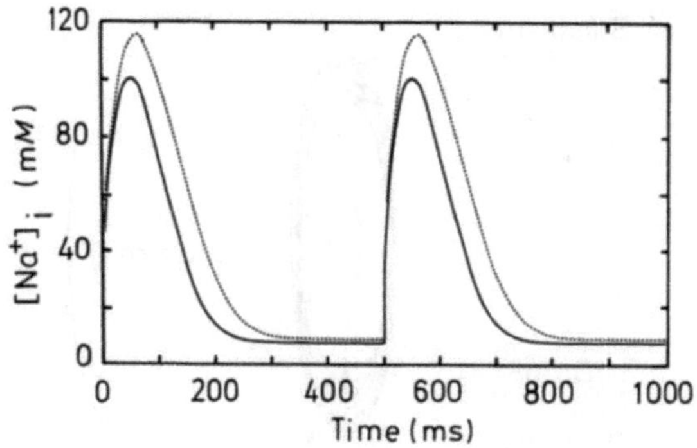

Fig. 6. Changes in concentration of Na$^+$ at inner surface of the cell membrane in cardiac muscle: computer simulation. Normal conditions *(solid line)* and with sodium pump inhibited 40% by positive inotropic concentration of ouabain *(broken line)*. A rapid Na$^+$ influx during the upstroke of action potential increases Na$^+$ concentration at inner surface of cell membrane (zone 1 of Fig. 7). Subsequently, Na$^+$ concentration in this space increases further because Na$^+$ influx rate through slow channels exceeds maximal capacity of the sodium pump. Approximately 70 ms after depolarization, decreasing Na$^+$ influx rate balances with sodium pump capacity (peak of sodium transients). Subsequently, $[Na^+]_i$ and sodium pump activity return to resting low level. With pump inhibited, peak Na$^+$ concentration is higher, and restoration of resting state is delayed. With moderate sodium pump inhibition, concentration of Na$^+$ at inner surface of sarcolemma is restored before next membrane excitation. Thus, 40% inhibition of sodium pump produces only minor increase in intracellular Na$^+$ concentration. (AKERA, 1977; see also AKERA et al., 1976a, for details of computer simulation)

concentration (5%–20%) and the degree of enzyme inhibition (40%–60%), indicating the presence of an apparent reserve capacity of the sodium pump in cardiac muscle (LANGER, 1968, 1972b).

Such a reserve capacity of the sodium pump, however, seems to be limited only to the later phase of each cycle of cardiac function (AKERA, et al., 1976a). Figure 6 is a representation of the intracellular Na$^+$ concentration at the inner surface of the cell membrane as it changes during cycles of cardiac function. Associated with membrane excitation, the Na$^+$ concentration at the inner surface of the cell membrane increases transiently (Na$^+$ transients), activating the sodium pump and the Na/Ca exchange reaction (Na efflux coupled with Ca influx). When the Na$^+$ concentration returns to a resting value of approximately 5 m*M* (LEE and FOZZARD, 1975), the sodium pump operates at a low level, owing to the low intracellular Na$^+$ concentration available to the pump. Inhibition of the sodium pump then enhances the Na$^+$ transients, and probably increases the amount of Na$^+$ which is extruded by the Na/Ca exchange mechanism, but would not markedly increase the average intracellular Na$^+$ concentration (AKERA, 1977).

Is it reasonable to assume that such an intracellular Na$^+$ transient exists? Apparently, the amount of Na$^+$ entering the cell associated with each membrane excitation is not sufficient to cause a marked increase in Na$^+$ concentration (BENTFELD et al., 1977). However, there may be a dynamic intracellular Na$^+$ gradient, and the intracellular Na$^+$ concentration may not be equilibrated throughout the cell at all times. We may divide the cardiac cell into a large number of small coaxial

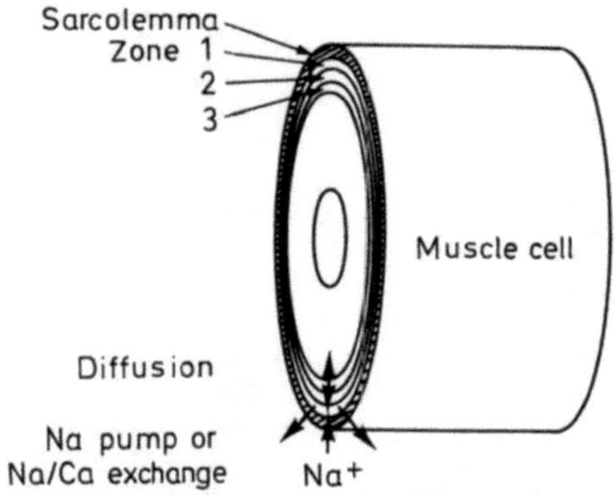

Fig. 7. Model for Na^+ movements across cell membrane. Na^+ which enters the cell during membrane excitation is extruded by Na/K exchange mechanism (sodium pump), is extruded by Na/Ca exchange mechanism, or enters into zone 2, etc. by diffusion. During later phase of cardiac cycle, last process reverses to maintain-steady state

zones, each having the thickness of the cell membrane (Fig. 7). Each zone may be numbered, starting with that closest to the cell membrane and working inward. Assuming a cylindrical shape of the myocardial cell and with its known diameter, there are probably several hundreds of these zones. During the action potential, Na^+ influx is markedly increased. Since Na^+ enters from the extracellular milieu via the cell membrane, it must enter zone 1 first. Subsequently, some Na^+ ions are extruded by the sodim pump or Na/Ca exchange reaction, while others may diffuse into the other zones. The time constant for equilibrium between two adjacent zones is probably very short. However, since there are several hundred zones, it is highly unlikely that equilibrium is established instantaneously. Thus, it is reasonable to assume that there is a dynamic Na^+ gradient within the cell. The marked fluctuation of Na^+ (Na^+ transient) may be limited to a few outer zones, and there may be no sodium transients in deeper sites as suggested by BENTFELD et al. (1977). It is, however, the concentration of Na^+ in zone 1, which regulates the activity of the sodium pump and the Na/Ca exchange reaction.

The concept of intracellular Na^+ transients predicts that a moderate sodium pump inhibition due to inotropic concentrations of cardiac glycosides does not normally cause marked changes in intracellular Na^+ or K^+ concentrations, but would cause an accumulation of myocardial Na^+ and a loss of K^+ when the frequency of membrane excitation is high, and that the time during which the sodium pump is idling is short. Such results have been observed by a number of investigators (VICK and KAHN, 1957; BROWN et al., 1962; TUTTLE et al., 1962; AKERA et al., 1976a; BENTFELD et al., 1977). This concept is also consistent with the observation of EBNER and REITER (1977) that the positive inotropic effect of dihydroouabain is smaller at higher stimulation frequencies.

Other consequences of glycoside binding to Na^+, K^+-ATPase have also been explored as possible mechanisms of the positive inotropic action of these agents, probably because of the apparent reserve capacity of the sodium pump which seemingly neutralizes a partial inhibition. One effect of cardiac glycosides is a slight membrane depolarization which may be observed before a marked cellular K^+

loss. This phenomenon may be due to a decrease in K^+ concentration in zone 1 illustrated in Fig. 7, resulting from sodium pump inhibition. Such a change may occur before a decrease in the average cellular K^+ concentration becomes evident. Alternatively, the glycoside may alter the electrogenicity of the sodium pump. The electrogenic sodium pump may be demonstrated in cardiac muscles under conditions in which the intracellular Na^+ available to the sodium pump is abundant (GLITSCH, 1972, 1973; see THOMAS, 1972). The sodium pump is electrogenic when it moves more Na^+ than K^+. The overall Na^+:K^+ ratio is largely determined by Na^+ influx and K^+ efflux during a cycle of myocardial function, and must be variable. In cardiac muscle, the rate of K^+ efflux is relatively unaffected by heart rate, whereas the rate of Na^+ influx is proportional to the frequency of membrane depolarizations (LANGER, 1968). Thus, the Na^+:K^+ ratio that the sodium pump has to move across the cell membrane in order to maintain a steady state is greater at higher heart rates. A slight inhibition of the sodium pump would make the remaining pump sites less electrogenic, if such an inhibition causes more Na^+ to be extruded by the nonelectrogenic Na/Ca exchange reaction, resulting in a lower net Na^+:K^+ ratio to be exchanged by the sodium pump. The relationship between the electrogenicity of sodium pump per se and the force of myocardial contractions, however, is not known.

Another action of the cardiac glycosides on Na^+, K^+-ATPase is to lock the enzyme in a particular conformation. Na^+, K^+-ATPase is an allosteric enzyme which is believed to take several different conformations during its cycle of ATP hydrolysis. Attempts to demonstrate ouabain-induced changes in enzyme conformation directly using a fluorescent probe, 1-anilino-8-naphthalenesulfonic acid, have been unsuccessful (MAYER and AVI-DOR, 1970; NAGAI et al., 1970). LINDENMAYER and SCHWARTZ (1970) and LÜLLMANN et al. (1975), however, demonstrated that glycosides alter the circular dichroism spectrum of the solubilized Na^+, K^+-ATPase or plasmalemmal fraction, related to their potency for positive inotropic effects. The glycoside–enzyme complex formed in vitro in the presence of Na^+, Mg^{2+}, and ATP is sensitive to K^+, which seems to alter the characteristics of the complex by inducing a conformational change (AKERA and BRODY, 1971; ALLEN et al., 1971a; AKERA et al., 1974b). After a K^+-induced conformational change and dephosphorylation, the ouabain-bound enzyme cannot enter the next cycle for further conformational changes. Since K^+ is available to the enzyme in situ, cardiac glycosides probably force the enzyme to assume a K^+-induced form, which may contribute to the inotropic action of cardiac glycosides.

Thus, the binding of cardiac glycosides to Na^+, K^+-ATPase results in an inactivation of the drug-bound enzyme unit and a stoichiometric inhibition of the sodium pump activity. In the presence of a nonelectrogenic Na/Ca exchange mechanism, the remaining sodium pump may become less electrogenic. In addition, the cardiac glycosides may lock the enzyme in a K^+-induced form. Available evidence indicate that Na^+, K^+-ATPase is inhibited 40%–60% at the time when the marked positive inotropic effect of the glycoside is observed. Such an extent of sodium pump inhibition is not likely to produce a progressive cellular Na^+ accumulation or K^+ loss except when the heart rate is high. A moderate inhibition of the sodium pump, generally produced by inotropic concentrations of cardiac glycosides, would enhance intracellular Na^+ transients.

H. A Mechanism of Positive Inotropic Action

The positive inotropic action of cardiac glycosides is due to an augmentation of an intracellular calcium transient which follows each membrane excitation (Allen and Blinks, 1978). Such an augmentation is caused by an increased efficacy of the mechanism or mechanisms which link membrane excitation (action potential) to calcium transients, because a glycoside-induced alteration in the action potential has been ruled out as a cause of the positive inotropic effect (see Lee and Klaus, 1971). Since the binding of cardiac glycosides to Na^+, K^+-ATPase is intimately related to the positive inotropic action (Table 2), possible mechanisms by which the binding of the glycosides to Na^+, K^+-ATPase affect calcium transients have been investigated.

Since the inotropic action of the cardiac glycosides is not accompanied by overt changes in cellular Na^+ or K^+ concentrations (Lee and Klaus, 1971), several investigators proposed that the inhibition of the enzyme or sodium pump activity per se is unrelated to the inotropic action of these agents, although the binding of the glycoside to the enzyme is essential. Dutta et al. (1968) postulate that Na^+, K^+-ATPase is a mechanism of transporting cardiac glycosides across the cell membrane to an unidentified intracellular site of action, based on the observation that a large amount of glycosides is found in the "microsomal" fraction after the glycoside perfusion of isolated guinea-pig heart. This view is supported by Park and Vincenzi (1975) who observed that the rate of development of the positive inotropic action of glycosides is temperature and beat dependent, whereas that of the highly lipid soluble aglycones is relatively insensitive to these factors. Since membrane depolarizations stimulate Na^+, K^+-ATPase via Na^+ influx, and temperature affects transport rates, it is suggested that the more polar glycosides are transported across the cell membrane by an active process, presumably involving Na^+, K^+-ATPase, whereas the lipid-soluble aglycones gain access to intracellular receptors by passively diffusing across the membrane. The observation that phloridzin and phloretin, compounds which block active sugar transport, inhibit the inotropic activity of ouabain is offered as evidence to support the hypothesis that glycosides are transported into the cell (Byrne and Dresel, 1969). Alternatively, however, phloridzin enhances the effect of K^+ on Na^+, K^+-ATPase and thereby inhibits ouabain binding (Akera et al., 1974b). Beat and temperature dependence of the action of glycosides is also explained from the dependence of the glycoside binding to Na^+, K^+-ATPase on intracellular Na^+ and temperature.

Fricke and Klaus (1977, 1978) support a similar hypothesis based on the observation that the affinity of isolated Na^+, K^+-ATPase for ^3H-ouabain is high at higher Na^+ concentrations and low at lower Na^+ concentrations (Inagaki et al., 1974; Fricke and Klaus, 1977). Whether the glycoside binds to the enzyme at a high affinity site (high extracellular Na^+ concentration) and moves to a low affinity site (low intracellular Na^+ concentration) from which the glycoside is released into the cell is unknown. Direct proof that Na^+, K^+-ATPase is capable of transporting cardiac glycosides is presently unavailable.

Implicit in the above hypothesis is that there is a receptor for cardiac glycosides responsible for the inotropic action but separate from Na^+, K^+-ATPase. Intracellular Ca^{2+} storage sites, such as mitochondria or sarcoplasmic reticulum, have

Table 2. Comparison of the interaction of digitalis with Na^+, K^+-ATPase and with the positive inotropic receptor

Binding to Na^+, K^+-ATPase or Na^+, K^+-ATPase inhibition	Positive inotropic action
Rate of binding or development	
Slow 1)	Slow 2)
Inhibited by K^+ 3)	Delayed by K^+ 2)
Stimulated by Na^+ 4)	Facilitated by Na^+ 5)
Dependent on $[Na^+]_i$ 6)	Dependent on depolarization 5)
No species differences 7)	No species differences 8)
Equilibrium	
Dependent on chemical structure 9)	Dependent on chemical structure 10)
Slightly decreased by K^+ (glycosides) 11)	Slightly decreased by K^+ (glycosides) 12)
Decreased by K^+ (aglycones) 11)	Decreased by K^+ (aglycones) 12)
Marked species differences 13)	Marked species differences, see reference cited in 13)
Rate of release or loss	
Slow 14)	Slow 15)
Dependent on chemical structure 16)	Dependent on chemical structure 17)
Delayed by K^+ 18)	Delayed by K^+ 12)
Marked species differences 7)	Marked species differences 15)

1) ALBERS et al. (1968), ALLEN and SCHWARTZ (1970), ALLEN et al. (1971a), AKERA (1971), AKERA and BRODY (1971)
2) PRINDLE et al. (1971), GOLDMAN et al. (1975)
3) MATSUI and SCHWARTZ (1968), BARNETT (1970), SCHÖN et al. (1970), AKERA and BRODY (1971), AKERA et al. (1973), CHOI and AKERA (1977)
4) BARNETT (1970), LINDENMAYER and SCHWARTZ (1973)
5) AKERA et al. (1977b)
6) YAMAMOTO et al. (1979)
7) TOBIN and BRODY (1972), TOBIN et al. (1972b)
8) AKERA et al. (1973), KU et al. (1976)
9) REPKE (1963), WILSON et al. (1970), ERDMANN and SCHONER (1974a, b), HAUSTEIN and HAUPTMANN (1974), FLASCH and HEINZ (1978)
10) HAUSTEIN (1974), HAUSTEIN and HAUPTMANN (1974), BEARD et al. (1975), CALDWELL and NASH (1977), FLASCH and HEINZ (1978)
11) AKERA et al. (1978)
12) AKERA et al. (1979c)
13) REPKE et al. (1965)
14) AKERA and BRODY (1971), ALLEN et al. (1971a)
15) AKERA et al. (1973)
16) YODA (1973, 1974), YODA and YODA (1974a, 1977), YODA et al. (1975)
17) FRICKE and KLAUS (1971b), KU et al. (1974)
18) AKERA and BRODY (1971), ALLEN et al. (1971a), AKERA et al. (1974b, 1976b)

been suggested to be involved in the inotropic action of the glycosides. However, cardiac glycosides do not have a direct effect on mitochondrial Ca^{2+} uptake or release (DRANSFELD et al., 1969). Cardiac sarcoplasmic reticulum fails to bind the glycosides (ALLEN et al., 1970a). The microsomal fraction used by DUTTA et al. (1968) apparently contains sarcolemma, since extracellular tracers are also found in this fraction (LÜLLMANN and PETERS, 1974). The binding observed in this fraction

(DUTTA et al., 1968) has been attributed to the large amount of contaminating sarcolemmal Na^+, K^+-ATPase (JONES et al., 1976). Consistent with these observations, the characteristics of bound ^3H-digoxin present in the microsomal fraction (KIM et al., 1972) resemble those of ouabain bound to Na^+, K^+-ATPase (AKERA et al., 1974a). The failure of ouabain to penetrate the cell membrane has been demonstrated autoradiographically after a long exposure of guinea-pig heart to a large concentration of ouabain (LÖHR et al., 1971). Electronmicroscopic studies with fractionated fragments also indicate that ^3H-ouabain is predominantly distributed in the sarcolemmal fraction which contains Na^+, K^+-ATPase and sparsely in the sarcoplasmic reticulum (LÜLLMANN and PETERS, 1976). Moreover, FRICKE et al. (1975) indicated that the glycoside receptor for the inotropic action in rat heart is in the sarcolemma, based on the observation that the onset of the positive inotropic effects of ouabain, digoxin, and digitoxin is extremely rapid. The slow onset of the inotropic action of cardiac glycosides in other species may be accounted for by the slow binding of these agents to cardiac Na^+, K^+-ATPase in the presence of K^+ (AKERA et al., 1973). Thus, substantive evidence to support the above hypothesis is presently unavailable.

Cardiac glycosides stimulate Na^+, K^+-ATPase or the sodium pump under certain conditions in extremely low concentrations (see LEE and KLAUS, 1971; also COHEN et al., 1976; GODFRAIND and GHYSEL-BURTON, 1977; see Chap. 16). Since such effects are observed at concentrations of the glycosides lower than those which cause enzyme inhibition, stimulation and inhibition of the enzyme activity have been implicated in the inotropic and toxic actions, respectively (LEE and YU, 1963; REPKE, 1963; COHEN et al., 1976; GODFRAIND and GHYSEL-BURTON, 1977). The mechanism which possibly links the stimulation of cardiac Na^+, K^+-ATPase to the enhancement of Ca^{2+} transients is unknown. Furthermore, cardiac Na^+, K^+-ATPase is inhibited, rather than stimulated, by concentrations of the glycosides which are in the therapeutic range, i.e., approximately 50% of those necessary to produce toxicity.

WILBRANDT (1955) proposed that digitalis increases intracellular Ca^{2+} (transients) indirectly by inhibiting Na and K transport. The possible relationship between Na^+, K^+-ATPase inhibition and the positive inotropic action of the cardiac glycosides, and the mechanisms by which enzyme inhibition may enhance the force of myocardial contraction, have been discussed by several investigators (GLYNN, 1964; REPKE, 1964; SCHWARTZ et al., 1969; LEE et al., 1970; LANGER, 1971, 1972a; SMITH and HABER, 1973; SCHWARTZ et al., 1975; AKERA and BRODY, 1976, 1977; AKERA, 1977). Although the inhibition of Na^+, K^+-ATPase by cardiac glycosides brings about inhibition of both Na^+ and K^+ transport, the loss of cellular K^+ has been ruled out as the cause of the glycoside-induced inotropic effect (POOLE-WILSON and LANGER,, 1975). The inhibition of Na^+ efflux by the cardiac glycosides is implicated in the positive inotropic action of these agents by REPKE (1964) who proposed the following: "during depolarization, Na^+ flows into the cell; return transport starts immediately. The speed, size, and duration of the net Na^+ gain depend on the activity of the sodium pump. The cardiac glycosides inhibit the sodium pump so that Na^+ is completely pumped out between contractions but the extrusion is delayed. This results in a greater and more rapid net growth of the intracellular Na^+ concentration, and a more rapid and greater ionization of intracellular

Ca^{2+} during ion exchange." Computer simulation studies and experimental results support this concept (AKERA et al., 1976a; AKERA, 1977). The Na^+ transient would be enhanced by a moderate Na^+, K^+-ATPase inhibition.

The concept of intracellular Na^+ transients and their enhancement by cardiac glycosides is consistent with the finding that a moderate Na^+, K^+-ATPase inhibition due to therapeutic concentrations of cardiac glycosides causes a relatively minor cellular ionic shift. There are alternative hypotheses, however, which may account for the inotropic action of the glycosides based on the binding of these agents to Na^+, K^+-ATPase and do not require an overt cellular ionic shift. LANGER (1971, 1977) REUTER (1974), and BLAUSTEIN (1976) suggest that a small increase in intracellular Na^+ concentration usually observed with therapeutic concentrations of the glycosides is sufficient to cause a marked enhancement of Ca^{2+} transients. Such hypotheses are based on the calculation that 2–3 mol Na^+ exchange with 1 mol Ca^{2+} across the cell membrane, and that the intracellular Ca^{2+} concentration required for the activation of contractile proteins is in a micromolar range, whereas the intracellular Na^+ concentration is at least 1,000-fold higher.

Other hypotheses generally assume that cardiac muscle has extra sodium pump or Na^+, K^+-ATPase units which, when combined with cardiac glycoside may become an extra Ca^{2+} pool (BESCH and SCHWARTZ, 1970). This hypothesis is developed later into a theory that the binding of cardiac glycosides to Na^+, K^+-ATPase alters the membrane lipids associated with the enzyme, altering their affinity for Ca^{2+} (SCHWARTZ, 1976; GERVAIS et al., 1977). An enhancement of the Ca^{2+} exchange reaction in subcellular particles has been reported in the presence of Na^+, K^+-ATPase and glycosides (ENTMAN et al., 1972). It is not known, however, whether such a change in membrane lipids enhances or diminishes Ca^{2+} transients. A similar hypothesis is proposed by LÜLLMANN and PETERS (1979), implicating membrane-bound phosphatidyl serine as the membrane Ca^{2+} pool (see Chap. 17). In support of this view, an intimate relationship between Na^+, K^+-ATPase and phosphatidyl serine has been reported, although this phospholipid does not seem to be essential for enzyme activity (DE PONT and BONTING, 1977). Phosphatidyl serine is localized at the inner aspect of the cell membrane (BERGELSON and BARSUKOV, 1977), where it may bind and release Ca^{2+} from and into the sarcoplasm. Whether the affinity of phosphatidyl serine for Ca^{2+} is adequate to control Ca^{2+} transients in submicromolar concentrations, and how the binding of the glycoside to Na^+, K^+-ATPase increases the size of Ca^{2+} released and rebound by this mechanism are not known.

Thus, there is presently no generally accepted theory which explains the mechanism of the action of cardiac glycosides. The binding of the glycoside to cardiac Na^+, K^+-ATPase seems to be the first step which leads to drug action. This is generally agreed upon. Several hypotheses have been advanced in attempts to explain the enhancement of Ca^{2+} transients resulting from the glycoside binding to Na^+, K^+-ATPase. These cannot be ruled out on the available evidence. It is also possible that the mechanism of the inotropic action of cardiac glycosides does not proceed as a single chain of events. Several of the above mechanisms and others presently unknown, may be simultaneously increasing the force of myocardial contraction in the presence of therapeutic concentrations of the cardiac glycosides as suggested by BLOOD and NOBLE (1977). Alternatively, the relationship between the binding of glycosides to Na^+, K^+-ATPase and the inotropic action may be coincidental, no matter how significant the observed correlation. In contrast to the therapeutic concentrations of cardiac glycosides, toxic concentrations cause marked Na^+, K^+-ATPase inhibition and the expected cellular Na^+ accumulation and K^+ loss. Thus, enzyme inhibition and ensuing sodium pump inhibition is generally accepted as the cause of digitalis toxicity (LEE and KLAUS, 1971; ZAVECZ and DUTTA, 1977). This aspect, however, has not been extensively studied.

J. Conclusions

Therapeutic concentrations of cardiac glycosides bind to isolated Na^+, K^+-ATPase under specific ligand conditions, and cause a moderate inhibition. In intact cells, the binding is promoted by intracellular Na^+ and inhibited by extracellular K^+. The binding of glycosides to isolated Na^+, K^+-ATPase observed in the presence of Na^+, Mg^{2+}, and ATP seems to be comparable to the binding of these agents to Na^+, K^+-ATPase in beating hearts. The binding of glycosides to Na^+, K^+-ATPase results in a stoichiometric inhibition of the enzyme activity, i.e., one glycoside molecule binds to one unit of the enzyme and makes the enzyme incapable of entering further cycles of the enzyme reaction. The release of glycosides from the enzyme results in its reactivation. The binding of cardiotonic steroids and glycosides to Na^+, K^+-ATPase is extensively influenced by ligand conditions, chemical structure of the compound, source of the enzyme, and other factors such as temperature and pH. These factors also influence the interaction between digitaloids and the positive inotropic receptor for these agents in a remarkably similar manner. Therapeutic concentrations of cardiac glycosides cause a moderate inhibition of Na^+, K^+-ATPase either in vitro in isolated enzyme preparations, in isolated heart preparations, or in vivo in anesthetized animals. Such an enzyme inhibition, however, does not cause a marked cellular Na^+ accumulation. Instead, a moderate Na^+ pump inhibition appears to enhance intracellular Na^+ transients which occur within a limited space of the cell, associated with membrane depolarizations. In addition, the binding of cardiac glycosides to sarcolemmal Na^+, K^+-ATPase may alter the electrogenicity of sodium pump, or force the enzyme to assume a particular conformation. Although several of these mechanisms may be responsible for the positive inotropic action of cardiac glycosides, the enhancement of Na^+ transients by moderate sodium pump inhibition appears sufficient to increase Ca^{2+} transient by the Na/Ca exchange mechanism, and to increase the force of myocardial contraction.

Acknowledgments: The author thanks Dr. Theodore M. Brody for encouragements and helpful suggestions, and Ms. Diane K. Hummel, and Ms. Theresa A. Blasen for help in preparation of the manuscript. The author's own studies cited herein were supported by USPHS Grants HL-16052, and HL-16055 from the National Heart, Lung, and Blood Institute, Grant BMS-74-19512 from the National Science Foundation, and grants-in-aid from the Michigan Heart Association.

References

Ahmed, K., Judah, J.D.: On the action of strophanthin G. Can. J. Biochem. *43*, 877–880 (1965)

Akera, T.: Quantitative aspects of the interaction between ouabain and ($Na^+ + K^+$)-activated ATPase in vitro. Biochim. Biophys. Acta (Amst.) *249*, 53–62 (1971)

Akera, T.: Membrane adenosinetriphosphatase: A digitalis receptor? Science *198*, 569–574 (1977)

Akera, T., Brody, T.M.: Membrane adenosine triphosphatase: The effect of potassium on the formation and dissociation of the ouabain-enzyme complex. J. Pharmacol. Exp. Ther. *176*, 545–557 (1971)

Akera, T., Brody, T.M.: Inotropic action of digitalis and ion transport. Life Sci. *18*, 135–142 (1976)

Akera, T., Brody, T.M.: The role of Na^+, K^+-ATPase in the inotropic action of digitalis. Pharmacol. Rev. *29*, 187–220 (1977)

Akera, T., Cheng, V.-J.K.: A simple method for the determination of affinity and binding site concentration in receptor binding studies. Biochim. Biophys. Acta (Amst.) *470*, 412–423 (1977)

Akera, T., Larsen, F.S., Brody, T.M.: The effect of ouabain on sodium- and potassium-activated adenosine triphosphatase from the hearts of several mammalian species. J. Pharmacol. Exp. Ther. *170*, 17–26 (1969)

Akera, T., Larsen, F.S., Brody, T.M.: Correlation of cardiac sodium- and potassium-activated adenosine triphosphatase activity with ouabain-induced inotropic stimulation. J. Pharmacol. Exp. Ther. *173*, 145–151 (1970)

Akera, T., Hook, J.B., Tobin, T., Brody, T.M.: Cardiac glycoside sensitivity of ($Na^+ + K^+$)-activated ATPase in new-born rats. Res. Commun. Chem. Pathol. Pharmacol. *4*, 699–706 (1972)

Akera, T., Baskin, S.I., Tobin, T., Brody, T.M.: Ouabain: Temporal relationship between the inotropic effect and the *in vitro* binding to, and dissociation from, ($Na^+ + K^+$)-activated ATPase. Naunyn-Schmiedebergs Arch. Pharmacol. *277*, 151–162 (1973)

Akera, T., Brody, T.M., So, R.H.-M., Tobin, T., Baskin, S.I.: Factors and agents that influence cardiac glycoside-Na^+, K^+-ATPase interaction. Ann. N.Y. Acad. Sci. *242*, 617–634 (1974a)

Akera, T., Tobin, T., Gatti, A., Shieh, I.-S., Brody, T.M.: Effect of potassium on the conformational state of the complex of ouabain with sodium- and potassium-dependent adenosine triphosphatase. Mol. Pharmacol. *10*, 509–518 (1974b)

Akera, T., Bennet, R.T., Olgaard, M.K., Brody, T.M.: Cardiac Na^+, K^+-adenosine triphosphatase inhibition by ouabain and myocardial sodium: A computer simulation. J. Pharmacol. Exp. Ther. *199*, 287–297 (1976a)

Akera, T., Ku, D., Tobin, T., Brody, T.M.: The complexes of ouabain with sodium and potassium activated adenosine triphosphatase formed with various ligands: Relationship to the complex formed in the beating heart. Mol. Pharmacol. *12*, 101–114 (1976b)

Akera, T., Ku, D., Brody, T.M.: Lack of effect on brain stem and cerebral cortex Na^+, K^+-ATPase during heart block produced by chronic digoxin treatment. Eur. J. Pharmacol. *45*, 243–249 (1977a)

Akera, T., Olgaard, M.K., Temma, K., Brody, T.M.: Development of the positive inotropic action of ouabain: Effects of transmembrane sodium movement. J. Pharmacol. Exp. Ther. *203*, 675–684 (1977b)

Akera, T., Temma, K., Wiest, S.A., Brody, T.M.: Reduction of the equilibrium binding of cardiac glycosides and related compounds to Na^+, K^+-ATPase as a possible mechanism for the potassium-induced reversal of their toxicity. Naunyn-Schmiedebergs Arch. Pharmacol. *304*, 157–165 (1978)

Akera, T., Brody, T.M., Wiest, S.A.: Saturable adenosine 5′-triphosphate-independent binding of (^{3}H)-ouabain to brain and cardiac tissue *in vitro*. Brit. J. Pharmacol. *65*, 403–409 (1979a)

Akera, T., Choi, Y.R., Yamamoto, S.: Potassium-induced changes in Na, K-ATPase: Influences on the interaction between cardiac glycosides and enzyme. In: Na, K-ATPase: Structure and Kinetics. Skou, J.C., Nørby, J.G. (eds.), pp. 405–420. London: Academic Press 1979b

Akera, T., Wiest, S.A., Brody, T.M.: Differential effect of potassium on the action of digoxin and digoxigenin in guinea-pig heart. Eur. J. Pharmacol. *57*, 343–351 (1979c)

Akera, T., Yamamoto, S., Chubb, J., McNish, R., Brody, T.M.: Biochemical basis for the low sensitivity of the rat heart to digitalis. Naunyn-Schmiedebergs Arch. Pharmacol. *308*, 81–88 (1979d)

Albers, R.W., Koval, G.J., Siegel, G.J.: Studies on the interaction of ouabain and other cardioactive steroids with sodium-potassium-activated adenosine triphosphatase. Mol. Pharmacol. *4*, 324–336 (1968)

Allen, D.G., Blinks, J.R.: Calcium transients in aequorin-injected frog cardiac muscle. Nature (Lond.) *273*, 509–513 (1978)

Allen, J.C., Schwartz, A.: A possible biochemical explanation for the insensitivity of the rat to cardiac glycosides. J. Pharmacol. Exp. Ther. *168*, 42–46 (1969)

Allen, J.C., Schwartz, A.: Effects of potassium, temperature, and time on ouabain interaction with the cardiac Na^+, K^+-ATPase: Further evidence supporting an allosteric site. J. Mol. Cell. Cardiol. *1*, 39–45 (1970)

Allen, J.C., Schwartz, A.: Human cardiac Na^+, K^+-ATPase: A pharmacologic receptor for cardiac glycosides. Cardiovasc. Res. Center Bull. *10*, 133–141 (1972)

Allen, J.C., Besch, H.R. Jr., Glick, G., Schwartz, A.: The binding of tritiated ouabain to sodium- and potassium-activated adenosine triphosphatase and cardiac relaxing system of perfused dog heart. Mol. Pharmacol. *6*, 441–443 (1970a)

Allen, J.C., Lindenmayer, G.E., Schwartz, A.: An allosteric explanation for ouabain-induced time-dependent inhibition of sodium, potassium-adenosine triphosphatase. Arch. Biochem. Biophys. *141*, 322–328 (1970b)

Allen, J.C., Harris, R.A., Schwartz, A.: The nature of the transport ATPase-digitalis complex. I. Formation and reversibility in the presence and absence of a phosphorylated enzyme. Biochem. Biophys. Res. Commun. *42*, 366–370 (1971a)

Allen, J.C., Harris, R.A., Schwartz, A.: The nature of the transport ATPase-digitalis complex. II. Some species differences and ouabain "exchange" characteristics. J. Mol. Cell. Cardiol. *3*, 297–300 (1971b)

Allen, J.C., Martines-Maldonado, M., Eknoyan, G., Suki, W.N., Schwartz, A.: Relation between digitalis binding in vivo and inhibition of sodium, potassium-adenosine triphosphatase in canine kidney. Biochem. Pharmacol. *20*, 73–80 (1971c)

Allen, J.C., Entman, M.L., Schwartz, A.: The nature of the transport adenosine triphosphatase-digitalis complex. VIII. The relationship between in vivo-formed (^{3}H-ouabain-Na^+, K^+-adenosine triphosphatase) complex and ouabain-induced positive inotropism. J. Pharmacol. Exp. Ther. *192*, 105–112 (1975)

Asano, Y., Liberman, U.A., Edelman, I.S.: Thyroid thermogenesis: Relationships between Na^+-dependent respiration and $Na^+ + K^+$-adenosine triphosphatase activity in rat skeletal muscle. J. Clin. Invest. *57*, 368–379 (1976)

Ash, A.S.F., Schwartz, A.: Sodium-plus-potassium ion-activated adenosine triphosphatase in a heavy-membrane fraction isolated from rat skeletal muscle. Biochem. J. *118*, 20P–21P (1970)

Baker, P.F., Willis, J.S.: On the number of sodium pumping sites in cell membranes. Biochim. Biophys. Acta (Amst.) *183*, 646–649 (1969)

Baker, P.F., Willis, J.S.: Potassium ions and the binding of cardiac glycosides to mammalian cells. Nature (Lond.) *226*, 521–523 (1970)

Baker, P.F., Willis, J.S.: Inhibition of the sodium pump in squid giant axons by cardiac glycosides: The dependence on extracellular ions and metabolism. J. Physiol. (Lond.) *224*, 463–475 (1972)

Balasubramanian, V., McNamara, D.B., Singh, J.N., Dhalla, N.S.: Biochemical basis of heart function. X. Reduction in the Na^+-K^+-stimulated ATPase activity in failing rat heart due to hypoxia. Can. J. Physiol. Pharmacol. *51*, 504–510 (1973)

Barnett, R.E.: Effect of monovalent cations on the ouabain inhibition of the sodium and potassium ion activated adenosine triphosphatase. Biochemistry *9*, 4644–4648 (1970)

Baskin, S.I., Akera, T., Puckett, C.R., Brody, S.L., Brody, T.M.: Effect of potassium canrenoate on cardiac functions and ($Na^+ + K^+$)-activated ATPase. Proc. Soc. Exp. Biol. Med. *143*, 495–498 (1973)
Beauge, L.A., Adragna, N.: The kinetics of ouabain inhibition and the partition of rubidium influx in human red blood cells. J. Gen. Physiol. *57*, 576–592 (1971)
Beard, N.A., Rouse, W., Somerville, A.R.: Cardiotonic steroids: Correlation of sodium-potassium adenosine triphosphate inhibition and ion transport in vitro with inotropic activity and toxicity in dogs. Br. J. Pharmacol. *54*, 65–74 (1975)
Beller, G.A., Conroy, J., Smith, T.W.: Ischemia-induced alterations in myocardial ($Na^+ + K^+$)-ATPase and cardiac glycoside binding. J. Clin. Invest. *57*, 341–350 (1976)
Bentfeld, M., Lüllmann, H., Peters, T., Proppe, D.: Interdependence of ion transport and the action of ouabain in heart muscle. Brit. J. Pharmacol. *61*, 19–27 (1977)
Bergelson, L.D., Barsukov, L.I.: Topological asymmetry of phospholipids in membranes. Science *197*, 224–230 (1977)
Besch, H.R. Jr., Schwartz, A.: On the mechanism of action of digitalis. J. Mol. Cell. Cardiol. *1*, 195–199 (1970)
Besch, H.R. Jr., Allen, J.C., Glick, G., Schwartz, A.: Correlation between the inotropic action of ouabain and its effects on subcellular enzyme systems from canine myocardium. J. Pharmacol. Exp. Ther. *171*, 1–12 (1970)
Besch, H.R. Jr., Jones, L.R., Watanabe, A.M.: Intact vesicles of canine cardiac sarcolemma: Evidence from vectorial properties of Na^+, K^+-ATPase. Circ. Res. *39*, 586–595 (1976)
Blaustein, M.P.: Sodium-calcium exchange and the regulation of cell calcium in muscle fibers. Physiologist *19*, 525–540 (1976)
Blood, B.E., Noble, D.: Glycoside induced inotropism of the heart – more than one mechanism? J. Physiol. (Lond.) *266*, 76P–78P (1977)
Bluschke, V., Bonn, R., Greeff, K.: Increase in the ($Na^+ + K^+$)-ATPase activity in heart muscle after chronic treatment with digitoxin or potassium defficient diet. Europ. J. Pharmacol. *37*, 189–191 (1976)
Bodemann, H.H., Hoffman, J.F.: Side-dependent effects of internal versus external Na and K on ouabain binding to reconstituted human red blood cell ghosts. J. Gen. Physiol. *67*, 497–525 (1976a)
Bodemann, H.H., Hoffman, J.F.: Comparison of the side-dependent effects of Na and K on orthophosphate-, UTP-, and ATP-promoted ouabain binding to reconstituted human red blood cell ghosts. J. Gen. Physiol. *67*, 527–545 (1976b)
Bonn, R., Greeff, K.: The effect of chronic administration of digitoxin on the activity of the myocardial (Na+K)-ATPase in guinea-pigs. Arch. Int. Pharmacodyn. *233*, 53–64 (1978)
Borsch-Galetke, E., Dransfeld, H., Greeff, K.: Specific activity and sensitivity of strophanthin of the $Na^+ + K^+$-activated ATPase in rats and guinea-pigs with hypoadrenalism. Naunyn-Schmiedebergs Arch. Pharmacol. *274*, 74–80 (1972)
Böttcher, H., Fischer, K., Proppe, D.: Untersuchungen über die Wirkung von Digoxigenin-mono-, -bis- und -tridigitoxosid am Herz-Lungen-Präparat der Katze. Basic. Res. Cardiol. *70*, 279–291 (1975)
Boutagy, J., Gelbart, A., Thomas, R.: Cardenolide analogues. IV. Inhibition of Na^+, K^+-ATPase. Aust. J. Pharmaceut. Sci. *NS2*, 41–46 (1973)
Brennan, F.J., McCans, J.L., Chiong, M.A., Parker, J.O.: Effects of ouabain on myocardial potassium and sodium balance in man. Circulation *45*, 107–113 (1972)
Brody, T.M., Akera, T.: Relations among Na^+, K^+-ATPase activity, sodium pump activity, transmembrane sodium movement, and cardiac contractility. Fed. Proc. *36*, 2219–2224 (1977)
Brown, T.E., Acheson, G.H., Grupp, G.: The saturated-lactone glycoside dihydro-ouabain: Effects on potassium balance of the dog heart. J. Pharmacol. Exp. Ther. *136*, 107–113 (1962)
Byrne, J.E., Dresel, P.E.: The effect of temperature and calcium concentration on the action of ouabain in quiescent rabbit atria. J. Pharmacol. Exp. Ther. *166*, 354–363 (1969)
Caldwell, R.W., Nash, C.B.: Comparison of effects of aminosugar cardiac glycosides with ouabain and digoxin on Na^+, K^+-adenosine triphosphatase and cardiac contractile force. J. Pharmacol. Exp. Ther. *204*, 141–148 (1977)

Chang, C.C., Trosko, J.E., Akera, T.: Characterization of ultraviolet light-induced ouabain-resistant mutations in Chinese hamster cells. Mutation Res. *51*, 85–98 (1978)
Charnock, J.S., Potter, H.A.: The effect of Mg^{2+} and ouabain on the incorporation of P^{32} from ATP^{32} into Na^+- and K^+-activated adenosine triphosphatase. Arch. Biochem. Biophys. *134*, 42–47 (1969)
Charnock, J.S., Rosenthal, A.S., Post, R.L.: Studies of the mechanism of cation transport. II. A phosphorylated intermediate in the cation stimulated enzymic hydrolysis of adenosine triphosphate. Aust. J. Exp. Biol. *41*, 675–686 (1963)
Charnock, J.S., Almeida, A.F., To, R.: Temperature-activity relationships of cation activation and ouabain inhibition of $(Na^+ + K^+)$-ATPase. Arch. Biochem. Biophys. *167*, 480–487 (1975)
Charnock, J.S., Simonson, L.P., Almeida, A.F.: Variation in sensitivity of the cardiac glycoside receptor characteristics of $(Na^+ + K^+)$-ATPase to lipolysis and temperature. Biochim. Biophys. Acta (Amst.) *465*, 77–92 (1977)
Choi, Y.R., Akera, T.: Kinetic studies on the interaction between ouabain and (Na^+, K^+)-ATPase. Biochim. Biophys. Acta (Amst.) *481*, 648–659 (1977)
Choi, Y.R., Akera, T.: Membrane $(Na^+ + K^+)$-ATPase of canine brain, heart, and kidney: Tissue dependent differences in kinetic properties and the influence of purification procedures. Biochim. Biophys. Acta (Amst.) *508*, 313–327 (1978)
Clark, A.F., Swanson, P.D., Stahl, W.L.: Increase in dissociation rate constants of cardiotonic steroid-brain $(Na^+ + K^+)$-ATPase complexes by reduction of unsaturated lactone. J. Biol. Chem. *250*, 9355–9359 (1975)
Clausen, T., Hansen, O.: Ouabain binding and Na^+-K^+ transport in rat muscle cells and adipocytes. Biochim. Biophys. Acta (Amst.) *345*, 387–404 (1974)
Cohen, I., Daut, J., Noble, D.: An analysis of the actions of low concentrations of ouabain on membrane currents in Purkinje fibres. J. Physiol. (Lond.) *260*, 75–103 (1976)
Curfman, G.D., Crowley, T.J., Smith, T.W.: Thyroid-induced alterations in myocardial sodium- and potassium-activated adenosine triphosphatase, monovalent cation active transport, and cardiac glycoside binding. J. Clin. Invest. *59*, 586–590 (1977)
DePont, J.J.H.H.M., Bonting, S.L.: The role of phospholipids in Na-K-ATPase. In: Bazan, N.G., Brenner, R.R., Guisto, N.M. (eds.). Function and biosynthesis of lipids, pp. 219–224. New York: Plenum Publ. Corp. 1977
DePover, A., Godfraind, T.: Sensitivity to cardiac glycosides of (Na+K) ATPase prepared from human heart, guinea-pig heart, and guinea-pig brain. Arch. Int. Pharmacodyn. Ther. *221*, 339–341 (1976)
Dhalla, N.S., Singh, J.N., Fedelesova, M., Balasubramanian, V., McNamara, D.B.: Biochemical basis of heart function. XII. Sodium-potassium stimulated adenosine triphosphatase activity in the perfused heart made to fail by substrate-lack. Cardiovasc. Res. *8*, 227–236 (1974)
Dransfeld, H., Greeff, K., Berger, H., Cautius, V.: Die verschiedene Empfindlichkeit der $Na^+ + K^+$-aktivierten ATPase des Herz- und Skeletmuskels gegen k-Strophanthin. Naunyn-Schmiedebergs Arch. Pharmacol. *254*, 225–234 (1966)
Dransfeld, H., Greeff, K., Schorn, A., Ting, B.T.: Calcium uptake in mitochondria and vesicles of heart and skeletal muscle in presence of potassium, sodium, k-strophanthin, and pentobarbital. Biochem. Pharmacol. *18*, 1335–1345 (1969)
Dudding, W.F., Winter, C.G.: On the reaction sequence of the K^+-dependent acetyl phosphatase activity of the Na^+ pump. Biochim. Biophys. Acta (Amst.) *241*, 650–660 (1971)
Dunham, E.T., Glynn, I.M.: Adenosine triphosphatase activity and the active movements of alkali metal ions. J. Physiol. (Lond.) *156*, 274–293 (1961)
Dutta, S., Goswami, S., Datta, D.K., Lindower, J.O., Marks, B.H.: The uptake and binding of six radiolabeled cardiac glycosides by guinea-pig hearts and by isolated sarcoplasmic reticulum. J. Pharmacol. Exp. Ther. *164*, 10–21 (1968)
Ebner, F., Reiter, M.: The dependence on contraction frequency of the positive inotropic effect of dihydro-ouabain. Naunyn-Schmiedebergs Arch. Pharmacol. *300*, 1–9 (1977)
Edelman, I.S.: Thyroidal regulation of renal energy metabolism and $(Na^+ + K^+)$-activated adenosine triphosphatase activity. Med. Clin. N. Amer. *59*, 605–614 (1975)

Entman, M.L., Allen, J.C., Schwartz, A.: Calcium-ouabain interaction in a "microsomal" membrane fraction containing Na^+, K^+-ATPase activity and calcium binding activity. J. Mol. Cell. Cardiol. *4*, 435–441 (1972)

Erdmann, E., Hasse, W.: Quantitative aspects of ouabain binding to human erythrocyte and cardiac membranes. J. Physiol. (Lond.) *251*, 671–682 (1975)

Erdmann, E., Schoner, W.: Ouabain-receptor interactions in $(Na^+ + K^+)$-ATPase preparations from different tissues and species. Determination of kinetic constants and dissociation constants. Biochim. Biophys. Acta (Amst.) *307*, 386–398 (1973a)

Erdmann, E., Schoner, W.: Ouabain-receptor interactions in $(Na^+ + K^+)$-ATPase preparations. II. Effects of cations and nucleotides on rate constants and dissociation constants. Biochim. Biophys. Acta (Amst.) *330*, 302–315 (1973b)

Erdmann, E., Schoner, W.: Ouabain-receptor interactions in $(Na^+ + K^+)$-ATPase preparations. III. On the stability of the ouabain receptor against physical treatment, hydrolases, and SH reagents. Biochim. Biophys. Acta (Amst.) *330*, 316–324 (1973c)

Erdmann, E., Schoner, W.: Ouabain-receptor interactions in $(Na^+ + K^+)$-ATPase preparations. IV. The molecular structure of different cardioactive steroids and other substances and their affinity to the glycoside receptor. Naunyn-Schmiedebergs Arch. Pharmacol. *283*, 335–356 (1974a)

Erdmann, E., Schoner, W.: Eigenschaften des Rezeptors für Herzglykoside. Klin. Wochenschr. *52*, 705–718 (1974b)

Erdmann, E., Bolte, H.-D., Lüderitz, B.: The $(Na^+ + K^+)$-ATPase activity of guinea pig heart muscle in potassium deficiency. Arch. Biochem. Biophys. *145*, 121–125 (1971)

Erdmann, E., Bolte, H.-D., Schoner, W.: Cardiac glycoside receptor in potassium depletion. Recent Adv. in studies on cardiac structure and metabolism. Basic functions of cations in myocardial activity. Fleckenstein, A., Dhalla, N.S. (eds.), Vol. 5, pp. 351–358. Baltimore: University Park Press 1975

Erdmann, E., Patzelt, R., Schoner, W.: The cardiac glycoside receptor: Its properties and its correlation to nucleotide binding sites, phosphointermediate, and $(Na^+ + K^+)$-ATPase activity. Recent Adv. in studies on cardiac structure and metabolism. The Sarcolemma. Roy, P-E., Dhalla, N.S. (eds.), Vol. 9, pp. 329–335. Baltimore: University Park Press 1976a

Erdmann, E., Philipp, G., Tanner, G.: Ouabain-receptor interactions in $(Na^+ + K^+)$-ATPase preparations. A contribution to the problem of nonlinear Scatchard plots. Biochim. Biophys. Acta (Amst.) *455*, 287–296 (1976b)

Erlij, D., Elizalde, A.: Rapidly reversible inhibition of frog muscle sodium pump caused by cardiotonic steroids with modified lactone rings. Biochim. Biophys. Acta (Amst.) *345*, 49–54 (1974)

Flasch, H., Heinz, N.: Correlation between inhibition of $(Na^+ + K^+)$-membrane-ATPase and positive inotropic activity of cardenolides in isolated papillary muscles of guinea pig. Naunyn-Schmiedebergs Arch. Pharmacol. *304*, 37–44 (1978)

Forbush, B., Kaplan, J.H., Hoffman, J.F.: Characterization of a new photoaffinity derivative of ouabain: Labeling of the large polypeptide and of a proteolipid component of the Na, K-ATPase. Biochemistry *17*, 3667–3676 (1978)

Fricke, U.: Lack of interaction of spironolactone with ouabain in guinea pig isolated heart muscle praparations. Eur. J. Pharmacol. *49*, 363–371 (1978)

Fricke, U., Klaus, W.: Die Reversibilität der Wirkung von Digitoxin, Strophanthidin und Strophanthidin-3-bromazetat am Papillarmuskel und einer mikrosomalen Na^+-K^+-aktivierbaren ATPase des Meerschweinchens. Experientia *25*, 685–686 (1969)

Fricke, U., Klaus, W.: Comparative studies of the influence of various K^+ concentrations on the action of k-strophanthidin, digitoxin, and strophanthidin-3-bromoacetate on papillary muscle and on membrane-ATPase of guinea pig hearts. Eur. J. Pharmacol. *15*, 1–7 (1971a)

Fricke, U., Klaus, W.: Über die Wirkung von Strophanthidin-3-bromacetat am Papillarmuskel des Meerschweinchens. Naunyn-Schmiedebergs Arch. Pharmacol. *268*, 192–199 (1971b)

Fricke, U., Klaus, W.: Die Haftung verschiedener Cardenolide am Papillarmuskel und einer mikrosomalen ATPase des Meerschweinchenherzens. Naunyn-Schmiedebergs Arch. Pharmacol. *268*, 200–209 (1971c)
Fricke, U., Klaus, W.: Evidence for two different Na^+-dependent (^3H)-ouabain binding sites of a Na^+-K^+-ATPase of guinea-pig hearts. Brit. J. Pharmacol. *61*, 423–428 (1977)
Fricke, U., Klaus, W.: Sodium-dependent cardiac glycoside binding: Experimental evidence and hypothesis. Brit. J. Pharmacol. *62*, 255–257 (1978)
Fricke, U., Hollborn, U., Klaus, W.: Inotropic action, myocardial uptake, and subcellular distribution of ouabain, digoxin, and digitoxin in isolated rat hearts. Naunyn-Schmiedebergs Arch. Pharmacol. *288*, 195–214 (1975)
Fujino, S., Kawagishi, S., Eguchi, N., Tanaka, M.: Binding site of ouabain in cardiac muscle cell and its positive inotropic effect in cat. Jpn. J. Pharmacol. *21*, 423–425 (1971)
Gardner, J.D., Kiino, D.R.: Ouabain binding and cation transport in human erythrocytes. J. Clin. Invest. *52*, 1845–1851 (1973)
Gervais, A., Lane, L.K., Anner, B.M., Lindenmayer, G.E., Schwartz, A.: A possible molecular mechanism of the action of digitalis. Ouabain action on calcium binding to sites associated with a purified sodium-potassium-activated adenosine triphosphatase from kidney. Circ. Res. *40*, 8–14 (1977)
Glitsch, H.G.: Hemmung der elektrogenen Na-Pumpe am Meerschweinchenvorhof durch Digitoxigenin. Pflügers Arch. *335*, 243–251 (1972)
Glitsch, H.G.: An effect of the electrogenic sodium pump on the membrane potential in beating guinea-pig atria. Pflügers Arch. *344*, 169–180 (1973)
Glynn, I.M.: The action of cardiac glycosides on ion movements. Pharmacol. Rev. *16*, 381–407 (1964)
Godfraind, T.: The therapeutic mode of action of cardiac glycosides. Arch. Int. Pharmacodyn. Ther. *206*, 384–388 (1973)
Godfraind, T., Ghysel-Burton, J.: Binding sites related to ouabain-induced stimulation or inhibition of the sodium pump. Nature (Lond.) *265*, 165–166 (1977)
Goodman, S.L., Wheeler, K.P.: Ouabain binding to phospholipid-dependent adenosine triphosphatase. Biochem. J. *169*, 313–320 (1977)
Goldman, R.H., Coltart, D.J., Friedman, J.P., Nola, G.T., Berke, D.K., Schweizer, E., Harrison, D.C.: The inotropic effects of digoxin in hyperkalemia. Relation to (Na^+, K^+)-ATPase inhibition in the intact animal. Circulation *48*, 830–838 (1973)
Goldman, R.H., Coltart, D.J., Schweizer, E., Snidow, G., Harrison, D.C.: Dose response in vivo to digoxin in normo- and hyperkalaemia: Associated biochemical changes. Cardiovasc. Res. *9*, 515–523 (1975)
Grobecker, V.H., Piechowski, U., Greeff, K.: Die Wirkung des k-Strophanthins und Digitonins auf den Ionentransport und die Membran-ATPase der Erythrocyten von Menschen, Meerschweinchen und Ratten. Med. Exp. *9*, 273–282 (1963)
Gubitz, R.H., Akera, T., Brody, T.M.: Control of brain slice respiration by ($Na^+ + K^+$)-activated adenosine triphosphatase and the effects of enzyme inhibitors. Biochim. Biophys. Acta (Amst.) *459*, 263–277 (1977)
Hajdu, S., Leonard, E.: The cellular basis of cardiac glycoside action. Pharmacol. Rev. *11*, 173–209 (1959)
Hall, R.J., Gelbart, A., Silverman, M., Goldman, R.H.: Studies on digitalis-induced arrhythmias in glucose- and insulin-induced hypokalemia. J. Pharmacol. Exp. Ther. *201*, 711–722 (1977)
Hansen, O.: The relationship between g-strophanthin-binding capacity and ATPase activity in plasma-membrane fragments from ox brain. Biochim. Biophys. Acta (Amst.) *233*, 122–132 (1971)
Hansen, O.: Non-uniform populations of g-strophanthin binding sites of ($Na^+ + K^+$)-activated ATPase: Apparent conversion to uniformity by K^+. Biochim. Biophys. Acta (Amst.) *433*, 383–392 (1976)
Hansen, O.: The effect of sodium on inorganic phosphate- and paranitrophenyl phosphate-facilitated ouabain binding to ($Na^+ + K^+$)-activated ATPase. Biochim. Biophys. Acta (Amst.) *511*, 10–22 (1978)
Hansen, O., Jensen, J., Nørby, J.G.: Mutual exclusion of ATP, ADP, and g-strophanthin binding to NaK-ATPase. Nature New Biol. *234*, 122–124 (1971)

Harris, W.E., Swanson, P.D., Stahl, W.L.: Ouabain binding sites and the (Na^+, K^+)-ATPase of brain microsomal membranes. Biochim. Biophys. Acta (Amst.) *298*, 680–689 (1973)
Haustein, K.-O.: Studies on cardioactive steroids. I. Structure-activity relationships on the isolated atrium. Pharmacology *11*, 117–126 (1974)
Haustein, K.-O., Hauptmann, J.: Studies on cardioactive steroids. Pharmacology *11*, 129–138 (1974)
Hegyvary, C.: Covalent labeling of the digitalis-binding component of plasma membranes. Mol. Pharmacol. *11*, 588–594 (1975)
Hegyvary, C.: Ouabain binding and phosphorylation of ($Na^+ + K^+$)-ATPase treated with N-ethylmaleimide or oligomycin. Biochim. Biophys. Acta (Amst.) *422*, 365–379 (1976)
Hegyvary, C.: Alterations of cardiac NaK-ATPase by the thyroid state in the rat. Res. Commun. Chem. Path. Pharmacol. *17*, 689–702 (1977)
Hegyvary, C., Post, R.L.: Binding of adenosine triphosphate to sodium and potassium ion-stimulated adenosine triphosphatase. J. Biol. Chem. *246*, 5234–5240 (1971)
Heinen, E., Noack, E.: Effects of k-strophanthin and digitoxigenin on contractile force, calcium content and exchange in guinea-pig isolated atria. Naunyn-Schmiedebergs Arch. Pharmacol. *275*, 359–371 (1972)
Hess, H.H., Pope, A.: Effect of metal cations on adenosine triphosphatase activity of rat brain. Fed. Proc. *16*, 196 (1957)
Hilden, S., Rhee, H.M., Hokin, L.E.: Sodium transport by phospholipid vesicles containing purified sodium and potassium ion-activated adenosine triphosphatase. J. Biol. Chem. *249*, 7432–7440 (1974)
Hoffman, J.F.: The red cell membrane and the transport of sodium and potassium. Am. J. Med. *41*, 666–680 (1966)
Hoffman, J.F.: The interaction between tritiated ouabain and the Na-K pump in red blood cells. J. Gen. Physiol. *54*, 343S–350S (1969)
Hokin, L.E., Dahl, J.L., Deupree, J.D., Dixon, J.F., Hackney, J.F., Perdue, J.F.: Studies on the characterization of the sodium-potassium transport adenosine triphosphatase. X. Purification of the enzyme from the rectal gland of Squalus acanthias. J. Biol. Chem. *248*, 2593–2605 (1973)
Hougen, T.J., Smith, T.W.: Inhibition of myocardial monovalent cation active transport by subtoxic doses of ouabain in the dog. Circ. Res. *42*, 856–863 (1978)
Huang, W., Askari, A.: Transport ATPase of erythrocyte membrane: Sensitivities of Na^+, K^+-ATPase and K^+-phosphatase activities to ouabain. Arch. Biochem. Biophys. *175*, 185–189 (1976)
Inagaki, C., Lindenmayer, G.E., Schwartz, A.: Effects of sodium and potassium on binding of ouabain to the transport adenosine triphosphatase. J. Biol. Chem. *249*, 5135–5140 (1974)
Jones, L.R., Besch, H.R. Jr., Watanabe, A.M.: Significance of a cardiac microsomal Na^+ plus K^+-stimulated ATPase apparently insensitive to ouabain. Fed. Proc. *35*, 833 (1976)
Jørgensen, P.L.: Purification of Na^+, K^+-ATPase: Active site determination and criteria of purity. Ann. N.Y. Acad. Sci. *242:* 36–52 (1974)
Jørgensen, P.L.: Purification and characterization of (Na^+, K^+)-ATPase. V. Conformational changes in the enzyme transitions between the Na-form and the K-form studied with tryptic digestion as a tool. Biochim. Biophys. Acta (Amst.) *401*, 399–415 (1975)
Jørgensen, P.L., Skou, J.C.: Purification and characterization of ($Na^+ + K^+$)-ATPase. I. The influence of detergents on the activity of ($Na^+ + K^+$)-ATPase in preparations from the outer medulla of rabbit kidney. Biochim. Biophys. Acta (Amst.) *233*, 366–380 (1971)
Jørgensen, P.L., Hansen, O., Glynn, I.M., Cavieres, J.D.: Antibodies to pig kidney ($Na^+ + K^+$)-ATPase inhibit the Na^+ pump in human red cells provided they have access to the inner surface of the cell membrane. Biochim. Biophys. Acta (Amst.) *291*, 795–800 (1973)
Kim, N.D., Bailey, L.E., Dresel, P.E.: Correlation of the subcellular distribution of digoxin with the positive inotropic effect. J. Pharmacol. Exp. Ther. *181*, 377–385 (1972)
Köhler, E., Greeff, K.: Der Einfluß des Blut-pH auf die Toxicität herzwirksamer Glycoside. Res. Exp. Med. *159*, 65–74 (1972)

Ku, D., Akera, T., Pew, C.L., Brody, T.M.: Cardiac glycosides: Correlation among Na^+, K^+-ATPase, sodium pump and contractility in the guinea pig heart. Naunyn-Schmiedebergs Arch. Pharmacol. *285*, 185–200 (1974)

Ku, D.D., Akera, T., Tobin, T., Brody, T.M.: Comparative species studies on the effect of monovalent cations and ouabain on cardiac Na^+, K^+-ATPase and contractile force. J. Pharmacol. Exp. Ther. *197*, 458–469 (1976)

Ku, D.D., Akera, T., Brody, T.M., Weaver, L.C.: Chronic digoxin treatment on canine myocardial Na^+, K^+-ATPase. Naunyn-Schmiedebergs Arch. Pharmacol. *301*, 39–47 (1977)

Kyte, J.: Phosphorylation of a purified $(Na^+ + K^+)$ adenosine triphosphatase. Biochem. Biophys. Res. Commun. *43*, 1259–1265 (1971)

Kyte, J.: The titration of the cardiac glycoside binding site of the $(Na^+ + K^+)$-adenosine triphosphatase. J. Biol. Chem. *247*, 7634–7641 (1972)

Kyte, J.: The reactions of sodium and potassium ion-activated adenosine triphosphatase with specific antibodies. J. Biol. Chem. *249*, 3652–3660 (1974)

Lane, L.K., Copenhaver, J.H. Jr., Lindenmayer, G.E., Schwartz, A.: Purification and characterization of and (^{3}H)-ouabain binding to the transport adenosine triphosphatase from outer medulla of canine kidney. J. Biol. Chem. *248*, 7197–7200 (1973)

Lane, L.K., Anner, B.M., Wallick, E.T., Ray, M.V., Schwartz, A.: Effect of phospholipase A treatment on the partial reactions of and ouabain binding to a purified sodium and potassium activated adenosine triphosphatase. Biochem. Pharmacol. *27*, 225–231 (1977)

Langer, G.A.: Ion fluxes in cardiac excitation and contraction and their relation to myocardial contractility. Physiol. Rev. *48*, 708–757 (1968)

Langer, G.A.: The intrinsic control of myocardial contraction – ionic factors. New Engl. J. Med. *285*, 1065–1071 (1971)

Langer, G.A.: Effects of digitalis on myocardial ionic exchange. Circulation *46*, 180–187 (1972a)

Langer, G.A.: Myocardial K^+ loss and contraction frequency. J. Mol. Cell. Cardiol. *4*, 85–86 (1972b)

Langer, G.A.: Relationship between myocardial contractility and the effects of digitalis on ionic exchange. Fed. Proc. *36*, 2231–2234 (1977)

Langer, G.A., Serena, S.D.: Effects of strophanthidin upon contraction and ionic exchange in rabbit ventricular myocardium: Relative to control of active state. J. Mol. Cell. Cardiol. *1*, 65–90 (1970)

Lee, C.O., Fozzard, H.A.: Activities of potassium and sodium ions in rabbit heart muscle. J. Gen. Physiol. *65*, 695–708 (1975)

Lee, K.S., Klaus, W.: The subcellular basis for the mechanism of inotropic action of cardiac glycosides. Pharmacol. Rev. *23*, 193–261 (1971)

Lee, K.S., Yu, D.H.: A study of the sodium- and potassium-activated adenosinetriphosphatase activity of heart microsomal fraction. Biochem. Pharmacol. *12*, 1253–1264 (1963)

Lee, K.S., Shin, M.R., Kang, D.H., Chen, K.K.: Studies on the mechanism of cardiac glycoside action. Biochem. Pharmacol. *19*, 1055–1069 (1970)

Lelievre, L., Charlemagne, D., Paraf, A.: Plasma membrane studies on drug-sensitive and -resistant cell lines. II. Ouabain sensitivity of $(Na^+ + K^+)$stimulated Mg^{2+}-ATPase. Biochim. Biophys. Acta (Amst.) *455*, 277–286 (1976)

Lin, M.H., Akera, T.: Increased (Na^+, K^+)-ATPase concentrations in various tissues of rats caused by thyroid hormone treatment. J. Biol. Chem. *253*, 723–726 (1978)

Lin, M.H., Romsos, D.R., Akera, T., Leveille, G.A.: Na^+, K^+-ATPase enzyme units in skeletal muscle from lean and obese mice. Biochem. Biophys. Res. Commun. *80*, 398–404 (1978)

Lindenmayer, G.E., Schwartz, A.: Conformational changes induced in Na^+, K^+-ATPase by ouabain through a K^+-sensitive reaction: Kinetic and spectroscopic studies. Arch. Biochem. Biophys. *140*, 371–378 (1970)

Lindenmayer, G.E., Schwartz, A.: Nature of the transport adenosine triphosphatase digitalis complex. IV. Evidence that sodium-potassium competition modulates the rate of ouabain interaction with ($Na^+ + K^+$) adenosine triphosphatase during enzyme catalysis. J. Biol. Chem. *248*, 1291–1300 (1973)
Lindenmayer, G.E., Laughter, A.H., Schwartz, A.: Incorporation of inorganic phosphate-32 into a $Na^+ + K^+$-ATPase preparation: Stimulation by ouabain. Arch. Biochem. Biophys. *127*, 187–192 (1968)
Lishko, V.K., Malysheva, M.K., Grevisirskaya, T.I.: The interaction of the (Na^+, K^+)-ATPase of erythrocyte ghosts with ouabain. Biochim. Biophys. Acta (Amst.) *288*, 103–106 (1972)
Lo, C.S., Edelman, I.S.: Effect of triiodothyronine on the synthesis and degradation of renal cortical ($Na^+ + K^+$)-adenosine triphosphatase. J. Biol. Chem. *251*, 7834–7840 (1976)
Lo, C.S., August, T.R., Liberman, U.A., Edelman, I.S.: Dependence of renal ($Na^+ + K^+$)-adenosine triphosphatase activity on thyroid status. J. Biol. Chem. *251*, 7826–7833 (1976)
Löhr, E., Makoski, H.B., Gobbeler, T., Strötges, M.W.: Beitrag zu der Membranpermeabilität von Cardiaca (g-Strophanthin, Digoxin, Oxyfedrin) auf Grund von Mikroautoradiographien am Meerschweinchenherzen. Arzneim. Forsch. (Drug Res.) *21*, 921–927 (1971)
Lüllmann, H., Peters, T.: Action of cardiac glycosides on the excitation-contraction coupling in heart muscle. Prog. Pharmacol. *2*, 5–57 (1979)
Lüllmann, H., Peters, T.: Studies on the site of action of cardiac glycosides. In: Digitalis. Storstein, O. (ed.), p. 125. Oslo: Gyldendal Norsk Forlag 1974
Lüllmann, H., Peters, T.: On the sarcolemmal site of action of cardiac glycosides. Recent Adv. in studies on cardiac structure and metabolism. The sarcolemma. Roy, P.-E., Dhalla, N.S. (eds.), Vol. 9, pp. 311–328. Baltimore: University Park Press 1976
Lüllmann, H., Peters, T., Preuner, J., Rüther, T.: Influence of ouabain and dihydroouabain on the circular dichroism of cardiac plasmalemmal microsomes. Naunyn-Schmiedebergs Arch. Pharmacol. *290*, 1–19 (1975)
Malur, J., Repke, K.R.H.: Modelluntersuchungen über die Beteiligung einer Wasserstoff-Brücken-Bindung an der Komplexbildung zwischen Cardenolidverbindungen und $Na^+ + K^+$-aktivierter, Mg^{2+}-abhängiger Adenosintriphosphat-Phosphohydrolase. Acta Biol. Med. Ger. *24*, K67–K72 (1970)
Matsui, H., Schwartz, A.: Kinetic analysis of ouabain-K^+ and Na^+ interaction on a Na^+, K^+-dependent adenosinetriphosphatase from cardiac tissue. Biochem. Biophys. Res. Commun. *25*, 147–152 (1966)
Matsui, H., Schwartz, A.: Mechanism of cardiac glycoside inhibition of the (Na^+-K^+)-dependent ATPase from cardiac tissue. Biochim. Biophys. Acta (Amst.) *151*, 655–663 (1968)
Matsui, H., Hayashi, Y., Homareda, H., Kimimura, M.: Ouabain-sensitive ^{42}K binding to Na^+,K^+-ATPase purified from canine kidney outer medulla. Biochem. Biophys. Res. Commun. *75*, 373–380 (1977)
Mayer, M., Avi-Dor, Y.: Fluorescence of 8-anilino-1-naphthalene-sulfonate bound to ox-brain Na^+- and K^+-stimulated adenosine triphosphatase. Isr. J. Med. Sci. *6*, 726–731 (1970)
McCans, J.L., Lane, L.K., Lindenmayer, G.E., Butler, V.P. Jr., Schwartz, A.: Effects of an antibody to a highly purified Na^+, K^+-ATPase from canine renal medulla: Separation of the "holoenzyme antibody" into catalytic and cardiac glycoside receptor-specific components. Proc. Natl. Acad. Sci. U.S.A. *71*, 2449–2452 (1974)
Michael, L., Wallick, E.T., Schwartz, A.: Modification of (Na^+, K^+)-ATPase function by purified antibodies to the holoenzyme – effects on enzyme activity and ^3H ouabain binding. J. Biol. Chem. *252*, 8476–8480 (1977)
Michael, L., Pitts, B.J.R., Schwartz, A.: Is pump stimulation associated with positive inotropy of the heart? Science *200*, 1287–1289 (1978)
Murthy, R.V., Kidwai, A.M., Daniel, E.E.: Dissociation of contractile effect and binding and inhibition of Na^+-K^+-adenosine triphosphatase by cardiac glycosides in rabbit myometrium. J. Pharmacol. Exp. Ther. *188*, 575–581 (1974)

Nagai, K., Lindenmayer, G.E., Schwartz, A.: Direct evidence for the conformational nature of the Na^+, K^+-ATPase system: Fluorescence and circular dichroism studies. Arch. Biochem. Biophys. *139*, 252–254 (1970)

Nagatomo, T., Jarmakani, J.M., Philipson, K.D., Nakazawa, M.: Effect of anoxia on membrane-bound ATPase and K^+-p-nitrophenyl phosphatase activities in rabbit heart. J. Mol. Cell. Cardiol. *10*, 981–989 (1978)

Naidoo, B.K., Witty, T.R., Remers, W.A., Besch, H.R. Jr.: Cardiotonic steroids: I Importance of 14β-hydroxy group in digitoxigenin. J. Pharm. Sci. *63*, 1391–1394 (1974)

Nayler, W.G., Stone, J., Carson, V., Chipperfield, D.: Effect of ischemia on cardiac contractility and calcium exchangeability. J. Mol. Cell. Cardiol. *2*, 125–143 (1971)

Okarma, T.B., Tramell, P., Kalman, S.M.: Inhibition of sodium- and potassium-dependent adenosine triphosphatase by digoxin covalently bound to sepharose. Mol. Pharmacol. *8*, 476–480 (1972)

Okita, G.T.: Dissociation of Na^+, K^+-ATPase inhibition from digitalis inotropy. Fed. Proc. *36*, 2225–2230 (1977)

Okita, G.T., Richardson, F., Roth-Schechter, B.F.: Dissociation of the positive inotropic action of digitalis from inhibition of sodium- and potassium-activated adenosine triphosphatase. J. Pharmacol. Exp. Ther. *185*, 1–11 (1973)

Park, M.K., Vincenzi, F.F.: Rate of onset of cardiotonic steroid-induced inotropism: Influence of temperature and beat interval. J. Pharmacol. Exp. Ther. *195*, 140–150 (1975)

Perrone, J.R., Blostein, R.: Asymmetric interaction of inside-out and right-side-out erythrocyte membrane vesicles with ouabain. Biochim. Biophys. Acta (Amst.) *291*, 680–689 (1973)

Pert, C., Snyder, S.: Opiate receptor binding of agonists and antagonists affected differentially by sodium. Mol. Pharmacol. *10*, 868–879 (1974)

Peters, T., Raben, R.H., Wassermann, O.: Evidence for a dissociation between positive inotropic effect and inhibition of the Na^+-K^+-ATPase by ouabain, cassaine, and their alkylating derivatives. Eur. J. Pharmacol. *26*, 166–174 (1974)

Pitts, B.J.R., Wallick, E.T., Van Winkle, W.B., Allen, J.C., Schwartz, A.: On the lack of inotropy of cardiac glycosides on skeletal muscle: A comparison of Na^+, K^+-ATPases from skeletal and cardiac muscle. Arch. Biochem. Biophys. *184*, 431–440 (1977)

Poole-Wilson, P.A., Langer, G.A.: Glycoside inotropy in the absence of an increase in potassium efflux in the rabbit heart. Circ. Res. *37*, 390–395 (1975)

Post, R.L., Merritt, C.R., Kinsolving, C.R., Albright, C.D.: Membrane adenosine triphosphatase as a participant in the active transport of sodium and potassium in the human erythrocyte. J. Biol. Chem. *235*, 1796–1802 (1960)

Post, R.L., Kume, S., Tobin, T., Orcutt, B., Sen, A.K.: Flexibility of an active center in sodium-plus-potassium adenosine triphosphatase. J. Gen. Physiol. *54*, 306S–326S (1969)

Prindle, K.H. Jr., Skelton, C.L., Epstein, S.E., Marcus, F.I.: Influence of extracellular potassium concentration on myocardial uptake and inotropic effect of tritiated digoxin. Circ. Res. *28*, 337–345 (1971)

Repke, K.: Metabolism of cardiac glycosides. In: Proc. 1 st International Pharmacol. Meeting, vol. 3, pp. 47–73. Oxford: Pergamon Press 1963

Repke, K.: Über den biochemischen Wirkungsmodus von Digitalis. Klin. Wochenschr. *42*, 157–165 (1964)

Repke, K.: Effect of digitalis on membrane adenosine triphosphatase of cardiac muscle. In: Drugs and enzymes. Proc. 2 nd International Pharmacol. Meeting, Prague. Brodie, B.B., Gillette, J. (eds.), vol. 4, pp. 65–87. Oxford: Pergamon, Prague: Chechoslovak Medical Press 1965

Repke, K., Portius, H.J.: Über die Identität der Ionenpumpen-ATPase in der Zellmembran des Herzmuskels mit einem Digitalis-Rezeptorenzym. Experientia *19*, 452–458 (1963)

Repke, K.R.H., Schön, R.: Flip-flop model of (NaK)-ATPase function. Acta Biol. Med. Germ. *31*, K19–K39 (1973)

Repke, K., Est, M., Portius, H.J.: Über die Ursache der Speciesunterschiede in der Digitalisempfindlichkeit. Biochem. Pharmacol. *14*, 1785–1802 (1965)

Reuter, H.: Exchange of calcium ions in the mammalian myocardium: Mechanisms and physiological significance. Circ. Res. *34*, 599–605 (1974)

Rhee, H.M., Dutta, S., Marks, B.H.: Cardiac NaK ATPase activity during positive inotropic and toxic action of ouabain. Eur. J. Pharmacol. *37*, 141–153 (1976)
Ross, C.R., Pessah, N.I.: Reversible inhibition of ($Na^+ + K^+$)-ATPase with a cardiac glycoside. Eur. J. Pharmacol. *33*, 223–226 (1975)
Roth-Schechter, B.F., Okita, G.T., Thomas, R.E., Richardson, F.F.: On the positive inotropic action of alkylating bromoacetates of strophanthidin and strophanthidol-(19-H^3). J. Pharmacol. Exp. Ther. *171*, 13–19 (1970)
Ruoho, A., Kyte, J.: Photoaffinity labeling of the ouabain-binding site on ($Na^+ + K^+$)-adenosinetriphosphatase. Proc. Natl. Acad. Sci. U.S.A. *71*, 2352–2356 (1974)
Ruoho, A.E., Hokin, L.E., Hemingway, R.J., Kupchan, S.M.: Hellebrigenin 3-haloacetates: Potent site-directed alkylators of transport adenosinetriphosphatase. Science *159*, 1354–1355 (1968)
Sachs, J.R.: Interaction of external K, Na, and cardioactive steroids with the Na-K pump of the human red blood cell. J. Gen. Physiol. *63*, 123–143 (1974)
Schatzmann, H.J.: Herzglycoside als Hemmstoffe für den aktiven Kalium- und Natriumtransport durch die Erythrocytenmembran. Helv. Physiol. Acta *11*, 346–354 (1953)
Scholz, H.: Effect of a "therapeutic" concentration of digitoxigenine on myocardial potassium and sodium content in Ca-poor media. Naunyn-Schmiedebergs Arch. Pharmacol. *273*, 434–437 (1972)
Schön, R., Schönfeld, W., Repke, K.R.H.: Zur Charakterisierung des Ouabain-bindenden Konformationszustandes der ($Na^+ + K^+$)-aktivierten ATPase. Acta Biol. Med. Germ. *24*, K61–K65 (1970)
Schön, R., Schönfeld, W., Menke, K.-H., Repke, K.R.H.: Mechanism and role of Na^+/Ca^{++} competition in (NaK)-ATPase. Acta Biol. Med. Germ. *29*, 643–659 (1972)
Schönfeld, W., Schön, R., Menke, K.-H., Repke, K.R.H.: Identification of conformational states of transport ATPase by kinetic analysis of ouabain binding. Acta Biol. Med. Germ. *28*, 935–956 (1972)
Schuurmans Stekhoven, F.M.A.H., DePont, J.J.H.H.M., Bonting, S.L.: Studies on ($Na^+ + K^+$)-activated ATPase. XXXVII. Stabilization by cations of the enzyme-ouabain complex formed with Mg^{2+} and inorganic phosphate. Biochim. Biophys. Acta (Amst.) *419*, 137–149 (1976)
Schwartz, A.: Is the cell membrane Na^+, K^+-ATPase enzyme system the pharmacological receptor for digitalis? Circ. Res. *39*, 2–7 (1976)
Schwartz, A., Matsui, H., Laughter, A.H.: Tritiated digoxin binding to ($Na^+ + K^+$)-activated adenosine triphosphatase: Possible allosteric site. Science *160*, 323–325 (1968)
Schwartz, A., Allen, J.C., Harigaya, S.: Possible involvement of cardiac Na^+, K^+-adenosine triphosphatase in the mechanism of action of cardiac glycosides. J. Pharmacol. Exp. Ther. *168*, 31–41 (1969)
Schwartz, A., Wood, J.M., Allen, J.C., Bornet, E.P., Entman, M.L., Goldstein, M.A., Sordahl, L.A., Suzuki, M.: Biochemical and morphologic correlates of cardiac ischemia. I. Membrane system. Am. J. Cardiol. *32*, 46–61 (1973)
Schwartz, A., Allen, J.C., Van Winkle, W.B., Munson, R.: Further studies on the correlation between the inotropic action of ouabain and its interaction with the Na^+, K^+-adenosine triphosphatase: Isolated perfused rabbit and cat hearts. J. Pharmacol. Exp. Ther. *191*, 119–127 (1974)
Schwartz, A., Lindenmayer, G.E., Allen, J.C.: The sodium-potassium adenosine triphosphatase: Pharmacological, physiological, and biochemical aspects. Pharmacol. Rev. *27*, 3–134 (1975)
Sen, A.K., Tobin, T., Post, R.L.: A cycle for ouabain inhibition of sodium- and potassium-dependent adenosine triphosphatase. J. Biol. Chem. *244*, 6596–6604 (1969)
Sharma, V.K., Banerjee, S.P.: Specific [^{3}H]ouabain binding to rat heart and skeletal muscle: Effects of thyroidectomy. Mol. Pharmacol. *14*, 122–129 (1978)
Siegel, G.J., Josephson, J.: Ouabain reaction with microsomal (sodium-plus-potassium)-activated adenosine-triphosphatase: Characteristics of substrate and ion dependencies. Eur. J. Biochem. *25*, 323–335 (1972)
Siegel, G.J., Koval, G.J., Albers, R.W.: Sodium-potassium-activated adenosine triphosphatase. VI. Characterization of the phosphoprotein formed from orthophosphate in the presence of ouabain. J. Biol. Chem. *244*, 3264–3269 (1969)

Singh, C.M., Flear, C.T.G., Nandra, A., Ross, D.N.: Electrolyte changes in the human myocardium after anoxic arrest. Cardiology *56*, 128–135 (1971)

Skou, J.C.: The influence of some cations on an adenosine triphosphatase from peripheral nerves. Biochim. Biophys. Acta (Amst.) *23*, 394–401 (1957)

Skou, J.C.: Further investigations on a $Mg^{++}+Na^{+}$-activated adenosinetriphosphatase, possibly related to the active, linked transport of Na^{+} and K^{+} across the nerve membrane. Biochim. Biophys. Acta (Amst.) *42*, 6–23 (1960)

Skou, J.C.: Enzymatic basis for active transport of Na and K across cell membrane. Physiol. Rev. *45*, 596–617 (1965)

Skou, J.C.: Study on the influence of the concentration of Mg^{2+}, Pi, K^{+}, Na^{+}, and Tris on (Mg^{2+}+Pi)-supported G-strophanthin binding to ($Na^{+}+K^{+}$)-activated ATPase from ox brain. Biochim. Biophys. Acta (Amst.) *311*, 51–66 (1973)

Skou, J.C., Hilberg, C.: The effect of cations, G-strophanthin and oligomycin on the labeling from (^{32}P) ATP of the ($Na^{+}+K^{+}$)-activated enzyme system and the effect of cations and G-strophanthin on the labeling from (^{32}P) ITP and ^{32}Pi. Biochim. Biophys. Acta (Amst.) *185*, 198–219 (1969)

Skou, J.C., Butler, K.W., Hansen, O.: The effect of magnesium, ATP, Pi, and sodium on the inhibition of the ($Na^{+}+K^{+}$)-activated enzyme system by g-strophanthin. Biochim. Biophys. Acta (Amst.) *241*, 443–461 (1971)

Smith, T.W., Haber, E.: Digitalis. New Engl. J. Med. *289*, 1125–1129 (1973)

Smith, T.W., Wagner, H. Jr.: Effects of ($Na^{+}+K^{+}$)-ATPase-specific antibodies on enzymatic activity and monovalent cation transport. J. Membrane Biol. *25*, 341–360 (1975)

Smith, T.W., Wagner, H. Jr., Markis, J.E., Young, M.: Studies on the localization of the cardiac glycoside receptor. J. Clin. Invest. *51*, 1777–1789 (1972)

Smith, T.W., Wagner, H. Jr., Young, M.: Cardiac glycoside interaction with solubilized myocardial sodium- and potassium-dependent adenosine triphosphatase. Mol. Pharmacol. *10*, 626–633 (1974)

Stickney, J.L.: Inhibition of ^{3}H-1-norepinephrine uptake by ouabain is species dependent. Res. Commun. Chem. Pathol. Pharmacol. *14*, 227–236 (1976)

Swanson, P.D.: Temperature dependence of sodium ion activation of the cerebral microsomal adenosine triphosphatase. J. Neurochem. *13*, 229–236 (1966)

Sybers, H.D., Helmer, P.R., Murphy, Q.R.: Effect of hypoxia on myocardial potassium balance. Am. J. Physiol. *220*, 2047–2050 (1971)

Taniguchi, K., Iida, S.: The binding of ouabain to Na^{+}-K^{+}-dependent ATPase treated with phospholipase. Biochim. Biophys. Acta (Amst.) *233*, 831–833 (1971)

Taniguchi, K., Iida, S.: Two apparently different ouabain binding sites of ($Na^{+}+K^{+}$)-ATPase. Biochim. Biophys. Acta (Amst.) *288*, 98–102 (1972)

Taniguchi, K., Iida, S.: The role of phospholipids in the binding of ouabain to sodium- and potassium-dependent adenosine triphosphatase. Mol. Pharmacol. *9*, 350–359 (1973)

Thomas, R.C.: Electrogenic sodium pump in nerve and muscle cells. Physiol. Rev. *52*, 563–594 (1972)

Thomas, R., Boutagy, J., Gelbart, A.: Cardenolide analogs. V. Cardiotonic activity of semisynthetic analogs of digitoxigenin. J. Pharmacol. Exp. Ther. *191*, 219–231 (1974)

Tobin, T., Brody, T.M.: Rates of dissociation of enzyme-ouabain complexes and $K_{0.5}$ values in ($Na^{+}+K^{+}$) adenosine triphosphatase from different species. Biochem. Pharmacol. *21*, 1553–1560 (1972)

Tobin, T., Sen, A.K.: Stability and ligand sensitivity of (^{3}H) ouabain binding to ($Na^{+}+K^{+}$)-ATPase. Biochim. Biophys. Acta (Amst.) *198*, 120–131 (1970)

Tobin, T., Baskin, S.I., Akera, T., Brody, T.M.: Nucleotide specificity of the Na^{+}-stimulated phosphorylation and (^{3}H) ouabain-binding reactions of ($Na^{+}+K^{+}$)-dependent adenosine triphosphatase. Mol. Pharmacol. *8*, 256–263 (1972a)

Tobin, T., Henderson, R., Sen, A.K.: Species and tissue differences in the rate of dissociation of ouabain from ($Na^{+}+K^{+}$)-ATPase. Biochim. Biophys. Acta (Amst.) *274*, 551–555 (1972b)

Tobin, T., Akera, T., Baskin, S.I., Brody, T.M.: Calcium ion and sodium- and potassium-dependent adenosine triphosphatase: Its mechanism of inhibition and identification of the E_1-P intermediate. Mol. Pharmacol. *9*, 336–349 (1973a)

Tobin, T., Akera, T., Hogg, R.E., Brody, T.M.: Ouabain binding to sodium- and potassium-dependent adenosine triphosphatase: Inhibition by the β-γ-methylene analogue of adenosine triphosphate. Mol. Pharmacol. *9*, 278–281 (1973b)

Tobin, T., Akera, T., Ku, D.: Reversibility of the interaction of strophanthidin bromoacetate with the cardiotonic steroid binding site of sodium- and potassium-dependent adenosine triphosphatase. Mol. Pharmacol. *9*, 676–685 (1973c)

Tobin, T., Akera, T., Brody, T.M.: Studies on the two phosphoenzyme conformations of $Na^+ + K^+$-ATPase. Ann. N.Y. Acad. Sci. *242*, 120–132 (1974a)

Tobin, T., Akera, T., Lee, C.Y., Brody, T.M.: Ouabain binding to $(Na^+ + K^+)$-ATPase: Effects of nucleotide analogues and ethacrynic acid. Biochim. Biophys. Acta (Amst.) *345*, 102–117 (1974b)

Tuttle, R.S., Witt, P.N., Farah, A.: Therapeutic and toxic effects of ouabain on K^+ fluxes in rabbit atria. J. Pharmacol. Exp. Ther. *137*, 24–30 (1962)

Vick, R.L., Kahn, J.B.,Jr.: The effects of ouabain and veratridine on potassium movement in the isolated guinea pig heart. J. Pharmacol. Exp. Ther. *121*, 389–401 (1957)

Wallick, E.T., Schwartz, A.: Thermodynamics of the rate of binding of ouabain to the sodium, potassium adenosine triphosphatase. J. Biol. Chem. *249*, 5141–5147 (1974)

Wallick, E.T., Dowd, F., Allen, J.C., Schwartz, A.: The nature of the transport adenosine triphosphatase-digitalis complex. VII. Characteristics of ouabagenin-Na^+, K^+-adenosine triphosphatase interaction. J. Pharmacol. Exp. Ther. *189*, 434–444 (1974)

Wallick, E.T., Lindenmayer, G.E., Lane, L.K., Allen, J.C., Pitts, B.J.R., Schwartz, A.: Recent advances in cardiac glycoside – Na^+, K^+-ATPase interaction. Fed. Proc. *36*, 2214–2218 (1977)

Weaver, L.C., Akera, T., Brody, T.M.: Digitalis toxicity: Lack of marked effect on brain Na^+, K^+-adenosine triphosphatase in the cat. J. Pharmacol. Exp. Ther. *200*, 638–646 (1977)

Whittam, R., Wheeler, K.P., Blake, A.: Oligomycin and active transport reactions in cell membranes. Nature (Lond.) *203*, 720–724 (1964)

Whittam, R., Hallam, C., Wattam, D.G.: Observations on ouabain binding and membrane phosphorylation by the sodium pump. Proc. Roy. Soc. London B *193*, 217–234 (1976)

Wilbrandt, W.: Zum Wirkungsmechanismus der Herzglykoside. Schweiz. Med. Wochenschr. *85*, 315–320 (1955)

Willerson, J.T., Scales, F., Mukherjee, A., Platt, M., Templeton, G.H., Fink, G.S., Buja, L.M.: Abnormal myocardial fluid retention as an early manifestation of ischemic injury. Am. J. Pathol. *87*, 159–181 (1977)

Wilson, W.E., Sivitz, W.I., Hanna, L.T.: Inhibition of calf brain membranal sodium- and potassium-dependent adenosine triphosphatase by cardioactive sterols. A binding site model. Mol. Pharmacol. *6*, 449–459 (1970)

Yamamoto, S., Akera, T., Brody, T.M.: Sodium influx rate and ouabain-sensitive rubidium uptake in isolated guinea pig atria. Biochim. Biophys. Acta (Amst.) *555*, 270–284 (1979)

Yoda, A.: Structure-activity relationships of cardiotonic steroids for the inhibition of sodium- and potassium-dependent adenosine triphosphatase. I. Dissociation rate constants of various enzyme-cardiac glycoside complexes formed in the presence of magnesium and phosphate. Mol. Pharmacol. *9*, 51–60 (1973)

Yoda, A.: Association and dissociation rate constants of the complexes between various cardiac monoglycosides and Na, K-ATPase. Ann. N.Y. Acad. Sci. *242*, 598–616 (1974)

Yoda, A.: Binding of digoxigenin to sodium- and potassium-dependent adenosine triphosphatase. Mol. Pharmacol. *12*, 399–408 (1976)

Yoda, A., Hokin, L.E.: On the reversibility of binding of cardiotonic steroids to a partially purified (Na + K)-activated adenosinetriphosphatase from beef brain. Biochem. Biophys. Res. Commun. *40*, 880–886 (1970)

Yoda, A., Hokin, L.E.: Studies on the characterization of the sodium-potassium transport adenosine triphosphatase. VIII. Effects of ligands on fluorescence due to interaction of the enzyme with a fluorescent derivative of Hellebrigenin. Mol. Pharmacol. *8*, 30–40 (1972)

Yoda, A., Yoda, S.: Structure-activity relationships of cardiotonic steroids for the inhibition of sodium- and potassium-dependent adenosine triphosphatase. III. Dissociation rate constants of various enzyme-cardiac glycoside complexes formed in the presence of sodium, magnesium, and adenosine triphosphate. Mol. Pharmacol. *10*, 494–500 (1974a)

Yoda, A., Yoda, S.: Influence of certain ligands on the dissociation rate constants of cardiac glycoside complexes with sodium- and potassium-dependent adenosine triphosphatase. Mol. Pharmacol. *10*, 810–819 (1974b)

Yoda, A., Yoda, S.: Structure-activity relationships of cardiotonic steroids for the inhibition of sodium- and potassium-dependent adenosine triphosphatase. V. Dissociation rate constants of digitoxin acetates. Mol. Pharmacol. *11*, 653–662 (1975)

Yoda, A., Yoda, S.: Association and dissociation rate constants of the complexes between various cardiac aglycones and sodium- and potassium-dependent adenosine triphosphatase formed in the presence of magnesium and phosphate. Mol. Pharmacol. *13*, 352–361 (1977)

Yoda, A., Yoda, S.: Influence of pH on the interaction of cardiotonic steroids with sodium- and potassium-dependent adenosine triphosphatase. Mol. Pharmacol. *14*, 624–632 (1978)

Yoda, A., Yoda, S., Sarrif, A.M.: Structure-activity relationships of cardiotonic steroids for the inhibition of sodium- and potassium-dependent adenosine triphosphatase. II. Association rate constants of various enzyme-cardiac glycoside complexes. Mol. Pharmacol. *9*, 766–773 (1973)

Yoda, S., Sarrif, A.M., Yoda, A.: Structure-activity relationships of cardiotonic steroids for the inhibition of sodium- and potassium-dependent adenosine triphosphatase. IV. Dissociation rate constants for complexes of the enzyme with cardiac oligodigitoxides. Mol. Pharmacol. *11*, 647–652 (1975)

Zavecz, J.H., Dutta, S.: The relationship between Na^+, K^+-ATPase inhibition and cardiac glycoside-induced arrhythmia in dogs. Naunyn-Schmiedebergs Arch. Pharmacol. *297*, 91–98 (1977)

CHAPTER 15

Influence of Cardiac Glycosides on their Receptor

E. ERDMANN

A. Introduction

Although the pharmacologic effects of cardiac glycosides are well known today, their mechanism of action is still subject to speculation. It is, however, generally accepted that their effects are directly on the cardiac cell (for reviews see LEE and KLAUS, 1971; LÜLLMANN and PETERS, 1979). A great variety of cellular and subcellular systems has been studied as putative primary points of interaction such as: polymerization of cardiac actin (HORVATH et al., 1949), myosin (OLSON et al., 1961), myosin-ATPase (JACOBSEN, 1968), contractile properties of actomyosin (WASER and VOLKART, 1954), sarcoplasmic reticulum (DUTTA et al., 1968), and several others (LEE and KLAUS, 1971).

The only system found to be affected reproducibly and by low concentrations ($<10^{-7}M$) of this group of drugs is Na^+, K^+-ATPase (EC 3.6.1.3). This enzyme system is supposed to represent the biochemical basis for the active Na^+ and K^+ transport of the cell membrane (SKOU, 1957, 1965; GLYNN, 1964; SCHONER, 1971; SCHWARTZ et al., 1975). Specific inhibition of this enzyme system as well as sodium and potassium transport (SCHATZMANN, 1953) by cardiac glycosides has been reported by numerous investigators (for references see SKOU, 1965, 1973; DUNHAM and GUNN, 1972; DAHL and HOKIN, 1974; GLYNN and KARLISH, 1975; AKERA and BRODY, 1978; see also Chap. 14), so much so, it has been proposed as the "receptor enzyme" for cardiac glycosides (REPKE and PORTIUS, 1963).

Experimentally, cardiac glycosides inhibit the active transport of sodium and potassium across erythrocyte membranes only when present on the outside (HOFFMANN, 1966; PERRONE and BLOSTEIN, 1973). The same seems to apply to the squid giant axon (CALDWELL and KEYNES, 1959) and cardiac cells (OKARMA et al., 1972). Hence, the drug molecule must interact with the cell membrane on the outside and thus assert its pharmacologic effects (among others, the increase in force of contraction).

Binding sites for cardiac glycosides on isolated cell membranes or isolated heart preparations have been measured using radioactively labeled digoxin (MATSUI and SCHWARTZ, 1967, 1968; SCHWARTZ et al., 1968; BELLER et al., 1975) or ouabain (DUTTA and MARKS, 1969; ALLEN and SCHWARTZ, 1970; ALLEN et al., 1970; WHITTAM and CHIPPERFIELD, 1973; ERDMANN and SCHONER, 1973a). In fact, the first specific pharmacological radioligand binding experiment was carried out by using ^3H-digoxin reported by MATSUI and SCHWARTZ in 1967 and 1968. Other radioligand drug studies such as β-adrenoceptor agonists and antagonists and opiates were introduced later. Such binding sites for cardiac glycosides could also be demonstrated in a variety of intact cells or tissue (BAKER and WILLIS, 1972; CLAUSEN and HANSEN, 1974, 1977; see also this Handbook, Vol. 56/II, Chap. 7). These membrane-bound binding sites closely associated with Na^+, K^+-ATPase activity have been postulated as pharmacologic receptors for cardiac glycosides (GODFRAIND, 1975; SCHWARTZ, 1976a, b; AKERA, 1977).

This chapter discusses some of the known properties of these glycoside-binding sites and their interaction with that class of drugs whilst reviewing the experimental evidence for the existence of a specific receptor for cardiac glycosides. Since a number of excellent reviews exist on Na^+, K^+-ATPase and ouabain binding to this enzyme (ASKARI, 1974; SCHWARTZ et al., 1975; WHITTAM and CHIPPERFIELD, 1975; GLYNN and KARLISH, 1975; NAKAO, 1975; AKERA and BRODY, 1978; WALLICK et al., 1979) this chapter is restricted to some aspects supporting the above concept (for further information see Chap. 14, AKERA, 1977).

I. Definition of Receptors

Specific binding of a drug to a biologically occurring molecule does not define a receptor completely, not even, when this binding is of high affinity. There are several proteins for instance in plasma that bind drugs or hormones tightly and very specifically and yet may not be called receptors. When EHRLICH (1913) wrote: "corpora non agunt nisi fixata," he evidently meant that structurally specific chemical substances, before initiating their effects, must be bound to certain macromolecules of the target tissue. Thus, a physicochemical interaction of drugs at definite sites constitutes only the first part of a "receptor" as it is a locus of recognition and discrimination. Secondly, the information of the drug or hormone has to be transmitted (for instance, from the outside of the plasma membrane into the cell) and a response must be produced (the biologic effect). This may be expressed in a simple reaction sequence (SMYTHIES and BRADLEY, 1978)

$$D+R \rightleftharpoons DR \xrightarrow[\text{steps}]{\text{intermediate}} \text{Effect},$$

where D is the drug molecule, R the receptor, and DR the drug–receptor complex. Of course there could be several intermediate steps (for instance elevation of cyclic AMP, increased Ca^{2+}, etc.) before such a response is elicited.

If the receptor is part of the cell membrane, a great number of environmental factors (membrane fluidity, ions etc.) may influence the drug–receptor interaction and hence the effect. Usually a membrane-bound receptor is in some way coupled to an enzyme, the stimulation or inhibition of which will then pass on the information into the interior of the cell (GOLDSTEIN et al., 1974; ROBERTS, 1977; STRAUB and BOLIS, 1978; YAMAMURA et al., 1978). Thus, a pharmacologic receptor for drugs and hormones may be defined as a macromolecule (usually a protein), which with high affinity and selectivity binds certain drugs or hormones eliciting (for instance by a conformational change of the macromolecule) their pharmacologic effects.

Such receptors have been demonstrated to exist for several hormones (insulin, glucagon, acetylcholine, β-adrenergic agents, oestrogen, progesterone, neurotransmitters, etc. for references see BLECHER, 1976; SMYTHIES and BRADLEY, 1978) and drugs (opiates, β-adrenergic antagonists, dopamine, bromocryptine, etc., for references see HARPER and SIMMONDS, 1977; STRAUB and BOLIS, 1978). Recently CHANG and SNYDER (1978) demonstrated that half-maximal receptor binding of benzodiazepines corresponded directly to the half-maximal effects of these drugs in intact animals. Similar experimental evidence of a direct correlation between receptor binding of cardiac glycosides and their effect remains to be produced. However, there exists a great number of valid arguments for this hypothesis (ALLEN et al., 1971; KU et al., 1974, 1975; AKERA et al., 1976; SCHWARTZ et al., 1975; BRODY and AKERA, 1977); these are discussed later.

II. Binding Sites for Cardiac Glycosides

Binding sites for cardiac glycosides should fulfill certain criteria as that class of drugs exerts its unique action at very low serum concentrations: they should have, besides saturability and reversibility of binding, a rather high affinity and selectivity for the drug. In fact, this has been demonstrated (for references see SCHWARTZ et al., 1975).

Specific and Nonspecific Binding of Cardiac Glycosides

With the easy availability of radioactively labeled compounds the number of published "binding experiments" with various substances has increased greatly. Quite often the specificity of such ligand–receptor binding is, however, not properly defined. Most ligands are bound by a great variety of subcellular particles, by the laboratory glassware, by glass fiber filters etc. (CUATRECASAS and HOLLENBERG, 1975). For reasons of convenience the rapid filtration method is usually employed in binding experiments to separate unbound from receptor-bound drug. With this procedure the tissue components (cell membranes etc.) containing the bound radioactivity are kept on the filters and are subject to more or less intensive washing with buffered solution. During this washing equilibrium conditions no longer exist. Hence the results from the rapid filtration method need to be checked by careful control experiments performed by different means (ultrafiltration, centrifugation, equilibrium dialysis etc.). Even if pitfalls in binding experiments can be avoided it remains a problem to define "specific" binding, i.e., drug–receptor binding. In an attempt to give a clear experimental definition the following experiment was performed (ERDMANN, 1978b) *(Fig. 1)*:

Isolated human cardiac cell membranes were incubated with a buffered medium known to support ^3H-ouabain binding to the cell membranes and 4 nM ^3H-ouabain (=40,000 cpm). After various incubation times the reaction was interrupted by rapid centrifugation (30 min, 80,000 g, 0 °C). The radioactivity in the sedimented membranes measured as total bound ^3H-ouabain was maximally 17,000 cpm. For an absolute measure of nonspecific ^3H-ouabain binding they were heated for 60 min at 60 °C. At that time there was no measurable ATPase activity. The ^3H-ouabain bound to this heat-denatured tissue preparation under identical conditions amounted to merely 250 cpm. The same low radioactivity level was found when the experiment was performed with native membranes in the presence of 10^{-4} M unlabeled ouabain or 150 mM NaCl, (the latter of which completely de-

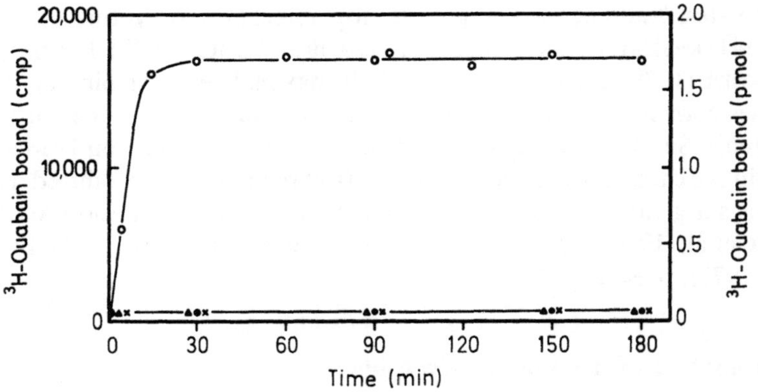

Fig. 1. Specific and nonspecific binding of ^3H-ouabain to human cardiac cell membranes. 0.3 mg membrane protein incubated in 50 m*M* imidazole/HCl buffer, pH 7.25, 3 m*M* $MgCl_2$, 3 m*M* imidazole/PO_4, and 4 n*M* ^{3}H-ouabain at 37 °C. Reaction interrupted by rapid centrifugation (80,000 *g*) every 30 min and radioactivity determined *(open circles* total radioactivity in pellet; *(solid circles)* with the addition of 0.1 m*M* unlabeled ouabain; *(crosses)* after heating membranes for 60 min at 60 °C before incubation; *(open triangles)* with the addition of 150 m*M* NaCl. (ERDMANN, 1978 b)

creases the affinity of the receptor for the drug in the presence of magnesium and phosphate without ATP).

Thus, three methods to measure unspecific ouabain–receptor binding agree: destruction of membrane proteins, occupation of specific binding sites by the unlabeled drug, and maximal reduction of affinity (Fig. 1). Ouabain is in fact almost exclusively bound by specific binding sites (~97%) in isolated cell membranes and even in a cardiac homogenate (~80%) (ERDMANN, 1977; MICHAEL et al., 1979).

The more lipophilic cardiac glycosides are bound to a greater extent by membrane components unrelated to cardiac glycoside receptors and they may penetrate intact cells, which ouabain reportedly is not supposed to do (for references see LÜLLMANN and PETERS, 1979). This is certainly one of the reasons why ouabain is usually used for such binding experiments. Besides this, ouabain (strophanthin G) is quite convenient to handle in the laboratory as its solubility in water is sufficient to make 10 m*M* solutions.

The addition of large amounts of unlabeled ligands to the incubation medium is often used to measure nonspecific binding. This is, however, by no means sufficient to define specific binding as some drugs or hormones do inhibit or displace even nonspecific binding to cell membranes, too, if present in high concentrations (KRAWIETZ and ERDMANN, 1979). Nonspecifically bound drugs (that is drugs bound to non-receptor material of cells or cell membranes) may in fact be displaced from their binding sites easily if competing drugs are present in very high concentrations. Therefore, the term "specific binding" should be used with caution.

B. Quantitative Aspects of Cardiac Glycoside–Receptor Interaction

Having established the specificity of ouabain binding to cell membranes one would like to know the maximal number of such binding sites, their tissue distribution,

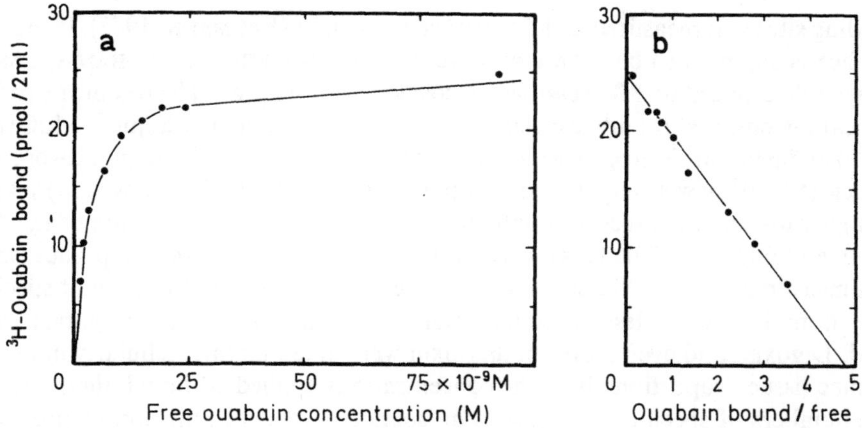

Fig. 2a, b. Quantitation of ouabain-binding sites in human heart at equilibrium. 2.3 mg membrane protein (Na^+, K^+-ATPase activity = 0.08 U/mg protein) incubated in 50 m*M* imidazole/HCl buffer pH 7.25, 3 m*M* $MgCl_2$, 3 m*M* imidazole/PO_4, and increasing amounts of ^{3}H-ouabain (lowest: 4 n*M* ^{3}H-ouabain) for 120 min at 37 °C. The experimental data **a** are plotted **b** according to SCATCHARD (1949). The dissociation constant (K_D) of the ouabain–receptor-complex was calculated as 2.5 n*M*. (ERDMANN, 1978b)

and their relation to Na^+, K^+-ATPase activity or to Na^+ and K^+ transport capacity. Ouabain serves as a tool to label and characterize a certain membrane component (CHIPPERFIELD and WHITTAM, 1973; TITUS, 1975; HANSEN, 1978 a, b, c).

I. Quantitation of Ouabain-Binding Sites

When purified cell membrane preparations are incubated with increasing amounts of ^3H-ouabain until steady-state conditions are established, the amount of specifically membrane-bound ouabain increases up to an equilibrium (Fig. 2). Thus, the maximal number of binding sites in a cell membrane preparation can be determined quite easily. If the experimental data are plotted according to SCATCHARD (1949), the dissociation constant (K_D) calculated from the slope of the plot gives an important value for the drug–receptor interaction. In human heart K_D is about 2.5 n*M*, i.e., at a free ouabain concentration of 2.5 n*M* 50% of the receptors are occupied by a drug molecule under the given optimal binding conditions (ERDMANN and HASSE, 1975; ERDMANN et al., 1976 a).

The ouabain concentration in human serum 2 h after the intravenous administration of 0.5 mg is about 0.8 ng/ml ($\sim 10^{-9}$ *M*) (GREEFF, 1974; GREEFF et al., 1975). Assuming steady state and nearly optimal drug–receptor binding conditions, some 25% of all cardiac glycoside receptors could then maximally be occupied. Probably it is much less, as the ionic environment and other factors in vivo tend to decrease the receptor affinity (HANSEN and SKOU, 1973; WHITTAM and CHIPPERFIELD, 1973; ERDMANN and SCHONER, 1973 b; HANSEN, 1976; AKERA et al., 1978; YODA and YODA, 1978).

In a similar equilibrium binding experiment with a crude homogenate of human cardiac tissue (left ventricle), the maximal number of specific ^3H-ouabain-binding sites was measured as 1.5×10^{14}/g wet weight (ERDMANN, 1977). If the cell surface is assumed to be 1,000 cm^2/g wet weight (WINEGRAD and SHANES, 1962), this would amount to 1,000 cardiac glycoside receptors/μm^2. The receptor density in beating, cultured rat myocardial cells has been measured as 720/μm^2 (MCCALL, 1979) and 656/μm^2 in frog skeletal muscle (MANERY et al., 1977). In guinea-pig left auricle (5×10^{14} g wet weight) and in papillary muscle (8×10^{14} g wet weight) slightly higher maximal numbers of ouabain-binding sites have been determined (LÜLLMANN and PETERS, 1979). This is certainly due to the experimental setup since what was measured was the "tissue uptake" of the drug rather than the amount specifically bound. Nevertheless, these data were in the same range when ouabain was used. Digoxin, and even more so, digitoxin were taken up intracellularly in intact cardiac tissue proportionally to the concentrations applied, although their respective numbers of specific binding sites in isolated cardiac cell membrane preparations equalled that of ouabain (ERDMANN et al., 1977; BOSSALLER and SCHMOLDT, 1979). Thus, the known accumulation of digoxin for instance in the human heart has posed some problems of interpretation. According to several reports the myocardial digoxin concentration after therapeutic doses may be accepted as about 60–80 ng/g wet weight (COLTART et al., 1974; GÜLLNER et al., 1974; THOMPSON et al., 1974; DOHERTY and PERKINS, 1967; HÄRTEL et al., 1976; MALCOLM and COLTART, 1977; HAASIS et al., 1977; LICHEY et al., 1978). As such there are about 0.4–0.5×10^{14} molecules of digoxin per g wet weight containing about 1.5×10^{14} specific cardiac glycoside receptors (ERDMANN, 1977). These data only permit speculation on the amount of receptor-bound fraction, as long as the in vivo affinity of the receptor for the drug is not known. However, it transpires that less than 30% of all cardiac glycoside receptors are occupied by a drug molecule under therapeutic conditions. This agrees with some theoretical reflections based on the supposed transport rate of Na^+ in the presence of ouabain (AKERA et al., 1976).

Unfortunately it is not possible to measure the receptor-bound ouabain molecules in the heart. This can be accomplished in erythrocytes, however, which have a well-defined surface area (see Sect. B.III).

II. Correlation of Ouabain Binding and Na^+, K^+-ATPase Inhibition

The positive inotropic effect of cardiac glycosides on heart muscle cells in vitro can be correlated temporally to the binding of ouabain to Na^+, K^+-activated ATPase and the opposite seems to be true for the dissociation of the ouabain–enzyme complex (AKERA et al., 1973). Because of these experiments an inhibition of Na^+, K^+-ATPase by ouabain has been accepted as necessary for the pharmacologic effects by most investigators (for references see SCHWARTZ et al., 1975). In fact, the simultaneous determination of ^3H-ouabain binding and Na^+, K^+-ATPase inhibition to isolated cell membrane preparations has revealed a coincidence of both effects when plotted against time as shown in Fig. 3 (ERDMANN and SCHONER, 1974a).

The extent of ouabain bound to the cardiac glycoside receptor correlates very well with the extent of inhibition of Na^+, K^+-ATPase activity (Fig. 4), in dog kidney (ALLEN et al., 1971), in bovine brain and heart as well as in human heart preparations (ERDMANN and SCHONER, 1974a, b; ERDMANN et al., 1976a, b). Such ex-

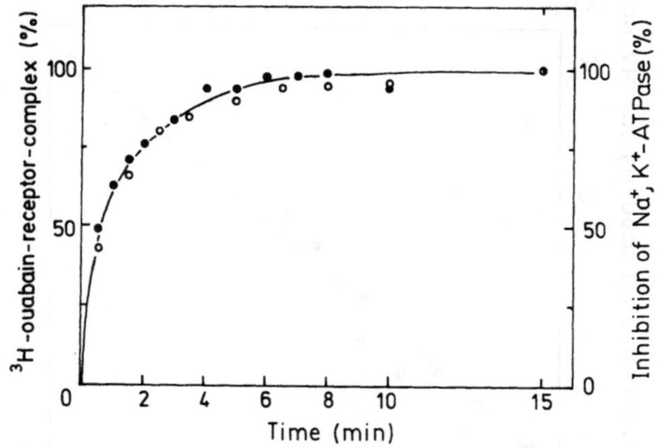

Fig. 3. Time course of ouabain binding *(solid circles)* and onset of Na^+, K^+-ATPase inhibition *(open circles)*. 0.7 mg enzyme protein (ox brain Na^+, K^+-ATPase activity = 3.0 U/mg) were incubated at 25 °C in 50 m*M* imidazole/HCl pH 7.25, 3 m*M* $MgCl_2$, 3 m*M* Na_2-ATP, 100 m*M* NaCl and 10^{-6} *M* ^{3}H-ouabain, total volume 2 ml. At the indicated time reaction was stopped either by rapid cooling in liquid air and successive centrifugation (30 min, 80,000 *g*) or by dilution of 0.1 ml aliquot in 1.5 ml of the reaction mixture of the coupled optical assay for Na^+, K^+-ATPase. Na^+, K^+-ATPase activity calculated from recorder reading of the first 2 min after start of Na^+, K^+-ATPase assay. (ERDMANN and SCHONER, 1974a)

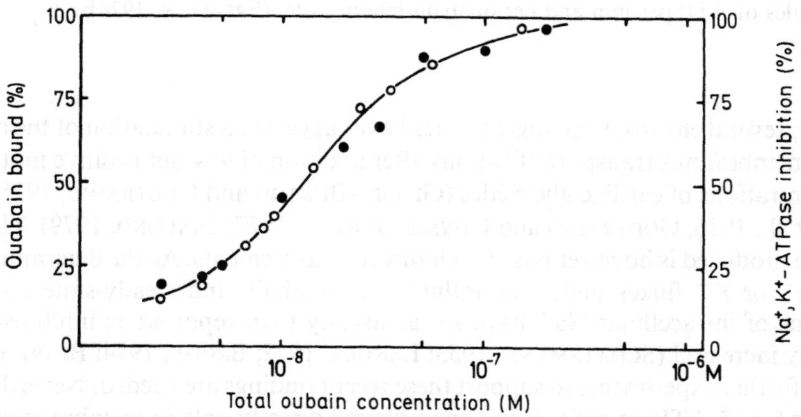

Fig. 4. Correlation between the occupation of ouabain receptor sites *(open circles)* and the inhibition of Na^+, K^+-ATPase *(solid circles)* in human cardiac cell membranes. Ouabain binding and Na^+, K^+-ATPase activity were determined after 2 h preincubation under the conditions of Fig. 2. (ERDMANN et al., 1976b)

periments have to be carried out under equilibrium conditions with identical receptor concentrations in both assays (ERDMANN and SCHONER, 1974a), otherwise slight inaccuracies will result.

Thus, several experiments (for references see SCHWARTZ et al., 1975) repeatedly confirmed an inhibition of Na^+, K^+-ATPase activity following the binding of cardiac glycosides (GODFRAIND and GHYSEL-BURTON, 1978). The rather inconsistent and irreproducible activation of this enzyme by low concentrations of these drugs seems to be an artifact (LEE and YU, 1963; REPKE, 1964; PETERS et al., 1974). Re-

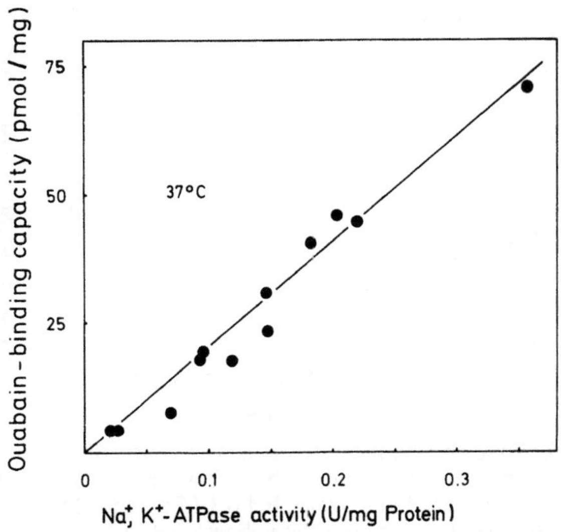

Fig. 5. Ouabain-binding capacity of different cardiac cell membrane preparations determined by equilibrium binding experiments as shown in Fig. 2 Na^+, K^+-ATPase activity determined according to SCHONER et al. (1977a, b). Figure demonstrates a linear relationship between number of ouabain receptors and enzyme activity. There are about 180×10^{-12} mol ouabain-binding sites per unit of Na^+, K^+-ATPase activity. This gives a turnover of 5,560 molecules of ATP per min and per ouabain-binding site. (ERDMANN, 1978b)

cently, several electrophysiologic studies have suggested a stimulation of the active transmembranous transport of cations after addition of low but positive inotropic concentrations of cardiac glycosides (GHYSEL-BURTON and GODFRAIND, 1975; COHEN et al., 1976; GODFRAIND and GHYSEL-BURTON, 1977; DIACONO, 1979). The evidence produced is however based on indirect measurements. As the determination of Na^+ or K^+-fluxes under the influence of ouabain and steady-state concentrations of intracellular Na^+ have so far usually been reported as inhibited and slightly increased (SCHATZMANN, 1953; LANGER, 1972; BRODY, 1974; BEARD et al., 1975) further experiments to support these recent findings are needed. Nevertheless the Na^+, K^+-ATPase activity can be measured directly only in membrane preparations and not in intact tissue. However, patients recieving cardiac glycosides in therapeutic concentrations have been found to have increased sodium concentrations in their erythrocytes (BURCK et al., 1975).

If the ouabain-binding site and the enzymatically active site are both located on one polypeptide chain of the enzyme (KYTE, 1972a, b, 1975; RUOHO and KYTE, 1974), there should be a constant relationship between the ouabain-binding capacity of the membranes and the Na^+, K^+-ATPase activity. This has been found to be true (ERDMANN and SCHONER, 1973a) (Fig. 5). It allows the calculation of the maximal turnover of the enzyme for ATP. In fact, values between 3,600 min^{-1} (KYTE, 1972a, b), 6,480 min^{-1} (LANE et al., 1973), and 8,550 min^{-1} (ERDMANN and SCHONER, 1973a) have been measured in different organs and species.

The above concept of a single protein extending the binding site for cardiac glycosides to the outside of the membrane and the enzymatically active site towards the inside has been confirmed by the finding of a constant one-to-one relationship between ouabain-binding capacity and nucleotide-binding capacity in cell membranes (HANSEN, 1971; ERDMANN et al., 1976b). Thus, the quantitation of specific ^{3}H-ouabain-binding sites in cell membranes rather seems to be a quantitation of Na^{+}, K^{+}-ATPase molecules (or rather parts thereof, ALEXANDER, 1974) or of "pump sites." These findings of course support REPKE's view of the Na^{+}, K^{+}-ATPase as the receptor enzyme for cardiac glycosides (REPKE and PORTIUS, 1963).

III. Quantitation of Ouabain-Binding Sites in Erythrocytes

SCHATZMANN (1953) showed that cardiac glycosides inhibit the active transport of Na^{+} and K^{+} across the human erythrocyte membrane. This inhibition is rather specific for cardioactive steroids and other lactones (KAHN and ACHESON, 1955) and is the result of a glycoside–receptor interaction at the outside of the membrane (GARDNER et al., 1973).

The specific ouabain or digoxin binding to this receptor in erythrocyte membranes involves a single class of binding sites; it is saturable and is a function of the extracellular glycoside concentration. It is reversible, temperature sensitive, dependent on the cation composition of the medium, and is inhibited in a competitive manner by other cardiac glycosides (GARDNER and CONLON, 1972; GARDNER and KIINO, 1973; JULIANO, 1973; GARDNER and FRANTZ, 1974; LAUF and JOINER, 1976). The number of these specific receptors in different reports ranges from 200 (DUNHAM and HOFFMANN, 1970), 650 (BAKER and WILLIS, 1972), to 1,200 (GARDNER and FRANTZ, 1974) per human red cell. In human erythrocyte membranes in which intracellular uptake of ^{3}H-ouabain could be completely excluded, the saturable binding at 37 °C under optimal binding conditions revealed some 228 ± 28 receptors per single red cell. The dissociation constant of the ouabain–receptor complex was determined as 2.8 nM (ERDMANN and HASSE, 1975). It is about the same as in human cardiac tissue. Neither the maximal number of ouabain receptors nor the affinity for the drug changed with the age (18–88 years) or sex of the blood donors.

LAUF and JOINER (1976) measured 450–500 receptors in intact human red cells. Under their experimental conditions with large concentrations of ^{3}H-ouabain they did not achieve true saturation kinetics, however, (personal communication). In this respect it has to be emphasized that nonspecific ^{3}H-ouabain binding to intact red cells may be as high as 30% or even more, if high concentrations of ^{3}H-ouabain are used. When using 2–4 nM ^{3}H-ouabain we repeatedly determined about 260 receptors in intact erythrocytes as shown in Fig. 6 (ERDMANN et al., 1979a).

These experimental results lead to some interesting reflections. The ouabain receptors, which are supposed to be located on the same polypeptide chain as the Na^{+}, K^{+}-ATPase (RUOHO and KYTE, 1974) and as such being a measure of the sodium pump sites (SACHS et al., 1974) might be diffusely distributed on the cell surface. Then one receptor site – equal to one sodium pumping site – corresponds to about 1–2 μm^{2} of the membrane. As we have seen, there are considerably more ouabain receptors in human myocardium (1,000/μm^{2}). Recently, MICHAEL et al. (1979) have reported interesting qualitative ^{3}H-ouabain binding studies in cat ventricular muscle. They have measured the biologically active number of cardiac glycoside receptors as 760/μm^{2} and about 5.2×10^{6} sites per cell. Thus, ouabain as a

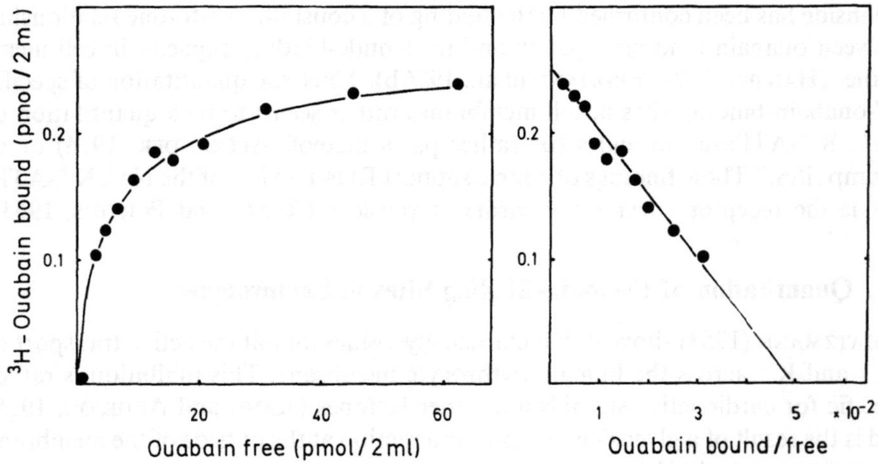

Fig. 6. ^{3}H-ouabain binding to human erythrocytes. Washed human red cells incubated for 300 min at 37 °C in 1 mM HEPES buffer pH 7.4, 3 mM $MgCl_2$, 150 mM NaCl, 3 mM ATP, 5 mM glucose, and increasing amounts of ^{3}H-ouabain. Radioactivity on the filters was determined after rapid filtration. Nonspecific ouabain binding was measured as in Fig. 1. *Solid circles* indicate results from 5.5×10^8 erythrocytes of control subject. One red cell contained 270 ^3H-ouabain-binding sites. $K_D = 2.5$ nM. (ERDMANN et al., 1979 a)

monolayer would cover only 0.08% of the membrane surface. Although, these measurements were carried out by completely different methods, MCCALL (1979) confirmed these data in cultured rat heart cells finding 720 ^3H-ouabain binding sites per μm^2. According to CUATRECASAS (1973) a single normal fat cell of a rat contains about 10 insulin receptors per μm^2 and 10^5 concanavalin-A-binding sites per μm^2. There are 10^5–10^6 cytochalasin-B high affinity binding sites per human red cell (LIN et al., 1978). The quantitation and determination of different hormone- or drug-binding sites on the surface of the cell (erythrocyte) membrane and the subsequent analysis of the binding structures might lead to a “mapping” of the membrane.

Na^+, K^+-ATPase activity is directly related to the quantity of nucleotide- and ouabain-binding sites per mg of membrane protein (KANIIKE et al., 1974; ERDMANN and SCHONER, 1973 c). There are about 180×10^{-12} mol ouabain-binding sites per Na^+, K^+-ATPase unit of activity in human cardiac membranes (see Sect. B.II). The ouabain-binding capacity of erythrocyte membranes being of the same order (about 135×10^{-12} mol per enzyme unit) indicates a turnover of about 8,000 molecules ATP hydrolyzed per minute and per enzymatic site. This value agrees well with previous data obtained for different tissues and species (LANE et al., 1973; ERDMANN and SCHONER, 1973 a).

If the sodium pump of the erythrocyte membrane moves sodium and potassium ions with a stoichiometry of $3Na^+:2K^+$ per molecule of ATP hydrolyzed (SEN and POST, 1964; SCHONER, 1971; SKOU, 1973), some 5.5×10^6 molecules sodium per erythrocyte could maximally be pumped out actively per minute or about 5–6 mequiv. $l^{-1}h^{-1}$. According to SOLOMON (1952), human erythrocytes have to pump out 3.1 equiv. l^{-1} h^{-1} to maintain the measured intra–extracellular cation difference.

Another reflection involves the identical dissociation constant of the ouabain–receptor complex in human myocardium and in human erythrocyte membranes. The same result was obtained for several other species (ERDMANN and SCHONER, 1973a). The membrane-bound receptor, having the same affinity for the drug in both tissues, might be composed of the same protein structures in some species. This is underlined by the fact that there seems to be only one class of binding sites according to our data (ERDMANN and HASSE, 1975) and those of others (GARDNER and KIINO, 1973). Recently, however, HELLER and BECK (1978) claim to have produced evidence for the existence of two types of binding sites for ouabain in human red cells. One receptor site designated as "type I" was detected at ouabain concentrations as low as $10^{-9} M$, but could not be saturated even at $10^{-3} M$. As the "type I ouabain" complex only approached equilibrium over a period of 24 h, it may consist of ouabain binding to some sites not associated with the Na^+, K^+-ATPase. All the evidence shows that binding to the saturable ouabain-binding sites is directly correlated with inhibition of Na^+ and K^+ transport inhibition, in these experiments only one class of receptors is found (GARDNER et al., 1973; GARDNER and KIINO, 1973; SACHS, 1974; see however Sect. B. V).

It is quite interesting that the number of ouabain-binding sites in erythrocytes of lambs developing low intracellular K^+ decreases within the first 40 days of life from 135 per red cell to about 40–50 per red cell along with the decreasing K^+ concentration (LAUF et al., 1978; JOINER and LAUF, 1978). In these animals the maturation of the cells involves the inactivation of most of the Na^+, K^+-pump activity (DUNHAM and BLOSTEIN, 1976). Such quantitative changes of the ouabain-binding sites have been found in human red cells in chronic hypokalemia (ERDMANN and KRAWIETZ, 1977; ERDMANN et al., 1979a), which is known to lead to increased Na^+, K^+-ATPase activity, too (CHAN and SANSLONE, 1969). Apparently the number of Na^+, K^+-ATPase molecules (measured by the ouabain-binding sites) is subject to regulative changes. Recently, increases in cardiac glycoside receptors of red cells were found in thalassemia minor (ERDMANN et al., 1979a), where cell membrane defects have been reported leading to increased Na^+, K^+-pump activity possibly due to leaky membranes (NATHAN and GUNN, 1966). Thus the quantitation of the maximal number of ouabain-binding sites provides a tool for probing the cell membrane composition.

In this respect, successful attempts have been made to label the digitalis-binding component of plasma membranes covalently with oxidized ^3H-ouabain (HEGYVARY, 1975). The successive identification of the labeled membrane protein revealed a single protein with an apparent molecular weight of 89,000 daltons as estimated by polyacrylamide gel electrophoresis, and could be identified as part of Na^+, K^+-ATPase in sheep kidney membranes.

Anthroylouabain, which has been used as a specific fluorescent probe for the cardiac glycoside receptor, allows the exact quantitation and possibly the location of receptors in the membrane (FORTES, 1977). Its application in rabbit kidney membranes has suggested that it may be useful for the study of orientation and mobility of the receptor molecule within the membrane.

The use of antibodies directed separately against the ouabain-binding site or the catalytic site of the enzyme has also proved a useful tool to support the above concept (MCCANS et al., 1975).

IV. Kinetics of Specific Ouabain–Receptor Binding

According to HANSEN (1971) the ouabain–receptor interaction follows the law of mass action

$$\mathrm{Ou}+\mathrm{R}\underset{k_{-1}}{\overset{k_{+1}}{\rightleftharpoons}}\mathrm{Ou\,R}, \quad (1)$$

where Ou is ouabain, R the receptor, OuR the ouabain–receptor complex, k_{+1} the association rate constant and k_{-1} the dissociation rate constant. Further one may accept

$$K_D=\frac{k_{-1}}{k_{+1}} \quad (2)$$

as the dissociation constant K_D calculated from the respective rate constants. These simple equations have been supported by experimental data in several cell membrane preparations from different species (BARNETT, 1970; LINDENMAYER et al., 1974; ERDMANN and SCHONER, 1973a; ERDMANN and HASSE, 1975). K_D as determined from equilibrium experiments and from the initial association and dissociation rates (Eq. 2) agrees well in (Mg^{2+}, Pi) or (Mg^{2+}, ATP, Na^+) incubation media for several species and organs (see Table 1). In the absence of Pi or at low Mg^{2+} concentrations the kinetics seem to be different (see Sect. B. V). According to Eq. 1 the binding process follows second-order kinetics while the dissociation is a unimolecular reaction. The association rate constant k_{+1}, can be calculated, if the initial receptor concentration b and the initial ouabain concentration a are known according to the second-order equation (Fig. 7)

$$k_{+1}=\frac{1}{(a-b)\,t}\times \ln\frac{b\,(a-x)}{a\,(b-x)}, \quad (3)$$

where x represents the amount of ouabain bound to the receptor in time t. The association rate according to this equation can be measured if a and b are in the same concentration range but $a \neq b$. If a is much larger than b, the reaction velocity is extremely rapid. This therefore has been termed "pseudo-first-order kinetics" (YODA et al., 1973; LINDENMAYER et al., 1974). YODA (1976) reported that aglycone binding, such as the binding of ^3H-digoxigenin to the receptor apparently does not follow this simple equation, although the maximal number of digoxigenin sites agreed well with the number of ouabain sites. In these experiments, however, the total number of binding sites was not constant but was dependent on the ligand concentrations posing some experimental difficulties. Later YODA and YODA (1977) used a different method to obtain results agreeing with the simple equation.

The association rates of the formation of the ouabain–receptor complexes of several species and tissues have been measured (Table 1). They are basically all in the same range, even that of the rat heart (1.1–$5.9\times 10^4\,M^{-1}\,s^{-1}$). The dissociation rates, however, vary considerably (0.6×10^{-2}–$6.6\times 10^{-5}\,s^{-1}$) (Table 1). Apparently the sensitivity to ouabain of the respective species is determined mainly by the dissociation rate constant (TOBIN et al., 1972). Saturation of the lactone ring of the cardiac glycoside molecule results in an increased dissociation rate, too, and thus in a decrease in binding affinity (CLARK et al., 1975). The same seems to be valid for aglycones (digitoxigenin, digoxigenin, strophanthidin, and ouabagenin) in

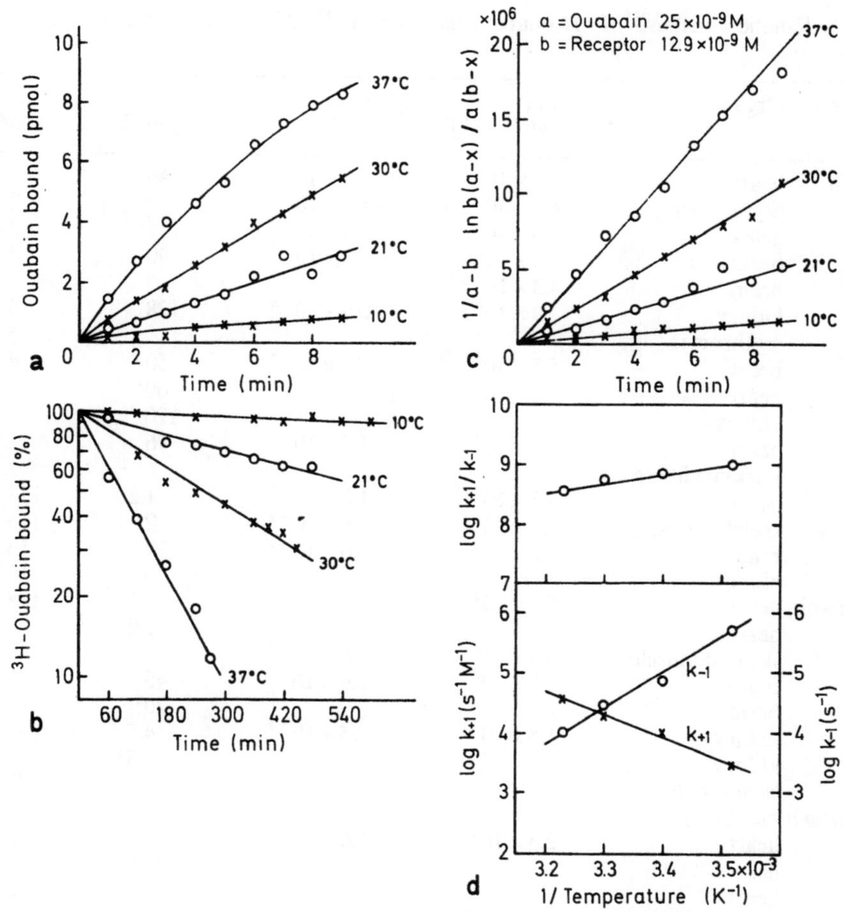

Fig. 7a–d. Kinetic analysis of ouabain binding to Na^+, K^+-ATPase from beef heart. Details of the determination of kinetic data are described elsewhere. Thermodynamic data for the enzyme-ouabain complex are tabulated below.

Temperature (°C)	K_D (*M*)
10	0.06×10^{-8}
21	0.14×10^{-8}
30	0.18×10^{-8}
37	0.28×10^{-8}

$\Delta G° = -11$ kcal/mol; $\Delta H° = -5$ kcal/mol; $\Delta S° = 21$ col °C^{-1} mol^{-1} (ERDMANN and SCHONER, 1973a)

which the dissociation rates were all the same, the association rates not following simple second-order kinetics (YODA and YODA, 1977). The sugar moiety seems to be very important for the binding process (YODA et al., 1975). The number of binding sites for aglycones equals that for ouabain. Scatchard plots of the aglycone–receptor interaction determined from equilibrium binding are linear despite a complex association reaction (see Sect. B. V; YODA and YODA, 1977).

Table 1. Kinetic constants of ^3H-ouabain-receptor binding in various organs of different species

Species /	Organ	k_{+1} $[M^{-1} s^{-1}]$	k_{-1} $[s^{-1}]$	$T/_2$ [min]	K_D [nM]
Human	heart	1.2×10^5	3.0×10^{-4}	40	2.5
	erythrocyte	4.6×10^4	1.4×10^{-4}	82	2.5
Beef	heart	3.6×10^4	1.0×10^{-4}	115	2.8
	brain	4.6×10^4	1.3×10^{-4}	90	2.8
	brain	4.3×10^4 [b]			
	kidney	1.9×10^4	0.9×10^{-4}	128	5.1
	erythrocyte				5.6
Dog	heart	5.5×10^4	2.3×10^{-4}	50	4.3
	heart			60[a]	
	kidney			120[a]	
Sheep	heart	1.1×10^4	1.2×10^{-4}	96	1.1
	Purkinje fibers				1.1
Rat	heart	4.3×10^4	1.0×10^{-2}	1.2	233
	skeletal muscle	3.3×10^4	0.6×10^{-2}	2	175
	brain	1.8×10^4	3.0×10^{-4}	39	16
	kidney				10^4
Guinea-pig	heart	5.5×10^4	1.1×10^{-2}	1.1	198
	heart			2.0[a]	
	skeletal muscle	5.9×10^4	2.6×10^{-3}	4.5	43
	brain	1.2×10^4	2.6×10^{-4}	45	22
	brain			20[a]	
	kidney	1.5×10^4	2.5×10^{-3}	4.5	166
	kidney			2.5[a]	
	erythrocyte				170
Frog (*Rana esculenta*)					
	heart	2.3×10^4	6.6×10^{-5}	175	3.3
Cat	heart			40[a]	
	heart				6.3

[a] TOBIN et al. (1972)
[b] LINDENMAYER and SCHWARTZ (1973)

Recently, the dissociation reaction of the ^3H-ouabain–receptor complex has been reported not to follow true first-order kinetics (SCHONER et al., 1977 a, b; HANSEN, 1978 a, c). Small amounts of K^+ or Tris buffer in the incubation medium, however, may transfer the dissociation curve to a "nearly monoexponential" (i.e., linear) curve (HANSEN, 1978 a, b, c). If the dissociation rate is not monoexponential but displays a two or more component course, then there should be two or more distinct types of cardiac glycoside receptors or possibly even negative cooperativity among the receptors (DE MEYTS, 1976). This in fact has been postulated by several authors (TANIGUCHI and IIDA, 1972; REPKE and SCHÖN, 1973; FRICKE and KLAUS, 1977; SCHONER et al., 1977 a; HANSEN, 1978 a, c). This merits detailed discussion.

V. Uniformity and Nonuniformity of Cardiac Glycoside Receptors

The interaction of ^3H-ouabain with its membrane-binding site under equilibrium conditions has been reported to follow the law of mass action (HANSEN, 1971; HAN-

SEN and SKOU, 1973; ERDMANN and SCHONER, 1973a). This ouabain–receptor binding has been shown to be reversible and saturable (BARNETT, 1970; TOBIN and BRODY, 1972). Thus, the data of equilibrium binding experiments may be plotted according to SCATCHARD (1949). In this plot linearity indicates the existence of one class of binding sites whereas curvilinear plots are obtained in the presence of two or more binding sites (BARTELS and HESCH, 1973; WEDER et al., 1974; CHAMNESS and MCGUIRE, 1975; HOLLEMANS and BERTINA, 1975; BOEYNAEMS and DUMONT, 1975). A curvilinear Scatchard plot, however, may have multiple meanings: heterogenity in binding affinity of labeled and unlabeled ligand (HOLLEMANS and BERTINA, 1975), cooperative effects of receptors (SCAF, 1975; BOEYNAEMS and DUMONT, 1975; DE MEYTS, 1976), multiple classes of independent binding sites (DE MEYTS and ROTH, 1975), nonspecific interaction or inaccuracies in the estimation of the free ligand concentration (CHANG et al., 1975), radioactive contaminants (REIMANN and SOLOFF, 1978), a two-step ligand–receptor interaction (BOEYNAEMS and DUMONT, 1977), ligand–ligand interaction, e.g., self-aggregation, polymerization, isomerization (JACOBS et al., 1975), instability of the binding site (ERDMANN et al., 1976c) and probably several more. Thus, curved Scatchard plots are extremely hard to interpret and may not be taken as the sole argument for the existence of more than one type of binding site. A different way of plotting the same experimental data (Hill, Schieldplots etc.) does not usually render more information, as they are merely graphical analyses.

TANIGUCHI and IIDA (1972) reported two apparently different ouabain-binding sites of Na^+, K^+-ATPase in ox brain microsomes at pH 6.1 in the presence of Mg^{2+}, Na^+, K^+, ATP, EDTA, and high concentrations of ^3H-ouabain (1–100 μM). They measured dissociation constants from Scatchard plots (high affinity $K_D = 0.18\,\mu M$, low affinity $K_D = 20\,\mu M$) which however did not agree with the values (3–12 nM) reported for one (high affinity) binding site by other groups, (HANSEN and SKOU, 1973; WHITTAM and CHIPPERFIELD, 1973; ERDMANN and SCHONER, 1973a). Possibly the low pH value (6.1) may have caused some alteration of the membranes or some conformational change.

In the presence of low phosphate concentrations (without ATP) several authors have demonstrated a nonuniformity of the ouabain–binding sites (HANSEN and SKOU, 1973; ERDMANN and SCHONER, 1973b; HANSEN, 1976). The same seems to be true for low Mg^{2+} concentrations (ERDMANN and SCHONER, 1973b). Low concentrations of K^+ in (Mg^{2++}, Pi)-supported ouabain binding cause a return to uniformity of binding sites (HANSEN, 1976).

Hence, the overall reaction scheme $Ou + R \rightleftharpoons OuR$ only seems to apply in special conditions, that is in the presence of Pi and/or K^+. HANSEN (1976) furthermore reported, that falsely linear Scatchard plots may have been measured using impure ^3H-ouabain (REIMANN and SOLOFF, 1978). He prepared pure ^3H-ouabain by chromatography on Na^+, K^+-ATPase by taking advantage of the stability of the enzyme–ouabain complex at 0 °C. These data have been reproduced (SCHONER et al., 1977a; ERDMANN et al., 1977) and may indicate under certain conditions a "half-of-sites reactivity" or a "flip-flop" model (REPKE and SCHÖN, 1973) of the Na^+, K^+-ATPase. The ouabain receptor under certain conditions might exist in two different conditions ($R_\alpha \rightleftharpoons R_\beta$) (ERDMANN and SCHONER, 1973b, INAGAKI et al., 1974). This concept is furthermore supported by the different ouabain binding kinetics in

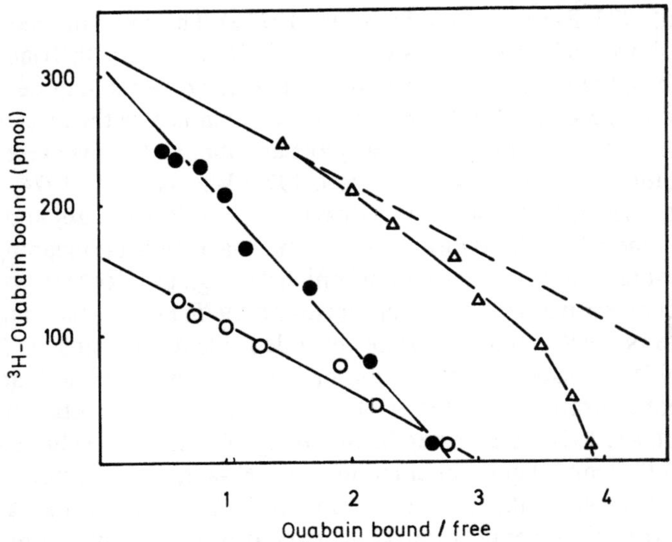

Fig. 8. Effect of protein concentration on ouabain–receptor binding data. Ox brain cell membranes Na^+, K^+-ATPase activity = 2.4 U/mg protein, 3.1 mg protein/ml) incubated in 50 m*M* imidazole/HCl pH 7.25, 3 m*M* $MgCl_2$, 1 m*M* imidazole/HCl, 3 n*M* ^{3}H-ouabain, and increasing amounts of unlabeled ouabain at 37 °C for 60 min. Data plotted according to SCATCHARD (1949). *(Open circles)* 0.31 mg membrane protein, total volume of incubation 2 ml; *(solid circles)* 0.62 mg membrane protein, total volume of incubation 4 ml; *(open triangles)* 0.62 mg membrane protein, total volume of incubation 2 ml. Higher protein concentrations caused increasingly concave plots leading to even further "false" binding data. (ERDMANN, unpublished)

the presence of different nucleotide analogs, some of which cannot phosphorylate Na^+, K^+-ATPase and decrease the affinity of the cardiac glycoside receptor (TOBIN et al., 1973; ERDMANN and SCHONER, 1973c; PATZELT-WENCZLER and SCHONER, 1975). Further experiments, however, will be necessary to understand this mechanism fully.

In ox brain microsomes large amounts of membrane protein in the usual incubation medium will cause curvilinear Scatchard plots of binding data with upward concavity (Fig. 8). This pitfall of artificial "positive cooperativity" can be avoided by using low protein concentrations.

Recently, FRICKE and KLAUS (1977) presented data suggestive of two different Na^+-dependent ^{3}*H*-ouabain-binding sites in guinea-pig heart. In the presence of Mg^{2+} and ATP, as the authors suppose, one type of ouabain-binding site is predominant at low and one at high Na^+-concentrations. From these data they speculate that the Na^+, K^+-ATPase actually represents a "carrier-mediated translocation system for cardiac glycosides operative in the heart muscle cell membrane." These experimental results have not yet been supported by others. SWEADNER (1979) has, however, produced convincing evidence that brain cell membranes, in contrast to other organs, contain two molecular forms of Na^+, K^+-ATPase. These two distinct forms have detectable structural and kinetic differences particularly in their different affinities for strophanthidin. Earlier electrophoretic analyses of brain enzymes did not detect the presence of the two forms because low resolution gel electrophoresis systems were used (SWEADNER, 1979).

In cell membrane preparations of rat and guinea-pig skeletal muscle we have consistently obtained nonlinear Scatchard plots when the reaction was allowed to proceed for more than 3–10 min (ERDMANN et al., 1976c). From this we might suggest, that an alteration of the membrane composition during incubation does mimic the appearance of nonuniformity of ouabain receptors or represents partly affected and native binding sites. This is supported by the fact that the kinetic data of ouabain binding to the unchanged membranes (3–10 min incubation) agree with those values obtained by other methods and other investigators (CLAUSEN and HANSEN, 1974, 1977). Furthermore, the Na^+, K^+-ATPase activity of the membranes decreases during the incubation procedure. These rather conflicting results caused us to investigate this problem of nonuniformity of ouabain receptors in membrane preparations more closely.

Ouabain-Binding Sites in a Cell Membrane Preparation and in Contracting Ventricular Strips of Rat Heart

When analyzing the experimental data of a concentration dependent ^3H-ouabain-binding experiment in a rat heart *cell membrane preparation* according to SCATCHARD (1949) a nonlinear plot was repeatedly obtained (Fig. 9). The data agreed

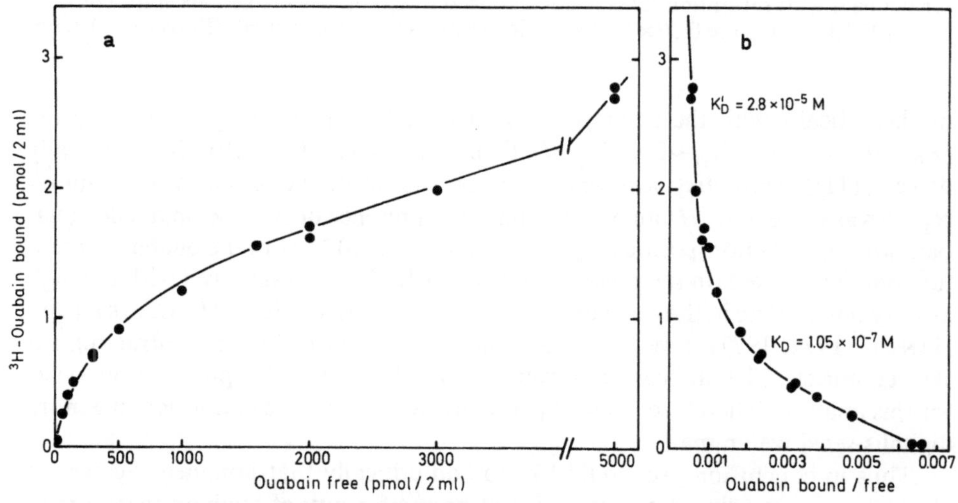

Fig. 9 a, b. ^{3}H-ouabain binding to isolated rat heart cell membranes. Effect of increasing ^{3}H-ouabain concentrations. Cardiac cell membranes (0.3 mg protein, Na^+, K^+-ATPase activity 0.25 U/mg protein) incubated at 37 °C for 50 m*M* imidazole/HCl pH 6.5, 3 m*M* $MgCl_2$, 3 m*M* imidazole/PO_4, 4 n*M* ^{3}H-ouabain, and increasing ouabain concentrations ($10^{-9}-10^{-3}$ *M*). Nonspecific ^{3}H-ouabain binding (in the presence of 10^{-3} *M* ouabain) was subtracted. After rapid filtration (Whatman GF/C glass fiber filters) the radioactivity on the filters was determined. **a** ouabain binding to the membranes increases steadily up to very high free ouabain concentrations. **b** Scatchard analysis demonstrates at least two components of ouabain-binding sites. A high affinity/low capacity receptor ($K_D = 1.05 \times 10^{-7}$ *M*/10% of binding sites) and a low affinity/high capacity receptor ($K'_D = 2.8 \times 10^{-5}$ *M*/90% of binding sites). Mathematical analysis according WEIDMANN et al. (1970). (ERDMANN et al., 1980)

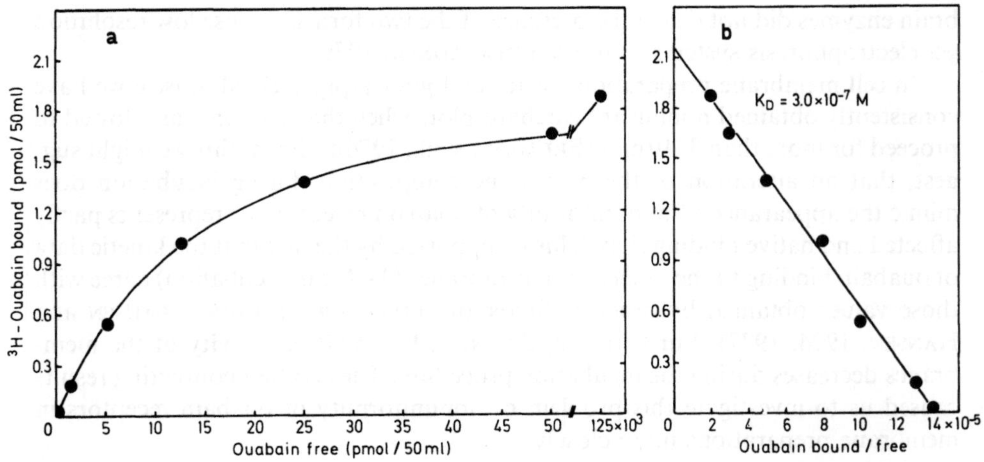

Fig. 10a, b. ^{3}H-ouabain binding to contracting rat heart ventricular strips (20 mg wet weight, for details see ERDMANN et al., 1980) incubated at 37 °C in Tyrode solution containing ^{3}H-ouabain (4×10^{-9} *M*, 2.5×10^{-8} *M*, 1×10^{-7} *M*, 2.5×10^{-7} *M*, 5×10^{-7} *M*, 1×10^{-6} *M*, 2.5×10^{-6} *M*; 1.5×10^{-4} *M* = nonspecific binding value) and electrically stimulated (1 Hz). Incubation volume 50 ml. Incubation time interrupted at time of peak contractile force (10–20 min), strips were blotted and assayed for radioactivity. Each point is the mean of four experiments. Radioactivity calculated per mg of cardiac tissue. **a** amount of ^{3}*H*-ouabain bound to the muscles plotted versus free ^{3}*H*-ouabain (in the incubation medium). **b** same data plotted according to SCATCHARD (1949); only one type of binding site for ouabain with saturation. Dissociation constant (K_D) of ^{3}*H*-ouabain–receptor complex is 3×10^{-7} *M*. There are 0.66×10^{14} specific receptors per g wet weight. (ERDMANN, 1978 b)

mathematically with the assumption of at least two different types of receptors ($K_D \sim 10^{-7}$ *M* and $K'_D \sim 3 \times 10^{-5}$ *M*) (WEIDEMANN et al., 1970). In electrically driven (1 Hz) *contracting ventricular strips*, however, only one type of ouabain receptor was present, if ^{3}*H*-ouabain bound in the presence of 10^{-4} *M* unlabelled ouabain was taken as nonspecific (Fig. 10). The $K_D \sim 3 \times 10^{-7}$ *M* of the ouabain–receptor complex formed (in the presence of 5.4 m*M* K^+) agrees very well with the high affinity binding site in the cell membrane preparation ($K_D \sim 10^{-7}$ *M*). CLAUSEN and HANSEN (1974, 1977) have also shown linear Scatchard plots in contracting rat skeletal muscle. The rat heart, however, may not be the suitable preparation to use for this analysis. Therefore, these experiments will have to be confirmed in a more digitalis-sensitive animal.

Thus in conclusion, we would like to add critically that nonlinear Scatchard plots or corresponding data suggestive of nonuniformity of ouabain receptors in cell membrane preparations should be considered with great caution. Some of the reasons for this have been quoted in this and in the preceding Chap. 14).

VI. Correlation of Ouabain Binding and Increase in Force of Contraction

A number of investigators have produced evidence demonstrating a correlation between ouabain binding and positive inotropic effect (AKERA et al., 1969; BESCH et al., 1970; AKERA et al., 1973; LÜLLMANN and PETERS, 1974; FRICKE et al., 1975; KU

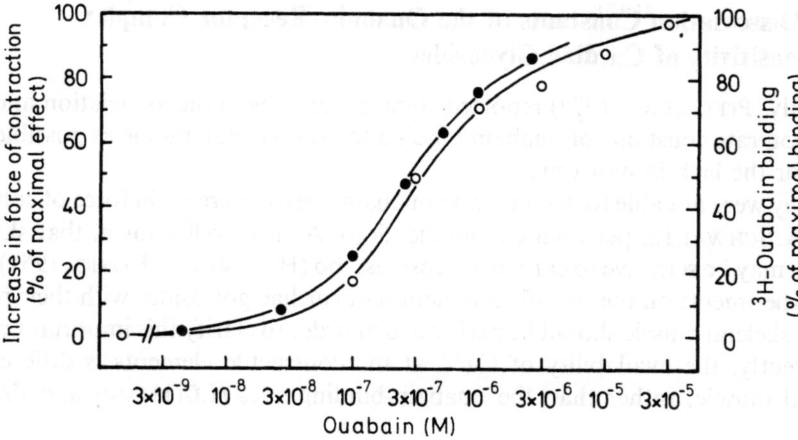

Fig. 11. Concentration-dependent ^{3}H-ouabain binding and effect of ouabain on force of contraction. Data from Fig. 10 plotted in a concentration-dependent manner. Maximal increase in force of contraction at the indicated ouabain concentration *(open circles)*, was measured in the very same cardiac preparations (for details see ERDMANN et al., 1980). The *solid circles* indicate ^{3}H-ouabain binding as a percentage of maximal binding. A close correlation exists between ouabain–receptor binding and its pharmacologic effect. (ERDMANN, 1978 b)

et al., 1975; PFLEGER et al., 1975; ALLEN et al., 1975; BRODY and AKERA, 1977; WALLICK et al., 1977; for references see AKERA, 1977). Usually, however, heart or isolated cardiac preparations were perfused with cardiac glycosides and the tissue-bound ^3H-glycoside was correlated with increased contractility. Thereby it is not possible to quantitate specific ouabain binding with its effect. Hence, several authors have questioned a relationship between the effect of cardiac glycosides on Na^+, K^+-ATPase and force of contraction (OKITA et al., 1973; TEN EICK et al., 1973; MURTHY et al., 1974; OKITA, 1975, 1977). As our experiments (Fig. 10) had shown that it is quite possible to measure specific ouabain binding to isolated ventricular strips, we recorded, in the very same paced heart tissue, the time course of increase in force of contraction. As soon as steady-state conditions were reached, usually after 10–20 min of incubation with the respective concentration of ouabain, the amount of receptor-bound ouabain was analyzed. The results are shown in Fig. 10 and 11 in detail.

There is a good quantitative correlation between ouabain–receptor binding and increased force of contraction even in the rat, which had deliberately been used as being rather insensitive to cardiac glycosides (DETWEILER, 1967). These results have been confirmed recently by MICHAEL et al. (1979) in contracting cat papillary muscles. These authors furthermore could demonstrate, that the number of occupied receptors decreased to zero, when the inotropic effect was washed out. Thus, ouabain binding to its specific receptors causes a positive inotropic effect of the same extent in cardiac muscle, i.e., if 50% of the cardiac glycoside receptors are occupied by a drug molecule, 50% of the maximal increase in force of contraction is observed. Toxic effects under these experimental conditions were regularly noted as arrhythmia at high ouabain concentrations.

VII. Dissociation Constants of the Ouabain–Receptor Complex and Sensitivity of Cardiac Glycosides

Recently, Pitts et al. (1977) reported some differences in the association and dissociation rate constants of ouabain binding to dog skeletal muscle as possible reasons for the lack in inotropy.

They were not able to demonstrate ouabain-caused increase in force of contraction (A. Schwartz, personal communication). As it is well known, that skeletal muscle may be sensitive to cardiac glycosides, too (Hotovy and König, 1951), further experiments on the specific interaction of cardiac glycosides with their receptors in skeletal muscle should be performed in order to clarify this important point. Apparently, the availability of Ca^{2+} at the contractile elements is different in skeletal muscle, rather than the ouabain-binding sites (Lüllmann and Peters, 1976).

The relative insensitivity of the rat to cardiac glycosides has its origins in a low affinity of the receptor for ouabain (Table 1). This low affinity is mainly caused by a rapid dissociation of the ouabain–receptor complex. Previous reports have either demonstrated irreproducible specific binding of the drug (Allen and Schwartz, 1969) because of insufficient methods at that time or a rather insensitive Na^+, K^+-ATPase activity (Repke et al., 1965; Dransfeld et al., 1966; Wallick et al., 1974). The same applies to other species. The higher the affinity of the receptor for the cardiac glycoside, the greater is the sensitivity of that species to this group of drugs. With the exception of sheep, humans are apparently most sensitive to cardiac glycosides (Table 1).

The properties of cardiac glycoside receptors from human right and left ventricle were the same (Fig. 12). Identical affinity to ouabain was measured in the atria as well. In sheep Purkinje fibers the apparent dissociation constant K_D was identical to that of ventricular cell membranes. Thus, differences in sensitivity or rapidity of onset of effects in various parts of the heart do not seem to be caused by higher or lower affinities of the receptors.

It is easy to understand, though, that the rat after injection of large doses of cardiac glycosides at first exhibits neurologic signs of intoxication rather than cardiac symptoms. The dissociation constant of the ouabain–receptor complex in brain Na^+, K^+-ATPase preparation was 16 nM whereas that of heart was measured as 233 nM under identical conditions (Table 1).

VIII. Changes in Ouabain Receptor Density

As already mentioned briefly (Sect. B.III) the number of ouabain receptors in erythrocytes may increase under certain circumstances, for instance in diseases such as chronic hypokalemia or thalassemia. After successful therapy of the hypokalemia the increased receptor density returned to normal (Erdmann and Krawietz, 1977) indicating a regulation of Na^+, K^+-ATPase molecules in the red cell membrane.

It is well known that some receptors may be regulated by substrates, hormones, or other effects (Hollenberg and Cuatrecasas, 1978). The insulin receptor concentration is low in patients with hyperinsulinemic diabetes, in obesity, or in

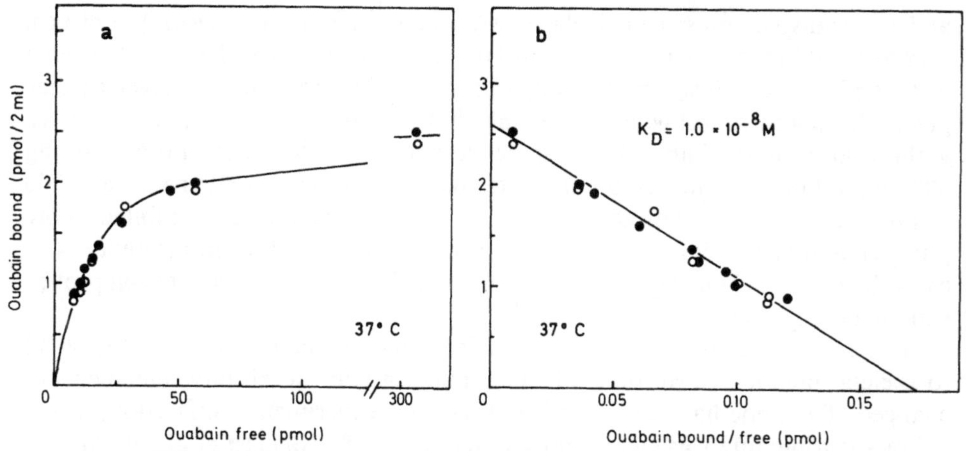

Fig. 12 a, b. Binding of ^3H-ouabain to a crude homogenate from human left and right ventricle. Experiment similar to that of Fig. 2. Dissociation constant was measured as 10 n*M* (instead of 2.5 n*M* at optimal binding conditions). Cardiac glycoside receptor has same affinity for ouabain in left and right ventricle, indicating the same protein conformation. (ERDMANN, unpublished)

growth hormone excess (FLIER et al., 1979). The number of β-adrenergic receptors apparently decreases after application of high doses of β-adrenergic drugs (LEFKOWITZ, 1976). Similar variations in cardiac glycoside receptors are controversial. After chronic treatment with subcutaneous digitoxin (0.3 mg/kg for up to 24 days) Na^+, K^+-ATPase activity was increased in guinea-pig hearts by about 30% (BLUSCHKE et al., 1976). Unfortunately the number of ouabain-binding sites was not measured in this study. Although kinetic properties of the enzyme were unchanged, an increase in enzyme molecules seems to have been measured. In dogs treated with 8–12 ng/kg digoxin twice daily for up to 4 weeks Na^+, K^+-ATPase and digoxin-binding sites were unchanged (KU et al., 1977). As tachyphylaxis is not known in treatment with digitalis drugs, a change in receptor number or affinity by chronic treatment is not to be expected. However, the extremely high doses used by BLUSCHKE et al. (1976) might have caused the observed increases in Na^+, K^+-ATPase activity, which were not seen after only 1–5 days of treatment (BONN and GREEFF, 1978).

Several other substances or conditions may change the number of Na^+, K^+-ATPase molecules in cell membranes as well: chronic administration of $CoCl_2$ (DRANSFELD et al., 1971), hypoadrenalism (BORSCH-GALETKE et al., 1972), hypo- and hyperthyroidism (KATZ and LINDHEIMER, 1973; CURFMAN et al., 1977; PHILIPSON and EDELMAN, 1977). The observed increased number of ouabain-binding sites (by 17%–19%) in guinea-pig heart in animals treated with triiodothryonine (T_3) may be one of the causes of relative digitalis resistence in hyperthyroidism (CURFMAN et al., 1977). The affinity of the cardiac receptor for ouabain was unchanged by T_3 treatment (LIN and AKERA, 1978). In order to understand this proposed mechanism of relative digitalis resistence when the number of cardiac glycoside receptors is increased, one has however to make several assumptions. First, the inhibition of the Na^+, K^+-ATPase activity and hence the inhibition of active Na^+

and K^+ transport must be correlated with the drug response. Second, a critical number of inhibited sodium pump sites must be necessary for that effect. Then a 50% inhibition of 1,000 sodium pump sites per μm^2 for instance at a given cardiac glycoside concentration would leave only 500 sodium pump sites active whereas, without an affinity change at the same cardiac glycoside concentration causing 50% inhibition of some 1,400 pump sites per μm^2 in the hyperthyroid state, 700 sodium pump sites are left active. Thus, although more sites are inhibited, more pump sites are left uninhibited, too, in hyperthyroidism, and the intracellular Na^+ can still be pumped out rapidly. This hypothetical mechanism has not been proved experimentally, however.

Recently, LIN et al. (1978) observed a decreased ^3H-ouabain binding (by 36%) to skeletal muscle preparations of obese mice. Again the affinity remained unchanged. The same has been found in dogs after experimental infarction, where ^3H-ouabain binding was significantly reduced to 58% (BELLER et al., 1976).

Apparently, some known cases of increased sensitivities to cardiac glycosides are caused by a reduced number of binding sites resulting in a decreased volume of distribution and higher serum levels of the drug at the sites of action as in hypothyroidism for instance (DOHERTY and PERKINS, 1966) – and vice versa (hyperthyroidism).

Critically, however, we would like to add that in most studies the specific ^3H-ouabain binding has not been defined exactly. The result is very much dependent on the reference (mg membrane protein, kg wet weight, dry weight, etc.) used. Changes of the tissue composition (water content, lipid metabolism, etc.) may distort such quantitative measurements greatly. We have consistently measured 0.9×10^{14} cardiac glycoside receptors in right and 1.5×10^{11}/g wet weight in left human ventricular tissue. The difference corresponded well with the known differences in myocardial digoxin content between right and left ventricular tissue (28%) after treatment (HAASIS et al., 1977). These differences are not the expression of a decreased receptor density but of a different water content of right ventricular tissue. The receptor density can be determined exactly only in erythrocytes (ERDMANN and HASSE, 1975) or in cell cultures (MCCALL, 1977).

C. Qualitative Aspects of Cardiac Glycoside–Receptor Interaction

The "receptors" defined by specific binding of ouabain or other cardiac glycosides represent the "recognition" site of the Na^+, K^+-ATPase molecule. That enzyme system, being firmly embedded in the cell membrane (DEGUCHI et al., 1977), is supposed to consist of several large (~100,000 daltons) and small (~50,000 daltons) polypeptide chains to give an overall molecular weight of about 250,000 daltons (JØRGENSEN, 1974; DIXON and HOKIN, 1974; GIOTTA, 1975; PERRONE et al., 1975). This large protein is surrounded by several phospholipid molecules necessary for full enzyme activity (CHARNOCK et al., 1973; STAHL, 1973; LANE et al., 1978). Changes of membrane fluidity, of protein conformation, or direct interaction at the receptor will therefore modify the binding of cardiac glycosides and thus its pharmacologic effects (ENGEL et al., 1977; ZWAAL, 1977). As the number of known effects of drugs, ions, or changes of membrane composition on specific ouabain

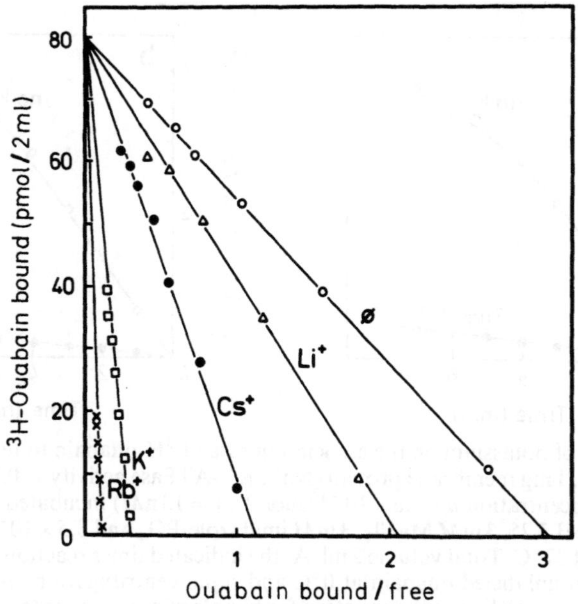

Fig. 13. Effect of alkali metal ions on ^{3}H-ouabain–receptor interaction. Scatchard plots of ouabain binding in the presence of monovalent cations. 0.3 mg of enzyme protein (ox brain, specific activity 2.05 U/mg protein $= 266 \times 10^{-12}$ mol receptor sites per mg protein) were incubated in 50 m*M* imidazole/HCl pH 7.25, 3 m*M* $MgCl_2$, 3 m*M* Tris-phosphate and increasing amounts of labeled ouabain at 37 °C for 120 min. Total volume 2 ml. Addition of 20 m*M* RbCl *(crosses)*, 10 m*M* KCl *(squares)*, 20 m*M* CsCl *(solid circles)*, 20 m*M* LiCl *(triangles)*, without any addition *(open circles)*. At identical cation concentration (20 m*M*) the following values were obtained: K_D: Li^+ 18.8 n*M*, Na^+ 28 n*M*, K^+ 205 n*M*, NH_4^+ 82.5 n*M*, Rb^+ 197.5 n*M*, Cs^+ 34.6 n*M*. (Erdmann and Schoner, 1973a)

binding to its receptor is very large (Allen and Schwartz, 1970; Hansen and Skou, 1973; Erdmann and Schoner, 1973c; Gardner and Frantz, 1974; Yoda and Yoda, 1978) we would like to concentrate on a few and discuss some interesting aspects.

I. Effects of Cations on Ouabain–Receptor Interaction

Clinicians have known for a long time that a rapid decrease of serum potassium levels will cause severe digitalis-induced arrhythmias in treated patients. On the other hand digitalis intoxication is often treated successfully by the application of potassium. Potassium, in fact, changes the conformation of the ouabain receptor towards a decreased affinity for cardiac glycosides (Allen and Schwartz, 1970; Whittam and Chipperfield, 1973; Hansen and Skou, 1973; Erdmann and Schoner, 1973b; Erdmann et al., 1976a; Akera et al., 1978). This is accomplished by all alkali metal ions but to a different extent (Han et al., 1976) (Fig. 13).

In the presence of Mg^{2+} and Pi in the incubation medium, potassium merely decreases the association rate but does not affect the dissociation rate in human cardiac cell membranes (Akera and Brody, 1971; Erdmann et al., 1976a). Recent-

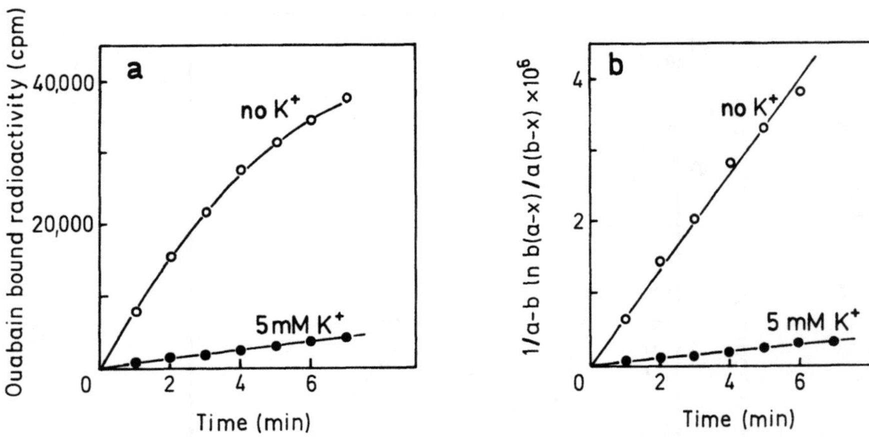

Fig. 14a, b. Effect of potassium on the association rate of ^{3}H-ouabain to its receptor in human cardiac tissue. 1 mg membrane protein (Na^+, K^+-ATPase activity = 0.2 U/mg protein; initial receptor concentration $a = 26 \times 10^{-12}$ mol/2 ml = 13 n*M*) incubated in 50 m*M* imidazole/HCl buffer pH 7.25, 3 m*M* $MgCl_2$, 3 m*M* imidazole/PO_4 and 8.5×10^{-12} mol ^{3}H-ouabain $b = 4.25$ n*M* at 37 °C. Total volume 2 ml. At the indicated times reaction was interrupted by addition of 2 ml unlabeled ouabain at 0 °C and rapid centrifugation (30 min, 80,000 *g*). Data from **a** plotted according to the equation for bimolecular reactions (see Sect. B.IV) in **b**. Association rate constant (k_{+1}) calculated as 1.2×10^5 M^{-1} s^{-1} and 8.2×10^5 M^{-1} s^{-1} (5 m*M* K^+). (ERDMANN et al., 1976a)

ly, AKERA et al. (1978) observed in the presence of Mg^{2+}, ATP, Na^+, and K^+ a less rapid dissociation (k_{-1}) of the ouabain–receptor complex. Because of the pronounced decreased association rate, the dissociation constant K_D was nevertheless increased. Up to about 5 m*M* K^+, equilibrium binding of ouabain decreases linearly (AKERA, 1971). In the presence of EDTA (used as chelator of Mg^{2+}) K^+ stabilizes the dissociation significantly (ALLEN et al., 1971). Thus, the experimental conditions may alter the influence of K^+ on ouabain binding to its receptor profoundly. Under nearly physiologic conditions (Mg^{2+}, Pi) or (Mg^{++}, ATP, Na^+) however, potassium does reduce the receptor affinity in a concentration-dependent way, mainly by slowing the rate of association (ERDMANN et al., 1976a; CHOI and AKERA, 1977; AKERA et al., 1978). Figure 14 shows this effect of potassium. These in vitro measurements of the ouabain–receptor interactions agree with clinical observations. In canine Purkinje fibers a rapid elevation of K^+ from 2.7 to 5.4 m*M* resulted in a release of formerly bound ^{3}*H*-ouabain and reversed digitalis toxicity (ANDERSON et al., 1976).

Sodium in the presence of (Mg^{2+}, Pi) inhibits ouabain binding whereas it stimulates it in a (Mg^{2+}, ATP)-containing medium (LINDENMAYER and SCHWARTZ, 1973; HARRIS et al., 1973; HANSEN and SKOU, 1973; ERDMANN and SCHONER, 1973b). Again, Na^+ does not change the number of binding sites but rather the affinity of the receptor for the drug. Ca^{2+} increases the affinity quite significantly (ERDMANN and SCHONER, 1973b). This effect is pronounced in the presence of K^+, whereas Ca^{2+} may reverse the K^+ effect completely. The action of calcium on the toxicity of cardiac glycosides is well known clinically. BOWER and MENGLE (1936) were the first to report two deaths because of the "additive effect of calcium and digitalis." In dogs this "synergistic" relationship between calcium and digitalis has been experimentally proved (NOLA et al., 1970). Recently an insensitivity to digoxin associated with hypocalcemia has been reported (CHOPRA et al., 1977). This does not, however, prove the relevance of the observed direct interaction of Ca^{2+} at the cardiac glycoside receptor site in vivo. Further investigations to elucidate this mechanism are necessary.

II. Effect of Vanadate on Ouabain Binding

Quite recently it has been observed, that vanadate (Na_3VO_4 or $NaVO_3$) may substitute for phosphate and stimulate ^3H-ouabain binding to its receptor (HANSEN, 1978c; ERDMANN et al., 1979c). This is, however, not caused by a direct interaction with the receptor but with the ATP-binding sites of the enzyme, where vanadate displaces ATP or Pi (CANTLEY et al., 1978) and thereby presumably changes the conformation of the enzyme to a state with higher affinity for the cardiac glycoside (ERDMANN et al., 1979b). The positive inotropic effect of vanadate on cat ventricular muscle (HACKBARTH et al., 1978) thus may be brought about by an interaction with Na^+, K^+-ATPase but not with the ouabain receptor. Both, cardiac glycosides and vanadate inhibit this enzyme, though at different sites.

In short, the affinity of the cardiac glycoside receptor for this group of cardioactive drugs is greatly influenced by the ionic composition of the incubation medium. Probably the conformation of the receptor molecule is changed by ions. Certainly the same is true for other agents indirectly interacting with the membrane fluidity (ZWAAL, 1977). This aspect of membrane biophysics is however, beyond the limits of this chapter.

III. Influence of pH and Temperature on Ouabain Binding

The pH dependence of the dissociation constant of the ouabain–receptor complex has been demonstrated to be different from that of the Na^+, K^+-ATPase activity (ERDMANN and SCHONER, 1973b). Above pH 7.25 the affinity is decreased steeply and steadily, the highest affinity being observed at or below pH 7.25. At pH 8 the dissociation constant in ox brain membranes is raised to about 100 nM. This behavior coincides with the clinically known increased sensitivity to cardiac glycosides in cor pulmonale and acidosis. The exact pH dependence of the ouabain–receptor complex seems to vary with the species tested (ALBERS et al., 1968).

In human cardiac cell membranes the dissociation constant of the ^3H-ouabain–receptor complex is increased with temperature (Fig. 15). This means a high affinity of the receptor at low temperatures and vice versa. A diminished response to digoxin in isolated heart muscle as a result of fever has been demonstrated experimentally (KOKENGE and VAN ZWIETEN, 1971).

D. Specificity of the Cardiac Glycoside Receptor

Cardiac glycosides inhibit Na^+, K^+-ATPase activity. Of course, not all inhibitors of this enzyme system are bound by the cardiac glycoside receptor. The same is true for positive inotropic agents. For some inhibitors the sites of action on Na^+,

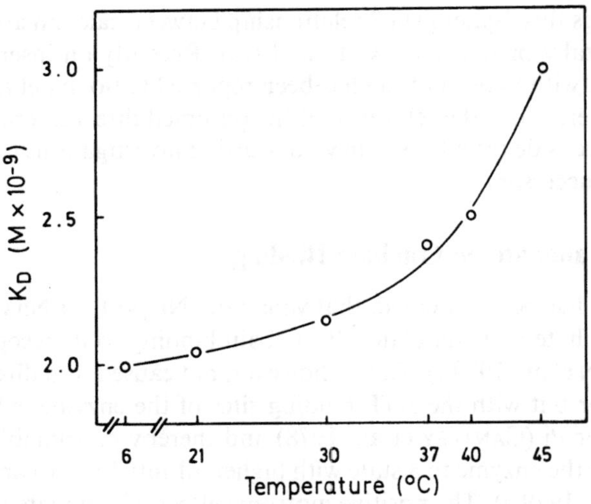

Fig. 15. Influence of temperature on ouabain binding to human cardiac cell membranes. Experiment performed according to Fig. 2. Equilibrium conditions had been established previously (26 h at 6 °C). Dissociation constants calculated from the respective Scatchard plots. Curve demonstrates an increased affinity (decreased dissociation constant) at low temperatures. (ERDMANN, unpublished)

K^+-ATPase are known, e.g., for vanadate (see Sect. C.II); several inhibiting agents (*N*-ethylmaleimide, p-chloromercuribenzoate, p-chloromercuribenzenesulfonic acid etc.) have however not been investigated that closely (FRICKE, 1978; TEMMA et al., 1978). In this chapter a few substances with known mode of action at the cardiac glycoside receptor are discussed.

I. Receptor Specificity for Cardiac Glycosides

The demonstration of a direct binding of cardiac glycosides or other substances to the receptor is, because of experimental reasons, only possible with the use of radioactively labeled drugs. With certain reservations the competitive displacement of labeled cardiac glycosides by unlabeled drugs may be taken as evidence of drug–receptor binding. In high concentrations, however, a great number of substances decreases bound ^3H-ouabain on the receptor, probably by an indirect action on adjacent membrane components. Thus, great caution has to be used when drugs are said to act by binding to the glycoside receptor if the data are taken only from indirect measurements (displacement) and reveal rather high concentrations necessary to displace ^3H-ouabain, for instance. Besides this, one should differentiate between true displacement, meaning the dissociation of the drug–receptor complex and inhibition of binding, meaning the association of the drug–receptor complex in the presence of another substance with inhibitory effect on that reaction. If a drug is bound specifically by that receptor, both experiments (displacement and inhibition) will render identical results, if assayed at equilibrium. On the other hand, substances that change the receptor conformation without binding to the same site (like K^+) will displace and inhibit ^3H-ouabain binding, too. This should

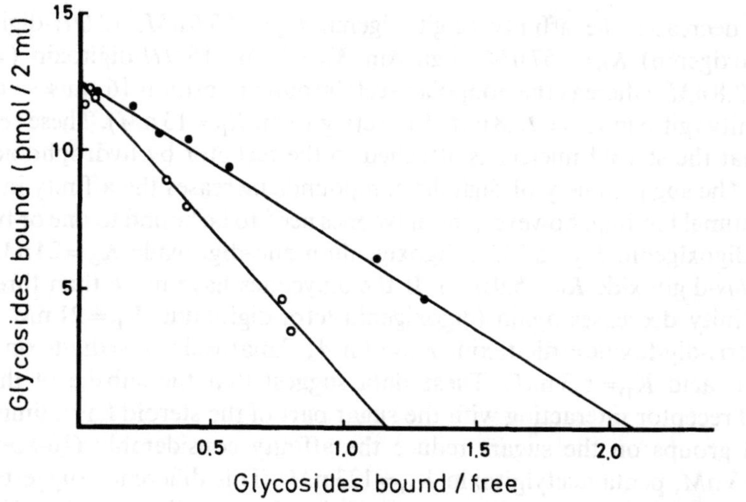

Fig. 16. Scatchard plots of glycosides binding to Na^+, K^+-ATPase of beef heart. Increasing amounts of ^{3}H-ouabain *(solid circles)* (4 n*M*–0.5 µ*M*, 98 nCi/2 ml) or ^{3}H-digoxin *(open circles)* (3 n*M*–0.5 µ*M*, 95 nCi/2 ml) incubated with the enzyme preparation for 2 h. Determination of enzyme-bound ^{3}*H*-activity by filtration procedure. Values are means of two experiments. (BOSSALLER and SCHMOLDT, 1979)

be kept in mind when analyzing such experiments on drug–receptor binding.

In human cardiac cell membranes as well as in other membrane preparations (ERDMANN and SCHONER, 1974a, b; BOSSALLER and SCHMOLDT, 1979) specific binding of labeled cardiac glycosides with different affinities but identical binding capacities can be demonstrated (Fig. 16). Displacement curves give identical values, if specific binding is measured exactly. Usually the lipophilic drugs (for instance digitoxin) have slightly higher unspecific binding (i.e., to membrane components unrelated to the cardiac glycoside receptor; see Sect. A.II.1) than ouabain.

If ^{3}H-ouabain and any other unlabeled drug compete for the same receptor, the dissociation constant K_D may be calculated from the amount of ^{3}*H*-ouabain displaced from the binding site at equilibrium conditions (ERDMANN and SCHONER, 1974a). An analysis of the reliability of such determinations has established a good agreement when compared with the direct evaluation of the data obtained with the labeled compounds (Fig. 16). It must be kept in mind, however, that this method is only applicable when competition for the same binding sites has been proved. Employing this method in ox brain cell membranes, it has been shown that the receptor probably has three points of attachment, interacting with:

1) The lactone group. The unsaturated lactone group is most important for the binding to the receptor. If the C=C bond is saturated, the affinity decreases sharply (digitoxin K_D= 3 n*M*, dihydrodigitoxin K_D= 37.8 n*M*; digoxin K_D= 10 n*M*, dihydrodigoxin K_D= 57.4 n*M*).

2) The steroid component. The *cis* configuration of the A–B ring junction is an essential structural requirement. Uzarigenin (K_D= 554 n*M*), which has an A–B *trans* configuration instead of the A–B *cis* configuration, is bound about 30 times less than digitoxigenin (K_D=15.6 n*M*). Introduction of polar groups in rings B, C

and D decreases the affinity (digitoxigenin K_D = 15.6 nM, 12OH-digitoxigenin (=digoxigenin) K_D = 57 nM, digitoxin K_D = 3 nM, 16OH-digitoxin (=gitoxin) K_D = 22.8 nM) whereas the nonpolar acetylgroup in position 16 causes an increase in affinity (gitoxin K_D = 22.8 nM, 16-acetylgitoxin K_D = 13 nM). These results suggest that the steroid nucleus is attached to the receptor by hydrophobic bonds.

3) The sugar moiety of digitalis compounds increases the affinity in all cases. For optimal binding, however, the aglycones need to be bound to one or two sugars only (digoxigenin K_D = 57 nM, digoxigenin-mono-digitoxide K_D = 24 nM, digoxigenin-*bis*-digitoxide K_D = 5.9 nM). If the aglycones have more than three sugars, the affinity decreases again (digoxigenin-tetra-digitoxide K_D = 21 nM; digitoxigenin-tris-digitoxide = digitoxin K_D = 3 nM; lanatoside A = digitoxin + glucose + acetic acid K_D = 6.3 nM). These data suggest that the subsite of the cardiac steroid receptor interacting with the sugar part of the steroid has a limited space. Acetyl groups on the sugars reduce the affinity considerably (16-acetylgitoxin K_D = 13 nM, penta-acetylgitoxin K_D = 123 nM). This difference suggests that the attachment of the sugar component of the drug to the receptor depends on hydrogen bonds.

It is interesting that the Gibbs free energy of binding ($\Delta G°$) of the cardioactive steroids used for therapeutical purposes is between −10 and −12 kcal/mol (ERDMANN and SCHONER, 1974a).

For further information on the affinity and specificity of cardiac glycoside–receptor binding see YODA (1973, 1976), YODA and YODA (1975, 1977), BOSSALLER and SCHMOLDT (1979). In respect to the inhibition of Na^+, K^+-ATPase there exists a great number of papers (see AKERA and BRODY, 1978).

The reported affinities for cardiac glycoside–receptor interaction agree well with the inhibitory action on Na^+, K^+-ATPase (REPKE and PORTIUS, 1966; ERDMANN and SCHONER, 1974a) and the increase in force on contraction – as far as that parameter has been tested (REITER, 1967; AKERA and BRODY, 1978; SCHOLZ et al., 1979).

II. Other Substances that Bind to the Cardiac Glycoside Receptor

In general, the specificity of the receptor is extremely high and thus there are only a very few compounds that are bound specifically. As these substances are not usually available in the radioactively labeled form and they often have moreover several effects and a variety of binding sites at the membrane, the evaluation of the present information is rather inconclusive.

Diphenylhydantoin, which has been said to have a distinct anti-arrhythmic effect in digitalis intoxication (LÜLLMANN and WEBER, 1968; HAGEN, 1971), does displace bound ^3H-ouabain from its receptor, though in very high concentrations, (ERDMANN and SCHONER, 1974b). The dissociation constant was determined indirectly, by the displacement of ^3H-ouabain, as $\sim 10^{-4} M$. Ro 2-2,985 ($K_D \sim 35\,\mu M$), an ionophore with positive inotropic action, as well as canrenone ($K_D \sim 50\,\mu M$), an aldosterone antagonist, do displace ^3H-ouabain and inhibit Na^+, K^+-ATPase. The necessary concentrations are, however, extremely high, when compared with cardiac glycosides (Fig. 17). Therefore, direct binding studies with these compounds need to be performed to prove the site of binding to the mem-

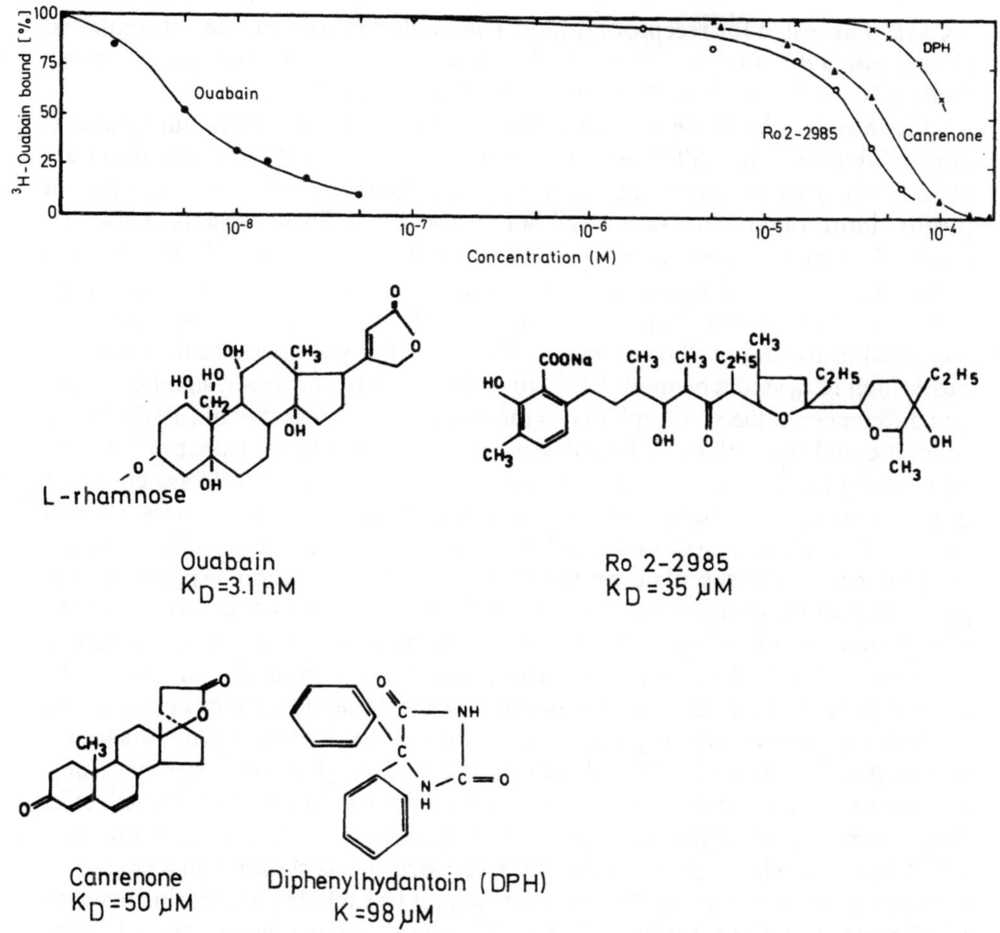

Fig. 17. Inhibition of ^3H-ouabain binding to a calf heart membrane preparation by ouabain, Ro 2-2985, canrenone, and diphenylhydantoin. 0.17 mg membrane protein incubated for 120 min at 37 °C in 50 mM imidazole/HCl buffer pH 7.25, 3 mM $MgCl_2$, 3 mM imidazole/PO_4, 8.5×10^{-9} M ^3H-ouabain and the indicated concentration of ouabain, Ro 2-2985, canrenone and diphenylhydantoin. Dissociation constants K_D calculated according to ERDMANN and SCHONER (1974). Note the striking concentration difference of the effects of ouabain and the other substances. (ERDMANN and SCHONER, 1974b)

brane. In a recent report LA BELLA et al. (1979) demonstrate that some progesterone derivatives (hydroxyprogesterone, chlormadinone acetate etc.) inhibit Na^+, K^+-ATPase, displace ^3H-ouabain from its specific binding sites and inhibit Na^+ efflux in cardiac tissue. These substances, however, do not increase the force of contraction. Possibly the measured inhibition of ^3H-ouabain–receptor binding to a dog heart membrane preparation is not caused by a direct interaction at the cardiac glycoside receptor but by an attack at another site of the Na^+, K^+-ATPase molecule.

As cardiac glycosides contain the lactone structure and lactones are known to inhibit active sodium transport in erythrocytes (KAHN and ACHESON, 1955), ascorbic acid was tested for receptor affinity. This could however not be substantiated (ERDMANN and SCHONER, 1974a). Probably, ascorbic acid exerts its inhibitory effect on Na^+, K^+-ATPase (FREY et al., 1973) at a different site.

Prednisone-3,20-*bis*-guanylhydrazone and prednisolon-3,20-*bis*-guanylhydrazone inhibit Na^+, K^+-ATPase (DRANSFELD and GREEFF, 1964; YAMAMOTO et al., 1978a), the active cation transport in red cells (GREEFF et al., 1964) and have a positive inotropic action (GREEFF and SCHLIEPER, 1967). These substances also displace ^3H-ouabain from its receptor in ox brain membranes (ERDMANN and SCHONER, 1974a) and guinea-pig heart (YAMAMOTO et al., 1978b). Dissociation constants of 7.8 μM and 8.7 μM respectively for the drug–receptor complexes were measured in ox brain cell membranes. This low affinity is in accordance with the rather high IC_{50} values of about 10^{-6}–10^{-5} M. There are, however, significant species differences in the susceptability to these drugs (DRANSFELD et al., 1967). In rat brain it could be established by indirect measurements (displacement of ^3H-ouabain, antagonism of K^+ on the inhibitory effect on Na^+, K^+-ATPase etc.) that prednisolon-3,20-*bis*-guanylhydrazone binding to the receptor is competitive to that of ^3H-ouabain (YAMAMOTO, 1978). Because of the rather low affinity to the receptor one might speculate on the structure–activity relation. Apparently the guanylhydrazone group seems not to be sufficient in replacing the missing lactone ring. Moreover the C–D ring junction of the prednison-3,20-*bis*-guanylhydrazone and prednisolon-3,20-*bis*-guanylhydrazone has a *trans* configuration instead of the *cis* configuration of the cardiac glycosides. It therefore cannot now be agreed upon, whether the missing lactone group or the different shape of the steroid nucleus is responsible for the rather low affinity to the cardiac glycoside receptor. Compounds without a lactone ring have not been investigated (to our knowledge) in drug–receptor binding studies. The inhibitory potency of such substances on Na^+, K^+-ATPase activity might be caused by secondary effects resulting in a change of the membrane composition. This, at least, cannot be excluded in experiments with a 17β-pyridazine ring substituted for the 17β-lactone ring to digoxin (TONA LUTETE et al., 1977).

Cassaine, an Erythrophleum alkaloid shows many of the pharmacological actions of cardiac glycosides but lacks the structural characteristics typical of cardiac glycosides. It displaces specifically bound ^3H-ouabain from the cardiac glycoside receptor, it has a very rapid dissociation from Na^+, K^+-ATPase, paralleling the rapid offset of cassaine-induced inotropy in several species. The inhibition of Na^+, K^+-ATPase activity is not competitive with respect to K^+ (TOBIN et al., 1975). Because of its high affinity to the receptor (25% of that ouabain), it is probably a specific agent at the cardiac glycoside-binding site.

III. Possible Application of Drug–Receptor Binding Studies in Experimental Pharmacology

Compounds with high affinity to the cardiac glycoside receptor are bound in spite of their low concentrations in the medium (incubation fluid, serum). They exhibit their pharmacologic effects at low concentrations. As a general rule it is thought that substances acting in low concentrations because of their high affinity to spe-

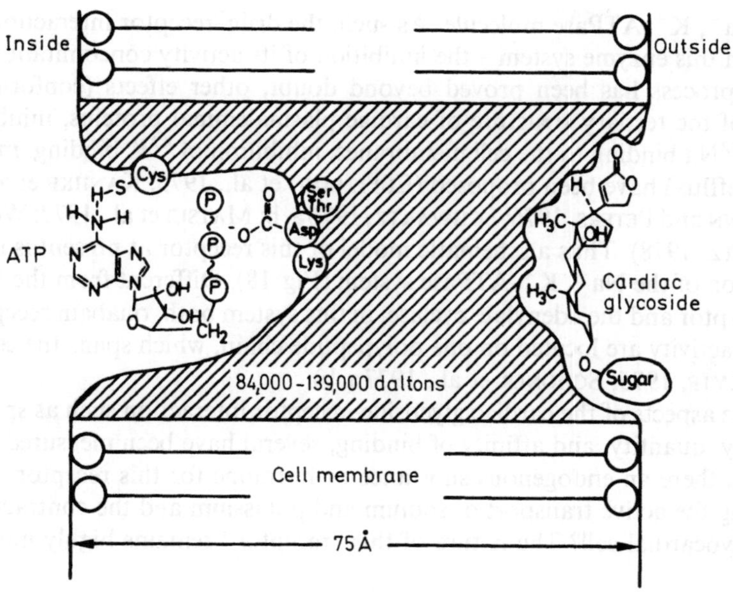

Fig. 18. Hypothetical molecular model of cardiac glycoside receptor and Na^+, K^+-ATPase. One of the membrane proteins spanning the cell membrane contains the binding site for cardiac glycosides on the outside and the enzymic activity (ATPase) on the inside. (SCHONER et al., 1977b)

cific receptors are bound to a lesser extent by nonspecific binding sites and therefore have less side effects.

In this respect it would be desirable to synthesize drugs with high affinity. On the other hand the low therapeutic range of cardiac glycosides finds its expression in the rather steep binding curves (Figs. 11 and 16) reflecting the specific and concentration-dependent drug–receptor interaction. The development of molecules exhibiting a more oblique binding curve (stretching over a larger concentration range) might reveal a broader therapeutic range. A screening for these or other properties with chemically modified molecules is possible by the use of binding studies with highly purified cardiac cell membranes of the desired species. These experiments are rather simple to perform, though some characteristic properties of drugs such as serum protein binding, metabolic fate, absorption, elimination etc., which greatly determine and modify their effects (SCHOLZ et al., 1979) cannot be measured by these experiments – this has advantages and disadvantages. A drug with high affinity to its specific receptor when tested in an intact animal might be metabolized too rapidly to be useful for therapy. Drug–receptor-binding studies will be important for the determination of the specificity and biologic potency of pharmacologically active substances in the future.

E. Conclusions

Although the exact mechanism of the action of cardiac glycosides remains controversial (AKERA and BRODY, 1978), it is established that these potent drugs interact with specific binding sites located in the cell membrane. These receptors are part of the Na^+, K^+-ATPase molecule. As such, the drug–receptor interaction has its effects on this enzyme system – the inhibition of its activity concomitant with the binding process has been proved beyond doubt, other effects (conformational change of the tertiary structure of cardiac plasmalemmal proteins, inhibition of specific ^{22}Na binding to the cell membranes, inhibition of ^{42}K binding, inhibition of ^{45}Ca efflux) have been proposed (LÜLLMANN et al., 1975; KANIIKE et al., 1976; LÜLLMANN and PETERS, 1976; SCHWARTZ, 1976a, b; MATSUI et al., 1977; WOOD and SCHWARTZ, 1978). Thus a schematic model of this receptor at present is mainly a description of the Na^+, K^+-ATPase system (Fig. 18). Different from the β-adrenergic receptor and the adenylate cyclase, in this system both, ouabain receptor and enzymic activity are located on one polypeptide chain, which spans the cell membrane (KYTE, 1975; SCHONER et al., 1977a, b).

Of the aspects of the cardiac glycoside–receptor interaction such as specificity, selectivity, quantity, and affinity of binding, several have been measured to some extent. Is there an endogenous substance or hormone for this receptor, possibly regulating the active transport of sodium and potassium and the contractile state of the myocardial cell? The nature of this "receptor" remains highly uncertain.

References

Akera, T., Larsen, F.S., Brody, T.M.: The effect of ouabain on sodium- and potassium-activated adenosine triphosphatase from the hearts of several mammalian species. J. Pharmacol. Exp. Ther. *170*, 17–26 (1969)

Akera, T.: Quantitative aspects of the interaction between ouabain and ($Na^+ + K^+$)-activated ATPase in vitro. Biochim. Biophys. Acta (Amst.) *249*, 53–62 (1971)

Akera, T., Brody, T.M.: Membrane adenosine triphosphatase: the effect of potassium on the formation and dissociation of the ouabain-enzyme complex. J. Pharmacol. Exp. Ther. *176*, 545–557 (1971)

Akera, T., Baskin, S.I., Tobin, T., Brody, T.M.: Ouabain: temporal relationship between the inotropic effect and the in vitro binding to, and dissociation from, ($Na^+ + K^+$)-activated ATPase. Naunyn-Schmiedeberg's Arch. Pharmacol. *277*, 151–162 (1973)

Akera, T., Bennett, R.T., Olgaard, M.K., Brody, T.M.: Cardiac Na^+, K^+-adenosine triphosphatase inhibition by ouabain and myocardial sodium: a computer simulation. J. Pharmacol. Exp. Ther. *199*, 287–297 (1976)

Akera, T.: Membrane adenosinetriphosphatase: a digitalis receptor? Science *198*, 569–574 (1977)

Akera, T., Brody, T.M.: The role of Na^+, K^+-ATPase in the inotropic action of digitalis. Pharmacol. Rev. *29*, 187–220 (1978)

Akera, T., Temma, K., Wiest, S.A., Brody, T.M.: Reduction of the equilibrium binding of cardiac glycosides and related compounds to Na^+, K^+-ATPase as a possible mechanism for the potassium-induced reversal of their toxicity. Naunyn-Schmiedebergs Arch. Pharmacol. *304*, 157–165 (1978)

Albers, R.W., Koval, G.J., Siegel, G.J.: Studies on the interaction of ouabain and other cardio-active steroids with sodium-potassium-activated adenosine triphosphatase. Mol. Pharmacol. *4*, 324–336 (1968)

Alexander, D.R.: Isolation of a digitoxin-bound protein from a brain membrane preparation containing Na^+, K^+-activated ATPase. FEBS Letters *45*, 150–154 (1974)

Allen, J.C., Schwartz, A.: A possible biochemical explanation for the insensitivity of the rat to cardiac glycosides. J. Pharmacol. Exp. Ther. *168*, 42–46 (1969)

Allen, J.C., Schwartz, A.: Effects of potassium, temperature, and time on ouabain interaction with the cardiac Na, K^+-ATPase: further evidence supporting an allosteric site. J. Mol. Cell. Cardiol. *1*, 39–45 (1970)

Allen, J.C., Besch, H.R. Jr., Glick, G., Schwartz, A.: The binding of tritiated ouabain to sodium- and potassium-activated adenosine triphosphatase and cardiac relaxing system of perfused dog heart. Mol. Pharmacol. *6*, 441–443 (1970)

Allen, J.C., Martinez-Maldonado, M., Eknoyan, G., Suki, W.N., Schwartz, A.: Relation between digitalis binding in vivo and inhibition of sodium, potassium-adenosine triphosphatase in canine kidney. Biochem. Pharmacol. *20*, 73–80 (1971)

Allen, J.C., Harris, R.A., Schwartz, A.: The nature of the transport ATPase-digitalis complex. I. Formation and reversibility in the presence and absence of a phosphorylated enzyme. Biochem. Biophys. Res. Commun. *42*, 366–370 (1971)

Allen, J.C., Entman, M.L., Schwartz, A.: The nature of the transport adenosine triphosphatase-digitalis complex. VIII. The relationship between in vivo-formed (^{3}*H*ouabain-Na^+, K^+-adenosine triphosphatase) complex and ouabain-induced positive inotropism. J. Pharmacol. Exp. Ther. *192*, 105–112 (1975)

Anderson, G.I., Bailey, J.C., Reiser, J., Freeman, A.: Electrophysiological observations on the digitalis-potassium interaction in canine purkinje fibers. Circ. Res. *39*, 717–723 (1976)

Askari, A. editor: Properties and functions of ($Na^+ + K^+$)-activated adenosinetriphosphatase. Ann. N.Y. Acad. Sci. *242*, 1–741 (1974)

Baker, P.F., Willis, J.S.: Binding of the cardiac glycoside ouabain to intact cells. J. Physiol. (Lond.) *224*, 441–462 (1972)

Barnett, R.E.: Effect of monovalent cations on the ouabain inhibition of the sodium and potassium ion activated adenosine triphosphatase. Biochemistry *9*, 4644–4648 (1970)

Bartels, H.-J., Hesch, R.-D.: Homotrope kooperative Effekte und aufsteigende B/F-Kurven bei Hormon-Antikörperreaktionen. Z. Klin. Chem. Klin. Biochem. *11*, 311–318 (1973)

Beard, N.A., Rouse, W., Somerville, A.R.: Cardiotonic steroids: correlation of sodium-potassium adenosine triphosphate inhibition and ion transport in vitro with inotropic activity and toxicity in dogs. Br. J. Pharmacol. *54*, 65–74 (1975)

Beller, G.A., Smith, Th.W., Hood, W.B.: Effects of ischemia and coronary reperfusion on myocardial digoxin uptake. Am. J. Cardiol. *36*, 902–907 (1975)

Beller, G.A., Conroy, J., Smith, Th.W.: Ischemia-induced alterations in myocardial ($Na^+ + K^+$)-ATPase and cardiac glycoside binding. J. Clin. Invest. *57*, 341–350 (1976)

Besch, H.R., Allen, J.C., Glick, G., Schwartz, A.: Correlation between the inotropic action of ouabain and its effects on subcellular enzyme systems from canine myocardium. J. Pharmacol. Exp. Ther. *171*, 1–12 (1970)

Blecher, M.: Methods in receptor research. I and II, pp. 1–763. New York: Marcel Dekker 1976

Bluschke, V., Bonn, R., Greeff, K.: Increase in the ($Na^+ + K^+$)-ATPase activity in heart muscle after chronic treatment with digitoxin or potassium deficient diet. Eur. J. Pharmacol. *37*, 189–191 (1976)

Boeynaems, J.M., Dumont, J.E.: Quantitative analysis of the binding of ligands to their receptors. J. Cyclic Nucleotide Res. *1*, 123–142 (1975)

Boeynaems, J.M., Dumont, J.E.: The two-step model of ligand-receptor interaction. Mol. Cell. Endocrinol. *7*, 33–47 (1977)

Bonn, R., Greeff, K.: The effect of chronic administration of digitoxin on the activity of the myocardial (Na+K)-ATPase in guinea-pigs. Arch. Int. Pharmacodyn. Ther. *233*, 53–64 (1978)

Borsch-Galetke, E., Dransfeld, H., Greeff, K.: Specific activity and sensitivity to strophanthin of the $Na^+ + K^+$-activated ATPase in rats and guinea-pigs with hypoadrenalism. Naunyn-Schmiedebergs Arch. Pharmacol. *274*, 74–80 (1972)

Bossaller, C., Schmoldt, A.: Dehydro-digitoxosides of digitoxigenin and digoxigenin: binding to beef heart ($Na^+ + K^+$)-ATPase in relation to unchanged digitoxosides. Naunyn-Schmiedebergs Arch. Pharmacol. *306*, 11–15 (1979)

Bower, J.O., Mengle, H.A.K.: The additive effect of calcium and digitalis. Am. Med. Assoc. *106*, 1151–1153 (1936)

Brody, T.M.: Discussion paper: ouabain-induced inhibition of cardiac ($Na^+ + K^+$)-ATPase and the positive inotropic response. Ann. N.Y., Acad. Sci. *242*, 684–687 (1974)

Brody, T.M., Akera, T.: Relations among Na^+, K^+-ATPase activity, sodium pump activity, transmembrane sodium movement, and cardiac contractility. Fed. Proc. *36*, 2219–2224 (1977)

Burck, H.C., Haasis, R., Larbig, D.: Beeinflussung der Erythrocyten-Elektrolyte durch β-Methyl-Digoxin bei Gesunden. Klin. Wochenschr. *53*, 125–128 (1975)

Caldwell, P.C., Keynes, R.D.: The effect of ouabain on the efflux of sodium from a squid giant axon. J. Physiol. (Lond.) *148*, 8P–9P (1959)

Cantley, L.C. Jr., Cantley, L.G., Josephson, L.: A characterization of vanadate interactions with the (Na, K)-ATPase. J. Biol. Chem. *253*, 7361–7368 (1978)

Chamness, G.C., McGuire, W.L.: Scatchard plots: common errors in correction and interpretation. Steroids *26*, 538–542 (1975)

Chan, P.C., Sanslone, W.R.: The influence of a low-potassium diet on rat-erythrocyten-membrane adenosine triphosphatase. Arch. Biochem. Biophys. *134*, 48–52 (1969)

Chang, K.-L., Jacobs, S., Cuatrecasas, P.: Quantitative aspects of hormone-receptor interactions of high affinity. Effect of receptor concentration and measurement of dissociation constants of labeled and unlabeled hormones. Biochim. Biophys. Acta (Amst.) *406* 294–303 (1975)

Chang, R.S.L., Snyder, S.H.: Benzodiazepine receptors: labeling in intact animals with ^3H-flunitrazepam. Eur. J. Pharmacol. *48*, 213–218 (1978)

Charnock, J.S., Cook, D.A., Almeida, A.F., To, R.: Activation energy and phospholipid requirements of membrane-bound adenosine triphosphatases. Arch. Biochem. Biophys. *159*, 393–399 (1973)

Chipperfield, A.R., Whittam, R.: Ouabain binding to the sodium pump. Nature *242*, 62–63 (1973)

Choi, Y.R., Akera, T.: Kinetics studies on the interaction between ouabain and (Na^+, K^+)-ATPase. Biochim. Biophys. Acta (Amst.) *481*, 648–659 (1977)

Chopra, D., Janson, P., Sawin, C.T.: Insensitivity to digoxin associated with hypocalcemia. New Engl. J. Med. *296*, 917–918 (1977)

Clark, A.F., Swanson, P.D., Stahl, W.L.: Increase in dissociation rate constants of cardiotonic steroid-brain ($Na^+ + K^+$)-ATPase complexes by reduction of the unsaturated lactone. J. Biol. Chem. *250*, 9355–9359 (1975)

Clausen, T., Hansen, O.: Ouabain binding and Na^+-K^+transport in rat muscle cells and adipocytes. Biochim. Biophys. Acta (Amst.) *345*, 387–404 (1974)

Clausen, T., Hansen, O.: Active Na-K transport and the rate of ouabain binding. The effect of insulin and other stimuli on skeletal muscle and adipocytes. J. Physiol. (Lond.) *270*, 415–430 (1977)

Cohen, I., Daut, J., Noble, D.: The influence of extracellular potassium ions on the action of ouabain on membrane currents in sheep purkinje fibers. J. Physiol. (Lond.) *251*, 42P–43P (1976)

Coltart, D.J., Güllner, H.G., Billingham, A., Goldman, R.H., Stinson, E.B., Kalman, S.M., Harrison, D.C.: Physiological distribution of digoxin in human heart. Br. Med. J. *28*, 733–736 (1974)

Cuatrecasas, P.: Insulin receptor of liver and fat cell membranes. Fed. Proc. *32*, 1836–1846 (1973)

Cuatrecasas, P., Hollenberg, M.D.: Binding of insulin and other hormones to non-receptor materials: saturability, specificity, and apparent "negative cooperativity." Biochem. Biophys. Res. Commun. *62*, 31–40 (1975)

Curfman, G.D., Crowley, T.J., Smith, Th.W.: Thyroid-induced alterations in myocardial sodium- and potassium-activated adenosine triphosphatase, monovalent cation active transport, and cardiac glycoside binding. J. Clin. Invest. *59*, 586–590 (1977)

Dahl, J.L., Hokin, L.E.: The sodium-potassium adenosinetriphosphatase. Annu. Rev. Biochem. *43*, 327–356 (1974)

Deguchi, N., Jørgensen, P.L., Maunsbach, A.B.: Ultrastructure of the sodium pump. Comparison of thin sectioning, negative staining, and freeze-fracture of purified, membrane bound ($Na^+ + K^+$)-ATPase. J. Cell. Biol. *75*, 619–634 (1977)

De Meyts, P., Roth, J.: Cooperativity in ligand binding: a new graphic analysis. Biochem. Biophys. Res. Commun. *66*, 1118–1126 (1975)

De Meyts, P.: Cooperative properties of hormone receptors in cell membranes. J. Supramol. Struct. *4*, 241–258 (1976)
Detweiler, D.K.: Comparative pharmacology of cardiac glycosides. Fed. Proc. *26*, 1119–1124 (1967)
Diacono, J.: Suggestive evidence for the activation of an electrogenic sodium pump in stimulated rat atria: apparent discrepancy between the pump inhibition and the positive inotropic response induced by ouabain. J. Mol. Cell. Cardiol. *11*, 5–30 (1979)
Dixon, J.F., Hokin, L.E.: Studies on the characterization of the sodium-potassium transport adenosine triphosphatase. Purification and properties of the enzyme from the electric organ of Electrophorus electricus. Arch. Biochem. Biophys. *163*, 749–758 (1974)
Doherty, J.E., Perkins, W.H.: Digoxin metabolism in hypo- and hyperthyroidism. Studies with tritiated digoxin in thyroid disease. Ann. Intern. Med. *64*, 489–507 (1966)
Doherty, J.E., Perkins, W.H.: The distribution and concentration of tritiated digoxin in human tissues. Ann. Int. Med. *65*, 116–124 (1967)
Dransfeld, H., Greeff, K.: Der Einfluß des Prednison- und Prednisolonbisguanylhydrazons auf die $Na^+ + K^+$-stimulierte Membran-ATPase des Meerschweinchenherzens. Naunyn-Schmiedebergs Arch. Exp. Path. Pharmak. *249*, 425–431 (1964)
Dransfeld, H., Greeff, K., Berger, H., Cautius, V.: Die verschiedene Empfindlichkeit der $Na^+ + K^+$-aktivierten ATPase des Herz- und Skeletmuskels gegen k-Strophanthin. Naunyn-Schmiedebergs Arch. Exp. Path. Pharmak. *254*, 225–234 (1966)
Dransfeld, H., Galetke, E., Greeff, K.: Die Wirkung des Prednisolonbisguanylhydrazons auf die $Na^+ + K^+$-aktivierte Membran-ATPase des Herz- und Skeletmuskels. Arch. Int. Pharmacodyn. *166*, 342–349 (1967)
Dransfeld, H., Lipinski, J., Borosch-Galetke, E.: Die $Na^+ + K^+$-aktivierte Transport-ATPase bei experimenteller Herzinsuffizienz durch Kobaltchlorid. Naunyn-Schmiedebergs Arch. Pharmacol. *270*, 335–342 (1971)
Dunham, P.B., Hoffman, J.F.: Partial purification of the ouabain-binding component and of Na, K-ATPase from human red cell membranes. Proc. Natl. Acad. Sci. U.S.A. *66*, 936–943 (1970)
Dunham, P.B., Gunn, R.B.: Adenosine triphosphatase and active cation transport in red blood cell membranes. Arch. Intern. Med. *129*, 241–247 (1972)
Dunham, P.B., Blostein, R.: Active potassium transport in reticulocytes of high-K^+ and low-K^+ sheep. Biochim. Biophys. Acta (Amst.) *455*, 749–758 (1976)
Dutta, S., Goswamin, S., Lindower, J.O., Marks, B.H.: Subcellular distribution of digoxin-H^3 in isolated guinea-pig and rat hearts. J. Pharmacol. Exp. Ther. *159*, 324–334 (1968)
Dutta, S., Marks, B.H.: Factors that regulate ouabain-H^3 accumulation by the isolated guinea-pig heart. J. Pharmacol. Exp. Ther. *170*, 318–325 (1969)
Engel, H., Proppe, D., Wassermann, O.: Influence of highly unsaturated phosphatidylcholine on the effects of ouabain and some cardioactive drugs on cardiac contractile force and Na^+, K^+-ATPase activity. Biochem. Pharmacol. *26*, 381–388 (1977)
Ehrlich, P.: Chemotherapeutics: scientific principles, methods and results. Lancet 1913 II, 445–451
Erdmann, E., Schoner, W.: Ouabain-receptor interactions in $(Na^+ + K^+)$-ATPase preparations from different tissues and species. Determination of kinetic constants and dissociation constans. Biochim. Biophys. Acta (Amst.) *307*, 386–398 (1973a)
Erdmann, E., Schoner, W.: Ouabain-receptor interactions in $(Na^+ + K^+)$-ATPase preparations. II. Effect of cations and nucleotides on rate constants and dissociation constants. Biochim. Biophys. Acta (Amst.) *330*, 302–315 (1973b)
Erdmann, E., Schoner, W.: Ouabain-receptor interactions in $(Na^+ + K^+)$-ATPase preparations. III. On the stability of the ouabain receptor against physical treatment, hydrolases, and SH reagents. Biochim. Biophys. Acta (Amst.) *330*, 316–324 (1973c)
Erdmann, E., Schoner, W.: Ouabain-receptor interactions in $(Na^+ + K^+)$-ATPase preparations. IV. The molecular structure of different cardioactive steroids and other substances and their affinity to the glycoside receptor. Naunyn-Schmiedebergs Arch. Pharmacol. *283*, 335–356 (1974a)
Erdmann, E., Schoner, W.: Die Affinität verschieden strukturierter Herzglykoside sowie DPH und Ro 2-2985 zum Herzglykosidrezeptor. Verh. Dtsch. Ges. Kreislaufforsch. *40*, 309–314 (1974b)

Erdmann, E., Hasse, W.: Quantitative aspects of ouabain binding to human erythrocyte and cardiac membranes. J. Physiol. (Lond.) *251*, 671–682 (1975)

Erdmann, E., Presek, P., Swozil. R.: Über den Einfluß von Kalium auf die Bindung von Strophanthin an menschliche Herzmuskelzellmembranen. Klin. Wochenschr. *54*, 383–387 (1976a)

Erdmann, E., Patzelt, R., Schoner, W.: The cardiac glycoside receptor: its properties and its correlation to nucleotide binding sites, phosphointermediate, and $(Na^+ + K^+)$-ATPase activity. Rec. Adv. Stud. Cardiac Struct. Metab. *9*, 329–335 (1976b)

Erdmann, E., Philipp, G., Tanner, G.: Ouabain-receptor interactions in $(Na^+ + K^+)$-ATPase preparations. A contribution to the problem of nonlinear Scatchard plots. Biochim. Biophys. Acta (Amst.) *455*, 278–296 (1976c)

Erdmann, E.: Cell membrane receptors for cardiac glycosides in the heart. Membrangebundene Herzglykosidrezeptoren der Herzmuskelzelle. Basic Res. Cardiol. *72*, 315–325 (1977)

Erdmann, E., Krawietz, W.: Increased number of ouabain binding sites in human erythrocyte membranes in chronic hypokalaemia. Acta Biol. Med. Germ. *36*, 879–883 (1977)

Erdmann, E., Krawietz, W., Presek, P.: Receptor for cardiac glycosides. In: Riecker, G., Weber, A., Goodwin, J. (eds.), pp. 120–131. Myocardial failure. Berlin, Heidelberg, New York: Springer 1977

Erdmann, E.: Vergleichende Messungen der Herzglykosid-Rezeptoraffinität und der Hemmung der $(Na^+ + K^+)$-ATPase durch Digitoxin, Digoxin, Methyldigoxin, Strophanthin, Proscillaridin und Meproscillarin an isolierten menschlichen Herzmuskelzellmembranen. Arzneim. Forsch. *28*, 531–535 (1978a)

Erdmann, E.: Quantitative Aspekte der spezifischen Bindung von Herzglykosiden an Membranrezeptoren. Habilitationsschrift, München, 1978b

Erdmann, E., Krawietz, W., Koch, M.: Cardiac glycoside receptors in disease. The number of ouabain binding sites in human erythrocytes is subject to regulation. In: $Na^+ + K^+$-ATPase structure, and kinetics. Skou, J.C., Nørby, J.G. (eds.), pp. 517–524. London: Academic Press 1979

Erdmann, E., Werdan, K., Krawietz, W., Koch, M.: Effect of vanadate on $(Na^+ + K^+)$-ATPase and on ouabain binding in mammalian cardiac muscle. Naunyn-Schmiedebergs Arch. Pharmacol. *307*, R37 (1979b)

Erdmann, E., Krawietz, W., Philipp, G., Hackbarth, I., Schmitz, W., Scholz, H.: Stimulatory effect of vanadate on $(Na^+ + K^+)$-ATPase activity and on ^3H-ouabain-binding in a cat heart cell membrane preparation. Nature *278*, 459–461 (1979c)

Erdmann, E., Philipp, G., Scholz, H.: Cardiac glycoside receptor, $(Na^+ + K^+)$-ATPase activity and force of concentration in rat heart. Biochem. Pharmacol. *29*, 3219–3229 (1980)

Flier, J., Kahn, C.R., Roth, J.: Receptors, antireceptor antibodies, and mechanism of insulin resistance. New Engl. J. Med. *300*, 413–419 (1979)

Fortes, G.: Anthroylouabain: a specific fluorescent probe for the cardiac glycoside receptor of the Na-K-ATPase. Biochemistry *16*, 531–540 (1977)

Frey, M., Pitts, B.J.R., Askari, A.: Vitamin C-effects on the Na^+, K^+ adenosine triphosphate phosphohydrolase complexes of several tissues. Biochem. Pharmacol. *22*, 9–15 (1973)

Fricke, U., Hollborn, U., Klaus, W.: Inotropic action, myocardial uptake and subcellular distribution of ouabain, digoxin and digitoxin in isolated rat hearts. Naunyn-Schmiedebergs Arch. Pharmacol. *288*, 195–214 (1975)

Fricke, U., Klaus, W.: Evidence for two different Na^+-dependent ^3H-ouabain binding sites of a Na^+-K^+-ATPase of guinea-pig hearts. Br. J. Pharmacol. *61*, 423–428 (1977)

Fricke, U.: Myocardial activity of inhibitors of the Na^+-K^+-ATPase: differences in the mode of action and subcellular distribution pattern of N-ethylmaleimide and ouabain. Naunyn-Schmiedebergs Arch. Pharmacol. *303*, 197–204 (1978)

Gardner, J.D., Conlon, Th.P.: The effects of sodium and potassium on ouabain binding by human erythrocytes. J. Gen. Physiol. *60*, 609–629 (1972)

Gardner, J.D., Kiino, D.R.: Ouabain binding and cation transport in human erythrocytes. J. Clin. Invest. *52*, 1845–1851 (1973)

Gardner, J.D., Kiino, D.R., Swartz, T.J., Butler, V.P. Jr.: Effects of digoxin-specific antibodies on accumulation and binding of digoxin by human erythrocytes. J. Clin. Invest. *52*, 1820–1833 (1973)
Gardner, J.D., Frantz, C.: Effects of cations on ouabain binding by intact human erythrocytes. J. Membrane Biol. *16*, 43–64 (1974)
Ghysel-Burton, J., Godfraind, T.: Stimulation and inhibition by ouabain of the sodium pump in guinea-pig atria. Br. J. Pharmacol. *55*, 249P (1975)
Giotta, G.J.: Native ($Na^+ + K^+$)-dependent adenosine triphosphatase has two trypsin-sensitive sites. J. Biol. Chem. *250*, 5159–5164 (1975)
Glynn, I.M.: The action of cardiac glycosides on ion movements. Pharmacol. Rev. *16*, 381–407 (1964)
Glynn, I.M., Karlish, S.J.D.: The sodium pump. Annu. Rev. Physiol. *37*, 13–55 (1975)
Godfraind, T.: Cardiac glycoside receptors in the heart. Biochem. Pharmacol. *24*, 823–827 (1975)
Godfraind, T., Ghysel-Burton, J.: Binding sites related to ouabain-induced stimulation or inhibition of the sodium pump. Nature *265*, 165–166 (1977)
Godfraind, T., Ghysel-Burton, J.: The action of digoxin and digoxigenin-monodigitoxoside on the sodium pump and on the contractility in isolated guinea-pig atria. Arch. Int. Pharmacodyn. Ther. *234*, 340–341 (1978)
Goldstein, A., Arrow, L., Kalman, S.M.: Principles of drug action. New York, London, Sydney, Toronto: J. Wiley and Sons 1974
Greeff, K., Meng, K., Schwarzmann, D.: Digitalis-ähnliche Eigenschaften des Prednison- und Prednisolonbisguanylhydrazons. Ihre Wirkung auf die Kaliumbilanz isolierter Herzpräparate und den Na/K-Transport an Erythrocyten. Naunyn-Schmiedebergs Arch. Exp. Path. Pharmak. *249*, 416–424 (1964)
Greeff, K., Schlieper, E.: Artspezifische Wirkungsunterschiede des k-Strophanthins und Prednisolonbisguanylhydrazons: Untersuchungen an isolierten Vorhofpräparaten und Erythrocyten des Menschen, Meerschweinchens, Kaninchens und der Ratte. Arch. Int. Pharmacodyn. *166;* 350–361 (1967)
Greeff, K.: Bestimmungen des Blutspiegels von Digoxin, Digitoxin und g-Strophanthin mit Hilfe radioimmunologischer Methoden. Herz. Kreisl. *6*, 145–149 (1974)
Greeff, K., Strobach, H., Verspohl, E.: Ergebnisse radioimmunologischer Bestimmungen von Digitoxin, Digoxin und g-Strophanthin am Menschen. In: Digitalistherapie, Beiträge zur Pharmakologie und Klinik. Jahrmärker, H. (Hrsg), S. 52–61. Berlin, Heidelberg, New York: Springer 1975
Güllner, H.-G., Stinson, E.B., Harrison, D.C., Kalman, S.M.: Correlation of serum concentrations with heart concentrations of digoxin in human subjects. Circulation *50*, 653–655 (1974)
Haasis, R., Larbig, D., Stunkat, R., Bader, H., Seboldt, H.: Radioimmunologische Bestimmung der Glykosidkonzentration im menschlichen Gewebe. Klin. Wochenschr. *55*, 23–30 (1977)
Hackbarth, I., Schmitz, W., Scholz, H., Erdmann, E., Krawietz, W., Philipp, G.: Positive inotropism of vanadate in cat papillary muscle. Nature *275*, 67 (1978)
Hagen, H.: Behandlung von Herzrhythmusstörungen mit Diphenylhydantoin. Dtsch. Med. Wochenschr. *96*, 380–384 (1971)
Han, C.S., Tobin, T., Brody, T.M.: Effects of alkali metal cations on phospho-enzyme levels and ^{3}H-ouabain binding to ($Na^+ + K^+$)-ATPase. Biochim. Biophys. Acta (Amst.) *429*, 993–1005 (1976)
Hansen, O.: The relationship between g-strophanthin binding capacity and ATPase-activity in plasma membrane fragments of ox brain. Biochim. Biophys. Acta (Amst.) *233*, 122–132 (1971)
Hansen, O., Jensen, J., Nørby, J.G.: Mutual exclusion of ATP, ADP, and g-strophanthin binding to NaK-ATPase. Nature New Biol. *234*, 122–124 (1971)
Hansen, O., Skou, J.C.: A study on the influence of the concentration of Mg^{2+}, Pi, K^+, Na^+, and tris on (Mg^{2+} + Pi)-supported g-strophanthin binding to ($Na^+ + K^+$)-activated ATPase from ox brain. Biochim. Biophys. Acta (Amst.) *311*, 51–66 (1973)

Hansen, O.: Non-uniform populations of g-strophanthin binding sites of $(Na^+ + K^+)$-activated ATPase. Apparent conversion to uniformity by K^+. Biochim. Biophys. Acta (Amst.) *433*, 383–392 (1976)
Hansen, O.: The effect of sodium on inorganic phosphate- and p-nitrophenyl phosphate-facilitated ouabain binding to $(Na^+ + K^+)$-activated ATPase. Biochim. Biophys. Acta (Amst.) *511*, 10–22 (1978 a)
Hansen, O.: Ouabain used as a tool for trapping and characterizing phosphorylation products of NaK-ATPase. FEBS, 11th Meeting 1977 Copenhagen. Oxford, New York: Pergamon Press 1978 b A-4-2-602
Hansen, O.: Reactive states of the Na^+, K^+-ATPase demonstrated by the stability of the enzyme-ouabain complex. In: Na^+, K^+-ATPase, structure, and kinetics. Skou, J.C., Nørby, J.G. (eds.), pp. 169–180. London, New York: Academic Press 1979
Harper, N.J., Simmonds, A.B.: Advances in drug research. London: New York: Academic Press 1977
Harris, W.E., Swanson, P.D., Stahl, W.L.: Ouabain binding sites and the $(Na^+$, $K^+)$-ATPase of brain microsomal membranes. Biochim. Biophys. Acta (Amst.) *298*, 680–689 (1973)
Härtel, G., Kyllönen, K., Merikallio, E., Ojala, K., Manninen, V., Reissell, P.: Human serum and myocardium digoxin. Clin. Pharmacol. Ther. *19*, 153–157 (1976)
Hegyvary, C.: Covalent labeling of the digitalis-binding component of plasma membranes. Mol. Pharmacol. *11*, 588–594 (1975)
Heller, M., Beck, S.: Interactions of cardiac glycosides with cells and membranes. Properties and structural aspects of two receptor sites for ouabain in erythrocytes. Biochim. Biophys. Acta (Amst.) *514*, 332–347 (1978)
Hoffman, J.F.: The red cell membrane and the transport of sodium and potassium. Am. J. Med. *41*, 666–680 (1966)
Hollemans, H.J.G., Bertina, R.M.: Scatchard plot and heterogeneity in binding affinity of labeled and unlabeled ligand. Clin. Chem. *21*, 1769–1773 (1975)
Hollenberg, M.D., Cuatrecasas, P.: Membrane receptors and hormone action: recent developments. Prog. Neuro-Psychopharmacol. *2*, 287–302 (1978)
Horvath, I., Kiraly, C., Szerb, J.: Action of cardiac glycosides on the polymerisation of actin. Nature *165*, 792 (1949)
Hotovy, R., König, W.: Weitere pharmakologische Studien an der Kaumuskulatur der Ratte. Arch. Exp. Path. Pharmakol. *213*, 175–184 (1951)
Inagaki, C., Lindenmayer, G., Schwartz, A.: Effects of sodium and potassium on binding of ouabain to the transport adenosine triphosphatase. J. Biol. Chem. *249*, 5135–5140 (1974)
Jacobs, S., Chang, K.-J., Cuatrecasas, P.: Estimation of hormone receptor affinity by competitive displacement of labeled ligand: effect of concentration of receptor and of labeled ligand. Biochem. Biophys. Res. Commun. *66*, 687–692 (1975)
Jacobsen, A.L.: Effect of ouabain on the ATPase of cardiac myosin R at high ionic strength. Circ. Res. *22*, 625–632 (1968)
Jopiner, C.H., Lauf, P.K.: The correlation between ouabain binding and potassium pump inhibition in human and sheep erythrocytes. J. Physiol. (Lond.) *283*, 155–175 (1978)
Jørgensen, P.L.: Purification and characterization of $(Na^+ + K^+)$-ATPase. IV. Estimation of the purity and of the molecular weight and polypeptide content per enzyme unit in preparations from the outer medulla of rabbit kidney. Biochim. Biophys. Acta (Amst.) *356*, 53–67 (1974)
Juliano, R.L.: The proteins of the erythrocyte membrane. Biochim. Biophys. Acta (Amst.) *300*, 341–378 (1973)
Kahn, J.B. Jr., Acheson, G.H.: Effects of cardiac glycosides and other lactones, and of certain other compounds, on cation transfer in human erythrocytes. J. Pharmacol. Exp. Ther. *115*, 305–318 (1955)
Kaniike, K., Erdmann, E., Schoner, W.: Study on the differential modifications of $(Na^+ + K^+)$-ATPase and its partial reactions by dimethylsulfoxide. Biochim. Biophys. Acta (Amst.) *352*, 275–286 (1974)

Kaniike, K., Lindenmayer, G.E., Wallick, E.T., Lane, L.K., Schwartz, A.: Specific sodium-22 binding to a purified sodium + potassium adenosine triphosphatase. J. Biol. Chem. *251*, 4794–4795 (1976)

Katz, A.I., Lindheimer, M.D.: Renal sodium- and potassium-activated adenosine triphosphatase and sodium reabsorption in the hypothyroid rat. J. Clin. Invest. *52*, 796–804 (1973)

Kokenge, F., Van Zwieten, P.A.: A diminished response to digoxin in isolated heart muscle as a result of fever. Klin. Wschr. *49*, 1236–1237 (1971)

Krawietz, W., Erdmann, E.: Specific and unspecific binding of ^{3}H-dihydroalprenolol to cardiac tissue. Biochem. Pharmacol. *28*, 1283–1288 (1979)

Ku, D., Akera, T., Pew, C.L., Brody, T.M.: Cardiac glycosides: correlations among Na^+, K^+-ATPase, sodium pump and contractility in the guinea pig heart. Naunyn-Schmiedebergs Arch. Pharmacol. *285*, 185–200 (1974)

Ku, D., Akera, T., Tobin, T., Brody, T.M.: Effects of monovalent cations on cardiac Na^+, K^+-ATPase activity and on contractile force. Naunyn-Schmiedebergs Arch. Pharmacol. *290*, 113–131 (1975)

Ku, D., Akera, T., Brody, T.M., Weaver, L.C.: Chronic digoxin treatment on canine myocardial Na^+, K^+-ATPase. Naunyn-Schmiedebergs Arch. Pharmacol. *301*, 39–47 (1977)

Kyte, J.: Properties of the two polypeptides of sodium- and potassium-dependent adenosine triphosphatase. J. Biol. Chem. *247*, 7642–7649 (1972a)

Kyte, J.: The titration of the cardiac glycoside binding site of the ($Na^+ + K^+$)-adenosine triphosphatase. J. Biol. Chem. *247*, 7634–7641 (1972b)

Kyte, J.: Structural studies of sodium and potassium ion-activated adenosine triphosphatase. The relationship between molecular structure and mechanism of active transport. J. Biol. Chem. *250*, 7443–7449 (1975)

La Bella, F.S., Bihler, I., Kim, R.S.: Progesterone derivative binds to cardiac ouabain receptor and shows dissociation between sodium pump inhibition and increased contractile force. Nature *278*, 571–573 (1979)

Lane, L.K., Copenhaver, J.H., Lindenmayer, G.E., Schwartz, A.: Purification and characterization of and ^{3}H-ouabain binding to the transport adenosine triphosphatase from outer medulla of canine kidney. J. Biol. Chem. *248*, 7197–7200 (1973)

Lane, L.K., Anner, B.M., Wallick, E.T., Ray, M.V., Schwartz, A.: Effect of phospholipase a treatment on the partial reactions of and ouabain binding to a purified sodium and potassium activated adenosine triphosphatase. Biochem. Pharmacol. *27*, 225–231 (1978)

Langer, G.A.: Effects of digitalis on myocardial ionic exchange. Circulation *46*, 180–187 (1972)

Lauf, P.K., Joiner, C.H.: Increased potassium transport and ouabain binding in human Rh_{null} red blood cells. Blood *48*, 457–468 (1976)

Lauf, P.K., Shoemaker, D.G., Joiner, C.H.: Changes in K^+ pump transport and ouabain binding sites in erythrocytes of genetically low K^+ lambs. Biochim. Biophys. Acta (Amst.) *507*, 544–548 (1978)

Lee, K.S., Yu, D.H.: A study of the sodium- and potassium- activated adenosine trophosphatase activity of heart microsomal fraction. Biochem. Pharmacol. *12*, 1254–1264 (1963)

Lee, K.S., Klaus, W.: The subcellular basis for the mechanism of inotropic action of cardiac glycosides. Pharmacol. Rev. *23*, 193–261 (1971)

Lefkowitz, R.J.: β-adrenergic receptors: recognition and regulation. New Engl. J. Med. *295*, 323–328 (1976)

Lichey, J., Havestatt, Ch., Weinmann, J., Hasford, J., Rietbrock, N.: Human myocardium and plasma digoxin concentration in patients on long-term digoxin treatment. Int. J. Clin. Pharmacol. *16*, 460–462 (1978)

Lin, M.H., Akera, T.: Increased (Na^+, K^+)-ATPase concentrations in various tissues of rats caused by thyroid hormone treatment. J. Biol. Chem. *253*, 723–726 (1978)

Lin, M.H., Romsos, D.R., Akera, T., Leveille, G.A.: Na^+, K^+-ATPase enzyme units in skeletal muscle from lean and obese mice. Biochem. Biophys. Res. Commun. *80*, 398–404 (1978)

Lindenmayer, G.E., Schwartz, A.: Nature of the transport adenosine triphosphatase digitalis complex. IV. Evidence that sodium-potassium competition modulates the rate of ouabain interaction with ($Na^+ + K^+$) adenosine triphosphatase during enzyme catalysis. J. Biol. Chem. *248*, 1291–1300 (1973)

Lindenmayer, G.E., Schwartz, A., Thompson, H.K. Jr.: A kinetic description for sodium and potassium effects on ($Na^+ + K^+$)-adenosine triphosphatase: a model for a two-non-equivalent site potassium activation and an analysis of multiequivalent site models for sodium activation. J. Physiol. (Lond.) *236*, 1–28 (1974)

Lüllmann, H., Weber, R.: Über die Wirkung von Phenytoin auf Digitalis-bedingte Arrhythmien. Ärztl. Forsch. *22*, 49–55 (1968)

Lüllmann, H., Peters, T.: Cardiac glycosides and contractility. The myocardium. Adv. Cardiol. *12*, 174–182 (1974)

Lüllmann, H., Peters, T., Preuner, J., Rüther, T.: Influence of ouabain and dihydroouabain on the circular dichroism of cardiac plasmalemmal microsomes. Naunyn-Schmiedebergs Arch. Pharmacol. *290*, 1–19 (1975)

Lüllmann, H., Peters, T.: On the sarcolemmal site of action of cardiac glycosides. Recent Adv. Stud. Cardiac Struct. Metab. *9*, 311–327 (1976)

Lüllmann, H., Peters, T.: Action of cardiac glycosides. Prog. Pharmacol. *2*, 5–57 (1979)

Malcolm, A., Coltart, J.: Relation between concentrations of digoxin in the myocardium and in the plasma. Br. Heart J. *39*, 935–938 (1977)

Manery, J.F., Dryden, E.E., Still, J.S., Madapallimattam, G.: Enhancement (by ATP, insulin, and lack of divalent cations) of ouabain inhibition of cation transport and ouabain binding in frog skeletal muscle; effect of insulin and ouabain on sarcolemmal (Na + K)MgATPase. Can. J. Physiol. Pharmacol. *55*, 21–33 (1977)

Matsui, H., Schwartz, A.: ATP-dependent binding of H^3-digoxin to a Na^+, K^+-ATPase from cardiac muscle. Fed. Proc. *26*, 398 (1967)

Matsui, H., Schwartz, A.: Mechanism of cardiac glycoside inhibition of the ($Na^+ + K^+$)-dependent ATPase from cardiac tissue. Biochim. Biophys. Acta (Amst.) *151*, 655–663 (1968)

Matsui, H., Hayashi, Y., Homareda, H., Kimimura, M.: Ouabain-sensitive ^{42}K binding to Na^+, K^+-ATPase purified from canine kidney outer medulla. Biochem. Biophys. Res. Commun. *75*, 373–379 (1977)

McCall, D.: Cation exchange and glycoside binding in cultured rat heart cells. Am. J. Physiol. *236*, C87–C95 (1979)

McCans, J.L., Lindenmayer, G.E., Pitts, B.J.R., Ray, M.V., Raynor, B.D., Butler, V.P., Schwartz, A.: Antigenic differences in (Na^+, K^+)-ATPase preparations isolated from various organs and species. J. Biol. Chem. *250*, 7257–7265 (1975)

Michael, L., Schwartz, A., Wallick, E.: Nature of the transport adenosine triphosphatase-digitalis complex: XIV. Inotropy and cardiac glycoside interaction with Na^+, K^+-ATPase of isolated cat papillary muscles. Molec. Pharmacol. *16*, 135–146 (1979)

Murthy, R.V., Kidwai, A.M., Daniel, E.E.: Dissociation of contractile effect and binding and inhibition of Na^+-K^+-adenosine triphosphatase by cardiac glycosides in rabbit myometrium. J. Pharmacol. Exp. Ther. *188*, 575–581 (1974)

Nakao, M.: Several topics concerning Na, K-ATPase. Life Sci. *15*, 1849–1859 (1975)

Nathan, D.G., Gunn, R.B.: Thalassaemia: the consequences of unbalanced hemoglobin synthesis. Am. J. Med. *41*, 815–830 (1966)

Nola, G.T., Pope, S., Harrison, D.C.: Assessment of the synergistic relationship between serum calcium and digitalis. Am. Heart J. *79*, 499–507 (1970)

Okarma, T.B., Tramell, P., Kalman, S.M.: The surface interaction between digoxin and cultured heart cells. J. Pharmacol. Exp. Ther. *183*, 559–576 (1972)

Okita, G.T., Richardson, F., Roth-Schechter, B.F.: Dissociation of the positive inotropic action of digitalis from inhibition of sodium and potassium-activated adenosine triphosphatase. J. Pharmacol. Exp. Ther. *185*, 1–11 (1973)

Okita, G.T.: Dissociation of the positive inotropic effects from the cardiotoxic effects of digitalis. Proc. West. Pharmacol. Soc. *18*, 14–19 (1975)

Okita, G.T.: Dissociation of Na^+, K^+-ATPase inhibition from digitalis inotropy. Fed. Proc. *56*, 2225–2230 (1977)

Olson, R.E., Ellenbogen, E., Iyengar, R.: Cardiac myosin and congestive heart failure in the dog. Circulation *24*, 475–482 (1961)
Patzelt-Wenczler, R., Schoner, W.: Die Herzglykosid-Bindungsstelle der (Na^+, K^+)-ATPase nach Affinitätsmarkierung der ATP-Bindungsstelle mit (sITP). Verh. Dtsch. Ges. Kreislaufforsch. *41*, 311–314 (1975)
Perrone, J.P., Blostein, R.: Asymmetric interaction of inside-out and rightside-out erythrocyte membrane vesicles with ouabain. Biochim. Biophys. Acta (Amst.) *291*, 680–689 (1973)
Perrone, J.R., Hackney, J.F., Dixon, J.F., Hokin, L.E.: Molecular properties of purified (sodium + potassium)-activated adenosine triphosphatases and their subunits from the rectal gland of Squalus acanthias and the electric organ of Electrophorus electricus. J. Biol. Chem. *250*, 4178–4184 (1975)
Peters, T., Raben, R.-H., Wassermann, O.: Evidence for a dissociation between positive inotropic effect and inhibition of the ($Na^+ + K^+$)-ATPase by ouabain, cassaine, and their alkylating derivatives. Eur. J. Pharmacol. *26*, 166–174 (1974)
Pfleger, K., Kolassa, N., Heinrich, W., Schneider, M.: Pharmakokinetik und Wirkung von Digitoxin und Ouabain am isolierten Herzen von Meerschweinchen und Ratte. Arch. Int. Pharmacodyn. Ther. *216*, 130–143 (1975)
Philipson, K.D., Edelman, I.S.: Characteristics of thyroid-stimulated Na^+-K^+-ATPase of rat heart. Am. J. Physiol. *232*, 202–206 (1977)
Pitts, B.J.R., Schwartz, A.: Improved purification and partial characterization of (Na^+, K^+)-ATPase from cardiac muscle. Biochim. Biophys. Acta (Amst.) *401*, 184–195 (1975)
Pitts, B., Wallick, E.T., van Winkle, W.B., Allen, J.C., Schwartz, A.: On the lack of inotropy of cardiac glycosides on skeletal muscle: a comparison of Na^+, K^+-ATPase from skeletal and cardiac muscle. Arch. Biochem. Biophys. *184*, 431–440 (1977)
Reimann, E.M., Soloff, M.S.: The effect of radioactive contaminants on the estimation of binding parameters by Scatchard analysis. Biochim. Biophys. Acta (Amst.) *533*, 130–139 (1978)
Reiter, M.: Die Wertbestimmung inotrop wirkender Arzneimittel am isolierten Papillarmuskel. Arzneim. Forsch. *17*, 1240–1253 (1967)
Repke, K.: Metabolism of cardiac glycosides. New aspects of cardiac glycosides. III, pp. 47–73. Oxford: Pergamon Press 1964
Repke, K., Est, M., Portius, H.J.: Über die Ursache der Speciesunterschiede in der Digitalisempfindlichkeit. Biochem. Pharmacol. *14*, 1785–1802 (1965)
Repke, K., Portius, H.J.: Analysis of structure activity relationship in cardioactive action on the molecular level. Sci. Pharmaceut. *1*, 39–57 (1966)
Repke, K.R.H., Schön, R.: Flip-flop model of (NaK)-ATPase function. Acta Biol. Med. Germ. *31*, 19–30 (1973)
Roberts, G.C.K. (edit.): Drug action at the molecular level. London: Macmillan Press 1977
Ruoho, A., Kyte, J.: Photoaffinity labeling of the ouabain-binding site on ($Na^+ + K^+$) adenosinetriphosphatase. Proc. Natl. Acad. Sci. USA *71*, 2352–2356 (1974)
Sachs, J.R.: Interaction of external K, Na, and cardioactive steroids with the Na-K pump of the human red blood cell. J. Gen. Physiol. *63*, 123–143 (1974)
Sachs, J.R., Ellory, J.C., Kropp, D.L., Dunham, P.B., Hoffman, J.F.: Antibody-induced alterations in the kinetic characteristics of the Na : K pump in goat red blood cells. J. Gen. Physiol. *63*, 389–414 (1974)
Scaf, A.H.J.: Cooperativity in classical receptor theory. Arch. Int. Pharmacodyn. Ther. *215*, 4–12 (1975)
Scatchard, G.: The attractions of proteins for small molecules and ions. N.Y. Acad. Sci. *51*, 660–672 (1949)
Schatzmann, H.-J.: Herzglykoside als Hemmstoffe für den aktiven Kalium- und Natriumtransport durch die Erythrocytenmembran. Helv. Physiol. Acta *11*, 346–354 (1953)
Scholz, H., Hackbarth, I., Schmitz, W.: Intensität und zeitlicher Verlauf der Digitoxinwirkung im Vergleich zu anderen herzwirksamen Glykosiden am isolierten Warmblüterherzen. In: Digitoxin als Alternative in der Therapie der Herzinsuffizienz. Greeff, K., Rietbrock, N. (Hrsg.), pp. 141–147. Stuttgart, New York: Schattauer 1979
Schoner, W.: Zum aktiven Na^+, K^+-Transport durch die Membran tierischer Zellen. Angew. Chem. *83*, 947–955 (1971)

Schoner, W., von Ilberg, C., Seubert, W.: On the mechanism of Na^+- and K^+-stimulated hydrolysis of adenosine triphosphate. 1. Purification and properties of a Na^+- and K^+-activated ATPase from ox brain. Eur. J. Biochem. *1*, 334–343 (1967)

Schoner, W., Pauls, H., Patzelt-Wenczler, R.: Biochemical characteristics of the sodium pump: indications for a half-of-the-sites reactivity of $(Na^+ + K^+)$-ATPase. In: Myocardial failure. Riecker, G., Weber, A., Goodwin, J. (eds.), pp. 104–119. Berlin, Heidelberg, New York: Springer 1977a

Schoner, W., Pauls, H., Patzelt-Wenczler, R., Erdmann, E., Stahl, I.: Some structural and functional aspects of the sodium pump: interrelation between the ATP-binding site and the ouabain receptor site. In: Diuretics in research and clinics. Siegenthaler, W., Beckerhoff, R., Vetter, W. (eds.), pp. 91–101. Stuttgart: Georg Thieme 1977b

Schwartz, A., Matsui, H., Laughter, A.H.: Tritiated digoxin binding to $(Na^+ + K^+)$-activated adenosine triphosphatase: possible allosteric site. Science *159*, 323–325 (1968)

Schwartz, A., Lindenmayer, G., Allen, J.C.: The sodium-potassium adenosine triphosphatase: pharmacological, physiological, and biochemical aspects. Pharmacol. Rev. *27*, 3–134 (1975)

Schwartz, A.: Sodium-potassium adenosine triphosphatase – a receptor for digitalis? Biochem. Pharmacol. *25*, 237–239 (1976a)

Schwartz, A.: Is the cell membrane Na^+, K^+-ATPase enzyme system the pharmacological receptor for digitalis? Circ. Res. *39*, 2–7 (1976b)

Sen, A.K., Post, R.L.: Stoichiometry and localization of adenosine triphosphate-dependent sodium and potassium transport in the erythrocyte. J. Biol. Chem. *239*, 345–352 (1964)

Skou, J.C.: The influence of some cations on an adenosine triphosphatase from peripheral nerves. Biochim. Biophys. Acta (Amst.) *23*, 349–401 (1957)

Skou, J.C.: Enzymatic basis for active transport of Na^+ and K^+ across cell membranes. Physiol. Rev. *45*, 596–617 (1965)

Skou, J.C.: The relationship of the $(Na^+ + K^+)$-activated enzyme system to transport of sodium and potassium across the cell membranes. Bioenergetics *4*, 1–30 (1973)

Smythies, J.R., Bradley, R.J.: Receptors in pharmacology. New York, Basel: Marcel Dekker 1978

Solomon, A.K.: The permeability of the human erythrocyte to sodium and potassium. J. Gen. Physiol. *36*, 57–110 (1952)

Stahl, W.L.: Role of phospholipids in the Na^+, K^+-stimulated adenosine triphosphatase system of brain microsomes. Arch. Biochem. Biophys. *154*, 56–67 (1973)

Straub, R.W., Bolis, L.: Cell membrane receptors for drugs and hormones. New York: Raven Press 1978

Sweadner, K.J.: Two molecular forms of $(Na^+ + K^+)$-stimulated ATPase in brain. J. Biol. Chem. *254*, 6060–6067 (1979)

Taniguchi, K., Iida, S.: Two apparently different ouabain binding sites of $(Na^+ + K^+)$-ATPase. Biochim. Biophys. Acta (Amst.) *288*, 98–102 (1972)

Temma, K., Akera, T., Ku, D.D., Brody, T.M.: Sodium pump inhibition by sulfhydryl inhibitors and myocardial contractility. Naunyn-Schmiedebergs Arch. Pharmacol. *302*, 63–71 (1978)

Ten Eick, R.E., Bassett, A.L., Okita, G.T.: Dissociation of electrophysiological and inotropic actions of strophanthidin-3-bromoacetate: possible role of adenosine triphosphatase in the maintenance of the myocardial transmembrane Na^+ and K^+ gradients. J. Pharmacol. Exp. Ther. *185*, 12–23 (1973)

Thompson, A.J., Hargis, J., Murphy, M.L., Doherty, J.E.: Tritiated digoxin. XX. Tissue distribution in experimental myocardial infarction. Am. Heart J. *88*, 319–324 (1974)

Titus, E.O.: Characterization of pharmacological receptors. Naunyn-Schmiedebergs Arch. Pharmacol. *288*, 269–281 (1975)

Tobin, Th., Brody, T.M.: Rates of dissociation of enzyme-ouabain complexes and $K_{0.5}$ values in $(Na^+ + K^+)$-adenosine triphosphatase from different species. Biochem. Pharmacol. *21*, 1553–1560 (1972)

Tobin, Th., Henderson, R., Sen, A.K.: Species and tissue differences in the rate of dissociation of ouabain from $(Na^+ + K^+$-ATPase. Biochim. Biophys. Acta (Amst.) *274*, 551–555 (1972)

Tobin, Th., Akera, T., Hogg, R.E., Brody, T.M.: Ouabain binding to sodium- and potassium-dependent adenosine triphosphatase: inhibition by the β,γ-methylene analogue. Mol. Pharmacol. *6*, 278–281 (1973)
Tobin, Th., Akera, T., Brody, S.L., Ku, D., Brody, T.M.: Cassaine: mechanism of inhibition of $Na^+ + K^+$-ATPase and relationship of this inhibition to cardiotonic actions. Eur. J. Pharmacol. *32*, 135–145 (1975)
Tona Lutete, N., Noel, F., de Pover, A., Godfraind, T.: The inhibition of human heart (Na+K)ATPase by semisynthetic digitalis glycosides. Arch. Int. Pharmacodyn. Ther. *227*, 166–167 (1977)
Wallick, E., Dowd, F., Allen, J., Schwartz, A.: The nature of the transport adenosine triphosphatase-digitalis complex. VII. Characteristics of ouabagenin-Na^+, K^+-adenosine triphosphatase interaction. J. Pharmacol. Exp. Ther. *189*, 434–444 (1974)
Wallick, E.T., Lindenmayer, G.E., Lane, L.K., Allen, J.C., Pitts, B.J.R., Schwartz, A.: Recent advances in cardiac glycoside-Na^+, K^+-ATPase interaction. Fed. Proc. *36*, 2214–2218 (1977)
Wallick, E.T., Lane, L.K., Schwartz, A.: Biochemical mechanism of the sodium pump. Ann. Rev. Physiol. *41*, 397–412 (1979)
Waser, P.G., Volkart, O.: Wirkung von Herzglykosiden auf Aktomyosin. Helv. Physiol. Acta *12*, 12–22 (1954)
Weder, H.G., Schildknecht, J., Lutz, R.A., Kesselring, P.: Determination of binding parameters from Scatchard plots. Eur. J. Biochem. *42*, 475–481 (1974)
Weidemann, M.J., Erdelt, H., Klingenberg, M.: Adenine nucleotide translocation of mitochondria. Eur. J. Biochem. *16*, 313–335 (1970)
Whittam, R., Chipperfield, A.R.: Ouabain binding to the sodium pump in plasma membranes isolated from ox brain. Biochim. Biophys. Acta (Amst.) *307*, 563–577 (1973)
Whittam, R., Chipperfield, A.R.: The reaction mechanism of the sodium pump. Biochim. Biophys. Acta (Amst.) *415*, 149–171 (1975)
Winegrad, S., Shanes, A.M.: Calcium flux and contractility in guinea pig atria. J. Gen. Physiol. *45*, 371–394 (1962)
Wood, J.M., Schwartz, A.: Effects of ouabain on calcium-45 flux in guinea pig cardiac tissue. J. Mol. Cell. Cardiol. *10*, 137–144 (1978)
Yamamoto, S.: Prednisolone-3,20-bis-guanylhydrazone: the mode of interaction with rat brain sodium and potassium-activated adenosine triphosphatase. Eur. J. Pharmacol. *50*, 409–418 (1978)
Yamamoto, S., Akera, T., Brody, T.M.: Prednisolone-3,20-bis-guanylhydrazone: Na^+, K^+-ATPase inhibition and positive inotropic action. Eur. J. Pharmacol. *49*, 121–132 (1978a)
Yamamoto, S., Akera, T., Brody, T.M.: Prednisolone-3,20-bis-guanylhydrazone: binding in vitro to sodium-and-potassium-activated adenosine triphosphatase of guinea pig heart ventricular muscle. Eur. J. Pharmacol. *51*, 63–69 (1978b)
Yamamura, H.I., Enna, S.J., Kuhar, M.J.: Neurotransmitter-receptor-binding. New York: Raven Press 1978
Yoda, A.: Structure-activity relationships of cardiotonic steroids for the inhibition of sodium- and potassium-dependent adenosine triphosphatase. 1. Dissociation rate constants of various enzyme-cardiac glycoside complexes formed in the presence of magnesium and phosphate. Mol. Pharmacol. *9*, 51–60 (1973)
Yoda, A., Yoda, S., Sarrif, A.M.: Structure-activity relationships of cardiotonic steroids for the inhibition of sodium- and potassium-dependent adenosine triphosphatase. 2. Association rate constants of various enzyme-cardiac glycoside complexes. Mol. Pharmacol. *9*, 766–773 (1973)
Yoda, S., Sarrif, A.M., Yoda, A.: Structure-activity relationships of cardiotonic steroids for the inhibition of sodium- and potassium-dependent adenosine triphosphatase. 4. Dissociation rate constants for complexes of the enzyme with cardiac oligodigitoxides. Mol. Pharmacol. *11*, 647–652 (1975)
Yoda, A., Yoda, S.: Structure-activity relationships of cardiotonic steroids for the inhibition of sodium- and potassium-dependent adenosine triphosphatase. 5. Dissociation rate constants of digitoxin acetates. Mol. Pharmacol. *11*, 653–662 (1975)

Yoda, A.: Binding of digoxigenin to sodium- and potassium-dependent adenosine triphosphatase. Mol. Pharmacol. *12*, 399–408 (1976)
Yoda, A., Yoda, S.: Association and dissociation rate constants of the complexes between various cardiac aglycones and sodium- and potassium-dependent adenosine triphosphatase formed in the presence of magnesium and phosphate. Mol. Pharmacol. *13*, 352–361 (1977)
Yoda, A., Yoda, S.: Influence of pH on the interaction of cardiotonic steroids with sodium- and potassium-dependent adenosine triphosphatase. Mol. Pharmacol. *14*, 624–632 (1978)
Zwaal, R.F.A.: Some aspects of structure-functions relationships in biological membranes. First European Symposium on Hormones and Cell Regulation, Dumont and Nunez (eds.), pp. 1–14. Elsevier/North-Holland: Biomedical Press 1977

CHAPTER 16

Stimulation and Inhibition of the Na^+, K^+-Pump by Cardiac Glycosides

T. GODFRAIND

A. Introduction

The inhibition of the Na^+, K^+-pump by cardiac glycosides is very well documented. Since the earlier work of SCHATZMANN (1953) and of GLYNN (1964) the analysis of the mechanism of this inhibition which paralleled the study of the properties of Na^+, K^+-ATPase (initially described by SKOU, 1957) has allowed us to describe the biochemical mechanism of the Na^+, K^+-pump. Furthermore, the presence of a Na^+, K^+-pump in most of the cells and the multiple functions controlled by its activity have been demonstrated by using ouabain as a tool. The sensitivity of a physiologic process to a high dose of ouabain is generally considered as a convincing argument for its relation to the Na^+, K^+-pump. Several cellular processes are coupled to the activity of the Na^+, K^+-pump; some appear to be very different one from the other, e.g., the membrane potential or the uptake of amino acids.

Owing to the great attention devoted to the study of the inhibition of the Na^+, K^+-pump, the earlier observations of an action of cardiac glycosides on the ionic composition of the heart have received less attention. HAGEN (1939) and BOYER and POINTDEXTER (1940) first reported that low doses of cardiac glycosides may increase the total K content of cardiac tissue. This effect has been re-examined in some detail in recent years. The concept of a biphasic action of cardiac glycosides on the activity of the Na^+, K^+-pump has been proposed in order to take into account most of the experimental observations. In this brief report, I will examine the various factors responsible for the stimulation and inhibition of the Na^+, K^+-pump by cardiac glycosides.

B. Dose–Response Relationship

The therapeutic concentration of free glycosides in the blood ranges between 10^{-9} and 5×10^{-9} *M*. A two-fold or three-fold increase is characteristic of toxic concentrations since in such conditions, the patients present symptoms of intoxication (ARONSON et al., 1977; BELPAIRE and BOGAERT, 1977; GODFRAIND, 1972; LESNE, 1977). Studies on human heart slices have shown that the uptake of ^{42}K was inhibited by concentrations equivalent to those found in the blood of intoxicated patients, whereas therapeutic concentrations enhanced ^{42}K uptake. Such observations suggested that besides their well-documented action as inhibitors of the Na^+, K^+-pump, cardiac glycosides showed the property of stimulating active Na/K transport across the cardiac cell membrane (GODFRAIND, 1972, 1973).

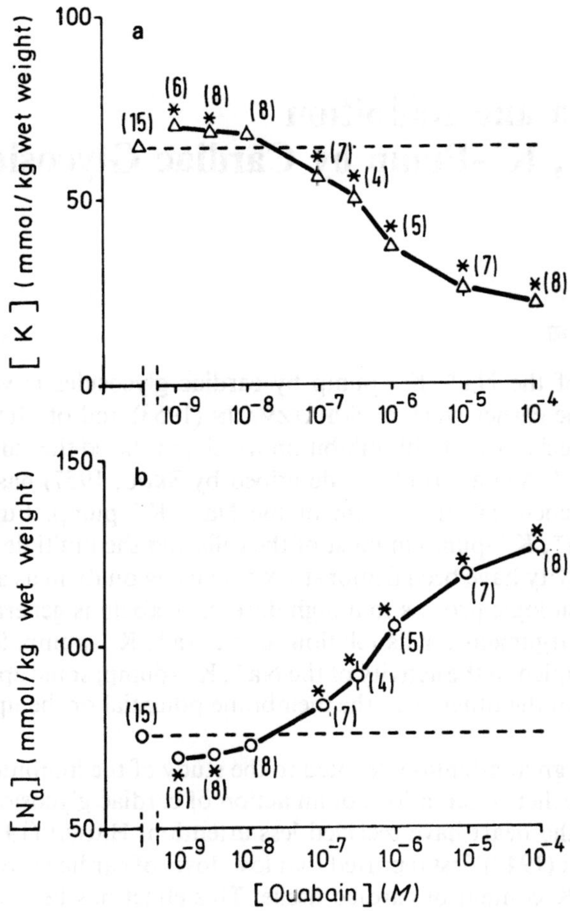

Fig. 1a, b. Potassium **a** and sodium **b** content of stimulated guinea pig left atria incubated at 30 °C for 3 h in various concentrations of ouabain. The number of determinations at each point is indicated in parentheses. Note increased gradient in the low range (10^{-9}–10^{-8} *M*) and decreased gradient in the higher range ($>10^{-8}$ *M*). (GODFRAIND and GHYSEL-BURTON, 1977)

These observations were confirmed in slices of guinea pig heart ventricle; after an incubation of 4 h in the presence of either ouabain 3×10^{-9} *M* or digitoxin 3×10^{-9} *M*, K influx was 120% of controls; higher concentrations reduced K influx (GODFRAIND and LESNE, 1972). A similar observation was made in squid axons where low doses of cardiac glycosides stimulated the Na^+, K^+-pump (BAKER and WILLIS, 1972).

This action of low doses of cardiac glycosides opposite to the action of high doses has been observed in different experimental designs where different techniques were used to estimate the activity of the Na^+, K^+-pump. In isolated guinea pig left atria electrically stimulated at a frequency of 3.3 Hz, GODFRAIND and GHYSEL-BURTON (1977) and GHYSEL-BURTON and GODFRAIND (1979) have observed that low doses of ouabain evoked a net decrease of intracellular sodium concentration $[Na^+]_i$ associated with an equivalent intracellular potassium concentration $[K^+]_i$ increase, when the ionic content was measured after an incubation of 3 h (Fig. 1).

In sheep cardiac Purkinje fibers, COHEN et al. (1976) have examined the influence of low concentrations of ouabain on the K gradient by measuring changes in the reversal potential for a K-specific current, iK_2, and by measuring total steady-state current–voltage relations. They observed that low doses of ouabain produced changes in the K gradient that reflects stimulation of the Na^+, K^+-pump. ELLIS (1977) has recorded the intracellular Na activity $a^i_{N_a}$ of sheep heart Purkinje fibers using recessed-tip Na^+-sensitive glass microelectrodes. He has observed that ouabain evoked an increase of $a^i_{N_a}$. In some experimental conditions, there was an indication of a small decrease in the $a^i_{N_a}$ in ouabain concentrations between 10^{-7} and 10^{-8} *M*.

In vivo experiments have also allowed us to observe an action of cardiac glycosides related to a stimulation of the Na^+, K^+-pump. STEINESS and VALENTIN (1976) have measured heart contractility and myocardial K content after intravenous (IV) infusion of 0.05 mg/kg digoxin in dogs. They observed an initial increase in K content, a decrease was noticed later.

Some reports do not fit with the above observations. LAMB and MCGALL (1972) have studied the effect of prolonged ouabain treatment on Na, K, Cl, and Ca concentrations and fluxes in Girardi and Hela cells (derived from human heart anc cervix respectively) grown as monolayer cultures. They examined a range of ouabain concentrations between 5×10^{-9} and 10^{-3} *M*. They observed only an inhibition of the Na^+, K^+-pump. In patients treated with digoxin, ASTRUP (1974) has measured the Na content of red blood cells. He observed an increase of this content which may be attributed to Na^+, K^+-pump inhibition by therapeutic concentrations of digoxin. This observation has been repeated by others. In rabbit and human myocardium, FRY et al. (1978) have reported that the only detectable change in K exchange was a ^{42}K loss associated with a positive inotropic effect. Such a ^{42}K loss could be attributed to an increased K conductance related to intracellular calcium concentration $[Ca^+]_i$ increase. It has indeed been reported that intracellular calcium controls membrane permeability for potassium (LEW and FERREIRA, 1978) and HEINEN and NOACK (1972) have shown that cellular calcium exchangeability was increased by cardiac glycosides.

More recently, HOUGEN et al. (1979) have measured ^{82}Rb uptake in dog heart for various digoxin doses ranging between 10^{-9} and 10^{-3} *M*. They have shown that digoxin only inhibited the pump. This observation appears to be in contradiction to the recent report that low digoxin concentrations produce a two-fold increase in Na^+, K^+-pump activity (Fig. 2) (GODFRAIND and GHYSEL-BURTON, 1979). As discussed below, these discrepancies may be accounted for by a difference in methodology.

C. Role of Duration of Treatment with Glycoside

Most of the experimental observations just summarized are consistent with the concept that cardiac glycosides do exert a biphasic action on the heart Na^+, K^+-pump: a stimulation and an inhibition. Although the stimulatory effect has

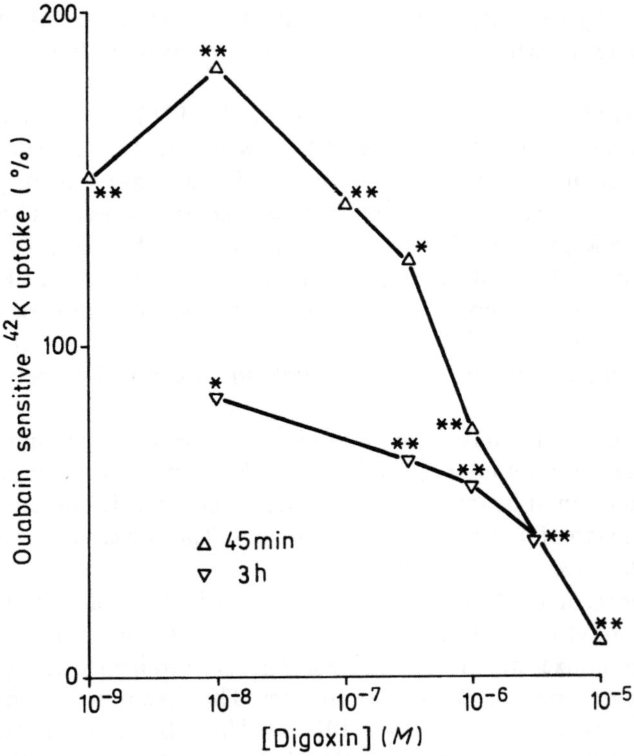

Fig. 2. The action of various digoxin dosages on ^{42}K uptake by stimulated guinea pig left atria incubated at 30 °C for 45 min *(upper curve)* and for 3 h *(lower curve)* in the presence of the glycoside. Note that the stimulation of the Na^+, K^+-pump by 10^{-8} *M* digoxin waned after a prolonged incubation. (After GODFRAIND and GHYSEL-BURTON, 1979)

been observed with prolonged incubations at low concentrations (near 10^{-9} *M*), this action could be only transient at high concentrations.

In squid giant axon, BAKER and WILLIS (1972) have followed the time course of ouabain action on the Na^+, K^+-pump. They reported that the inhibition was preceeded by a transient stimulation of Na efflux. In Purkinje fibers, an a^i_{Na} decrease preceeded the a^i_{Na} increase evoked by moderate concentrations of ouabain (ELLIS, 1977; DEITMER and ELLIS, 1978). Furthermore, electrophysiologic measurements of COHEN et al. (1976) indicated that the pump stimulation was transient.

We have recently measured ^{42}K uptake by isolated guinea pig left atria incubated at 30 °C in physiologic solution containing 6 m*M* KCl. After 45 min in the presence of 10^{-8} *M* digoxin, we observed a two-fold increase in the rate of K influx; whereas a slight inhibition of K influx was observed after an incubation of 3 h (Fig. 2) (GODFRAIND and GHYSEL-BURTON, 1979). These observations indicate that the stimulatory sites are inactivated during a prolonged incubation in the presence of cardiac glycosides. It is therefore not suprising that the Na^+, K^+-pump stimulation has not been always observed since it depends upon the duration of the glycoside treatment. Furthermore, it might be that in tissue such as red blood cells, stimulatory sites are not functional.

D. Molecular Requirements for Stimulation of the Pump

GHYSEL-BURTON and GODFRAIND (1975, 1979) have compared the action of ouabain, ouabagenin and, dihydroouabain on the ouabain-sensitive K content of guinea pig atria. They have observed that ouabagenin inhibited the Na^+, K^+-pump at concentrations 30 times higher than ouabain but that the stimulation occurred for similar concentrations of the two cardioactive steroids. Furthermore, dihydroouabain was 80 times less potent than ouabain for inhibition but did not stimulate the pump. DEITMER and ELLIS (1978) have confirmed this observation as they have not observed that dihydroouabain decreased intracellular Na activity in Purkinje fibers.

These observations show that the unsaturation of the lactone ring of ouabain is required for the stimulation of the pump. DEITMER and ELLIS have also reported that actodigin behaved like dihydroouabain. Actodigin is a semisynthetic cardiac glycoside differing from digitoxin in two respects : the butenolide ring is attached to the steroid through the carbon atom that is α to the carboxyl group (instead of β in the natural compounds) and the sugar residue at C3 is glucose and not digitoxose. Since the sugar chain does not seem necessary for the stimulation of the pump as shown by the comparison of heterosides and genins (GHYSEL-BURTON and GODFRAIND, 1977, 1979), the properties of actodigin are probably related to the configuration of the isomeric lactone. THOMAS et al. (1979) have proposed that this α-lactone will not allow the molecule to be properly bound to the receptor. According to the model they have proposed, the lactone would bind to a two-point attachment: an electrostatic bound between a receptor negative charge and the electron deficient C20 and a hydrogen bond between C24 (=, i.e., between the oxygen of the lactone) and another receptor site. This would allow the binding of the steroids followed, according to YODA and YODA (1974), by the binding of the sugar. The electrostatic binding is impaired in the dihydro derivative as well as in actodigin as proposed by THOMAS et al. (1979). If this model is correct, it would suggest that the stimulation of the pump might be related to the electrostatic binding of C20 and the inhibition to the hydrogen binding of the carbonyl. This is consistent with the observation that SC4453 which is a semisynthetic analog of digoxin, in which the lactone is substituted by a C17 pyridazine ring, is as potent as digoxin for stimulation of the Na^+, K^+-pump (GODFRAIND and GHYSEL-BURTON, 1979). The C20 in the pyridazine ring is also electron deficient through a charge separation due to resonance. The question as to whether the high affinity binding sites are represented by these negative sites, allowing a transient interaction between the glycoside and the receptors, could be answered with compounds in which the electron deficiency in C20 is increased by substitution on the lactone or on the pyridazine ring.

E. Influence of Extracellular KCl on Stimulation and Inhibition of the Pump by Ouabain

It has been shown that the increase of extracellular potassium $[K^+]_o$ raised the concentration of ouabain providing a 50% inhibition of the Na^+, K^+-pump. GHYSEL-BURTON and GODFRAIND (1977, 1979) have reported that the concentration of oua-

Table 1. Parameters describing the interaction of ouabain with the high affinity sites in guinea pig atria. The experiments were performed at 30 °C with isolated left atria stimulated at a rate of 3.3 Hz. The data were calculated from GHYSEL-BURTON and GODFRAIND (1979)

KCl concentration	Ouabain concentration (*mM*)		Capacity of the high affinity sites (% total specific uptake)	Hill coefficient
	Upper limit for pump stimulation	50% Saturation of the high affinity sites		
2.7	10^{-9}	3.7×10^{-10}	0.7	2.4
6	10^{-8}	2.9×10^{-9}	2.4	2.1
12	2.5×10^{-8}	1.2×10^{-8}	4	1.4

bain producing the pump stimulation was increased by increasing $[K^+]_o$: the upper limit for KCl, 2.7 m*M* was found with 10^{-9} *M* ouabain; with 6 m*M* KCl, it increased to 10^{-8} *M* ouabain and approximately to 3×10^{-8} *M* ouabain with 12 m*M* KCl. These limits corresponded to the concentrations required for binding to the high affinity sites (Table 1). These observations indicate that digitalis receptor sites responsible for pump stimulation might present an affinity for K^+ of the same order as the inhibitory sites. The K shift of the ouabain dose–response curve might explain the discrepancies between the observations of HOUGEN et al. (1979) on one hand and GODFRAIND and GHYSEL-BURTON (1979) on the other hand. This possibility has been reanalyzed recently by measuring the activity of the Na^+, K^+-pump in isolated guinea pig atria bathed in various $[K^+]_o$ and treated with various concentrations of ouabain. The pump activity was estimated by measuring ^{42}K uptake after 30 min incubation in the presence of the glycoside (GHYSEL-BURTON and GODFRAIND, unpublished). The biphasic action of ouabain was seen with ouabain concentrations between 10^{-10} and 10^{-8} *M*. With 2 m*M* KCl, ouabain stimulated the Na^+, K^+-pump maximally at concentrations $<10^{-9}$ *M* and higher concentrations became inhibitory. Raising the concentration of KCl raised the ouabain concentration at which its action was reversed.

Thus at low KCl, the range of concentrations over which ouabain stimulates the Na^+, K^+-pump is reduced and the range of concentrations over which it inhibits it is increased. At 2 m*M* KCl, 10^{-9} *M* ouabain was not stimulatory. Since dog heart is more sensitive than guinea pig heart to the inhibitory effect of cardiac glycoside, this is consistent with the observations of HOUGEN et al. (1979) that at 2 m*M* KCl, 10^{-9} *M* digoxin slightly depressed ^{86}Rb uptake by dog heart biopsies.

A further analysis of the K effect will undoubtly give interesting information on the relation between stimulation and inhibition of the Na^+, K^+-pump.

F. Changes in Pump Activity and Inotropic Effect

It is widely accepted that muscle contractility is controlled by cytosolic calcium activity. Muscle relaxation is caused by a reduction of free calcium concentration down to 10^{-7} *M*. This is achieved by several processes including a Ca outward flux coupled to the inward Na gradient. This Na/Ca exchange might occur by means of a hypothetical membrane carrier (JUNDT et al., 1975). It has been postulated that increased $[Na^+]_i$ would impair the binding of Ca to the carrier and therefore would allow $[Ca^{2+}]_i$ to increase. It has also been proposed that the $[Na^+]_i$ increase will stimulate the activity of an electroneutral Na/Ca carrier system allowing the stimulation of an outward Na and an inward Ca movement (LANGER, 1977). Further experimental evidence is provided by the demonstration of a correlation between the inhibition of the Na^+, K^+-pump and the force of contraction. KU et al. (1974) have measured the pump activity by the determination of the ^{86}Rb uptake into the cardiac cell after preloading the cell with Na^+. In ventricular slices prepared during the inotropic action of digitoxin, the uptake of ^{86}Rb was decreased.

The degree of inhibition was related to the magnitude of the inotropic effect. When the heart was perfused with drug-free solutions after 20 min exposure to digitoxin, the inotropic effect decreased with a half-life of about 40 min; concurrently, the activity of the Na^+, K^+-pump was restored. The amount of inhibited Na^+, K^+-ATPase was also estimated in the same hearts by determining the initial velocity of ATP-dependent ^3H-ouabain binding. The authors have observed that the proportion of enzyme inhibition paralleled the reduction of the sodium pump activity.

We have recently re-examined the correlation between the inhibition of the Na^+, K^+-pump and the increase in contractility (GODFRAIND and GHYSEL-BURTON, 1980).

The exposure of isolated atria to low K media caused the systolic tension to increase. This inotropic effect was reversible and it was inversely related to the KCl concentration. The time to peak tension was 10–20 min, after which the contractility was sustained for at least 40 min. The diastolic tension remained unchanged.

^{42}K was added to the bath after an incubation of 30 min in low K solution or a similar period of time in standard physiological solution containing 6 m*M* KCl. The net ^{42}K uptake, i.e., the sum of passive and active K exchanges was dependent on $[K^+]_o$. The passive exchange was estimated in atria treated with 10^{-4} *M* ouabain which blocked the Na^+, K^+-pump. The active exchange was obtained by subtracting ^{42}K passive exchange from ^{42}K net uptake. The turnover of the Na^+, K^+-pump in low K solution was expressed, as a percentage, from the ratio between the active (ouabain-sensitive) ^{42}K uptake at a given KCl concentration of the medium and the active ^{42}K uptake at 6 m*M* KCl. As Fig. 3 illustrates a linear relation was found between the inhibition of active ^{42}K uptake and the increase in the systolic tension of isolated left atria immersed in low K solutions. In other experiments, the atria bathed in 6 m*M* KCl were treated by various concentrations of ouabain or dihydroouabain sufficient to inhibit the Na^+, K^+-pump. With the lowest active concentrations, the time to maximum inotropic effect was 15–20 min after which the tension was sustained for at least 45 min. ^{42}K was added to the bath 30 min after the addition of the glycosides in order to obtain an estimate of the active K uptake. Within the range of concentrations which did not alter the diastolic tension, the increase in systolic tension was dependent on the inhibition of the Na^+, K^+-pump, but the dependence was not the same with ouabain or dihydroouabain (Fig. 3). As far as dihydroouabain was concerned, the relation between the inhibition of the Na^+, K^+-pump was not different from that found in the low K solutions. With ouabain, the slope was steeper than with dihydroouabain.

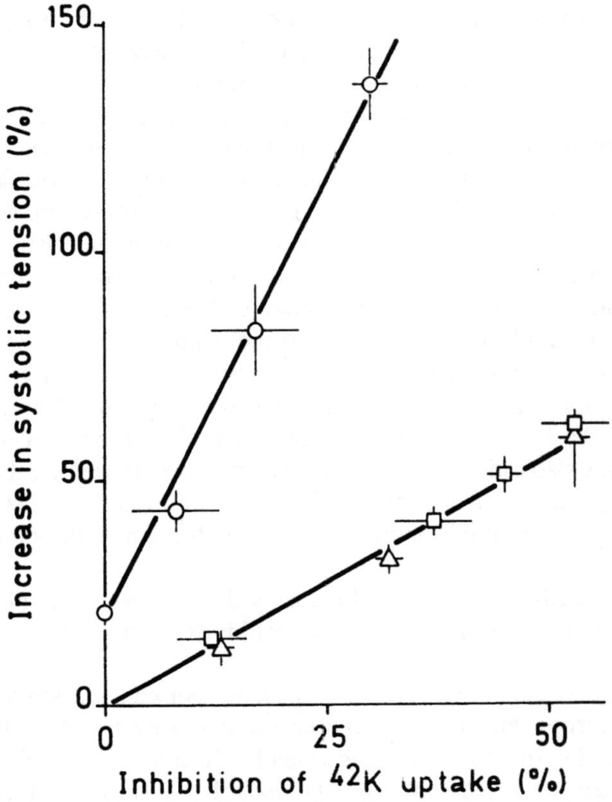

Fig. 3. The relation between the inhibition of ^{42}K uptake and the increase in systolic tension of isolated guinea pig atria treated by ouabain *(circles)*, dihydroouabain *(triangles)* or low K solutions *(squares)*. Note the similarity of the relation for dihydroouabain and low K solution whereas for a similar inhibition of the Na^+, K^+-pump, ouabain evokes a much higher inotropic effect. (GODFRAIND and GHYSEL-BURTON, 1980)

Some other data are summarized in Table 2 which compares, for a 25% inhibition of the Na pump, the increase in contractility evoked by low K and by several cardioactive glycosides. The only glycoside producing an increase in contractility similar to low K was dihydroouabain whereas it was higher with the other compounds.

In other experiments ED_{50} values (i.e., dose producing 50 percent of the maximum inotropic effect) for the positive inotropic effect of ouabain were found to be lower than I_{50} values (i.e., dose producing 50 percent of the inhibition of the pump) for the pump and were not similarly affected by an increase in KCl concentration. For ouabagenin, ED_{50} values for the two effects were in the ratio 1:20 at 6 m*M* whereas the two values were nearly equal with dihydroouabain (Table 3). In Table 3 are also reported ED_{50} values of digoxin, digoxigenin-mono-digitoxoside and SC4453. This comparison confirms the dissociation between inhibition of the pump and inotropic effect. The observation that ED_{50} for contractility are lower than I_{50} for pump inhibition may be the result of the existence of two groups of sites with different affinities for the glycosides.

Table 2. The increase in systolic tension evoked by hypokalemia and by a dose of cardioactive steroid producing a 25% inhibition of the sodium pump in guinea pig isolated atria. The experiments were carried out at 30 °C on isolated guinea pig left atria stimulated at 3.3 Hz in physiologic solution containing 6 m*M* KCl. Hypokalemia consisted in the replacement of this solution by the same solution but containing 3 m*M* KCl. ^{42}K uptake was measured at the peak of the inotropic effect (GODFRAIND and GHYSEL-BURTON, unpublished)

Treatment	Increase in contractility (% of control before treatment)	Cardiactive steroid concentration (*M*)
Hypokaliemia	28	
Dihydroouabain	28	2×10^{-6}
Ouabain	127	5×10^{-7}
Digoxigenin	49	3×10^{-6}
Digoxin	69	10^{-6}
Digoxigenin-mono-digitoxoside	107	10^{-6}
SC4453	130	2×10^{-6}

Table 3. Comparison of cardioactive steroid concentration parameters describing the inhibition of the Na pump (I_{50}), and the inotropic effect (ED_{50})

Drug	[KCl] *M*	I_{50} (*M*)	ED_{50} (*M*)
Ouabain	2.7	2×10^{-7}	10^{-7}
	6	5.4×10^{-7}	3.4×10^{-7}
Ouabagenin	12	10^{-6}	5×10^{-6}
	6	1.4×10^{-5}	3×10^{-6}
Dihydroouabain	6	4.2×10^{-5}	2.2×10^{-5}
Digoxin	6	2.5×10^{-6}	10^{-6}
SC4453	6	4.2×10^{-6}	10^{-6}
Digoxigenin-mono-digitoxoside	6	2.5×10^{-6}	6×10^{-7}

These experiments were undertaken in order to analyze the relation existing between the inhibition of the Na^+, K^+-pump and the contractility of isolated guinea pig atria. For a similar action on the pump, low K solutions and dihydroouabain evoked a similar inotropic effect. This strongly supports the hypothesis that the contractility of the heart is dependent on the turnover of the Na^+, K^+-pump which regulates indirectly the transmembrane Na/Ca exchange.

The quantitative relationship between the inhibition of the Na^+, K^+-pump and the increase in contractility was different, considering dihydroouabain and low K on one hand and ouabain and the other cardiac glycosides with a saturated lactone or a pyridazine ring on the other hand. As far as ouabain was concerned, the observed increase in contractility was much higher than might have been expected from the experiments with dihydroouabain and low K. It is therefore likely that the inotropic effect of ouabain is the sum of more than one process; one related to the inhibition of the Na^+, K^+-pump and another to a still unknown mechanism. It has been shown that ouabain stimulates the Na^+, K^+-pump through an interaction with high affinity binding sites different from the low affinity binding sites responsible for the inhibition of the pump (GODFRAIND and GHYSEL-BURTON, 1977, 1979; GHYSEL-BURTON and GODFRAIND, 1979). The interaction with these sites also evokes a positive inotropic effect. As dihydroouabain does not evoke the stimulation of the Na^+, K^+-pump, it is likely that the inotropic effect of ouabain which is not accounted for by the inhibition of the Na^+, K^+-pump is due to interaction with these high affinity sites.

G. Receptor Sites Responsible for Stimulation and Inhibition of the Pump

Most of the observations on the uptake of cardiac glycosides by the whole heart and on their binding to microsomes are reviewed in other chapters (Chap. 14 and this Handbook, Vol. 56/II, Chap. 7). Since the pump stimulation was observed at low concentration of glycosides, recent binding studies were conducted not only at high but also at these low concentrations both on the whole heart and on microsomes (GODFRAIND and GHYSEL-BURTON, 1977; GHYSEL-CURTON and GODFRAIND, 1979; DE POVER and GODFRAIND, 1979; GODFRAIND et al., 1980). These experiments have allowed the identification of two classes of ouabain-specific binding sites in guinea pig heart.

High affinity sites (S sites) are occupied within the dose range for which a stimulation of the Na^+, K^+-pump is observed and low affinity sites (I sites) within the dose range which for which pump inhibition is observed.

Increase of the KCl concentration in the medium reduces the affinity of both groups of sites for ouabain, but increase the binding capacity of S sites (from 1% at 2.7 mM KCl to 4% at 12 mM KCl). The binding capacity of I sites was not affected by $[K^+]_o$. changes.

In microsomal preparations enriched in Na^+, K^+-ATPase, two classes of ouabain-specific binding sites have also been identified. The dissociation constant K_D for the binding of ouabain to low affinity sites was close to $[K^+]_i$ for ouabain inhibition of Na^+, K^+-ATPase. The high affinity sites are apparently inactive in fragmented membranes (DE POVER and GODFRAIND, 1979).

H. Concluding Remarks

The experimental data reported here show that the action of cardiac glycosides on the active transport of Na^+ and K^+ in the heart is biphasic. The stimulation of this active transport seems to be related to the interaction with high affinity binding sites. The inhibition of the Na^+, K^+-pump is due to the interaction with sites which exhibit a lower affinity for the glycosides. It is likely that the low affinity sites are part of the Na^+, K^+-ATPase molecule as a great many experimental results have demonstrated that inhibition of the Na^+, K^+-pump is due to inhibition of this enzyme (see Chap. 14). As far as the high affinity sites are concerned, there is at present no direct biochemical indication that they constitute one of the reactive sites of this enzyme. A few biochemical studies have reported stimulation of Na^+,

K^+-ATPase by cardiac glycosides or by other drugs, but is appears to be an indirect process (GODFRAIND et al., 1974).

The structural requirements for binding to the two groups of sites are not identical. Indeed, the presence of a sugar moeity in $C_3\beta$ of the steroid ring increases the affinity for the inhibitory sites but is not necessary for an interaction with the stimulatory sites as shown by the action of ouabagenin. Furthermore, the electron deficiency of C20 favors the interaction with the stimulatory sites which is absent with dihydroouabain.

Although several questions are still unanswered, it becomes more and more evident that the concentration parameters describing the inhibition of the Na^+, K^+-pump (I_{50}) and the inotropic effect (ED_{50}) are not quantitatively similar, the latter being lower than the former. Furthermore, affinity changes obtained by modifying $[K^+]_o$ and by using cardenolides with different structures have not produced an identical shift of these parameters. This indicates that the inhibition of the Na^+, K^+-pump is not the only mechanism responsible for the positive inotropic effect. It is likely that the additional and still unknown process is activated through interaction of cardiac glycosides with the high affinity sites. The identification of the biochemical nature of the additional mechanism will undoubtedly improve the further development of pharmacology and pharmacotherapy of congestive heart failure. It will also open new areas in the understanding of the physiologic regulation of the Na^+, K^+-pump, since the stimulatory S sites could constitute receptors for endogenous ligands.

Acknowledgments. The author's own studies cited herein were supported by grants from the Fonds de la Recherche Scientifique Médicale and from the Fonds de Developpement Scientifique de l'Université Catholique de Louvain.

References

Aronson, J.K., Grahame-Smith, D.G., Hallis, K.F., Hibble, A., Wigley, F.: Monitoring digoxin therapy: I plasma concentrations and an in vitro assay of tissue response. Br. J. Clin. Pharmacol. *4*, 213–222 (1977)

Astrup, J.: Changes in sodium and potassium in human red cells in vivo related to treatment with digoxin, to hypokalemia, and to cell density. Ann. N.Y. Acad. Sci. *242*, 683–702 (1974)

Baker, P.F., Willis, J.S.: Inhibition of the sodium pump in squid axons by cardiac glycosides: The dependence on extracellular ions and metabolism. J. Physiol. (Lond.) *224*, 463–475 (1972)

Belpaire, F.M., Bogaert, M.G.: Methods for assay of cardiac glycosides. In: Plasma digitalis concentration and digitalis therapy. Godfraind, T. (ed.). Bruxelles: Arscia 1977

Blood, B.E., Noble, D.: Glycoside induced inotropism of the heart – more than one mechanism: J. Physiol. (Lond.) *266*, 76P–78P (1977)

Boyer, P.K., Pointdexter, C.A.: The influence of digitalis on the electrolyte and water balance of heart muscle. Am. Heart J. *20*, 586–191 (1940)

Cohen, I., Daut, J., Noble, D.: An analysis of the actions of low concentrations of ouabain on membrane currents in Purkinje fibres. J. Physiol. (Lond.) *260*, 75–103 (1976)

Deitmer, J.W., Ellis, D.: The intracellular sodium activity of cardiac Purkinje fibres during inhibition and reactivation of the Na-K pump. J. Physiol. (Lond.) *284*, 241–259 (1978)

De Pover, A., Godfraind, T.: Interaction of ouabain with ($Na^+ + K^+$)-ATPase from human heart and from guinea-pig heart. Biochem. Pharmacol. *28*, 3051–3056 (1979)

Ellis, D.: The effects of external cations and ouabain on the intracellular sodium activity of sheep heart Purkinje fibres. J. Physiol. (Lond.) *273*, 211–240 (1977)

Fry, C.H., Galindez, E., Poole-Wilson, P.A.: Potassium exchange and the positive inotropic effect of ouabain in rabbit and human myocardium. J. Physiol. (Lond.) *280*, 72–73P (1978)

Ghysel-Burton, J., Godfraind, T.: Stimulation and inhibition by ouabain of the sodium pump in guinea-pig atria. Br. J. Pharmacol. *55*, 249P (1975)

Ghysel-Burton, J., Godfraind, T.: Importance of the lactone ring for the action of therapeutic doses of ouabain in guinea-pig atria. J. Physiol. (Lond.) *266*, 75–76 (1976)

Ghysel-Burton, J., Godfraind, T.: Stimulation and inhibition of the sodium pump by cardioactive steroids in relation to their binding sites and their inotropic effect on guinea-pig isolated atria. Brit. J. Pharmacol. *66*, 175–184 (1979)

Glitsch, H.G., Reuter, H., Scholz, H.: The effect of the internal sodium concentration on calcium fluxes in isolated guinea-pig auricles. J. Physiol. (Lond.) *209*, 25–43 (1970)

Glynn, I.M.: The action of cardiac glycosides on ion movements. Pharmacol. Rev. *16*, 381–407 (1964)

Godfraind, T.: Pharmacologie des récepteurs digitaliques. Bull. Acad. Roy. Med. Belg. *12*, 403–443 (1972)

Godfraind, T.: The therapeutic mode of action of cardiac glycosides. Arch. Int. Pharmacodyn. Ther. *206*, 384–388 (1973)

Godfraind, T., De Pover, A.: La pharmacologie de la $(Na^+ + K^+)$-ATPase du muscle cardiaque humain. Bull. Acad. Roy. Belg. *131*, 295–311 (1976)

Godfraind, T., Ghysel-Burton, J.: Binding sites related to ouabain-induced stimulation or inhibition of the sodium pump. Nature (Lond.) *265*, 165–166 (1977)

Godfraind, T., Ghysel-Burton, J.: The action of digoxin and digoxigenin-monodigitoxoside on the sodium pump and on the contractility in isolated guinea-pig atria. Arch. Int. Pharmacodyn. Ther. *234*, 340 (1978)

Godfraind, T., Ghysel-Burton, J.: The cardioactive properties of SC4453, a digoxin analogue with a C_{17} pyridazine ring. Eur. J. Pharmacol. *60*, 337–344 (1979)

Godfraind, T., Ghysel-Burton, J.: Independence of the positive inotropic of ouabain from the inhibition of the heart Na-K pump. Proc. Natl. Acad. Sci. USA *77*, 3067–3069 (1980)

Godfraind, T., Godfraind-de Becker, A.: The action of ouabain on the response of the isolated guinea-pig auricles to catecholamines in relation with its inotropic and chronotropic effects. Arch. Int. Pharmacodyn. Ther. *158*, 453–465 (1965)

Godfraind, T., Lesne, M.: Estimation of digitoxin uptake by isolated cardiac and smooth muscle preparations. Arch. Int. Pharmacodyn. Ther. *166*, 195–199 (1967)

Godfraind, T., Lesne, M.: The uptake of cardiac glycosides in relation to their actions in isolated cardiac muscle. Br. J. Pharmacol. *46*, 488–497 (1972)

Godfraind, T., Tona Lutete, N.: Inhibition of digoxin and SC4453 of $(Na^+ + K^+)$-ATPase prepared from human heart, guinea-pig heart, and guinea-pig brain. Eur. J. Pharmacol. *60*, 329–336 (1979)

Godfraind, T., Koch, M.C., Verbeke, N.: The action of EGTA on the catecholamines stimulation of rat brain Na-K-ATPase. Biochem. Pharmacol. *23*, 3505–3511 (1974)

Godfraind, T., De Pover, A., Verbeke, N.: Influence of pH and sodium on the inhibition of guinea-pig heart $(Na^+ + K^+)$-ATPase by calcium. Biochim. Biophys. Acta *451*, 202–211 (1977)

Godfraind, T., De Pover, A., Tona Lutete, N.: Identification with potassium and vanadate of two classes of specific ouabain binding sites in a $(Na^+ + K^+)$-ATPase preparation from the guinea-pig heart. Biochem. Pharmacol. *29*, 1195–1199 (1980)

Hagen, P.S.: The effects of digilanid C in varying dosages upon the potassium and water content of rabbit heart muscle. J. Pharm. Exp. Ther. *67*, 50–55 (1939)

Heinen, E., Noack, E.: Effects of k-strophanthin and digitoxigenin on contractile force, calcium content and exchange in guinea-pig isolated atria. Naunyn-Schmiedebergs Arch. Pharmacol. *275*, 359–371 (1972)

Heller, M., Beck, S.: Interactions of cardiac glycosides with cells and membranes. Properties and structural aspects of two receptor sites for ouabain in erythrocytes. Biochim. Biophys. Acta *514*, 332–347 (1978)

Jundt, H., Porzig, H., Reuter, H., Stucki, J.W.: The effect of substances releasing intracellular calcium ions on sodium – dependent calcium efflux from guinea-pig auricles. J. Physiol. (Lond.) *246*, 229–253 (1975)
Ku, D., Akera, T., Pew, C.L., Brody, T.M.: Cardiac glycosides: Correlation among Na^+, K^+-ATPase, sodium pump and contractility in the guinea-pig heart. Naunyn-Schmiedebergs Arch. Pharmacol. *285*, 185–200 (1974)
Kuschinsky, K., Lahrtz, H.G., Lüllmann, H., van Zwieten, P.A.: Accumulation and release of ^{3}H-digoxin by guinea-pig heart muscle. Br. J. Pharmacol. *30*, 317–328 (1967)
Lamb, J.F., McCall, D.: Effect of prolonged ouabain treatment and fluxes in cultured human cells. J. Physiol. (Lond.) *225*, 599–617 (1972)
Landowne, D., Ritchie, J.M.: The binding of tritiated ouabain to non-myelinated nerve fibres. J. Physiol. (Lond.) *207*, 529–537 (1970)
Langer, G.A.: Ion fluxes in cardiac excitation and contraction and their relation to myocardial contractility. Physiol. Rev. *48*, 708–757 (1968)
Langer, G.A.: Relationship between myocardial contractility and the effects of digitalis on ionic exchange. Fed. Proc. *36*, 2231–2234 (1977)
Lesne, M.: Principe et utilisation du dosage radioimmunologique des digitaliques. In: Plasma digitalis concentration and digitalis therapy. Godfraind, T. (ed.). Bruxelles: Arscia 1977
Lew, V.L., Ferreira, H.G.: Calcium transport and the properties of a calcium-activated potassium channel in red cell membranes. In: Current topics in membranes and transport, Vol. 10, pp. 218–270. New York: Academic Press 1978
Lüllmann, H., Peters, T.: On the sarcolemmal site of action of cardiac glycosides. In: Recent advances in studies on cardiac structure and metabolism, Vol. 9: The sarcolemma. Roy and Dhalla (eds.), pp. 311–328. Baltimore: University Park Press 1976
Müller, P.: Ouabain effects on cardiac contractions, action potential and cellular potassium. Circ. Res. *17*, 46–56 (1965)
Noack, E., Felgenträger, J., Zettner, B.: Changes in myocardial Na and K content during the development of cardiac glycoside inotropy. J. Mol. Cell Cardiol. *11*, 1189–1194 (1979)
Reiter, M.: Cardioactive steroids with special reference to calcium. In: Calcium and cellular function. Cuthbert (ed.), pp. 270–279. London: Macmillan 1970
Schatzmann, H.J.: Herzglycoside als Hemmstoffe für den aktiven Kalium- und Natriumtransport durch die Erythrocytenmembran. Helv. Physiol. Acta *11*, 346–354 (1953)
Skou, J.C.: The influence of some cations on an adenosine triphosphatase from peripheral nerves. Biochim. Biophys. Acta *23*, 394–401 (1957)
Steiness, E., Valentin, N.: Myocardial digoxin uptake: dissociation between digitalis-induced inotropism and myocardial loss of potassium. Brit. J. Pharmacol. *58*, 183–188 (1976)
Thomas, R., Allen, J.C., Pitts, B. Jr., Schwartz, A.: Cardenolide analogs. An explanation for the unusual properties of AY 22241. Eur. J. Pharmacol. *53*, 227–237 (1979)
Thomas, R., Boutagy, J., Gelbart, A.: Synthesis and biological activity of semisynthetic digitalis analogs. J. Pharm. Sci. *11*, 1645–1683 (1974a)
Thomas, R., Boutagy, J., Gelbart, A.: Cardenolide analogs. V. Cardiotonic activity of semisynthetic analogs of digitoxigenin. J. Pharmacol. Exp. Ther. *191*, 219–231 (1974b)
Yoda, A., Yoda, S.: Structure-activity relationships of cardiotonic steroids for the inhibition of sodium- and potassium-dependent adenosine triphosphatase. III. Dissociation rate constants of various enzyme-cardiac glycoside complexes formed in the presence of sodium, magnesium, and adenosine triphosphate. Mol. Pharmacol. *10*, 494–500 (1974)

CHAPTER 17

Influence of Cardiac Glycosides on Cell Membrane*

H. LÜLLMANN and T. PETERS

A. Function of Na^+, K^+-ATPase in Heart Muscle Cells

Na^+, K^+-ATPase is a proteolipidic enzyme integrated into the plasmalemma. Under in vivo conditions Na^+, K^+-ATPase faces two different ionic milieus, i.e., the intra- and the extracellular milieu. Considering the working conditions of Na^+, K^+-ATPase in both isolated and membrane-integrated forms the following should be kept in mind: the *isolated* enzyme works at constant ion conditions since, in spite of continuous pump activity, the Na and K concentrations remain unaltered. The ATP-hydrolysis proceeds as long as substrate is available. In contrast, *the membrane-integrated* Na^+, K^+-ATPase of *intact cells* effectively transports Na and K against gradients until the physiologic intracellular ion concentrations have been attained, i.e., the pump works only upon demand. The demand is given by the Na load (influx per unit time) imposed upon the cell according to the prevailing conditions (e.g., resting permeability, frequency of excitations, Na flux per action potential). Under physiologic conditions, the main determinant will be the beat frequency. Each action potential adds to the basic Na load resulting from the comparatively small resting flux.

The membrane-bound Na^+, K^+-ATPase in cardiac tissue is not only responsible for the ion transport but may also serve as an essential Ca-binding structure. The lipid moiety of the enzyme consists of phospholipids surrounding the protein portion at the intracellular surface. The lipids are predominantly composed of acidic phospholipids such as phosphatidylserines (REDWOOD et al., 1973; COLEMAN, 1973; KNAUF et al., 1974; ZWAAL et al., 1973; DE PONT and BONTING, 1977). An essential part of the Na^+, K^+-ATPase molecule consists of the internal annulus of acidic phospholipids, in particular of phosphatidylserines. In this context it is of importance that the protein of the ATPase undergoes conformational alterations during transport activity (ALLEN and SCHWARTZ, 1970; SCHÖNFELD et al., 1972; DITTRICH et al., 1974). The lipids participate in conformational changes occurring during each transport cycle of the protein portion of the Na^+, K^+-ATPase molecule. In consequence each cycle will be accompanied by alterations of the spatial arrangement of the lipids. The acidic phospholipids display a high affinity for Ca, provided that the plasmalemma is polarized. Along with a depolarization the high affinity is more or less lost, resulting in a release of Ca^{2+} into the cytosol (LÜLLMANN and PETERS, 1977, 1979). The extent of Ca release induced by a depolarization will critically depend upon the spatial arrangement of the complexing lipids. The remaining affinity of Ca to the phospholipids during the depolarized

* The manuscript was submitted in May 1979

state *(residual affinity)* will thus determine the actual amount of Ca^{2+} released per excitation. The actual residual affinity is responsible for the fraction of the complexed Ca which becomes released and hence effective for initiating a contraction. The residual affinity, determined by the spatial arrangement of the acidic lipids, is influenced by the activity of Na^+, K^+-ATPase: with increasing ATPase activity the residual affinity will decline and consequently increasingly more Ca^{2+} will be released per excitation. The driving force for the ATPase cycling frequency is given by the actual Na load. Conditions which are associated with an enhanced Na load should, therefore, be accompanied by an increment of contractile force, e.g., the mechanisms underlying the force–frequency relationship, and the treppe (staircase) phenomenon.

B. Interaction of Cardiac Glycosides with Na^+, K^+-ATPase of Cardiac Tissue

It is well documented that different cardiac glycosides are accumulated by heart muscles to a variable extent. Hydrophobic compounds such as digitoxin display high accumulation rates which predominantly reflect nonspecific intracellular uptake, as can be expected from the physicochemical properties (DUTTA et al., 1968; KUSCHINSKY et al., 1967, 1968). In contrast, a hydrophilic glycoside such as ouabain cannot penetrate into the intracellular space, and is, therefore, suited as a tool to determine specific binding sites (GODFRAIND and LESNE, 1972; LÜLLMANN and PETERS, 1979). Even ouabain becomes to a minor extent nonspecifically bound, as revealed by an analysis of concentration-dependent binding curves: at pure inotropically acting concentrations, the nonspecific binding maximally attains 10% of the total binding, with increasing concentrations this proportion grows (GODFRAIND and LESNE, 1972). This is why only ouabain should be used to determine the specific binding capacity in cardiac tissue since, with increasing hydrophobicity, the binding kinetics of the specific sites become obscured by nonspecific uptake processes.

For the understanding of the binding kinetics of cardiac glycosides to specific binding sites it is of the utmost importance that only one particular conformation of Na^+, K^+-ATPase displays high affinity towards the drug. This conformation only occurs transiently during a pump cycle (SCHWARTZ et al., 1975). A situation like this implies that the description of drug–receptor interaction has to account for two variables, namely the concentration of the glycoside and the actual concentration of the susceptible conformation of the ATPase, the latter being subject to the functional state of cardiac cells (LÜLLMANN and PETERS, 1979).

The cycling rate of one particular ATPase molecule will depend statistically upon the hit probability of Na^+ provided that ATP and Mg are adequately supplied and that K^+ is present in the extracellular space (Fig. 1). The rate of the conformational cycling will be influenced by the temperature, which should be kept in mind when working at different experimental temperatures. From the foregoing it becomes obvious that the actual concentration of $ATPase_2$, which is of particular interest with respect to binding of cardiac glycosides, varies over a wide range according to the prevailing conditions: in resting heart muscles, only the resting Na flux will burden the ATPase and can be counteracted by as little as 5% of the total

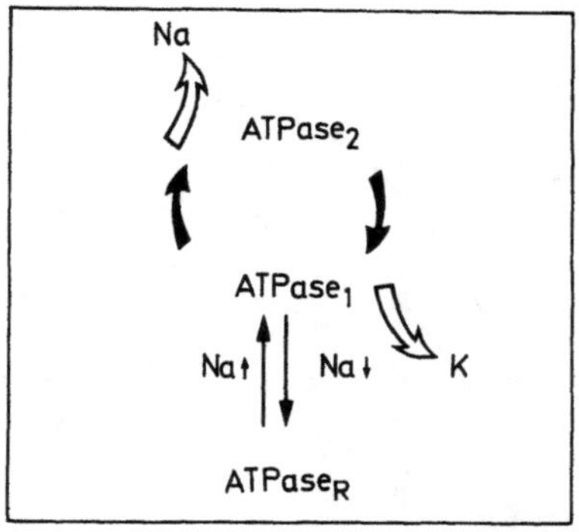

Fig. 1. Schematic presentation of functional states of Na^+, K^+-ATPase in intact heart muscle cells. $ATPase_R$ = ATPase in resting conformation; $ATPase_1$ = phosphorylated intermediate, ready to transport Na; $ATPase_2$ = briefly occurring intermediate which, prior to K-induced dephosphorylation, can bind cardiac glycosides; *upward arrow* = increase in Na load; *downward arrow* = decrease in Na load. The *open arrows* indicate transport of Na to the extra- and of K to the intracellular space

ATPase molecules (LÜLLMANN and PETERS, 1979). Each action potential will add to the basic Na flux and thus, with increasing beat frequencies, more and more ATPase molecules are required to maintain the Na^+, K^+-homeostasis. There will be an upper limit at which the total amount of ATPase molecules is no longer able to counterbalance the Na load. At this critical beat frequency the ionic balance escapes the control. Under physiologic conditions of beat frequency and temperature the Na^+, K^+-ATPase of heart muscle cells by far exceeds the actual requirement (AKERA and BRODY, 1977). This surplus of ATPase molecules allows the adaptation to conditions imposing a higher Na load upon the cells within the physiologic range.

An experimental example of the beat-frequency-dependent binding of ouabain to cardiac muscle is given in Fig. 2. Atria of guinea-pigs were exposed to identical concentrations of ouabain at different frequencies. As can be seen, both the rate of uptake and the final equilibria are determined by the beat frequency which imposes varying Na loads upon the cells. This phenomenon can, however, only be observed within a limited frequency range in which a pure inotropic response is evoked (LÜLLMANN et al., 1979). Whenever the first signs of intoxication occur the binding kinetics are altered by additional factors such as increase of cellular Na^+ and Ca^{2+} concentrations, decrease of K^+ concentration, and arrhythmia (BENTFELD et al., 1977). Even in the pure inotropic range, a quantitative description of the dependence of the rate of ouabain uptake and the final binding equilibrium on the beat frequency cannot be given since the system appears to be too complex. The basic features of ouabain binding are shown in simplified form in Fig. 3.

The scheme shows that the actual concentration of ouabain–ATPase complexes ($OuATPase_2$) is determined both by the concentration of ouabain and by the cycling frequency of the Na pump, which is responsible for the actual concentration of the susceptible conformation, $ATPase_2$ (LÜLLMANN and PETERS, 1979; BENTFELD et al., 1977; BUSSE et al., 1979; LÜLLMANN et al., 1979). In experiments

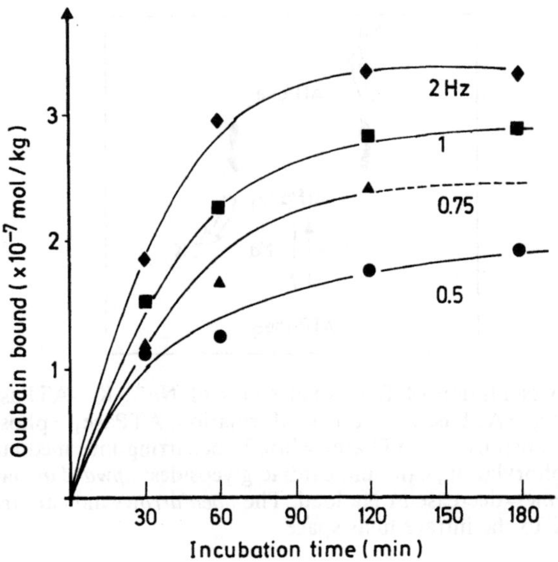

Fig. 2. Frequency-dependent binding of ouabain to isolated left guinea-pig atria at a constant drug concentration of 1×10^{-7} *M*. Guinea-pig left atria electrically driven at different frequencies as indicated. (Data from LÜLLMANN et al., 1979)

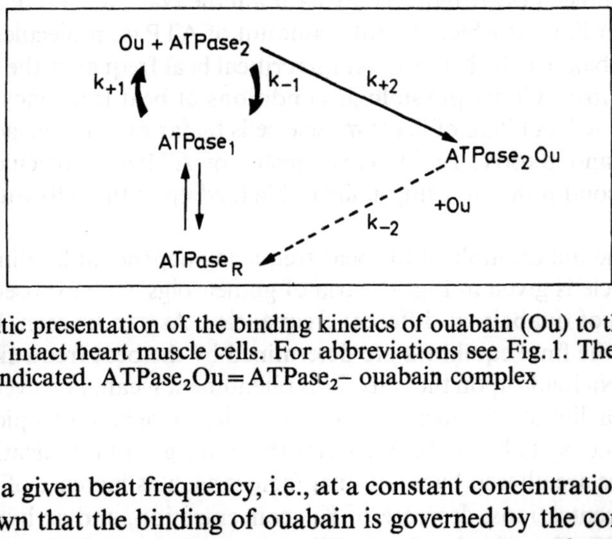

Fig. 3. Schematic presentation of the binding kinetics of ouabain (Ou) to the cycling Na^+, K^+-ATPase in intact heart muscle cells. For abbreviations see Fig. 1. The respective rate constants are indicated. $ATPase_2Ou = ATPase_2$– ouabain complex

conducted at a given beat frequency, i.e., at a constant concentration of $ATPase_2$, it can be shown that the binding of ouabain is governed by the concentration of the drug, as expected. Fig. 4 gives an experimental example performed on guinea-pig atria with varying concentrations of ouabain but at constant beat frequency (BUSSE et al., 1979). With increasing concentrations of ouabain the rate of uptake and the amount bound at equilibrium rise. Also in this case, the relation holds true only for a pure inotropic concentration range.

The kinetic model shows that only a certain proportion of the total ATPase can be occupied by ouabain since, depending on the conditions, more or less ATPase molecules are required to maintain the ionic homeostasis. The higher the Na load the smaller is the proportion of surplus ATPase which can be withdrawn from the

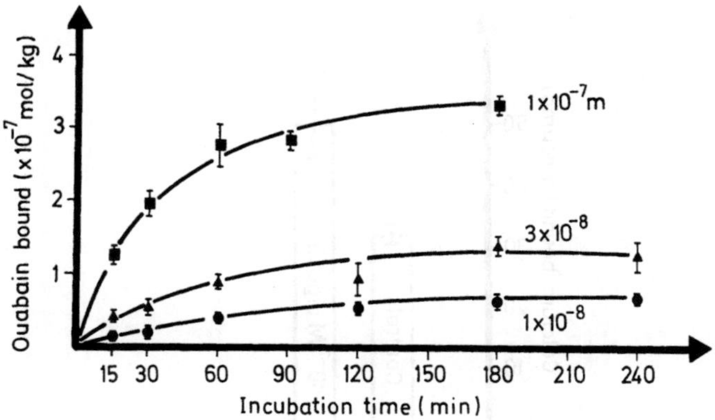

Fig. 4. Concentration-dependent binding of ouabain to isolated guinea-pig atria at a constant beat frequency of 1 Hz. The different concentrations of ouabain applied are indicated. (Data from BUSSE et al., 1979)

ion pump system. For the determination of specific binding sites it consequently follows that only part of the total binding sites are estimated as long as the homeostasis is kept intact. The lower the Na load, the higher this estimate will be. An estimation of the total number of specific binding sites is practically impossible in intact tissue since an impairment of the ion transport is inevitable and in turn leads to alterations of the binding characteristics.

Intoxication of cardiac cells will occur as soon as the transport capacity for Na fails to meet the actual demand. Consequently, the intracellular Na concentration but also the Ca concentration (see below) will rise and accordingly the K concentration declines. Particularly the rise of the cellular Ca^{2+} concentration will by itself inhibit Na^+, K^+-ATPase activity, thus supporting the glycoside-induced inactivation of the ion pump (DE POVER and GODFRAIND, 1976; BENTFELD et al., 1977; LÜLLMANN and PETERS, 1979). Additionally other essential cellular processes become impaired, e.g., ATP synthesis, membrane stability. Even if the experimental conditions are such that the glycoside is removed from the tissue the intoxication is irreversible and self-sustaining (BENTFELD et al., 1977).

I. Na^+-K^+-ATPase as a Calcium Binding Partner

So far, the discussion has dealt with the role of the ATPase as an ion pump. To understand the inotropic action of cardiac glycosides, the *second function* of Na^+, K^+-ATPase in heart muscle is of particular interest. As outlined above the lipid portion of ATPase is able to bind and release Ca; the extent to which Ca can be released or become bound is critically influenced by the conformational state, namely the cycling frequency of the individual ATPase molecule. The binding of a cardiac glycoside molecule to one particular conformation of Na^+, K^+-ATPase results in a preservation of this conformation and has two consequences: (1) exclusion of this complex from transport activity as long as the glycoside molecule remains bound, and (2) transient sustaining of the "labilization" of Ca binding inherent to this conformation. The labilization is brought about by a lowering of the

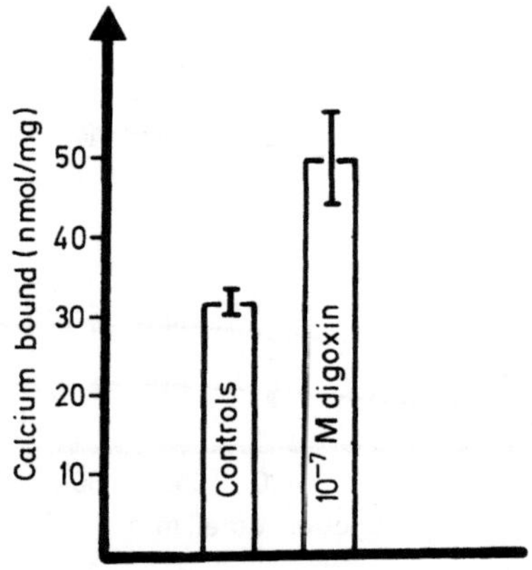

Fig. 5. Binding of calcium to isolated plasmalemmal microsomes from guinea-pig ventricular heart muscle. The columns indicate the amount of calcium bound to the microsomes after 150 min incubation in the absence *(left)* and the presence *(right)* of 1×10^{-7} *M* digoxin. The data reflect binding (and not Ca transport). The incubation medium contained 10^{-4} *M* Ca containing negligible traces of Ca^{45}. Ordinate: nmol Ca bound per mg protein. (Data from BRADE et al., 1974)

residual affinity. An additional consequence might be that the amount of Ca bound to the phospholipids during polarization increases when glycosides are attached to $ATPase_2$. Probably both processes will contribute to the improved release of Ca^{2+} during excitation after binding of ouabain.

We now discuss the experimental basis for the concept that cardiac glycosides evoke the inotropic response via an alteration of the Ca-binding and Ca-release characteristics of cardiac plasmalemma. By using purified Na^+, K^+-ATPase from cardiac plasmalemma. ALLEN and SCHWARTZ (1970) were able to demonstrate by means of circular dichroism (CD) measurements that binding of ouabain induces conformational alterations of the protein part of the preparation. A similar effect was observed upon binding of ouabain to cardiac plasmalemmal microsomes; a quantitative analysis of the spectra suggested that not only the protein part of the enzyme but also adjacent membrane proteins contributed to the observed CD alterations (LÜLLMANN et al., 1975). Since the CD measurement detects only changes of the protein structure, the findings of REDWOOD et al. (1973) are of particular interest, because a participation of the lipid portion of the Na^+, K^+-ATPase molecule in structural changes could be demonstrated upon binding of ouabain. It is this mutual interdependence between protein conformation and lipid array which forms the link between the ATPase activity on the one hand and the Ca binding properties on the other. A mechanism like this probably underlies the following observations. The Ca-binding capacity was found to be raised by digoxin in microsomes prepared from cardiac plasmalemma as shown in Fig. 5 (BRADE et al., 1974). A more detailed investigation (PREUNER, 1979) revealed that the binding of Ca is particularly enhanced when intact hearts were perfused prior to the preparation of the microsomes with ouabain and when the conditions preserved the ouabain–ATPase complex. In studies on purified kidney ATPase GERVAIS et al. (1977) demonstrated that, in this preparation also, the Ca-binding affinity was enhanced in the presence of ouabain.

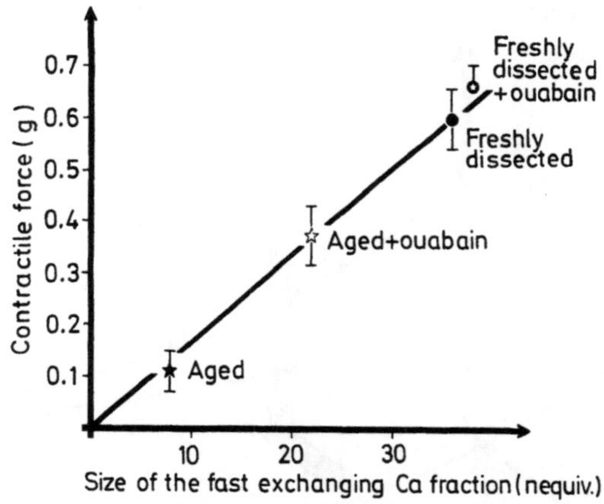

Fig. 6. Correlation between size of fast exchangeable calcium fraction and contractile force in guinea-pig atria. While "aging" of the muscles and pentobarbital depress the fast exchanging fraction and contractile force as compared with the respective conditions in freshly dissected muscle preparations, ouabain is able to augment the depressed exchangeable fraction and concomitantly to restore contractile force. (Data from CARRIER et al., 1974)

These results obtained from isolated plasmalemma or ATPase preparations can be correlated to findings in intact cardiac tissue. Already in the early papers dealing with the influence of cardiac glycosides upon the Ca metabolism of atrial tissue it has been reported that the exchange rate of Ca was increased at a constant concentration of total tissue Ca (HOLLAND and SEKUL, 1961; LÜLLMANN and HOLLAND, 1962; KLAUS and KUSCHINSKY, 1962; WINEGRAD and SHANES, 1962; GERSMEYER and HOLLAND, 1963). This finding has thereafter repeatedly been confirmed. From the more recent results it can be concluded that cardiac glycosides enlarge the size of the fast exchangeable fraction of the tissue Ca (NAYLER, 1973; CARRIER et al., 1974). The size of this particular fraction depends also on physiologic conditions: with increasing beat frequencies the rate of exchange rises, whereas with "aging" of an isolated heart preparation the rate of Ca exchange declines. Depending on the control state, cardiac glycosides will more or less accelerate the Ca exchange rate (CARRIER et al., 1974). The effect of ouabain will become the more pronounced the lower the exchange rate of the atrial muscle under control conditions. In conclusion it appears that the size of the fast exchangeable fraction is closely related to the contractile force (BAILEY and DRESEL, 1968; CARRIER et al., 1974). An example is given in Fig. 6.

A different experimental approach to check on the "lability" of a superficial Ca fraction of ventricular muscle consists of a rapid lowering of the extracellular Ca^{2+} concentration and determination of the Ca released per unit time with simultaneous recording of the contractile force. As demonstrated in Fig. 7, there exist identical time courses for the decline of contractile force and loss of Ca from superficial sites. In the presence of ouabain, the size of the superficial Ca fraction as well as the contractility are increased, the time course, however, of the adaptation process to the lowered extracellular Ca concentration remain similar to each other and correspond to the time courses under control conditions. Comparable results obtained

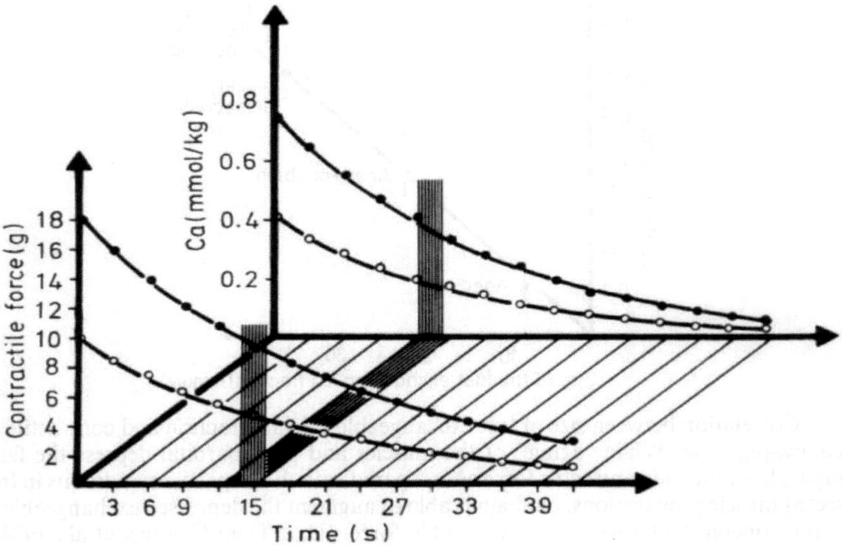

Fig. 7. Influence of a reduction of the extracellular calcium concentration on contractile force and calcium release of isolated coronary perfused hearts of guinea-pigs in the absence *(open circles)* and presence *(solid circles)* of 1.5×10^{-7} *M* ouabain. Abscissa indicates the time after reduction of external calcium concentration from 1.8 to 0.18 m*M* and is corrected for the time required for washout of the extracellular space. Ordinates indicate contractile force and fast removable calcium respectively. *Hatched bars* mark the range of half-lives of all processes. (Data from LENSCHOW, 1978)

on cat hearts have been reported by BAILEY and DRESEL (1968), ONG and BAILEY (1972). By means of the lanthanum-displacement method NAYLER (1973) observed similar effects.

II. Action of Cardiac Glycosides on Resting Cardiac Muscle

So far, experimental data obtained from beating heart muscle have been discussed. It seems of particular interest to investigate the Ca-binding characteristics of resting heart muscle cells in the presence of ouabain since it has been claimed that the inotropic action of cardiac glycosides is based on an excitation Na influx, Na/Ca exchange hypothesis (REPKE, 1964; REUTER, 1974; AKERA and BRODY, 1977). In 1963 GERSMEYER and HOLLAND investigated the influence of ouabain upon the Ca exchange rate of isolated guinea-pig atria at rest and at different beat frequencies. Upon exposure to ouabain (10^{-6} *M*) for 75 min, the exchangeable Ca fraction was found to be increased by 25% in resting preparation (see the Table in the paper by GERSMEYER and HOLLAND, 1963). VINCENZI (1967) clearly demonstrated that exposure of resting atrial tissues to ouabain (4×10^{-7} *M*) for 50 min induced alterations which led, even after washout of the drug immediately prior to recommencement of stimulation, to enhanced contractily. VINCENZI concluded that the inotropic effect of cardiac glycosides can be preformed in the resting state. Obviously the

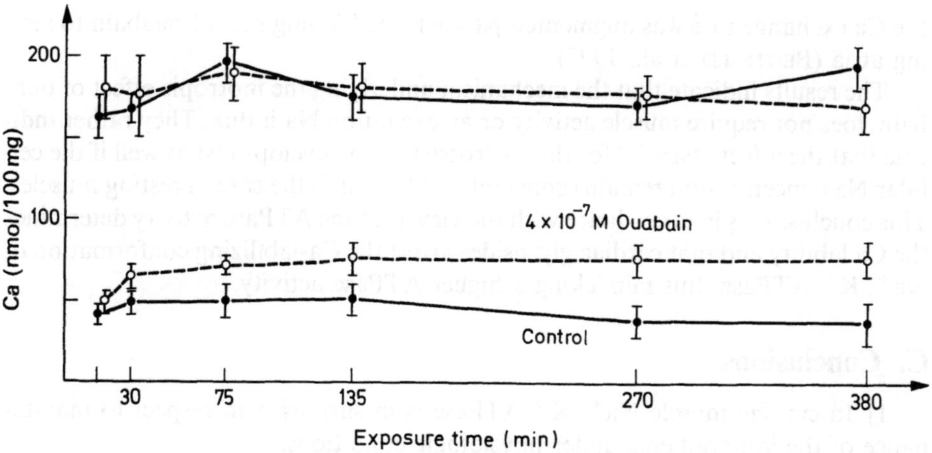

Fig. 8. Influence of ouabain (4×10^{-7} *M*) on total tissue calcium and ^{45}Ca exchange rate of isolated resting guinea-pig atria. ^{45}Ca was offered 15 min prior to the time points of determination. *Closed circles* = control conditions. *Open circles* = presence of 4×10^{-7} *M* ouabain. Points on the curves indicate mean values; vertical bars indicate standard errors of mean. (Data from KASPAREK et al., 1981)

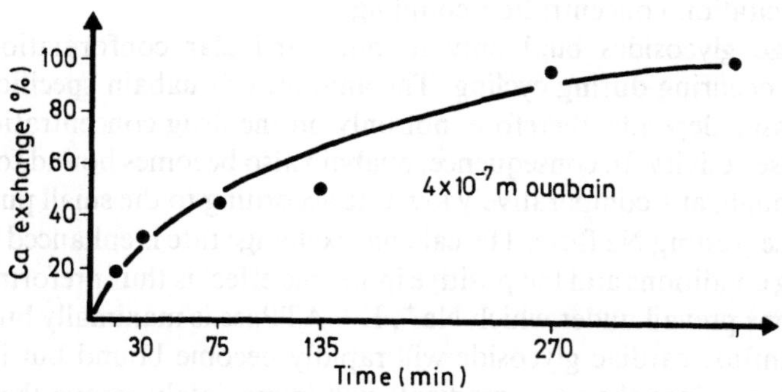

Fig. 9. Influence of ouabain on fractional calcium exchange. Points on the curves are corrected for the time-dependent decrease of calcium exchange rate under control conditions. For details see text. (Data from KASPAREK et al., 1981)

development of the underlying change proceeds more slowly in resting than in contracting muscles. This is in accordance with the rate of binding of ouabain. An attempt to demonstrate directly the action of ouabain upon the Ca lability of resting guinea-pig atria yielded the following results (KASPAREK et al.,1981), see Fig. 8. In control muscles the Ca fraction, exchanged within 15 min, amounted to almost 50 nmol/100 mg w.w. and slightly declined within about 6 h of observation, reflecting the reduced exchangeability with "aging" of the preparations. In the presence of ouabain (4×10^{-7} *M*) the exchanged fraction rose within 75 min to an apparent equilibrium approaching 75 nmol/100 mg w.w. Taking into account, however, the decline of the exchangeability of the controls due to "aging", ouabain induced a continuous increase of the exchanged fraction as shown in Fig. 9. The total Ca content of the tissue remained constant in these experiments. The time course at which the Ca exchange rate was augmented parallels the binding rate of ouabain to resting atria (BENTFELD et al., 1977).

The results indicate that the mechanism underlying the inotropic effect of ouabain does not require muscle activity or an excitation Na influx. They rather indicate that the effect essential for the inotropic action develops just as well if the cellular Na concentration remains constant and low, as is the case in resting muscles. This conclusion is in accordance with the view that the ATPase activity determines the Ca lability and that cardiac glycosides arrest the Ca-labilizing conformation of Na^+, K^+-ATPase thus mimicking a higher ATPase activity.

C. Conclusions

1) In cardiac muscle Na^+, K^+-ATPase is in surplus with respect to maintenance of the ion gradients under physiologic conditions.

2) Beside the pump function which is taken care of by the protein moiety of the ATPase molecule, the lipid portion of the ATPase molecule provides binding sites for Ca, some of which becomes released upon membrane depolarization. With increasing Na load and hence enhanced cycling frequency the lability is augmented, rendering the action potential more effective in releasing Ca^{2+} for contraction, i.e., improved excitation-concentration coupling.

3) Cardiac glycosides bind only to one particular conformation of Na^+, K^+-ATPase occuring during cycling. The amount of ouabain specifically bound to cardiac tissue depends, therefore, not only on the drug concentration but also on the ATPase activity. In consequence, ouabain also becomes bound to the resting muscle, although, at a comparatively low rate according to the small pump activity of the ATPase (resting Na flux). The calcium exchange rate is enhanced by ouabain under resting conditions and the positive inotropic effect is thus preformed. If however conditions prevail under which Na^+, K^+-ATPase is maximally burdened due to high Na influx, cardiac glycoside will rapidly become bound but induce only toxic symptoms, since the ionic gradients will immediately escape the control.

4) The binding of ouabain to the susceptible ATPase conformation has two consequences: (a) transport inhibition of the occupied ATPase molecules, and (b) transient preservation of that particular conformation resulting in a labilization of the complexed Ca.

5) Although the attachment of a cardiac glycoside molecule to an individual ATPase molecule inhibits its pump function, the inotropic response can only be evoked and maintained as long as sufficient functioning ATPase molecules are present to counteract the Na influx. It is not the reduced Na pump efficiency which produces the inotropic response, but rather an alteration of Ca-binding characteristics of the plasmalemma which is controlled by the ATPase activity.

References

Akera, T., Brody, T.M.: The role of Na^+, K^+-ATPase in the inotropic action of digitalis. Pharmacol. Rev. *29*, 187–220 (1977)

Allen, J.C., Schwartz, A.: Effects of potassium, temperature, and time on ouabain interaction with the cardiac Na^+, K^+-ATPase: Further evidence supporting an allosteric site. J. Mol. Cell. Cardiol. *1*, 39–45 (1970)

Bailey, L.E., Dresel, P.E.: Correlation of contractile force with a calcium pool in the isolated cat heart. J. Gen. Physiol. *52*, 969–982 (1968)

Bentfeld, M., Lüllmann, H., Peters, T., Proppe, D.: Interdependence of ion transport and the action of ouabain in heart muscle. J. Pharmacol. *61*, 19–27 (1977)

Brade, H., Lübbers, H., Peters, T.: Isolation and properties of a microsomal plasmamembrane fraction from guinea-pig ventricular muscle. Naunyn-Schmiedebergs Arch. Pharmacol. *282*, R 11 (1974)

Busse, F., Lüllmann, H., Peters, T.: Concentration dependence of the binding of ouabain to isolated guinea-pig atria. J. Cardiovasc. Pharmacol. *1*, 687–698 (1979)

Carrier, G.O., Lüllmann, H., Neubauer, L., Peters, T.: The significance of a fast exchanging superficial calcium fraction for the regulation of contractile force in heart muscle. J. Mol. Cell. Cardiol. *6*, 333–347 (1974)

Coleman, R.: Membrane bound enzymes and membrane ultrastructure. Biochim. Biophys. Acta *300*, 1–30 (1973)

De Pont, J.J.H.H.M., Bonting, S.L.: The role of phospholipids in Na-K-ATPase. In: Function and biosynthesis of lipids. Bazan, N.G., Brenner, R.R., Guisto, N.M. (eds.), pp. 219–224. New York: Plenum 1977

De Pover, A., Godfraind, T.: Sensitivity to cardiac glycosides of (Na+K) ATPase prepared from human heart, guinea-pig heart, and guinea-pig brain. Arch. Int. Pharmacodyn. Ther. *221*, 339–341 (1976)

Dittrich, F., Schön, R., Repke, K.R.H.: Mechanism of ATP free-energy transfer and utilization in (Na, K)-ATPase transport function. Acta Biol. Med. Ger. *33*, K17–K25 (1974)

Dutta, S., Goswami, S., Lindower, J.O., Marks, B.H.: Subcellular distribution of digoxin-H^3 in isolated guinea-pig and rat hearts. J. Pharmacol. Exp. Ther. *170*, 318–325 (1968)

Gersmeyer, E.F., Holland, W.C.: Effect of heart rate on action of ouabain on Ca-exchange in guinea-pig left atria. Am. J. Physiol. *205*, 795–798 (1963)

Gervais, A., Lane, L.K., Anner, B.M., Lindenmayer, G.E., Schwartz, A.: A possible molecular mechanism of the action of digitalis. Ouabain action on calcium binding to sites associated with a purified sodium-potassium-activated adenosine triphosphatase from kidney. Circ. Res. *40*, 9–14 (1977)

Godfraind, T., Lesne, M.: The uptake of cardiac glycosides in relation to their actions in isolated cardiac muscle. Br. J. Pharmacol. *46*, 488–497 (1972)

Holland, W.C., Sekul, A.A.: Influence of K^+ and Ca^{2+} on the effect of ouabain on Ca^{45} entry and contracture in rabbit atria. J. Pharmacol. Exp. Ther. *133*, 288–294 (1961)

Kasparek, R., Lüllmann, H., Peters, T.: Influence of ouabain on the Ca-exchangeability in resting atria of guinea-pig. Europ. J. Pharmac. in press 1981

Klaus, W., Kuschinsky, G.: Über die Wirkung von Digitoxigenin auf den zellulären Ca-Umsatz im Herzmuskelgewebe. Naunyn Schmiedebergs Arch. Pharmacol. *244*, 237–253 (1962)

Knauf, P.A., Proverbio, F., Hoffman, J.S.: Chemical characterization and pronase susceptibility of the Na, K pump-associated phosphoprotein of human red blood cells. J. Gen. Physiol. *63*, 305–323 (1974)

Kuschinsky, K., Lahrtz, H.G., Lüllmann, H., van Zwieten, P.A.: Accumulation and release of ^{3}H-digoxin by guinea-pig heart muscle. Br. J. Pharmacol. *30*, 317–328 (1967)

Kuschinsky, K., Lüllmann, H., van Zwieten, P.A.: A comparison of the accumulation and release of ^{3}H-ouabain and ^{3}H-digitoxin by guinea-pig heart muscle. Br. J. Pharmacol. *32*, 598–608 (1968)

Lenschow, V.: Zum Einfluß der Herzglykoside auf den Calciumumsatz im Herzmuskel. Thesis, Med. Fac. Kiel (1978)

Lüllmann, H., Holland, W.C.: Influence of ouabain on an exchangeable calcium fraction, contractile force and resting tension of guinea-pig atria. J. Pharmacol. Exp. Ther. *137*, 186–192 (1962)

Lüllmann, H., Peters, T.: Plasmalemmal calcium in cardiac excitation-contraction coupling. Clin. Exp. Pharmacol. Physiol. *4*, 49–57 (1977)

Lüllmann, H., Peters, T.: Action of cardiac glycosides on the excitation-contraction coupling in heart muscle. Prog. Pharmacol. *2*, 3–57 (1979)

Lüllmann, H., Peters, T., Preuner, J., Rüther, T.: Influence of ouabain and dihydro-ouabain on the circular dichroism of cardiac plasmalemmal microsomes. Naunyn-Schmiedebergs Arch. Pharmacol. *290*, 1–19 (1975)
Lüllmann, H., Peters, T., Ziegler, A.: Kinetic events determining the effects of cardiac glycosides. TIPS, 102–106 (1979)
Nayler, W.G.: An effect of ouabain on the superficially-located stores of calcium in cardiac muscle cells. J. Mol. Cell. Cardiol. *5*, 101–110 (1973)
Ong, S.D., Bailey, L.E.: Two functionally distinct calcium pools in the excitation-contraction coupling process. Experientia *28*, 1446–1447 (1972)
Preuner, J.: Cardiac glycoside-induced changes of the Ca-binding in guinea-pig plasmalemma. Naunyn-Schmiedebergs Arch. Pharmacol. *307*, R 36 (1979)
Redwood, W.R., Gibbes, D.C., Thompson, T.E.: Interaction of solubilized membrane ATPase with lipid bilayer membranes. Biochim. Biophys. Acta *31*, 10–22 (1973)
Repke, K.R.H.: Über den biochemischen Wirkungsmechanismus von Digitalis. Klin. Wochenschr. *42*, 157–165 (1964)
Reuter, H.: Exchange of calcium ions in the mammalian myocardium. Mechanism and physiological significance. Circ. Res. *34*, 599–605 (1974)
Schönfeld, W., Schön, R., Menke, K.-H., Repke, K.R.H.: Identification of conformational states of transport ATPase by kinetic analysis of ouabain binding. Acta Biol. Med. Ger. *28*, 935–956 (1972)
Schwartz, A., Lindenmayer, G.E., Allen, J.C.: The sodium-potassium adenosine triphosphatase: Pharmacological, physiological and biochemical aspects. Pharmacol. Rev. *27*, 3–134 (1975)
Vincenzi, F.F.: Influence of myocardial activity on the rate of onset of ouabain action. J. Pharmacol. Exp. Ther. *155*, 279–287 (1967)
Winegrad, S., Shanes, A.M.: Calcium flux and contractility of guinea-pig atria. J. Gen. Physiol. *45*, 371–394 (1962)
Zwaal, R.F.A., Roelofsen, B., Colley, C.M.: Localization of red cell membrane constituents. Biochim. Biophys. Acta *300*, 159–182 (1973)

CHAPTER 18

Influence of Cardiac Glycosides on Electrolyte Exchange and Content in Cardiac Muscle Cells

W. NAYLER and E. A. NOACK

A. Introduction

Whether the positive inotropism of the cardiac glycosides (CG) can be accounted for in terms of an altered cellular distribution, concentration or exchangeability of certain ions has attracted widespread attention. As a result several review articles dealing with this subject have already appeared (WOLLENBERGER, 1949; GLYNN, 1964; LEE and KLAUS, 1971). In this present chapter, therefore, we intend concentrating on the experimental findings which have been published during the past decade (1970–1979).

There is widespread agreement that, at the molecular level, the positive inotropism of these drugs does not involve their direct interaction with the contractile proteins (see Chap. 19). Nor can their positive inotropism be explained in terms of a direct effect on myocardial energy metabolism (WOLLENBERGER, 1949). By contrast, there are many observations which link the positive inotropism of these drugs with their effect on electrolyte metabolism and exchangeability. Thus, as early as 1931, CALHOUN and HARRISON noted that the administration of digitals caused a loss of K^+ from canine heart. Other investigators have shown a parallelism between the positive inotropism of the glycosides and an increase in intracellular Ca^{2+} exchangeability (THOMAS et al., 1958; HOLLAND and SEKUL 1959; GROSSMAN and FURCHGOTT, 1964; HEINEN and NOACK, 1972; NAYLER, 1973b). Whilst comparatively recent experiments indicate that the positive inotropism of the glycosides can be dissociated from their effect on potassium and sodium movements (BESCH and WATANABE, 1978; POOLE-WILSON and LANGER, 1975), the possibility that calcium ions may be involved remains strong. If this is the case, however, then these drugs must evoke a positive inotropic response by increasing the intracellular availability of Ca^{2+} so that more calcium ions are available to interact with the contractile and regulatory proteins. There is no evidence of any change in the sensitivity of the contractile proteins to calcium (KATZ, 1970) or in the amount of calcium needed to develop peak tension.

Other investigators believe that the positive inotropic activity of the glycosides is closely linked with, or even dependent upon, their ability to bind to and inhibit the activity of the Na^+, K^+-ATPase enzyme. Although first identified in crab nerve (SKOU, 1957) a glycoside-sensitive, Na^+, K^+-activated, Mg^{2+}-dependent-ATPase enzyme is also present in heart muscle (see Chaps. 14 and 15). Some investigators (LANGER, 1972) have argued that because this particular ATPase enzyme is responsible for maintaining the differential distribution of Na^+ and K^+ across cardiac cell membranes its inhibition would result in a raised intracellular concentration of Na^+, particularly in the immediate vicinity of the cytosolic surface of the cell membrane. Under these conditions and particularly because of the ability of cardiac muscle cells to exchange Ca^{2+} for Na^+ (REUTER, 1974) the intracellular availability of Ca^{2+} may increase (Fig. 1). This increase in intracellular Ca^{2+} may be insufficient by itself to activate or influence tension development directly (SOLARO et al., 1974). It is conceivable, however, that it may supplement the supply of "trigger Ca^{2+}", thereby promoting an additional release (Fig. 2) of "activator Ca^{2+}" (NAYLER, 1966) and hence a positive inotropic response. If these or closely related events are involved in the positive inotropism of the glycosides then it might be possible to detect changes in electrolyte content or exchangeability associated with the increase in tension development. There are, however, methodological difficulties with such an approach.

B. Critical Evaluation of Factors Influencing the Validity of Myocardial Electrolyte Determinations

Attempts to establish the biochemical basis for the mode of action of the cardiac glycosides by determining total tissue electrolyte content have generally been unrewarding, partly because of methodological problems and in part because of difficulties in interpretation.

The results obtained from skeletal muscle cannot be translated directly to the myocardium. In skeletal muscle, excitation–contraction (E–C) coupling depends entirely upon the mobilization of "activator calcium" (NAYLER, 1966) from internal storage sites. By contrast the electrolyte composition of the fluid which bathes cardiac muscle cells has a profound effect on their inotropic state and responsiveness to drugs (LANGER, 1972). Calcium, sodium, and potassium (FARAH, 1969) are all involved and to some extent their activities are interdependent. For example, there is competition between Ca^{2+} and Na^+ at the level of the cell membrane (NIEDERGERKE, 1963). Although this competition is usually accounted for in terms of the Na^+/Ca^{2+} exchange system (REUTER, 1974), recent electrophysiological studies have shown that this competition may extend to the voltage-activated, ion-selective channels which are responsible (Fig. 1) for the inward transport of Ca^{2+} during the plateau phase of the action potential, because these channels can also transport Na^+ (KASS and TSIEN, 1975). The movements of Ca^{2+} and K^+ are also closely linked, a raised concentration of Ca^{2+} at the cytosolic face of the cell membrane triggering an efflux of K^+ (ISENBERG, 1977).

There are other difficulties – for example if a raised tissue concentration of Ca^{2+} is detected (LANGER and SERENA, 1970) it cannot be concluded that the additional Ca^{2+} is available for interaction with the myofilaments. It might be entrapped in either the sarcoplasmic reticulum or the mitochondria.

Moreover the cardiac glycosides increase force and velocity of contraction in the normal as well as in the failing heart, but owing to the fact that the tissue electrolyte content of failing heart muscle differs by an increased sodium and a decreased potassium content from that of nonfailing muscle, the glycoside may produce opposite effects on their electrolyte composition (BOYER and POINDEXTER, 1940; HOCHREIN, 1965). This altered electrolyte composition also lowers the con-

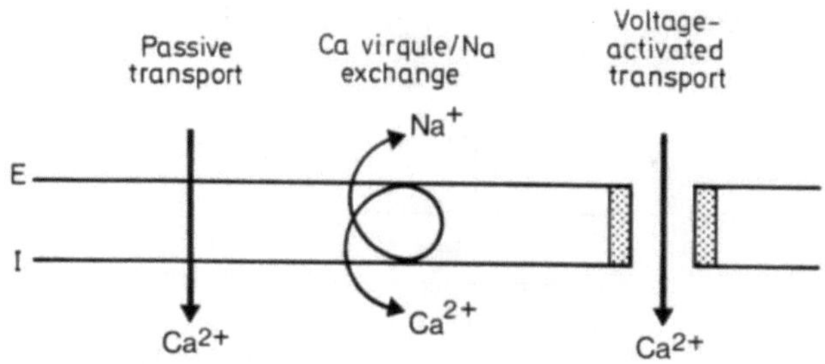

Fig. 1. Schematic representation of the three processes whereby Ca^{2+} can penetrate cardiac cell membrane: by passive diffusion, in exchange for Na^+ and through voltage-activated channels. E = extracellular; I = intracellular

centration of glycoside needed to achieve a given response, making the evaluation and comparison of results even more difficult.

It might be appropriate to question now whether the concentrations of glycosides which are used to achieve a positive inotropic response under experimental conditions bear any resemblance to or are comparable with those used therapeutically. For instance the reported plasma levels of CG in patients with congestive heart failure is about 1.5×10^{-9} mol/l for digoxin and 2.5×10^{-8} mol/l for digitoxin (BELLER et al., 1971; SMITH, 1970), whilst higher concentrations are used to provoke a positive inotropic response in isolated heart muscle preparations.

There are also species-dependent variations in sensitivity, and these differences further complicate the interpretation of results and the comparison of results obtained for different species. Age must also be taken into account (BOERTH et al., 1975; WOLLENBERGER et al., 1953; SHERIDAN, 1978), perhaps because of an effect of age on electrolyte content and the mechanisms responsible for regulating intracellular Ca^{2+} homeostasis (NAYLER et al., 1979). Certainly species, heart rate, and the prior contractile state of the heart muscle must be taken into account; for example, depending on the frequency of stimulation ouabain may evoke either a positive inotropic or a toxic response from guinea-pig atrial muscle (BENTFELD et al., 1977; GREEFF et al., 1971). Plasma potassium and calcium content (FORRESTER and MAINWOOD, 1974), plasma acidity and oxygen partial pressure (SMITH and HABER, 1973) also help to determine whether a particular CG concentration will be therapeutic or toxic. The hormonal state of the animal also seems to be of importance – particularly if hypo- or hyperthyroidism is present (see Chap. 24). Even the different regions of the heart respond differently. Hence the results of studies on Purkinje fibres, for example, (ELLIS, 1977) cannot be extrapolated directly to the myocardium. The same holds true for atrial preparations (CARSLAKE and WEATHERALL, 1962; REITER, 1972).

Even if all the factors which have been discussed so far have been adequately controlled there are other reasons for conflicting results being obtained for the effect of the glycosides on myocardial electrolyte metabolism. These reasons are largely methodological. The detection of small changes in electrolyte content is essentially impossible in whole myocardium unless special techniques are simultaneously applied to assess the electrolyte content of the extracellular space. For example, a relatively high extracellular concentration of Na^+ will obscure small

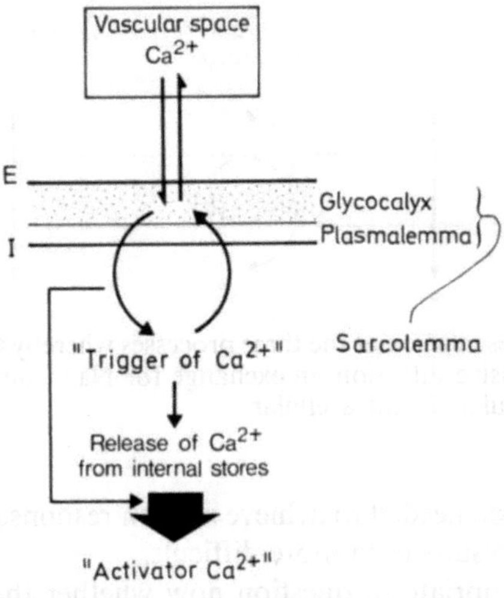

Fig. 2. Schematic representation of the mobilization of "activator Ca^{2+}" by the arrival of a small amount of "trigger Ca^{2+}"

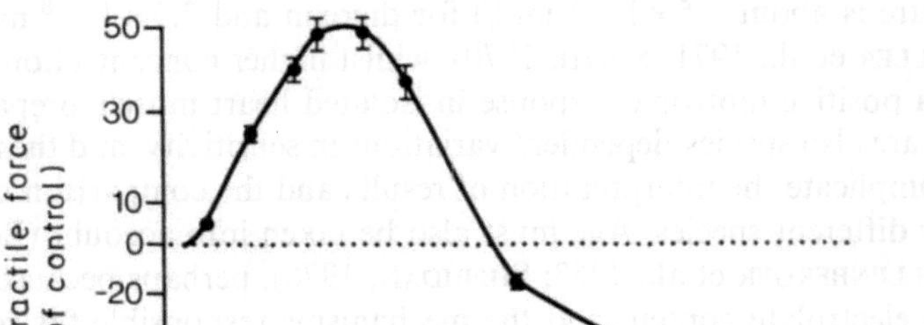

Fig. 3. The influence of a toxic concentration of ouabain (2.5 μmol/l) on the contractile force and the size of the extracellular space in guinea-pig isolated atria. The means ± standard errors from 8 to 38 (contractile force) or 8 (extracellular space volume) separate experiments are shown. Control value of contractile force (at zero time) 2.44 ± 0.09 g. (NOACK, unpublished)

changes in intracellular Na^+. Because of this, studies in isolated muscle preparations which involve the detection of ions by flame photometry or atomic absorption spectrometry are useless unless they include the determination of extracellular volume and electrolyte content. There are many factors which influence the size of the extracellular space: it increases with the wet weight of atrial preparations, decreasing extracellular calcium concentrations or increasing frequency of stimulation (NOACK, 1979). GODFRAIND and GHYSEL-BURTON (1977) and NOACK et al. (1979) found that the inulin diffusion space was never modified in the presence of therapeutic concentrations of CG. In contrast it decreases rapidly with toxic concentrations (Fig. 3). It should be stressed, however, that whichever method is used (inulin, SO_4^{2-}, Cl^- etc.) to measure extracellular space it will represent only a compromise for the estimation of the exact size of the extracellular space. Therefore, determinations of the cellular bulk electrolyte content of the myocardial cell should be regarded as providing only a semiquantitative measure. Nevertheless in combination with isotope flux studies substantial information may be derived about the basic changes in electrolyte composition.

Rapid advances in electrophysiological and biochemical methods, such as the use of ion-selective microelectrodes, have revealed new limitations for the evidence obtained from total electrolyte studies. LEE and FOZZARD (1975) have demonstrated that a major portion of the intracellularly located sodium and potassium is not free in cytoplasm but rather compartmentalized, possibly in subcellular structures like mitochondria (PAGE et al., 1971), and sarcoplasmic reticulum (ROGUS and ZIERLER, 1973) which are not directly linked to the extracellular space or in the central elements of T systems (CALDWELL, 1968). In addition the cytosolic sodium activity may be unevenly distributed, showing perhaps an accumulation in a space close to the sarcolemma (BLOOD, 1975; DEITMER and ELLIS, 1978a, b). Such intracellular compartmentalization of electrolytes, and the exchange and translocation of ions between these compartments, will be overlooked in studies which rely completely on the classical methods of measuring total tissue electrolytes.

Before considering whether the effect of the CG on the electrolyte content and ionic exchangeability of heart muscle correlates with their positive inotropic or toxic responses it may be useful to examine briefly whether the electrolyte composition of the extracellular fluid affects their activity.

C. Influence of Extracellular Electrolyte Composition on the Pharmacological Effect of Cardiac Glycosides

The many theories which have been formulated to account for the positive inotropism of the cardiac glycosides have one common feature: they are all based on the assumption that these drugs evoke an inotropic response by increasing the intracellular free Ca^{2+} concentration, and that this increase in intracellular Ca^{2+} is secondary to an interaction of these drugs with the cell membrane. According to one popular theory (LANGER, 1972), by inhibiting the Na^+, K^+-ATPase enzyme these drugs evoke a raised intracellular Na^+ which in turn stimulates Ca^{2+} influx via the Ca/Na exchange system (REUTER, 1974). Another although less popular theory is that by interacting with the Na^+, K^+-ATPase enzyme the glycosides effect a conformational change in the enzyme thereby affecting its ion pumping activity to favour an exchange of K^+ for Ca^{2+} (LINDENMAYER and SCHWARTZ, 1975). Another possibility is that the interaction of the CG with either Na^+, K^+-ATPase or some other component of the cell membrane affects the exchange between K^+ and Ca^{2+} which takes place during the action potential (MORAD and GREENSPAN, 1973). Later in this chapter we will consider the viability of these three possibilities. For the moment, however, it may be useful to consider whether the extracellular availability of the three ions which are concerned – K^+, Na^+, and Ca^{2+} – affect or are necessary for the positive inotropic activity of the CG.

I. Potassium

Studies in which a parallelism has been observed between plasma and extracellular K^+ concentrations and the rate at which the CG are accumulated by the myocardium (DUTTA and MARKS, 1972) add little to our understanding of how these drugs achieve their positive inotropism, because of the likelihood of nonspecific binding and accumulation.

Using isolated rabbit, guinea-pig, and cat myocardium many investigators (GARB and VENTURI, 1954; CAPRIO and FARAH, 1967; PRINDLE et al., 1971) have shown that a raised extracellular concentration of K^+ delays the onset of the positive inotropic effect of the glycosides. Raised extracellular concentrations of K^+ have also been shown to reduce the rate at which the myocardium accumulates these drugs (MARKS, 1972; DUTTA and MARKS, 1972). Conversely COHN et al. (1967) found that in guinea-pig heart muscle a reduced extracellular K^+ shows a reduced rate of CG accumulation. This "antagonism" between K^+ and the glycosides is of therapeutic interest, because it provides a basis for the clinical management of digitalis-induced toxicity (see this Handbook, Vol. 56/II, Chap. 13). Glycoside-induced toxicity in humans initially manifests itself in the form of arrhythmias and conduction abnormalities, long before end-diastolic volume decreases. The arrhythmias and conduction defects are almost certainly due to or associated with a loss of tissue K^+ (see Sect. E). By increasing plasma K^+ the ionic gradient for K^+ will be reduced and hence any further loss minimized, without necessarily reversing the positive inotropic efficacy of the CG (WILLIAMS et al., 1966).

Some investigators have used the parallelism which exists between the effect of an ionic substitution such as a raised extracellular K^+ on the rate of development of inotropic response (PRINDLE et al., 1971) and their rate of binding to Na^+, K^+-ATPase (ALLEN et al., 1970; AKERA and BRODY, 1971) to support the hypothesis that the Na^+, K^+-ATPase is involved in at least some aspect of the glycoside's inotropic activity (see Chap. 14).

II. Sodium

The positive inotropic effect of the CG is reduced and the onset of the response delayed when the extracellular concentration of Na^+ is reduced (CAPRIO and FARAH, 1967; REITER 1972). This Na^+ dependence of the positive inotropy of CG is usually explained in terms of the influence of Na^+ on the binding of CG to the Na^+, K^+-ATPase enzyme (SCHWARTZ et al., 1975).

Reducing the extracellular Na^+ has been shown to prevent the onset of ouabain-induced arrhythmias (TODA and WEST, 1966). If, as seems likely, the toxicity of the CG is linked with their ability to inhibit Na^+, K^+-ATPase, and if a reduction in the extracellular Na^+ impedes the binding of CG to the ATPase complex, it is not altogether surprising to find that a reduced extracellular Na^+ provides some protection against CG-induced toxicity.

The relationship between the extracellular concentration of Na^+ and the magnitude of the positive inotropic response induced by the CG is explicable irrespective of whether the inotropic response is explained in terms of an altered Ca/Na exchange secondary to the inhibition of Na^+, K^+-ATP transport ATPase (LANGER, 1972) or to a direct effect on the intracellular availability of Ca^{2+} (WEINGART et al., 1978). If the inhibition of the Na^+, K^+-ATPase enzyme is involved then it can be argued that the resultant increase in intracellular Na^+ will stimulate the Na/Ca exchange system. Under these conditions the Na^+ that has accumulated intracellularly will be exchanged for Ca^{2+}. This is known as the "Na-lag" hypothesis (LANGER and SERENA, 1970; LANGER, 1972). On the other hand, if a direct effect on Ca^{2+} metabolism is involved, and particularly if that effect involves the mobilization of Ca^{2+} from superficially located depots (NAYLER, 1973a, b; BAILEY, 1977; SHERIDAN, 1978) the relative change in the size of the "activator" pool of Ca^{2+} induced by the CG will be reduced if extracellular Na^+ is low. This follows logically from the well-documented competition displayed by cardiac cell membranes for Na^+ and Ca^{2+} (LÜTTGAU and NIEDERGERKE, 1958).

There is one other factor relevant to a discussion of the importance of extracellular Na^+ in determining the positive inotropism of the CG and the dependence of this inotropism on the inhibition of the Na^+, K^+-ATPase enzyme. All the studies which have been undertaken so far have been unanimous in showing that the Na^+-sensitive sites for this enzyme, which is vectorial, are located on the cytosolic surface of the cell membrane (for review see SCHWARTZ et al., 1975). Surely this means that the effect of extracellular Na^+ is indirect!

III. Calcium

The cardiotonic effect of the CG closely resembles that of Ca^{2+}. This is an old observation (CLARK, 1912, LOEWI, 1917) and it is, therefore, hardly suprising to find that many studies have been aimed at establishing whether the positive inotropism of the CG is influenced by the extracellular concentration of Ca^{2+}. Studies on cat, frog, guinea-pig heart muscle (CLARK, 1912; KONSCHEGG, 1913; LOEWI, 1917, 1918; REITER, 1963; CARRIER et al., 1974) have revealed evidence of a synergistic or additive effect between Ca^{2+} and the CG.

A raised extracellular Ca^{2+} also potentiates the cardiotoxicity of the CG, irrespective of whether that toxicity is assessed in terms of a raised end-diastolic tension (resulting from imperfect relaxation), decreased active tension development or the occurrence of extrasystoles and conduction abnormalities.

One of the most definitive of the earlier studies on this topic is that of SALTER et al. (1949, 1951). These investigators reduced the contractile force of isolated frog hearts by sequentially removing Ca^{2+} from the extracellular phase. They then established the amount of CG needed to restore the contractile force to a set starting point. Under the conditions of their experiments the positive inotropy of 1 mol ouabain was equivalent to that caused by 6,400 mol Ca^{2+}.

Some investigators (CAPRIO and FARAH, 1967) have reported that low concentrations of Ca^{2+} decrease the time necessary for a known concentration of CG to produce its maximum inotropic response. Others (PARK and VINCENZI, 1978) have found conditions – for example, in the presence of an excessively high concentration of Ca^{2+} – under which this dependence of the rate of onset of the response on the extracellular Ca^{2+} does not always apply. Since species, frequency of contraction (MORAN, 1967; VINCENZI, 1967; KOCH-WESER, 1971) and the concentration of other ions in the extracellular phase (see Sects. C.I.II) all affect the rate of onset of the response to CG it is difficult to compare the results of experiments.

The existing data which relate to the effect of Ca^{2+} on the rate of uptake of the CG are somewhat confusing. FRICKE and KLAUS (1975) found that increasing the Ca^{2+} from 0.45 to 7.2 mmol/l resulted in a decreased uptake of digoxin. Earlier KUSCHINSKY et al. (1967) had reported with isolated atria that reducing the extracellular Ca^{2+} from 1.2 to 0.36 mmol/l slowed the rate of uptake of ^3H-digoxin. These latter results are in good agreement with those obtained by DUTTA and MARKS (1969, 1972). Using guinea-pig and rat heart these investigators found that reducing the Ca^{2+} from 2.5 to 0 mmol/l gradually decreased the myocardial uptake of digoxin, ouabain, and digitoxin. The use of a Ca^{2+}-free medium may be questioned however, because under these conditions the myocardial cells will be quiescent, thereby possibly influencing the distribution of the isotope.

Although there is no doubt that the presence of Ca^{2+} in the extracellular phase is a prerequisite for the action of the CG, the relationship between the magnitude of the inotropic response and the extracellular concentration of Ca^{2+} is often misunderstood. The magnitude of the positive inotropic response depends upon the contractile force developed prior to adding the CG. A heart beating at its maximal force of contraction cannot further increase its tension development. Hence at high concentrations of Ca^{2+}, the addition of CG may fail to cause any increase in active tension development. This same failure to produce a positive inotropic response is observed at fast heart rates (GREEFF et al., 1971) in a wide variety of preparations including atrial and papilary muscle preparations from guinea-pigs, rabbits and chickens (KOCH-WESER, 1971). Under normal conditions, however, and at a normal beat frequency the ability of the CG to evoke a positive inotropic response is dependent upon the presence of Ca^{2+} in the extracellular phase; an abnormally high extracellular Ca^{2+} facilitates the development of the cardiotoxicity. At normal concentrations of Ca^{2+} there is a synergism between the inotropic activity of the CG, and that due to Ca^{2+} (see LEE and KLAUS, 1971 for references).

In conclusion, therefore, the concentration of Ca^{2+}, Na^+, and K^+ in the extracellular phase plays an important role in determining the rate of onset and the magnitude of the inotropic response to the CG. Whether or not the drugs elicit a toxic response is also determined, in part at least, by the extracellular concentration of these ions.

D. Effect of Cardiac Glycosides on Transmembrane Ion Movements

Because of the well-documented ability of the CG to inhibit the activity of Na^+, K^+-ATPase many investigators have suggested that Na^+, K^+-ATPase might contain the inotropic receptor for these drugs (see Chap. 14). There is, however, no general consensus of opinon as to how the interaction of the CG with this complex triggers the inotropic response. The only point of agreement is that ultimately the response must depend upon a raised intracellular availability of Ca^{2+}. The question to be answered, therefore, is whether this raised intracellular availability of Ca^{2+} occurs directly, or is it an indirect result of an altered Na^+ or K^+ flux.

I. K^+ and Na^+ Fluxes in the Presence of Positive Inotropic Response to Cardiac Glycosides

In Sects. E.I and E.II we have summarized the evidence which shows that concentrations of CG which are within the therapeutic range evoke a positive inotropic response without necessarily either depleting the myocardium of K^+ or overloading it with Na^+. There may, however, have been regional distributions in Na^+ and K^+ that would be missed by assays of total tissue Na^+ or K^+. Where flux studies have been undertaken, however, an efflux of K^+ occurring in association with CG-induced positive inotropic response has been relatively easy to demonstrate particularly in the isolated perfused interventricular septum preparation of the rabbit (LANGER and SERENA, 1970). The existence of an increased K^+ flux does not necessarily mean that it is functionally related to the accompanying inotropic response – although MORAD and GREENSPAN (1973) would like to argue that because of the K/Ca exchange efflux of K^+ triggers an enhanced Ca^{2+} influx. This possibility can be discarded, however, because the experiments of POOLE-WILSON and LANGER (1975) in which they used a respiratory acidosis to block K^+ efflux using the same experimental method, showed quite clearly that the CG can evoke a positive inotropic response without necessarily causing an increase in K^+ efflux. This does not mean, however, that K^+ loss is not associated with the toxic effects of these drugs.

The dependence of the development of the positive inotropic action of the CG on the number of contractions per unit time (SANYAL and SAUNDERS, 1957; MORAN, 1967; PARK and VINCENZI, 1978) has caused many investigators to believe that there is a close correlation between the positive inotropic effect of the CG and an enhanced Na^+ influx, dependent in turn upon the inhibition of Na^+, K^+-ATPase (AKERA et al., 1977). Recently AKERA et al. (1977) used paired stimulation and studied the effect of grayanotoxin I, an agent which increases Na^+ influx, in the hope of obtaining further evidence to support the theory that there is a causal link between the effect of the CG on Na^+ influx and the inotropic state of the myocardium, as suggested by LANGER (1974). Even more recently, however, BESCH and WATANABE (1978) have re-examinded this hypothesis. By using tetrodotoxin they set their experimental conditions so that the positive inotropic effect of the glycosides was maintained under conditions which prevented Na^+ influx during the rapid depolarization phase of the action potential (PEPER and TRAUTWEIN, 1965; PAPPANO, 1970; BOSTEELS and CARMELEIT, 1972) they seem to have demolished the argument that Na^+ influx, and the associated activation of the Na/Ca exchange plays an important role in the positive inotropism of the glycosides (REUTER, 1974). There is just one reservation with these experiments of BESCH and WATANABE, however – they did not block the influx of Na^+ which occurs during the plateau phase of the action potential (KASS and TSIEN, 1975). However the earlier experiments of LANGER and SERENA (1970) provided fairly conclusive evidence of the fact that the positive inotropism of the CG need not involve an enhanced influx of Na^+.

II. Ca^{2+} Fluxes in the Presence of Positive Inotropic Concentrations of Cardiac Glycosides

There is practically no doubt that the cardiotoxicity of the glycosides is due, in part at least, to raised tissue levels of Ca^{2+}; evidence supporting the hypothesis that their positive inotropism results from an enhanced intracellular availability of Ca^{2+} due to an enhanced Ca^{2+} influx has been much more difficult to obtain. This difficulty is due partly to the small amount of additional calcium that is needed to increase peak tension development substantially (10–20 μmol Ca^{2+}/kg wet weight) relative to the total calcium content of the tissue (2–3 mmol/kg wet weight). It could also be due to the rapid exchangeability of the particular component of cellular calcium (LÜLLMANN and PETERS, 1979) that participates in E–C coupling, or the prolonged period of time needed to establish a steady state with respect to labelling with either $^{45}Ca^{2+}$ or $^{47}Ca^{+}$ (LEE and KLAUS, 1971). In addition, because most interventions that are positively inotropic decrease the size of the extracellular space, this factor, together with any change in efflux rate which might occur, must be taken into account. Analysis of guinea-pig, rat and turtle heart muscle for total cellular calcium content has not shown any significant increase associated with the positive inotropic effect of the glycosides (GERSMEYER and HOLLAND, 1963; GOVIER and HOLLAND, 1964; KLAUS, 1967); in marked contrast to the results obtained when excessively high concentrations of glycosides are used (see Sect. D.III). Nevertheless, despite an apparently unchanged total calcium content of the tissue, studies using radioactively labelled calcium (KLAUS and KUSCHINSKY, 1962; LÜLLMANN and HOLLAND, 1962; HEINEN and NOACK, 1972) revealed evidence of an increase in the proportion of the total tissue Ca^{2+} which could be labelled. These experiments have been interpreted to mean that the glycosides increase the proportion of total tissue Ca^{2+} that is available for exchange and which therefore can be utilized for E–C coupling.

More recently, using sensitive techniques which enable the sequential monitoring of Ca^{2+} uptake into isolated strips of heart muscle LANGER and SERENA (1970) were able to detect a small but significant increase in Ca^{2+} uptake in response to the addition of acetylstrophanthidin. Unfortunately the authors monitored the response for a relatively short period of time, so that it is not possible to exclude a small increase in Ca^{2+} uptake associated with a very early phase of a toxic response. Even more recently, however, evidence has come from experiments implicating an increased intracellular Ca^{2+} exchange in association with the positive inotropism of the CG. ALLEN and BLINKS (1978) using thin trabeculae from frog *(Rana pipiens)* atrial muscle and intracellularly injected aequorin, found an increase in the aequorin-induced light emission following the injection of acetylstrophanthidin and, whilst their experiments provide no information concerning the site of origin of the Ca^{2+}, they substantiate the concept that an increased intracellular availability of ionized calcium occurs concurrently with the increase in developed tension.

The second line of evidence is provided by the voltage clamp studies of WEINGART et al. (1978). Using Purkinje fibres from calf hearts WEINGART et al. (1978) studied the effect of the aglycone strophanthidin and found a rapid enhancement of the slow inward current for which Ca^{2+} is the predominant charge carrier.

Why then, had GREENSPAN and MORAD (1975), and MCDONALD et al. (1975) failed to find such an effect? The answer is not yet clear but it is reassuring to find that others are able to confirm WEINGART'S results (BROWN, GILLIS, and NOBLE, personal communication) and even to extend them (BELARDINELLI et al., 1979) to vascular smooth muscle.

Despite the many difficulties already referred to the older literature contains repeated references to the fact that exposing heart muscle to nontoxic doses of CG results in a small increase in Ca^{2+} uptake and an increase in Ca^{2+} exchangeability. This has been shown for a variety of different species: guinea-pig (GERSMEYER and HOLLAND, 1963a; GROSSMANN and FURCHGOTT, 1964; KLAUS and KUSCHINSKY, 1962), rabbit (HOLLAND and SEKUL, 1961; KEETON and BRIGGS, 1965), turtle (GOVIER and HOLLAND, 1964), and rat (GERSMEYER and HOLLAND, 1963). To these must be added the more recent experiments of LANGER and SERENA (1970), although in these particular studies the concentration of CG used was sufficient ultimately to produce toxicity. This increase in Ca^{2+} exchangeability occurs in the absence of a diminished rate of efflux (LANGER and SERENA, 1970). Nor is there any increase in total Ca^2 content – indeed, as discussed later, total tissue Ca^{2+} may fall.

Probably the small increase in Ca^{2+} influx involves the transport of Ca^{2+} through the slow channels (WEINGART et al., 1978), but it seems unlikely that this is of sufficient magnitude to account for the positive inotropic response. There is, however, the possibility that this additional influx of Ca^{2+} concerns transport into a specialized compartment closely linked with "trigger Ca^{2+}" (Fig. 2). There is one other possibility that warrants attention, i.e. that the positive inotropism of the CG involves, in addition to a small increase in the slow inward current, a redistribution of Ca^{2+} normally associated with the glycocalyx or other superficially located depots NAYLER (1973a), BAYLEY (1977), and SHERIDAN (1978) have all shown that the interaction of the glycosides with the cell membrane increases the exchangeability of a "pool" of Ca^{2+} associated with the cell membrane. Under these conditions the CG may increase the intracellular availability of Ca^{2+} without increasing Ca^{2+} uptake or inhibiting efflux.

In summary, therefore, whilst the positive inotropism of the CG can be dissociated from their effect on Na^+ and K^+ exchange the available evidence indicates that there is a close correlation between their positive inotropism and their effect on the intracellular availability of Ca^{2+}. Because the increase in intracellular availability of Ca^{2+} involves a component due to a redistribution of cellular Ca^{2+}, measurement of changes in Ca^{2+} influx or slow Ca^{2+} current may provide a totally inaccurate estimate of the magnitude of the CG-induced increase in the amount of Ca^{2+} made available for interaction with the myofibrils.

III. K^+, Na^+, and Ca^{2+} Exchange in the Presence of Toxic Concentrations of Cardiac Glycosides

Numerous studies have shown that toxic doses of CG produce an apparent increase in K^+efflux. A loss of K^+ under these conditions has been described for human patients as well as for a wide variety of animals including dog, cat, rabbit, guinea-pig, rat, turtle, and pigeon (see LEE and KLAUS, 1971 for references). This loss of K^+ is due to an increased rate of K^+ efflux, and a diminished rate of influx

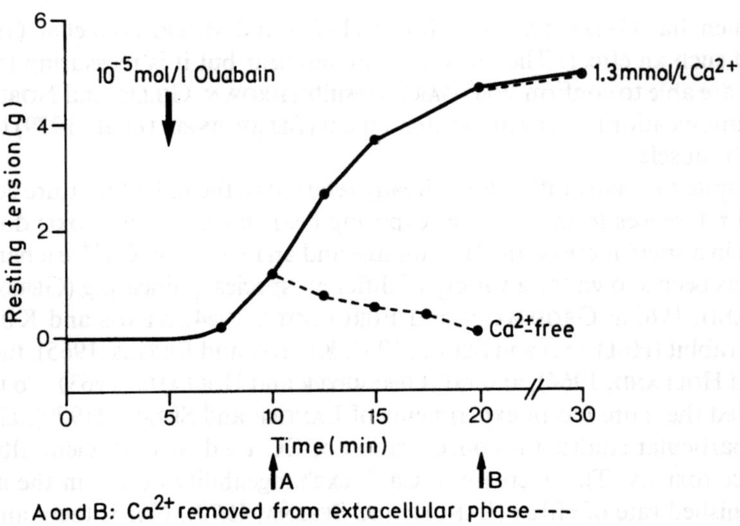

Fig. 4. Effect of a toxic concentration of ouabain (10 µmol/l) on resting tension in isolated, paced (180 beats/min) Langendorff-perfused rabbit heart perfused with a Krebs–Henseleit buffer solution containing 1.5 mol/l Ca^{2+}. When indicated *(broken line)* perfusion was continued in the absence of any added Ca^{2+} but in the presence of 10 µmol/l ouabain. Note that removing Ca^{2+} early but not late during the toxic response reversed this aspect of the glycoside-induced toxicity. (NAYLER, unpublished)

(TUTTLE and FARAH, 1962). An increase in K^+ efflux with high concentrations of CG has been reported by other investigators, including GRUPP and CHARLES (1964). Presumably this enhanced K^+ efflux is due to the inhibition of Na^+, K^+-ATPase by toxic doses of CG. In this case we should expect to find an increase in Na^+ influx, or a diminished Na^+ efflux (LANGER and SERENA, 1970; DEITMER and ELLIS, 1978a, b) and hence an increase in intracellular Na^+ (see Sect. E).

If toxic concentrations of CG cause a diminished K^+ influx and an enhanced Na^+ influx and K^+ efflux, it is not difficult to understand why these same toxic concentrations cause a marked increase in Ca^{2+} influx (LANGER andSERENA, 1970; LÜLLMANN and PETERS, 1979) for Ca^{2+} will enter via the two pathways we have previously described, i.e. in exchange for either Na^+ (REUTER, 1974) or K^+ (MORAD and GREENSPAN, 1973). Once intracellular Ca^{2+} rises we can predict the following sequence of events: raised cytosolic Ca^{2+} → increased usage of endogenous ATP → reduced ATP availability → failure of ATP-dependent mechanisms for maintenance of intracellular homeostasis → raised cytosolic Ca^{2+} → activation of Ca^{2+}-sensitive phospholipases and mitochondrial Ca^{2+} overload → impaired ATP production → imperfect relaxation → toxicity. There is a wide variety of evidence (NAYLER and WILLIAMS, 1978; GERVAIS et al., 1978) which indicates that this sequence of events may be correct. In this context it is interesting to note that whereas the removal of Ca^{2+} from the extracellular phase in the early stages of the toxic response to the glycosides allows relaxation to occur, its late removal is ineffectual (Fig. 4).

The arrhythmias and conduction defects, which are associated with the cardiotoxicity of the CG are certainly due to the loss of K^+ and the accompanying changes in Ca^{2+} and K^+ across cardiac cell membranes; the effect of the CG on the action potential is complex. Initially the action potential is prolonged, probably because of the enhancement of the inward Ca^{2+} current (see Chap. 12) and perhaps a diminished K^+ efflux (BROWN et al., 1962). But as the concentration of the CG increases and Na^+, K^+-ATPase is inactivated, Na^+ efflux is impaired and K^+ efflux stimulated. The resultant gain in tissue Na^+ and loss of K^+ stimulates an exchange of these ions for Ca^{2+} such that the concentration of Ca^{2+} at the inner surface of the sarcolemma rises. Under these conditions the outward K^+ current is further stimulated (ISENBERG, 1975) leading to a progressive shortening of the action potential.

In summary, therefore, nontoxic doses of CG may cause a small increase in Ca^{2+} influx, with probably no change in Na^+ or K^+. Toxic doses stimulate K^+ efflux and Na^+ influx, probably by inhibiting Na^+, K^+-ATPase; this Na^+ and K^+ then exchanges for Ca^{2+}, resulting in a marked increase in Ca^{2+} influx and perhaps Ca^{2+} overloading. In marked contrast to the situation which is encountered in hypoxic heart muscle (NAYLER and WILLIAMS, 1978) Ca^{2+} overloading caused by the administration of toxic doses of CG precedes the decline in the tissue stores of ATP.

E. Effect of Cardiac Glycosides on Intracellular Electrolyte Content

I. Influence of Cardiac Glycosides on Myocardial K^+ and Na^+

If Na^+, K^+-ATPase is the specific pharmacological receptor for the CG and since the CG inactivate it in a dose-dependent manner we could expect that the administration of these drugs would cause an intracellular accumulation of sodium and a loss of potassium. This has not always been observed for concentrations which produce only a positive inotropic and not a toxic response. Despite this, the well-documented augmentation of the intracellular exchangeable fraction of Ca^{2+} is consistently attributed to an increase in intracellular Na^+ (LANGER and SERENA, 1970).

In some of the older studies there is even evidence of a CG-induced decrease in myocardial potassium and an increase in sodium (ORDOBONA and MANGANELLI, 1954; RAYNER and WEATHERALL, 1957; SANYAL and SAUNDERS, 1957; AIKAWA and RHOADES, 1955; BLACKMAN et al., 1960; BOYER and POINDEXTER, 1940; BROWN et al., 1962; CARSLAKE and WEATHERALL, 1962; HARVEY, 1954; HELLEMS et al., 1955; KLAUS et al., 1962; LANGER and SERENA, 1970; LEE et al., 1961; SHERROD, 1957; TUTTLE et al., 1962; WITT and TUTTLE, 1960). Using isolated Langendorff preparations BRODY and AKERA (1977) showed that the gain in sodium may be frequency dependent. Thus after 25 min perfusion with 0.4 μmol/l ouabain at 120 beats/min there was no significant change in sodium content while at 240 beats/min an intracellular gain of sodium from 0.36 to 0.48 μmol/mg protein was observed, without cardiac arrhythmias. LANGER and SERENA (1970) added 5×10^{-6} mol/l of acetylstrophanthidin to the arterially perfused rabbit interventricular septum and obtained a net uptake of cellular calcium and sodium together with a net loss of potassium. There was a coupling of approximately 2.5 mol sodium to 1 mol potassium.

Another technique that has been widely used to study the effect of the CG on tissue electrolytes indirectly involves the measurement of arteriovenous differences in sodium and potassium content. With this method a transient loss of potassium was registered by several authors, using the isolated, perfused, guinea-pig heart, the heart–lung preparation of the dog or the intact dog or human heart in vivo. Thus VICK and KAHN (1957) applying 0.05–0.2 µmol/l ouabain to the isolated, perfused guinea-pig heart observed only small losses of potassium in the range of the IC_{50} for the inotropic effect which were rising continously as the concentration exceeded 0.24 µmol/l. The time of maximal potassium loss preceded the maximal positive inotropic effect. With the same experimental model GREEFF et al. (1962) found a correlation between the inotropic effect of strophanthoside K, digitoxin and digitoxigenin and a potassium release of up to 4% of the total tissue content. An even more pronounced increase of myocardial potassium loss was seen in the presence of toxic concentrations, becoming obvious with > 12 nmol strophanthoside K, > 2 nmol digitoxin, and > 8 nmol digitoxigenin. GÜTTLER et al. (1979), also using the isolated, perfused guinea-pig heart, showed a concentration-dependent increase in myocardial potassium loss with increasing ouabain concentrations (6×10^{-7} mol/l ouabain: 2.76 ± 0.22 µmol g^{-1} min^{-1} K^{+}).

With the heart–lung preparation of the dog ARESKOG (1962) described a net loss of potassium from the myocardium induced by convollatoxin. No significant changes of the sodium balance due to the CG action was found. After injecting strophanthoside K into the intact dog heart in vivo HARRIS et al. (1955) observed a significant release of myocardial potassium by measuring the arteriovenous differences of sodium and potassium. Comparative results were obtained by REGAN et al. (1956) on eight mongrel dogs. With 0.05–0.1 mg acetylstrophanthidin/kg body weight a negative potassium (0.86 ± 0.44 mmol/l) and a positive sodium (6 ± 4 mmol/l) arteriovenous difference was induced, the maximum change occurring in an average of 6 min. Comparable results were obtained by BLACKMAN et al. (1960) with intact dog heart. While there was a transient loss of myocardial potassium with low doses a more pronounced loss was seen with toxic doses. In addition a complete dissociation between the positive inotropic effect of CG and the disturbance in potassium balance generally became obvious. REGAN et al. (1969) noted that in the closed chest dog a nontoxic dose of acetylstrophanthidin produced a 78% increase in maximum dP/dt accompanied by an average increase in potassium efflux of 1.2 mmol/kg.

In humans BRENNAN et al. (1972) observed a loss of myocardial potassium which occured prior to positive inotropism. This finding was confirmed by McCANS et al. (1973) in a study in humans with ouabain. SLANY and MÖSSLACHER (1976) administered ouabain to patients by an intracardiac catheter and registered a slight transient potassium loss after 3–8 min; there was no change after application of a placebo. But, such an affect on K^{+} loss has not always been observed (for review see GLYNN, 1964). As discussed by LEE and KLAUS (1971) caution should be exercised in assuming that there is a causal relationship between myocardial potassium loss and changes in contractility, even if only because numerous studies show that a continuous loss or gain of potassium may occur without any concomitant change in contractile force (BROWN et al., 1960; MORALES and ACHESON, 1961). Moreover it cannot be fully excluded that the effect on potassium content may be produced by establishing transient toxic concentrations in the myocardium especially if CG are administered as a bolus or if rather large concentrations are used. It is also possible that the tendency of the heart to lose potassium in the presence of CG is very much dependent on its prior contractile state. In heart failure there is already a reduction in muscle potassium and an increase in sodium and water content (HARRISON et al., 1930; CULLEN et al., 1933; MANGUN and MEYERS, 1936). With isolated, perfused, guinea-pig heart in which a cardiac failure was provoked by pretreatment with phenylbutazone, HOCHREIN (1965) made the interesting finding that 4×10^{-7} g/ml ouabain or digitoxin induced a significant potassium uptake and sodium release in contrast to normal hearts showing an opposite effect. On the other hand GONLUBOL et al. (1956), applying 0.9–1.5 mg Cedilamid to ten patients with cardiac failure, did not see any change in the coronary arteriovenous difference of potassium and sodium. But there was a significant increase in the potassium concentration of arterial and coronary sinus blood which indicates that potassium liberation by CG may primarily originate from extracardiac tissue.

In contrast, some investigators have described an increase in myocardial potassium and a decrease in sodium content in the presence of positive inotropic concentrations of CG. Thus in the dog heart–lung exposed to positive inotropic concentrations of ouabain, VICK and KAHN (1957) showed an accumulation of potassium from the perfusing medium when the heart was driven slowly but a net loss at higher heart rates. HAGEN (1939) found a significant increase in the potassium content of isolated rabbit heart in the presence of therapeutic concentrations. Similar results were obtained by TUTTLE et al. (1961) and CARSLAKE and WEATHERALL (1962) for isolated atria and by HOLLAND et al. (1954) for isolated perfused guinea-pig heart. An overview of the literature indicates that the findings are inconsistent. It may, however, be confirmed that if the CG concentrations are within the lower therapeutic range there need be no significant change in potassium or sodium content. This was found in vivo in experiments with rabbits (GERTLER et al., 1956) and in vitro with isolated heart muscle preparations of the guinea-pig (HOLLAND et al., 1954; KLAUS et al., 1962; NOACK et al., 1978, 1979), of cats (LEE et al., 1960), and of rabbits (TUTTLE et al., 1961; WITT and TUTTLE, 1960).

OKITA et al. (1978) studied the effect of 2.6×10^{-6} mol/l strophanthidin on ^{22}Na efflux and electrolyte levels in left atrial preparations during the initial stages of inotropy and found no change in ^{22}Na efflux rate or in sodium and potassium content. The authors concluded, therefore, that the reserve capacity of the sodium pump can maintain normal sodium efflux and sodium and potassium levels in the presence of a moderate inhibition of the sodium pump since the CG concentration used was sufficient to inhibit rabbit heart ATPase slightly. This is also the basis of AKERA's concept (see Chap. 14 and AKERA et al., 1976a, b). AKERA postulated that there is a relative transient increase in sodium content, but this does not result in a measurable change. From this transient increase in sodium concentration a net accumulation of sodium should not occur until the inhibition exceeds a critical point.

It is interesting to note that different results are obtained, depending upon whether it is the steady state or a transient effect of the CG that is being linked to changes in sodium and potassium content. During the early stages of the response many investigators have described a stimulation of the sodium pump, followed by an inhibition. In studies with ion-selective electrodes COHEN et al. (1976) demonstrated that if the extracellular potassium concentration $[K^+]_o$ is relatively high or ouabain dosage is relatively low there is stimulation of the Na^+, K^+-pump, which can be accounted for by a depletion of $[K^+]_o$ in a restricted extracellular space adjacent to the plasmalemmal membrane. Possibly this involves a shift of potassium to sites within the membrane. Even in those experiments in which an inhibitory effect of ouabain (10^{-7} mol/l) was recorded during the first 5 min there was a small, but probably significant stimulatory effect. Moreover the effect of positive inotropic concentrations of ouabain (in the range 5×10^{-10}–10^{-7} mol/l) was completely reversible while at higher doses changes attributable to pump blockade were not. From these results the authors were inclined to reject the hypothesis that positive inotropic action of CG is causally related to net pump inhibition. Their findings are consistent with studies of BAKER and WILLIS (1972) in squid nerve where they noticed that concentrations of ouabain (10^{-6} mol/l and greater) that produce inhibition of sodium efflux may first produce a transient stimulation of efflux.

Digitalis-treated cardiac tissue of the tunicate *Ciona intestinalis* showed significant losses of intracellular sodium in contractile, mitochondrial and basal cell membranes, characteristics of a stimulation of the sodium pump (KEEFNER and AKERS, 1971). Using an ion-sensitive electrode ELLIS (1977) has detected that concentrations of between 10^{-8} and 10^{-7} mol/l ouabain cause a slight decrease in the intracellular sodium activity in Purkinje fibres. This decrease lasted for only 10 min. The concentrations of ouabain used by ELLIS are those which reversibly increase the contractility of sheep heart Purkinje fibres (BLOOD, 1975). Concentrations of ouabain higher than 5×10^{-7} mol/l induced a gain of intracellular sodium activity. Under comparable experimental conditions DEITMER and ELLIS (1978 b) found that very low strophanthidin concentrations (10^{-8}–10^{-7} mol/l) can actually decrease the intracellular Na concentration while at higher concentrations of CG (dihydroouabain, actodigin, acetylstrophanthidin, strophanthidin) above a threshold concentration of about 10^{-7} mol/l, intracellular Na increases after a transient decrease, the rate of rise depending on the CG concentration. Half-maximal effect was seen with 8.1×10^{-7} mol/l acetylstrophanthidin or 8.4×10^{-7} mol/l strophanthidin. The maximum rate that could be achieved was 0.75 mmol l^{-1} min^{-1} sodium, by 10^{-5} mol/l acetylstrophanthidin and strophanthidin, indicating a net passive Na influx into the Purkinje fibre cells of approximately 4.3 pmol cm^{-2} s^{-1}.

COHEN et al. (1976) also found that concentrations of CG in the therapeutic range can produce changes in the potassium gradient that reflect stimulation of the Na^+, K^+-pump. After 5×10^{-7} mol/l ouabain, hyperpolarization occured, with a shift of the membrane potential from -64 to -72 mV. The authors attribute this effect to either an increase in intracellular potassium, a decrease in $[K^+]_o$ or a combination of these two possibilities. Since the immediate extracellular potassium concentration is rarely equal to $[K^+]_o$ of the bathing solution and the effect is closely mimicked by those produced by decreasing $[K^+]_o$ in the absence of ouabain, the mechanism of action might be due to a reduced potassium concentration in the clefts of the Purkinje cells. Hopefully these results can be extrapolated to the myocardial cell, where studies with ion-selective microelectrodes have not yet been performed, for technical reasons. GODFRAIND and GHYSEL-BURTON (1977) incubated isolated guinea-pig atria for 3 h in the presence of 10^{-9}–10^{-8} mol/l ouabain. After that time they found a significant increase in intracellular potassium and loss of sodium. With higher concentrations an opposite effect was always obtained. Simultaneous binding studies with ^{3}H-ouabain revealed high and low affinity binding sites for ouabain and the authors concluded that obviously the saturable binding sites and pumping sites are the same.

In another study GHYSEL-BURTON and GODFRAIND (1975) reported that low concentrations of ouabain (10^{-9}–10^{-8} mol/l) stimulated ^{42}K uptake into the myocardium in correlation with its inotropic effect. Comparing several CG (ouabain, ouabagenin, and dihydroouabain) GHYSEL-BURTON and GODFRAIND (1977) studied the molecular requirements for the biphasic action of CG on the sodium pump. It was shown that in contrast to ouabain, dihydroouabain produced only an inhibitory effect, so that unsaturation of the lactone ring is required for stimulation.

NOACK et al. (1978, 1979) investigated the effect of positive inotropic concentrations of ouabain and digitoxigenin on the cellular potassium and sodium content in isolated guinea-pig atria. In the first three to five minutes after the application of 0.2–0.1 μmol/l ouabain or 0.6 μmol/l digitoxigenin they always obtained a significant gain in cellular potassium, of maximally 17.4 mmol/kg tissue weight and a simultaneous loss of sodium, of up to 6.2 mmol/kg tissue weight. At the time of maximal inotropic response the values no longer differed from the control values. There was a direct relationship between the extent of positive inotropic response and the magnitude of the increased intracellular $K^+:Na^+$ ratio. In the presence of a reduced extracellular calcium concentration the development of the inotropic response as well as the appearance of intracellular potassium gain and sodium loss were markedly delayed. The effect was shown to be CG specific since a positive inotropic concentration of isoproterenol (18 nmol/l) had no such effect. Figure 5 shows an experiment in the presence of a toxic concentration of digitoxigenin (3.75 μmol/l). It is obvious that even if the digitoxigenin concentration is high it is possible to see the transient effect of pump stimulation soon after drug administration. It may be concluded therefore that a stimulatory effect can be established as long as the concentration of CG at the receptor site is low. This could mean that the rapid binding of the drug to one site – perhaps a high affinity site (GHYSEL-BURTON and GODFRAIND, 1975, 1979) – on the enzyme stimulates the pump whereas binding to another site inhibits. Thus the transient stimulation could represent different rates of binding to two different sites.

At present it is not known whether the net stimulation of the sodium pump which finds a parallel in the occasionally observed in vitro stimulation of Na^+, K^+-ATPase, is in any way linked with the positive inotropic action of CG. Until now the stimulation of the sodium pump by CG was regarded as an unorthodox hypothesis, raised for the first time by CARSLAKE and WEATHERALL (1962). Under normal positive inotropic concentrations its transient character may be another reason for the conflicting results reported in the literature relating to changes in cellular potassium and sodium content.

In summary, on the basis of recent studies there is some evidence that in the presence of positive inotropic concentrations of ouabain no changes in intracellular potassium and sodium content occur at the time of maximal inotropy. By contrast, there are indications that the positive inotropic action of CG is preceded by

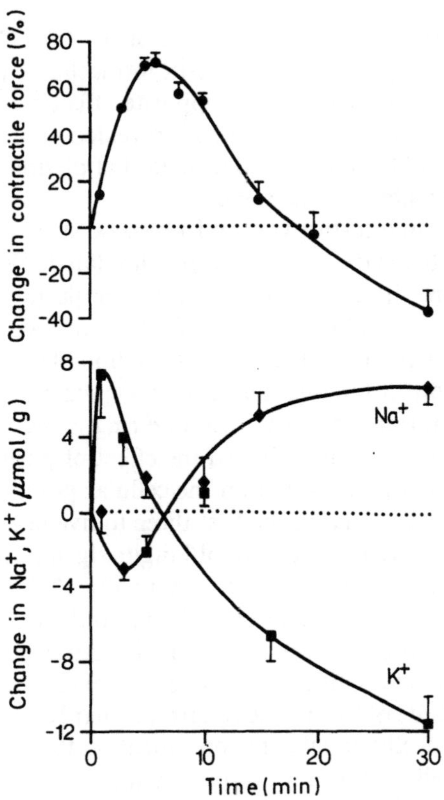

Fig. 5. Changes in contractile force (*circles;* number of samples $n = 24$–80) and intracellular total potassium (*squares;* $n = 10$–23) and sodium (*diamonds;* $n = 10$–23) content in isolated guinea-pig atria exposed to a toxic concentration of digitoxigenin (3.75 μmol/l). Results are expressed as mean $\pm$ standard errors. Control value for contractile force 2.27 ± 0.07 g; for cellular potassium 106 ± 0.6 mmol/kg cellular weight; for cellular sodium: 39.8 ± 0.4 mmol/kg cellular weight. (NOACK, unpublished)

a transient stimulation of the sodium pump, the significance of which is still obscure. In contrast to the effect of positive inotropic concentrations of CG on cellular electrolytes their toxic effect is well established, showing a gain in tissue sodium and a loss of potassium. This is probably due to a more or less pronounced blockade of Na^+, K^+-ATPase or the sodium pump which is then operating beyond its reserve capacity for repumping sodium ions from the cell during the diastolic phase of the cardiac cycle. Figure 5 shows an experiment in which the potassium and sodium content was determined in isolated guinea-pig atria in the presence of a toxic concentration of digitoxigenin. From the time when the negative inotropic effect develops there is a continued increase in sodium and decrease in potassium content.

In rabbit ventricular myocardium isotopic studies of sodium exchange (LANGER and SERENA, 1970) following CG administration demonstrate that the sodium that is gained is sequestered or bound in a special cellular compartment and may thus produce a very large change in content within some restricted region of the cell. Through an exchange of intracellular sodium against extracellular calcium the intracellular "free" and total calcium concentration rises to unphysiological and contracture-inducing levels. The amount and extent of potassium loss determines whether or not an arrhythmic toxicity will occur owing to the lowered transmembrane potential and shortened action potential. According to LANGER and SERENA (1970) inhibition of ^{42}K uptake by 5×10^{-6} mol/l acetylstrophanthidin decreases at the time where systolic tension and dP/dt decreases and diastolic tension increases. This finding is not so evident with ^{42}K effluent measurement where the increase in ^{42}K starts some 5 min prior to development of a raised diastolic tension.

The study of MIURA and ROSEN (1978) with ion-selective electrodes demonstrates that after 30 min exposure to a toxic concentration of ouabain there is a decrease in the intracellular potassium activity in Purkinje fibres. The authors observed a decrease in potassium from 130.0 to 112.3 mmol (−13.7%) and this occured concomitantly with a decrease in transmembrane potential. Potassium activity was also measured in the extracellular space. The mean value before superfusion was 4.6 mmol/l. After 30 min exposure it was 5 mmol/l. The potassium can obviously accumulate at the cell surface and temporarily reach concentrations far in excess of the bulk phase. These perhaps locally restricted movements of electrolytes may manifest themselves as surface charge effects at the cell membrane; they could, therefore, influence excitation–contraction coupling.

In a whole animal trial WATSON and WOODBURY (1973) administered a toxic dose of 0.35 mg/kg ouabain to guinea-pigs. They observed an increased plasma potassium and myocardial sodium and chloride and a decreased intracellular potassium content. Potassium loss from the heart muscle following an infusion of 1.1 mg acetylstrophanthidin into the right atrium was also confirmed in patients. HELLEMS et al. (1956) observed, in seven patients suffering from a low output left heart failure, a myocardial potassium loss, attaining a maximum negative arteriovenous difference of 0.7 ± 0.4 mmol/l ($P < 0.001$) at 4 min, with a restoration to control values within 20 min while ventricular stroke work increased not earlier than 5 min after drug infusion. There was no significant myocardial sodium uptake. In a study on 18 patients with left-sided heart failure REGAN et al. (1959) usually registered a maximum negative potassium arteriovenous difference at 4 min after infusion of 1.1 mg acetylstrophanthidin ($P < 0.001$). The theoretically calculated decrement of potassium ranged from 3% to 7% of the cellular potassium content. There occurred no significant changes in sodium balance. Likewise no correlation between the haemodynamic response and the potassium loss was obtained.

II. Influence of Cardiac Glycosides on Myocardial Calcium Content

Considerable effort has been expended on measuring the effect of CG on the exchange and distribution of calcium in heart muscle, presumably because it is generally agreed that the mechanical performance of the heart is determined by the "free" intracellular concentration of calcium (LANGER, 1968; WOOD et al., 1969).

As described earlier in this chapter CG increase the amount of exchangeable calcium and the intracellular concentration of ionized calcium, at the expense of intracellularly bound calcium. Since the intracellular electrolytes of the myocardial cell are in equilibrium with the extracellular electrolytes it is conceivable that this

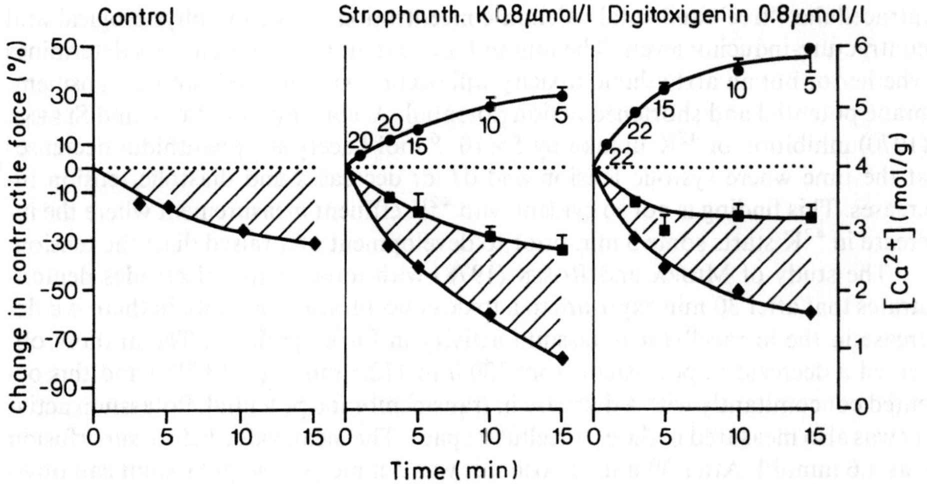

Fig. 6. Changes in contractile force *(circles)*, tissue Ca^{2+} *(squares;* number of samples $n=12$) and in the exchanged and unexchanged calcium fraction (*diamonds;* $n=5$–10) in the presence of positive inotropic concentrations of strophanthoside K or digitoxigenin: changes were plotted as a function of the incubation time. The mean values of the force of contraction at time 0 are: control 720 ± 70 mg ($n=27$); 0.8 µmol/l strophanthoside K 680 ± 60 mg ($n=24$); and 0.8 µmol/l digitoxigenin 680 ± 70 mg ($n=27$). Means ($\pm$ standard errors). (HEINEN and NOACK, 1972)

action is associated with intracellular calcium ion shifts or even in a change in the net calcium content. The literature provides no evidence of the fact that the positive inotropic action of CG is associated with a significant increase in total calcium. On the other hand there are numerous, older reports (for references see LEE and KLAUS, 1971) showing that the total cellular calcium content does not change after CG administration. Indeed KASPAREK (1976) demonstrated with ouabain that in quiescent atria the rate of calcium exchange can increase by as much as 200% without any accompanying change in total calcium.

On the other hand there are several studies reporting a significant decrease in total tissue calcium (LANGER and BRADY, 1963; LANGER, 1964; KLAUS et al., 1962; LÜLLMANN and HOLLAND, 1962; GERSMEYER and HOLLAND, 1963). BAILEY and KRIP (1972) investigated the compartmentalization of calcium by washout experiments in gas-perfused cat heart and found no effect of positive inotropic concentrations of ouabain, but the quantity of Ca remaining in the tissue after two washouts was significantly lower in ouabain-treated than in control heart. In studies with isolated guinea-pig atria HEINEN and NOACK (1972) showed that the cellular calcium content declined in the presence of inotropically effective concentrations of strophanthoside-K (0.8 µmol/l) and digitoxigenin (0.8 µmol/l). After 15 min incubation it was reduced by 33.5% and 20.7%. This finding was recently confirmed by NOACK et al. (1978) in studies which showed that the net calcium content was reduced in isolated atria in the presence of positively inotropic concentrations of ouabain (0.2 µmol/l) or digitoxigenin (0.8 µmol/l). The reduction was between 30% and 20%. The time course of cellular calcium loss paralleled the development of positive inotropy (Fig. 6).

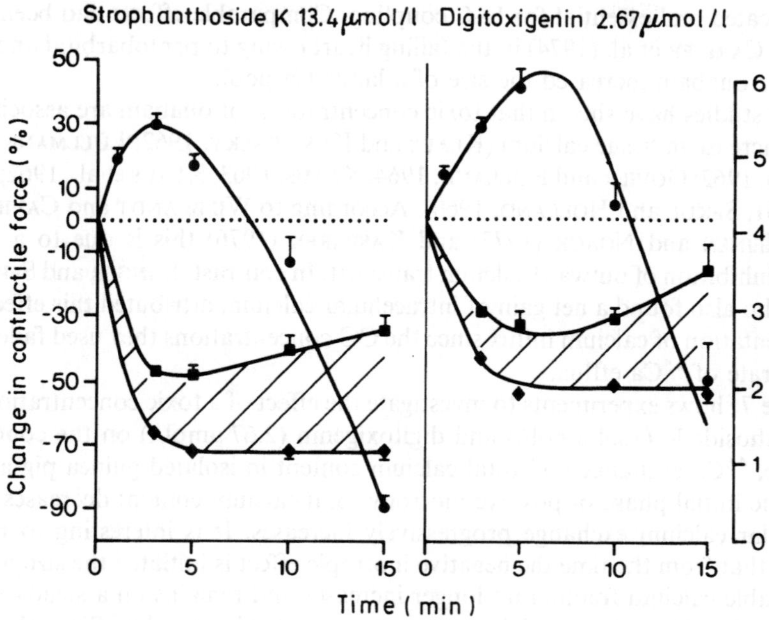

Fig. 7. Changes in contractile force *(circles)*; number of samples $n=5$–20), in cellular calcium (*squares*; $n=12$–24), and in the exchanged (45Ca) or unexchanged calcium fraction *(diamonds)*; $n=5$–10) in isolated guinea-pig atria exposed to toxic concentrations of strophanthoside K (13.4 μmol/l) or digitoxigenin (2.67 μmol/l). Means ± standard errors. (HEINEN and NOACK, 1972)

Some authors have attempted a quantitative analysis of myocardial calcium metabolism. KLAUS and KREBS (1974) studied the content of the calcium compartments in isolated, isovolumic perfused guinea-pig heart under equilibrium conditions in the presence of a positive inotropic concentration (1.5×10^{-7} mol/l) of ouabain. Myocardial total calcium content decreased significantly (by 21%) compared with control hearts. In addition there was an increase in calcium content in a slowly exchangeable calcium compartment from 0.30 to 0.46 mmol/kg wet weight.

There is growing evidence that at least one effect of positive inotropic concentrations of CG may be to alter the intracellular sequestration of calcium and to increase the quantity of calcium contained in the superficially located calcium pools within the sarcolemma, bringing about an increased availability of calcium for inward displacement during excitation (NAYLER, 1973a). In heart muscle "coupling calcium" seems to be much more superficially located (sarcolemmal surface and T tubular system) than in skeletal muscle.

In studies with cell membrane preparation GERVAIS et al. (1977) noted an ouabain-induced increase in the proportion of bound calcium which potentially may become available for E–C coupling. In the same year BAILEY (1977) showed that the administration of 5×10^{-8} g/ml ouabain to hypodynamic Langendorff-perfused kitten heart caused a significant increase in the quantity of calcium taken up by a calcium pool which contained relatively labile calcium thought to be superficially located and essential for E–C coupling. Comparable effects had been obtained by CARRIER et al. (1974) in the failing heart owing to pentobarbital and aging, where ouabain increased the size of a labile Ca pool.

Many studies have shown that toxic concentrations of ouabain are associated with an increase in tissue calcium (KLAUS and KUSCHINSKY, 1962; LÜLLMANN and HOLLAND, 1962; GOVIER and HOLLAND, 1964; KLAUS, 1963; KLAUS et al., 1962; LEE et al., 1961; SEKUL and HOLLAND, 1960). According to WILBRANDT and CAVIEZEL (1954), HEINEN and NOACK (1972) and KASPAREK (1976) this is due to a pronounced inhibition of outward calcium transport. In contrast, LANGER and SERENA (1970), who also found a net gain of intracellular calcium, attributed this effect to an augmentation of calcium influx since the CG concentrations they used failed to alter the rate of ^{45}Ca efflux.

Figure 7 shows experiments to investigate the effect of a toxic concentration of strophanthoside K (13.4 μmol/l) and digitoxigenin (2.67 μmol/l) on the contraction force, ^{45}Ca exchange and total calcium content in isolated guinea-pig atria. During the initial phase of positive inotropy total calcium content decreases and intracellular calcium exchange progressively increases. It is interesting to note, however, that from the time the negative inotropic effect is initiated the size of the exchangeable calcium fraction no longer increases and remains on a steady-state level while the total cellular calcium content begins to rise steadily. Since the size of the exchangeable calcium fraction was found to be smaller than in the presence of positive inotropic concentrations of CG (HEINEN and NOACK, 1972) it may be concluded that an appreciable outward movement of calcium ion does not take place during the negative inotropic phase of toxic CG concentrations.

F. Effect of Cardiac Glycosides on Subcellular Calcium Storage Sites

I. Sarcoplasmic Reticulum

There can be little doubt now that the CG have no effect on the Ca^{2+}-accumulating and releasing activity of cardiac isolated sarcoplasmic reticulum (SR). The reports which did indicate a possible site of action at the sarcoplasmic reticulum (ENTMAN et al., 1969) were probably based on the use of plasmalemmal-contaminated preparations. PRETORIUS et al. (1969), BESCH et al. (1970), WORSFOLD and PETER (1970) and others now consistently fail to find any change in the Ca^{2+}-transporting activity of the SR even when large doses of CG are administered. If however, the toxicity of the CG is accompanied by Ca^{2+} overloading to such an extent that ATP production fails, insufficient ATP might remain available under in vivo conditions to support the Ca^{2+}-accumulating activity of the SR, i.e. its Ca^{2+}-accumulating activity could decline as an indirect result of the administration of toxic doses.

II. Mitochondria

Although the glycosides have no direct effect on the Ca^{2+}-accumulating activity of isolated cardiac mitochondria (DRANSFELD et al., 1967; NOACK and DRANSFELD, 1976) it is quite possible that changes in cytosolic Na^+ and K^+ caused by the administration of toxic doses of CG affect the Ca^{2+}-accumulating activity of these organelles so that they release Ca^{2+} into the cytosol (NOACK and GREEFF, 1975). Hence, as was noted for the SR, the effect of CG on mitochondrial Ca^{2+} metabolism is indirect. Nevertheless it could contribute to the myocardial action of CG.

In conclusion, therefore, there seems to be no good reason for arguing that the positive inotropism of the CG involves their interaction with either the SR or the mitochondria. Both these subcellular organelles, however, may be indirectly involved in the CG action, especially in the presence of toxic concentrations.

G. Conclusions

There is little doubt now that toxic concentrations of CG cause marked and easily detectable changes in the electrolyte content of heart muscle. The ions that are involved are Na^+, K^+, and Ca^{2+}; toxicity due to CG administration involves a gain in Na^+ and Ca^{2+} and a loss of K^+. The most likely explanation for this altered tissue electrolyte content is the inhibition of the Na^+, K^+-pump and the resultant activation of the Na/Ca and Ca/K exchange mechanisms. It seems unlikely, however, that this same mechanism provides the basis for the positive inotropic effect of the CG. Nor does it seem likely that the positive inotropism of the CG involves an increase in total intracellular Na^+, K^+ or Ca^{2+}. It is possible, however, that there is an altered cellular distribution and exchangeability of one or more of these cations, resulting in a regional accumulation that favours E–C coupling. The most likely candidate for this altered exchangeability and regional accumulation is Ca^{2+}, and there is some evidence that the site of accumulation is closely associated with the cell membrane. It is impossible at the present time to distinguish between two possibilities – whether the altered regional distribution of Ca^{2+}, and hence its increased availability for activation of the myofibrils, results from a direct effect of the CG at the cell membrane, or whether it is related to a changed activity of the Na^+, K^+-pump. There is ample evidence for discarding the K/Ca exchange hypothesis as a mechanism underlying the inotropism of the CG; the evidence against Na^+ is less convincing. In some more recent experiments, however, the positive inotropism of the CG is accompanied by an activation of the Na^+, K^+-pump.

New techniques, probably involving ion-sensitive electrodes are needed to differentiate between these possibilities. Measurements of total tissue electrolytes, or even the measurement of their rate of accumulation using radioactively labelled electrolytes, lack the precision needed to solve the basic problem: is the increased intracellular availability of Ca^{2+} a direct effect or does it occur because of a changed regional distribution of Na^+?

References

Aikawa, J.K., Rhoades, E.L.: Effect of digitoxin on exchangeable and tissue potassium content. Proc. Soc. Exp. Biol. Med. *90*, 332–335 (1955)

Akera, T., Bennett, R.T., Olgaard, M.K., Brody, T.M.: Cardiac Na^+, K^+-adenosine triphosphatase inhibition by ouabain and myocardial sodium: a computer simulation. J. Pharmacol. Exp. Ther. *199*, 287–297 (1976a)

Akera, T., Brody, T.M.: Membrane adenosine triphosphate. The effect of potassium on the formation and dissociation of the ouabain-enzyme complex. J. Pharmacol. Exp. Therap. *176*, 545–557 (1971)

Akera, T., Larson, F.S., Brody, T.M.: Correlation of cardiac sodium- and potassium-activated adenosine triphosphatase activity with ouabain-induced inotropic stimulation. J. Pharm. Exp. Ther. *173*, 145–151 (1970)
Akera, T., Olgaard, M.K., Brody, T.M.: Effect of ouabain on sodium movement in cardiac cells. Rec. Adv. Stud. Cardiac Struct. Met. *11*, 401–405 (1976b)
Akera, T., Olgaard, M.K., Temma, K., Brody, T.M.: Development of the positive inotropic action of ouabain: effects of transmembrane sodium movement. J. Pharm. Exp. Ther. *203*, 675–684 (1977)
Allen, D.G., Blinks, J.R.: Calcium transients in aequrorin-injected frog cardiac muscle. Nature *213*, 509–513 (1978)
Allen, J.C., Lindenmayer, G.E., Schwartz, A.: An allosteric explanation for ouabain induced time-dependent inhibition of sodium, potassium-adenosine triphosphatase. Arch. Biochem. Biophys. *141*, 322–328 (1970)
Allen, J.C., Schwartz, A.: Na^+, K^+-ATPase, the transport enzyme: evidents for its proposed role as pharmacologic receptor for CG. Ann. N.Y. Acad. Sci. *242*, 646–657 (1974)
Areskog, N.A.: Electrolyte effect of cardiac glycosides on dog's heart-lung preparation. Acta Physiol. Scand. *55*, 264 (1962)
Bailey, L.E.: Changes in myocardial calcium and E–C coupling associated with failure and ouabain treatment. Arch. Int. Pharmacodyn. *226*, 178–181 (1977)
Bailey, L.E., Krip, G.: The effect of ouabain on the distribution of calcium in the cat heart. Arch. Int. Pharmacodyn. Ther. *196*, 36–45 (1972)
Baker, P.F., Willis, J.S.: Inhibition of the sodium pump in squid giant axons by cardiac glycosides: dependence on extracellular ions and metabolism. J. Physiol. (Lond.) *224*, 463–475 (1972)
Belardinelli, L., Harder, D., Sperelakis, N., Rubio, R., Berne, R.M.: Cardiac glycoside stimulation of inward Ca^{2+} current in vascular smooth muscle of canine coronary artery. J. Pharmacol. Exp. Therap. *209*, 62–66 (1979)
Beller, G.A., Smith, T.W., Abelmann, W.H., Haber, E., Hood, W.B.: Digitalis intoxication. A prospective clinical study with serum level correlations. N. Engl. J. Med. *284*, 989 (1971)
Bentfeld, M., Lüllmann, H., Peters, T., Proppe, D.: Interdependence of ion transport and the action of ouabain in heart muscle. Br. J. Pharmacol. *61*, 19–27 (1977)
Besch, H.R., Allen, J.C., Glick, G., Schwartz, A.: Correlation between the inotropic action of ouabain and its effect on subcellular enzyme system from canine myocardium. J. Pharm. Exp. Ther. *171*, 1–12 (1970)
Besch, H.R., Watanabe, A.M.: The positive inotropic effect of digitoxin: independence from sodium accumulation. J. Pharm. Exp. Ther. *207*, 958–965 (1978)
Billingheimer, E.: Vergleichende Untersuchungen über die Wirkung des Calcium und der Digitalis. Z. Klin. Med. *100*, 411–457 (1924)
Blackman, J.R., Hellenstein, J.K., Gillespie, L., Berne, R.M.: Effect of digitalis glycosides on the myocardial sodium and potassium balance. Circ. Res. *8*, 1003–1012 (1960)
Blood, B.E.: The influence of low doses of ouabain and potassium ions on sheep Purkinje fibre contractility. J. Physiol. (Lond.) *251*, 69P–70P (1975)
Boerth, R.C.: Decreased sensitivity of newborn myocardium to the positive inotropic effects of ouabain. In: Basic and therapeutical aspects of perinatal pharmacology. Morselli, P.L., Garattini, S., Sereni, F. (eds.). New York: Raven Press 1975
Bosteels, S., Carmeleit, E.: Estimation of intracellular sodium concentration and transmembrane sodium flux in cardiac Purkinje fibres. Pflügers Arch. Eur. J. Physiol. *376*, 35–47 (1972)
Boyer, P.K., Poindexter, C.A.: The influence of digitalis on the electrolyte and water balance of heart muscle. Am. Heart J. *1940*, 586–590
Brennan, F.J., McCans, J.L., Chiong, M.A., Parker, J.O.: Effects of ouabain on myocardial K and Na balance in man. Circulation *45*, 107–113 (1972)
Brody, T.M., Akera, T.: Relations among Na^+, K^+-ATPase activity, sodium pump activity, transmembrane sodium movement, and cardiac contractility. Fed. Proc. *36*, 2219–2224 (1977)
Brown, T.E., Acheson, G.H., Grupp, G.: The saturated lactone glycoside dihydro-ouabain: effects on potassium balance of the dog heart. J. Pharmacol. Exp. Ther. *136*, 107–113 (1962)

Brown, T.E., Grupp, G., Acheson, G.H.: Potassium balance of the dog heart: Effects of increasing heart rate and of pentobarbital and dihydro-ouabain. J. Pharmacol. Exp. Ther. *129*, 42–48 (1960)
Caldwell, P.C.: Factors governing movement and distribution of inorganic ions in nerve and muscle. Physiol. Rev. *48*, 1–64 (1968)
Calhoun, J.A., Harrison, T.R.: Studies in congestive heart failure. IX. The effect of digitalis on the potassium content of the cardiac muscle of dogs. J. Clin. Invest. *10*, 139–144 (1931)
Caprio, A., Farah, A.: The effect of the ionic milieu on the response of rabbit cardiac muscle to ouabain. J. Pharmacol. Exp. Ther. *155*, 403–414 (1967)
Carrier, G.A., Lüllmann, H., Neubauer, L., Peters, T.: The significance of a fast exchanging superficial calcium fraction for the regulation of contractile force in heart muscle. J. Mol. Cell. Cardiol. *6*, 333–347 (1974)
Carslake, M.C., Weatherall, M.: Changes in the sodium, potassium, and chloride of rabbit auricles treated with ouabain. J. Physiol. (Lond.) *163*, 347–361 (1962)
Clark, A.J.: The influence of ions upon the action of digitalis. Proc. R. Soc. Med. *5*, 181–199 (1912)
Cohen, I., Daut, J., Noble, D.: An analysis of the actions of low concentration of ouabain on membrane currents in Purkinje fibers. J. Physiol. (Lond.) *260*, 75–103 (1976)
Cohn, K.E., Keliger, R.E., Harrison, D.C.: Influence of potassium depletion on myocardial concentration of tritiated digoxin. Circ. Res. *20*, 473–476 (1967)
Cullen, G.E., Wilkins, W., Harrison, T.E.: Electrolytes in humans tissue. II. The electrolyte content of hearts and other tissues from cases with various diseases. J. Biol. Chem. *102*, 415–423 (1933)
Deitmer, J.W., Ellis, D.: Comparison of the action of various cardiac glycosides on the intracellular sodium activity of sheep heart Purkinje fibres. J. Physiol. (Lond.) *276*, 26–27P (1978a)
Deitmer, J.W., Ellis, D.: The intracellular sodium activity of cardiac Purkinje fibres during inhibition and re-activation of the Na-K pump. J. Physiol. (Lond.) *284*, 241–259 (1978b)
Dransfeld, H., Greeff, K., Hess, D., Schorn, A.: Die Abhängigkeit der Ca^{2+}-Aufnahme isolierter Mitochondrien des Herzmuskels von der Na^{+}- und K^{+}-Konzentration als mögliche Ursache der inotropen Digitaliswirkung. Experientia *23*, 375–377 (1967)
Dransfeld, H., Greeff, K., Schorn, A., Ting, B.T.: Calcium uptake in mitochondria and vesicles of heart and skeletal muscle in presence of potassium, sodium, k-strophanthin, and pentobarbital. Biochem. Pharmacol. *18*, 1335–1345 (1969)
Dutta, S., Marks, B.H.: Factors that regulate ouabain H^3-uptake by the isolated guinea pig heart. J. Pharm. Exp. Ther. *170*, 318–325 (1969)
Dutta, S., Marks, B.H.: Species and ionic influence on the accumulation of digitalis glycosides by isolated perfused hearts. Br. J. Pharmacol. *46*, 401–408 (1972)
Ellis, D.: The effects of external cations and ouabain on the intracellular sodium activity of sheep heart Purkinje fibres. J. Physiol. (Lond.) *273*, 211–240 (1977)
Entman, J., Cook, W.J., Bressler, R.: The influence of ouabain and alpha angelica lactone on calcium metabolism in dog cardiac microsomes. J. Clin. Invest. *48*, 229–234 (1969)
Farah, A.: The effect of the ionic milieu on the response of cardiac muscle to cardiac glycosides. In: Digitalis. Fisch, C., Surawiez, B. (eds.), pp. 55–64. New York: Grune & Stratton 1969
Felgenträger, J., Zettner, B.: Effect of positive inotropic concentrations of cardiac glycosides on myocardial electrolyte metabolism. Thesis, Düsseldorf 1979
Forester, G.V., Mainwood, G.W.: Ouabain sensitivity in the rat myocardium. Correlation with a model of subcellular calcium movement. Recent Adv. Stud. Cardiac Struct. Metab. *4*, 273–279 (1974)
Fricke, U., Klaus, W.: Dependence of the cardiac uptake of digitalis glycoside on the extracellular calcium concentration in guinea pig isolated hearts. Eur. J. Pharmacol. *30*, 182–187 (1975)
Garb, S., Venturi, V.: The differential actions of potassium on the therapeutic and toxic effect of ouabain. J. Pharm. Exp. Ther. *112*, 94–98 (1954)
Gersmeyer, E.F., Holland, W.C.: Effect of heart rate on action of ouabain on Ca exchange in guinea-pig left atria. Am. J. Physiol. *205*, 795–798 (1963)

Gertler, M.M., Kream, J., Hylin, J.W., Robinson, H., Neidle, E.G.: Effect of digitoxin and quinidine on intracellular electrolytes of the rabbit heart. Proc. Soc. Exp. Biol. Med. *92*, 629–632 (1956)

Gervais, A., Lane, L.K., Amer, B.M., Lindenmayer, G.E., Schwartz, A.: A possible molecular mechanism of the action of digitalis. Circ. Res. *40*, 8–19 (1977)

Ghysel-Burton, J., Godfraind, T.: Stimulation and inhibition by ouabain of the sodium pump in guinea-pig atria. Br. J. Pharmacol. *55*, 249P (1975)

Ghysel-Burton, J., Godfraind, T.: Importance of the lactone ring for the action of therapeutic doses of ouabain in guinea-pig atria. J. Physiol. (Lond.) *266*, 75P–76P (1977)

Glitsch, H.G., Reuter, H., Scholz, H.: The effect of the internal sodium concentration on calcium fluxes in isolated guinea-pig auricles. J. Physiol. (Lond.) *209*, 25–43 (1970)

Glynn, I.M.: The action of cardiac glycosides on ion movements. Pharmacol. Rev. *16*, 381–407 (1964)

Godfraind, T.: The therapeutic mode of action of cardiac glycosides. Arch. Int. Pharmacodyn. Ther. *206*, 384–388 (1973)

Godfraind, T., Ghysel-Burton, J.: Binding sites related to ouabain-induced stimulation or inhibition of the sodium pump. Nature *265*, 165–166 (1977)

Gonlubol, F., Siegel, A., Bing, R.J.: Effect of a cardiac glycoside (Cedilanid) on the sodium and potassium balance of the human heart. Circ. Res. *4*, 298–301 (1956)

Govier, W.C., Holland, W.C.: Effects of ouabain on tissue calcium exchange in pacemaker of turtle heart. Am. J. Physiol. *207*, 195–198 (1964)

Grahame-Smith, D.G., Everest, M.S.: Measurement of digoxin in plasma and its use in diagnosis of digoxin intoxication. Br. Med. J. *I*, 286–289 (1969)

Greeff, K., Meng, K., Moog, E.: Der Einfluß nichttoxischer und toxischer Konzentrationen herzwirksamer Glykoside auf die Kaliumbilanz isolierter Herzpräparate. Naunyn-Schmiedebergs Arch. Exp. Path. Pharm. *244*, 270–282 (1962)

Greeff, K., Pereira, E., Wagner, J.: Die Wirkung des Strophanthins bei Änderung der Schlagfrequenz und der extrazellulären K^+- und Ca^{2+}-Konzentration. Arch. Int. Pharmacodyn. Ther. *190*, 219–228 (1971)

Greenspan, A.M., Morad, M.: Electromechanical studies on the inotropic effects of acetylstrophanthidin in ventricular muscle. J. Physiol. (Lond.) *253*, 357–384 (1975)

Grossman, A., Furchgott, R.F.: The effects of various drugs on calcium exchange in the isolated guinea-pig left auricle. J. Pharmacol. Exp. Ther. *145*, 162–172 (1964)

Grupp, G., Charles, A.: Effect of ouabain and 3-acetyl strophanthidin on potassium exchange in dog heart in situ. J. Pharm. Exp. Ther. *143*, 356–365 (1964)

Güttler, K., Klaus, W., Land, E.: Antagonistic effect of triamterene to ouabain toxicity. Arzneim. Forsch. Drug Res. *29*, 623–628 (1979)

Hagen, P.S.: The effects of digilanid C in varying dosage upon the K and water content of rabbit heart muscle. J. Pharm. Exp. Ther. *67*, 50–55 (1939)

Harris, A.S., Firestone, J.E., Liptak, R.A.: Relation of potassium ions in extracellular fluid to ectopic ventricular arrhythmias by k-strophanthoside. Circ. Res. *12*, 718–719 (1955)

Harrison, T.R., Pilcher, C., Ewing, G.: Studies in congestive heart failure. IV. The potassium content of skeletal and cardiac muscle. J. Clin. Invest. *8*, 325–335 (1930)

Harvey, S.C.: Effect of digitoxin on cardiac water and electrolytes. Fed. Proc. *13*, 364 (Abstr. 1195) (1954)

Heinen, E., Noack, E.: Effects of k-strophanthidin and digitoxigenin on contractile force, calcium content, and exchange in guinea-pig isolated atria. Naunyn-Schmiedebergs Arch. Exp. Path. Pharm. *275*, 359–371 (1972)

Hellems, H.K., Regan, T.J., Talmers, F.N.: Influence of acetyl strophanthidin on myocardial electrolyte exchange. J. Clin. Invest. *34*, 915 (1955)

Hellems, H.K., Regan, T.J., Talmers, F.N., Christensen, R.C., Wada, T.: The mode of action of acetyl strophanthidin on the failing human heart. J. Clin. Invest. *35*, 710 (1956)

Hochrein, H.: VIII. Metabolischer Digitalis-Effekt am suffizienten und insuffizienten Myokard. In: Herzinsuffizienz und Myokardstoffwechsel. Hochrein, H. (ed), pp. 59–62. Aulendorf i. Württ.: Editio Cantor K.G. 1965

Holland, W.C., Greig, M.E., Dunn, C.W.: Factor affecting the action of lanatoside C on the potassium content of isolated perfused guinea-pig hearts. Am. J. Physiol. *176*, 227–231 (1954)

Holland, W.C., Sekul, A.A.: Effect of ouabain on ^{45}Ca and ^{38}Cl exchange in isolated rabbit atria. Am. J. Physiol. *197*, 757–760 (1959)

Holland, W.C., Sekul, A.A.: Influence of K^+ and Ca^{2+} on the effect of ouabain on Ca^{45} entry and contracture in rabbit atria. J. Pharmacol. Exp. Ther. *133*, 288–294 (1961)

Isenberg, G.: Is potassium conductance of cardiac Purkinje fibres controlled by Ca^{2+}? Nature *253*, 273–274 (1975)

Jenny, E., Turina, M., Waser, P.G.: Über den Einfluß herzaktiver Glykoside auf physikochemische Eigenschaften von natürlichen und rekonstruiertem Actomyosin aus Kalbherzmuskel. Helv. Physiol. Pharmacol. Acta *25*, 147–155 (1967)

Kasparek, R.: Effects of ouabain on rate of calcium exchange under different stimulation frequencies in guinea-pig left auricles. Naunyn-Schmiedebergs Arch. Exp. Path. Pharm. *293*, R25 (1976)

Kass, R.S., Tsien, R.W.: Multiple effects of calcium antagonists on plateau currents in cardiac Purkinje fibres. J. Gen. Physiol. *66*, 169–192 (1975)

Katz, A.M.: Contractile proteins of the heart. Physiol. Rev. *50*, 63–158 (1970)

Keefner, K.R., Akers, T.K.: Sodium localization in digitalis- and epinephrine-treated tunicate myocardium. Comp. Gen. Pharmacol. *2*, 415–422 (1971)

Keeton, W.F., Briggs, A.H.: In vico effects of ouabain on calcium metabolism in rabbit hearts and plasma. Proc. Soc. Exp. Biol. Med. *118*, 1127–1129 (1965)

Klaus, W.: Vergleichende Untersuchungen über die Wirkung verschiedener Digitoxigeninderivate auf die Kontraktionskraft und den Ca-Austausch isolierter Meerschweinchenvorhöfe. Arch. Exp. Path. Pharmakol. *246*, 226–239 (1963)

Klaus, W., Krebs, R.: Analysis of the calcium compartments in guinea-pig hearts under control conditions and under the influence of ouabain. Naunyn-Schmiedebergs Arch. Pharmacol. *283*, 277–292 (1974)

Klaus,W., Kuschinsky, G.: Über die Wirkung von Digitoxigenin auf den cellulären Calcium-Umsatz im Herzmuskelgewebe. Arch. Exp. Path. Pharmakol. *244*, 237–253 (1962)

Klaus, W., Kuschinsky, G., Lüllmann, H.: Über den Zusammenhang zwischen positiv inotroper Wirkung von Digitoxigenin, Kaliumflux und intrazellulären Ionenkonzentrationen im Herzmuskel. Arch. Exp. Path. Pharmakol. *242*, 480–496 (1962)

Koch-Weser, J.: Myocardial contraction frequency and onset of glycoside action. Circ. Res. *28*, 34–49 (1971)

Konschegg, A.: Über die Beziehung zwischen Herzmittel- und physiologischer Kationenwirkung. Arch. Exp. Pathol. Pharmakol. *71*, 251–260 (1913)

Kuschinsky, K., Lahrtz, H., Lüllmann, H., van Zwieten, P.A.: Accumulation and release of ^3H digoxin by guinea-pig heart muscle. Brit. J. Pharmacol. *30*, 317–328 (1967)

Langer, G.A.: A kinetic study of calcium distribution in ventricular muscle of the dog. Circ. Res. *15*, 393–405 (1964)

Langer, G.A.: Ion fluxes in cardiac excitation and contraction and their relation to myocardial contractility. Physiol. Rev. *48*, 708–757 (1968)

Langer, G.A.: The intrinsic control of myocardial contraction-ionic factors. N. Engl. J. Med. *285*, 1065–1071 (1971)

Langer, G.A.: "Effects of digitalis on myocardial ionic exchange." Circulation *46* (42), 180–187 (1972)

Langer, G.A.: Ionic movements and the control of contraction. In: The mammalian myocardium. Langer, G.A., Brady, A.J. (eds.), pp. 193–217. New York: John Wiley & Sons 1974

Langer, G.A., Brady, A.J.: Calcium flux in the mammalian ventricular myocardium. J. Gen. Physiol. *46*, 703–720 (1963)

Langer, G.A., Serena, S.D.: Effects of strophanthidin upon contraction and ionic exchange in rabbit ventricular myocardium: relation to control of active state. J. Mol. Cell. Cardiol. *1*, 65–90 (1970)

Lee, C.O., Fozzard, H.A.: Activities of potassium and sodium ions in rabbit heart muscle. J. Gen. Physiol. *65*, 695–708 (1975)

Lee, K.S., Klaus, W.: The subcellular basis for the mechanism of inotropic action of CG. Pharmacol. Rev. *23*, 193–261 (1971)

Lee, K.S., Yu, D.H., Burnstein, R.: The effect of ouabain on the oxygen consumption, the high energy phosphates and the contractily of the cat papillary muscle. J. Pharmacol. Exp. Ther. *129*, 115–122 (1960)
Lee, K.S., Yu, D.H., Lee, D.I., Burnstein, R.: The influence of potassium and calcium on the effect of ouabain on cat papillary muscle. J. Pharmacol. Exp. Ther. *132*, 139–148 (1961)
Lindenmayer, G.E., Schwartz, A.: A kinetic characterization of calcium on Na^+, K^+-ATPase and its potential role as a link between extracellular and intracellular events; hypotheses for digitalis-induced inotropism. J. Mol. Cell. Cardiol. *7*, 591–612 (1975)
Loewi, O.: Über den Zusammenhang zwischen Digitalis- und Kalziumwirkung. Arch. Exp. Path. Pharmakol. *82*, 131–158 (1917)
Loewi, O.: Über den Zusammenhang zwischen Digitalis- und Kalziumwirkung. Arch. Exp. Path. Pharmakol. *83*, 366–380 (1918)
Lüllmann, H., Holland, W.C.: Influence of ouabain on an exchangeable calium fraction, contractile force, and resting tension of guinea-pig atria. J. Pharmacol. Exp. Ther. *137*, 186–192 (1962)
Lüllmann, H., Peters, T.: Action of cardiac glycosides on the excitation-contraction coupling in heart muscle. Prog. Pharmacol. *2*, 1–57 (1979)
Lüttgau, H.C., Niedergerke, R.: The antagonism between Ca and Na ions in the frog's heart. J. Physiol. (Lond.) *143*, 486–505 (1958)
Mangun, G.H., Myers, V.C.: Creatine, potassium, and phosphorus content of cardiac and voluntary muscle. Proc. Soc. Exp. Biol. Med. *35*, 455–456 (1936)
Marks, B.H.: Factors that affect the accumulation of digitalis glycosides by the heart. Basic and clinical pharmacology of digitalis. Marks, B.H., Weissler, A.M. (eds.), pp. 69–93. Springfield, Ill.: Charles C. Thomas 1972
Matsui, H., Schwartz, A.: Mechanism of cardiac glycoside inhibition of the Na^+, K^+-dependent ATPase from cardiac tissues. Biochim. Biophys. Acta *151*, 655–663 (1968)
McCans, J.L., Brennan, F.J., Chiong, M.A., Parker, J.O.: Effects of ouabain and DHT on myocardial K balance in man. Am. J. Cardiol. *31*, 320–326 (1973)
McDonald, T.F., Nawrath, H., Trautwein, W.: Membrane currents and tension in cat ventricular muscle treated with cardiac glycosides. Circ. Res. *37*, 674–685 (1975)
Miura, D.S., Rosen, M.R.: The effect of ouabain on the transmembrane potentials and intracellular K activity of canine cardiac Purkinje fibres. Circ. Res. *42*, 333–338 (1978)
Morad, M., Greenspan, A.M.: Excitation-contraction coupling as a possible site for the action of digitalis on heart muscle. In: Cardiac arrhythmias. Driefus, L.S., Likoff, W. (eds.), pp. 479–489. New York: Grune & Stratton 1973
Morales, A.G., Acheson, G.H.: Effect of veratridine on the potassium balance of the dog heart-lung preparation. J. Pharmacol. Exp. Ther. *134*, 230–244 (1961)
Moran, N.C.: Contraction dependency of the positive inotropic action of cardiac glycosides. Circ. Res. *21*, 727–740 (1967)
Nayler, W.G.: Influx and efflux of calcium in the physiology of muscle contraction. J. Clin. Orthopaed. *46*, 151–182 (1966)
Nayler, W.G.: An effect of ouabain on the superficially located stores of calcium in cardiac muscle cells. J. Mol. Cell. Cardiol. *5* (11) 101–110 (1973a)
Nayler, W.G.: Positive inotropic action of ouabain. Br. Heart. J. *35* (5), 561 (1973b)
Nayler, W.G., Poole-Wilson, P.A., Williams, A.: Hypoxia and calcium. J. Mol. Cell. Cardiol. *11*, 683–706 (1979)
Nayler, W.G., Williams, A.: Relaxation in heart muscle: some morphological and biochemical considerations. Eur. J. Cardiol. 7/Suppl., 35–50 (1978)
Niedergerke, R.: Movements of Ca in frog heart ventricles at rest and during contractures. J. Physiol. (Lond.) *167*, 515–550 (1963)
Noack, E.: Factors influencing the size of inulin space in isolated atria. J. Mol. Cell. Cardiol. *11*, Suppl. 2, 42 (1979)
Noack, E., Dransfeld, H.: The uptake of potassium ions into isolated heart mitochondria in the presence of calcium ions and k-strophanthin. Arzneim. Forsch. (Drug Res.) *26*, 1538–1542 (1976)
Noack, E., Felgenträger, J., Zettner, B.: Kinetics of changes in K^+, Na^+, and Ca^{2+} content in heart tissue in the presence of cardiac glycosides. Naunyn Schmiedebergs Arch. Pharmacol. *302*, R34 (1978)

Noack, E., Felgenträger, J., Zettner, B.: Changes in myocardial Na and K content during the development of cardiac glycoside inotropy. J. Mol. Cell. Cardiol. *11*, 1189–1194 (1979)

Noack, E., Greeff, K.: Calcium uptake and storage in isolated heart mitochondria influenced by sodium and potassium ions. Recent. Adv. Stud. Cardiac. Struct. Metab. *5*, 165–170 (1975)

Okita, G.T., Matlack, M.A., Johnson, C.: Inability of strophanthidin to alter ^{22}Na efflux and myocardial Na and K levels during initial stages of digitalis inotropy. Pharmakolog. Kongr. Paris, p. 732, Abstr. 2286, 1978

Ordabona, M.L., Manganelli, G.: Modificazione del K^+ nei tessuti di ratti sottoposti a sforza ad in ratti tratti con digitale. Nota II. Boll. Soc. Ital. Biol. Sper. *30*, 10–12 (1954)

Page, E., McAllister, L.P., Power, B.: Stereological measurements of cardiac ultrastructures implicated in excitation-contraction coupling. Proc. Natl. Acad. Sci. USA *68*, 1465 (1971)

Pappano, A.J.: Calcium dependent action potentials produced by catecholamines in guinea-pig atrial muscle fibres depolarised by potassium. Circ. Res. *27*, 379–390 (1970)

Park, M.K., Vincenzi, F.F.: Influence of calcium concentration on the rate of onset of cardiac glycoside and aglycone inotropism. J. Pharm. Exp. Ther. *198*, 680–686 (1978)

Peper, K., Trautwein, W.: A note on the pacemaker current in Purkinje fibres. Pflügers Arch. Europ. J. Physiol. (Lond.) *309*, 356–361 (1965)

Poole-Wilson, P.A., Langer, G.A.: Glycoside inotropy in the absence of an increase in potassium efflux in rabbit heart. Circ. Res. *34*, 599 (1975)

Pretrorius, P.J., Pohl, W.G., Smithen, C.S., Inesi, G.: Structural and functional characteristics of dog heart microsomes. Clin. Res. *25*, 487–499 (1969)

Prindle, K.H., Skelton, C.L., Epstein, S.E. et al.: Influence of extracellular potassium concentration on myocardial uptake and inotropic effect of tritiated digoxin. Circ. Res. *28*, 337–345 (1971)

Rayner, B.X., Weatherall, M.: Digoxin, ouabain, and potassium movements in rabbit auricles. Brit. J. Pharmacol. *12*, 371–381 (1957)

Regan, T.J., Christensen, R.C., Wada, T., Talmers, F.N., Hellems, H.K.: Myocardial response to acetyl strophanthidin in congestive heart failure: A study of electrolytes and carbohydrate substrate. J. Clin. Invest. *38*, 306–316 (1959)

Regan, T.J., Markov, A., Oldewurtel, H.A., Harman, M.A.: Myocardial K^+ loss after countershock and the relation to ventricular arrhythmias after nontoxic doses of acetylstrophanthidin. Am. Heart J. *77*, 367–371 (1969)

Regan, T.J., Talmers, F.N., Hellems, H.K.: Myocardial transfer of sodium and potassium: Effect of acetylstrophanthidin in normal dogs. J. Clin. Invest. *35*, 1220–1228 (1956)

Reiter, M.: Die Beziehung von Calcium und Natrium zur inotropen Glykosidwirkung. Naunyn Schmiedebergs Arch. Pharmakol. *245*, 487–499 (1963)

Reiter, M.: Drugs and heart muscle. Ann. Rev. Pharmacol. *12*, 111–124 (1972)

Repke, K.R.H.: The biochemical action of digitalis. Klin. Wochenschr. *42*, 157–165 (1964)

Repke, K.R.H.: Effect of digitalis on membrane adenosine triphosphatase. In: Second international pharmacological meeting. Drugs and enzymes, Vol. 4, pp. 65–88. New York: Pergamon Press 1965

Reuter, H.: Exchange of calcium ions in mammalian myocardium. Mechanism and physiological significance. Circ. Res. *34*, 599–605 (1974)

Rogus, E., Zierler, K.L.: Sodium and water contents of sarcoplasm and sarcoplasmic reticulum in rat skeletal muscle: effects of anisotonic media, ouabain, and external sodium. J. Physiol. (Lond.) *233*, 227 (1973)

Salter, W.T., Runels, E.A.: A monogram for cardiac contractility involving calcium, potassium, and digitalis-like drugs. Am. J. Physiol. *165*, 520–526 (1951)

Salter, W.T., Sciarini, T.J., Gemmel, J.: Inotropic synergism of cardiac glycosides with calcium action on the frog's heart in artificial media. J. Pharm. Exp. Ther. *96*, 272–279 (1949)

Sanyal, P.N., Saunders, P.R.: Action of ouabain upon normal and hypodynamic myocardium. Proc. Soc. Exp. Biol. Med. *95*, 156–157 (1957)

Schwartz, A., Lindenmayer, G.E., Allen, J.C.: The sodium-potassium adenosine triphosphatase: pharmacological, physiological and biochemical aspects. Pharm. Rev. *27*, 1–134 (1975)

Sekul, A.A., Holland, W.C.: Effect of ouabain on ^{45}Ca entry in quiescent and electrically driven rabbit atria. Am. J. Physiol. *199*, 457–459 (1960)

Shapiro, W., Taubert, K., Narahara, K.: Nonradioactive serum digoxin and digitoxin levels. Arch. Int. Med. *130*, 31–36 (1972)

Sheridan, D.J.: Effects of ouabain on concentracture and superficially bound calcium in neonatal and adult cardiac muscle. J. Mol. Cell. Cardiol. *12*, 1123–1130 (1978)

Sherrod, T.R.: Effects of digitalis on electrolytes of heart muscle. Proc. Soc. Exp. Biol. Med. *65*, 89–90 (1947)

Skou, J.C.: The influence of some cations on an adenosinetriphosphatase from peripheral nerves. Biochim. Biophys. Acta *23*, 394–401 (1957)

Slany, J., Mösslacher, H.: Einfluß von Spironolactone auf die myokardiale Kaliumbilanz unter Stroph. beim Menschen. Klin. Wochenschr. *54*, 671–676 (1976)

Smith, T.W.: Radioimmunoassay for serum digoxin concentration: methodology and clinical experience. J. Pharmacol. Exp. Ther. *175*, 352–360 (1970)

Smith, T.W., Haber, E.: Digitalis: N. Engl. J. Med. *289*, 945–1011 (1973)

Solaro, R.J., Wise, R.M., Shiner, J.S. et al.: Calcium requirement for cardiac myofibrillar activation. Circ. Res. *34*, 525–530 (1974)

Thomas, L.J., Weldon, B.J., Grechman, R.: Effect of potassium lack and ouabain on ^{45}calcium uptake in frog's heart. Fed. Am. Soc. Exp. Biol. *17*, 162 (1958)

Toda, N., West, T.C.: Modification by sodium and calcium of the cardiotoxicity induced by ouabain. J. Pharm. Exp. Ther. *154*, 239–249 (1966)

Tuttle, R.S., Farah, A.: The effect of ouabain on the frequency-force relationship and on post-stimulation potentiation in isolated atrial and ventricular muscle. J. Pharmacol. Exp. Ther. *135*, 142–150 (1962)

Tuttle, R.S., Witt, P.N., Farah, A.: The influence of ouabain on intracellular Na and K concentrations in the rabbit myocardium. J. Pharmacol. Exp. Ther. *133*, 281–287 (1961)

Tuttle, R.S., Witt, P.N., Farah, A.: Therapeutic and toxic effects of ouabain on K^+ fluxes in rabbit atria. J. Pharmacol. Exp. Ther. *137*, 24–30 (1962)

Vick, R.L., Kahn, J.B.: The effects of ouabain and veratridine on potassium movement in the isolated guinea-pig heart. J. Pharmacol. Exp. Ther. *121*, 389–401 (1957)

Vincenzi, F.F.: Influence of myocardial activity on the rate of onset of ouabain action. J. Pharmacol. Exp. Therap. *155*, 279–287 (1967)

Watson, E.L., Woodbury, D.M.: The effect of DHT and ouabain, alone, and in combination on the electrocardiogram and on cellular electrolytes of guinea-pig heart and sarcoplasmic reticulum. Arch. Int. Pharmacodyn. Ther. *201*, 389–399 (1973)

Wedd, A.M.: The influence of digoxin on the potassium content of heart muscle. J. Pharmacol. Exp. Ther. *65*, 268–274 (1939)

Weingart, R., Kass, R.S., Tsien, R.W.: Is digitalis inotropy associated with enhanced slow inward calcium current? Nature *273*, 389–392 (1978)

Wilbrandt, W., Caviezel, R.: Die Beeinflussung des Austritts von Calcium aus dem Herzmuskel durch Herzglykosid. Helv. Physiol. Acta *12*, C40–42 (1954)

Williams, J.F., Klocke, F.J., Braunwald, E.: Studies on digitalis. XIII. A comparison of the effects of potassium on the inotropic and arrhythmia producing actions of ouabain. J. Clin. Invest. *45*, 346–352 (1965)

Witt, P.N., Tuttle, R.S.: Effects of ouabain on K and Na concentrations in resting rabbit auricles. Fed. Proc. *19*, 123 (1960)

Wollenberger, A.: The energy metabolism of the failing heart and the metabolic action of the cardiac glycosides. Pharmacol. Rev. *1*, 311–352 (1949)

Wollenberger, A., Jehl, Jo'ann, Karsh, M.L.: Influence of age on the sensitivity of the guinea-pig and its myocardium to ouabain. J. Pharmacol. Exp. Ther. *108*, 52–60 (1953)

Wood, E.H., Heppner, R.L., Weidmann, S.: Inotropic effects of electric currents. II. Hypotheses: Calcium movements, excitation contraction coupling, and inotropic effects. Circ. Res. *24*, 409–443 (1969)

Worsfold, M., Peters, J.B.: Kinetics of calcium transport by fragmented sarcoplasmic reticulum. J. Biol. Chem. *245*, 5545–5552 (1970)

CHAPTER 19

Effects of Cardiac Glycosides on Myofibrils

P. G. WASER and M. C. SCHAUB

A. Introduction

Different sites in the cell may be involved in the control of myocardial function, but it seems unlikely that any one of these is the sole controlling factor under physiologic and pathologic conditions. Ever since it was realized that cardiac glycosides (CG) directly affect the myocardium (GOLD and CATTEL, 1940), all major areas of cellular and subcellular activity in the myocardium have been subjected to investigation (LEE and KLAUS, 1971). The conclusions drawn from the large body of data, however, remain contradictory in part. In the past, it was thought that the inhibition of the membrane Na^+, K^+-pump was correlated with the positive inotropic action of CG. Recent developments, however, would seem to require modification of such a simple mechanism, as a growing number of reports point to dissociation of these two effects and to changes in the intracellular environment as the cause of inotropic effects (MURTHY et al., 1974; BEELER, 1977; WEINGART et al., 1978; DIACONO, 1979; LA BELLA et al., 1979). In fact, small but therapeutically effective doses of CG are even reported to activate the Na^+, K^+-pump system (BLOOM et al., 1974; PETERS et al., 1974; COHEN et al., 1976). Other results indicate that the toxic effects of CG may instead be due to alterations in the intracellular ionic composition produced by the inhibition of the Na^+, K^+-pump (OKITA, 1977; LÜLLMANN and PETERS, 1979). Thus, today, we see some of the attention paid in the past to sites located on the outer side of the cellular membrane being diverted to intracellular sites.

Indeed, in species with high sensitivity to CG, such as the guinea pig, cat, dog, and man, the largest amount of CG is recovered after tissue fractionation both in the so-called nuclear membrane fraction, which also contains the bulk of the myofibrils, and in the microsomal fraction (KIM et al., 1972; FRICKE et al., 1975; FRICKE, 1978; COLTART, 1978). These findings give more prominence to possible target sites of CG action at the control points of ionic conditions within the cell itself.

The level of free Ca^{2+} ion concentration is indeed important for the control of cellular activity in general. In the heart cell in particular, these ions serve as the messenger for regulating the excitation-contraction coupling as well as the level of the concomitant metabolic activity needed to meet the energy requirements (CARAFOLI and CROMPTON, 1978; DRUMMOND and SEVERSON, 1979; FABIATO and FABIATO, 1979). As the intracellular Ca^{2+} messenger, which is mainly compartmentalized between the membranous structures of the sarcolemma, sarcoplasmic reticulum, and mitochondria, represents the last link in the stimulus transfer chain, its target site on the contractile machinery itself may be a possible location for pharmacologic interference. The last few years have seen considerable advances in the understanding of the elementary events in the mechanism involved in contraction and its regulation in both skeletal and cardiac muscles. These advances have opened the way for research into the effects of CG at the molecular level of the contractile apparatus.

B. Molecular Aspects of Myofibrillar Proteins

I. Contractile Mechanism

In the myocardium every cell contracts in each cardiac cycle, and the heart contractility cannot be modulated as in skeletal muscle by varying the number of active cells, i.e., by summation of so-called functional motor units. Hence, each cardiac cell must be capable of a broad range of contractile states. Nevertheless, it has become clear that on the molecular level, the contractile mechanism and its regulation function in the same way in skeletal and cardiac muscles. In both types of muscle tissues, the histochemical architecture of the myofibrillar arrangement looks very much alike (HUDDART, 1975). The sarcomere consists of the two sets of interdigitating filaments containing myosin on the one hand and actin on the other. Myosin is a large (470,000 daltons mol. wt.), highly asymmetric duplex molecule comprising six polypeptide subunits, two heavy chains (200,000 daltons mol. wt. each), and four light chains in the molecular weight range of 20,000 daltons (LOWEY and RISBY, 1971). At one end the two heavy chains that form a long double α-helical tail structure curl up to form a couple of elongated globular head portions to which are attached two light chains each. These heads protrude from the myosin filaments, which are located in the middle of the sarcomere, and when muscle is activated, extend to the actin filaments where the two types of filamentous structures overlap. Then the so-called cross-bridges, by tilting of the myosin heads through an angle of around 45°, pull the actin filaments, which are attached to the Z disks at both sarcomere ends, toward the sarcomere center (HUXLEY, 1969). Finally, the cross-bridges detach from the actin filaments, swing back to their original rectangular position, and are ready to reach out for the next movement again (Fig. 1). From several up to about a dozen such repeated cross-bridge cycles are necessary to produce a macroscopic shortening of the muscle of 5%–10% in a fraction of a second. In the isometric case, the rotatory movement of the myosin heads produces tension. Unlike depicted in Fig. 1, however, the cross-bridges do not cycle synchronously in active muscle. If that were the case, all bridges would be interlocked at one time with actin, and at another time all bridges would be detached allowing the muscle to slacken instantaneously. In contrast, the cross-bridges work in an asynchronous manner to ensure that sufficient bridges are always attached to sustain the resistance against which the muscle is working. In biochemical terms, such definite directional movements coupled with force generation of the cross-bridges require definite conformational changes in the proteins involved. The energy is derived ultimately from the terminal phosphate bond in ATP, which is split at the active sites located in both myosin heads. The chemical-mechanical trans-

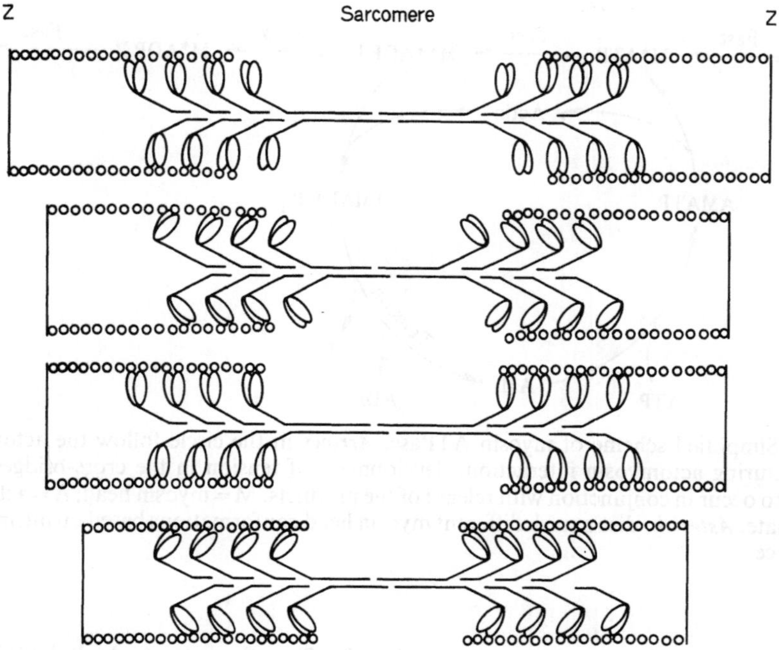

Fig. 1. Sarcomere shortening in muscle contraction. Shortening is brought about by a circular movement of the attached cross-bridges toward the center of the sarcomere, thus pulling the actin filaments inward. The cross-bridges then detach from actin and revert to their rectangular position, ready to attach to actin again and repeat the cycle. Each cycle produces further shortening. One cross-bridge is considered to consist of the two heads of one myosin molecule in this scheme. Z = Z line

duction process will only be fully understood when the sequence of protein conformational changes can be related to the elementary steps in the hydrolytic cycle.

In resting muscle when the myosin heads are detached from actin, ATP is turned over at a very slow steady-state rate. Among over half a dozen intermediary steps that have been recognized in the enzymatic cycle, the actual hydrolysis step of ATP proceeds fast with a rate constant of several hundreds per second. On the other hand, the subsequent hydrolytically induced conformation represents a long-lived intermediate which has the products ADP and inorganic phosphate bound at the active center (Fig. 2). The half-life of this species lasts for over 30 s and is then followed by dissociation of the products (TRENTHAM et al., 1976). This represents an energy-saving enzyme mechanism in resting muscle that keeps the conversion of ATP to ADP far from equilibrium in favor of ATP. It is this long-lived intermediary conformation that becomes affected by the interaction with actin when muscle is activated. The interaction with actin accelerates the product dissociation, once it has been formed at the active sites over 2,000 times. This actin activation during muscle contraction elevates the steady-state turnover of ATP, close to the maximal speed of the actual hydrolytic intermediary step on the active site (TAYLOR, 1979). It has to be borne in mind that the hydrolytic step takes place on the myosin heads when the cross-bridges are released from actin and return to their

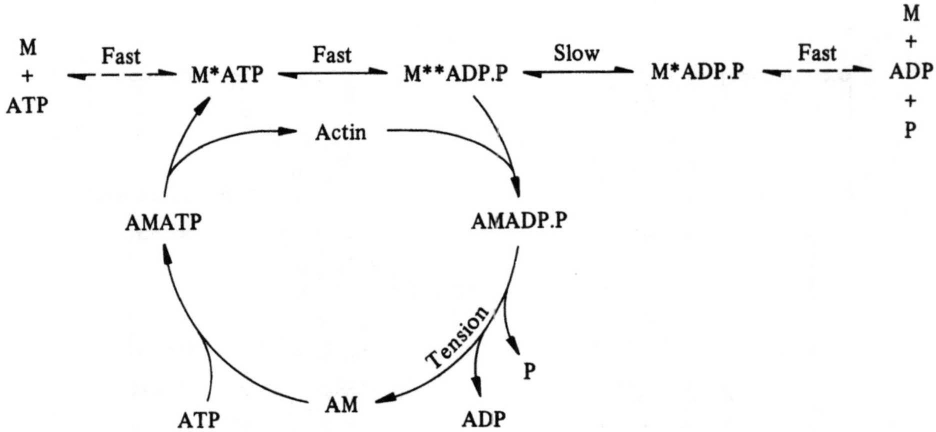

Fig. 2. Simplified scheme of myosin ATPase. *Arrows* in the circle follow the actomyosin cycle. During actomyosin interaction, development of tension in the cross-bridges is assumed to occur in conjunction with release of the products. M = myosin head; A = actin; P = phosphate. *Asterisks* distinguish different myosin head conformations based on intrinsic fluorescence

starting position for the next mechanical cycle. Thus the rate at which ATP is used during repetitive cross-bridge cycles in contraction is tightly coupled to the interaction with actin. Conceptually it is now understandable that just one ATP molecule is split per one interaction event of one myosin head with an individual actin molecule in the actin filament (Fig. 3). In this way, the chemical and mechanical cycles can, at least in part, be brought together (EISENBERG and HILL, 1978). The largest changes in free energy arise from those steps involving the binding of the substrate ATP and desorption of the products ADP and inorganic phosphate from the active sites, while the hydrolytic step itself yields only about 15% of the negative free energy in the overall degradation of ATP to ADP (BOYER et al., 1975). Recent evidence indicates that allosteric protein conformational changes connected with the hydrolytic cycle at the active sites in the myosin heads do indeed take place in a distant part of the molecule, namely, in the neck region where the heads that form the cross-bridges are joined together to the rod tail and where flexibility is assumed to occur during active cross-bridge cycling (SCHAUB et al., 1978; SHUKLA et al., 1979). Furthermore, though the resolution achieved at present in mechanical measurements as well as in X-ray diffraction studies does not allow the two heads of a single myosin molecule to be distinguished, biochemical approaches indicate that in cardiac and skeletal muscle myosin, the two heads (probably forming one cross-bridge) seem to operate together in a concerted way (SCHAUB et al., 1977; KUNZ et al., 1980; SCHAUB and WATTERSON, 1981).

II. Regulation of Contraction

With activation of muscle involving myofibrillar cross-bridge cycling, within a fraction of a second, the rate of ATP hydrolysis is accelerated some 2,000—fold as compared to the relaxed state. This clearly requires a very precise control mech-

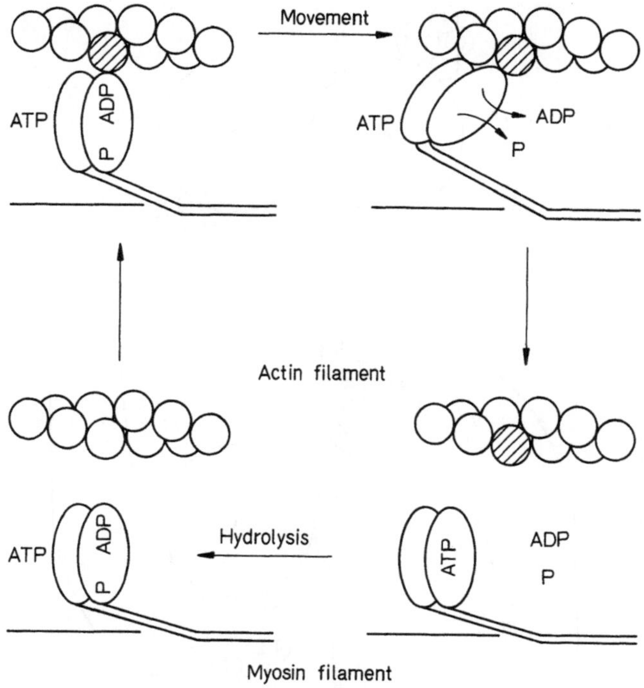

Fig. 3. Cross-bridge cycle coupled to ATP hydrolysis. The second myosin head is assumed to follow suit through the same cycle, fractionally lagging behind the first one. P = phosphate. Adapted from LYMN and TAYLOR (1971), MANNHERZ et al. (1973), and SCHAUB et al. (1977)

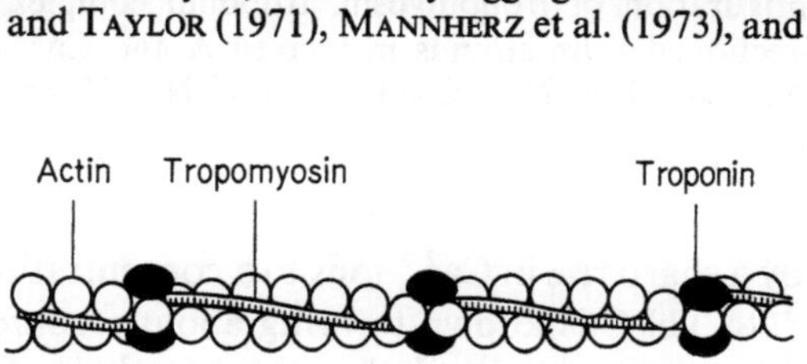

Fig. 4. Localization of the regulatory proteins in the actin filament. Adapted from EBASHI and ENDO (1968)

anism, which is governed by the free cytosolic Ca^{2+} ion concentration. This control is exerted on the reaction partner of myosin, the actin filament, in such a way that at low Ca^{2+} ion concentration ($Ca^{2+} < 10^{-7}$ *M*) at rest, the actin molecules exhibit low affinity toward the energized myosin cross-bridges. This is effected by steric blockage of the actin domains, which in the active state interact with the myosin heads (WAKABAYASHI et al., 1975). The rodlike tropomyosin molecules connected by end-to-end contact run as two threads along the two grooves that are formed by the double-stranded actin filament. Each tropomyosin molecule covers seven globular actins. At 40-nm intervals, the troponin complex consisting of three components holds the tropomyosin threads in place (Fig. 4). The troponin-I (TN-I) alone inhibits actomyosin interaction, troponin-C (TN-C) specifically binds Ca^{2+} ions, and via troponin-T (TN-T) the complex is attached to tropomyosin. When

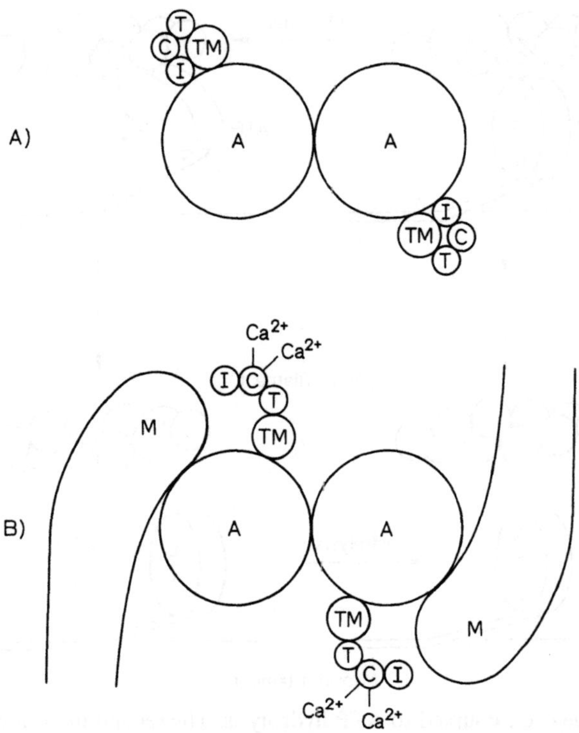

Fig. 5A, B. Steric configuration of tropomyosin, troponin complex, and myosin heads relative to the cross-sectioned actin strands in relaxed **A** and Ca^{2+}-activated muscle **B**. A = actin; TM = tropomyosin; T = TN-T; C = TN-C; I = TN-I; M = myosin head. Adapted from GERGELY (1976)

muscle is activated by a sharp rise in Ca^{2+} ions to a concentration around 10^{-5} *M*, the binding of Ca^{2+} to TN-C is known to bring about a conformational change that is believed to be transmitted through the other members of the troponin complex to tropomyosin. Consequently, the tropomyosin threads change position along the grooves of the double-stranded actin filament, now allowing the energized myosin cross-bridges to interact with individual actin molecules (Fig. 5).

This molecular regulatory mechanism had been deduced from structural studies using X-ray diffraction techniques and was recently confirmed in an elegant way by biochemical methods (MIKAWA, 1979). By cross-linking isolated actin filaments containing the regulatory proteins with glutaraldehyde in the presence or absence of Ca^{2+} ions, it was possible to "freeze" the two conformational states of the actin strands, the switched-on and the switched-off states. In subsequent recombination of "frozen" actin filaments with myosin, the ATPase activity was either high or low according to the corresponding "frozen" state; but in both cases neither presence nor absence of Ca^{2+} exerted any regulatory influence. In the cross-linked states, the troponin complex was obviously no longer able to transmit its Ca^{2+}-dependent conformational changes to the tropomyosin threads along the actin filaments.

This elaborate regulatory protein system has led to a well-defined stoichiometric structural arrangement both in cardiac and skeletal muscle (Table 1). In both

Table 1. Protein composition of rabbit skeletal muscle myofibrils. Adapted from Potter (1974)

Protein	Molecular weight	% of total protein	Molar ratio	Role
Myosin	470,000	47	1	Contraction
Actin	42,000	28	7	Contraction
Tropomyosin	65,000	6.3	1	Regulation
Troponin-T	30,500	3.1	1	Regulation
Troponin-I	21,000	2.0	1	Regulation
Troponin-C	18,000	1.6	1	Regulation
Others		12		Structure (?)

types of muscle, the contractile and regulatory proteins of the myofibrils account for over 60% of total cell proteins. Their state of activity is governed in a subtle way by the Ca^{2+} ions, which act as allosteric effectors, and in addition by reversible covalent protein modification by phosphorylation reactions (see following paragraphs).

To understand the particularities in regulation by Ca^{2+} and phosphorylation, detailed knowledge of all the proteins involved is essential. The amino acid sequences of all regulatory proteins and of actin and myosin light chains from fast contracting skeletal muscle are now known. Comparative sequence studies on these proteins from different types of muscle have shown that none of them is common to more than one type of muscle (COLLINS, 1976; ELZINGA and LU, 1976; WILKINSON and GRAND, 1978; SMILLIE, 1979). Thus, in cardiac and skeletal muscle cells different structural genes are expressed in these myofibrillar proteins. There are the amino acid side chains that dictate the properties of the proteins they comprise. In actin, which during evolution is the most conservative among these proteins, so far only one amino acid replacement has been found between cardiac and skeletal muscles, and no functional differences. In contrast, the larger differences in primary structure of the other proteins are also accompanied by differences in their functional properties. Although the physicochemical properties of myosin in cardiac muscle and in fast contracting skeletal muscle are identical (LEGER et al., 1975), the most obvious differences concern their light chain complement (LOWEY and RISBY, 1971) and the lower steady-state ATPase activity of the cardiac myosins (KATZ, 1970). It has recently been shown, however, that the light chain complement is not responsible for the different enzymatic activities (WAGNER and WEEDS, 1977), but that sequence differences exist in a peptide that is closely related to the active site and could be located in the heavy chain in the myosin head portion, i.e., within the cross-bridge proper (ELZINGA and COLLINS, 1977; FLINK et al., 1977; KUNZ et al., 1977).

Despite a lower enzymatic activity of cardiac myosin, which must be due to its difference in primary structure of its heavy chains, conformational changes are induced by the hydrolytic cycle similar to those observed with skeletal myosin (PFISTER et al., 1975; TAYLOR and WEEDS, 1976). Assuming the actins are indistinguishable in both tissues, maximal activation in heart can never induce the cross-bridges to cycle as rapidly as in fast contracting skeletal muscle (MARSTON and TAYLOR,

1980). It should be mentioned here that differences in peptides from the heavy chains were observed when myosin was isolated from hypothyroid as well as from thyrotoxic rabbit hearts as compared to cardiac myosin from control animals (HOH et al., 1979; MORKIN, 1979). Similar peptide differences are also apparent between atrial and ventricular myosins from calf heart (FLINK et al., 1978; DALLA LIBERA et al., 1979). All these myosin isoenzymes vary in their primary structure and display different ATPase activities. Also in hearts with chronic pressure overload a myosin with significantly lower ATPase activity and differences in the primary structure of its heavy chains has been found (SWYNGHEDAUW et al., 1976; SCHEUER and BHAN, 1979; HOH et al., 1977; HOH et al., 1979). Thus the heart muscle cell has obviously retained the ability to produce a number of myosin isoenzymes. One cannot consider this versatility in gene expression to be regulation but rather an adaptation process according to the functional state.

On the other hand, the myosin light chains, in particular the so-called light chain-2 (LC-2), which can be phosphorylated and can bind divalent metal ions, Ca^{2+} or Mg^{2+}, seem to modulate the cross-bridge interaction with actin (PEMRICK, 1977; MALHOTRA et al., 1979). In muscles where no regulatory proteins are associated with the actin filament, myofibrillar ATPase activity was found to be regulated by binding of Ca^{2+} to the homologous light chain in molluscan myosin or by phosphorylation of the corresponding light chain in vertebrate smooth muscle myosin (KENDRICK-JONES and JAKES, 1977; SCHOLEY et al., 1981). Each of the two LC-2 of skeletal myosin binds one divalent metal ion with high affinity. Mg^{2+} and Ca^{2+} compete for the same site; thus, in relaxed muscle when the Ca^{2+} ion concentration is low, Mg^{2+} is bound. On isolated myosin it has been shown that at near physiologic Mg^{2+} ion concentrations (in the millimolar range according to POLIMENI and PAGE, 1973; BRINLEY et al., 1977) raising the Ca^{2+} to 10^{-5} *M*, which is likely to occur in active muscle, leads to binding of just 1 mol of Ca^{2+} per myosin while the second LC-2 probably retains its bound Mg^{2+} (WATTERSON et al., 1979). Thus, a dual Ca^{2+} regulatory system operating simultaneously on the myosin and actin filaments may well be anticipated in skeletal and cardiac muscles (LEHMAN, 1978).

C. Effect of Cardiac Glycosides on Myofibrillar Function

I. Calcium

The positive inotropic agent within the muscle cell is unquestionably the Ca^{2+} ion. In skeletal muscle, sufficient Ca^{2+} is stored in the sarcoplasmic reticulum, which upon stimulation is released and fully activates the contractile apparatus. In contrast, the myocardium depends from beat to beat on transsarcolemmal Ca^{2+} entry into the cell during the action potential (FABIATO and FABIATO, 1977; MARBAN et al., 1980). Thus, in the heart muscle cell the Ca^{2+} release from the sarcoplasmic reticulum is supplemented by the Ca^{2+} set free during activation from binding sites associated with the sarcolemma (LANGER et al., 1976; LÜLLMANN and PETERS, 1979). This provides several locations where the inotropic agent may be pharmacologically influenced, as discussed in other parts of this volume. The ultimate target sites for the inotropic effect of the Ca^{2+} ions are certainly the regulatory proteins of the contractile machinery. The intracellular occurrence of CG at therapeutically effective doses has been demonstrated by various techniques, such as cell fractionation (DUTTA et al., 1968; BASKIN et al., 1973) and earlier autoradiographic studies (WASER, 1962). However, these results have met with criticism on technical grounds (LÜLLMANN and PETERS, 1979). On the other hand, it has long been known that CG do associate with high affinity to isolated contractile proteins displaying an equilibrium constant around $10^6\ M^{-1}$ (WASER, 1956). Subsequently, conflicting results of the effects CG might have on the physicochemical and enzymatic properties of isolated contractile proteins have been reported (WASER, 1963; LEE and KLAUS, 1971). On balance, it seems that the purer the isolated contractile proteins actin and myosin were, the lesser the effects of CG observed, while in more crude preparations that obviously still contained regulatory proteins, some of their effects persisted (JENNY et al., 1967). Therefore, it needs to be clarified whether CG may affect the contractile apparatus by interfering with the subtle interplay between the regulatory and contractile proteins.

Skeletal TN-C binds four Ca^{2+} ions per molecule, two with high affinity (equilibrium constant K_1 around $10^7\ M^{-1}$) and two with about 100 times lower affinity (K_2 around $10^5\ M^{-1}$). In the intact troponin complex, these affinities are about 10 times higher (POTTER and GERGELY, 1975). Since Mg^{2+} ions compete with Ca^{2+} in binding to the high affinity sites, in muscle the apparent affinity for Ca^{2+} to these sites is somewhat lower. On structural grounds, it has been suggested that the high affinity sites must always be occupied by either Ca^{2+} or Mg^{2+} (LEVINE et al., 1977). Kinetic measurements, on the other hand, have indicated that Ca^{2+} ions bind and dissociate sufficiently fast only to those lower affinity sites that are specific for Ca^{2+} and do not bind Mg^{2+}, thus accounting for the fast response to the trigger of contraction (JOHNSON et al., 1979). Both Mg^{2+} and Ca^{2+} dissociate from the so-called Ca^{2+}-Mg^{2+} mixed sites about 100 times more slowly, and since at rest they are always occupied by Mg^{2+}, this would not allow contraction to be triggered by Ca^{2+} binding to them.

These findings led to the concept that the low affinity Ca^{2+}-specific sites are the "trigger sites" while the mixed Ca^{2+}-Mg^{2+} sites may be considered "relaxing sites." The mechanism envisaged depicts the Ca^{2+} changing location slowly from the "trigger sites" to the "relaxing sites" because of the higher affinity of the latter. This may initiate the relaxation process before Ca^{2+} is returned to the Ca^{2+}-storing membrane systems (HAIECH et al., 1979).

In cardiac TN-C, only three Ca^{2+} ions were found to bind, two with high affinity to the mixed Ca^{2+}-Mg^{2+} sites and one with lower affinity to a Ca^{2+}-specific site (POTTER et al., 1977; LEAVIS and KRAFT, 1978). Thus for triggering contraction in heart muscle the binding of only one Ca^{2+} to TN-C is thought to be sufficient (JOHNSON et al., 1980). VAN EERD and TAKAHASHI (1975) have shown that the amino acid sequence of bovine cardiac TN-C differs by 35% from that of rabbit fast muscle TN-C and have ascribed this to tissue difference rather than species difference. They have also shown that one of the specific Ca^{2+} binding sites contains seven amino acid replacements out of 14 residues while the other three binding sites exhibit a maximum of two replacements only when compared with the homologous sites in rabbit skeletal muscle TN-C. From the nature of the amino acid replacements, one could predict that this particular site in cardiac TN-C will no longer be able to bind Ca^{2+}. Nevertheless, under identical conditions, half-maximal activation of tension and ATPase activity in skinned fibers and myofibrillar preparations from skeletal and cardiac muscles is induced by a free Ca^{2+} ion concentration of around 10^{-6} *M* in the additional presence of $MgCl_2$ in the millimolar range (STULL and BUSS, 1978; ALPERT et al., 1979).

This indicates that in the integrated filamentous structure, the regulatory Ca^{2+} ions exhibit similar apparent affinity to troponin in both muscle types. Ca^{2+} binding to myofibrils is not only affected by the free Mg^{2+} ion concentration (DONALDSON et al., 1978) but also by the myosin cross-bridge interaction with actin as well as by the degree of overlap between the myosin and actin filaments (FUCHS, 1977). Careful analysis of Ca^{2+} binding data seems to support the hypothesis that the affinity for this ion to troponin varies during the course of contraction (HONIG and REDDY, 1975; RUPP, 1980).

Despite the molecular differences between the regulatory components TN-C in heart and skeletal muscles, the similar characteristics in the Ca^{2+} activation let one speculate that CG, if they do affect the contractile process via the troponin-linked Ca^{2+} regulation, should also display their action on skeletal myofibrils. In the past the results of CG action on myofibrils, whether from heart or from skeletal muscle, were inconclusive, but this might have been due to inappropriate experimental conditions (LEE and KLAUS, 1971). Until the present time, no detailed studies seem to have been done on the effect of CG on myofibrillar ATPase activity at free Ca^{2+} concentrations, which allow the system to be only partially activated. Our studies confirm older results that at low ionic strength CG do not affect the Mg^{2+}-ATPase of myofibrils, i.e., the actin-activated ATPase involved in contraction, even at various Ca^{2+} concentrations (GORDON and BROWN, 1966). However, in these studies the free Ca^{2+} ion concentrations were all above 10^{-5} *M* so that the system was fully activated. CG also did not affect the enzyme reaction when the Ca^{2+} concentration was around 10^{-8} *M* or lower and the regulatory proteins exerted their full inhibition. However, at intermediate degrees of activation (controlled by a Ca^{2+} ion buffer system) application of k-strophanthin, albeit at rather high concentrations of 10^{-4}–10^{-3} *M*, raises the enzyme activity to 100%–130% of the maximum obtainable (Fig. 6). The effect was by far largest at small degrees of Ca^{2+} activation. The use of a Ca^{2+} chelating agent and the determination of Ca^{2+} impurities in the drug preparations ensured that the level of free Ca^{2+} ion concentration could always be calculated (SCHAUB et al., 1979, unpublished results). This marked activation of the partially inhibited ATPase activity could be demonstrated equally well on myofibrils from fast contracting skeletal as well as from cardiac muscles. Cardiac myofibrils previously had to be treated with detergents to remove Ca^{2+}-binding membrane systems (SOLARO et al., 1971). These results then indicate that CG may affect the myofibrillar activity via the regulatory components only if the system is at an intermediate level of activation. If so, they then display their effect on myofibrils from cardiac or skeletal muscles.

II. Phosphorylation

As the mode of action of CG on myofibrils is unknown at present, a further aspect should be considered, which is also involved in myofibrillar regulation. In the

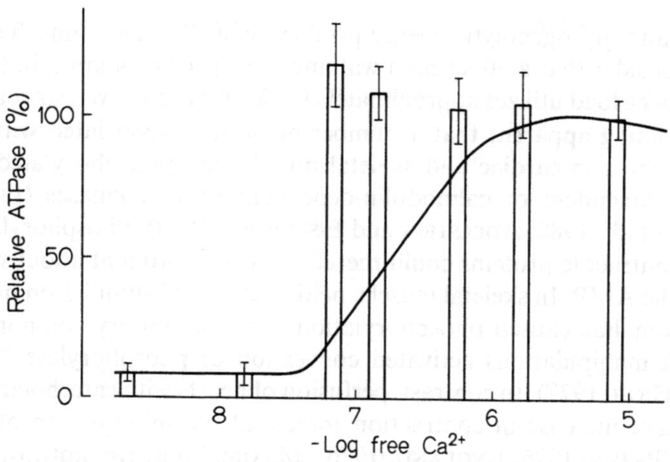

Fig. 6. Effect of k-strophanthin on myofibrillar Mg^{2+}-ATPase activity at different free Ca^{2+} ion concentrations. Mg^{2+}-ATPase activities *(curve)* were measured after incubation of myofibrils from rapidly contracting rabbit skeletal muscle at 25 °C and pH 7.6 in a medium containing 2 m*M* EGTA and various amounts of Ca^{2+} with a total ionic strength of 0.08–0.10. Columns give maximum enzymatic activities in the additional presence of 10^{-3} *M* k-strophanthin (average and range of three to six preparations). ATPase activity at 100% mark was on average 0.43 μmol of liberated phosphate per minute and milligram of protein

heart, β-adrenergic receptor activation by catecholamines initiates a sequence of reactions that alter contractility, ion transport, and metabolic responses in a concerted way (NAYLER, 1977). Adenosine 3′,5′-monophosphate (cyclic AMP) becomes increased by receptor-linked activation of adenylate cyclase. Cyclic AMP then precipitates the well-known cascade of chain reactions involving several phosphorylation steps resulting in glycogenolytic energy production. At several stages, Ca^{2+} is required in concentrations that also induce positive inotropism. It exerts its regulatory function by binding to calmodulin (16,700 daltons mol. wt.). Calmodulin (formerly called Ca^{2+}-dependent regulatory protein) is an ubiquitous target protein for Ca^{2+}-regulation, and homologous to TN-C. It is able to regulate a number of different enzyme systems including cyclic nucleotide phosphodiesterase, adenylate cyclase, phosphorylase kinase, and certain other protein kinases (MEANS and DEDMAN, 1980). Thus calmodulin is involved in the heart cell in Ca^{2+}-regulated processes of cyclic AMP metabolism, Ca^{2+}-fluxes, glycogen metabolism and heart muscle contraction (COHEN et al., 1978; WALSH et al., 1980). The temporal relationship between the formation of cyclic AMP, activation of protein kinases, transformation of phosphorylase "b" into "a" and increase in contractility as observed in the intact rat heart in the open chest preparation supports the suggestion that β-adrenergic stimulation initiates glycogenolysis and positive inotropism by the same mechanism (DOBSON, 1978). The concomitant activation of the process that furnishes new energy seems to be a prerequisite for mechanical force production in view of the fact that the entire readily available energy in the form of ATP and creatine phosphate would be depleted in a few minutes. This certainly also holds true for skeletal muscles where depolarization-induced Ca^{2+} release triggers contraction and glycogenolytic energy production at the same time. To illustrate the point, consider that a 70-kg man walking at a speed of 4 km/h in loose snow carrying a 20-kg load utilizes approximately 0.5 kg ATP/min (WILSON et al., 1978).

It is becoming apparent that a number of proteins associated with the contractile apparatus in cardiac and skeletal muscles are phosphorylated either by cyclic AMP-dependent or calmodulin-dependent protein kinases (PIRES et al., 1974; STULL et al., 1980; ADELSTEIN and EISENBERG, 1980). Phosphorylation of at least some contractile proteins could therefore also be brought about without increase in cyclic AMP. In skeletal muscle, neither electrical stimulation nor application of isoprenaline caused phosphorylation of the inhibitory component TN-I, though both manipulations activated conversion of phosphorylase "b" to "a" (STULL and HIGH, 1977). In contrast, perfusion of hearts with catecholamines leads to concomitant increase in contraction force and phosphorylation of the TN-I component (PERRY, 1976; ENGLAND, 1976). Maximal positive inotropy is accompanied by incorporation of about one phosphate group per mole of TN-I. Thereby, both the chemical and physiologic effects are temporally linked and level off together. From the study of peptide fragments isolated from TN-I, two parts of the primary sequence can be identified as being associated with biologic activity. The TN-C binding site is localized near the NH_2 terminal, and the region involved in the inhibitory action and binding with actin is close to the other end of the polypeptide chain toward the COOH terminal. Both regions contain a predominance of basic amino acids and possess net positive charges. The major site of the protein kinase-catalyzed phosphorylation has been identified as the serine-20 residue closely related to the TN-C binding site (MOIR and PERRY, 1977; PERRY, 1979). It follows that phosphorylation of this amino acid residue reduces the net positive charge at the interacting site and therefore would probably weaken the binding to the negatively charged TN-C. Interaction between these two components has been shown to be essential to relieve the TN-I-induced inhibition of the actomyosin ATPase (SCHAUB et al., 1972). Moreover, as this relief of inhibition is governed by trace amounts of Ca^{2+} ions, it is to be expected that the phosphorylation of TN-I affects cardiac contractility. In particular, one may speculate that the degree of phosphorylation of TN-I affects the binding of Ca^{2+} to the TN-C component. It has indeed been shown convincingly that the higher the degree of TN-I phosphorylation, the lower the Ca^{2+} sensitivity of isolated cardiac actomyosin, myofibrils, and even of intact skinned cardiac muscle fibers (SOLARO and SHINER, 1976; ENGLAND et al., 1979). These findings are corroborated by a recent observation that the higher the degree of TN-I phosphorylation, the lower the specific myofibrillar ATPase activity (WYBORNY and REDDY, 1978). In the heart, then, the catecholamine-induced increase in contraction force is accompanied by a decrease in the interaction of myosin cross-bridges with actin due to the phosphorylation of TN-I (Fig. 7). Therefore, the hormonal activation must at the same time ensure a sufficiently high cytosolic-free Ca^{2+} concentration to overcome the adverse effect of TN-I phosphorylation. Thus, this phosphorylation reaction seems to act as a negative feedback system. Its significance may be that the lower affinity of the regulatory complex troponin for Ca^{2+} facilitates the relaxation process. This effect can then be seen as one of several, all initiated by β-adrenergic stimulation and operating together in a concerted way: cyclic AMP-dependent phosphorylation of sar-

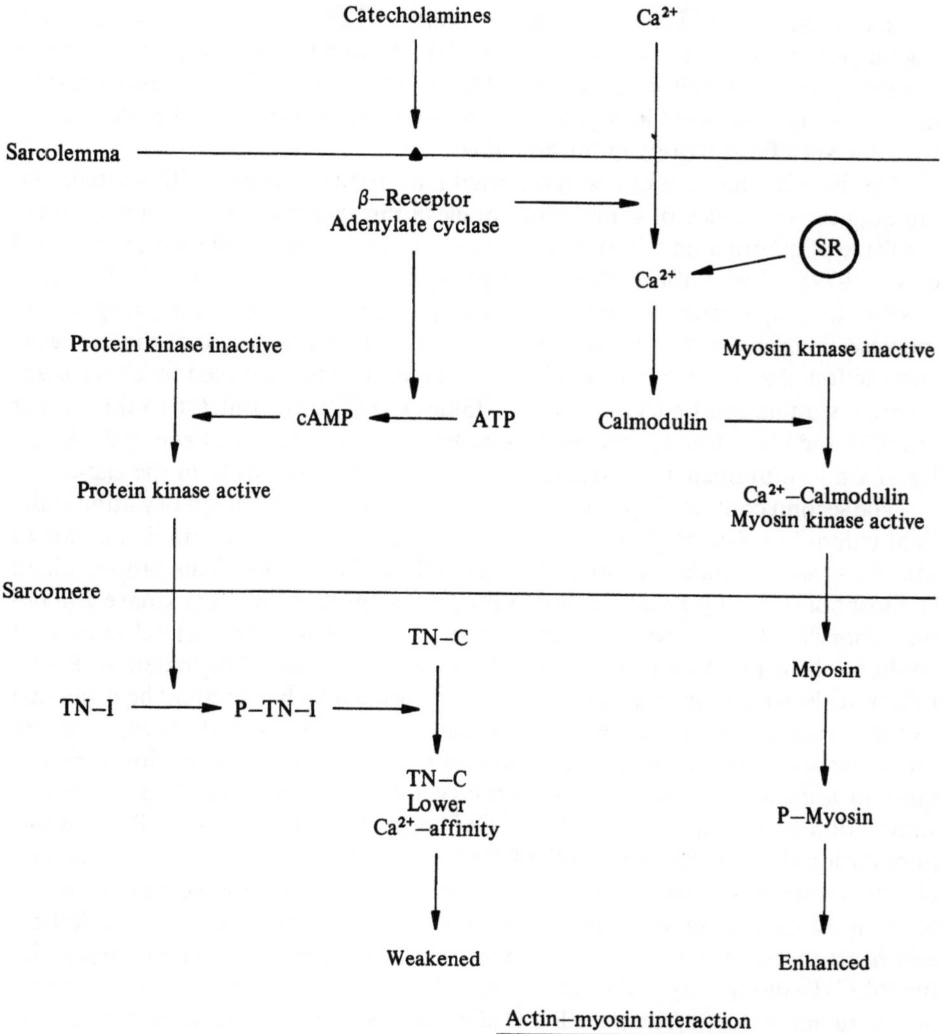

Fig. 7. Phosphorylation pathways for modulation of cardiac muscle contraction. Stimulation of β-adrenergic receptors promotes cyclic AMP-dependent phosphorylation of the regulatory component TN-I in the actin filament. Increased cytosolic free Ca^{2+} induces Ca^{2+}-calmodulin-dependent phosphorylation of the P-LC of myosin. Cardiac glycosides in addition to their effects on the sarcolemma, may interfere intracellularly with these modulatory pathways at several stages (for explanations see text). SR stands for sarcoplasmic resiculum

colemmal and sarcoplasmic reticulum membrane proteins altering the kinetics of Ca^{2+}-movements, so that increased amounts of Ca^{2+} are injected into the cytosol at faster rates but are also removed at faster rates. All effects together lead to a systolic contraction of larger amplitude and shorter duration (KATZ, 1979). It remains to be explained why no catecholamine-induced phosphorylation of skeletal TN-I is observed. Comparative amino acid sequence studies have revealed that cardiac TN-I differs by 55% from the skeletal TN-I (WILKINSON and GRAND, 1978). Furthermore, the cardiac TN-I comprises an additional 28 amino acid residues including the phosphorylatable serine-20 near its NH_2 terminal. This brings its molecular weight up to 23,500 daltons as compared to 21,000 of skeletal TN-I. This represents an interesting example of how phylogenetic variation in homologous proteins leads to tissue specific reactions in hormonal control.

On the other hand, it has been reported that perfusing hearts with ouabain, increasing the frequency of stimulation, or elevating external Ca^{2+} almost doubled the force of contraction without an increase in TN-I phosphorylation (EZRAILSON et al., 1977). This confirms that TN-I phosphorylation is not causally linked to positive inotropic interventions but acts only as a negative regulatory response to increases in cyclic AMP. In this context, it is of interest to note that the isometric contraction characteristics also differ between inotropism induced by either β-adrenergic stimulation or CG and Ca^{2+} (GREEFF, 1977). In contrast to the former case CG and Ca^{2+} lead to increased force development without affecting the kinetics of Ca^{2+}-movements, the duration of contraction remains then the same.

The second contractile protein that undergoes reversible phosphorylation is the light cahin-2 or P-light chain (P-LC) of myosin. Each myosin contains a pair of this P-LC with a molecular weight of around 19,000 daltons. This protein chain is phosphorylated by a Ca^{2+}-calmodulin-dependent myosin P-LC kinase and dephosphorylated by a specific phosphatase (PERRY, 1976). In skeletal muscle, it could be shown that phosphorylation (at maximum 1 mol phosphate/mol P-LC) followed electrically-induced short tetani with some time lag or could be provoked by increasing the cytosolic free Ca^{2+}-concentration (BARANY et al., 1979). Thereby the increased degree of phosphorylation correlated with larger tension development in immediately following isometric twitches. In heart muscle β-adrenergic stimulation or increased extracellular Ca^{2+} increased the degree of P-LC phosphorylation (Fig. 7). The degree of P-LC phosphorylation during diastole has been postulated to pre-determine directly the force of active tension generated by the heart in the ensuing contraction (KOPP and BARANY, 1979). Thus in both skeletal and heart muscle the mode of actomyosin interaction seems to be monitored by the P-LC (PEMRICK, 1977; MALHOTRA et al., 1979). Together with the troponin-tropomyosin regulatory system, the P-LC of myosin would then constitute a dual control mechanism for cross-bridge activity in vertebrate sarcomeric muscles (LEHMAN, 1978; WATTERSON et al., 1979). The P-LC homologs in myosins from smooth muscle and non-muscle cells are known to be involved in the regulation of the actomyosin interaction (SCHOLEY et al., 1981). One may then ask whether phosphorylation of P-LC affects its interaction with Ca^{2+} ions. Although the degree of phosphorylation was found to influence the Ca^{2+} binding to the isolated P-LC, this does not seem to be the case in intact myosin (HOLROYDE et al., 1979).

The third contractile protein that has been shown in skeletal and cardiac muscles to be reversibly phosphorylated depending on the state of activity is tropomyosin. In this case, the phosphorylation occurs at the penultimate amino acid residue (serine-283) at the COOH terminal of the molecule (MAK et al., 1978). This region of the molecule is thought to be involved in the head-to-tail interaction of tropomyosin forming the threads that run along the actin filament. Its reversible covalent modification may also represent an additional mechanism for the fine tuning of the regulatory protein system (SMILLIE, 1979).

D. Conclusions and Outlook

Though CG have now been in use for over 200 years for man's benefit, their precise mode of action is still unclear. Based on their distribution and ensemble of actions, intracellular sites also seem to be involved in addition to the better known effects on membranous structures. Great advances in basic research over the last few years have led to a detailed understanding of contraction and its regulation on the molecular level.

In view of the considerable differences in the mode of regulation between skeletal and cardiac muscles, one wonders whether the mechanical properties are comparable in the two tissues. Tension measurements on live skeletal muscle indicate that activation by Ca^{2+} ions is always paralleled by a proportionate increase in series elastic stiffness (HUXLEY and SIMMONS, 1971). This has been interpreted to mean that force development is directly related to the number of activated cross-bridges at any one moment. The proportionate relationship between tension and stiffness is maintained in cardiac muscle strips at various extracellular Ca^{2+} concentrations as well as under conditions of positive inotropism induced by the application of adrenaline or noradrenaline (HERZIG, 1978). Systolic force development and positive inotropism are therefore due to the same process: a recruitment of cross-bridges so that on average a larger number of them are attached to actin. At low free Ca^{2+} ion concentrations below half-maximal tension development, the cross-bridge kinetics are apparently accelerated with small increments of Ca^{2+}. Higher Ca^{2+} concentrations, as may occur during inotropic intervention, have no further effect on the cross-bridge kinetics; hence, contraction speed is not affected though isometric tension and stiffness still increase greatly. Thus, the combination of enzyme kinetics of ATP hydrolysis with the mechanical cycle comprising the force-generating step reveals that increased contractility, i.e., positive inotropism, results from elevated cross-bridge activity in both skeletal and cardiac muscle. This cross-bridge activity is primarily regulated in a direct way by Ca^{2+} ions acting as allosteric effectors on the regulatory proteins.

As we have seen above, however, in both types of muscle there are additional enzyme systems, specific protein kinases, and phosphatases that by covalent protein modification in rapid response to the functional state, further affect the regulation of contraction. Phosphorylation may exert its effect either by changing the affinity for Ca^{2+} binding to the myosin P-LC and to the troponin complex or by directly influencing the protein conformational changes occurring during cross-bridge activity. This second indirect regulatory mechanism shows a clear distinction between skeletal and cardiac muscle. The negative feedback induced by phosphorylation of TN-I upon β-adrenergic stimulation represents an evolutionary specialization of the cardiac tissue because skeletal TN-I lacks this phosphorylatable site in its primary structure.

In intact muscle, inotropic manipulations are more effective in cardiac than skeletal muscles, as the latter type usually works near its maximum performance when stimulated. However, CG still cause an increase in contraction force even in skeletal muscles when applied in situ at relatively high doses (SMULYAN and EICH, 1976). In the normally working cardiac muscle, between systole and diastole, the degree of activation oscillates but never reaches maximal activation or full relaxation. Therefore, in cardiac muscle, a much larger residual potential than in skeletal muscle always remains, which may respond to inotropic interventions. Furthermore, it is known that chronic pressure overload results in cardiac hypertrophy associated with defective mechanical performance and a myosin type of lower ATPase activity. In experiments on cats with chronic pressure overload, prophylactic treatment with digoxin did not impede the hypertrophic process but significantly attenuated the decline of mechanical function. In addition, the cardiac myosin isolated from the digitalis-treated animals exhibited normal ATPase activity comparable to that from nonhypertrophic control animals, while the enzyme activity of myosin from hearts of untreated hypertrophic cats was still 20% lower (Carey et al., 1978). These results are in agreement with the observation that the myosin may change its properties according to the functional state. Since the application of CG immediately elicits positive inotropism on the mechanically impaired hypertrophic hearts, one wonders whether myofibrillar preparations from such cardiac muscles would respond even more readily to the action of CG than those from normal striated muscles. The finding that the cardiac regulatory protein system seems to be more labile and more readily influenced by ions and cardioactive drugs would then indicate that these proteins may indeed play an important role in positive inotropism (Reddy and Honig, 1972).

As CG may act by more than one mechanism, it is attractive to speculate that they do affect the phosphorylation and Ca^{2+} binding properties of the regulatory proteins in the actin filament. However, since the degree of phosphorylation of myosin has also been shown to be in some way correlated with the state of muscle activity, the myosin filaments themselves might also be susceptible to the action of CG (Fig. 7). The present stage in fundamental research on contraction opens up the possibility for such questions to be examined. Hypotheses are, after all, to be proved or disproved.

Acknowledgements. We are indebted to Dr. J. G. Watterson for helpful discussions and reading of the manuscript. Financial support came from the Swiss National Science Foundation (grants 3.675.75 and 3.230.77).

References

Adelstein, R.S., Eisenberg, E.: Regulation and kinetics of the actin-myosin-ATP interaction. Ann. Rev. Biochem. *49*, 921–956 (1980)

Alpert, N.R., Hamrell, B.B., Mulieri, L.A.: Heart muscle mechanics. Ann. Rev. Physiol. *41*, 521–537 (1979)

Barany, K., Barany, M., Gillis, J.M., Kushmerick, M.J.: Phosphorylation-dephosphorylation of the 18,000-dalton light chain of myosin during the contraction-relaxation cycle of frog muscle. J. Biol. Chem. *254*, 3617–3623 (1979)

Baskin, S.I., Dutta, S., Marks, B.H.: The effects of diphenylhydantoin and potassium on the biological activity of ouabain in the guinea-pig heart. Br. J. Pharmacol. *47*, 85–96 (1973)

Beeler, G.W.: Ionic currents in cardiac muscle; a fromework for glycoside action. Fed. Proc. *36*, 2209–2213 (1977)

Bloom, S., Brady, A.J., Langer, G.A.: Calcium metabolism and active tension in mechanically disaggregated heart muscle. J. Mol. Cell. Cardiol. *6*, 137–148 (1974)

Boyer, P.D., Stokes, B.O., Wolcott, R.G., Degani, C.: Coupling of "high-energy" phosphate bond to energy transductions. Fed. Proc. *34*, 1711–1717 (1975)

Brinley, F.J., Scarpa, A., Tiffert, T.: The concentration of ionized magnesium in barnacle muscle fibres. J. Physiol. (Lond.) *266*, 545–565 (1977)

Carafoli, E., Crompton, M.: The regulation of intracellular calcium. Curr. Top. Membr. Transp. *10*, 151–216 (1978)

Carey, R.A., Bove, A.A., Coulson, R.L., Spann, J.F.: Normal cardiac myosin ATPase and mechanics in pressure overload with digitalis treatment. Am. J. Physiol. *234*, H253–H259 (1978)

Cohen, I., Daunt, J., Noble, D.: An analysis of the actions of low concentrations of ouabain on membrane currents in Purkinje fibers. J. Physiol. (Lond.) *260*, 75–103 (1976)

Cohen, P., Burchell, A., Foulkes, J.G., Cohen, T.W., Vanaman, T.C., Naim, A.C.: Identification of the Ca^{2+}-dependent modulator protein as the fourth subunit of rabbit skeletal muscle phosphorylase kinase. FEBS Lett. *92*, 287–293 (1978)

Collins, J.H.: Structure and evolution of troponin-C and related proteins. Symp. Soc. Exp. Biol. *30*, 303–334 (1976)

Coltart, J.: Significance of plasma concentration of digoxin in relation to the myocardial concentration of the drug. In: Cardiac glycosides. Bodem, G., Dengler, H.J. (eds.), pp. 135–158. Berlin, Heidelberg, New York: Springer 1978

Dalla Libera, L., Sartore, S., Schiaffino, S.: Comparative analysis of chicken atrial and ventricular myosins. Biochim. Biophys. Acta *581*, 283–294 (1979)

Diacono, J.: Suggestive evidence for the activation of an electrogenic sodium pump in stimulated rat atria: Apparent discrepancy between the pump inhibition and the positive inotropic response induced by ouabain. J. Mol. Cell. Cardiol. *11*, 5–30 (1979)

Dobson, J.G.: Protein kinase regulation of cardiac phosphorylase activity and contractility. Am. J. Physiol. *234*, H638–H645 (1978)

Donaldson, S.K.B., Best, P.M., Kerrick, W.G.: Characterization of the effects of Mg^{2+} on Ca^{2+}- and Sr^{2+}-activated tension generation of skinned rat cardiac fibers. J. Gen. Physiol. *71*, 645–655 (1978)

Drummond, G.I., Severson, D.L.: Cyclic nucleotides and cardiac function. Circ. Res. *44*, 145–153 (1979)

Dutta, S., Goswami, S., Lindower, J.O., Marks, B.H.: Subcellular distribution of digoxin-H^3 in isolated guinea-pig and rat hearts. J. Pharmacol. Exp. Ther. *170*, 318–325 (1968)

Ebashi, S., Endo, M.: Calcium ion and muscle contraction. Prog. Biophys. Mol. Biol. *18*, 123–183 (1968)

Eisenberg, E., Hill, T.L.: A cross-bridge model of muscle contraction. Prog. Biophys. Mol. Biol. *33*, 55–82 (1978)

Elzinga, M., Collins, J.H.: Amino acid sequence of a myosin fragment that contains SH-1, SH-2, and N-methylhistidine. Proc. Natl. Acad. Sci. USA *74*, 4281–4284 (1977)

Elzinga, M., Lu, R.C.: Comparative amino acid sequence studies on actins. In: Contractile systems in non-muscle tissues. Perry, S.V., Margreth, A., Adelstein, R.S. (eds.), pp. 23–37. Amsterdam, Oxford, New York: North-Holland 1976

England, P.J.: Studies on the phosphorylation of the inhibitory subunit of troponin during modification of contraction in perfused rat heart. Biochem. J. *160*, 295–304 (1976)

England, P.J., Ray, K.P., Hibberd, M.G., Jeacocke, S.A., Murray, K.J., Hollingworth, D.N.: The control of cardiac contractility by protein phosphorylation. Horm. Cell Regul. *3*, 99–114 (1979)

Ezrailson, E.G., Potter, J.D., Michael, L., Schwartz, A.: Positive inotropy induced by ouabain, by increased frequency, by X537A (RO2-2985), by calcium and by isoproterenol; the lack of correlation with phosphorylation of TN-I. J. Mol. Cell. Cardiol. *9*, 693–698 (1977)

Fabiato, A., Fabiato, F.: Calcium release from the sarcoplasmic reticulum. Circ. Res. *40*, 119–129 (1977)

Fabiato, A., Fabiato, F.: Calcium and cardiac excitation-contraction coupling. Ann. Rev. Physiol. *41*, 473–484 (1979)

Flink, I.L., Morkin, E., Elzinga, M.: Cyanogen bromide peptide from bovine cardiac myosin containing two essential thiols; evidence for sequence homology with skeletal myosin in the region of the active site. FEBS Lett. *84*, 261–265 (1977)

Flink, I.L., Rader, J.H., Banerjee, S.K., Morkin, E.: Atrial and ventricular cardiac myosins contain different heavy chain species. FEBS Lett. *94*, 125–130 (1978)
Fricke, U.: Myocardial activity of inhibitors of the Na^+-K^+-ATPase; differences in the mode of action and subcellular distribution pattern of N-ethylmaleimide and ouabain. Naunyn-Schmiedebergs Arch. Pharmacol. *303*, 197–204 (1978)
Fricke, U., Hollborn, U., Klaus, W.: Inotropic action, myocardial uptake, and subcellular distribution of ouabain, digoxin, and digitoxin in isolated rat hearts. Naunyn-Schmiedebergs Arch. Pharmacol. *288*, 195–214 (1975)
Fuchs, F.: Cooperative interactions between calcium-binding sites on glycerinated muscle fibers; the influence of cross-bridge attachment. Biochim. Biophys. Acta *462*, 314–322 (1977)
Gergely, J.: Excitation-contraction coupling: Cardiac muscle events in the myofilament. Fed. Proc. *35*, 1283–1287 (1976)
Gold, H., Cattel, M.: Mechanism of digitalis action in abolishing heart failure. Arch. Intern. Med. *65*, 263–278 (1940)
Gordon, M.S., Brown, A.L.: Myofibrillar adenosine triphosphatase activity of human heart tissue in congestive failure: Effects of ouabain and calcium. Circ. Res. *18*, 534–542 (1966)
Greeff, K.: Contraction and relaxation of heart muscle as influenced by cAMP, isoproterenol, glucagon, ouabain, and calcium. In: Myocardial failure. Riecker, G., Weber, A., Goodwin, J. (eds.), pp. 293–297. Berlin, Heidelberg, New York: Springer 1977
Haiech, J., Derancourt, J., Pechere, J.F., Demaille, J.G.: Magnesium and calcium binding to parvalbumins: Evidence for differences between parvalbumins and an explanation of their relaxing function. Biochemistry *18*, 2752–2758 (1979)
Herzig, J.W.: A cross-bridge model for inotropism as revealed by stiffness measurements in cardiac muscle. Basic Res. Cardiol. *73*, 273–286 (1978)
Hoh, J.F.Y., McGrath, P.A., Hale, P.T.: Electrophoretic analysis of multiple forms of rat cardiac myosin. Effects of hypophysectomy and thyroxine replacement. J. Mol. Cell. Cardiol. *10*, 1053–1076 (1977)
Hoh, J.F.Y., Yeoh, G.P.S., Thomas, M.A.W., Higginbottom, L.: Structural differences in the heavy chains of rat ventricular myosin isoenzymes. FEBS Lett. *97*, 330–334 (1979)
Holroyde, M.J., Potter, J.D., Solaro, R.J.: The calcium binding properties of phosphorylated and unphosphorylated cardiac and skeletal myosins. J. Biol. Chem. *254*, 6478–6483 (1979)
Honig, C.R., Reddy, Y.S.: Interaction of Ca^{2+} binding sites of troponin; significance for Ca^{2+} movements. Am. J. Physiol. *228*, 172–178 (1975)
Huddart, H.: The comparative structure and function of muscle. Oxford: Pergamon 1975
Huxley, A.F., Simmons, R.M.: Proposed mechanism of force generation in striated muscle. Nature *233*, 533–538 (1971)
Huxley, H.E.: The mechanism of muscular contraction. Science *164*, 1356–1366 (1969)
Jenny, E., Turina, M., Waser, P.G.: Über den Einfluß herzaktiver Glykoside auf physicochemische Eigenschaften von natürlichem und rekonstituiertem Actomyosin aus Kalbsherzmuskel. Helv. Physiol. Pharmacol. Acta *25*, 147–155 (1967)
Johnson, J.D., Charlton, S.C., Potter, J.D.: A fluorescence stopped flow analysis of Ca^{2+} exchange with troponin C. J. Biol. Chem. *254*, 3479–3502 (1979)
Johnson, J.D., Collins, J.H., Robertson, S.P., Potter, J.D.: A fluorescent probe study of Ca^{2+} binding to the Ca^{2+}-specific sites of cardiac and skeletal troponin C. J. Biol. Chem. *255*, 9635–9640 (1980)
Katz, A.M.: Contractile proteins of the heart. Physiol. Rev. *50*, 63–158 (1970)
Katz, A.M.: Role of the contractile proteins and sarcoplasmic reticulum in the response of the heart to catecholamines. An historical review. Adv. Cyclic Nucleotide Res. *11*, 303–343 (1979)
Kendrick-Jones, J., Jakes, R.: Myosin-linked regulation; a chemical approach. In: Myocardial failure. Piecker, G., Weber, A., Goodwin, J. (eds.), pp. 28–40. Berlin, Heidelberg, New York: Springer 1977
Kim, N.D., Bailey, L.E., Dresel, P.E.: Correlation of the subcellular distribution of digoxin with the positive inotropic effect. J. Pharmacol. Exp. Ther. *181*, 377–385 (1972)

Kopp, S.J., Barany, M.: Phosphorylation of the 19,000-dalton light chain of myosin in perfused rat heart under the influence of negative and positive inotropic agents. J. Biol. Chem. *254*, 12007–12012 (1979)
Kunz, P.A., Loth, K., Watterson, J.G., Schaub, M.C.: Nucleotide induced head-head interaction in myosin. J. Muscle Res. Cell. Mot. *1*, 15–30 (1980)
Kunz, P.A., Walser, J.T., Watterson, J.G., Schaub, M.C.: Isolation of cyanogen bromide and tryptic peptides containing the essential thiol groups from isolated myosin heads. FEBS Lett. *83*, 137–140 (1977)
LaBella, F.S., Bihler, I., Kim, R.S.: Progesterone derivative binds to cardiac ouabain receptor and shows dissociation between sodium pump inhibition and increased contractile force. Nature *278*, 571–573 (1979)
Langer, G.A., Frank, J.S., Nudd, L.M., Seraydarian, K.: Sialic acid; effect of removal on calcium exchangeability of cultured heart cells. Science *193*, 1013–1015 (1976)
Leavis, P.C., Kraft, E.L.: Calcium binding to cardiac troponin-C. Arch. Biochem. Biophys. *186*, 411–415 (1978)
Lee, K.S., Klaus, W.: The subcellular basis for the mechanism of inotropic action of cardiac glycosides. Pharmacol. Rev. *28*, 193–261 (1971)
Leger, J.J., Berson, G., Delcayre, C., Klotz, C., Schwartz, K., Leger, J., Stephens, M., Swynghedauw, B.: Heart contractile proteins. Biochimie *57*, 1249–1273 (1975)
Lehman, W.: Thick-filament-linked calcium regulation in vertebrate striated muscle. Nature *274*, 80–81 (1978)
Levine, B.A., Mercola, D., Thornton, J.M., Coffman, D.: Calcium binding by troponin-C; a protein magnetic resonance study. J. Mol. Biol. *115*, 743–760 (1977)
Lowey, S., Risby, D.: Light chains from fast and slow muscle myosins. Nature *234*, 81–85 (1971)
Lüllmann, H., Peters, T.: Action of cardiac glycosides on the excitation-contraction coupling in heart muscle. Prog. Pharmacol. *2*, 1–53 (1979)
Lymn, R.W., Taylor, E.W.: Mechanism of adenosine triphosphate hydrolysis by actomyosin. Biochemistry *10*, 4617–4624 (1971)
Mak, A., Smillie, L.B., Barany, M.: Specific phosphorylation at serine-283 of alpha-tropomyosin from rabbit skeletal and cardiac muscle. Proc. Natl. Acad. Sci. USA *75*, 3588–3592 (1978)
Malhotra, A., Huang, S., Bhan, A.: Subunit function in cardiac myosin: Effect of removal of LC-2 (18,000) on enzymic properties. Biochemistry *18*, 461–467 (1979)
Mannherz, H.G., Barrington-Leigh, J., Holmes, K.C., Rosenbaum, G.: Identification of the transitory complex myosin-ATP by the use of α,β-methylene-ATP. Nature [New Biol.] *241*, 226–229 (1973)
Marban, E., Rink, T.J., Tsien, R.W., Tsien, R.Y.: Free calcium in heart muscle at rest and during contraction measured with Ca^{2+}-sensitive microelectrodes. Nature *286*, 845–850 (1980)
Marston, S.B., Taylor, E.W.: Comparison of the myosin and actomyosin ATPase mechanisms of the four types of vertebrate muscles. J. Mol. Biol. *139*, 573–600 (1980)
Means, A.R., Dedman, J.R.: Calmodulin, an intracellular calcium receptor. Nature *285*, 73–77 (1980)
Mikawa, T.: "Freezing" of Ca-regulated conformation of reconstituted thin filament of skeletal muscle by glutaraldehyde. Nature *278*, 473–474 (1979)
Moir, A.J.G., Perry, S.V.: The sites of phosphorylation of rabbit cardiac troponin-I by adenosine 3′,5′-cyclic monophosphate-dependent protein kinase; effect of interaction with troponin-C. Biochem. J. *167*, 333–343 (1977)
Morkin, E.: Stimulation of cardiac adenosine triphosphatase in thyrotoxicosis. Circ. Res. *44*, 1–7 (1979)
Murthy, P.V., Kidwai, A.M., Daniel, E.E.: Dissociation of contractile effect and binding and inhibition of Na^+, K^+-ATPase by cardiac glycosides in rabbit myometrium. J. Pharmacol. Exp. Ther. *188*, 575–581 (1974)
Nayler, W.G.: Cyclic nucleotides and the heart. Adv. Drug Res. *12*, 39–51 (1977)
Okita, G.T.: Dissociation of Na^+, K^+-ATPase inhibition from digitalis inotropy. Fed. Proc. *36*, 2225–2230 (1977)

Pemrick, S.M.: Comparison of the calcium sensitivity of actomyosin from native and L2-deficient myosin. Biochemistry *16*, 4047–4054 (1977)

Perry, S.V.: Regulation of contraction in muscle. In: Contractile systems in non-muscle tissues. Perry, S.V., Margreth, A., Adelstein, R.S. (eds.), pp. 141–151. Amsterdam, Oxford, New York: North-Holland 1976

Perry, S.V.: The regulation of contractile activity in muscle. Biochem. Soc. Trans. *7*, 593–617 (1979)

Peters, T., Raben, R.H., Wassermann, O.: Evidence for a dissociation between positive inotropic effect and inhibition of the Na^+-K^--ATPase by ouabain, cassaine, and their alkylating derivatives. Eur. J. Pharmacol. *26*, 166–174 (1974)

Pfister, M., Schaub, M.C., Watterson, J.G., Knecht, M., Waser, P.G.: Radioactive labelling and location of specific thiol groups in myosin from fast, slow, and cardiac muscles. Biochim. Biophys. Acta *410*, 193–209 (1975)

Pires, E., Perry, S.V., Thomas, M.A.W.: Myosin light chain kinase, a new enzyme from striated muscle. FEBS Lett. *41*, 292–296 (1974)

Polimeni, P., Page, E.: Magnesium in heart muscle. Circ. Res. *33*, 367–374 (1973)

Potter, J.D., Gergely, J.: The calcium and magnesium binding sites on troponin and their role in the regulation of myofibrillar adenosine triphosphatase. J. Biol. Chem. *250*, 4628–4633 (1975)

Potter, J.D., Johnson, J.D., Dedman, J.R., Schreiber, W.E., Mandel, F., Jackson, R.L., Means, A.R.: Calcium-binding proteins; relationships of binding, structure, conformation, and biological function. In: Calcium-binding proteins and calcium function. Wasserman, R.H., Corradino, R.A., Carafoli, E., Kretsinger, R.H., Mac Lennan, D.H., Siegel, F.L. (eds.), pp. 239–250. New York: North-Holland 1977

Reddy, Y.S., Honig, C.R.: Ca^{2+}-binding and Ca^{2+}-sensitizing functions of cardiac native tropomyosin, troponin, and tropomyosin. Biochim. Biophys. Acta *275*, 453–463 (1972)

Rupp, H.: Modulation of tension generation at the myofibrillar level. An analysis of the effect of magnesium adenosine triphosphate, magnesium, pH, sarcomere length, and state of phosphorylation. Basic Res. Cardiol. *75*, 295–317 (1980)

Schaub, M.C., Perry, S.V., Häcker, W.: The regulatory proteins of the myofibril; characterization and biological activity of the calcium-sensitizing factor. Biochem. J. *126*, 237–249 (1972)

Schaub, M.C., Watterson, J.G.: Symmetry and asymmetry in the contractile protein myosin. Biochimie *63*, 291–299 (1981)

Schaub, M.C., Watterson, J.G., Waser, P.G.: Evidence for head-head interaction in myosin from cardiac and skeletal muscles. Basic Res. Cardiol. *72*, 124–132 (1977)

Schaub, M.C., Watterson, J.C., Walser, J.T., Waser, P.G.: Hydrolytically induced allosteric change in the heavy chain of intact myosin involving nonessential thiol groups. Biochemistry *17*, 246–253 (1978)

Scheuer, J., Bhan, A.K.: Cardiac contractile proteins. Adenosine triphosphatase activity and physiological function. Circ. Res. *45*, 1–12 (1979)

Scholey, M.J., Taylor, K.A., Kendrick-Jones, J.: The role of myosin light chains in regulating actin-myosin interaction. Biochimie *63*, 255–271 (1981)

Shukla, K.K., Ramirez, F., Marecek, J.F., Levy, H.M.: A mechanism for the hydrolysis of MgATP by actomyosin of skeletal muscle. J. Theor. Biol. *76*, 359–367 (1979)

Smillie, L.B.: Structure and function of tropomyosins from muscle and non-muscle sources. Trends Biochem. Sci. *4*, 151–155 (1979)

Smulyan, H., Eich, R.H.: Effect of digitalis on skeletal muscle in man. Am. J. Cardiol. *37*, 716–723 (1976)

Solaro, R.J., Pang, D.C., Briggs, F.N.: The purification of cardiac myofibrils with triton X-100. Biochim. Biophys. Acta *245*, 259–262 (1971)

Solaro, R.J., Shiner, J.S.: Modulation of Ca^{2+} control of dog and rabbit cardiac myofibrils by Mg^{2+}; comparison with rabbit skeletal myofibrils. Circ. Res. *39*, 8–14 (1976)

Stull, J.T., Blumenthal, D.K., Cooke, R.: Regulation of contraction by myosin phosphorylation. A comparison between smooth and skeletal muscles. Biochem. Pharmacol. *29*, 2537–2543 (1980)

Stull, J.T., Buss, J.E.: Calcium binding of beef cardiac troponin. J. Biol. Chem. *253*, 5932–5938 (1978)

Stull, J.T., High, C.W.: Phosphorylation of skeletal muscle contractile proteins in vivo. Biochem. Biophys. Res. Commun. *77*, 1078–1083 (1977)

Stull, J.T., Manning, D.R., High, C.W., Blumenthal, D.K.: Phosphorylation of contractile proteins in heart and skeletal muscle. Fed. Proc. *39*, 1552–1557 (1980)

Swinghedauw, B., Leger, J.J., Schwartz, K.: The myosin isozyme hypothesis in chronic heart overloading. J. Mol. Cell. Cardiol. *8*, 915–924 (1976)

Taylor, E.W.: Mechanism of actomyosin ATPase and the problem of muscle contraction. CRC Crit. Rev. Biochem. *6*, 103–164 (1979)

Taylor, R.S., Weeds, A.G.: The magnesium-ion-dependent adenosine triphosphatase of bovine cardiac myosin and its subfragment-1. Biochem. J. *159*, 301–315 (1976)

Trentham, D.R., Eccleston, J.F., Bagshaw, C.R.: Kinetic analysis of ATPase mechanisms.Q. Rev. Biophys. *9*, 217–281 (1976)

Van Eerd, J.P., Takahashi, K.: The amino acid sequence of bovine cardiac troponin-C; comparison with rabbit skeletal troponin-C. Biochem. Biophys. Res. Commun. *64*, 122–127 (1975)

Wagner, P.D., Weeds, A.G.: Studies on the role of myosin alkali light chains. J. Mol. Biol. *109*, 455–473 (1977)

Wakabayashi, T., Huxley, H.E., Amos, L.A., Klug, A.: Threedimensional image reconstruction of actin-tropomyosin complex and actin-tropomyosin-troponin T-troponin I complex. J. Mol. Biol. *93*, 477–497 (1975)

Walsh, M.P., Le Peuch, C.J., Vallet, B., Cavadore, J.C., Demaille, J.G.: Cardiac calmodulin and its role in the regulation of metabolism and contraction. J. Mol. Cell. Cardiol. *12*, 1091–1101 (1980)

Waser, P.G.: Über die Wirkung von Herzglykosiden auf Actomyosin. Cardiologia *29*, 214–229 (1956)

Waser, P.G.: Nachweis von ^{3}H-digitoxin in der Herzmuskelzelle. Experientia *18*, 35–36 (1962)

Waser, P.G.: Cardiac glycosides and actomyosin. In: New aspects of cardiac glycosides. Wilbrandt, W. (ed.), Vol. 3, pp. 173–184. Oxford, London, New York, Paris: Pergamon 1963

Watterson, J.G., Kohler, L., Schaub, M.C.: Evidence for two distinct affinities in the binding of divalent metal ions to myosin. J. Biol. Chem. *254*, 6470–6477 (1979)

Weingart, R., Kass, R.S., Tsien, R.W.: Is digitalis inotropy associated with enhanced slow inward calcium current? Nature *273*, 389–392 1978)

Wilkinson, J.M., Grand, R.J.A.: Comparison of amino acid sequence of troponin-I from different striated muscles. Nature *271*, 31–35 (1978)

Wilson, D.F., Brecinska, M., Sussman, I.: Control of energy flux in biological systems. Mosbach Colloq. *29*, 255–263 (1978)

Wyborny, L.E., Reddy, Y.S.: Phosphorylated cardiac myofibrils and their effect on ATPase activity. Biochem. Biophys. Res. Commun. *81*, 1175–1179 (1978)

CHAPTER 20

Substances Possessing Inotropic Properties Similar to Cardiac Glycosides

T. AKERA, A. L. FOX, and K. GREEFF

A. Introduction

The positive inotropic actions of the cardiotonic steroids and their glycosides can be distinguished from those of other positive inotropic agents, such as catecholamines or xanthine derivatives (see Chap. 10 and 11). The differences in the inotropic actions of these substances suggest differences in their mechanisms of action. Several substances, however, possess positive inotropic properties similar to those of cardiac glycosides. Comparative studies on these glycosides and aglycones indicate strict structural requirements for their cardiotonic actions (see Chap. 2). Considering that only a slight modification of the chemical structure results in a marked reduction in the inotropic potency, it is quite remarkable that substances of different chemical structure (Fig. 1) exhibit positive inotropic actions quite similar to those of cardiac glycosides. Some of these compounds inhibit, as does digitalis, cardiac Na^+, K^+-ATPase or the sodium pump, whereas others have no direct effect on this enzyme system. The relationships between effects of these substances on Na^+, K^+-ATPase activity, sodium pump activity, or transmembrane sodium fluxes and the force of myocardial contraction merit consideration.

B. Na^+, K^+-ATPase Inhibitors

The inhibition of the sarcolemmal Na^+, K^+-ATPase system has been implicated by several investigators as the mechanism for the positive inotropic action of cardiac glycosides (see Chap. 14). In support of this hypothesis, various Na^+, K^+-ATPase inhibitors have been shown to possess positive inotropic properties. It follows that all inhibitors of cardiac Na^+, K^+-ATPase should increase the force of myocardial contraction, unless this effect is offset by other actions of these agents.

I. Cassaine

Among various Na^+, K^+-ATPase inhibitors not belonging to the cardiotonic steroids or their derivatives, the erythrophleum alkaloids, particularly cassaine (Fig. 1), have been studied most extensively. Their digitalis-like cardiotonic and cardiotoxic actions have been noted for some time (MALING and KRAYER, 1946; KAHN et al., 1963; NIX et al., 1971). These alkaloids inhibit potassium transport in human erythrocytes (KAHN, 1962) and Na^+, K^+-ATPase activity in homogenates and partially purified enzyme preparations obtained from various animal tissues (REPKE and PORTIUS, 1963; BONTING et al., 1964; PETERS et al., 1974). The con-

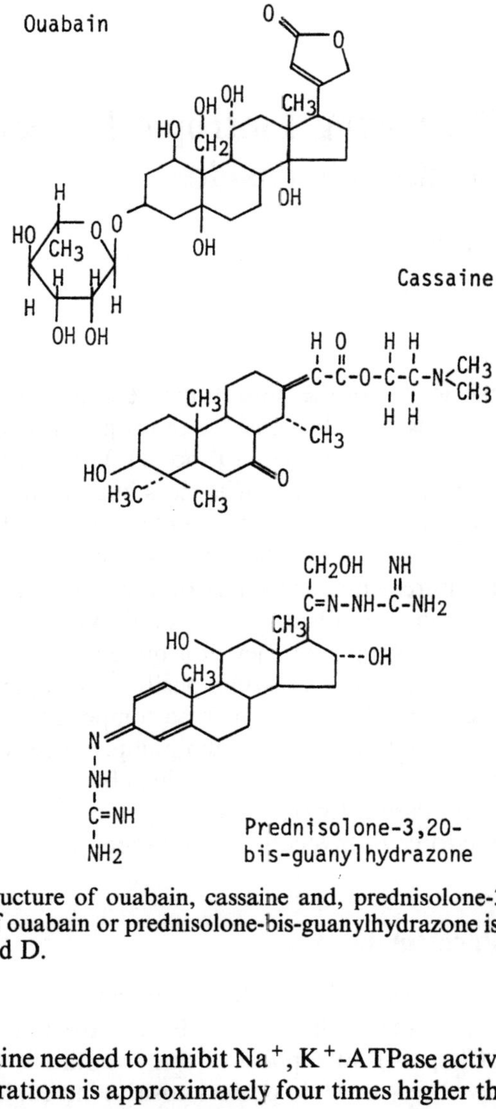

Fig. 1. Chemical structure of ouabain, cassaine and, prednisolone-3,20-bis-guanylhydrazone. Steroid ring of ouabain or prednisolone-bis-guanylhydrazone is labelled from the left to right; A, B, C and D.

centration of cassaine needed to inhibit Na^+, K^+-ATPase activity in partially purified enzyme preparations is approximately four times higher than that of ouabain; the apparent affinity of cassaine is therefore about one-quarter of that of ouabain (PETERS et al., 1974; TOBIN et al., 1975). In isolated guinea-pig atria, the positive inotropic action of cassaine develops and dissipates more rapidly than that of ouabain (PETERS et al., 1974; TOBIN et al., 1975). As such, the action of cassaine on Na^+, K^+-ATPase and the force of contraction resembles that of cardiotonic aglycones rather than glycosides (see Chap. 14).

In extremely high concentrations (greater than 0.1 m*M*), cassaine causes a poorly reversible inhibition of Na^+, K^+-ATPase isolated from rat brain (TOBIN et al., 1976). The development of this inhibition does not require the presence of ATP, unlike that by the cardiac glycosides or low concentrations of cassaine. The inhibition of Na^+, K^+-ATPase activity is associated with the inhibition of phosphorylation and ^3H-ouabain binding. It is not known, however, if the stable inhibition of Na^+, K^+-ATPase by high concentrations of cassaine is due to its binding to low affinity binding sites on the enzyme, or due to contaminants in cassaine preparations. This poorly reversible inhibition of the Na^+, K^+-ATPase by cassaine requires drug concentrations at least 100-fold higher than those needed for positive inotropic effects, and thus is unlikely to be involved in the cardiotonic actions of this drug.

At low concentrations (0.1–10 μ*M*), the inhibition of Na^+, K^+-ATPase by cassaine is readily reversible (TOBIN et al., 1975). The simultaneous presence of Mg^{2+} and inorganic phosphate seems to support the binding of cassaine to Na^+, K^+-ATPase in a manner similar to the binding of the cardiac glycosides to this enzyme, because inorganic phosphate enhances the inhibitory action of cassaine (TOBIN et al., 1975) or ouabain (see Chap. 14). With the exception of lower affinity for, faster binding to, and faster release from Na^+, K^+-ATPase cassaine appears to have characteristics indistinguishable from ouabain. These include: antagonism of binding by Na^+ in the presence of Mg^{2+} and Pi; antagonism of binding by K^+ in the presence of Na^+, Mg^{2+}, and ATP; stabilization of the phosphorylated intermediate of Na^+, K^+-ATPase; displacement of ATP-dependent ^{3}H-ouabain binding, and species-dependent differences in the affinity of Na^+, K^+-ATPase obtained from rat, guinea-pig, and dog heart (TOBIN et al., 1975).

Species-dependent differences in the affinity of cardiac Na^+, K^+-ATPase for ouabain result from differences in the stability of the ouabain–enzyme complex (see Chap. 14). Just as with cardiac glycosides, the loss of the positive inotropic effect of cassaine was faster in guinea-pig heart (Langendorff preparations) than in dog heart preparations (TOBIN et al., 1975). When the rate of washout of the positive inotropic effect was compared in the same species, it was faster with cassaine than with ouabain, reflecting a faster release of cassaine from Na^+, K^+-ATPase. These results indicate that cassaine binds to the cardiotonic steroid-binding site on Na^+, K^+-ATPase in micromolar concentrations, and inhibits the enzyme activity in a manner similar to ouabain. The binding of cassaine to Na^+, K^+-ATPase is hence associated with the positive inotropic action of this agent.

II. Prednisolone-bis-Guanylhydrazone (PBGH)

Corticosteroids which have a transconfiguration between the C and D rings of the steroid nucleus (Fig. 1) lack the digitalis-like positive inotropic action. When a positive inotropic effect is produced by corticosteroids, it apparently resembles that of catecholamines (PENEFSKY and KAHN, 1971).The addition of one or more guanylhydrazone groups to prednisone, prednisolone, or progesterone, however, makes the corticosteroid with a *trans* configuration more digitalis-like (cardiotonic) with a concomitant loss of the corticoid activity. These guanylhydrazone steroids produce digitalis-like positive inotropic effects in frog or guinea-pig heart preparations (EHMER et al., 1964; KRONEBERG and STOEPEL, 1964). In isolated perfused guinea-pig hearts (Langendorff preparations), the positive inotropic effect of these compounds is accompanied by a loss of potassium from the tissue, similar to strophanthin K (GREEFF et al., 1964). The active Na^+ and K^+ transport of erythrocytes (EHMER et al., 1964; GREEFF et al., 1964), and Na^+, K^+-ATPase activity of membrane fractions obtained from guinea-pig hearts (DRANSFELD and GREEFF, 1964) can be inhibited by these compounds. PBGH is more potent than prednisone-bis-guanylhydrazone or strophanthin K in its actions on Na^+, K^+-ATPase and the force of contraction, suggesting a close relationship between the inhibition of Na^+, K^+-ATPase or active Na^+ and K^+ transport and positive inotropic effects.

In cold-stored human erythrocytes, the concentration of strophanthin K needed to inhibit active Na^+ and K^+ transport is markedly lower than that required to inhibit ion transport in guinea-pig erythrocytes (GREEFF et al., 1964). In contrast, the concentrations of prednisolone and prednisone-bis-guanylhydrazone required to inhibit active Na^+ and K^+ transport are markedly higher in human erythrocytes than in guinea-pig erythrocytes. The potency difference between bis-guanylhydrazone derivatives of prednisolone and prednisone is maintained within each species (GREEFF et al., 1964), emphasizing the difference in the spectrum of species sensitivity of Na^+, K^+-ATPase to digitalis and guanylhydrazone steroids. Consistent with this, atrial preparations obtained from rat, guinea-pig, rabbit, dog, and human heart have remarkably different spectra of species sensitivity to the positive inotropic action of PBGH compared with digitalis (GREEFF and SCHLIEPER, 1967; YAMAMOTO et al., 1978 a). Guinea-pig heart was uniquely sensitive to PBGH, and there was no significant difference in sensitivities between rat, rabbit, and dog heart, although they do exhibit remarkable differences in sensitivity to the digitalis glycosides.

Guinea-pigs are uniquely sensitive to histamine (LEVI et al., 1976). The positive inotropic action of PBGH is, however, not mediated by a release of histamine or by the activation of the H_1 receptor, because a blocking concentration (1 μM) of tripelennamine failed to affect the positive inotropic effect of the prednisolone derivative (YAMAMOTO et al., 1978 a). Metiamide (a histamine H_2-antagonist), propranolol, or reserpine pretreatment also failed to affect the positive inotropic effect. The positive inotropic action of PBGH seems to involve Na^+, K^+-ATPase inhibition, since there is a good correlation between its potency to inhibit isolated cardiac Na^+, K^+-ATPase and that needed to produce a positive inotropic effect in several species (GREEFF and SCHLIEPER, 1967; YAMAMOTO et al., 1978 a). The high sensitivity of guinea-pig Na^+, K^+-ATPase to PBGH seems to be confined to cardiac muscle. Na^+, K^+-ATPase obtained from skeletal muscle exhibits a similar sensitivity to this compound in guinea-pig, rabbit, or rat (DRANSFELD et al., 1967). In rabbits, the Na^+, K^+-ATPase prepared from skeletal muscle is more sensitive to PBGH than that of heart muscle, which may explain the relaxation of skeletal muscle by PBGH at lower doses than those necessary to produce cardiac arrhythmias (BADURA, 1966).

The inhibition of Na^+, K^+-ATPase by PBGH is the result of its binding to the enzyme. It inhibits ^3H-ouabain binding to the enzyme and ouabain-sensitive ^{86}Rb uptake by thin slices of ventricular muscle (YAMAMOTO et al., 1978 a). The binding of PBGH and ouabain is mutually exclusive (YAMAMOTO, 1978; YAMAMOTO et al., 1978 a). Therefore, the binding sites for ouabain and PBGH have a strong interaction, although they do not appear to be identical, since (as we have already seen) the spectra of species sensitivities are markedly different. The binding of PBGH to Na^+, K^+-ATPase requires ATP in the presence of 100 mM NaCl and 5 mM $MgCl_2$ (Yamamoto et al., 1978 b), similar to the binding of the cardiac glycosides to this enzyme (see Chap. 14). Moreover, the binding of both cardiac glycosides and PBGH to Na^+, K^+-ATPase observed in vitro is inhibited in the presence of

K^+ (YAMAMOTO, 1978). These results suggest that these compounds probably bind to the same phosphorylated intermediate form of Na^+, K^+-ATPase.

Thus, PBGH increases the force of myocardial contraction in various animal species. The mechanism for the positive inotropic action of this compound seems to be similar to that of cardiac glycosides, although the binding sites on Na^+, K^+-ATPase are probably not identical.

III. Benzylaminodihydrodimethoxyimidazoisoquinoline Hydrochloride (BIIA)

An isoquinoline derivative, benzylaminodihydrodimethoxyimidazoisoquinoline hydrochloride (BIIA) (Fig. 2), was first described by SZEKERES et al. (1974) as an anti-arrhythmic agent with a positive inotropic activity.

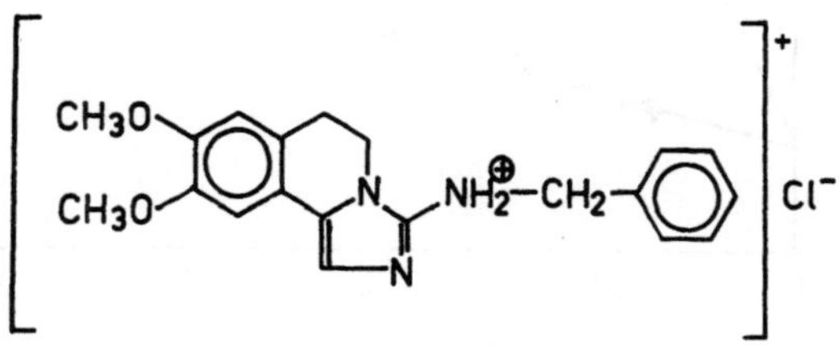

Fig. 2. Chemical structure of BIIA

BIIA has recently been reported to inhibit isolated Na^+, K^+-ATPase at relatively low concentrations (FOX, 1979; FOX and GREEFF, 1981). This compound possesses intriguing pharmacologic actions; it has a marked positive inotropic action on isolated guinea-pig atrial and papillary muscles in concentrations one order of magnitude lower than those for the inhibition of cardiac Na^+, K^+-ATPase isolated from the same species. BIIA decreases the frequency of spontaneously beating right atria, increases the effective refractory period and the threshold for arrhythmia and fibrillation. Higher doses of BIIA cause digitalis-like toxicity, i.e., a decrease in the S–T segment in the ECG in the anesthetized cat, followed by arrhythmias (BORCHARD et al., 1980). The intrinsic activity of BIIA is similar to that of cardiac glycosides, although its dose–response curve lies over an order of magnitude higher on the concentration scale. The positive inotropic action of BIIA does not appear to be mediated by catecholamine release. Broadening of the action potential and increase in resting potential in guinea-pig papillary muscle observed in the presence of BIIA seem to explain the anti-arrhythmic properties of this substance, whereas Na^+, K^+-ATPase inhibition might account for its digitalis-like toxicity (BORCHARD et al., 1980). There is, as yet, no electrophysiologic evidence, however, explaining the positive inotropic action of BIIA. The Na^+, K^+-ATPase inhibition is competitive with Na for enzyme phosphorylation (see Fig. 3), noncompetitive with ATP and potentiated by K^+, suggesting an intracellular site of action, where a low Na^+ concentration could potentiate binding and perhaps mediate the positive inotropic effect.

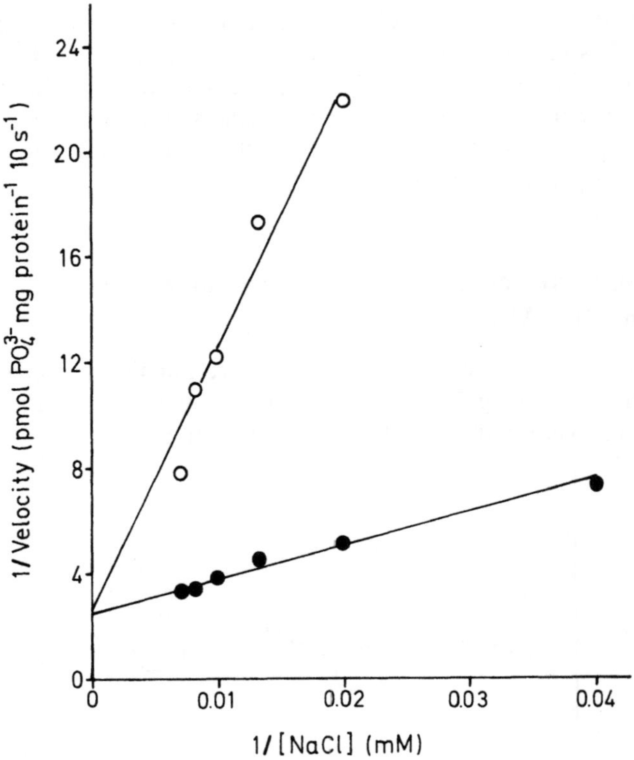

Fig. 3. Inhibition of the Na^+-dependent phosphorylation of Na^+, K^+-ATPase by BIIA; competitive with sodium ions. Guinea-pig kidney Na^+, K^+-ATPase phosphorylated at 0 °C in a medium containing (m*M*): Tris-HCl 50, Pi 0.5, and Na_2 EDTA 1, for 10 s with *(open circles)* and without *(solid circles)* 5.7 μ*M* BIIA and varying Na^+ concentrations. Results, plotted in the form of a Lineweaver–Burk plot, demonstrate competitive interaction of Na^+ ions and BIIA for this form of the enzyme. (FOX, 1979a)

IV. Sulfhydryl Blocking Agents

Isolated Na^+, K^+-ATPase can be inhibited by several sulfhydryl blocking agents, such as *N*-ethylmaleimide, p-chloromercuribenzoate, and p-chloromercuribenzenesulfonic acid (GLYNN, 1963; REPKE and PORTIUS, 1963; SKOU, 1963; SCHWARTZ and LASETER, 1964; SKOU and HILBERG, 1965; FAHN et al., 1966; FUJITA et al., 1968; JEAN and BADER, 1970). The positive inotropic action of *N*-ethylmaleimide was demonstrated with cat papillary muscles (BENNETT et al., 1958) and subsequently in guinea-pig heart, either pretreated with reserpine or in the presence of propranolol (FROM, 1970; YAMAMOTO et al., 1973; FRICKE, 1976, 1978). A time course study revealed the effect to have two phases with an initial transient phase reaching a peak at 2–3 min and a later one with a peak at 10–15 min (TEMMA et al., 1978). The initial phase is abolished by either reserpine pretreatment or in the presence of propranolol, and thus involves the drug-induced release of catecholamines from the sympathetic nerve terminals (Fig. 4). The second slower phase, however, is independent of this mechanism, and is, in fact, associated with an in-

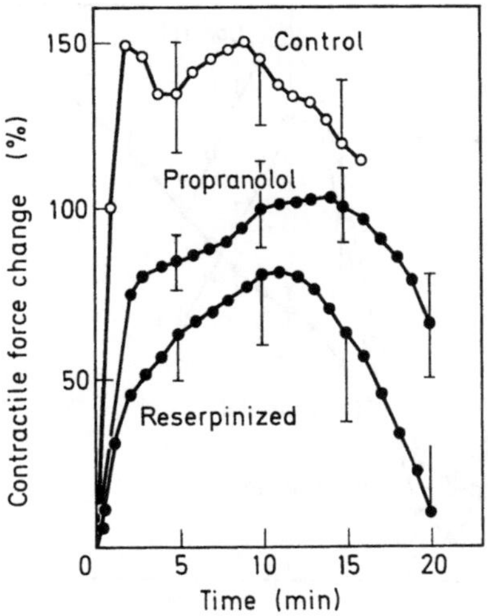

Fig. 4. Positive inotropic action of *N*-ethylmaleimide. Left atrial preparations of guinea-pig heart suspended in Krebs–Henseleit bicarbonate buffer solution (pH 7.4) at 30 °C and electrically stimulated at 2 Hz. After equilibration, 100 μ *M* *N*-ethylmaleimide added at time zero (control). P: 10 μ*M* (±)-propranolol added to the incubation mixture 20 min before addition of *N*-ethylmaleimide. R: animals pretreated with subcutaneous reserpine (5 mg/kg, 24 h before sacrifice). Diastolic tension 1.0 g. Developed tension prior to addition of *N*-ethylmaleimide: control, 1.00±0.06 g; P, 0.73±0.06 g; R, 0.92±0.11 g (mean ± standard error). Each point represents mean of 5 experiments. Vertical line indicates representative standard error of the mean. (TEMMA et al., 1978)

hibition of the ouabain-sensitive ^{86}Rb uptake, an estimate of sodium pump activity (TEMMA et al., 1978).

There is some disagreement as to whether *N*-ethylmaleimide produces its positive inotropic effects by inhibiting Na^+, K^+-ATPase. An early study, on isolated rabbit atrial preparations, indicated that *N*-ethylmaleimide had no effect on the force of contraction at 0.01 m*M* and inhibited it at 0.1 m*M* at 20 and 40 min after the drug addition (TODA and KONISHI, 1969). The reason these observations differ from those of others is not known. It should be noted that the inotropic action of *N*-ethylmaleimide is biphasic in guinea-pig heart; the initial positive effect being followed by a negative one, which becomes apparent after 20 min, probably due to the action of this compound on intracellular organelles (Fig. 4).

FRICKE (1978) observed a positive inotropic action of *N*-ethylmaleimide in papillary muscle preparations of guinea-pig heart. A half-maximal inotropic effect was observed at 9 μ*M*, whereas the half-maximal inhibition of Na^+, K^+-ATPase by this agent occurred at 1 m*M* in enzyme preparations isolated from the ventricular muscle. FRICKE concluded that the inhibition of Na^+, K^+-ATPase by *N*-ethylmaleimide does not account for its positive inotropic action. The magnitude of enzyme inhibition by this irreversible inhibitor, however, is affected by the length of incubation (TOBIN and AKERA, 1975) and the relative concentrations of sulfhydryl groups and inhibitor in the incubation mixture. The observation of FRICKE is possibly on that fraction of the inotropic effect produced by catecholamine release

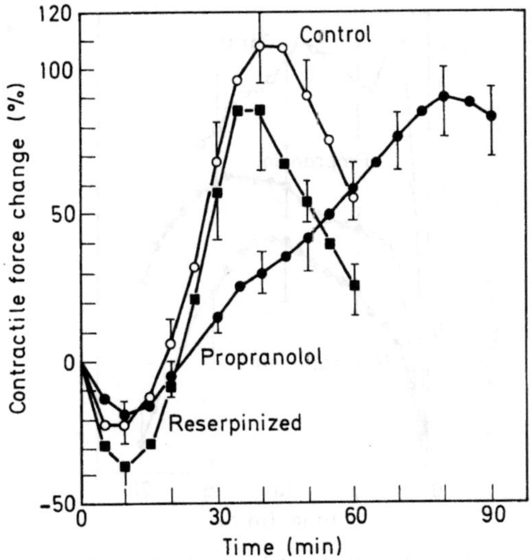

Fig. 5. Positive inotropic action of p-chloromercuribenzoate. See legend to Fig. 3. p-Chloromercuribenzoate (final concentration, 20 μ*M*) added at time zero instead of *N*-ethylmaleimide. Developed tension prior to addition of p-chloromercuribenzoate: control, 1.09 ± 0.10 g; P, 0.80 ± 0.11 g; R, 0.98 ± 0.13 g (mean ± standard error). (TEMMA et al., 1978)

(half-life of development of the inotropic effect, 2 min). The half-maximal concentration for the inotropic action may be further underestimated because of the negative inotropic action of *N*-ethylmaleimide at high concentrations (TEMMA et al., 1978). Thus, it is uncertain whether there is actually a marked difference between the concentration of *N*-ethylmaleimide causing Na^+, K^+-ATPase inhibition and that producing a positive inotropic effect. A moderate inhibition of ouabain-sensitive ^{86}Rb uptake was observed in thin slices of ventricular muscle prepared at a time when the positive inotropic effect was observed in isolated, perfused guinea-pig heart (TEMMA et al., 1978). As such, the sodium pump is inhibited at the time of the positive inotropic action of *N*-ethylmaleimide to a similar degree to that caused by inotropic concentrations of the cardiac glycosides.

Other sulfhydryl blocking agents, p-chloromercuribenzoate and p-chloromercuribenzenesulfonic acid, also inhibit Na^+, K^+-ATPase (GLYNN, 1963; FAHN et al., 1966; AKERA, 1971) and produce positive inotropic effects (YAMAMOTO, 1967; FROM and PROBSTFIELD, 1971; TEMMA et al., 1978). The positive inotropic effect is unaffected by reserpine pretreatment, indicating minimal involvement of an adrenergic mechanism (Fig. 5). Onset of the positive inotropic action is fastest with *N*-ethylmaleimide, slower with p-chloromercuribenzoate and slowest with p-chloromercuribenzenesulfonic acid (TEMMA et al., 1978), which seems to relate to the rate of penetration through the cell membranes (JACOB and JANDEL, 1962). In contrast the inorganic mercurial diuretic and sulfhydryl-reagent, mercuric chloride, has been found to induce a rapidly developing negative inotropic effect in guinea-pig papillary muscle in a concentration range of 0.1–4 m*M* (HALBACH, 1975), and also causes a strong inhibition of isolated Na^+, K^+-ATPase (TAYLOR, 1963). This negative inotropic effect is, however, interrupted by a transient increase in force which is apparently due to catecholamine release, since propranolol or reserpine pretreatment abolishes the transient positive inotropic effect without affecting the negative effect.

Propranolol does not reduce the magnitude of the positive inotropic effect of p-chloromercuribenzoate or p-chloromercuribenzenesulfonic acid. The development of the positive inotropic effect of these compounds, however, is markedly delayed by (±)-propranolol (Fig. 5) and also by (+)-propranolol (TEMMA et al., 1978). The latter form does not block β-adrenergic receptors but has a potency similar to the (−) isomer for a local anesthetic action (BARRETT and CULLUM, 1968). Thus, this delay seems to result from stabilization of the cell membrane, indicating that the access of sulfhydryl blocking agents to the receptor sites involves membrane penetration.

Several investigators demonstrated that chlorpromazine and related phenothiazine derivatives in relatively high concentrations inhibit isolated Na^+, K^+-ATPase (JÄRNEFELT, 1962; JUDAH and AHMED, 1964; SQUIRES, 1965; DAVIS and BRODY, 1966; ROBINSON et al., 1968; PALATINI, 1978). A semiquinone free-radical form of these compounds, produced by either photo-oxidation or chemical or enzymatic oxidation, is a more potent inhibitor of the enzyme (AKERA and BRODY, 1968, 1969, 1970). The inhibition involves sulfhydryl groups on the enzyme. Although the chlorpromazine free-radical is extremely unstable, dihydroxy derivatives of chlorpromazine and related compounds form stable free-radicals, and inhibit Na^+, K^+-ATPase in low concentrations (BRODY et al., 1974). These compounds increase the force of myocardial contraction in isolated left atrial preparations of guinea-pig heart (AKERA et al., 1978). Detailed analyses, however, revealed that the characteristics of the positive inotropic actions of hydroxylated derivatives of chlorpromazine are similar to those induced by catecholamines rather than those induced by digitalis (TEMMA et al., 1977). The positive inotropic effect of 7,8-dihydroxychlorpromazine is blocked by reserpine pretreatment or by propranolol. It is accompanied by an increased maximal rate of depolarization, a prolongation of the action potential duration, and an increase in cellular cyclic AMP concentrations. Inotropic concentrations of 7,8-dihydroxychlorpromazine increase the release of preloaded metaraminol from sympathetic nerve terminals in vitro, but fail to inhibit ouabain-sensitive ^{86}Rb uptake in ventricular slices. Thus, this potent sulfhydryl inhibitor cannot penetrate the cell membrane and gain access to the essential sulfhydryl groups on Na^+, K^+-ATPase when applied extracellularly.

Ethacrynic acid, a diuretic which is capable of inhibiting sulfhydryl groups, inhibits isolated Na^+, K^+-ATPase (DUGGAN and NOLL, 1965; CHARNOCK et al., 1970; DAVIS, 1970) and active Na^+ and K^+ transport in resealed human erythrocyte ghosts (ASKARI and RAO, 1970) and in isolated frog muscle (RIORDAN et al., 1972). The inhibition of isolated Na^+, K^+-ATPase by this agent is somewhat different from that by other sulfhydryl inhibitors, because the inhibitory effect of ethacrynic acid is apparently enhanced, rather than reduced, by Na^+ and ATP (BANERJEE et al., 1971). Nevertheless, this compound produces positive inotropic effects in isolated guinea-pig atrial preparations (FROM et al., 1975). This effect is partially reduced by reserpine pretreatment or in the presence of propranolol, but a significant component of the inotropic effect is independent of the β-adrenergic mechanism. Although ethacrynic acid inhibits Na^+, K^+-ATPase irreversibly

(BANERJEE et al., 1970), inhibition of enzyme activity at the time of its inotropic action has not been reported. The action of ethacrynic acid to inhibit Na^+, K^+-ATPase and to increase the force of myocardial contraction might be complicated by the effect of this agent on passive transmembrane sodium flux (HERRERA, 1975; LAW, 1976).

V. Monovalent Cations

The rate of ATP hydrolysis by Na^+, K^+-ATPase observed in vitro in the presence of Mg^{2+}, Na^+, and ATP is stimulated by NH_4^+, K^+, Rb^+, or Cs^+ (SKOU, 1960). However, Rb^+ causes a partial inhibition of the enzyme activity when the enzyme is fully activated in the presence of Mg^{2+}, Na^+, K^+, and ATP (POST et al., 1972), because Rb^+ has a lower intrinsic activity than K^+ to activate the enzyme in the presence of Na^+, Mg^{2+}, and ATP (REPKE and PORTIUS, 1963). Positive inotropic effects of this Na^+, K^+-ATPase inhibitor have been reported (PRASAD and MIDHA, 1972; KU et al., 1974, 1975). Thallous ion also substitutes for K^+ in stimulating ATPase activity observed in the presence of Mg^{2+}, Na^+, and ATP (GEHRING and HAMMOND, 1967; BRITTEN and BLANK, 1968). In human erythrocytes, Tl^+ inhibits Na^+, K^+-ATPase and ouabain-sensitive sodium transport in the presence of Na^+ and K^+ (SKULSKII et al., 1975). In Na^+, K^+-ATPase preparations obtained from guinea-pig heart, Rb^+ and Tl^+ both inhibit the enzyme activity in concentrations less than 5 mM, whereas Cs^+, NH_4^+, and Li^+ have no effect or may slightly stimulate enzyme activity, when assays are performed in the presence of Na^+ and K^+ (KU et al., 1975). In isolated left atrial preparations, these concentrations of Rb^+ and Tl^+ produce sustained positive inotropic effects, whereas Cs^+ fails to alter the force of contraction significantly and NH_4^+ produced a transient inotropic effect probably due to a catecholamine release (Fig. 6). The positive inotropic effects produced by Rb^+ or Tl^+ are insensitive to propranolol, as is that of digitalis. Concentrations of Tl^+ and cardiac glycosides which produce similar inotropic effects cause a similar degree of sodium pump inhibition (KU et al., 1975). In isolated atrial preparations, concentrations of Rb^+ or Tl^+ needed to produce equivalent positive inotropic effects and the same degree of cardiac Na^+, K^+-ATPase inhibition in vitro are comparable in rats and guinea-pigs (KU et al., 1976). Neither of these cations shows the species difference exhibited by digitalis with respect to either enzyme inhibition or positive inotropic effect. Thus, among monovalent cations, only Na^+, K^+-ATPase inhibitors produce a sustained positive inotropic effect (Fig. 6). The inotropic effect seems to be related to the enzyme inhibition.

The effects of Li^+ on Na^+, K^+-ATPase are distinctly different from those of other monovalent cations. Seemingly, Li^+ is capable of substituting for both Na^+ and K^+ (WILLIS and FANG, 1970), although Li^+ alone cannot fully activate ATPase activity in the presence of Mg^{2+} and ATP (SKOU, 1960). This cation stimulates ATPase activity in vitro in the presence of Na^+ and K^+ by stimulating the dephosphorylation reaction and dissociating rapidly from the enzyme to allow rephosphorylation and hence increasing turnover of the enzyme (TOBIN et al., 1974b; HAN et al., 1976). Despite its ability to stimulate isolated Na^+, K^+-ATPase (WILLIS and FANG, 1970; GUTMAN et al., 1973), Li^+ inhibits the sodium pump in renal cortical slices of the ground squirrel (WILLIS and FANG, 1970), desheathed rat

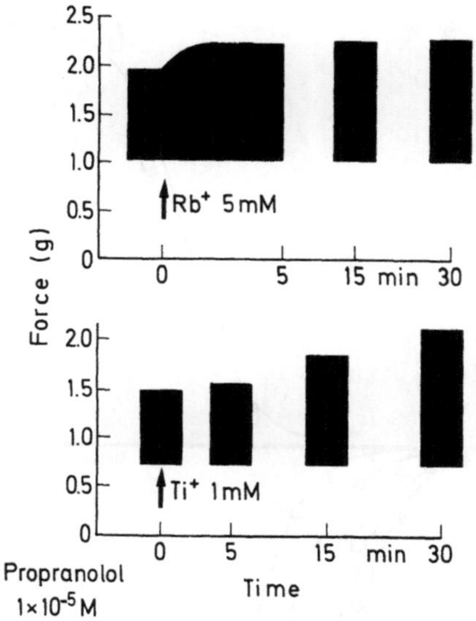

Fig. 6. Effects of rubidium and thallous ions on the force of myocardial contraction. Left atrial preparations of guinea-pig heart suspended in Krebs–Henseleit bicarbonate buffer solution (pH, 7.4) at 30 °C and electrically stimulated at 1 Hz. The indicated final ionic concentrations of various ions were added to the incubation medium. For Tl^+, thallous acetate was used since TlCl is relatively insoluble in the buffer solution. Medium for Tl^+ experiments contains 10 μ*M* (±)-propranolol. (AKERA et al., 1976a)

vagus nerve (PLOEGER, 1974), and in isolated guinea-pig heart (KU et al., 1978). In the isolated papillary muscle of the cat heart (PRINDLE et al., 1970), or the atrial preparation of guinea-pig heart (KU et al., 1975), relatively high concentrations (10–30 m*M*) of Li^+ produce positive inotropic effects, associated with inhibition of sodium pump activity (KU et al., 1975, 1978). The positive inotropic effect is observed in the presence of propranolol, indicating that the effect is independent of β-adrenergic mechanism. Therefore, it appears that sodium pump inhibition, rather than Na^+, K^+-ATPase inhibition per se, determines the force of myocardial contraction. Alternatively, Li^+ may actually be inhibiting Na^+, K^+-ATPase at high concentrations (GUTMAN et al., 1973).

VI. Vanadate

Vanadate (VO_3^-) has recently been shown to be a potent inhibitor of isolated Na^+, K^+-ATPase (CANTLEY et al., 1977). The inhibition is enhanced by Mg^{2+} and K^+, and antagonized by Na^+ and ATP, suggesting that vanadate interacts with the dephosphoenzyme (BEAUGE and GLYNN, 1978; NECHAY and SAUNDERS, 1978; QUIST and HOKIN, 1978). This essential trace nutrient (HOPKINS and MOHR, 1974) reacts with Na^+, K^+-ATPase in erythrocyte ghosts from the inner side of the membrane (CANTLEY et al., 1978b) and is present in muscles in sufficiently high concentrations to cause a substantial enzyme inhibition if it is in active form and in contact with the enzyme (CANTLEY et al., 1977). This inhibitor is of particular interest since

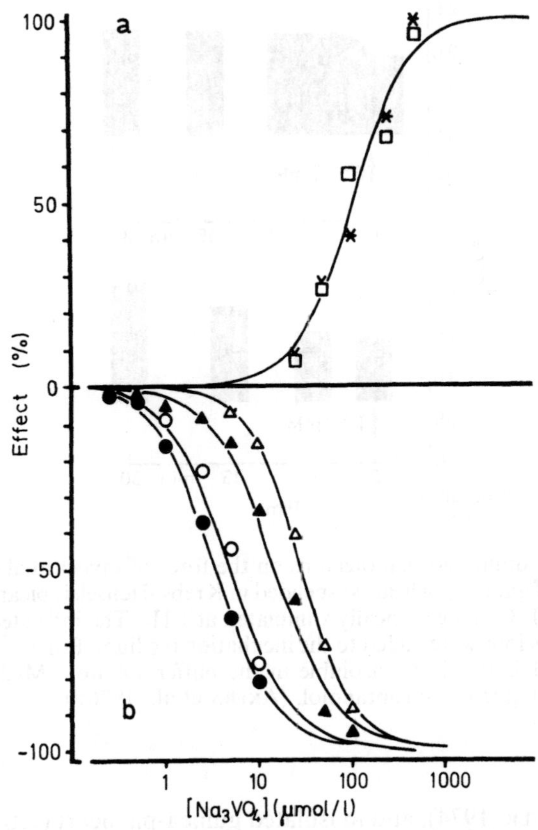

Fig. 7a, b. Change in force of contraction in **a** papillary muscle and **b** atria. Median effective doses (µmol/l): papillary muscle, guinea-pig *(squares)* 110.3 ± 1.9, cat *(asterisks)* 124.1 ± 2.6; left atria, guinea-pig *(open circles)* 4.73 ± 0.10, cat *(open triangles)* 28.1 ± 0.47; right atria, guinea-pig *(solid circles)* 3.29 ± 0.06, cat *(solid triangles)* 14.8 ± 0.28. (BORCHARD et al., 1979)

it may have a physiologic role in regulating the enzyme activity (CANTLEY et al., 1977; HUDGINS and BOND, 1977; BEAUGE and GLYNN, 1977).

Vanadate has been shown to produce a positive inotropic effect in cat papillary muscle preparations (HACKBARTH et al., 1978). Subsequently, however, BORCHARD et al. (1979) and GRUPP et al. (1979a), demonstrated that vanadate exerts, paradoxically, a negative inotropic effect on isolated guinea-pig and bovine atria, although it has a positive inotropic effect on papillary muscle preparations from the same animals (Fig. 7). These actions correlate better with electrophysiologic changes, i.e., shortening of the action potential in atrial tissue and broadening of that in papillary muscle, rather than with an inhibition of Na^+, K^+-ATPase; isolated enzymes of atrial and ventricular origin being equally sensitive to vanadate (ID_{50} in μM: bovine atria 0.59, ventricle 0.60; guinea-pig atria 0.75, ventricle 0.62) (BORCHARD et al., 1979). It is therefore once again evident that the positive inotropic action of a Na^+, K^+-ATPase inhibitor, in this case vanadate, is not necessarily related to the inhibition of this enzyme.

Concentrations of vanadate necessary to produce a positive inotropic effect in bovine ventricular papillary muscle (50–500 μ*M*) are two to three orders of magnitude higher than those required to inhibit isolated Na^+, K^+-ATPase (BORCHARD et al., 1979), but correspond to 55%–80% inhibition of ^{86}Rb uptake in erythrocytes (CANTLEY et al., 1978 b). AKERA et al. (1979) reported that vanadate failed to inhibit ouabain-sensitive ^{86}Rb uptake in electrically stimulated guinea-pig atria in concentrations that would inhibit isolated Na^+, K^+-ATPase. It was suggested therefore that the inhibitory sites, probably located at the internal surface of the cell membrane, are normally inaccessible to vanadate in intact tissue. Radioactive isotope studies using ^{48}V, however, demonstrate that vanadate equilibrates rapidly across erythrocyte membranes, followed by an accumulation process with a slower kinetic (CANTLEY et al., 1978 b). WERDAN et al. (1979) also reported vanadate accumulation in cultivated rat heart muscle cells. Thus, it is suggested that vanadate accumulates in the cells in a form incapable of inhibiting Na^+, K^+-ATPase (CANTLEY et al., 1978 b; POST, 1979).

SCHWABE et al. (1979) reported that vanadate stimulates adenylate cyclase in membrane preparations isolated from rat fat cells, producing maximal activation even in the absence of catecholamine. In preparations obtained from guinea-pig and rat heart, 100 μ*M* vanadate caused 100% and 50% activation of adenylate cyclase, respectively (GRUPP et al., 1979 b). There was no difference, however, between the atrial and ventricular preparations. Moreover, subsequent investigations by SCHMITZ et al. (1979) on intracellular cyclic AMP concentrations during the positive inotropic action of 300 μ*M* vanadate indicated only a 20% increase in cat papillary muscle. Thus, the role of adenylate cyclase stimulation in the positive inotropic action of vanadate is presently unknown. None of the effects of vanadate were significantly diminished in the presence of propranolol or after reserpine pretreatment, excluding the involvement of β-adrenergic mechanism.

The inhibitory action of vanadate on the Na^+, K^+-ATPase is reversed in vitro by high concentrations of certain catecholamines which complex and reduce vanadium (CANTLEY et al., 1978 a; HUDGINS and BOND, 1979). The significance of this interaction in vivo has not yet been investigated; a possible regulatory function of vanadate in catecholamine metabolism and vice versa could exist and contribute to the pharmacologic actions of these substances.

VII. Other Na^+, K^+-ATPase Inhibitor

Among various anions, fluoride (F^-) has been shown to inhibit isolated Na^+, K^+-ATPase (OPIT et al., 1966; YOSHIDA et al., 1968; LAHIRI and WILSON, 1971; PENZOTTI and TITUS, 1972). The inhibition is enhanced by K^+ and antagonized by Na^+ and ATP. In the presence of physiologic concentrations of Na^+ and ATP, however, F^- is still capable of inhibiting Na^+, K^+-ATPase (TOBIN et al., 1974 a). A 50% inhibition of isolated Na^+, K^+-ATPase occurs in the presence of 1–4 m*M* F^-. These concentrations of F^- produce positive inotropic effects in isolated heart preparations (LOEWI, 1955; KATZUNG et al., 1957; COVIN and BERMAN, 1959). It is not known, however, whether the inotropic action of F^- involves sarcolemmal Na^+, K^+-ATPase inhibition, or some other mechanism especially since, quite unlike cardiac glycosides, F^- has been shown to increase the time to peak tension

(REITER, 1965). Neither is it clear whether F^- gains access to the site of action on Na^+, K^+-ATPase in intact cells. Despite a relatively stable inhibition of Na^+, K^+-ATPase by F^- (YOSHIDA et al., 1968; TOBIN et al., 1974a), no attempt has been made to determine whether sarcolemmal Na^+, K^+-ATPase is inhibited by this anion when the positive inotropic effect is observed.

Relatively low concentrations of fluoride-containing metabolic inhibitors, such as fluoroacetate, methyl fluoroacetate, and fluorobutylate, produce positive inotropic effects in isolated hearts (CHENOWETH and PENGSRITONG, 1950; BENNETT and CHENOWETH, 1951; REILLY et al., 1953; KATZUNG et al., 1957), although higher concentrations of these agents and other metabolic inhibitors decrease the force of myocardial contraction, presumably owing to an energy depletion (LEE, 1954; KATZUNG et al., 1957). In the presence of a high concentration of pyruvate to overcome the metabolic inhibition, the positive inotropic effect of methyl fluoroacetate is enhanced, indicating that the effect is not a consequence of inhibition of cellular metabolism (REILLY et al., 1953). After a long exposure to fluoroacetate (0.1–10 m*M*), the force of contraction of guinea-pig papillary muscle declines below the control value as expected with a metabolic inhibitor (KORTH et al., 1978). However, a marked positive inotropic effect is observed with these concentrations of fluoroacetate prior to the development of the negative inotropic effect; the latter effect becoming apparent after 50–120 min exposure. The positive inotropic effect of fluoroacetate resembles that of the cardiac glycosides, and seems to result from an inhibition of a sodium extrusion mechanism (KORTH et al., 1978), although it is not known whether Na^+, K^+-ATPase or sodium pump activity is actually inhibited in papillary muscle preparations when the inotropic effect is observed.

Doxorubicin (Adriamycin) is a potent antitumor antibiotic which has limited clinical value because of its unusual and potentially lethal cardiac toxicity. The toxic action may involve lipid peroxidation (MYERS et al., 1977). This agent produces positive inotropic effects in isolated rabbit heart (KOBAYASHI et al., 1972; VAN BOXTEL et al., 1978; GOSALVEZ et al., 1979). The inotropic effect is characterized by increases in the rate of tension development and peak developed tension. Since time to peak tension is increased, it is unlikely that the action of this compound involves adrenergic mechanisms. Doxorubicin fails to alter the myocardial response to isoproterenol, but significantly shifts the dose–response curve for ouabain to the right. VAN BOXTEL et al. (1978) suggest that doxorubicin and ouabain have similar mechanisms of action, since doxorubicin binds to and inhibits isolated Na^+, K^+-ATPase (GOSALVEZ et al., 1979). The properties of its positive inotropic effect, such as an increase in time to peak tension, however, are different from those of the effect of cardiac glycosides. Whether Na^+, K^+-ATPase or sodium pump activity is inhibited in isolated heart preparations exposed to doxorubicin is not known.

Sanguinarine, an alkaloid from *Sanguinaria canadensis*, produces a positive inotropic effect in isolated guinea-pig atria (SEIFEN and STRAUB, 1974), at concentrations between 1 and 100 μ*M*, similar to those which inhibit Na^+, K^+-ATPase isolated from guinea-pig heart or brain (I_{50} = 10 μ*M*). This inotropic effect is not modified by 1 μ*M* propranolol. Chelidione, a tertiary analog of sanguinarine (sanguinarine is a quarternary compound), produces a small negative inotropic effect and does not inhibit Na^+, K^+-ATPase. Thus, the positive inotropic action of sanguinarine has some features in common with cardiac glycosides. An irreversible inhibition of bovine cardiac Na^+, K^+-ATPase by sanguinarine is associated with a reduction in ouabain binding (PITTS and MEYERSON, 1978). Sanguinarine, however, has in contrast to cardiac glycosides, a positive chronotropic effect in isolated guinea-pig atria (SEIFEN and STRAUB, 1974).

Divalent cations such as Zn^{2+}, Cu^{2+}, Fe^{2+}, Mn^{2+}, and Pb^{2+} inhibit isolated Na^+, K^+-ATPase of various origin (DONALDSON et al., 1971; HEXUM, 1974; SIEGEL and FOGT, 1977). The relationship between divalent cation-induced inhibition of cardiac Na^+, K^+-ATPase and the force of contraction, or if these cations inhibit Na^+, K^+-ATPase in beating heart muscle at all, is presently unknown. Calcium ion is an inhibitor of isolated Na^+, K^+-ATPase (SKOU, 1957; REPKE and PORTIUS, 1963; SOMOGYI, 1964; EPSTEIN and WHITTAM, 1966; BLOSTEIN and BURT, 1971). The inhibition is due to a reduced rate of the phosphorylation reaction and the inhibition of the conversion of ATP-sensitive phosphoenzyme to the K^+-sensitive phosphoenzyme (TOBIN et al., 1973). A 50% inhibition of Na^+, K^+-ATPase by Ca^{2+} is observed in the presence of 0.5 m*M* Ca^{2+}. Whether sarcolemmal Na^+, K^+-ATPase is inhibited by Ca^{2+} in functioning cardiac muscle is an interesting but as yet unresolved question. Since the inotropic effect of digitalis is diminished in the presence of high extracellular Ca^{2+} concentrations (LEE et al., 1961; LÜLLMANN and HOLLAND, 1962), REPKE (1964) proposed that an increased extracellular Ca^{2+} concentration causes an inhibition of the sodium pump in a manner similar to the cardiac glycosides, thus making the pump insensitive to further actions of the inotropic agents. Alternatively, the force of myocardial contraction is already at the limit in the presence of high concentrations of Ca^{2+} and therefore might be insensitive to most positive inotropic interventions.

Several agents and factors that inhibit Na^+, K^+-ATPase fail to increase the force of myocardial contraction. Among them is oligomycin, which inhibits Na^+, K^+-ATPase obtained from various sources including cardiac muscle (JÄRNEFELT, 1962; GLYNN, 1963; JÖBSIS and VREMAN, 1963; VAN GRONINGEN and SLATER, 1963; WHITTAM et al., 1964; MATSUI and SCHWARTZ, 1966). This compound, owing to its high affinity for the mitochondrial oxidative phosphorylation mechanism, may deplete an energy source required for contraction before significant consequences of Na^+, K^+-ATPase inhibition on the force of myocardial contraction can be observed.

Though not all Na^+, K^+-ATPase inhibitors produce positive inotropic effects, most however, have biochemical effects other than Na^+, K^+-ATPase inhibition. These effects may offset the influence of enzyme inhibition on the force of myocardial contraction. Several other Na^+, K^+-ATPase inhibitors such as mercurial diuretics, ethacrynic acid, diphenylhydantoin, oligomycin, hydroxylamine, chlorpromazine, fusidic acid, dimethylsulfoxide, phlorizin, ethanol, sulfhydryl reagents, suramin, and adrenal steroids are reviewed by SCHWARTZ et al. (1975). Na^+, K^+-ATPase inhibition by short-chain fatty acids (DAHL, 1968), long-chain fatty acids (MILLER and WOODHOUSE, 1977), snake venom cardiotoxin (VINCENT et al., 1976), dichlorodiphenyltrichloroethane, and polychlorinated biphenyls (SHARP et al., 1974), anesthetics (SONG and SCHEUER, 1968), rubratoxin B (PHILLIPS et al., 1978), the anti-anginal drug amidarone (BROEKJUYSEN et al., 1972), quinidine (KENNEDY and NAYLER, 1965; LOWRY et al., 1973; BESCH and WATANABE, 1977), and Δ^9-tetrahydrocannabinol (TORO-GOYCO et al., 1978) has been reported. Among them, cardiotoxin has been shown to produce a transient positive inotropic effect in frog heart and rat atria (LEE et al., 1968). Some cardiovascular effects of amidarone (see CHARLIER, 1971) have been described suggesting positive inotropic properties of this compound. The positive inotropic action of dimethylsulfoxide in isolated guinea-pig atria has been reported (SHLAFER et al., 1974). The relationship between enzyme inhibition and positive inotropic action of these latter agents is poorly understood.

Na^+, K^+-ATPase is an enzyme system which has a relatively high temperature coefficient, presumably because conformational transitions between at least two forms are required for the transport of Na^+ and K^+, and a concomitant hydrolysis of ATP (REPKE and PORTIUS, 1963; SWANSON, 1966; ALBERS et al., 1968; POST et al., 1969; CHARNOCK et al., 1971). Consequently, the enzyme activity decreases precipitously as the temperature is lowered. Since the force of myocardial contraction is greater and the inotropic action of cardiac glycosides is reduced at lower temperatures (SAUNDERS and SANYAL, 1958; TUTTLE et al., 1962), lower Na^+, K^+-ATPase activity is implicated in the increased force of contraction (REPKE, 1964). Calcium pump activity and the rate of sodium influx, however, may also be altered at low temperatures. The balance between sodium influx and the capacity of the sodium pump is probably more important than the sodium pump activity per se in determining the force of myocardial contraction (AKERA, 1977). Thus, when these parameters are possibly affected as the result of drug treatment or alterations in factors such as temperature, one should carefully evaluate these parameters in beating hearts, in order to assess the importance of these changes in the altered inotropic state of the cardiac muscle.

C. Substances that Enhance Sodium Influx

The rate of sodium entry into the myocardial cells can be increased by several agents which appear to act through different mechanisms. A veratrum alkaloid, veratridine, opens the sodium channel which is sensitive to relatively high concentrations of tetrodotoxin in squid and crayfish giant axons (OHTA et al., 1973). This and other veratrum alkaloids including germitrine appear to delay the inactivation of the fast sodium channel, and enhance sodium influx associated with membrane excitation in frog and guinea-pig hearts (HORACKOVA and VASSORT, 1974; HONERJÄGER and REITER, 1975, 1977a). Grayanotoxins (andromedotoxins) increase the rate of sodium influx in squid axons (SEYAMA and NARAHASHI, 1973; NARAHASHI and SEYAMA, 1974) and in guinea-pig cardiac muscle (KU et al., 1977). Although the effect of grayanotoxins on the electrical properties of rat skeletal muscle fibers, i.e., sustained after depolarization (SEYAMA, 1970), is quite similar to that of veratridine or germitrine, and although the effect of grayanotoxins on sodium permeability of squid axons is reversed by relatively high concentrations of tetrodotoxin (SEYAMA and NARAHASHI, 1973), it has been proposed that grayanotoxins increase the rate of sodium influx by increasing resting sodium permeability in squid axons (SEYAMA and NARAHASHI, 1973; NARAHASHI and SEYAMA, 1974). A steroidal alkaloid, batrachotoxin, extracted from the skin of the Columbian arrow poison frog, also increases sodium permeability in squid giant axons (NARAHASHI et al., 1971; ALBUQUERQUE et al., 1973), dog Purkinje fiber (HOGAN and ALBUQUERQUE, 1971), and in cat papillary muscle (SHOTZBERGER et al., 1976). According to studies performed by HONERJÄGER and REITER (1977a) there is evidence to suggest that, in isolated guinea-pig papillary muscles, an increased sodium permeability induced by batrachotoxin occurs secondarily to a delay in repolarization resulting in prolonged sodium influx rather than to an alteration in resting membrane permeability. The effects of this toxin were found to increase in intensity from beat to beat and were suggested to be associated with the activity of the fast sodium channels. This action of batrachotoxin is also antagonized by relatively high concentrations of tetrodotoxin. All these agents are capable of producing positive inotropic effects in vivo and in isolated heart preparations in a number of species including dog, guinea-pig, and frog. Positive inotropic effects of veratridine (KRAYER and ACHESON, 1946; BENFORADO, 1957, 1968; HORACKOVA and VASSORT, 1974; HONERJÄGER and REITER, 1975), germitrine (HONERJÄGER and REITER, 1977b), grayanotoxins (MORAN et al., 1954; COTTEN et al., 1956; AKERA et al., 1976b; KU et al., 1977), and batrachotoxin (HOGAN and ALBUQUERQUE, 1971; SHOTZBERGER et al., 1976) have been reported.

The positive inotropic action of grayanotoxins (Fig. 8) has many features in common with cardiac glycosides. These include: that a substantial part of the effect is not abolished by β-adrenergic blockade, that the inotropic action is not related to the changes in transmembrane potential or action potential configurations, and that higher concentrations produce arrhythmias in isolated guinea-pig heart preparations (AKERA et al., 1976b; KU et al., 1977). Grayanotoxins, however, do not inhibit Na^+,K^+-ATPase activity. Ouabain-sensitive ^{86}Rb uptake, an estimate of sodium pump activity, by ventricular slices obtained from grayanotoxin-treated guinea-pig hearts is increased, suggesting that sodium pump activity is increased, probably as a result of the grayanotoxin-induced increase in intracellular sodium ions available to the pump (KU et al., 1977).

Several other agents seem to possess similar properties. A toxin isolated from scorpion venom increases the force of contraction in rat heart (CORABOEUF et al., 1975). The inotropic effect is observed in reserpine-pretreated heart, and is associated with a marked increase in amplitude and duration of the action potential plateau. In this species, the inotropic effect is suggested to be associated with a delayed or incomplete inactivation of the sodium channel. In guinea-pig and rabbit heart, however, the toxin fails to alter the action potential, but increases the force of contraction. The mechanism of the action of this toxin is presently unknown.

An organic cation, triaminopyrimidine, increases the force of contraction and prolongs the action potential duration in isolated guinea-pig atrial preparations (FRANK and FLOM, 1978). The inotropic effect can be observed in the presence of propranolol. The authors suggest that a slower activation of the potassium channel, rather than slower inactivation of sodium channel, causes these changes.

Polypeptides obtained from the sea anemone produce positive inotropic effects (RAVENS, 1976; SHIBATA et al., 1976, 1978). The inotropic effect is not altered by propranolol, and is not associated with an inhibition of Na^+,K^+-ATPase or the increase in cellular cyclic AMP content (SHIBATA et al., 1976). The effect is abolished by the calcium antagonist, ryanodine which also diminishes the positive ino-

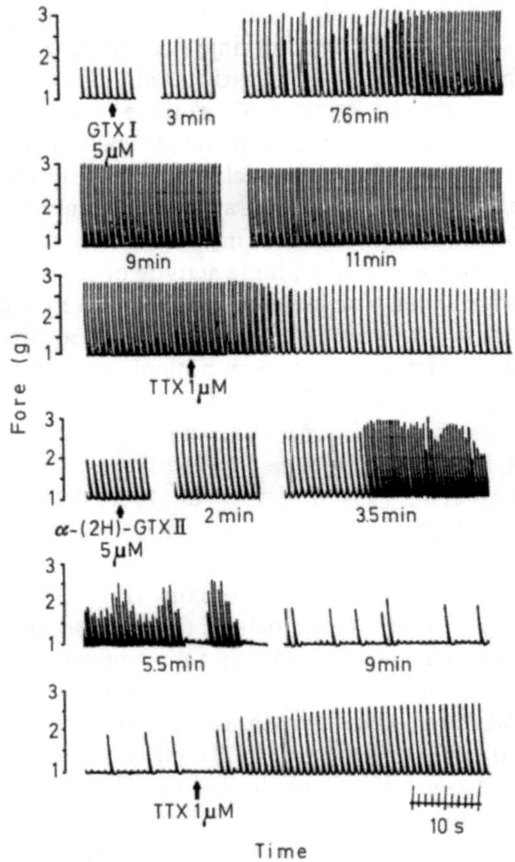

Fig. 8. Positive inotropic and arrhythmogenic actions of grayanotoxins. Left atrial preparations of guinea-pig heart suspended in Krebs–Henseleit bicarbonate buffer solution (pH 7.4) at 30 °C and electrically stimulated at 1 Hz. After equilibration grayanotoxin (GTX I) or α-dihydrograyanotoxin II [α-(2*H*)-GTX II] were added to the incubation medium. After the indicated time interval tetrodotoxin (TTX) was added. (KU et al., 1977)

tropic effect of ouabain and isoproterenol but not antagonized by verapamil or tetrodotoxin (SHIBATA et al., 1978). The polypeptides obtained from the Mediterranean (ATX II, WUNDERER et al., 1976) and Hawaiian Seas (anthopleurin A, TANAKA et al., 1977) have similar amino acid sequences, differing by only a few amino acids. These peptides are suggested to slow inactivation of the sodium channel in cardiac muscle (RAVENS, 1976), crayfish giant axons (LOWE et al., 1979), and frog myelinated fibers (ROMEY et al., 1976). The relationship between such an action of the polypeptides and inotropic effects is not firmly established.

The compounds that have been discussed increase membrane sodium permeability by affecting existing mechanisms for sodium movement. There is one group of compounds, the cation-specific ionophores, which increase membrane sodium permeability by an entirely different mechanism. These compounds form lipid-soluble complexes with certain cations, and by virtue of their lipid solubility, allow the cation to cross the lipid bilayer membrane following a concentration gradient. Among them, the carboxylic antibiotic ionophore, monensin, has been shown to be highly specific for Na^+ (PRESSMAN and HEEB, 1972) with a negligible ability to transport Ca^{2+} (PRESSMAN et al., 1975). Monensin in low concentrations (0.4–30 μM) produces positive inotropic effects in isolated heart preparations of guinea-pig, rabbit, and cat (SUTKO et al., 1977; SHLAFER et al., 1978). In higher concentrations and after a long exposure, monensin produces negative inotropic effects associated with an elevation of the resting tension. Part of the positive inotropic effect of monensin is apparently due to a release of catecholamines from sympathetic nerve terminals. There is, however, a component of the positive inotropic effect which is insensitive to the blockade of the adrenergic system either by 6-hydroxydopamine pretreatment or by propranolol (2 μM) plus phentolamine (2 μM) (SUTKO et al., 1977). In canine Purkinje fibers, the positive inotropic concentration of monensin shortens the action potential duration, but fails to alter the resting potential. These results indicate that the sodium-specific ionophore, monensin, is capable of producing positive inotropic effects similar to the cardiac glycosides.

Studies cited in this section do not establish a cause and effect relationship between the increase in sodium influx rate or intracellular sodium, and the observed positive inotropic effect. Various toxins, "sodium-specific" ionophores, or "sodium loading" (GLITSCH et al., 1970) may alter other cellular functions, in addition to increasing the rate of sodium influx or intracellular sodium concentration. However, the common feature of these agents or treatment is a positive inotropic effect associated with an increase in sodium ion available to the sodium pump and to Na^+/Ca^{2+} exchange mechanism. These results thus strongly suggest that an increase in sodium ions available to the Na^+/Ca^{2+} exchange mechanism increases the force of myocardial contraction, probably facilitating the calcium influx associated with membrane excitation.

D. Conclusions

There are a number of compounds that produce positive inotropic effects similar to cardiac glycosides. Most of them are either sodium pump inhibitors or compounds that enhance sodium influx. These two groups of compounds should enhance intracellular sodium transients, and thus may enhance intracellular calcium transients via Na^+/Ca^{2+} exchange reaction, similar to the cardiac glycosides.

Among Na^+, K^+-ATPase inhibitors, only a few compounds fail to produce positive inotropic effects. Some of these compounds affect biochemical systems other than Na^+,K^+-ATPase, precluding the manifestation of the inotropic action, while the other potent inhibitors of isolated Na^+,K^+-ATPase fail to inhibit the enzyme or sodium pump in intact cells under the conditions of inotropic studies.

It should be noted that the Na^+, K^+-ATPase inhibition may not be the sole mechanism for the inotropic action of these compounds. Similarly, the Na^+/Ca^{2+} exchange reaction triggered by sodium transients may not be the mainstream of the excitation–contraction coupling mechanism under the normal conditions, i.e., in the absence of the inotropic interventions. Nevertheless, we can conclude that either the inhibition of the sodium pump or the enhancement of sodium influx may be sufficient to cause positive inotropic effects.

Acknowledgments. Studies by one of the authors (T.A.) cited herein were supported by Grants HL-16052 and HL-16055 from the National Heart, Lung, and Blood Institute, and DA-01881 from the National Institute on Drug Abuse. U.S. Public Health Service, Grant BMS-74-19512 from the National Science Foundation, and grants-in-aid from the Michigan Heart Association. Help by Ms. Diane K. Hummel in preparation of the manuscript is gratefully acknowledged. The authors thank Dr. Theodore M. Brody for critically reading the manuscript. Studies by the other authors (A.L.F. and K.G.) were supported by a grant of the Deutsche Forschungsgemeinschaft SFB 30.

References

Akera, T.: Quantitative aspects of the interaction between ouabain and (Na^+, K^+)-activated ATPase in vitro. Biochim. Biophys. Acta *249*, 53–62 (1971)

Akera, T.: Membrane adenosinetriphosphatase: A digitalis receptor? Science *198*, 569–574 (1977)

Akera, T., Brody, T.M.: Inhibition of brain sodium- and potassium-stimulated adenosine triphosphatase activity by chlorpromazine free radical. Mol. Pharmacol. *4*, 600–612 (1968)

Akera, T., Brody, T.M.: The interaction between chlorpromazine free radical and microsomal sodium- and potassium-activated adenosine triphosphatase from rat brain. Mol. Pharmacol. *5*, 605–614 (1969)

Akera, T., Brody, T.M.: Inhibitory sites on sodium- and potassium-activated adenosine triphosphatase for chlorpromazine free radical and ouabain. Mol. Pharmacol. *6*, 557–566 (1970)

Akera, T., Ku, D.D., Brody, T.M.: Alterations of ion movements as a mechanism of drug-induced arrhythmias and inotropic responses. In: Taurine, Huxtable, R., Barbeau, A. (eds.) pp. 121–134, New York: Raven Press 1976a

Akera, T., Ku, D.D., Frank, M., Brody, T.M., Iwasa, J.: Effects of grayanotoxin I on cardiac Na^+,K^+-adenosine triphosphatase activity, transmembrane potential, and myocardial contractile force. J. Pharmacol. Exp. Ther. *199*, 247–254 (1976b)

Akera, T., Ku, D.D., Brody, T.M., Manian, A.A.: Inotropic action of hydroxylated chlorpromazine metabolites and related compounds. Biochem. Pharmacol. *27*, 995–998 (1978)

Akera, T., Takeda, K., Yamamoto, S., Brody, T.M.: Effects of vanadate on Na, K-ATPase and on the force of contraction in guinea-pig hearts. Life Sci. *25*, 1803–1812 (1979)

Albers, R.W., Koval, G.J., Siegel, G.J.: Studies on the interaction of ouabain and other cardioactive steroids with sodium-potassium activated adenosine triphosphatase. Mol. Pharmacol. *4*, 324–336 (1968)

Albuquerque, E.X., Seyama, I., Narahashi, T.: Characterization of batrachotoxin-induced depolarization of the squid giant axons. J. Pharmacol. Exp. Ther. *184*, 308–314 (1973)

Askari, A., Rao, S.N.: Drugs affecting sodium transport in human erythrocyte ghosts. J. Pharmacol. Exp. Ther. *172*, 211–223 (1970)

Badura, H.O.: Die Wirkung des Prednisolonbisguanylhydrazons auf die Kontraktionskraft des Skeletmuskels. Dissertation, Düsseldorf 1966

Banerjee, S.P., Khanna, V.K., Sen, A.K.: Inhibition of sodium- and potassium-dependent adenosine triphosphatase by ethacrynic acid: Two modes of action. Mol. Pharmacol. *6*, 680–690 (1970)

Banerjee, S.P., Khanna, V.K., Sen, A.K.: Inhibition of sodium- and potassium-dependent adenosine triphosphatase by ethacrynic acid: Ligand-induced modifications. Biochem. Pharmacol. *20*, 1649–1660 (1971)

Barrett, A.M., Cullum, V.A.: The biological properties of the optical isomers of propranolol and their effects on cardiac arrhythmias. Brit. J. Pharmacol. *34*, 43–55 (1968)

Beauge, L.A., Glynn, I.M.: A modifier of ($Na^+ + K^+$)-ATPase in commercial ATP. Nature (Lond.) *268*, 355–356 (1977)

Beauge, L.A., Glynn, I.M.: Commerical ATP containing traces of vanadate alters response of ($Na^+ + K^+$)-ATPase to external potassium. Nature (Lond.) *272*, 551–552 (1978)

Benforado, J.M.: Studies on veratrum alkaloids. XXVI. Comparison of the cardiac action of various tertiary amine ester alkaloids. J. Pharmacol. Exp. Ther. *120*, 412–425 (1957)
Benforado, J.M.: The veratrum alkaloids. In: Physiological Pharmacology, Root, W.S., Hoffmann, F.G. (eds.), 2 ed., Vol. 4, part D., pp. 331–398. New York: Academic Press 1968
Bennett, D., Chenoweth, M.B.: Metabolism associated with positive inotropic action. Fed. Proc. *10*, 280 (1951)
Bennett, D.R., Andersen, K.S., Andersen, M.V., jr., Robertson, D.N., Chenoweth, M.B.: Structure-activity analysis of the positive inotropic action of conjugated carbonyl compounds on the cat papillary muscle. J. Pharmacol. Exp. Ther. *122*, 489–498 (1958)
Besch, H.R.; jr., Watanabe, A.M.: Binding and effect of tritiated quinidine on cardiac subcellular enzyme systems: Sarcoplasmic reticulum vesicles, mitochondria, and Na^+,K^+-adenosine triphosphatase. J. Pharmacol. Exp. Ther. *202*, 354–364 (1977)
Blostein, R., Burt, V.K.: Interaction of N-ethylmaleimide and Ca^{2+} with human erythrocyte membrane ATPase. Biochim. Biophys. Acta *241*, 68–74 (1971)
Bonting, S.L., Hawkins, N.M., Canady, M.R.: Studies of sodium-potassium activated adenosine triphosphatase. VII. Inhibition by erythrophleum alkaloids. Biochem. Pharmacol. *13*, 13–22 (1964)
Borchard, U., Fox, A.A.L., Greeff, K.: Positive inotropic and antiarrhythmic actions of the Na^+,K^+-ATPase inhibitor, isoquinoline derivative, BIIA. Naunyn-Schmiedebergs Arch. Pharmacol. *312*, 187–192 (1980)
Borchard, U., Fox, A.A.L., Greeff, K., Schlieper, P.: Negative and positive inotropic action of vanadate on atrial and ventricular myocardium. Nature (Lond.) *279*, 339–341 (1979)
Britten, J.S., Blank, M.: Thallium activation of the (Na^+-K^+)-activated ATPase of rabbit kidney. Biochim. Biophys. Acta *159*, 160–166 (1968)
Brody, T.M., Akera, T., Baskin, S.I., Gubitz, R.H., Lee, C.Y.: Interaction of Na, K-ATPase with chlorpromazine free radical and related compounds. Ann. N.Y. Acad. Sci. *242*, 527–542 (1974)
Broekjuysen, J., Clinet, M., Delisee, C.: Action of amidarone on guinea pig heart sodium and potassium activated adenosine triphosphatase: Comparison with ouabain. Biochem. Pharmacol. *21*, 2951–2960 (1972)
Cantley, L.C., jr., Josephson, L., Warner, R., Yanagisawa, M., Lechene, C., Guidotti, G.: Vanadate is a potent (Na, K)-ATPase inhibitor found in ATP derived from muscle. J. Biol. Chem. *252*, 7421–7423 (1977)
Cantley, L.C., Ferguson, J.H., Kustin, K.: Norepinephrine complexes and reduces vanadium (V) to reverse vanadate inhibition of the (Na, K)-ATPase. J. Am. Chem. Soc. *100*, 5210–5212 (1978 a)
Cantley, L.C., Resh, M.D., Guidotti, G.: Vanadate inhibits red-cell ($Na^+ + K^+$)-ATPase from cytoplasmic side. Nature (Lond.) *272*, 552–554 (1978 b)
Charlier, R.: Cordarone. In: Handbook of Exper. Pharmacol. *31*, 255–288 (1971)
Charnock, J.S., Potter, H.A., McKee, D.: Ethacrynic acid inhibition of ($Na^+ + K^+$)-activated adenosine triphosphatase. Biochem. Pharmacol. *19*, 1637–1641 (1970)
Charnock, J.S., Doty, D.M., Russell, J.C.: The effect of temperature on the activity of ($Na^+ + K^+$)-ATPase. Arch. Biochem. Biophys. *142*, 633–637 (1971)
Chenoweth, M.B., Pengsritong, K.: Positive inotropic and other actions of fluoroacetate. Fed. Proc. *9*, 263 (1950)
Coraboeuf, E., Deroubaix, E., Tazieff-Depierre, F.: Effect of Toxin II isolated from scorpion venom on action potential and contraction of mammalian heart. J. Mol. Cell. Cardiol. *7*, 643–653 (1975)
Cotten, M. deV., Maling, H.M., Moran, N.C.: Comparison of the direct and indirect effects of andromedotoxin on the contractile force of the heart. J. Pharmacol. Exp. Ther. *118*, 55–62 (1956)
Covin, J.M., Berman, D.A.: Metabolic aspects of the positive inotropic action of fluoride on rat ventricle. J. Pharmacol. Exp. Ther. *125*, 137–141 (1959)
Dahl, D.R.: Short chain fatty acid inhibition of rat brain Na-K adenosine triphosphatase. J. Neurochem. *15*, 815–820 (1968)
Davis, P.W.: Inhibition of renal Na^+, K^+-activated adenosine triphosphatase activity by ethacrynic acid. Biochem. Pharmacol. *19*, 1983–1989 (1970)

Davis, P.W., Brody, T.M.: Inhibition of Na^+K^+-activated adenosine triphosphatase activity in rat brain by substituted phenothiazines. Biochem. Pharmacol. *15*, 703–710 (1966)
Donaldson, J., St-Pierre, T., Minnich, J., Barbeau, A.: Seizures in rats associated with divalent cation inhibition of Na^+-K^+-ATPase. Can. J. Biochem. *49*, 1217–1224 (1971)
Dransfeld, H., Greeff, K.: Der Einfluß des Prednison- und Prednisolonbisguanylhydrazons auf die $Na^+ + K^+$-stimulierte Membran-ATPase des Meerschweinchenherzens. Naunyn-Schmiedebergs Arch. Pharmakol. *249*, 425–431 (1964)
Dransfeld, H., Galetke, E., Greeff, K.: Die Wirkung des Prednisolonbisguanylhydrazons auf die $Na^+ + K^+$-aktivierte Membran-ATPase des Herz- und Skeletmuskels. Arch. Int. Pharmacodyn. *166*, 342–349 (1967)
Duggan, D.E., Noll, R.M.: Effects of ethacrynic acid and cardiac glycosides upon a membrane adenosine triphosphatase of renal cortex. Arch. Biochem. Biophys. *109*, 388–396 (1965)
Ehmer, A., Jahr, K., Kuschinsky, G., Lüllmann, H., Reuter, H., Woller, U.: Über die Herzglykosid-artige Wirkung von Progesteronbisguanylhydrazon (Progesteronbiguazon). Naunyn-Schmiedebergs Arch. Pharmakol. *248*, 521–539 (1964)
Epstein, F.H., Whittam, R.: Mode of inhibition by calcium of cell-membrane adenosine-triphosphatase activity. Biochem. J. *99*, 232–238 (1966)
Fahn, S., Hurley, M.R., Koval, G.J., Albers, R.W.: Sodium-potassium-activated adenosine triphosphatase of electrophorus electric organ. II. Effects of N-ethylmaleimide and other sulfhydryl reagents. J. Biol. Chem. *241*, 1890–1895 (1966)
Fox, A.A.L.: Charakterisierung des Wirkungsmechanismus von Hemmstoffen der Na^+, K^+-aktivierbaren, Mg^{2+}-abhängigen adenosine triphosphatase. Dissertation, Düsseldorf 1979
Fox, A.A.L., Borchard, U., Greeff, K.: Digitalisähnliche and antiarrhythmische Wirkung des Isochinolinderivates BIIA. Z. Kardiol. *68*, 244 (1979)
Fox, A.A.L., Greeff, K.: Mechanism of inhibition of sodium- and potassium-dependent adenosine triphosphatase by the isoquinoline derivate BIIA: A specific interaction with sodium activation. Biochem. Pharmacol. *30*, 611–617 (1981)
Frank, M., Flom, L.L.: Effects of 2,4,6-triaminopyridine on the electromechanical properties of guinea-pig myocardium. J. Pharmacol. Exp. Ther. *204*, 175–182 (1978)
Fricke, U.: Neuere Erkenntnisse über den Wirkungsmechanismus der Herzglykoside. Fortschr. Med. *94*, 1839–1845 (1976)
Fricke, U.: Myocardial activity of inhibitors of the Na^+-K^+-ATPase: Differences in the mode of action and subcellular distribution pattern of N-ethylmaleimide and ouabain. Naunyn-Schmiedebergs Arch. Pharmacol. *303*, 197–204 (1978)
From, A.H.L.: N-ethylmaleimide (NEM) induced inotropism. Clin. Res. *18*, 306 (1970)
From, A.H.L., Probstfield, J.: P-Chloromercuribenzene sulfonic acid induced inotropism. Fed. Proc. *30*, 632 (1971)
From, A.H., Probstfield, J.L., Smith, T.R.: Ethacrynic acid induced inotropism. Proc. Soc. Exp. Biol. Med. *149*, 1059–1062 (1975)
Fujita, M., Nagano, K., Mizuno, N., Tashima, Y., Nakao, T., Nakao, M.: Comparison of some minor activities accompanying a preparation of sodium-plus-potassium ion-stimulated adenosine triphosphatase from pig brain. Biochem. J. *106*, 113–121 (1968)
Gehring, P.J., Hammond, P.B.: The interrelationship between thallium and potassium in animals. J. Pharmacol. Exp. Ther. *155*, 187–201 (1967)
Glitsch, H.G., Reuter, H., Scholz, H.: The effect of the internal sodium concentration on calcium fluxes in isolated guinea-pig auricles. J. Physiol. (Lond.) *209*, 25–43 (1970)
Glynn, I.M.: "Transport adenosinetriphosphatase" in electric organ. The relation between ion transport and oxidative phosphorylation. J. Physiol. (Lond.) *169*, 452–465 (1963)
Gosalvez, M., Van Rossum, G.D.V., Blanco, M.F.: Inhibition of sodium-potassium-activated adenosine 5′-triphosphatase and ion transport by adriamycin. Cancer Res. *39*, 257–261 (1979)
Greeff, K., Schlieper, E.: Artspezifische Wirkungsunterschiede des k-Strophanthins und Prednisolonbisguanylhydrazons: Untersuchungen an isolierten Vorhofpräparaten und Erythrocyten des Menschen, Meerschweinchens, Kaninchens und der Ratte. Arch. Int. Pharmacodyn. *166*, 350–361 (1967)

Greeff, K., Meng, K., Schwarzmann, D.: Digitalisähnliche Eigenschaften des Prednison- und Prednisolonbisguanylhydrazons: Ihre Wirkung auf die Kaliumbilanz isolierter Herzpräparate und den Na/K-transport an Erythrocyten. Naunyn-Schmiedebergs Arch. Pharmakol. *249*, 416–424 (1964)

Grupp, G., Grupp, I., Johnson, C.L., Schwartz, A.: Effects of vanadate and ouabain on contractile force of rat, guinea pig, rabbit, and cat atria and ventricles. Fed. Proc. *38*, 1042 (1979a)

Grupp, G., Grupp, I., Johnson, C.L., Wallick, E.T., Schwartz, A.: Effects of vanadate on cardiac contraction and adenylate cyclase. Biochem. Biophys. Res. Commun. *88*, 440–447 (1979b)

Gutman, Y., Hochman, S., Wald, H.: The differential effect of Li^+ on microsomal ATPase in cortex, medulla, and papilla of the rat kidney. Biochim. Biophys. Acta *298*, 284–290 (1973)

Hackbarth, I., Schmitz, W., Scholz, H., Erdmann, E., Krawietz, W., Philipp, G.: Positive inotropism of vanadate in cat papillary muscle. Nature (Lond.) *275*, 67 (1978)

Halbach, S.: Effect of mercuric chloride on contractility and transmembrane potential of the guinea-pig myocardium. Naunyn-Schmiedebergs Arch. Pharmacol. *289*, 137–148 (1975)

Han, C.S., Tobin, T., Akera, T., Brody, T.M.: Effects of alkali metal cations on phosphoenzyme levels and (^{3}H)-ouabain binding to ($Na^+ + K^+$)-ATPase. Biochim. Biophys. Acta *429*, 993–1005 (1976)

Herrera, F.C.: The role of the active and passive sodium pathways in the mechanism of action of ethacrynic acid. Gen. Pharmacol. *6*, 201–207 (1975)

Hexum, T.D.: Studies on reaction catalyzed by transport (Na, K) adenosine triphosphatase. I. Effects of divalent metals. Biochem. Pharmacol. *23*, 3441–3447 (1974)

Hogan, P.M., Albuquerque, E.X.: The pharmacology of batrachotoxin. III. Effect on the heart Purkinje fibers. J. Pharmacol. Exp. Ther. *176*, 529–537 (1971)

Honerjäger, P., Reiter, M.: The relation between the effects of veratridine on action potential and contraction in mammalian ventricular myocardium. Naunyn-Schmiedebergs Arch. Pharmacol. *289*, 1–28 (1975)

Honerjäger, P., Reiter, M.: Sarcolemmal sodium permeability and contractile force of guinea pig papillary muscle. Effects of germitrine. Circ. Res. *40*, 90–98 (1977a)

Honerjäger, P., Reiter, M.: The cardiotoxic effect of batrachotoxin. Naunyn-Schmiedebergs Arch. Pharmacol. *299*, 239–252 (1977b)

Hopkins, L.L., jr., Mohr, H.E.: Vanadium as an essential nutrient. Fed. Proc. *33*, 1773–1775 (1974)

Horackova, M., Vassort, G.: Excitation-contraction coupling in frog heart: Effect of veratrine. Pflügers Arch. Eur. J. Physiol. *352*, 291–302 (1974)

Hudgins, P.M., Bond, G.H.: ($Mg^{2+} + K^+$)-dependent inhibition of NaK-ATPase due to a contaminant in equine muscle ATP. Biochem. Biophys. Res. Commun. *77*, 1024–1029 (1977)

Hudgins, P.M., Bond, G.H.: Reversal of vanadate inhibition of Na, K-ATPase by catecholamines. Res. Commun. Chem. Pathol. Pharmacol. *23*, 313–326 (1979)

Jacob, H.S., Jandel, J.H.: Effects of sulfhydryl inhibition on red blood cells. I. Mechanism of hemolysis. J. Clin. Invest. *41*, 779–792 (1962)

Järnefelt, J.: Properties and possible mechanism of the Na^+ and K^+-stimulated microsomal adenosine triphosphatase. Biochim. Biophys. Acta *59*, 643–654 (1962)

Jean, D.H., Bader, H.: Activation and inhibition of ($Na^+ + K^+$)-ATPase by N-ethylmaleimide. Biochim. Biophys. Acta *212*, 198–200 (1970)

Jöbsis, F.F., Vreman, H.J.: Inhibition of a Na^+ and K^+-stimulated adenosine triphosphatase by oligomycin. Biochim. Biophys. Acta *73*, 346–348 (1963)

Judah, J.D., Ahmed, K.: Inhibitors of transport and cation activated ATPase. J. Cell. Comp. Physiol. *64*, 355–361 (1964)

Kahn, J.B., jr.: Effects of two erythrophleum alkaloids on potassium transfer in human erythrocytes. Proc. Soc. Exp. Biol. Med. *110*, 412–414 (1962)

Kahn, J.B., jr., Van Atta, R.A., jr., Johnson, G.L.: Some effects of cassaine on cardiovascular dynamics in the dog. J. Pharmacol. Exp. Ther. *142*, 215–222 (1963)

Katzung, B., Rosin, H., Scheider, F.: Frequency-force relationship in the rabbit auricle and its modification by some metabolic inhibitors. J. Pharmacol. Exp. Ther. *120*, 324–333 (1957)

Kennedy, K.G., Nayler, W.G.: The effect of quinidine on the activity of a sodium-potassium activated, magnesium-dependent ATPase enzyme isolated from toad cardiac muscle. Biochim. Biophys. Acta *110*, 174–180 (1965)

Kobayashi, T., Nakayama, R., Takatani, O., Kimura, K.: Positive chronotropic and inotropic actions of new antitumor agent adriamycin and its cardiotoxicity. Its special references to myocardial contractile force and the change of the transmembrane action potential. Jpn. Circ. J. *36*, 259–265 (1972)

Korth, M., Weger, N., Reiter, M.: The positive inotropic action of sodium fluoroacetate on guinea-pig ventricular myocardium. Naunyn-Schmiedebergs Arch. Pharmacol. *303*, 7–14 (1978)

Krayer, O., Acheson, G.H.: The pharmacology of the veratrum alkaloids. Physiol. Rev. *26*, 383–446 (1946)

Kroneberg, G., Stoepel, K.: Synthetische Verbindungen mit Digitaliswirkung. Naunyn-Schmiedebergs Arch. Pharmakol. *249*, 393–415 (1964)

Ku, D., Akera, T., Tobin, T., Brody, T.M.: Effects of rubidium on cardiac tissue: Inhibition of Na^+,K^+-ATPase and stimulation of contractile force. Res. Comm. Chem. Path. Pharmacol. *9*, 431–440 (1974)

Ku, D., Akera, T., Tobin, T., Brody, T.M.: Effects of monovalent cations on cardiac Na^+, K^+-ATPase activity and on contractile force. Naunyn-Schmiedebergs Arch. Pharmacol. *290*, 113–131 (1975)

Ku, D.D., Akera, T., Tobin, T., Brody, T.M.: Comparative species studies on the effect of monovalent cations and ouabain on cardiac Na^+,K^+-ATPase and contractile force. J. Pharmacol. Exp. Ther. *197*, 458–469 (1976)

Ku, D.D., Akera, T., Frank, M., Brody, T.M., Iwasa, J.: The effects of grayanotoxin I and α-dihydrograyanotoxin II on guinea pig myocardium. J. Pharmacol. Exp. Ther. *200*, 363–372 (1977)

Ku, D.D., Akera, T., Olgaard, M.K., Brody, T.M.: Effects of lithium and thallous ions on sodium pump activity in the guinea-pig heart and their relationship to the positive inotropic action. Naunyn-Schmiedebergs Arch. Pharmacol. *304*, 167–173 (1978)

Lahiri, A.K., Wilson, I.B.: On the inhibition of $(Na^+ + K^+)$-activated adenosine triphosphatase by diisopropyl fluorophosphate. Mol. Pharmacol. *7*, 46–51 (1971)

Law, R.O.: The effects of ouabain and ethacrynic acid on the intracellular sodium and potassium concentrations in renal medullary slices incubated in cold potassium-free Ringer solution and re-incubated at 37 °C in the presence of external potassium. J. Physiol. (Lond.) *254*, 743–758 (1976)

Lee, K.S.: The metabolism and contraction of cat heart muscles as affected by drugs. J. Pharmacol. Exp. Ther. *112*, 484–494 (1954)

Lee, K.S., Yu, D.H., Lee, D.I., Burstein, R.: The influence of potassium and calcium on the effect of ouabain on cat papillary muscles. J. Pharmacol. Exp. Ther. *132*, 139–148 (1961)

Lee, C.Y., Chang, C.C., Chiu, T.H., Chiu, P.J.S., Tseng, T.C., Lee, S.Y.: Pharmacological properties of cardiotoxin isolated from Formosan cobra venom. Naunyn-Schmiedebergs Arch. Pharmacol. *259*, 360–374 (1968)

Levi, R., Allan, G., Zavecz, J.H.: Cardiac histamine receptors. Fed. Proc. *35*, 1942–1947 (1976)

Loewi, O.: On the mechanism of the positive inotropic action of fluoride, oleate, and calcium on the frog's heart. J. Pharmacol. Exp. Ther. *114*, 90–99 (1955)

Lowe, P.A., Wu, C.H., Narahashi, T.: The effect of anthopleurin-A on crayfish giant axon. J. Pharmacol. Exp. Ther. *210*, 417–421 (1979)

Lowry, K., Rao, S.N., Pitts, B.J.R., Askari, A.: Effects of quinidine on some reactions and ion translocations catalyzed by the Na^+,K^+-ATPase complex. Biochem. Pharmacol. *22*, 1369–1377 (1973)

Lüllmann, H., Holland, W.: Influence of ouabain on an exchangeable calcium fraction, contractile force, and resting tension of guinea-pig atria. J. Pharmacol. Exp. Ther. *137*, 186–192 (1962)

Maling, H.M., Krayer, O.: The action of the erythrophleum alkaloids on the isolated mammalian heart. J. Pharmacol. Exp. Ther. *86*, 66–78 (1946)
Matsui, H., Schwartz, A.: Purification and properties of a highly active, ouabain sensitive Na^+-K^+-dependent adenosine-triphosphatase from cardiac tissue. Biochim. Biophys. Acta *128*, 380–390 (1966)
Miller, H.M., Woodhouse, S.P.: Long-chain fatty-acid inhibition of sodium plus potassium-activated adenosine-triphosphatase from rat heart. Aust. J. Exp. Biol. *55*, 741–752 (1977)
Moran, N.C., Dresel, P.E., Perkins, M.E., Richardson, A.P.: The pharmacological actions of andromedotoxin, an active principle from Rhododendron Maximum. J. Pharmacol. Exp. Ther. *110*, 415–432 (1954)
Myers, C.E., McGuire, W.P., Liss, R.H., Ifrim, I., Grotzinger, K., Young, R.C.: Adriamycin: The role of lipid peroxidation in cardiac toxicity and tumor response. Science *197*, 165–167 (1977)
Narahashi, T., Seyama, I.: Mechanism of nerve membrane depolarization caused by grayanotoxin I. J. Physiol. (Lond.) *242*, 471–487 (1974)
Narahashi, T., Albuquerque, E.X., Deguchi, T.: Effects of batrachotoxin on membrane potential and conductance of squid giant axons. J. Gen. Physiol. *58*, 54–70 (1971)
Nechay, B.R., Saunders, J.P.: Inhibition of $Na^+ + K^+$-ATPase by vanadium pentoxide (V_2O_5). Fed. Proc. *37*, 240 (1978)
Nix, C.R., Holland, W.C., Berry, R.A.: The action of cassaine and two of its alkylated derivatives on atrial contractions. Pharmacol. Res. Comm. *3*, 75–78 (1971)
Ohta, M., Narahashi, T., Keeler, R.F.: Effects of veratrum alkaloids on membrane potential and conductance of squid and crayfish giant axons. J. Pharmacol. Exp. Ther. *184*, 143–154 (1973)
Opit, L.J., Potter, H., Charnock, J.S.: The effect of anions on (Na^+, K^+)-activated ATPase. Biochim. Biophys. Acta *120*, 159–161 (1966)
Palatini, P.: The interaction of tricyclic antipsychotics with (Na^+, K^+)-ATPase. Gen. Pharmacol. *9*, 215–220 (1978)
Penefsky, Z.J., Kahn, M.: Inotropic effects of dexamethasone in mammalian heart muscle. Eur. J. Pharmacol. *15*, 259–266 (1971)
Penzotti, S.T., jr., Titus, E.: Evidence for two forms of fluoride-treated sodium- and potassium-dependent adenosine triphosphatase. Mol. Pharmacol. *8*, 149–158 (1972)
Peters, T., Raben, R., Wassermann, O.: Evidence for a dissociation between positive inotropic effect and inhibition of the Na-K-ATPase by ouabain, cassaine and their alkylating derivatives. Eur. J. Pharmacol. *26*, 166–174 (1974)
Phillips, T.D., Hayes, A.W., Ho, I.K., Desaiah, D.: Effects of rubratoxin-B on kinetics of cationic and substrate activation of (Na^+-K^+)-ATPase and paranitrophenyl phosphatase. J. Biol. Chem. *253*, 3487–3493 (1978)
Pitts, B.J.R., Meyerson, L.R.: Inhibition of ouabain binding to Na, K-ATPase by sanguinarine: Correlation with the inotropic effect. Fed. Proc. *37*, 240 (1978)
Ploeger, E.J.: The effects of lithium on excitable cell membranes. On the mechanism of inhibition of the sodium pump of non-myelinated nerve fibres of the rat. Eur. J. Pharmacol. *25*, 316–321 (1974)
Post, R.L.: A model for regulation of vanadate inhibition of (Na, K)-ATPase by reduction. Fed. Proc. *38*, 242 (1979)
Post, R.L., Kume, S., Tobin, T., Orcutt, B., Sen, A.K.: Flexibility of an active center in sodium plus potassium adenosine triphosphatase. J. Gen. Physiol. *54*, 306s–326s (1969)
Post, R.L., Hegyvary, C., Kume, S.: Activation by adenosine triphosphate in the phosphorylation kinetics of Na and K ion transport adenosine triphosphatase. J. Biol. Chem. *247*, 6530–6540 (1972)
Prasad, K., Midha, K.K.: Effect of rubidium on cardiac function. Jpn. Heart J. *13*, 317–324 (1972)
Pressman, B.C., Heeb, M.J.: Permeability studies on erythrocyte ghosts with ionophorous antibiotics. In: Munoz, E., Garcis-Fernandiz, F., Vasquez, D. (eds.), Molecular mechanisms of antibiotic action on protein biosynthesis and membranes, 3 ed., pp. 603–614. New York: Elsevier 1972

Pressman, B.C., DeGuzman, N.T., Somani, P.: Correlation of inotropic and transport properties of carboxylic ionophores. Pharmacologist *17*, 245 (1975)
Prindle, K.H., jr., Gold, H.K., Cardon, P.V., Epstein, S.E.: Effects of pschychopharmacologic agents on myocardial contractility. J. Pharmacol. Exp. Ther. *173*, 133–137 (1970)
Quist, E.E., Hokin, L.E.: The presence of two (Na^+, K^+)-ATPase inhibitors in equine muscle ATP: Vanadate and a dithioerythritol-dependent inhibitor. Biochim. Biophys. Acta *511*, 202–212 (1978)
Ravens, U.: Electromechanical studies of an anemonia sulcata toxin in mammalian cardiac muscle. Naunyn-Schmiedebergs Arch. Pharmacol. *296*, 73–78 (1976)
Reilly, J., Riker, W.F., jr., Whitehouse, W.C., Kuriaki, K.: The actions of methyl fluoroacetate on the papillary muscle of the cat with comparative experiments on the actions of methyl chloroacetate. J. Pharmacol. Exp. Ther. *108*, 393–409 (1953)
Reiter, M.: The effect of various amines on the contractility of the guinea-pig papillary muscle. Experientia *21*, 87–89 (1965)
Repke, K.: Über den biochemischen Wirkungsmodus von Digitalis. Klin. Wochenschr. *42*, 157–165 (1964)
Repke, K., Portius, H.J.: Über die Identität der Ionenpumpen-ATPase in der Zellmembran des Herzmuskels mit einem Digitalis-Rezeptorenzym. Experientia *19*, 452–458 (1963)
Riordan, J.R., Manery, J.F., Dryden, E.E., Still, J.S.: Influence of ethacrynic acid on muscle surface enzymes and of ethacrynic acid and ouabain on Na, K, and H_2O in frog muscle. Can. J. Physiol. Pharmacol. *50*, 432–444 (1972)
Robinson, J.D., Lowinger, J., Bettinger, B.: Chlorpromazine: Differential effects on membrane-bound enzyme from rat brain. Biochem. Pharmacol. *17*, 1113–1116 (1968)
Romey, G., Abita, J.P., Schweitz, H., Wunderer, G., Lazdunski, M.: Sea anemone toxin: A tool to study molecular mechanisms of nerve conduction and excitation-secretion coupling. Proc. Natl. Acad. Sci. USA *73*, 4055–4059 (1976)
Saunders, P.R., Sanyal, P.N.: Effect of temperature upon the positive inotropic action of ouabain. J. Pharmacol. Exp. Ther. *123*, 161–163 (1958)
Schmitz, W., Hackbarth, I., Scholz, H.: Effect of vanadate on force of contraction in mammalian cardiac muscle. Naunyn-Schmiedebergs Arch. Pharmacol. *307*, R37 (1979)
Schwabe, U., Puchstein, C., Hannemann, H., Söchtig, E.: Activation of adenylate cyclase by vanadate. Nature (Lond.) *277*, 143–145 (1979)
Schwartz, A., Laseter, A.H.: A sodium- and potassium-stimulated adenosine triphosphatase from cardiac tissues. II. The effects of ouabain and other agents that modify enzyme activity. Biochem. Pharmacol. *13*, 337–348 (1964)
Schwartz, A., Lindenmayer, G.E., Allen, J.C.: The sodium-potassium adenosine triphosphatase: Pharmacological, physiological, and biochemical aspects. Pharmacol. Rev. *27*, 3–134 (1975)
Seifen, E., Straub, K.D.: Effects of sanguinarine on the isolated mammalian heart. Pharmacologist *16*, 245 (1974)
Seyama, I.: Effect of grayanotoxin I on the electrical properties of rat skeletal muscle fibers. Jpn. J. Physiol. *20*, 381–393 (1970)
Seyama, I., Narahashi, T.: Increase in sodium permeability of squid axon membranes by α-dihydrograyanotoxin II. J. Pharmacol. Exp. Ther. *184*, 299–307 (1973)
Sharp, C.W., Hunt, H.G., Clements, S.T., Wilson, W.E.: The influence of dichlorodiphenyltrichloroethane, polychlorinated biphenyls, and anionic amphiphilic compounds on stabilization of sodium- and potassium-activated adenosine triphosphatase by acidic phospholipids. Mol. Pharmacol. *10*, 119–129 (1974)
Shibata, S., Norton, T.R., Izumi, T., Matsuo, T., Katsuki, S.: A polypeptide (AP-A) from sea anemone (Anthopleura xanthogrammica) with potent positive inotropic action. J. Pharmacol. Exp. Ther. *199*, 298–309 (1976)
Shibata, S., Izumi, T., Seriguchi, D.G., Norton, T.R.: Further studies on the positive inotropic effect of the polypeptide anthopleurin-A from a sea anemone. J. Pharmacol. Exp. Ther. *205*, 683–692 (1978)
Shlafer, M., Matheny, J.L., Karow, M. jr.: Cardiac inotropism of dimethyl sulfoxide: Osmotic effects and interactions with calcium ion. Eur. J. Pharmacol. *28*, 276–287 (1974)

Shlafer, M., Somani, P., Pressman, B.C., Palmer, R.F.: Effects of the carboxylic ionophore monensin on atrial contractility and Ca^{2+} regulation by isolated cardiac microsomes. J. Mol. Cell. Cardiol. *10*, 333–346 (1978)
Shotzberger, G.S., Albuquerque, E.X., Daly, J.W.: The effects of batrachotoxin on cat papillary muscle. J. Pharmacol. Exp. Ther. *196*, 433–444 (1976)
Siegel, G.J., Fogt, S.M.: Inhibition by lead ion of electrophorus electroplax ($Na^+ + K^+$)-adenosine triphosphatase and K^+-p-nitrophenylphosphatase. J. Biol. Chem. *252*, 5201–5205 (1977)
Skou, J.C.: The influence of some cations on an adenosine triphosphatase from peripheral nerves. Biochim. Biophys. Acta *23*, 394–401 (1957)
Skou, J.C.: Further investigations on a $Mg^{++} + Na^+$-activated adenosine-triphosphatase, possibly related to the active, linked transport of Na^+ and K^+ across the nerve membrane. Biochim. Biophys. Acta *42*, 6–23 (1960)
Skou, J.C.: Studies on the $Na^+ + K^+$ activated ATP hydrolyzing enzyme system. The role of SH groups. Biochem. Biophys. Res. Comm. *10*, 79–84 (1963)
Skou, J.C., Hilberg, C.: The effect of sulphydryl blocking reagents and of urea on the ($Na^+ + K^+$)-activated enzyme system. Biochim. Biophys. Acta *110*, 359–369 (1965)
Skulskii, I.A., Manninen, V., Järnefelt, J.: Thallium inhibition of ouabain-sensitive sodium transport and of the (Na^+, K^+)-ATPase in human erythrocytes. Biochim. Biophys. Acta *394*, 569–576 (1975)
Somogyi, J.: Über die Wirkung der Ca-ionen auf die durch Na^+ und K^+ aktivierbare Adenosintriphosphatase des Hirngewebes. Hoppe-Seylers Z. Physiol. Chem. *336*, 264–270 (1964)
Song, S.Y., Scheuer, J.: The effect of pharmacologic agents on myocardial sodium (Na^+) and potassium (K^+) stimulated ATPase. Pharmacology *1*, 209–217 (1968)
Squires, R.F.: On the interaction of Na^+, K^+, Mg^{++}, and ATP with the Na^+ plus K^+ activated ATPase from rat brain. Biochem. Biophys. Res. Comm. *19*, 27–32 (1965)
Sutko, J.L., Besch, J.R. jr., Bailey, J.C., Zimmerman, G., Watanabe, A.M.: Direct effects of the monovalent cation ionophores monensin and nigericin on myocardium. J. Pharmacol. Exp. Ther. *203*, 685–700 (1977)
Swanson, P.D.: Temperature dependence of sodium ion activation of the cerebral microsomal adenosine triphosphatase. J. Neurochem. *13*, 229–236 (1966)
Szekeres, L., Papp, J.G., Udvary, E.: On two new isoquinoline derivatives with marked antianginal and antiarrhythmic actions. Naunyn-Schmiedebergs Arch. Pharmacol. *284*, R79 (1974)
Tanaka, M., Hanin, M., Yasunobu, K.T., Norton, T.R.: Amino acid sequence of Anthopleura xanthogrammica heart stimulant, anthopleurin-A. Biochemistry *16*, 204–208 (1977)
Taylor, S.R.: The effect of mercurial diuretics on ATPase of rabbit kidney in vitro. Biochem. Pharmacol. *12*, 539–550 (1963)
Temma, K., Akera, T., Brody, T.M.: Hydroxylated chlorpromazine metabolites: Positive inotropic action and the release of catecholamines. Mol. Pharmacol. *13*, 1076–1085 (1977)
Temma, K., Akera, T., Ku, D.D., Brody, T.M.: Sodium pump inhibition by sulfhydryl inhibitors and myocardial contractility. Naunyn-Schmiedebergs Arch. Pharmacol. *302*, 63–71 (1978)
Tobin, T., Akera, T.: Showdomycin, a nucleotide-site-directed inhibitor of ($Na^+ + K^+$)-ATPase. Biochim. Biophys. Acta *389*, 126–136 (1975)
Tobin, T., Akera, T., Baskin, S.I., Brody, T.M.: Calcium ion and sodium- and potassium-dependent adenosine triphosphatase: Its mechanism of inhibition and identification of the E_1-P intermediate. Mol. Pharmacol. *9*, 336–349 (1973)
Tobin, T., Akera, T., Dworin, J.Z., Brody, T.M.: Fluoride nephropathy: Lack of direct involvement of renal ATPase. Can. J. Physiol. Pharmacol. *52*, 489–495 (1974a)
Tobin, T., Akera, T., Han, C.S., Brody, T-M.: Lithium and rubidium interactions with sodium- and potassium-dependent adenosine triphosphatase: A molecular basis for the pharmacological actions of these ions. Mol. Pharmacol. *10*, 501–508 (1974b)

Tobin, T., Akera, T., Brody, S.L., Ku, D., Brody, T.M.: Cassaine: Mechanism of inhibition of $Na^+ + K^+$-ATPase and relationship of this inhibition to cardiotonic actions. Eur. J. Pharmacol. *32*, 133–145 (1975)
Tobin, T., Akera, T., Han, C.S., Brody, T.M.: Studies on the stable inhibition of Na^+, K^+-ATPase by cassaine. Eur. J. Pharmacol. *35*, 59–68 (1976)
Toda, N., Konishi, N.: Effects of a sulfhydryl reagent, N-ethylmaleimide, on electrical and mechanical activities of isolated atria. Eur. J. Pharmacol. *7*, 5–13 (1969)
Toro-Goyco, E., Rodriguez, M.B., Preston, A.M.: On the action of Δ^9-tetrahydrocannabinol as an inhibitor of sodium- and potassium-dependent adenosine triphosphatase. Mol. Pharmacol. *14*, 130–137 (1978)
Tuttle, R.S., Witt, P.N., Farah, A.: Therapeutic and toxic effects of ouabain on K^+ fluxes in rabbit atria. J. Pharmacol. Exp. Ther. *137*, 24–30 (1962)
Van Boxtel, C.J., Olson, R.D., Boerth, R.C., Oates, J.A.: Doxorubicin: inotropic effects and inhibitory action on ouabain. J. Pharmacol. Exp. Ther. *207*, 277–283 (1978)
Van Groningen, H.E.M., Slater, E.C.: The effect of oligomycin on the Na^+, K^+-activated magnesium-ATPase of brain microsomes and erythrocyte membrane. Biochim. Biophys. Acta *73*, 527–530 (1963)
Vincent, J.P., Schweitz, H., Chicheportiche, R., Fosset, M., Balerna, M., Lenoir, M.C., Lazdunski, M.: Molecular mechanism of cardiotoxin action on axonal membranes. Biochemistry *15*, 3171–3175 (1976)
Werdan, K., Bozsik, M., Erdmann, E., Krawietz, W., Schmitz, W., Scholz, H.: Vanadataufnahme und Vanadatwirkung in kultivierten, kontrahierenden Rattenherzmuskelzellen. Z. Kardiol. *68*, 244 (1979)
Whittam, R., Wheeler, K.P., Blake, A.: Oligomycin and active transport reactions in cell membranes. Nature (Lond.) *203*, 720–724 (1964)
Willis, J.S., Fang, L.S.T.: Li^+ stimulation of ouabain-sensitive respiration and $(Na^+ + K^+)$-ATPase of kidney cortex of ground squirrels. Biochim. Biophys. Acta *219*, 486–489 (1970)
Wunderer, G., Fritz, H., Wachter, E., Machleidt, W.: Amino-acid sequence of a coelenterate toxin: Toxin II from Anemonia sulcata. Eur. J. Biochem. *68*, 193–198 (1976)
Yamamoto, H.: Pharmacological studies on thiamine triphosphate. Folia Pharmacol. Jpn. *63*, 134–152 (1967)
Yamamoto, H., Kitano, T., Nishino, H., Murano, T.: Studies on pharmacodynamic action of N-ethylmaleimide (NEM). I.: Influence of NEM on synaptic transmission of sympathetic nerves in guinea-pigs. Jpn. J. Pharmacol. *23*, 151–160 (1973)
Yamamoto, S.: Prednisolone-3,20-bisguanylhydrazone: The mode of interaction with rat brain sodium and potassium-activated adenosine triphosphatase. Eur. J. Pharmacol. *50*, 409–418 (1978)
Yamamoto, S., Akera, T., Brody, T.M.: Prednisolone-3,20-bisguanylhydrazone: Na^+, K^+-ATPase inhibition and positive inotropic action. Eur. J. Pharmacol. *49*, 121–132 (1978a)
Yamamoto, S., Akera, T., Brody, T.M.: Prednisolone-3,20-bisguanylhydrazone: Binding in vitro to sodium- and potassium-activated adenosine triphosphatase of guinea-pig heart ventricular muscle. Eur. J. Pharmacol. *51*, 63–69 (1978b)
Yoshida, H., Nagai, K., Kamei, M., Nakagawa, Y.: Irreversible inactivation of $(Na^+ + K^+)$-dependent ATPase and K^+-dependent phosphatase by fluoride. Biochim. Biophys. Acta *150*, 162–164 (1968)

Non-Cardiac Effects of Cardiac Glycosides

CHAPTER 21

Effects of Cardiac Glycosides on Central Nervous System

H. F. BENTHE

A. Introduction

As compared with their cardiac side effects, the central nervous system (CNS) side effects caused by the therapeutic use of cardiac glycosides are of relatively minor significance. This difference is partly due to the preferential concentration of the glycosides in the myocardium; the levels reached in the CNS are much lower. Furthermore, such neurologic side effects as do occur are more or less subjective in nature and are not measurable. This is true even of the commonest central nervous side effect, namely nausea and vomiting. Even in cases of unquestionable digitalis poisoning, where the diagnosis has been confirmed by serum assays, the frequency of these symptoms is the subject of widely varying estimates by different authors – from 61% by one group to 25% by another (BELLER et al., 1974; HAASIS et al., 1975).

There are other neurologic symptoms such as drowsiness, lack of initiative, motor inactivity, visual disorders, and confusional states, but it is almost impossible to substantiate their reality, especially as geriatric patients so often complain of such symptoms when not receiving digitalis. The discovery and elucidation of the central nervous system effects of digitalis has therefore been based mainly on work in experimental animals.

B. Central Vomiting

Oral administration of glycosides to unanesthetized cats is promptly followed by profuse salivation with gagging and vomiting reflexes, the response depending on the dose and the speed with which the preparation is absorbed from the intestine. At the same time, about 15–30 min after an oral dose, typical glycoside changes appear in the electrocardiogram (ECG). Glycosides which, because of poor absorbability or bioavailability, lack cardiac efficacy do not evoke vomiting. The central mechanism was worked out by BORISON and BRIZZEE (1951) and BORISON and FAIRBANKS (1952) in dogs and cats. In the medulla oblongata there is a chemoreceptive trigger zone situated in the dorsal part of the ala cinerea. Localized electrocoagulation of this area makes the animals unresponsive to an intravenous (i.v.) injection of apomorphine. Whereas lanatoside C, scillaren A, and ouabain produce vomiting within 15 min when given intravenously to intact dogs, the same glycosides have no effect on animals which have been successfully operated upon (as checked by giving apomorphine). This confirmed the earlier work of DRESBACH (1947) who found that even surgical ablation of all known visceral afferent pathways did not prevent digitalis vomiting. Cats in which the trigger zone has been surgically destroyed are also to a great extent protected against digitalis vomiting induced by administration of glycosides by mouth. Vomiting has nevertheless occasionally been observed several hours after oral or intravenous administration of glycosides to animals which had had successful operations. In these instances reflexes arising in the viscera must have activated the still intact vomiting center (BORISON and FAIRBANKS, 1952).

The statement that injection of glycosides (ouabain, desacetyllanatoside C, strophanthidin) into the fourth ventricle of conscious cats fails to produce vomiting has been denied, the results being explained as due to incorrect injection into another part of the ventricular system (GAITONDE et al., 1965; CHAI et al., 1973). For ouabain, injected correctly into the fourth ventricle (slightly above the area postrema), the emetic dose in cats was found to be 6–8 μg, while a dose of 20–30 μg produced convulsions without vomiting. Because of the greater size of the ventricles in dogs, vomiting can be produced by injecting glycosides into the fourth, the third, or either of the lateral ventricles (SHARE et al., 1965).

All in all, these studies confirm that the chemoreceptor trigger zone in the area postrema is the afferent center for digitalis-induced vomiting ("early vomiting"), but this does not exclude the possibility that reflex stimuli may also arise at the periphery ("late vomiting"). Catecholamines appear to play an intermediary role in the mechanism of centrally evoked vomiting, as their depletion by reserpine or by pretreatment with phenoxybenzamine has a partial protective effect (GAITONDÉ and JOGLEKAR, 1975). In conscious cats early vomiting induced by oral glycosides can be lessened by pretreatment with anti-emetic phenothiazines.

C. Respiration

One of the pharmacologic hallmarks of cardiac glycosides is the fact that when toxic doses are given to animals susceptible to the cardiac effects of digitalis (dogs, cats, guinea pigs) cardiac arrest invariably precedes respiratory arrest. During administration of cardiac glycosides spontaneous respiration remains unaffected unless and until arrhythmias become so severe as to interfere with cardiac output. This observation is consistent with clinical experience of the therapeutic use of cardiac glycosides in humans. However, during conventional testing of genins in experimental animals (Hatcher infusion) abnormalities of spontaneous respiration may be noted before the onset of cardiac arrest, and in such cases the lethal dose can be raised by artificial ventilation.

Careful measurements of respiration rate, CO_2 partial pressure, pH, and cerebrospinal fluid (CSF) potassium levels in anesthetized dogs receiving ouabain infusions have shown that when the infusion rate is raised high enough to produce cardiac arrhythmias there is an increase in respiration rate, a rise in pH in plasma and CSF, a fall in CO_2 partial pressure, and a rise in arterial K^+ concentration, while the CSF K^+ level remains unchanged. The cause of the hyperventilation is thought to be the rise in arterial K^+ (YEN and CHOW, 1974). Injection of ouabain into the cerebral ventricles also causes hyperventilation and this too is believed to be the result of a rise in extracellular K^+ concentration in the vicinity of the neurons which control respiration (CAMERON, 1967).

The results of respiratory measurements in humans are not altogether consistent. A dose of 1–2 mg of digitoxin was found to depress the CO_2 sensitivity of the respiratory center in healthy subjects, whereas normal respiration was unchanged (RIGGERT and SCHWAB, 1955). Tests with strophanthin K carried out by the same technique showed no change in respiratory values (SCHWAB and WAGNER, 1957).

Cardiac glycosides are known to produce a clearly defined pattern of central effects in species which are insensitive to their action on the heart. In such animals they depress the respiratory center, as shown by measurements of action potentials in rat phrenic nerve. There may also be peripheral respiratory paralysis due to blockage of neuromuscular transmission as the result of hyperkalemia caused by toxic doses of glycosides (GREEFF, 1957; GREEFF and WESTERMANN, 1955; DAL RI and SCHMIDT, 1960).

D. Central Parasympathetic Activity

In view of the general excitatory effect of digitalis glycosides on parasympathetic nuclear areas in the brain stem it is reasonable to expect that they will also stimulate central vagomimetic reactions in the cardiovascular system. Certain typical digitalis effects such as sinus bradycardia, delayed atrioventricular conduction, and suppression of ectopic atrial impulses are mainly attributable to central vagal stimulation. In patients with transplanted hearts digoxin does not reduce ventricular rate in cases of atrial flutter, and even in regular sinus tachycardia it does not have any slowing effect (LEACHMAN et al., 1971).

An increase in spontaneous electrical activity in efferent parasympathetic fibers (cat) after administration of digitalis was described some time ago (MCLAIN, 1969). The causes are general stimulation of autonomic centers and, secondly, peripheral evocation of reflex impulses from the receptors in the carotid sinus and the aortic arch, a phenomenon independent of changes in blood pressure. Surgical ablation of the nerves from the carotid sinus and aorta largely abolishes the parasympathetic effect of cardiac glycosides (GILLIS et al., 1972).

E. Central Sympathetic Activity

The excitation of autonomic nuclei in the CNS is accompanied by activation of the efferent sympathetic outflow by cardiac glycosides. This seems to be a direct central effect of glycosides, and is independent of peripheral receptors (carotid sinus, aortic arch). In contrast to the parasympathetic excitation, the activation of the sympathetic system appears only when the dose is high enough to provoke arrhythmias, and it is in no way dependent on blood pressure changes. A causal connection between the sympathetic activation and the arrhythmogenic effect of digoxin has been postulated. Sympathetic denervation by atlanto-occipital transection of the spinal cord raises the dose of digoxin necessary to induce arrhythmias. When clonidine is given to depress central sympathetic activity it will at the same time convert the arrhythmia into normal sinus rhythm (PACE and GILLIS, 1976), a fact which provides further evidence of a causal link. The same interpretation has been placed on the observation that the lethal dose of ouabain by infusion in the spinal cat is much higher than in the intact cat (LEVITT et al., 1973; SOMBERG et al., 1979).

Other workers using practically the same method (cat, digoxin i.v.) deduced that the enhanced sympathetic activity is of ganglionic origin, because it is not measurable at preganglionic level (splanchnic nerve). They concluded that the target site of intravenous digoxin is not located in the autonomic centers of the CNS, though they do not rule out the possibility that a more lipid-soluble glycoside might be able to evoke central autonomic effects of the kind under consideration (WEAVER et al., 1976).

This direct mode of action is certainly feasible if the glycoside is injected directly into the brain stem centers or the ventricular system. Injection of ouabain into the nuclei of the vagus nerve and of the tractus solitarius (cat) stimulates arrhythmias which can be prevented by vagotomy or spinal transection (BASU-RAY et al., 1972). When injected into the ventricular system of vagotomized dogs and cats, ouabain evokes typical sympathomimetic reactions involving blood pressure, heart rate (sometimes causing arrhythmia), and myocardial contraction. Central depletion of noradrenalin (by 6-hydroxydopamine) abolishes the ouabain effect, as does intraventricular injection of β-blockers. The absence of any cardiovascular effects after intravenous administration of the same dose is regarded as final proof that the ouabain effect is of central origin (SAXENA and BHARGAVA, 1975).

F. Central Excitation

When unanesthetized animals of species susceptible to the cardiac effects of digitalis are poisoned with the glycoside, the intoxication runs its course without any conspicuous motor excitation. After lethal doses of digitalis, conscious cats usually show no signs of central excitation; if such signs do occur, they do not appear until the terminal stage and are undoubtedly secondary to the circulatory changes. Digitoxigenin, on the other hand, when given in lethal doses to conscious cats produces marked central excitation with convulsions and respiratory arrest.

In species which are not susceptible to the cardiac effects of digitalis the picture is dominated by purely central manifestations. Lethal doses of digitoxigenin given intravenously to rats and mice cause paralyses as their chief effect, sometimes followed by fits and respiratory arrest. When cardiac glycosides are given intravenously the paralytic manifestations precede respiratory depression. However, doses of cardiac glycosides and their genins injected into the cerebral ventricles invariably produce convulsions, and in these circumstances the glycosides are always more potent than the genins. The fact that the genins are more potent by the intravenous route and the glycosides by the intraventricular route is thought to be due to the better penetration of the genins into the CNS (GREEFF and KASPERAT, 1961; REPKE, 1963; LAGE and SPRATT, 1966).

Further elucidation of the central toxic mechanism of digitoxigenin given intravenously has come from pretreatment experiments with reserpine, tetrabenazine, and syrosingopine. Pretreatment with these agents has a protective effect, its strength being related to their central monoamine-depleting action (BUTERBAUGH and SPRATT, 1970a).

By using different agents the CNS can be selectively depleted of either noradrenalin or serotonin. Such experiments have shown that only by lowering the CNS serotonin concentration can any substantial decrease in digitoxigenin toxicity be achieved; a fall in noradrenalin concentration is ineffectual (BUTERBAUGH and SPRATT, 1970b).

Similar studies in anesthetized cats (i.v. infusion of desacetyllanatoside C) have demonstrated the mediator function of serotonin in the brain, and even its role in the generation of cardiac arrhythmias. Selective depletion of central serotonin levels considerably increases the dose of the glycoside needed to induce arrhythmias, an effect which is even more pronounced after intravenous administration of the serotonin antagonist methysergide. The conclusion drawn from this work is that central serotonergic mechanisms are the cause of the heightened sympathetic activity (HELKE et al., 1976).

Similar results have been obtained from experiments with peruvoside and ouabain in conscious cats, though admittedly by intracerebroventricular (i.c.v.) injection. The toxic response was assessed by grading the neurologic and autonomic reactions. Bromlysergic acid diethylamide i.c.v., which is a serotonin antagonist and causes serotonin depletion, had a protective effect. Haloperidol (a dopamine antagonist, i.c.v.) and lithium carbonate i.c.v. and intraperitoneal (i.p.) had no protective effect. Furthermore, i.c.v. doses of peruvoside and ouabain raised the serotonin concentration in the ventricular perfusate, but after pretreatment with parachlorphenylalanine (serotonin depletion) no serotonin release was observed (GAITONDÉ and JOGLEKAR, 1977).

In mice given toxic doses of digitoxigenin i.v. the leading feature is attacks of motor convulsions, and here too the digitoxigenin convulsive threshold can be raised by previous monoamine depletion. In contrast to this observation, the thresholds for electroshock convulsions and pentylenetetrazol convulsions are lowered by monoamine depletion. In nonpretreated mice a dose of digitoxigenin insufficient to induce convulsions (half the median convulsive dose) will lower the electroshock convulsive threshold. However, this heightening of excitability by digitoxigenin is not significantly enhanced by supplementary depletion of noradrenalin and serotonin, though this by itself will lower the electroshock convulsive threshold (BUTERBAUGH and LONDON, 1977).

G. Visual Symptoms

Ophthalmic disorders are among the commoner extracardiac side effects of digitalis glycosides and have been known ever since the early days of digitalis therapy. The main feature is derangement of color vision (xanthopsia), a change which has been demonstrated in several clinical studies by the well-known color chart tests. Scotomata are less commonly noted, but are sometimes encountered during digitoxin treatment, and may be present even though the ECG does not show typical digitalis abnormalities (GELFAND, 1951; SCHLIACK et al., 1967; BELZ and SCHMIDT-VOIGT, 1971). All these changes are fully reversible when the cardiac glycoside is stopped. In the literature there are reports of impairment of visual acuity, chiefly in connection with digitoxin. The pathophysiologic basis of these abnormalities, especially those cases with a central scotoma, is thought to be retrobulbar optic neuritis. In the broader sense digitalis glycosides are classified under the category of retinotoxic drugs. Distribution studies have revealed high local concentrations in the optic nerve – several times higher than in the brain as a whole.

H. Neurotoxicity in Humans

Because of their subjective nature, central nervous system side effects occurring during treatment with digitalis are frequently overlooked. Even in case reports of digitalis poisoning (overdoses, attempted suicide) a direct causal link between the glycoside and the cerebral manifestations is not always obvious, and only in a few such cases have serum glycoside levels been measured. Furthermore, all the reported series are made up of elderly patients, so that any psychological abnormalities are likely to be masked by the mental changes of old age.

Nevertheless, unequivocal psychological abnormalities (psychoses, paranoia, delirium) have been described in patients receiving digitalis glycosides, digitoxin being the chief offender. In practically every case the manifestations subsided when the drug was discontinued, the rate of reversion to normal being consistent with the rate of removal of the drug from the body (CHURCH and MARRIOTT, 1959; MILLER and FORKER, 1974; SHEAR and SACHS, 1978; STORZ, 1972).

An outbreak of mass poisoning with digitoxin occurred in Holland in 1969 and gave an opportunity for sophisticated statistical analysis of the central nervous system side effects. In all, 179 patients were admitted to hospital and the following neurologic abnormalities were observed: sedation and apathy (95%), muscle weakness (82%), abnormal behavior (65%); 12 patients developed true psychoses, and 4 patients with delirium required treatment (LELY and ENTER, 1972).

References

Basu-Ray, B.N., Dutta, S.N., Pradhan, S.N.: Effects of microinjections of ouabain into certain medullary areas in cats. J. Pharmacol. Exp. Ther. *181*, 357–361 (1972)

Beller, G.A., Hood, W.B., jr., Smith, T.W., Abelman, W.H., Wacker, W.E.C.: Correlation of serum magnesium levels and cardiac digitalis intoxication. Am. J. Cardiol. *33*, 225–229 (1974)

Belz, G.G., Schmidt-Voigt, J.: Entoptische Farberscheinungen unter Digitalis-Therapie. Fortsch. Med. *89*, 93–96 (1971)

Borison, H.L., Brizzee, K.R.: Morphology of emetic chemoreceptor trigger zone in cat medulla. Proc. Soc. Exp. Biol. Med. *77*, 38–42 (1951)

Borison, H.L., Fairbanks, V.F.: Mechanism of veratrum-induced emesis in the cat. J. Pharmacol. Exp. Ther. *105*, 317–325 (1952)

Buterbaugh, G.G., Spratt, J.L.: Observations on the possible role of central mechanisms in acute digitoxigenin toxicity. Toxicol. Appl. Pharmacol. *17*, 387–399 (1970a)

Buterbaugh, G.G., Spratt, J.L.: The possible role of brain monoamines in the acute toxicity of digitoxigenin. J. Pharmacol. Exp. Ther. *175*, 121–130 (1970b)

Buterbaugh, G.G., London, E.D.: The relationship between magnitude of electroshock stimulation and the effects of digitoxigenin, pentylenetetrazol, and brain monoamine reduction on electroshock convulsive threshold. Neuropharmacol. *16*, 617–623 (1977)

Cameron, J.R.: The respiratory response to injection of ouabain into the cerebral ventricles. Resp. Physiol. *3*, 55 (1967)

Chai, C.Y., Hsu, P.L., Wang, S.C.: Central locus of emetic action of digitalis substances in cats. Neuropharmacol. *12*, 1187–1193 (1973)

Church, G., Marriott, H.J.L.: Digitalis delirium: A report on three cases. Circulation *20*, 549–553 (1959)

Dresbach, M.: Further experiments in an attempt to locate the site of the emetic action of the digitalis glycosides. J. Pharmacol. Exp. Ther. *91*, 307–316 (1947)

Gaitondé, B.B., McCarthy, L.E., Borison, H.L.: Central emetic action and toxic effects of digitalis in cats. J. Pharmacol. Exp. Ther. *147*, 409–415 (1965)

Gaitondé, B.B., Joglekar, S.N.: Role of catecholamines in the central mechanism of emetic response induced by peruvoside and ouabain in cats. Br. J. Pharmacol. *54*, 157–162 (1975)
Gaitondé, B.B., Joglekar, S.N.: Mechanism of neurotoxicity of cardiotonic glycosides. Br. J. Pharmacol. *59*, 223–229 (1977)
Gelfand, M.L.: Visual symptoms after digitoxin therapy. J. Am. Med. Assoc. *147*, 1231–1233 (1951)
Gillis, R., Raines, A., Sohn, Y.J., Levitt, B., Standaert, G.: Neural excitatory effects of digitalis and their role in the development of cardiac arrhythmias. J. Pharmacol. Exp. Ther. *183*, 154–168 (1972)
Greeff, K., Westermann, E.: Untersuchungen über die muskellähmende Wirkung des Strophanthins. Naunyn-Schmiedebergs Arch. Exp. Path. Pharmakol. *226*, 103–113 (1955)
Greeff, K.: Mineralstoffwechselwirkung und Toxicität des Strophanthins bei nebennierenlosen Ratten. Naunyn-Schmiedebergs Arch. Exp. Path. Pharmakol. *231*, 391–400 (1957)
Greeff, K., Kasperat, H.: Vergleich der neurotoxischen Wirkung von Digitalisglykosiden und Geninen bei intracerebraler und intravenöser Injektion an Mäusen, Ratten und Meerschweinchen. Naunyn-Schmiedebergs Arch. Exp. Path. Pharmakol. *242*, 76–89 (1961)
Haasis, R., Larbig, D.: Serumglykosidkonzentration und Digitalisintoxikation. Dtsch. Med. Wochenschr. *100*, 1768–1773 (1975)
Helke, C.J., Souza, J.D., Hamilton, B.L., Morgenroth, V.H., Gillis, R.A.: Evidence for a role of central serotonergic neurones in digitalis-induced cardiac arrhythmias. Nature *263*, 246–248 (1976)
Lage, G.L., Spratt, J.L.: Structure activity correlation of lethality and central effects of selected cardiac glycosides. J. Pharmacol. Exp. Ther. *152*, 501–508 (1966)
Leachman, R.D., Cokkinos, D.V.P., Cabrera, R., Leatherman, L.L., Rochelle, D.G.: Response of the transplanted denervated human heart to cardiovascular drugs. Am. J. Cardiol. *27*, 272–276 (1971)
Lely, A.H., Enter, C.H.J. van: Non-cardiac symptoms of digitalis intoxication. Am. Heart J. *83*, 149–152 (1972)
Levitt, B., Cagin, N.A., Somberg, J., Bounous, H., Mittag, Th., Raines, A.: Alteration of the effects and distribution of ouabain by spinal cord transection in the cat. J. Pharmacol. Exp. Ther. *185*, 24–28 (1973)
McLain, P.L.: Effects of cardiac glycosides on spontaneous efferent activity in vagus and sympathetic nerves of cats. Int. J. Neuropharmacol. *8*, 379–387 (1969)
Miller, St., Forker, A.D.: Digitalis toxicity. J. Kans. Med. Soc. *75*, 263–267 (1974)
Pace, D.G., Gillis, R.A.: Neuroexcitatory effects of digoxin in the cat. J. Pharmacol. Exp. Ther. *199*, 583–600 (1976)
Repke, K.: Metabolism of cardiac glycosides. In: Proceedings of the First International Pharmacological Meeting, Stockholm. Wilbrandt, W. (ed.), Vol. 3, pp. 47–73. New York: Pergamon Press 1963
Ri, H. Dal., Schmidt, G.: Zentrale und periphere Beeinflussung der Atmung durch toxische Herzglykosiddosen an der Ratte. Naunyn-Schmiedebergs Arch. Exp. Path. Pharmakol. *239*, 158–169 (1960)
Riggert, H., Schwab, M.: Der Einfluß von Digitoxin auf die Atmung des herzgesunden Menschen. Z. Ges. Exp. Med. *125*, 80–92 (1955)
Saxena, P.R., Bhargava, K.P.: The importance of a central adrenergic mechanism in the cardiovascular responses to ouabain. Eur. J. Pharmacol. *31*, 332–346 (1975)
Schliack, H., Fischer, G., Ruiz-Torres, A.: Bild einer doppelseitigen retrobulbären Opticusneuritis bei Digitalisüberdosierung. Dtsch. Med. Wochenschr. *92*, 973–977 (1967)
Schwab, M., Wagner, P.H.: Der Einfluß von k-Strophanthin auf die Atmung des herzgesunden Menschen. Z. Ges. Exp. Med. *129*, 28–32 (1957)
Share, N.N., Chai, C.Y., Wang, S.C.: Emesis induced by intracerebroventricular injections of apomorphine and deslanoside in normal and chemoceptive trigger zone ablated dogs. J. Pharmacol. Exp. Ther. *147*, 416–421 (1965)
Shear, M.K., Sachs, M.H.: Digitalis delirium: Report of two cases. Am. J. Psychiatry *135*, 109–110 (1978)

Somberg, J.C., Bounous, H., Levitt, B.: The antiarrhythmic effects of quinidine and propranolol in the ouabain-intoxicated spinally transected cat. Eur. J. Pharmacol. *54*, 161–166 (1979)
Storz, H.: Zur Erhaltungsdosis von β-Methyldigoxin. Herz-Kreislauf *11*, 396–399 (1972)
Weaver, L.C., Akera, T., Brody, Th.: Digoxin toxicity: Primary sites of drug action on the sympathetic nervous system. J. Pharmacol. Exp. Ther. *197*, 1–9 (1976)
Yen, M.H., Chow, S.Y.: Effects of intravenous infusion of ouabain on respiration. Eur. J. Pharmacol. *28*, 95–99 (1974)

CHAPTER 22

Effects of Cardiac Glycosides on Vascular System

D. T. MASON

A. Introduction

The finding that the digitalis-induced stimulation of the force of contraction of the normal heart is not translated into an increase in cardiac output, suggests possible extracardiac vascular effects of the glycosides. A substantial body of evidence has now been accumulated to indicate that the actions of the digitalis glycosides are not limited to the heart. Indeed, it was at one time postulated that the primary action of digitalis was extracardiac and that the direct effects on the vascular bed were responsible, in part, for its therapeutic action (DOCK and TAINTER, 1930; KATZ et al., 1938; MCMICHAEL and SHARPEY-SCHAFER, 1944; LENDLE and MERCKER, 1961). Further, it has been shown that the contractile response of excised arterial and venous strips is enhanced in the presence of digitalis (BRENDER et al., 1969; COW, 1911; FRANKLIN, 1925; LEONARD, 1957). More recently, considerable attention has been devoted to the clarification of the actions of digitalis on the peripheral vasculature both in animals and in humans. It is the purpose of this chapter to review these investigations in order to define the alterations produced by the drug on the resistance vessels (systemic arterioles) and on the capacitance beds (systemic veins) of the peripheral circulation.

B. Systemic Arterioles in Experimental Animals

In the past, it has been difficult to define the direct effects of the glycosides on the peripheral circulation alone because of alterations in the cardiac output which usually accompany digitalization. The hemodynamic dissociation of the heart from the peripheral circulation during cardiopulmonary bypass now makes it possible to separate the direct vascular actions from the cardiac effects of the drug. If the rate at which blood is delivered from the extracorporeal circuit into the systemic arterial bed is held constant, any change in arterial pressure will reflect a change in the resistance offered by the arterial bed.

I. Total Vascular Resistance

To ascertain whether or not digitalis preparations have any effect on the systemic arterial bed of experimental animals, these drugs were given to several dogs placed on total cardiopulmonary bypass (ROSS et al., 1960 b). In the representative experiment shown in Fig. 1, with the perfusion rate maintained constant, digitalis produced a substantial increase in arterial pressure, reflecting an identical percentage increase in peripheral vascular resistance. Ligation of the splanchnic arteries did

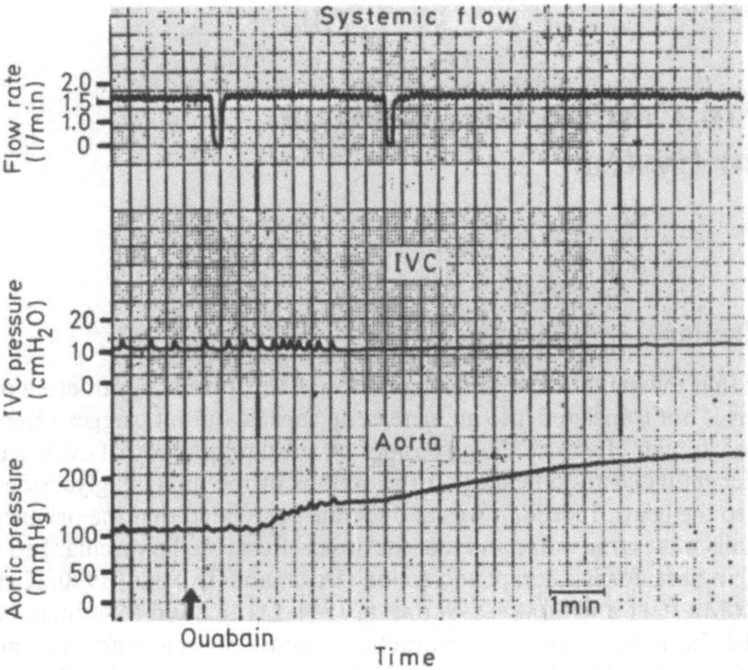

Fig. 1. Simultaneous recording of systemic flow, inferior vena caval (IVC), and aortic pressures following injection of ouabain, at time indicated by *arrow*. (Ross et al., 1960 b)

not appreciably modify the magnitude or duration of the pressor effect of digitalis. Similarly, bilateral adrenalectomy did not prevent the elevation of arterial pressure following administration of the glycosides. These findings, which were obtained with acetylstrophanthidin, ouabain, and lanatoside C, indicate that digitalis produced systemic arteriolar constriction. In several dogs, ganglionic blockade produced with hexamethonium and maintained with a constant infusion of this drug did not prevent the pressor effects of ouabain. Thus, following administration of the glycoside, arterial pressure rose, indicating an increased resistance of the arterial bed. These observations that the pressor action of digitalis is unaltered by adrenergic blockade, suggest that the drug exerts a direct stimulating effect on arteriolar smooth muscle, and that this effect is not mediated through the sympathetic nervous system.

II. Regional Vascular Resistance

In order to characterize further the peripheral arterial effects of digitalis, another preparation was devised (Ross et al., 1960 b). The aorta and inferior vena cava were ligated in the lower abdomen and their distal ends cannulated. When digitalis was injected into the lower circulatory segment, an increase in femoral arterial pressure occurred with no change in brachial arterial pressure. Since the perfusion rate of the lower segment was maintained constant, the change in femoral arterial pressure reflected a proportional increase in vascular resistance. In contrast, following the injection of digitalis into the upper circulatory segment, the pressor response was confined to this segment while in the femoral artery the pressure declined, presumably due to a reflex. These experiments support the view that the action of digitalis responsible for the increase in peripheral vascular resistance is a direct one on arteriolar smooth muscle. In additional studies in normal dogs, ouabain has been shown to constrict the mesenteric vasculature directly (HARRISON et al., 1969; PAWLIK and JACOBSON, 1974), as well as the renal and iliac vascular beds (HIGGINS et al., 1971). In monkeys SHANBOUR et al. (1971) observed after intravenous injection of a maximal therapeutic dose of 20 to 35 μg per kg a decrease in mesenteric blood flow and an increase of mesenteric vascular resistance.

Further evidence also comes from experiments with strips from coronary arteries wherein ouabain exerts a contraction antagonized by Ca^{2+}-antagonistic agents like verapamil (FLECKENSTEIN et al., 1975, 1976). Contrary reports exist too. ZIMPFER et al. (1977) observed no changes in coronary blood flow in anaesthetized dogs following intravenous injection of ouabain, resulting in plasma concentrations of 50–200 ng/ml. STEINESS et al. (1978) however, described a reduced myocardial blood flow in acute and chronic digitalization with digoxin (single dose of 0.05 mg/kg or maintenance dose of 0.02 mg/kg for 8 days). There are some indications that also in man cardiac glycosides in therapeutic or toxic amounts cause a decrease in splanchnic blood flow or even mesenteric infarction (SHANBOUR and JACOBSON, 1972; BYNUM et al., 1973; HESS and STUCKI, 1975; BROBMANN et al., 1980).

C. Systemic Veins in Experimental Animals

It has been repeatedly shown that the cardiac output declines following administration of digitalis to experimental animals (COHN and STEWART, 1928; COTTEN and STOPP, 1958; HARRISON and LEONARD, 1926; REGAN et al., 1956) and to normal human subjects (BING et al., 1950; BURWELL et al., 1927; DRESDALE et al., 1959; GERTZ et al., 1966; GOODYER et al., 1960; HARVEY et al., 1951; RODMAN et al., 1961; SELZER et al., 1959; STEWART and COHN, 1932). This finding is apparently inconsistent with the well-known positive inotropic action of the drug (BRAUNWALD et al., 1965; MASON and BRAUNWALD, 1963; SONNENBLICK et al., 1966). It has been suggested that this phenomenon may be explained by peripheral pooling of blood leading to a decreased venous return to the heart (DOCK and TAINTER, 1930; KATZ et al., 1938; MCMICHAEL and SHARPEY-SCHAFER, 1944). The use of the cardiopulmonary bypass technique made it possible to determine the actions of the digitalis preparations on the systemic and portal venous beds of a group of dogs (ROSS et al., 1960a). By maintaining the pump output constant, the effects of glycosides on venous return and venous tone could be studied by determining the changes in the volume of blood contained in the extracorporeal circuit. In this preparation alterations in extracorporeal blood volume, reflect reciprocal changes in intravascular blood volume. These relative shifts of blood volume permit study of venous tone.

I. Total Systemic Venous Tone

In dogs with intact portal ciculations, venous return decreased and intravascular pooling of blood occurred. In order to define further the actions of digitalis on the

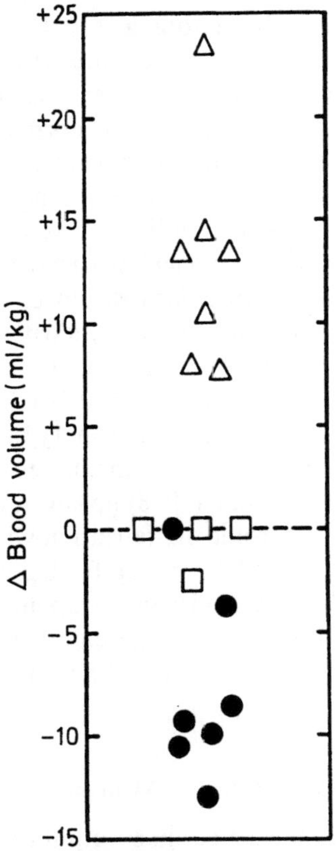

Fig. 2. Alterations in intravascular blood volume, resulting from acute digitalization in the dog. *Open triangles:* portal vein not vented. *Open squares:* portal vein vented, small elevation of portal venous pressure after digitalization. *Closed circles:* portal vein vented, portal venous pressure unchanged after digitalization. When portal circulation was intact, a decline in venous return and a consequent intravascular pooling of blood occured. When pooling of blood in splanchnic bed was prevented, venous return increased and intravascular volume declined. When portal venous bed was only partially decompressed, intravascular blood volume essentially unchanged. (ROSS et al., 1960b)

venous system, the portal veins were decompressed by means of a cannula which returned blood directly from the portal bed to the oxygenator (ROSS et al., 1960a). Following the administration of digitalis, venous return to the oxygenator increased, resulting in a diminution of intravascular blood volume. The pressure in the vena cava increased. These observations, made while portal venous pooling was prevented, indicate that contraction of the systemic venous system had occurred. When decompression of the portal bed was not carried out, the systemic venoconstrictor effect was masked by the more marked effect of portal venous pooling.

II. Hepatic Venous Tone

The alterations in intravascular blood volume resulting from the acute administration of digitalis to all animals studied are summarized in Fig. 2. The illustrated data are consistent with the hypothesis that in the dog, digitalis induces a generalized venoconstriction, most striking in the hepatic veins. Thus, in the dogs with intact portal circulations, the substantial decrease in venous return was attributed to pooling of blood in the portal venous system, presumably due to hepatic venous constriction. These observations offer a possible explanation for the decline in cardiac output that occurs following digitalis administration in the dog with a normal heart and an intact circulation. Although the difference in the smooth-muscle content of the hepatic veins in dog and humans is well recognized (BAUER et al., 1932), it has also been suggested that hepatic venoconstriction and splanchnic pooling of blood may occur in humans. Thus, an increased hepatic venous wedge pressure and an increased gradient between the hepatic venous wedge and inferior vena cava pressures have been observed after administration of digitalis to normal human subjects (BASCHIERI et al., 1957). In anesthetized dogs, previously subjected to endotoxic or oligemic shock, ouabain evoked significant increase in portal venous pressures (ULANO et al., 1971).

D. Systemic Arterioles in Patients

The observations which indicate that digitalis produces arteriolar constriction in experimental animals have been confirmed in a large number of patients studied in the course of open cardiac operations during cardiopulmonary bypass at a constant perfusion rate (BRAUNWALD et al., 1961). A dose of 0.5–1.5 mg of acetylstrophanthidin or 0.8–1.2 mg of lanatoside C resulted in a prompt increase in systemic vascular resistance, which averaged 23% above the control levels. No consistent change in arterial pressure occurred during perfusion in any of the patients who had not received glycosides. These observations indicate that digitalis constricted the systemic arterial bed in humans. The relationships between right ventricular contractile force, recorded with a Walton–Brodie strain guage arch (BONIFACE et al., 1953), and arterial pressure before and following the injection of acetylstrophanthidin are shown in Fig. 3. It should be noted that systemic arteriolar constriction was earlier both in onset and peak effect, with a shorter overall duration than the cardiotonic action of the drug. The arterial baroreceptors appear to attenuate considerably the vasopressor response produced by intravenous digitalis (MCRITCHIE and VATNER, 1976).

E. Systemic Arterioles and Veins in Normal Human Subjects

Little information has been available concerning the extracardiac actions of the usual therapeutic doses of digitalis glycosides administered to intact human subjects. In order to characterize the effects of digitalis on a specific vascular bed, the effects of the drug on both the arteriolar resistance and venous capacitance vessels of the forearm were examined in normal subjects and in patients with congestive heart failure (MASON and BRAUNWALD, 1964). After the vascular bed of the hand was eliminated from the measurements by inflating the sphygmomanometer cuff at the wrist to a suprasystolic pressure, venous occlusion of the forearm was produced by suddenly inflating the cuff on the upper arm to a pressure below the di-

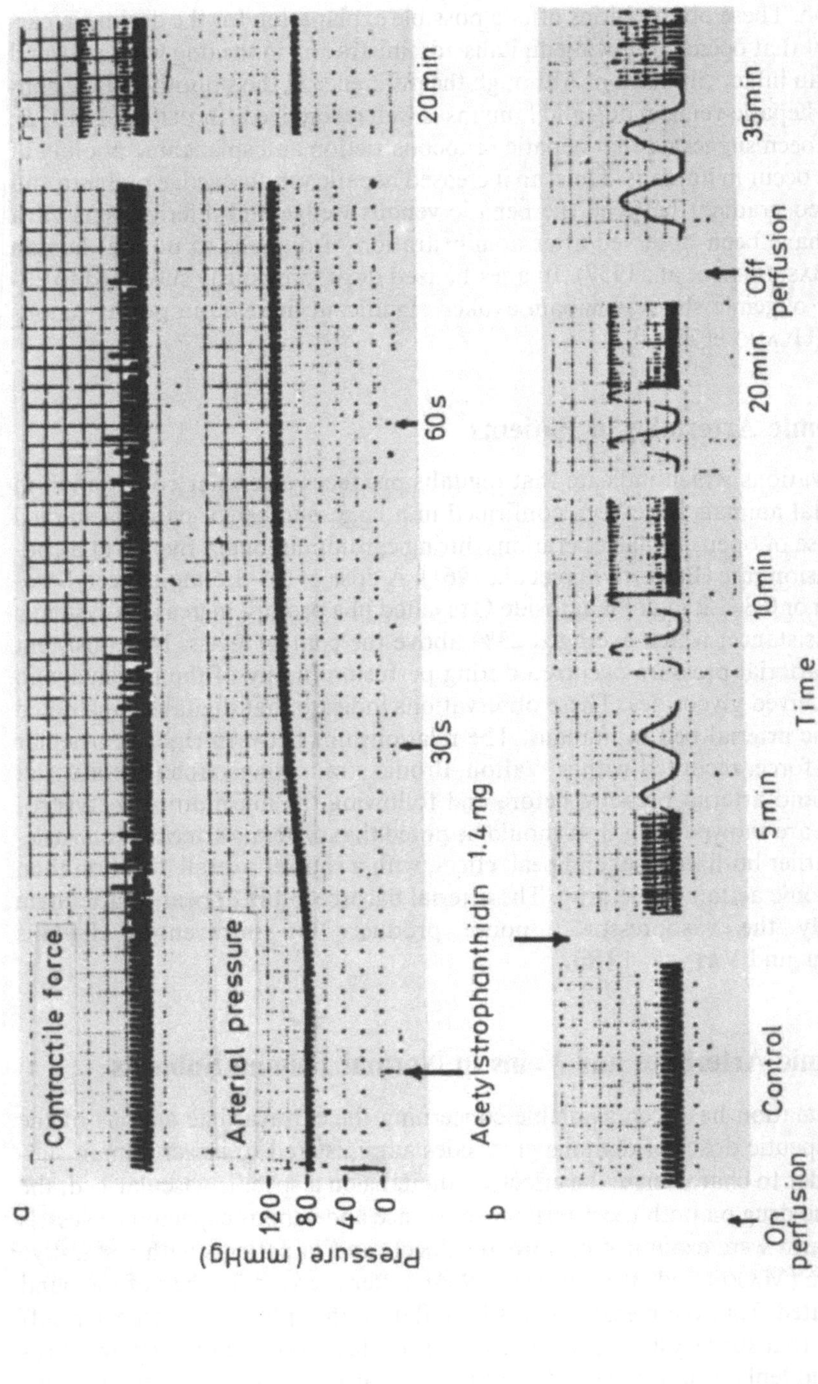

Fig. 3a, b. Acute digitalization during extracorporeal circulation. **a** Contractile force and arterial pressure recordings immediately and 20 min after injection of acetylstrophanthidin in patient with atrial septal defect. **b** Contractile force recordings before injection and at intervals after acetylstrophanthidin. Final recording obtained after bypass had been completed 35 min after administration of drug. (BRAUNWALD et al., 1961)

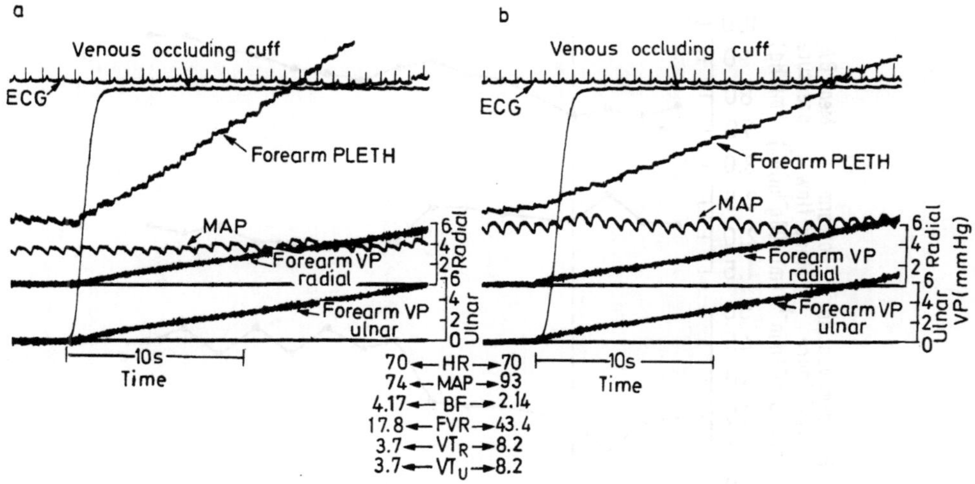

Fig. 4 a, b. Two segments of recording of venous occlusion curves in normal subject. **a** Tracing obtained during control period. *b* Tracing recorded 40 min following the injection of 0.5 mg ouabain. *MAP:* mean arterial pressure. *PLETH:* forearm phethysmographic tracing. *VP:* venous pressure, recorded simultaneously in two veins, one on the radial, the other on the ulnar side of the forearm. Figures below tracing are values of the variables (measured or calculated). *HR:* heart rate. *BF:* Forearm blood flow. *FVR:* forearm vascular resistance (mm Hg ml^{-1} min^{-1}). VT_R: venous tone, calculated using the vain on the radial aspect of the forearm. VT_U: venous tone, calculated using the vein on the ulnar aspect of the forearm. (MASON and BRAUNWALD, 1964)

astolic pressure. Since blood could freely flow into the forearm but was unable to leave the venous compartment, the forearm circumference increased and was measured by a rubber strain guage plethysmograph (Fig. 4). Forearm blood flow was directly proportional to the initial slope of the increase in forearm circumference. Forearm vascular resistance was calculated as the ratio of mean arterial pressure obtained via an indwelling arterial needle in the opposite arm to blood flow. Venous tone was determined by relating the increment in forearm venous pressure, obtained via a venous catheter, to the increment in forearm volume which occurred during venous occlusion. The patients were in a supine position with their forearms elevated in such a manner that the initial venous pressure was always nearly zero. Keeping the initial venous pressure near zero provided a constant reference point for the venous pressure–volume relationship and enabled the determinations to be made at a point where this relationship is known to be linear.

I. Total and Regional Vascular Effects

Following ouabain administration, forearm blood flow declined while arterial pressure rose, and thus forearm vascular resistance increased markedly. Further, there was a lesser increment in forearm volume and a greater increment in venous pressure, indicating that forearm venous tone had increased. The alterations of forearm vascular dynamics caused by ouabain in a normal subject are shown in sequential fashion in Fig. 5 (MASON and BRAUNWALD, 1964). The drug elevated arterial pressure, reduced forearm blood flow, and increased forearm vascular resis-

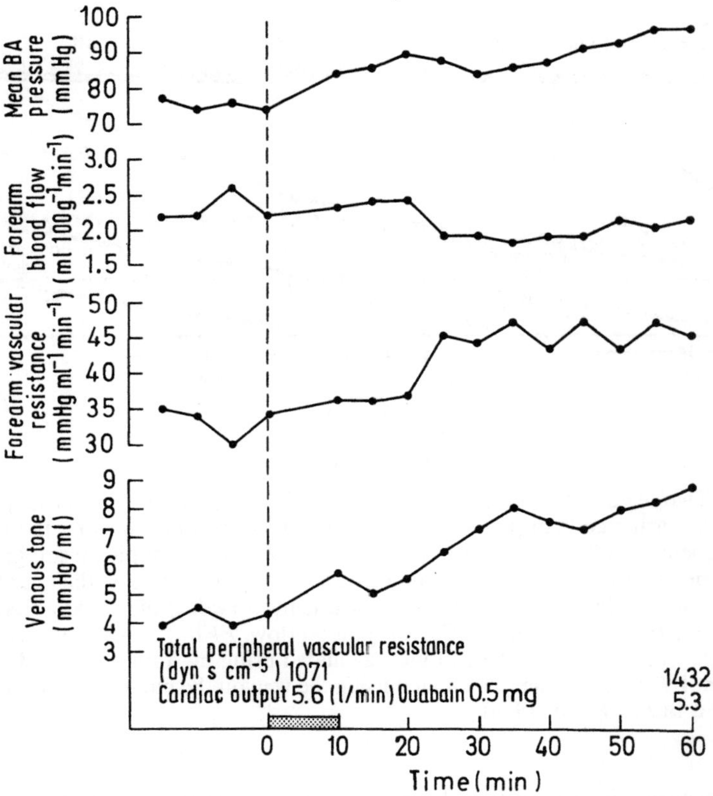

Fig. 5. Serial determinations of arterial pressure (BA), forearm blood flow, forearm vascular resistance, venous tone, cardiac output, and systemic vascular resistance before and after ouabain administration in normal subject. (MASON and BRAUNWALD, 1964)

tance and venous tone. In these normal subjects the peak rise of resistance averaged 38% above the average value during the control period (Fig. 6). The peak elevation of venous tone averaged 52% above the average control value (Fig. 7). Since the cardiac output remained essentially unchanged following ouabain administration, the calculated total systemic vascular resistance was elevated in all subjects. The pressure–volume relation for the veins of the forearm was shifted downward after clinically significant doses of ouabain (Fig. 8). Thus, following ouabain administration, at each level of venous pressure, the forearm volume was smaller, indicating that venoconstriction had occurred in the forearm.

II. Direct Vascular Actions

Since the changes in the resistance and capacitance vessels produced by ouabain occurred in the presence of a stable cardiac output and an elevation of systemic arterial pressure, the drug has a direct effect on vascular smooth muscle. The constriction of the vascular beds could not have resulted from reflexes mediated through the baroreceptor reflex arc, since the elevation of mean arterial pressure that occurred would have resulted in an inhibition, rather than an augmentation,

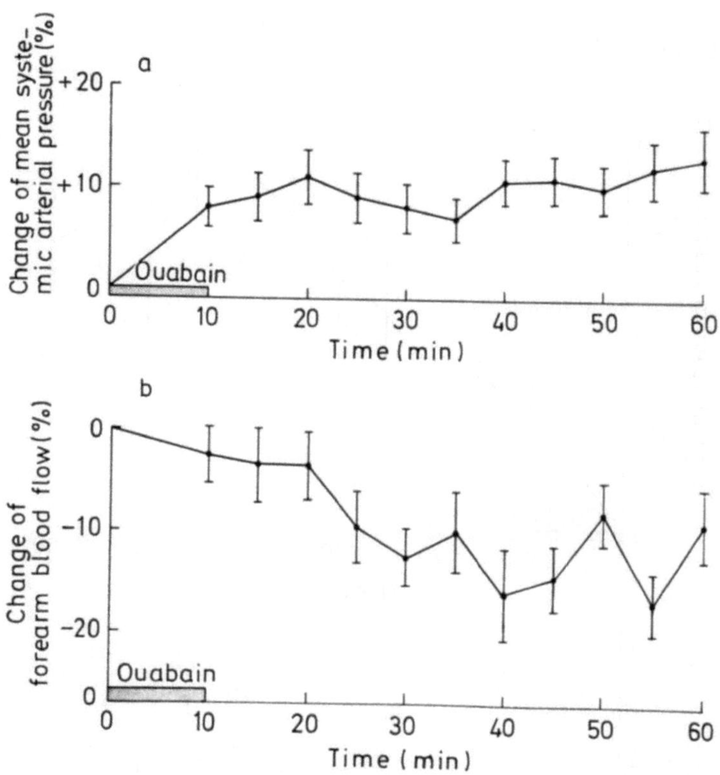

Fig. 6. Average change (± standard error of mean) of mean arterial pressure **a** and forearm blood flow **b** determined by plethysmography, expressed as percentage changes from control values in normal subjects. As arterial pressure was elevated forearm blood flow declined. (MASON and BRAUNWALD, 1964)

of both arterial and venous tone. That ouabain has a direct constricting effect on vascular smooth muscle was supported by the experiments in which the acute venous occlusion method was used to study normal subjects after prolonged treatment with the anti-adrenergic drug, guanethidine (MASON and BRAUNWALD, 1964). In these subjects, direct arteriolar constriction and venous constriction still occurred to the same degree following the administration of ouabain (Fig. 9).

F. Systemic Arterioles and Veins in Patients with Congestive Heart Failure

The actions of digitalis on the vascular dynamics of the forearm were also determined in patients with congestive heart failure (MASON and BRAUNWALD, 1964). During the control period, before the administration of ouabain, patients with congestive heart failure had lower forearm blood flow, and higher calculated forearm vascular resistance and venous tone than normal subjects. The effects of ouabain in the patients with heart failure differed strikingly from those observed in the normal subjects (Fig. 10).

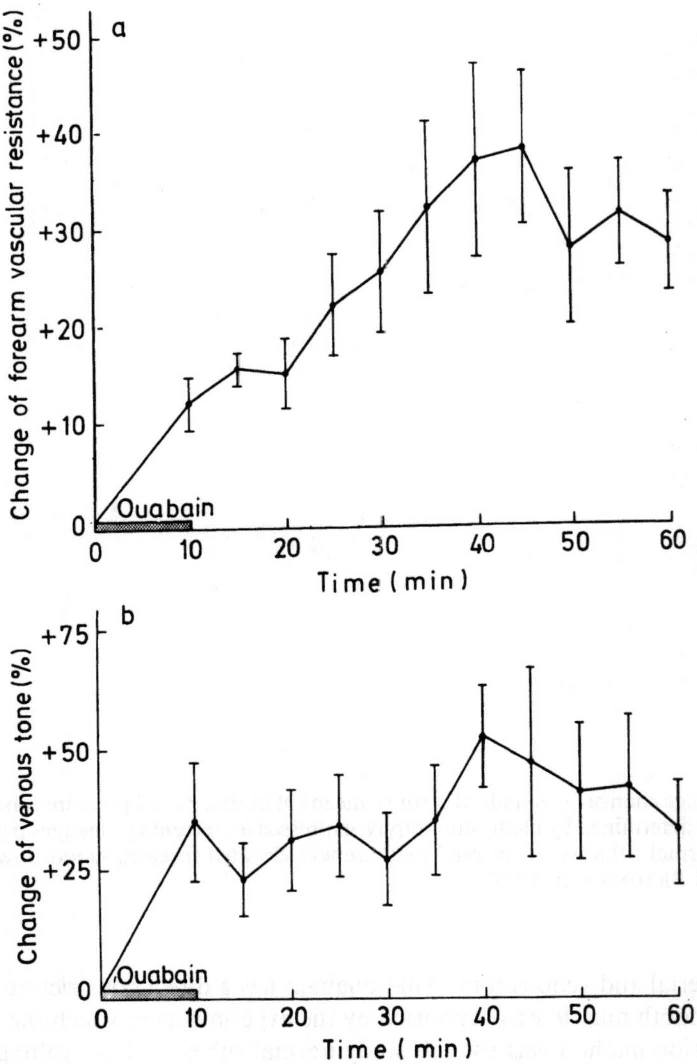

Fig. 7. Average change (± standard error of mean) of forearm vascular resistance **a** and forearm venous tone **b** expressed as percentage changes from control values in normal subjects. (MASON and BRAUNWALD, 1964)

I. Total and Regional Vascular Effects

In all patients with heart failure, the glycoside augmented forearm blood flow, while mean arterial pressure rose slightly or remained unchanged. Therefore, forearm vascular resistance declined. Venous tone, calculated by the acute venous occlusion technique, decreased after ouabain administration in every patient with congestive heart failure. Again in contrast to the normal subjects, ouabain produced a marked rise in cardiac index and a decline in heart rate, central venous pressure, and calculated systemic vascular resistance (Fig. 10). The effects of ouabain on the venous bed of the forearm in patients with congestive heart failure were also

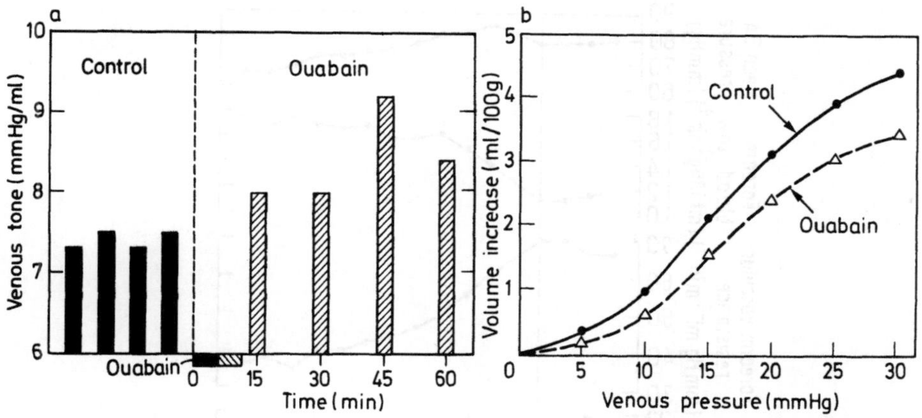

Fig. 8. **a** Venous tone measurements 2 min after inflation of occlusion cuff. *Solid bars* represent four values during control period. *Hatched bars* represent measurements made after ouabain administration. **b** Relationship between venous pressure and forearm volume increase determined by stepwise occlusion method during control period and after ouabain administration. (MASON and BRAUNWALD, 1964)

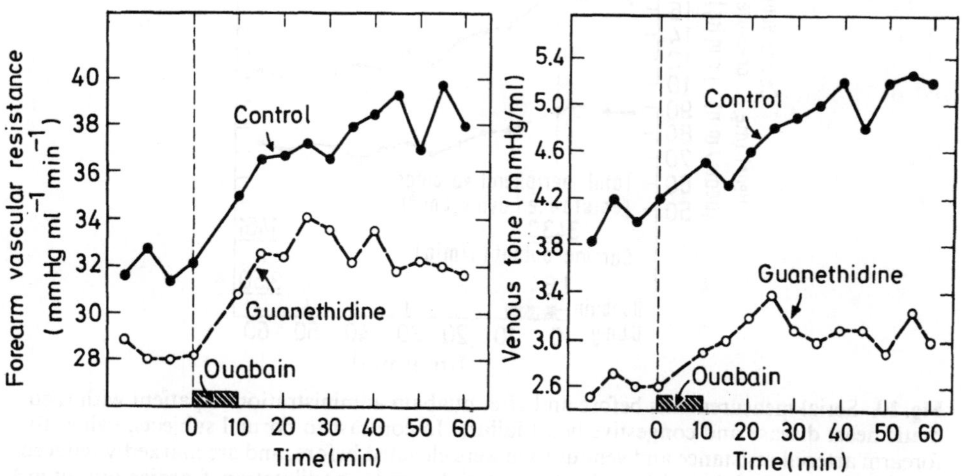

Fig. 9. Average values in four normal subjects showing the effects of ouabain on forearm vascular resistance **a** and venous tone **b** during a control period *(solid circles, solid line)* and after guanethidine administration *(open circles, broken line)*. (MASON and BRAUNWALD, 1964)

studied by the stepwise venous occlusion method in which forearm venous volume was allowed to equilibrate at serial increments of venous pressure (MASON and BRAUNWALD, 1964). Figure 11 illustrates the relationship between venous pressure and forearm volume determined by this equilibration method, before and after ouabain administration in a patient with heart failure. Following ouabain administration, at any given venous pressure, the increment in forearm volume was greater than before the drug was administered, signifying that a decrease in venous tone (venodilation) had occurred.

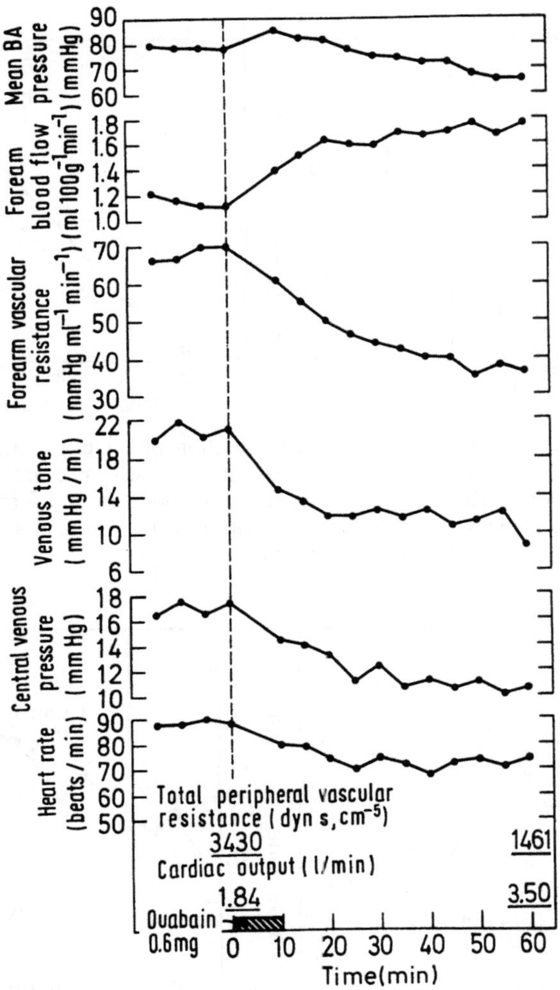

Fig. 10. Serial measurements before and after ouabain administration in patient with rheumatic heart disease and congestive heart failure. In contrast to normal subjects, values for forearm arterial resistance and venous tone were elevated before, and are markedly reduced following drug administration, indicating arteriolar and venodilatation. Cardiac output increased, while central venous pressure fell and total systemic vascular resistance was reduced. (Mason and Braunwald, 1964)

II. Mechanisms of Vascular Action

The effects of the digitalis glycosides on the vascular dynamics of the forearm depended, therefore, on the circulatory state of the patient at the time the drug was administered. These opposite effects of ouabain on the forearm vascular resistance and venous tone in patients with and those without heart failure are of great interest. There is now considerable evidence that the peripheral vasoconstriction in patients with congestive heart failure results from increased activity of the sympathetic nervous system (Braunwald et al., 1966; Burch, 1955; Kelley et al., 1953).

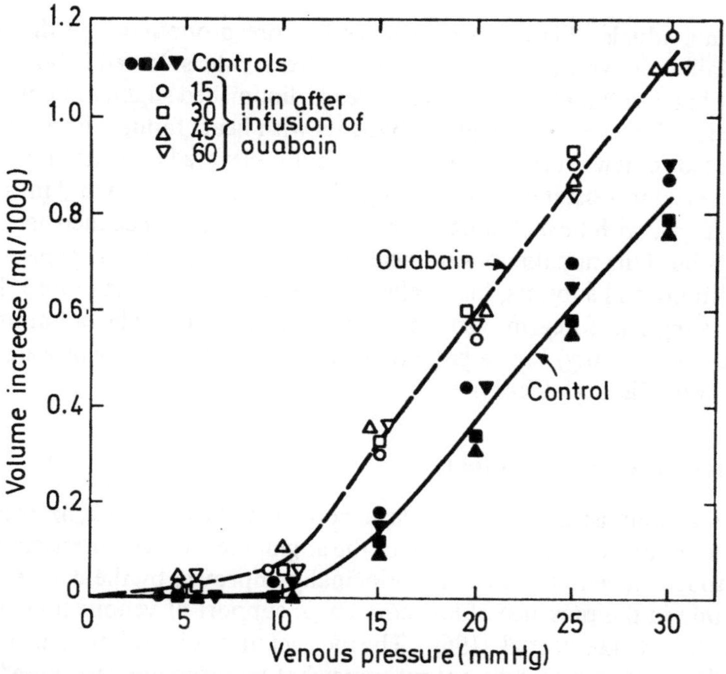

Fig. 11. Relationship between venous pressure and forearm volume increase determined by stepwise occlusion method during four control measurements and at 15 min intervals after ouabain administration in patient with congestive heart failure. (MASON and BRAUNWALD, 1964)

The elevation of cardiac output induced by ouabain in patients with heart failure may result in a diminution of this adrenergically induced arteriolar and venous constriction in the forearm. This interpretation is in agreement with the observation that the constriction of the forearm veins in patients with heart failure diminishes as the state of cardiac compensation improves (WOOD et al., 1956). This conclusion is also in agreement with the finding that a given level of muscular exercise produces a more intense forearm venoconstriction in patients with heart failure than in patients with compensated heart disease (WOOD, 1962). Thus, it is postulated that, as a consequence of the improvement in cardiac dynamics induced by ouabain in the patients with heart failure, the state of the resistance and capacitance vessels in their forearms changes towards normal. In the patients with congestive heart failure, this inhibition of vasoconstriction appears to override the direct vasoconstrictive action of the drug on vascular smooth muscle, observed consistently in the normal individuals.

G. Significance of Digitalis-Induced Vasomotor Changes

I. Normal Versus Heart Failure

The effects of digitalis on circulatory dynamics in normal subjects and patients with heart failure can be summarized in the following manner (MASON, 1974). Mean arterial pressure tends to rise following digitalis administration in both groups of individuals; this pressor response is more pronounced in the patients with heart failure if hypotension is present. The heart rate is slowed in both groups, particularly in heart failure. The cardiac index is diminished slightly in normal subjects, while it is increased strikingly in patients with heart failure. Forearm blood flow is reduced somewhat in normal subjects, but is elevated in heart failure. Total peripheral vascular resistance increases slightly in normal subjects and falls significantly in patients with heart failure. Forearm vascular resistance increases in normal subjects but falls in patients with heart failure. Central venous pressure tends to increase in normal subjects, but declines in patients with heart failure. Finally, in normal subjects, forearm venous tone rises after digitalis administration (venoconstriction occurs) but in patients with heart failure, forearm venous tone diminishes (venodilation occurs).

II. Influence on Cardiac Output

Following the administration of digitalis in congestive heart failure, arterial blood pressure is maintained with a reduction of the augmented arteriolar resistance, and cardiac output and venous return are principally improved by the drug-enhanced inotropic state in the presence of less adrenergic support of venous tone and cardiac contractility (MASON et al., 1969). The indirect arteriolar dilation also aids the elevation of the reduced cardiac output somewhat by providing less impedance to ventricular contraction (MASON, 1978). In addition, there is the possibility that the glycoside may release the hepatic venous sphincter by sympathetic withdrawal, an action which in itself might result in increased venous return despite the presence of generalized indirect venodilation.

III. Influence on Cardiac Preload and Energetics

The finding that digitalis causes systemic venodilation in patients with congestive heart failure provides an explanation for the clinical observation that rapid digitalization of such patients may result in a fall in elevated systemic venous pressure and in relief of pulmonary congestion before the onset of diuresis. In addition, the relaxing effects of the glycosides on the veins in heart failure confirm the suggestion originally made by MCMICHAEL and SHARPEY-SCHAFER (1944) that the drugs reduced elevated venous tone. Present evidence, however, indicates that this is an indirect action (MASON, 1974) and not as they suggested, the prime cause of the beneficial effects of digitalis on cardiac output and the force of contraction of the myocardium. The observation that digitalis produces systemic venodilation in patients with heart failure is in keeping with the findings that the elevation of cardiac output and fall in central venous pressure induced by digoxin in these patients is associated with a decline of splanchnic blood volume (BRADLEY, 1963). It is possible that digitalis eliminates the widespread systemic venoconstriction and that this action results in the displacement of blood from the splanchnic vascular bed to the systemic veins. In patients with heart failure, in spite of overall systemic venodilation and reduced central venous pressure following digitalis administration, venous return is augmented and there is a shift of blood in the venous reservoirs to the systemic bed. In addition, the digitalis-induced inotropic effects provide a more forceful ventricular contraction from a relatively lower filling pressure, resulting in marked enhancement of cardiac output with indirectly lowered myocardial oxygen requirements (COVELL et al., 1966; MASON, 1974).

IV. Coronary Heart Disease

Since digitalis possesses the action of direct arteriolar constriction, this property has been suggested to account for a decrease in cardiac output in some patients with acute myocardial infarction shock to whom the agent has been acutely administered (COHN et al., 1969). When diminished cardiac output is not increased by digitalis, as may occur in a severely damaged left ventricle unable to respond to the cardiotonic agent, the cardiac output may decline further because the moderate direct systemic arteriolar constrictor action of digitalis is not overcome by greater vasodilation due to sympathetic withdrawal (MASON and AWAN, 1979). Further, it is possible that pump output might be reduced by an abrupt increase of peripheral resistance caused by acute digitalization, occurring before the drug's positive inotropic action becomes manifest (BRAUNWALD et al., 1961). In addition, rapidly administered fast-acting glycoside preparations may have a direct constrictor effect on the coronary vascular bed (DEMOTS et al., 1975; VATNER et al., 1971).

V. Acute Pulmonary Edema

The unusual development of acute pulmonary edema has been reported following rapid digitalization in patients with severe cardiac disease (BAYLISS et al., 1950). This observation, which must be quite rare, could be explained by the moderate vasopressor effect of intravenous digitalis taking place prior to the cardiotonic actions (BRAUNWALD et al., 1961), or by the direct systemic arteriolar constriction of digitalis being unopposed by the drug's lack of improvement of ventricular dysfunction without consequent decrease of heightened adrenergic activity (MASON and AWAN, 1979). It is also possible that pulmonary congestion might be worsened by direct drug-effected venoconstriction in the absence of the ability of the diseased heart to improve its impaired pump performance to the inotropic stimulation provided by digitalis. In addition, it is pointed out that the degree of vasoconstriction produced by the glycosides is directly related to the speed of intravenous administration (DEMOTS et al., 1975).

VI. Compensated Heart Failure

In patients with heart disease without failure, the compensatory mechanisms of ventricular dilation, myocardial hypertrophy, and sympathetic stimulation are capable of maintaining the resting cardiac output at the normal level (MASON et al., 1970). Although sympathetic nervous activity is increased, the degree of elevated adrenergic stimulation of the cardiovascular system is less than in overt decompensated heart failure. Thus, vasoconstriction is not severe in the resting state in heart disease without decompensation, and both the opposing direct stimulatory effect and the indirect relaxing action of digitalis on blood vessels are equally balanced. As a result, there is no alteration of vascular tone following the administration of the drug (MASON et al., 1969). Therefore, the administration of digitalis in patients with heart disease without heart failure results in no change in cardiac output, since the effects of digitalis on cardiac contractility and ventricular impedance are equally balanced in terms of the drug's overall action on cardiac pump performance. In normal subjects in whom there is little sympathetic tone at rest, the cardiac output may fall slightly within the normal range after digitalis administration, since the glycoside may exert a greater effect on ventricular impedance than on contractile state in this setting (MASON, 1974).

H. Conclusions

It has been shown in experimental and clinical studies that peripheral and inotropic actions play a role in the hemodynamic alterations following digitalis administration. These findings indicate that in normal subjects, digitalis constricts the arterial and venous beds of the systemic circulations by a direct stimulating action. These extracardiac vascular actions of digitalis clarify the mechanisms of essentially unaltered normal cardiac output by the agent in patients without cardiac decompensation in spite of the fact that the glycoside clearly increases myocardial contractility. The drug-induced increased impedance to ventricular ejection offsets the augmented force of myocardial contraction. Further, the drug-induced systematic venoconstriction, being apparently more intense in the hepatic veins, may lead to portal venous pooling and ultimately diminish venous return to the heart. The evidence that this occurs in the dog is presented, and the possibility that this phenomenon might occur in humans as well is indicated. In contrast to normal subjects, in patients with heart failure, the glycoside increases the reduced forearm blood flow and lowers the elevated forearm vascular resistance and venous tone. It is postulated that this indirect vasodilator action, which overrides the direct stimulating effect of digitalis on the peripheral vascular beds, results from augmentation of the lowered cardiac output in heart failure. Thus, the effects of the glycoside on the peripheral vascular bed are dependent on the circulatory state of the patient at the time the drug is administered. These observations indicate that the digitalis glycosides have both cardiac and extracardiac actions and that these responses occur in normal subjects as well as in patients with congestive heart failure.

References

Baschieri, L., Ricci, P.D., Mazzuoli, G.F., Vassalle, M.: Studi su la portata epatica nell'uomo: Modificazioni del flusso epatico da digitale. Cuore e Circ *41*, 103–110 (1957)

Bauer, W., Dale, H.H., Poulsson, L.T., Richard, D.W.: The control of circulation through the liver. J. Physiol. (Lond.) *74*, 343–344 (1932)

Bayliss, H.G., McMichael, J., Read, E.A.S.: The effect of digoxin on the right ventricular pressure in hypertensive and ischaemic heart failure. Brit. Heart J. *12*, 317–321 (1950)

Bing, R.J., Maraist, F.M., Dammann, J.F., jr., Draper, A., jr., Heimbecker, R., Daley, R., Gerard, R., Calazel, P.: Effect of strophanthus on coronary blood flow and cardiac oxygen consumption of normal and failing human hearts. Circulation *2*, 513–520 (1950)

Boniface, K.J., Brodie, O.J., Walton, R.P.: Resistance strain guage arches for direct measurement of heart contractile force in animals. Proc. Soc. Exp. Biol. (N.Y.) *84*, 263–266 (1953)

Bradley, S.E.: Circulation and the liver. Gastroenterology *44*, 403–410 (1963)
Braunwald, E., Bloodwell, R.D., Goldberg, L.I., Morrow, A.G.: Studies on digitalis. IV. Observations in man on the effects of digitalis preparations on the contractility of the non-failing heart and on total vascular resistance. J. Clin. Invest. *40*, 52–60 (1961)
Braunwald, E., Chidsey, C.A., Pool, P.E., Sonnenblick, E.H., Ross, J., jr., Mason, D.T., Spann, J.F., Covell, J.W.: Congestive heart failure: Biochemical and physiological considerations. Ann. Intern. Med. *64*, 904–925 (1966)
Braunwald, E., Mason, D.T., Ross, J., jr.: Studies on the cardiocirculatory actions of digitalis. Medicine *44*, 233–245 (1965)
Brender, D., Vanhoutte, P.M., Shepherd, J.T.: Potentiation of adrenergic venomotor responses in dogs by cardiac glycosides. Circ. Res. *25*, 597–606 (1969)
Brobmann, G.F., Mayer, M., Grimm, W., Safer, A.: Aufhebung Herzglycosid-induzierter Spasmen der Mesenterialgefäße durch Calcium-Antagonisten im Hundexperiment. In: Calcium-Antagonismus. A. Fleckenstein and H. Roskamm (eds.), 230–242. Berlin-Heidelberg-New York: Springer-Verlag 1980
Burch, R.R.: The effects of intravenous hexamethonium on venous pressure of normotensive and hypertensive patients with and without congestive heart failure. Circulation *11*, 271–279 (1955)
Burwell, C.S., Neighbors, DeW, Regen, E.M.: The effect of digitalis upon the output of the heart in normal man. J. Clin. Invest. *5*, 125–132 (1927)
Bynum, T.E., Hanley, H.O., Cole, J.S.: Effect of digitalis glycosides on splanchnic blood flow in man. Clin. Res. *22*, 509 (1973)
Cohn, A.E., Stewart, J.J.: The relation between cardiac size and output per minute following the administration of digitalis in normal dogs. J. Clin. Invest. *6*, 53–62 (1928)
Cohn, J.N., Tristani, F.E., Khatri, I.M.: Cardiac and peripheral vascular effects of digitalis in clinical cardiogenic shock. Am. Heart J. *78*, 318–330 (1969)
Cotten, M.D., Stopp, P.E.: The action of digitalis on the non-failing heart of the dog. Am. J. Physiol. *192*, 114–122 (1958)
Covell, J.W., Braunwald, E., Ross, J., jr.: Studies on digitalis. XVI. Effects on myocardial oxygen consumption. J. Clin. Invest. *45*, 1535–1542 (1966)
Cow, D.: Some reactions of surviving arteries. J. Physiol. (Lond.) *42*, 125–132 (1911)
Demots, H., McAnulty, J., Porter, G., Rahimtoola, S.: Effects of rapid and slow infusion of ouabain on systemic and coronary vascular resistance in patients without heart failure. Circulation *52*, Suppl. II, 77 (1975)
Dock, W., Tainter, M.L.: The circulatory changes after full therapeutic doses of digitalis, with a critical discussion of views on cardiac output. J. Clin. Invest. *8*, 467–475 (1930)
Dresdale, D.T., Yuceoglu, Y.Z., Michtom, R.J., Schultz, M., Lunger, M.: Effects of lanatoside C on cardiovascular hemodynamics: Acute digitalizing doses in subjects with normal hearts and with heart disease without failure. Am. J. Cardiol. *4*, 88–97 (1959)
Fleckenstein, A., Nakayama, G., Fleckenstein-Grün, G., Byon, Y.K.: Interactions of vasoactive ions and drugs with Ca-dependent excitation-contraction coupling of vascular smooth muscle. In: Calcium transport in contraction and secretion. Carafoli, E. (ed.), pp. 555–566. Amsterdam: North-Holland Publ. Comp. 1975
Fleckenstein, A., Nakayama, K., Fleckenstein-Grün, G., Byon, Y.K.: Interactions of H ions, Ca-antagonistic drugs, and cardiac glycosides with excitation-contraction coupling of vascular smooth muscle. In: Ionic actions on vascular smooth muscle. Betz, E. (ed.), pp. 117–123. Berlin, Heidelberg, New York: Springer 1976
Franklin, K.J.: The pharmacology of the isolated vein ring. J. Pharmacol. Exp. Ther. *26*, 215–223 (1925)
Gertz, E.W., Hess, M.L., Briggs, F.N.: Augmentation of calcium uptake in cardiac vesicles from failed heart-lung preparations by G-Strophanthin. Pharmacologist *8*, 222–230 (1966)
Goodyer, A.V.N., Chetrick, A., Huvos, A.: The effect of Lanatoside-C on the response of the human cardiac output to walking exercise. Yale J. Biol. Med. *32*, 265–276 (1960)
Harrison, L.A., Blaschke, J., Phillips, R.S., Price, W.E., Cotten, M.DeV., Jacobson, E.D.: Effects of ouabain on the splanchnic circulation. J. Pharmacol. Exp. Ther. *169*, 321–327 (1969)

Harrison, T.R., Leonard, B.W.: The effect of digitalis on the cardiac output of dogs and its bearing on the action of the drug in heart disease. J. Clin. Invest. *3*, 1–15 (1926)
Harvey, R.M., Ferrer, M.I., Cathcart, R.T., Alexander, J.K.: Some effects of digoxin on the heart and circulation in man. Digoxin in enlarged hearts not in clinical congestive failure. Circulation *4*, 366–375 (1951)
Hess, T., Stucki, P.: Mesenterialinfarkt bei Digitalisintoxikation. Schweiz. med. Wschr. *105*, 1237–1240 (1975)
Higgins, C.B., Vatner, S.F., Franklin, D., Braunwald, E.: Effects of digitalis on regional vascular resistances in conscious dogs with and without experimental heart failure. Fed. Proc. *30*, 283 (1971)
Katz, L.N., Rodbard, S., Friend, M., Rottersman, W.: The effect of digitalis on the anesthetized dog. I. Action on the splanchnic bed. J. Pharm. Exp. Ther. *62*, 1–10 (1938)
Kelley, R.T., Freis, E.D., Higgins, R.F.: The effects of hexamethonium on certain manifestations of congestive heart failure. Circulation *7*, 169–178 (1953)
Lendle, L., Mercker, H.: Extrakardiale Digitaliswirkungen. Ergeb. Physiol. *51*, 200–298 (1961)
Leonard, E.: Alteration of contractile response of artery strips by a potassium-free solution, cardiac glycosides, and changes in stimulation frequency. Am. J. Physiol. *189*, 185–194 (1957)
McMichael, J., Sharpey-Schafer, E.P.: The action of intravenous digoxin in man. Quart. J. Med. *13*, 123–132 (1944)
McRitchie, R.J., Vatner, S.F.: The role of arterial baroreceptors in mediating the cardiovascular response to a cardiac glycoside in conscious dogs. Circ. Res. *38*, 321–326 (1976)
Mason, D.T.: Digitalis pharmacology and therapeutics: Recent advances. Ann. Int. Med. *80*, 520–530 (1974)
Mason, D.T.: Afterload reduction and cardiac performance. Am. J. Med. *65*, 106–125 (1978)
Mason, D.T., Awan, N.A.: Recent advances in digitalis research. Am. J. Cardiol. *43*, 1056–1059 (1979)
Mason, D.T., Braunwald, E.: Studies on digitalis. IX: Effects of ouabain on the nonfailing human heart. J. Clin. Invest. *42*, 1105–1115 (1963)
Mason, D.T., Braunwald, E.: Studies on digitalis. X. Effect of ouabain on forearm vascular resistance and venous tone in normal subjects and in patients in heart failure. J. Clin. Invest. *43*, 532–543 (1964)
Mason, D.T., Spann, J.F., jr., Zelis, R.: New developments in the understanding of the actions of the digitalis glycosides. Prog. Cardiovasc. Dis. *6*, 443–478 (1969)
Mason, D.T., Spann, J.F., jr., Zelis, R.: Alterations of hemodynamics and myocardial mechanics in patients with congestive heart failure: Pathophysiologic mechanisms and assessment of cardiac function and ventricular contractility. Prog. Cardiovasc. Dis. *12*, 507–557 (1970)
Pawlik, W., Jacobson, E.D.: Effects of digoxin on the mesenteric circulation. Cardiovas. Res. Cent. Bull. *12*, 80–84 (1974)
Regan, T.J., Talmers, F.N., Hellems, H.K.: Myocardial transfer of sodium and potassium: Effect of acetylstrophanthidin in normal dogs. J. Clin. Invest. *35*, 1220–1228 (1956)
Rodman, T., Gorczyca, C.A., Pastor, B.H.: The effect of digitalis on the cardiac output of the normal heart at rest and during exercise. Ann. Int. Med. *55*, 620–630 (1961)
Ross, J., jr., Braunwald, E., Waldhausen, J.A.: Studies on digitalis. II. Extracardiac effects on venous return and on the capacity of the peripheral vascular bed. J. Clin. Invest. *39*, 937–942 (1960a)
Ross, J., jr., Waldhausen, J.A., Braunwald, E.: Studies on digitalis. I. Direct effects on peripheral vascular resistance. J. Clin. Invest. *39*, 930–938 (1960b)
Selzer, A., Hultgren, H.N., Ebnother, C.L., Bradley, H.W., Stone, A.O.: Effect of digoxin on the circulation in normal man. Brit. Heart J. *21*, 335–342 (1959)
Shanbour, L.L., Jacobson, E.D., Brobmann, G.F., Hinshaw, L.B.: Effects of ouabain on splanchnic hemodynamics in the rhesus monkey. Am. Heart J. *81*, 511–515 (1971)

Shanbour, L.L., Jacobson, E.D.: Digitalis and the mesenteric circulation. Digestive Diseases *17*, 826–827 (1972)

Sonnenblick, E.H., Williams, J.R., jr., Glick, G., Mason, D.T., Braunwald, E.: Studies on digitalis. XV. Effects of cardiac glycosides on myocardial force-velocity relations in the nonfailing human heart. Circulation *34*, 532–540 (1966)

Steiness, E., Bille-Brahe, N.E., Hansen, J.F., Lomholt, N., Ring-Larsen, H.: Reduced myocardial blood flow in acute and chronic digitalization. Acta pharmacol. et toxicol. *43*, 29–35 (1978)

Stewart, H.J., Cohn, A.E.: Studies on the effect of the action of digitalis on the output of blood from the heart. III. The effect on the output in normal human hearts. J. Clin. Invest. *11*, 917–930 (1932)

Ulano, H.B., Treat, E., Chang, A.C.K., Jacobson, E.D.: Splanchnic circulatory responses to ouabain in shock. Surgery *70*, 678–684 (1971)

Vatner, S.F., Higgins, C.B., Franklin, D., Braunwald, E.: Effects of a digitalis glycoside on coronary and systemic dynamics in conscious dogs. Circ. Res. *28*, 470–479 (1971)

Wood, J.E.: The mechanism of the increased venous pressure with exercise in congestive heart failure. J. Clin. Invest. *41*, 2020–2032 (1962)

Wood, J.E., Litter, T., Wilkins, R.W.: Peripheral venoconstriction in human congestive heart failure. Circulation *13*, 524–534 (1956)

Zimpfer, M., Schütz, W., Raberger, R.: Haemodynamics and coronary actions of ouabain during coronary infusion. Naunyn-Schmiedebergs Arch. Pharmacol. *299*, 61–64 (1977)

CHAPTER 23

Effects of Cardiac Glycosides on Skeletal Muscle

B. DÉNES and K. GREEFF

A. Introduction

Cardiac glycosides, in therapeutic concentrations, are considered to influence the force of contraction of the heart, not only during an existing insufficiency but also in normal healthy conditions. A similar influence of these glycosides on the skeletal muscle is, however, not clear-cut since, only much higher concentrations produce an effect. Indeed, in this context, the basic differences between heart and skeletal muscle in their respective morphology (e.g., number of mitochondria, structure of the sarcoplasmic reticulum), electrophysiologic properties and contraction–relaxation velocity, activity of the membrane enzyme Na^+, K^+-ATPase and its specificity, and binding and accumulation characteristics, as well as basic distinctions between red and the white muscle, all assume significance and importance. An assessment and discussion of these various characteristics and their causal role in the differential sensitivity of the heart and skeletal muscle to known cardiac glycosides is, therefore, appropriate. For a discussion of differences in heart and skeletal muscle myofibrillar proteins, see Chap. 19.

B. Basic Differences Between Skeletal and Heart, and Red and White Muscle

I. Morphological Structure

The morphological differences between skeletal and heart muscle have been set out schematically in Fig. 1. Significantly, there are fewer mitochondria in skeletal, than in heart muscle.

The distribution of the transverse tubular system and of the sarcoplasmic reticulum is different in these two types of muscle. The transverse tubuli, representing inward extensions of the extracellular space are smaller, and the sarcoplasmic reticulum more prominent in skeletal muscle. The sarcoplasmic reticulum is assumed to be responsible for the release of calcium (Ca) during stimulation and, its rebinding during relaxation. This release and rebinding of Ca are principally intracellular processes in skeletal muscle (EBASHI, 1961, 1976; HASSELBACH, 1964; EBASHI and ENDO, 1968; NAYLER, 1973; INESI and MALAN, 1976; RICH and LANGER, 1975). Stimulation of cardiac muscle is coupled with a strong inward Ca current from the extracellular space and a release from binding sites in the plasmalemma, whereas storage in the sarcoplasmic reticulum or release from it have only a modifying influence on the extent of the intracellular concentration of Cations, and on the veloc-

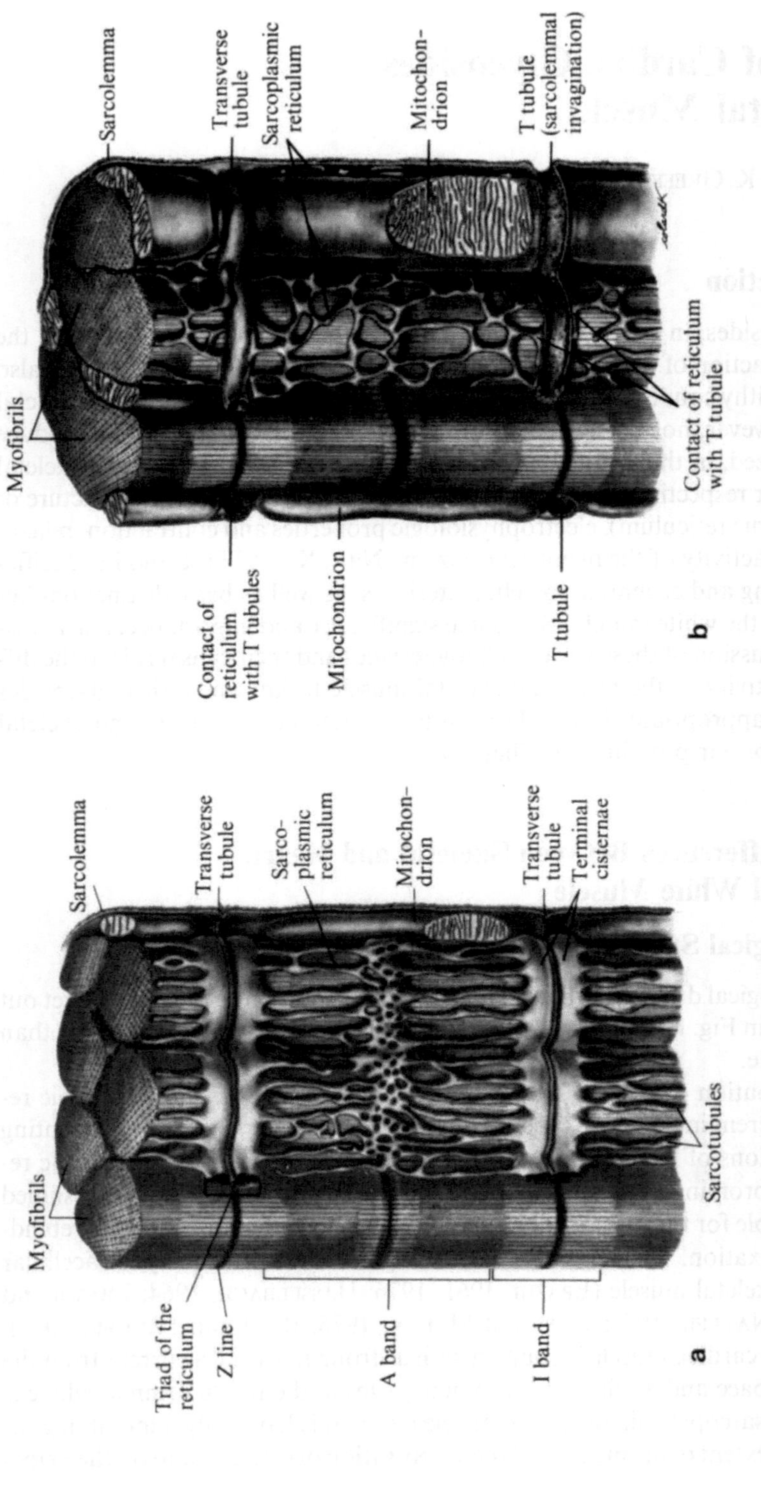

Fig. 1a, b. Schematic representation of differences in disposition of transverse tubules, sarcoplasmic reticulum, and mitochondria in skeletal **a** and cardiac **b** muscle. In skeletal muscle, the longitudinal sarcotubules are confluent with transverse elements called the terminal cisternae, flanking the transverse tubuli. Two terminal cisternae and one transverse tubule form the triads of the reticulum. In frog skeletal muscle, depicted here, the triads are at the Z line. In mammalian muscle there are two to each sarcomere, located at the A–I junctions. (FAWCETT and MCNUTT, 1969)

ity of muscle contraction and relaxation (see Chap. 18). This is in line with the observation that the force of contraction of skeletal muscle reflects extracellular Ca concentration less than in heart muscle, and may be reduced only after prolonged exposure to low extracellular Ca.

II. Electrophysiologic Properties

Distinct differences exist between the action potentials of skeletal and heart muscle; in skeletal muscle, the duration is less than 10 ms, while in ventricular muscle it is higher than 100 ms and includes a longer plateau phase, reflecting a slow inward Ca current, though with some exceptions for the cardiac muscle, as in the rat. Correspondingly, the duration of contraction of skeletal muscle is shorter than that of ventricular muscle. These differences are considered to originate from the release and binding of Ca ions at intracellular bindings sites in skeletal muscle.

III. The Membrane Enzyme Na^+, K^+-ATPase

This supposed receptor enzyme for digitalis in heart muscle is found in skeletal muscle too, and is inhibited by cardiac glycosides. Compared with what is seen in cardiac muscle, the specific activity of skeletal muscle Na^+, K^+-ATPase is lower, especially compared with magnesium-stimulated activity. SAMAHA and GERGELY (1965, 1966) were the first to purify a human, striated muscle Na^+, K^+-ATPase with a relatively high specific activity upon the addition of deoxycholate, and followed by aging for 10–14 days. This Na^+, K^+-ATPase was totally inhibited by 10^{-5} *M* ouabain. In comparative experiments with membrane preparations from both cardiac and skeletal muscle of the rabbit, the guinea pig, and the rat, DRANSFELD et al. (1966) found that the Na^+, K^+-ATPase from skeletal muscle was generally far less sensitive to strophanthoside K than from heart muscle. Prednisolone-bis-guanylhydrazone (for details, see Chap. 20), compared with strophanthoside K was more effective as an Na^+, K^+-ATPase inhibitor in skeletal than in cardiac muscle of the rabbit and the rat (DRANSFELD et al., 1967).

Several authors have described an ouabain-sensitive Na^+, K^+-ATPase preparation from skeletal muscle of the rat (ROGUS et al., 1969; PETER, 1970; ASH and SCHWARZ, 1970; KIDWAI et al., 1973; ANDREW and APPEL, 1973; FESTOFF and ENGEL, 1974; REDDY et al., 1976), hamster (McNAMARA et al., 1971), dog (AKERA et al., 1970), cat (BONTING et al., 1961), mouse (BRAY, 1973), and frog (BOEGMAN et al., 1970). All these different enzyme preparations differed in their specific activity and ouabain sensitivity, possibly owing to innate variations in their preparation. On the other hand, the comparative experiments of PITTS et al. (1977) showed only a slight difference in the ouabain sensitivities of the Na^+, K^+-ATPase from beef heart and dog skeletal muscles. The dissociation and association velocity of ouabain for Na^+, K^+-ATPase in skeletal muscle was, however, greater than for that in the cardiac muscle, which could be explained by differences in enzyme structures or preparations (for details, see Chap. 14 and 15). Furthermore, YAMAMOTO et al. (1981) have reported that the ouabain-sensitivities of Na^+, K^+-ATPase is rather higher in diaphragm than in cardiac tissue of guinea pig.

Table 1. Distribution of cardiac glycosides in blood, skeletal, and heart muscle. Mean values ± standard errors of mean. Number of determinations in parentheses

Species	Sampling (t)	Blood (ng/ml)	Skeletal muscle (ng/g wet weight)	Heart muscle (ng/g wet weight)	Ratio H/S	Ratio S/B	Reference	Method of determination
Digoxin								
Human	Post mortem (48–96h)		20 ± 3.5 (13)	Left ventricle 112 ± 19 (13)	5.6		KARJALAINEN et al. (1974)	RIA
Human	Post mortem (12–36 h)	Serum 2.3 ± 0.2 (7)	25 ± 2.4 (7)	Left ventricle 105 ± 10.3 (7)	4.2	10.8	HAASIS et al. (1977)	RIA
Human	Post mortem (2–4 days)	Plasma 1.3 ± 0.5 (7)	16 ± 3 (36)	Left ventricle 61 ± 15 (9)	3.8	12.3	WEINMANN et al. (1979)	RIA
Human	Surgically removed	Plasma 1.2 ± 0.3 (8)	11 ± 2 (8)	Left papillary muscle 78 ± 15 (8)	7.1	9.2	COLTART et al. (1972)	RIA
Human	Surgically removed (48–96 h)	Plasma 1.2 ± 0.1 (32)	13 ± 2 (32)	Papillary muscle 95 ± (11)	7.3	10.8	CARRUTHERS et al. (1975)	RIA
Human (premature neonates)	Post mortem (6–8 h)	Serum 1.9 ± 0.2 (5)	37 ± 12 (7)	Left ventricle 191 ± 26 (7)	5.2	19.5	KIM et al. (1975)	RIA
Human (full-term neonates)	Post mortem (6–8 h)	Serum 2.7 (1)	32 ± 8 (4)	Left ventricle 196 ± 20 (3)	6.1	11.9	KIM et al. (1975)	RIA
Human (children 1.5–7 years)	Post mortem (6–8 h)	Serum 0.6 ± 0.1 (4)	8 ± 3 (4)	Left ventricle 74 ± 21 (3)	9.2	13.3	KIM et al. (1975)	RIA
Human (infant)	Post mortem	Serum 3.5 ± 1.3 (5)	41.6 ± 18.6 (5)	Left ventricle 352 ± 115 (5)	8.4	11.9	GORODISCHER (1976)	RIA
Human (infant)	Post mortem (approx. 12 h)	Serum 4.0 (1)	52 (1)	Left ventricle 215 (1)	4.1	13	HAASIS et al. (1977)	RIA

Table 1. contiued

Species	Sampling (t)	Blood (ng/ml)	Skeletal muscle (ng/g wet weight)	Heart muscle (ng/g wet weight)	Ratio H/S	Ratio S/B	Reference	Method of determination
Digoxin								
Greyhound dog	Sacrificed 1 h after 0.033 mg/kg i.v.		22 ± 3 (10)	Left ventricle 129±8 (10)	5.9		Binnion and Morgan (1971)	^3H
Mongrel dog	Sacrificed 1 h after 0.038 mg/kg i.v.	Whole blood 9.1±0.7 (3)	25 ± 1.8 (6)	Left ventricle with septum 231±13 (3)	9.2		Francis et al. (1974)	^3H
Ouabain								
Guinea pig	Sacrificed 6 h after 50 µg/kg i.p.	Plasma 1.6±0.4 (8)	9.1± 1.1 (8)	Heart 12.0±1.5 (8)	1.3	5.7	Grope (1978)	RIA
Strophanthoside K								
Guinea pig	Sacrificed 5 h after 250 µg/kg i.v.	Whole blood 16±0.4 (7)	189± 1.1 (7)	Heart 95±0.8 (7)	0.5		Marzo et al. (1974)	^3H
Guinea pig	Sacrificed 6 h after 60 µg/kg i.p.	Plasma 0.9±0.1 (4)	18.2± 1.4 (4)	Heart 10±1.3 (4)	0.6	20.2	Merk (1979)	^3H
Guinea pig	Sacrificed 6 h after 60 µg/kg i.p.	Plasma 0.5±0.1 (4)	16.6± 4.3 (4)	Heart 9.1±0.7 (4)	0.6	33.2	Merk (1979)	RIA

H/S = heart/skeletal; S/B = skeletal/serum, or skeletal/plasma; RIA = radioimmunoassay; ^3H = radiochemical; (t) = time of last glycoside application.

Specific activity of the enzyme Na^+, K^+-ATPase apparently reflects the function of the musculature. SULAKHE et al. (1971) found a significantly higher activity for the Na^+, K^+-ATPase in the hind-leg skeletal muscle of the genetically dystrophic hamster, than for that of the healthy animal. Ouabain (0.4 m*M*) inhibited the enzyme in both states, though it is not evident if the degree of sensitivity to ouabain differed. SEVERIN et al. (1974) also found a significant increase in the Na^+, K^+-ATPase activity in denervated skeletal muscle; these studies were on the gastrocnemius muscle of rabbits, evaluated 20–25 days after denervation. On the other hand, BRAY (1973) reported significantly lower ouabain-sensitive Na^+, K^+-ATPase activity in dystrophic muscle preparations from mice compared with controls. Such varying observations are probably attributable to differences in preparative techniques employed, in addition to differences due to innate species characteristics.

IV. Binding and Accumulation

Compared with blood plasma, a higher glycoside concentration may be found not only in cardiac, but also in skeletal muscle. Binding to skeletal muscle represents a significant glycoside depot, especially since this muscle makes up a significant proportion of the body weight. The binding capacity of digoxin per unit weight of cardiac muscle is greater than that for skeletal muscle, the ratio for myocardial (left ventricle) to skeletal muscle in humans being in the range 4–9, for digoxin (Table 1). Ouabain and strophanthoside K, being more water-soluble than digoxin, exhibit in the guinea pig a binding capacity per unit weight approximately twice as high for skeletal muscle than for heart muscle 5–6 h following application.

V. Red and White Muscle

In an evaluation of the action of cardiac glycosides on skeletal muscle, the fact that skeletal muscle has various types of fibers assumes importance. Red, white, and mixed muscle have been differentiated on the basis of myoglobin content. The general classification into slow-twitch and fast-twitch red fibers, and fast-twitch white fibers is preferred. In addition to differing myoglobin content and velocity of contraction and relaxation, they further differ in Na^+, K^+-ATPase content, content of mitochondria, size of the sarcoplasmic reticulum, and content and nature of the myofibrillar proteins (PETER et al., 1972; PETER, 1973; PETTE and DÖLKEN, 1975; PETTE, 1977; LIBERA et al., 1978; VAN WINKLE and SCHWARTZ, 1978; VAN WINKLE et al., 1978). Slow-twitch red fibers possess a low myosim ATPase activity, low glycolytic capacity, moderately high respiratory capacity, and a low capacity for Ca-accumulation in the sarcoplasmic reticulum. Fast-twitch red and white fibers, on the other hand, have a high myosin ATPase activity, a high glycolytic capacity, a high capacity for Ca accumulation in the sarcoplasmic reticulum, and a high (fast-twitch red) or low (fast-twitch white) respiratory capacity. Examples of the different muscle types generally employed in pharmacologic work are the vastus lateralis (fast-twitch red), soleus (slow-twitch red), and the rectus femoris (fast-twitch white); the masseter muscle, also commonly employed is a fast-twitch red muscle type (SCHIAFFINO, 1974).

FESTOFF et al. (1977) reported that Na^+, K^+-ATPase activity in the soleus (slow-twitch red) was greater than in the extensor digitorum longus (fast-twitch white). The ^3H-ouabain-binding capacity was also higher in soleus than in the superficial portion of the gastrocnemius muscle (BRAY et al., 1977). Also in agreement are the results of SHARMA et al. (1977) who observed in the cat soleus muscle a higher binding capacity for ^3H-ouabain than in the anterior tibialis muscle. Ablation of the motor nerve increased Na^+, K^+-ATPase activity (SEVERIN et al., 1974; FESTOFF et al., 1977), and the binding capacity of ^3H-ouabain (SHARMA and BANERJEE, 1978). These observations support the idea that the motor nerve may be involved in the regulation of Na^+, K^+-ATPase activity in skeletal muscle.

Physical exercise too has an important influence on the structure and enzyme activity of skeletal muscle (BALDWIN et al., 1972; HOLLOSZY and BOOTH, 1976). It is also of significance that the fast- and slow-twitch muscles may, through experimental cross-innervation, be transformed morphologically and functionally (BULLER et al., 1960; GUTMANN, 1976). Thus, HEILMANN and PETTE (1979) found a decrease in Ca^{2+}-dependent ATPase and in Ca^{2+} uptake by sarcoplasmic reticulum vesicles in the fast-twitch muscle of the rabbit following prolonged electrical stimulation by implanted electrodes, indicating a transformation of the sarcoplasmic reticulum from the fast-twitch to the slow-twitch muscle type.

C. Influences on Force of Contraction of Skeletal Muscle

Reviews of the early literature by LENDLE (1935) and by LENDLE and MERCKER (1961) have stressed the contradictory nature of the data in this context. Specifically, in experiments with white or red skeletal muscle of the rabbit, increasing either the concentrations of cardiac glycosides or exposure time resulted in an increased force of contraction followed by contracture (ISHIKAWA, 1922). The glycosides further caused an increase or prevented a decrease in performance of fatigued muscle (NEUSCHLOSS, 1922) in experiments with the gastrocnemius muscle of *Rana temporaria*. They have also been attributed a curare-like paralytic effect in frog nerve–muscle preparations (LAPICQUE and LAPICQUE, 1923).

I. In Situ Experiments

For the investigation of extracardiac effects the rat is a suitable experimental animal, since its heart is quite insensitive to cardiac glycosides. A significant increase in the force of contraction was first observed in rat masticatory muscles, following intravenous (i.v.) or intraperitoneal (i.p.) injection of ouabain or strophanthoside K (0.3–0.5 mg/kg), (HOTOVY and ERDNISS, 1950; HOTOVY and KÖNIG, 1951). This effect is concentration dependent, the increment in force of contraction varying between 2% and 18% (LENDLE and OLDENBURG, 1950), and especially noticeable in fatigued or anoxic masticatory muscle (ANDERS et al., 1953). RIEKER (1953) used this preparation for a comparison of the extracardiac efficacy of several cardiac glycosides.

A distinct increase in the contractile force of the rat gastrocnemius muscle can be demonstrated by an i.v. injection of 1–3 mg/kg strophanthoside K (GREEFF and

WESTERMANN, 1955); this increase is subsequently followed by a neuromuscular inhibition, intensified by succinylcholine.

Physical performance, as measured by swimming capacity in the rat, is apparently increased by digitalis glycosides (FALKENHAHN et al., 1967). These observations are similar to those of MOSKOPF and SARRE (1953) in the guinea pig. Higher doses of strophanthoside K (1 mg/kg) in the rat and mouse can, however, bring about a transient paralysis, which to all outward appearance resembles that induced by curare (GREEFF and WESTERMANN, 1955; GREEFF and KASPERAT, 1961).

Recently, SMULYAN and EICH (1976) have reported an inotropic effect on the skeletal muscle of healthy volunteer subjects. Using thumb adduction by electrical nervous stimulation, they observed a significant increase in the contractile force upon infusion of 0.5 mg ouabain into the brachialis artery. Systemic digitalization [1.0 mg digoxin/day, given orally (p.o.) during the first day, followed by 0.75 mg/day for a further 9 days] using a double-blind approach, had, by contrast, no influence on muscular contractile force in the extremities of healthy volunteers (HOLLMANN et al., 1965). EL ZAYAT and KOURA (1971), on the other hand, demonstrated following digoxin administration (3.5 mg/day p.o. for 3 days, followed by 0.5 mg/day for a further 4 days) a beneficial effect on both contractility and motor unit action potential in patients suffering from progressive muscular dystrophy. Of 20 patients, 16 exhibited a quantitative increase in force of hand-grip, as measured by a dynamometer. These subjects further showed an increase in the duration and a decrease in the sharp nature of the motor unit action potential of the biceps muscle, apparently characteristic of muscular dystrophy, though, no increase was observed in the amplitude. None of the patients, however, demonstrated any subjective improvement in muscular force following digoxin therapy.

II. In Vitro Experiments

The observed effects of cardiac glycosides upon the force of contraction of skeletal muscle in vitro are also heterogeneous. WESTERMANN (1954) found no change in the contractile force of the isolated phrenic nerve–diaphragm preparation from the mouse, after administration of 1.35×10^{-5} *M* ouabain, under either direct or indirect electrical stimulation. Upon combining ouabain with other drugs, however, it was possible to detect an interaction with neuromuscular impulse transmission; the inhibitory effects induced by depolarizing substances (succinylcholine, acetylcholine, and decamethonium) were enhanced, the effects of stabilizing substances (curare and gallamin), lessened. Ouabain itself was inhibitory at 1.35×10^{-5} *M* after administration of prostigmine. HOFMANN and SHERROD (1963) found a dose-dependent potentiation in the isometric twitch tension of isolated phrenic nerve–diaphragm preparation from the rat by ouabain, in a concentration range of 7.0×10^{-7}–5.5×10^{-6} *M*. These authors, however, did not stimulate the diaphragm directly, and so this result does not preclude the possibility of an effect on neuromuscular transmission.

On increasing the K^+ concentration in the Tyrode solution from 2.5 to 7.5 m*M*, GREEFF and WESTERMANN (1955) noted an inhibition of the contractile force of the diaphragm of the rat, using direct stimulation and high concentrations of ouabain or strophanthoside K (1.35–2.0×10^{-5} *M* or 1.15–1.72×10^{-5} *M*). This effect was antagonized by $CaCl_2$ or epinephrine.

In guinea pigs, ouabain (3.0×10^{-6} *M*) caused a continual decrease in the contractile force of isolated phrenic nerve–diaphragm preparations, becoming unresponsive within 20 min, when stimulated at 1 Hz, directly or indirectly (HUGHES and WEATHERALL, 1970). At lower rates of stimulation, the paralytic response occurred more slowly. This has led to the conclusion that paralysis results from an intracellular accumulation of Na^+, sodium transport being impaired by the glycoside. In contrast, FAUST and SAUNDERS (1957) have reported earlier an increase in the contractile force of the isolated guinea-pig phrenic nerve–diaphragm preparation when either direct or indirect stimulation was used, a decrease resulting at high concentrations (0.7–4.0×10^{-6} *M*) of ouabain. YAMAMOTO et al. (1981) also observed a slight and transient increase in force of contraction by ouabain in isolated guinea-pig diaphragm preparations. They supposed that the lack of distinct positive inotropic effect of ouabain in the above tissue preparations is due neither to the difference in the ouabain-Na^+, K^+-ATPase interaction between diaphragm and cardiac tissues nor to the failure of the sodium pump inhibition by ouabain in diaphragm.

The actions of cardiac glycosides on directly stimulated frog sartorius muscle preparations are also contradictory. Whilst the isometric tension developed by this muscle appears unaltered after incubation with 1.5×10^{-4} *M* ouabain (NESTEROV, 1972), in a similar preparation from the toad, 10^{-9} *M* ouabain stimulated, and 10^{-6} *M* ouabain depressed contraction (AMARANTH and ANDERSEN, 1976). On individual muscle fibers from the carpopodite flexor of the marine crab, ouabain itself did not influence tension at up to 10 m*M*, but 1.5×10^{-7} *M* decreased the isometric muscle contraction caused by electrical stimulation, K^+, or caffeine (FUJINO et al., 1969). If, however, ouabain was applied intracellularly, muscle tension increased. These results seem to imply an intracellular receptor for ouabain and a normally low cellular penetration of the glycoside to reach the active site.

D. Influences on Skeletal Muscle Electrolyte Content

Potassium loss from skeletal muscle following administration of digitalis glycosides was first reported by CATTELL (1937) and CATTELL and GOODELL (1937), working with isolated, frog sartorius muscle. GREEFF and WESTERMANN (1955) found that the paralyzing effect in the rat of 1.5 mg/kg strophanthoside K (10% of the lethal dose) was accompanied by a decrease in K^+ and an increase in Na^+ content of the gastrocnemius and diaphragm muscles. These electrolyte changes in the skeletal muscle are accompanied by an exactly opposite pattern of changes in serum (Fig. 2), i.e., an increase in K^+ and a decrease in Na^+; simultaneously, an even larger urinary excretion of K^+ occurs (BODE and GREEFF, 1956; GREEFF, 1956a, b). The increased renal K^+ clearance after administration of strophanthoside K, ouabain, digitoxin, or other cardiac glycosides is dose dependent (GREEFF, 1958) and, like skeletal muscle K^+ loss and Na^+ accumulation, occurs in adrenalectomized rats (GREEFF, 1957). Such changes in electrolyte content of skeletal muscle occur also in animals with hearts generally more sensitive to cardiac glycosides; thus, in parallel, an infusion of strophanthoside K in the dog results in increased blood K^+, whilst in the guinea pig, following a subcutaneous (s.c.) injection of this glycoside, there is an increased K^+ excretion (GREEFF, 1956b).

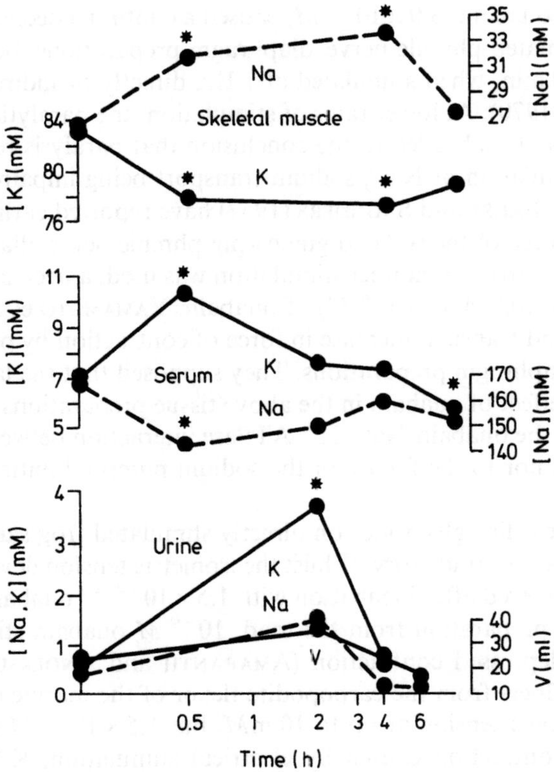

Fig. 2. Changes in potassium and sodium content of serum, skeletal muscle, and urine after 1 mg/kg s.c. injection of strophanthoside K. Each group consisted of six rats (150–200 g). *Asterisks* indicate $p < 0.05$ compared to control-values before the injection of strophanthoside K. (GREEFF, 1956b)

In the case of the isolated rat diaphragm, an increase in contractile force caused by 5.5×10^{-6} *M* ouabain was associated with a decrease in K^+ and an accumulation of Na^+ (HOFMANN and SHERROD, 1963). Similar electrolyte changes in the rat diaphragm muscle have been reported by GREEFF and WESTERMANN (1955) and BOLTE and LÜDERITZ (1969). The very early observation of CATTELL (1937), that ouabain increases the Na^+ and decreases the K^+ content in skeletal muscle, has been reproduced by several investigators in frog muscle (SCHATZMANN and WITT, 1954; JOHNSON, 1956; FLEAR et al., 1975; RIORDAN et al., 1972; NESTEROV, 1972). Inhibition of the recovery of K^+ to normal levels and a loss of Na^+ in preparations which were K^+ depleted and Na^+ loaded has been observed too (DUNKLEY et al., 1966). Similar observations were made by MANERY et al. (1977), who also found that the concentration of ouabain necessary to cause maximal inhibition of the recovery to normal electrolyte contents is the same (approximately 10^{-6} *M*) as that required to inhibit Na^+, K^+-ATPase. It is worth mentioning here that cardiac glycosides inhibit K^+ uptake and that Na^+ depletion through an inhibition of Na^+, K^+-ATPase has been often used in physiologic investigations of the mechanisms of ion transport and ion permeability (LING and PALMER, 1972; KEYNES and STEINHARDT, 1968; DAHL-HANSEN and CLAUSEN, 1973).

E. Influence on the Electrophysiologic Parameters of Skeletal Muscle

LOCKE and SOLOMON (1967) investigated the effect of ouabain on the resting membrane potential of innervated and denervated rat gastrocnemius and soleus muscles. They observed in vivo, a decrease in the innervated muscle, the effect of ouabain being even weaker in the denervated muscle. Similar results have been obtained with the rat diaphragm (BRAY et al., 1976) and the extensor digitorum longus muscle (SELLIN and MCARDLE, 1977) with 1.0–5.0 × 10^{-3} M, and 1.7 × 10^{-4} M ouabain respectively, decreasing the membrane resting potential. Denervation of skeletal muscle itself brings about a decrease in the membrane resting potential, but this decrease could be restored by application of catecholamines or dibutyryl-cyclic-AMP, in the presence of theophylline (BRAY et al., 1976). This treatment restored the decrement caused by ouabain too, though as stated earlier, the effect of ouabain on the denervated muscle is normally very weak.

Re-innervation, which restored the membrane potential, also restored ouabain sensitivity (SELLIN and MCARDLE, 1977). MCARDLE and ALBUQUERQUE (1975) investigated the effects of ouabain on normal and dystrophic mouse soleus muscle, both before and after denervation. Membrane potential was reduced to a similar extent in both the normal and dystrophic innervated muscle, although the latter already had a reduced resting potential. That of the denervated muscle in either physiologic state was not significantly affected by ouabain.

Ablation of the motor nerve increases the number of binding sites for ^3H-ouabain (FESTOFF et al., 1977; SHARMA et al., 1977; SHARMA and BANERJEE, 1978) and the Na^+, K^+-ATPase activity of skeletal muscle microsomal preparations (SEVERIN et al., 1974; FESTOFF et al., 1977). These observations are at variance with the theory of LOCKE and SOLOMON (1967) that, the contribution of the electrogenic Na^+ pump to membrane potential is directly dependent on innervation, and is reduced by denervation. Depolarization by ouabain has been inferred by these (LOCKE and SOLOMON, 1967) and by other investigators (MCARDLE and ALBUQUERQUE, 1975; BRAY et al., 1976) as evidence for an electrogenic contribution of the Na^+, K^+-ATPase to membrane potential. WAREHAM (1978) has stressed the same atypical response of the membrane potential to changes in external K^+ concentration, both after denervation and after treatment with ouabain. This has lent support to the views of ROBBINS (1977) which suggest an increased permeability to Na^+ as the direct response to both denervation and ouabain treatment, in addition to proposing that under these conditions the muscle membrane would behave as a mixed-ion electrode with a significant contribution through Na^+ permeability, hence the atypical response of the membrane potential to external K^+.

Additionally, because ouabain and denervation open common Na^+ channels, it has been suggested that ouabain had no effect on the denervated muscle (BRAY et al., 1976); moreover, following denervation these channels are already opened and ouabain is ineffective (ROBBINS, 1977).

It is appropriate to record what is known about the effects of cardiac glycosides on membrane potentials and electrolyte contents in the rat skeletal muscle. The effect of subcutaneously administered ouabain on the muscle action potential has been noted to be biphasic; there was an immediate shortening of depolarization and repolarization times, no serum or muscle electrolyte changes being demonstrable during this phase (HADDOW and KLEIN, 1970). After 15 min, repolarization became prolonged, and the prolongation of the action potential correlated in time only with a diminution of muscle K^+ content, although muscle Na^+ and urinary and plasma K^+ increased. Shortening of the action potential has been suggested as a consequence of therapeutic doses of ouabain.

The state of knowledge at present permits the conclusion that cardiac glycosides increase the force of contraction of skeletal muscle, this effect being reversed by high concentrations. A change in the force of contraction is accompanied by a loss of K^+ and an uptake of Na^+. Cardiac glycosides further seem to exercise an inhibitory effect on neuromuscular conduction, potentiating the effect of succinylcholine. While these statements are well supported by results from studies employing animal models, there is as yet no clear-cut evidence that human skeletal muscle is influenced in a like manner by cardiac glycosides. It is hoped that the progress that has been made by studying animals will stimulate further work on human subjects.

References

Akera, T., Larsen, F.S., Brody, T.M.: Correlation of cardiac sodium- and potassium-activated adenosine triphosphatase activity with ouabain-induced inotropic stimulation. J. Pharmacol. Exp. Ther. *173*, 145–151 (1970)

Amaranath, L., Andersen, N.B.: The effect of general anestethic agents, ouabain, and aldosterone on striated muscle contraction in toad. Anesth. Analg. (Cleve.) *55*, 409–414 (1976)

Anders, M., Nieschke, W., Dahm, H., Taugner, R.: Vergleichende pharmakologische Untersuchungen am ermüdeten bzw. anoxämischen und normalen Skeletmuskel. Naunyn Schmiedebergs Arch. Pharmacol. *217*, 406–412 (1953)

Andrew, C.G., Appel, S.H.: Macromolecular characterization of muscle membranes. J. Biol. Chem. *248*, 5156–5163 (1973)

Ash, A.S.F., Schwartz, A.: Sodium-plus-potassium ion-activated adenosine triphosphatase in a heavy-membrane fraction isolated from rat skeletal muscle. Biochem. J. *118*, 20P–21P (1970)

Baldwin, K.M., Klinkerfuss, G.H., Terjung, R.L., Molé, P.A., Holloszy, J.O.: Respiratory capacity of white, red, and intermediate muscle: adaptative response to exercise. Am. J. Physiol. *222*, 373–378 (1972)

Binnion, P.F., Morgan, L.M.: Effect of acute hypokalaemia on ^{3}H-digoxin metabolism. Cardiovasc. Res. *5*, 431–435 (1971)

Bode, H., Greeff, K.: Über die Wirkung des Strophanthins auf die renale Kalium- und Natriumausscheidung bei Ratten. Naunyn Schmiedebergs Arch. Pharmacol. *227*, 436–443 (1956)

Boegman, R.J., Manery, J.F., Pinteric, L.: The separation and partial purification of membrane-bound $(Na^+ + K^+)$-dependent Mg^{2+}-ATPase and $(Na^+ + K^+)$-independent Mg^2-ATPase from frog skeletal muscle. Biochim. Biophys. Acta *203*, 506–530 (1970)

Bolte, H.-D., Lüderitz, B.: Intra-extracellulärer Kaliumkonzentrationsgradient und g-Strophanthinwirkung (Ouabain) beim Kaliummangel. Untersuchungen an Skelettmuskelzellen. Klin. Wochenschr. *47*, 889–891 (1969)

Bonting, S.L., Simon, K.A., Hawkins, N.M.: Studies on sodium-potassium-activated adenosine triphosphatase. I. Quantitative distribution in several tissues of the cat. Arch. Biochem. Biophys. *95*, 416–423 (1961)

Bray, G.M.: A comparison of the ouabain-sensitive ($Na^+ + K^+$)-ATPase of normal and dystrophic skeletal muscle. Biochim. Biophys. Acta *298*, 239–245 (1973)
Bray, G.M., Wilcox, W., Aguayo, A.J.: Ouabain binding sites in skeletal muscle from normal and dystrophic mice. J. Neurol. Sci. *34*, 149–156 (1977)
Bray, J.J., Hawken, M.J., Hubbard, J.I., Pockett, S., Wilson, L.: The membrane potential of rat diaphragm muscle fibres and the effect of denervation. J. Physiol. (Lond.) *255*, 651–667 (1976)
Buller, A.J., Eccles, J.C., Eccles, R.M.: Interactions between motoneurones and muscles in respect of the characteristic speeds of their responses. J. Physiol. (Lond.) *150*, 417–439 (1960)
Carruthers, S.G., Cleland, J., Kelly, J.G., Lyons, S.M., McDevitt, D.G.: Plasma and tissue digoxin concentrations in patients undergoing cardiopulmonary bypass. Br. Heart J. *37*, 313–320 (1975)
Cattell, Mc.K.: The action of ouabain on skeletal muscle. J. Pharmacol. Exp. Ther. *60*, 101–102 (1937)
Cattell, Mc.K., Goodell, H.: On the mechanism of the action of digitalis glucosides on muscle. Science *86*, 106–107 (1937)
Coltart, J., Howard, M., Chamberlain, D.: Myocardial and skeletal muscle concentrations of digoxin in patients on long-term therapy. Br. Med. J. *1972 I*, 318–319
Dahl-Hansen, A.B., Clausen, T.: The effect of membrane stabilizers and ouabain on the transport of Na^+ and K^+ in rat soleus muscle. Biochim. Biophys. Acta *318*, 147–153 (1973)
Dransfeld, H., Greeff, K., Berger, H., Cautius, V.: Die verschiedene Empfindlichkeit der $Na^+ + K^+$-aktivierten ATPase des Herz- und Skelettmuskels gegen k-Strophanthin. Naunyn Schmiedebergs Arch. Pharmacol. *254*, 225–234 (1966)
Dransfeld, H., Galetke, E., Greeff, K.: Die Wirkung des Prednisolonbisguanylhydrazons auf die $Na^+ + K^+$-aktivierte Membran-ATPase des Herz- und Skelettmuskels. Arch. Int. Pharmacodyn. Ther. *166*, 342–349 (1967)
Dunkley, C.R., Manery, J.F., Dryden, E.E.: The conversion of ATP to IMP by muscle surface enzymes. J. Cell. Physiol. *68*, 241–247 (1966)
Ebashi, S.: Calcium binding activity of vesicular relaxing factor. J. Biochem. *50*, 236–244 (1961)
Ebashi, S.: Excitation-contraction coupling. Annu. Rev. Physiol. *38*, 293–313 (1976)
Ebashi, S., Endo, M.: Calcium ion and muscle contraction. Prog. Biophys. Mol. Biol. *18*, 123–183 (1968)
El Zayat, A., Koura, F.: Effect of digitalis glycosides on the myopathic skeletal muscle. J. Egypt. Med. Assoc. *54*, 535–547 (1971)
Falkenhahn, A., Hollmann, W., Kenter, H., Venrath, H., Bouchard, C.: Der Einfluß von Digitalis auf die Leistungsfähigkeit gesunder Ratten im Schwimmversuch. Arzneim. Forsch. *17*, 551–553 (1967)
Faust, R.M., Saunders, P.R.: Comparative effects of ouabain upon contractile force of guinea pig diaphragm and heart. Proc. Soc. Exp. Biol. Med. *94*, 351–356 (1957)
Fawcett, D.W., Mc Nutt, N.S.: The ultrastructure of the cat myocardium. I. Ventricular papillary muscle. J. Cell. Biol. *42*, 1–45 (1969)
Festoff, B.W., Engel, W.K.: In vitro analysis of the general properties and junctional receptor characteristics of skeletal muscle membranes. Isolation, purification, and partial characterization of sarcolemmal fragments. Proc. Natl. Acad. Sci. USA *71*, 2435–2439 (1974)
Festoff, B.W., Oliver, K.L., Reddy, N.B.: In vitro studies of skeletal muscle membranes. Effects of denervation on the macromolecular components of cation transport in red and white skeletal muscle. J. Membr. Biol. *32*, 345–360 (1977)
Flear, C.T.G., Greener, J.S., Bhattacharya, S.S.: Effects of ouabain on sodium uptake by frog heart and skeletal muscle. Recent Adv. Stud. Cardiac. Struct. Metab. *5*, 343–349 (1975)
Francis, D.J., Georoff, M.E., Jackson, B., Marcus, F.I.: The effect of insulin and glucose on the myocardial and skeletal muscle uptake of tritiated digoxin in acutely hypokalemic and normokalemic dogs. J. Pharmacol. Exp. Ther. *188*, 564–574 (1974)

Fujino, S., Tanaka, M., Fujino, M.: Microinjection of ouabain into crab muscle fibres. Nature *223*, 413–414 (1969)
Gorodischer, R., Jusko, W.J., Yaffe, S.J.: Tissue and erythrocyte distribution of digoxin in infants. Clin. Pharmacol. Ther. *19*, 256–263 (1976)
Greeff, K.: Über die Wirkung des Strophanthins auf den Elektrolythaushalt. Dtsch. Med. Wochenschr. *81*, 666–668/677–678 (1956a)
Greeff, K.: Tierexperimentelle Untersuchungen über den Einfluß herzwirksamer Glykoside auf den Natrium- und Kaliumhaushalt. Verh. Dtsch. Ges. Kreisl. Forsch. *22*, 312–318 (1956b)
Greeff, K.: Mineralstoffwechselwirkung und Toxizität des Strophanthins bei nebennierenlosen Ratten. Naunyn Schmiedebergs Arch. Pharmacol. *231*, 391–400 (1957)
Greeff, K.: Vergleich der kaliuretischen Wirkung verschiedener Glykoside des Strophanthins und Digitoxigenins bei parenteraler und enteraler Applikation. Naunyn Schmiedebergs Arch. Pharmacol. *233*, 468–483 (1958)
Greeff, K., Kasperat, H.: Vergleich der neurotoxischen Wirkung von Digitalisglykosiden und Geninen bei intracerebraler und intravenöser Injektion an Mäusen, Ratten und Meerschweinchen. Naunyn Schmiedebergs Arch. Pharmacol. *242*, 76–89 (1961)
Greeff, K., Westermann, E.: Untersuchungen über die muskellähmende Wirkung des Strophanthins. Naunyn Schmiedebergs Arch. Pharmacol. *226*, 103–113 (1955)
Grope, W.: Pharmakokinetik der Herzsteroide g-Strophanthin, Digoxin und Digitoxin im Organismus von Meerschweinchen. Dissertation, Düsseldorf 1978
Gutmann, E.: Neurotrophic relations. Annu. Rev. Physiol. *38*, 177–216 (1976)
Haasis, R., Larbig, D., Stunkat, R., Bader, H., Seboldt, H.: Radioimmunologische Bestimmung der Glykosidkonzentration im menschlichen Gewebe. Klin. Wochenschr. *55*, 23–30 (1977)
Haddow, J.E., Klein, R.: Cardiac glycoside effects on rat skeletal muscle potentials and electrolytes. Proc. Soc. Exp. Biol. Med. *133*, 138–143 (1970)
Hasselbach, W.: Relaxing factor and the relaxation of muscle. Prog. Biophys. Mol. Biol. *14*, 167–222 (1964)
Heilmann, C., Pette, D.: Molecular transformations in sarcoplasmic reticulum of fast-twitch muscle by electro-stimulation. Eur. J. Biochem. *93*, 437–446 (1979)
Hofmann, L.M., Sherrod, T.R.: Action of ouabain on isometric twitch tension in striated muscle. Proc. Soc. Exp. Biol. Med. *113*, 940–943 (1963)
Hollmann, W., Hettinger, Th., Venrath, H., Herkenrath, G.: Untersuchungen über den Einfluß von Digitalis auf die Skelettmuskelkraft beim Menschen. Münch. Med. Wochenschr. *107*, 1338–1340 (1965)
Holloszy, J.O., Booth, F.W.: Biochemical adaptations to endurance exercise in muscle. Annu. Rev. Physiol. *38*, 273–291 (1976)
Hotovy, R., Erdniss, H.: Pharmakologische Studien am Musculus masseter der Ratte. Naunyn Schmiedebergs Arch. Pharmacol. *209*, 204–234 (1950)
Hotovy, R., König, W.: Weitere pharmakologische Studien an der Kaumuskulatur der Ratte. Naunyn Schmiedebergs Arch. Pharmacol. *213*, 175–184 (1951)
Hughes, R., Weatherall, M.: The response of stimulated and quiescent phrenic nerve diaphragm preparations to digoxin and ouabain. Br. J. Pharmacol. *39*, 233P–234P (1970)
Inesi, G., Malan, N.: Mechanisms of calcium release in sarcoplasmic reticulum. Life Sci. *18*, 773–779 (1976)
Ishikawa, Y.: Pharmakologische Untersuchungen an den überlebenden roten und weißen Kaninchenmuskeln. Acta Scholae med. Kyoto *5*, 123–138 (1922)
Johnson, J.A.: Influence of ouabain, strophanthidin, and dihydrostrophanthidin on sodium and potassium transport in frog sartorii. Am. J. Physiol. *187*, 328–332 (1956)
Karjalainen, J., Ojala, K., Reissell, P.: Tissue concentrations of digoxin in an autopsy material. Acta Pharmacol. Toxicol. (Kbh.) *34*, 385–390 (1974)
Keynes, R.D., Steinhardt, R.A.: The components of the sodium efflux in frog muscle. J. Physiol. (Lond.) *198*, 581–599 (1968)
Kidwai, A.M., Radcliffe, M.A., Lee, E.Y., Daniel, E.E.: Isolation and properties of skeletal muscle plasma membrane. Biochim. Biophys. Acta *298*, 593–607 (1973)

Kim, P.W., Krasula, R.W., Soyka, L.F., Hastreiter, A.R.: Postmortem tissue digoxin concentrations in infants and children. Circulation *52*, 1128–1131 (1975)
Lapicque, L., Lapicque, M.: L'action de la strophantine sur le coeur et son action musculaire en général. C.R. Soc. Biol., Paris *89*, 315–317 (1923)
Lendle, L.: Digitaliskörper und verwandte herzwirksame Glykoside (Digitaloide). In: Handbuch der experimentellen Pharmakologie. Heubner, W., Schüller, J. (eds.), Ergänzungswerk, Vol. 1, pp. 11–241. Berlin: Springer 1935
Lendle, L., Mercker, H.: Extrakardiale Digitaliswirkungen. Ergeb. Physiol. *51*, 200–298 (1961)
Lendle, L., Oldenburg, D.: Prüfung extrakardialer Wirkungen an der digitalisunterempfindlichen Ratte. Naunyn Schmiedebergs Arch. Pharmacol. *211*, 243–263 (1950)
Libera, L.D., Margreth, A., Mussini, I., Cerri, C., Scarlato, G.: Myosin polymorphism in human skeletal muscles. Muscle Nerve *1*, 280–291 (1978)
Ling, G.N., Palmer, L.G.: Studies on ion permeability: IV. The mechanism of ouabain action on the Na^+-ion efflux in frog muscles. Physiol. Chem. Phys. *4*, 517–525 (1972)
Locke, S., Solomon, H.C.: Relation of resting potential of rat gastrocnemius and soleus muscles to innervation, activity, and the Na-K pump. J. Exp. Zool. *166*, 377–386 (1967)
Manery, J.F., Dryden, E.E., Still, J.S., Madapallimattam, G.: Enhancement (by ATP, insulin, and lack of divalent cations) of ouabain inhibition of cation transport and ouabain binding in frog skeletal muscle; effect of insulin and ouabain on sarcolemmal (Na+K)MgATPase. Can. J. Physiol. Pharmacol. *55*, 21–33 (1977)
Marzo, A., Ghirardi, P., Marchetti, G.: The absorption, distribution and excretion of k-strophanthoside-^{3}H in guinea pigs after parenteral administration. J. Pharmacol. Exp. Ther. *189*, 185–193 (1974)
McArdle, J.J., Albuquerque, E.X.: Effects of ouabain on denervated and dystrophic muscles of the mouse. Exp. Neurol. *47*, 353–356 (1975)
McNamara, D.B., Sulakhe, P.V., Dhalla, N.S.: Properties of the sarcolemmal calcium ion-stimulated adenosine triphosphatase of hamster skeletal muscle. Biochem. J. *125*, 525–530 (1971)
Merk, H.: Die Verteilung von Strophanthusglykosiden im Organismus des Meerschweinchens. Dissertation, Düsseldorf 1979
Moskopf, E., Sarre, H.: Die Wirkung von Herzglykosiden (Strophanthin, Digitoxin) auf die Leistungsfähigkeit des gesunden Herzens von Meerschweinchen im Schwimmversuch und auf Herz- und Nebennierengewicht. Verh. Dtsch. Ges. Kreislaufforsch. *19*, 283–289 (1953)
Nayler, W.G.: Regulation of myocardial function – a subcellular phenomenon. J. Mol. Cell. Cardiol. *5*, 213–219 (1973)
Nesterov, V.P.: Effect of strophanthin and ethacrinic acid on the functional properties and distribution of sodium and potassium ions in the sartorius muscle of the frog *Rana temporaria*. Zh. Evol. Biokhim. Fiziol. *8*, 592–597 (1972)
Neuschloss, S.M.: Beiträge zur Kenntnis der Wirkung der Herzglykoside auf den quergestreiften Skelettmuskel. Pflügers Arch. *197*, 235–256 (1922)
Peter, J.B.: A $(Na^+ + K^+)$ATPase of sarcolemma from skeletal muscle. Biochem. Biophys. Res. Commun. *40*, 1362–1367 (1970)
Peter, J.B.: Skeletal muscle: Diversity and mutability of its histochemical, electron microscopic, biochemical, and physiologic properties. In: The striated muscle. Pearson, C.M., Mostofi, F.K. (eds.), pp. 1–18. Baltimore: The Williams and Wilkins Company 1973
Peter, J.B., Barnard, R.J., Edgerton, V.R., Gillespie, C.A., Stempel, K.E.: Metabolic profiles of three fiber types of skeletal muscle in guinea pigs and rabbits. Biochemistry *11*, 2627–2633 (1972)
Pette, D.: Enzymmuster des energieliefernden Stoffwechsels im Skelettmuskel. In: Muskelstoffwechsel, körperliche Leistungsfähigkeit und Diabetes mellitus. Jahnke, K., Mehnert, H., Reis, H.E. (eds.), pp. 21–46. Stuttgart, New York: Schattauer 1977
Pette, D., Dölken, G.: Some aspects of regulation of enzyme levels in muscle energy-supplying metabolism. Adv. Enzyme Regul. *13*, 355–377 (1975)
Pitts, B.J.R., Wallick, E.T., Van Winkle, W.B., Allen, J.C., Schwartz, A.: On the lack of inotropy of cardiac glycosides on skeletal muscle: A comparison of Na^+, K^+-ATPases from skeletal and cardiac muscle. Arch. Biochem. Biophys. *184*, 431–440 (1977)

Reddy, N.B., Engel, W.K., Festoff, B.W.: In vitro studies of skeletal muscle membranes. Characterization of a phosphorylated intermediate of sarcolemmal ($Na^+ + K^+$) ATPase. Biochim. Biophys. Acta *433*, 365–382 (1976)

Rich, T.L., Langer, G.A.: A comparison of excitation-contraction coupling in heart and skeletal muscle: An examination of "calcium-induced calcium-release." J. Mol. Cell. Cardiol. *7*, 747–765 (1975)

Riecker, G.: Wirkung von Herzglykosiden auf den quergestreiften Muskel. Klin. Wochenschr. *31*, 191–192 (1953)

Riordan, J.R., Manery, J.F., Dryden, E.E., Still, J.S.: Influence of ethacrynic acid on muscle surface enzymes and of ethacrynic acid and ouabain on Na, K, and H_2O in frog muscle. Can. J. Physiol. Pharmacol. *50*, 432–444 (1972)

Robbins, N.: Cation movements in normal and short-term denervated rat fast twitch muscle. J. Physiol. (Lond.) *271*, 605–624 (1977)

Rogus, E., Price, T., Zierler, K.L.: Sodium plus potassium-activated, ouabain-inhibited adenosine triphosphatase from a fraction of rat skeletal muscle, and lack of insulin effect on it. J. Gen. Physiol. *54*, 188–202 (1969)

Samaha, F.J., Gergely, J.: Na^+- and K^+-stimulated ATPase in human striated muscle. Arch. Biochem. Biophys. *109*, 76–79 (1965)

Samaha, F.J., Gergely, J.: Studies on the Na^+- and K^+-activated adenosine triphosphatase in human striated muscle. Arch. Biochem. Biophys. *114*, 481–487 (1966)

Schatzmann, H.J., Witt, P.N.: Action of k-strophanthin on potassium leakage from frog sartorius muscle. J. Pharmacol. Exp. Ther. *112*, 501–508 (1954)

Schiaffino, S.: Histochemical enzyme profile of the masseter muscle in different mammalian species. Anat. Rec. *180*, 53–62 (1974)

Sellin, L.C., McArdle, J.J.: Effect of ouabain on reinnervating mammalian skeletal muscle. Eur. J. Pharmacol. *41*, 337–340 (1977)

Severin, S.E., Boldyrev, A.A., Tkachuk, V.A.: Some properties of Na^+, K^+-stimulated ATPase from normal and denervated muscles of rabbit with a special reference to its sensitivity to acetylcholine. Comp. Gen. Pharmacol. *5*, 181–185 (1974)

Sharma, V.K., Banerjee, S.P.: Denervation of cat fast- and slow-skeletal muscles: Effect on ouabain binding. Mol. Pharmacol. *14*, 1204–1211 (1978)

Sharma, V.K., Dasgupta, S.R., Banerjee, S.P.: [^{3}H]-ouabain binding to denervated and innervated skeletal muscle. Eur. J. Pharmacol. *43*, 273–276 (1977)

Smulyan, H., Eich, R.H.: Effect of digitalis on skeletal muscle in man. Am. J. Cardiol. *37*, 716–723 (1976)

Sulakhe, P.V., Fedelesova, M., McNamara, D.B., Dhalla, N.S.: Isolation of skeletal muscle membrane fragments containing active Na^+-K^+ stimulated ATPase: comparison of normal and dystrophic muscle sarcolemma. Biochem. Biophys. Res. Commun. *42*, 793–800 (1971)

Van Winkle, W.B., Schwartz, A.: Morphological and biochemical correlates of skeletal muscle contractility in the cat. I. Histochemical and electron microscopic studies. J. Cell. Physiol. *97*, 99–120 (1978)

Van Winkle, W.B., Entman, M.L., Bornet, E.P., Schwartz, A.: Morphological and biochemical correlates of skeletal muscle contractility in the cat. II. Physiological and biochemical studies. J. Cell. Physiol. *97*, 121–136 (1978)

Wareham, A.C.: Effect of denervation and ouabain on the response of the resting membrane potential of rat skeletal muscle to potassium. Pflügers Arch. *373*, 225–228 (1978)

Weinmann, J., Hasford, J., Kuhlmann, J., Bippus, P.H., Lichey, J., Rietbrock, N.: Digoxinkonzentrationen in Plasma und Gewebe. Eine Post-mortem-Untersuchung. Med. Klin. *74*, 613–619 (1979)

Westermann, E.: Zur Digitaliswirkung auf die neuromuskuläre Übertragung. Naunyn Schmiedebergs Arch. Pharmacol. *222*, 398–407 (1954)

Yamamoto, S., Fox, A.A.L., Greeff, K.: Inotropic effects and Na^+, K^+-ATPase inhibition of ouabain in isolated guinea-pig atria and diaphragm. Eur. J. Pharmacol. *71*, 437–446 (1981)

CHAPTER 24

Effects of Cardiac Glycosides on Autonomic Nervous System and Endocrine Glands

P. H. JOUBERT

A. Autonomic Effects of Cardiac Glycosides

It has been known for over a century that the cardiac glycosides affect the autonomic nervous system and that both the sympathetic and parasympathetic components can be involved. This review will look mainly at the evidence for the autonomic effects of peripherally administered cardiac glycosides as well as postulated mechanisms and will not consider in great detail work concerned with the effects of the cardiac glycosides on the central nervous system.

I. Effects on the Sympathetic Nervous System

Both sympathomimetic and sympatholytic effects have been attributed to the cardiac glycosides. There are two types of evidence for these effects. Many workers have observed the influence of modifying the actions of the sympathetic nervous system, either by drug administration or surgical intervention, on the therapeutic and toxic effects of digitalis, whereas some have studied the influence of cardiac glycosides on directly measured activity in sympathetic nerves.

The possible involvement of sympathetic nerves in the positive inotropic effects of digitalis has been investigated. Studies in the dog (ECKSTEIN et al., 1961; FAWAZ, 1967; KOCH-WESER, 1971), rabbit (KOCH-WESER, 1971), cat (SPANN et al., 1968; SWAMY et al., 1965), and guinea pig (KOCH-WESER, 1971) failed to show modification of the positive inotropic effect of various cardiac glycosides in the presence of sympathetic depletion by reserpine pretreatment. Another study (LEVY and RICHARDS, 1965a) failed to show any effect of β-adrenergic blockade on the positive inotropic effect of ouabain in rabbits. Work in normal dogs (DAGGET and WEISFELDT, 1965) suggested that acetylstrophanthidin is indirectly sympatholytic because of the failure to observe a positive inotropic effect in dogs, except when the cardiac sympathetic nerves were cut. This could be due to a compensatory lowering of sympathetic tone in the presence of digitalis and has a parallel in the inhibition of the reflex tachycardia seen clinically when digitalis is administered to patients with cardiac failure (JOUBERT, 1976).

Reserpine pretreatment inhibited the positive inotropic effect of digitalis in vitro in isolated rabbit left atrium (LEVY and RICHARDS, 1965b) and cat papillary muscle (TANZ, 1964). Similar findings were reported with β-blockade in isolated guinea pig atria (FÖRSTER and KALSOW, 1965). Based on the experimental approach discussed thus far, it therefore seems that in the intact animal, the sympathetic nervous system is not involved significantly in the therapeutic effects of digitalis as far as inotropy is concerned. Secondary adjustment of sympathetic tone might, however, result from the hemodynamic changes elicited by these drugs.

In contrast to the inotropic effects, there appears to be ample evidence for a sympathetic component in the tachyarrhythmias induced by digitalis as well as in the effect on atrioventricular conduction. The former is of relevance in digitalis toxicity, whereas the latter is important in both therapeutic and toxic situations.

Reserpine pretreatment appears to afford in vivo protection against the arrhythmogenic effects of digitalis, suggesting a sympathomimetic effect when digitalis is given in toxic doses. These observations have been made in species such as the rat (ATTREE et al., 1972; NADEAU and DE CHAMPLAIN, 1973), guinea pig (DOGGET and CASE, 1975), dog, (TAKAGI et al., 1965), and cat (LEVITT and ROBERTS, 1966; SWAMY et al., 1965). Failure of reserpine pretreatment to protect against ouabain-induced arrhythmias has been found in in vitro experiments with isolated rabbit atria (LEVY and RICHARDS, 1965 b), although another study reported protection in isolated atrial preparations of several species (KOCH-WESER, 1971). Protection against digitalis-induced arrhythmias by 6-hydroxydopamine has been reported in cats (KELLIHER and ROBERTS, 1972), but it had no effect in rats (NADEAU and DE CHAMPLAIN, 1973). The latter authors suggest that the fact that 6-hydroxydopamine does not cross the blood-brain barrier could be a possible explanation. It is interesting to note that one of the studies where protection with reserpine was produced (BOYAJY and NASH, 1965) showed a potentiation of the arrhythmogenic effect of digitalis when reserpine was given concomitantly. This has a parallel in the potentiation of digitalis arrhythmias by tricyclic antidepressants (ATTREE et al., 1972).

Some authors have examined the influence of digitalis on the effects produced by adrenaline administration or sympathetic nerve stimulation. In humans digitalis leaf had no effect on the heart rate response to adrenaline (MASSARO et al., 1960). A study in vagotomized dogs (NADEAU and JAMES, 1963) suggests a sympatholytic effect of acetylstrophanthidin. These workers found that digitalis produced a significant prolongation of atrioventricular conduction in vagotomized dogs. This effect could be abolished by thoracic sympathectomy. Contrasting observations suggesting sympathomimetic effects have been made in rats (JOUBERT and VAN DER MEER, 1978, 1979), where a dose of digitoxin that produced a positive inotropic effect of approximately 50% did not produce significant PR-interval prolongation. In the presence of propranolol, however, significant PR-interval prolongation was found. Similar observations were made in humans (LE WINTER et al., 1977), where digoxin or a β-adrenergic blocker alone failed to produce significant PR-interval prolongation, but when given concomitantly a significant prolongation was observed.

In sympathetic nerves it was found that recordings from preganglionic fibers to the stellate ganglion in the cat (GILLIS, 1969), showed that low doses of ouabain inhibited spontaneous sympathetic activity. Denervation of the reflexogenic areas in the carotid sinus and aortic arch prevented this. Larger doses of ouabain caused an increase in sympathetic nerve activity that was directly related to the development of ventricular tachycardia and could be antagonized by propranolol. Activation of carotid sinus nerves has been reported in the cat by PACE and THIBODEAUX (1973) and in the rabbit by MATHENY and AHLQUIST (1973).

The dilemma of establishing a mechanism for the sympathomimetic effects of digitalis is exemplified in a recent review (GILLIS et al., 1978), where contradictory literature allowed very few conclusions to be made. Studies on the mechanism of digitalis-induced sympathomimetic effects have focused mainly on the site of action and the possible neurotransmitters involved.

Many workers have reported arrhythmias after administering cardiac glycosides directly into the central nervous system (WEINBERG and HALEY, 1955; GREEFF and KASPERAT, 1961; STICKNEY and LUCCHESI, 1969; DOGGET and SPENCER, 1971). The most sensitive area in this regard appears to be the posterior hypothalamus (SAXENA and BHARGAVA, 1975). There appears to be no preferential concentration of cardiac glycosides in specific brain areas after systemic administration (DUTTA et al., 1977). It is difficult to ascertain from the literature whether sites of sympathetic stimulation within the central nervous system are of importance when digitalis is administered systemically. GILLIS (1978) maintains that large doses of digitalis administered systemically cause an activation of sympathetic preganglionic fibers. This is in keeping with a central mechanism of action. In contrast to this, WEAVER et al. (1976) found that toxic doses of digitalis only enhanced activity in postganglionic and not in preganglionic sympathetic nerves.

They explain their findings by the fact that they maintained blood pressure by volume expansion after the onset of arrhythmias and suggest that the findings of GILLIS were due to reflex activation of the sympathetic nervous system induced by hypovolemia. GILLIS (1969), however, described increased preganglionic activity when the carotid sinus and aortic arch areas were denervated and also reported a lack of correlation between changes in blood pressure and sympathetic activity. Both GILLIS and WEAVER used cats for their experiments, but the former used ouabain and the latter digoxin.

Considerable attention has been given to the effects of digitalis on central neurotransmitters which is dealt with in more detail in another section of this book. The role of adrenergic mechanisms (SAXENA and BHARGAVA, 1975), serotonergic mechanisms (HELKE et al., 1976), and dopaminergic mechanisms (GAITONDÉ and JOGLEKAR, 1975) have been investigated, but there appears to be no convincing evidence at present to implicate a particular central neurotransmitter in the sympathetic effects of systemically administered digitalis.

A vast literature exists on the effect of digitalis on the synthesis, storage, release, reuptake, and breakdown of catecholamines. Unfortunately, the results are conflicting and cover the whole range of possibilities, namely, enhancement, no effect, or inhibition. Ouabain has been found to decrease noradrenalin synthesis in the central nervous system (GOLDSTEIN et al., 1970; ANAGNOSTE and GOLDSTEIN, 1967). Most studies on storage have reported either a decrease (HARVEY, 1975; GÖTHERT, 1971; ANGELUCCI et al., 1966) or no change (KOCH-WESER, 1971; TANZ et al., 1968; SWAMY et al., 1965) in the content of catecholamines in organs such as the heart, adrenals, spleen, and brain. The majority of studies on catecholamine release have reported cardiac glycosides to cause enhanced release (HARVEY, 1975; SEIFEN, 1974; KIRPEKAR et al., 1970) though KOCH-WESER (1971) found no change in the isolated, perfused cat heart.

As far as effects on reuptake mechanisms are concerned, widely conflicting findings have been reported. The majority of workers have reported an in vitro inhibition of uptake (STICKNEY, 1976; LEITZ and STEFANO, 1970; DENGLER et al., 1962). In a recent study (EIKENBURG and STICKNEY, 1977), in vivo inhibition of reuptake was shown in the guinea pig. As far as breakdown is concerned, an inhibition of monoamine oxidase activity by cardiac glycosides has been reported (ROY and CHATTERJEE, 1970; POPOV and FORSTER, 1966).

It therefore appears reasonable to conclude that the cardiac glycosides in therapeutic concentrations and under in vivo conditions have little or no direct effects on the sympathetic nervous system. There does, however, appear to be evidence suggesting that due to hemodynamic changes under normal conditions and particularly in cardiac failure or due to a direct effect on baroreceptors and chemoreceptors in the carotid sinus and aortic arch, a reflex lowering of sympathetic tone may occur. In toxic concentrations, however, there is ample evidence for a sympathomimetic effect (Fig. 1). Very little firm evidence for the involvement of particular neurotransmitters exists.

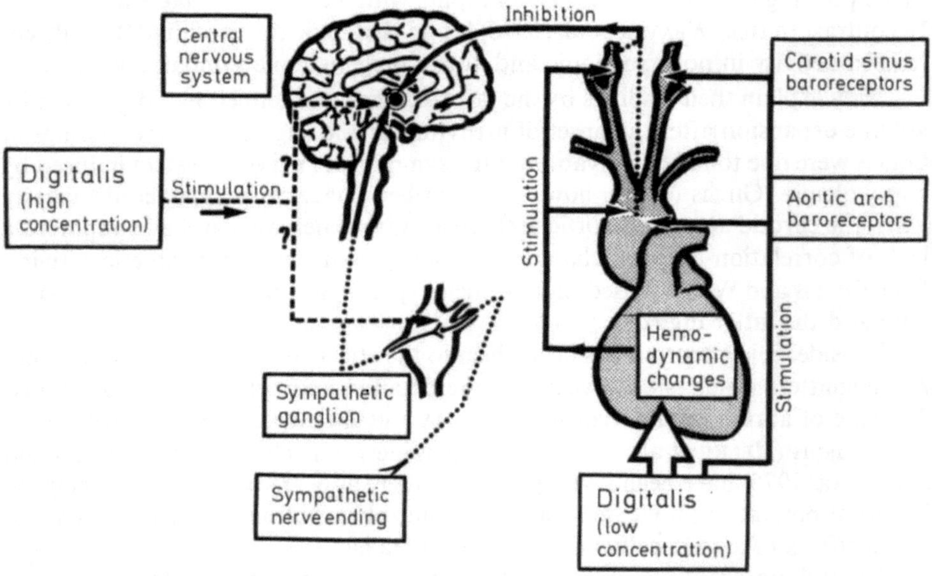

Fig. 1. Effects of digitalis on the sympathetic nervous system

II. Effects on the Parasympathetic Nervous System

Vagal stimulation by the cardiac glycosides has been implicated in the context of toxicity as well as in therapeutic effects. Many workers have investigated the vagal effects of digitalis, and there appears to be more agreement in this area than regarding the effects on the sympathetic nervous system. Various approaches have been used. The basic approach has been to administer digitalis by various routes to intact animals or to expose isolated organ preparations to digitalis. Subsequently, the effect of the vagus is monitored by measuring activity directly or by observing the effects produced by vagal stimulation, particularly in terms of heart rate and atrioventricular conduction. These initial observations are then usually extended by observing the modifying effect of destruction of parts of the nervous system (e.g., vagotomy, decerebration) or the administration of cholinergic and anticholinergic drugs on the vagal effects of digitalis.

Many studies of digitalis administration to a variety of animal species have reported bradycardia that could be abolished by vagotomy or atropine (KORTH et al., 1937; LENDLE, 1949; ABIKO et al., 1965; CHAI et al., 1967; FARGES et al., 1978). It has, however, been reported that a vagomimetic procedure (neostigmine administration) does not affect the lethal dose of strophanthidin G in guinea pigs (BUSSE et al., 1955) and that a vagolytic procedure (atropine) failed to influence the lethal dose of digitoxin in cats (MCLAIN et al., 1959).

Many investigators have examined the effects of cardiac glycosides on the heart rate response to vagal stimulation. Several workers found that cardiac glycosides produced an increased susceptibility to vagal stimulation (GAFFNEY et al., 1958; ZINNITZ, 1959; TODA and WEST, 1966). Some authors reported no effect (HEIM and BÄNDER, 1951; LENDLE, 1953) although the one group found a potentiation when the heart was damaged by pernocton anesthesia for 24 h. It appears as if digitalis only potentiates the effects of high frequency vagal stimulation on heart rate (more than 15/s) and not that of low frequency stimulation; the relevance of these findings to the effect of digitalis in the presence of normal vagal tone is not clear. Cardiac glycosides, however, potentiate the slowing of atrioventricular conduction produced by vagal stimulation in dogs (GREENSPAN and LORD, 1973; PACE and MARTIN, 1978) and appear to potentiate the effects of acetylcholine on the heart (GAFFNY et al., 1958; FARGES et al., 1978) both in terms of heart rate and atrioventricular conduction in intact dogs. These findings, however, are not confirmed by experiments with isolated frog and rat hearts (LENDLE and WIENKE, 1951).

A number of workers have looked at the effect of the administration of cardiac glycosides on activity in the vagus and other nerves. In 1965, ABIKO et al. reported that 20%–30% of the lethal dose of strospeside could produce bradycardia after vagotomy. This was associated with inhibition of afferent nerve impulses in postganglionic stellate fibers. When only the carotid sinus nerves were cut, the same observation was made, but when both vagi and carotid sinus nerves were cut, no bradycardia or inhibition of postganglionic stellate discharge were seen. It was therefore concluded that in addition to the carotid sinus nerves, afferent vagal fibers were involved in the inhibition of sympathetic discharge produced by cardiac glycosides. MCLAIN (1969) reported an increase in efferent vagal activity unrelated to heart rate after intravenous administration of digitoxin and ouabain to cats. GILLIS et al. (1972) also observed increased vagal activity, which could be abolished by propranolol or spinal cord transection after ouabain administration to cats. GILLIS et al. (1978) reported that denervation of the reflexogenic areas by section of the carotid sinus nerves and vagi prevented the increased efferent vagal discharge after digoxin administration. In the same experiments it was found that digoxin caused increased afferent discharge in the carotid sinus nerves and vagi. The authors consequently concluded that vagal effects are produced peripherally. This is supported by earlier work (QUEST and GILLIS, 1971), which showed increased carotid sinus nerve activity and vagal effects after intracarotid administration of acetylstrophanthidin. Other workers (CHAI et al., 1967) showed that acetylstrophanthidin, administered intravenously in small doses, produced no vagal effects, whereas intracarotid administration of the same dose produced marked effects. Intravertebral administration produced only slight vagal effects, which could be abolished by vagotomy. These findings tend to confirm the findings of GILLIS and his co-workers and tend to be against a central mechanism.

Although cardiac glycosides administered directly into the central nervous system produce vagal effects (KORTH et al., 1937; CHAI et al., 1967), there is little evidence for a central effect when administered systemically. Vagal stimulation by digitalis appears to be due to stimulation of carotid sinus and aortic arch receptors, but there is some evidence that certain cardiac sensory receptors are also involved (QUEST et al., 1973; OBERG and THOREN, 1972).

In humans, digitalis causes a decrease in heart rate in patients treated for cardiac failure or atrial fibrillation. GOLD et al. (1939) reported that the slowing seen after high doses could not be reversed by atropine. Based on this, he postulated a vagal and extravagal factor in the effect of digitalis on atrioventricular conduction. HOHENSEE and LENDLE (1949) postulated that in cardiac failure the decreased heart rate caused by digitalis results from improved cardiac output and not from vagal stimulation but that the bradycardia seen with digitalis toxicity was of vagal origin and can be reversed by atropine. This contention was also expressed in a later review (LENDLE and MERCKER, 1961). In the light of present knowledge, it seems clear that the decrease in heart rate seen in the treatment of cardiac failure is mainly due to a reversal of the reflex tachycardia produced by a reflex increase in sympathetic tone (JOUBERT, 1976).

The situation with regard to atrioventricular conduction appears to be different. In a study of patients with denervated hearts (cardiac transplants), it was found that the production of significant increases in the refractory period of the atrioventricular node by therapeutic doses of digitalis was dependent on an intact cardiac innervation (GOODMAN et al., 1975).

In conclusion, it therefore appears as if vagal effects are insignificant in the therapy of cardiac failure in terms of negative chronotropism but could be important in the context of supraventricular arrhythmias where the required therapeutic effect is a delay in atrioventricular conduction. Vagal effects in clinical situations appear to be mediated by a peripheral effect on the carotid sinus, aortic arch, and myocardial sensory receptors (Fig. 2).

III. Effects on the Gastrointestinal System

Cardiac glycosides can influence the gastrointestinal tract in two ways. The first and most important is the well-known emetic action about which very little controversy exists, and the second is the experimentally observed direct effect on gastrointestinal smooth muscle, which appears to be of little clinical relevance.

Since an extensive review on the physiology of vomiting due to cardiac glycosides (BORISON and WANG, 1953), a review on drug-induced vomiting (WANG and BORISON, 1952), and an exhaustive review of the extracardiac effects of cardiac glycosides (LENDLE and MERCKER, 1961) were published, very little of significance has been added to our knowledge about digitalis-induced vomiting.

As long ago as 1912 (HATCHER and EGGLESTON), it was demonstrated that emesis occurred after intravenous administration of digitalis to dogs after surgical re-

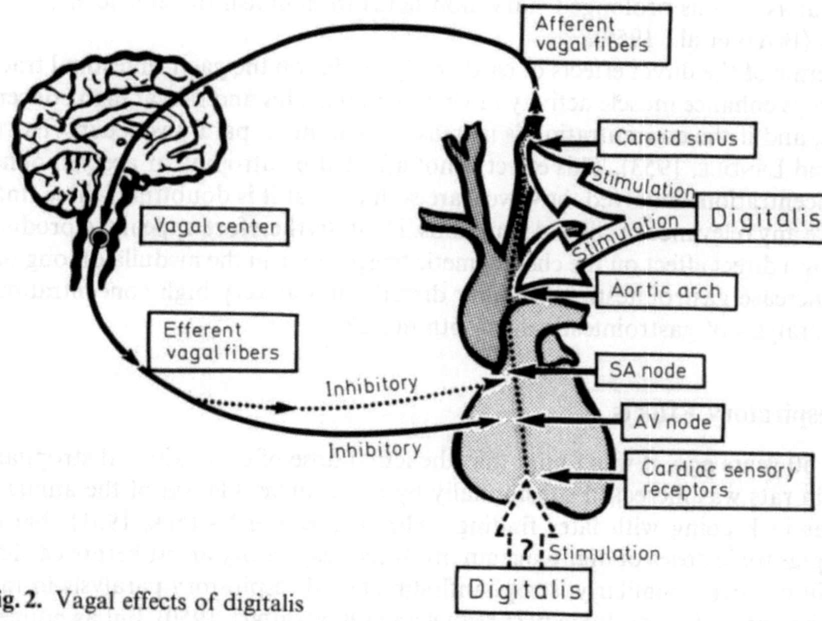

Fig. 2. Vagal effects of digitalis

moval of the gastrointestinal tract. Despite some claims for local irritation by digitalis as a basis for emesis in man (GOLD et al., 1952), another study in man led to the opposite conclusion (ÖSTLING, 1950). Several workers have shown that ablation of the chemoemetic trigger zone in the dorsal region of the medulla oblongata prevents digitalis-induced vomiting (BORISON and BRIZZEE, 1951; BORISON and WANG, 1951; WANG and BORISON, 1952), and it is generally accepted that this is the central site of action and the major factor in digitalis-induced vomiting.

Although cardiac and other peripheral receptors have been implicated by earlier workers of the emetic effects of digitalis (EGGLESTON, 1916; HATCHER and WEISS, 1923), it has been found subsequently that vagotomy (DRESBACH, 1939), cardiac denervation (DRESBACH and WADDELL, 1929), removal of carotid bodies and sinuses (PINSCHMIDT, 1946), and gut denervation (WANG and BORISON, 1952) do not prevent this effect. Nicotine in cats (HATCHER and WEISS, 1927) and diphenhydramine and chlorpromazine in dogs (BUSSE and LOENNECKEN, 1953) partially prevent cardiac glycoside emesis. The former can be related to a ganglionic effect and the latter to antidopaminergic effect in the chemoemetic trigger zone.

A puzzling finding is that although cats are susceptible to digitalis-induced vomiting by the oral and the intravenous route, neither application to the medulla oblongata (DRESBACH, 1947) nor intracerebroventricular administration (MELVILLE and SHISTER, 1957) produced emesis. GAITONDÉ et al. (1965) showed that the most effective route for producing emesis was administering cardiac glycosides into the vertebral artery and concluded that the central site for producing emesis can only be reached via the blood stream. In dogs, however, intracerebroventricular administration does elicit vomiting (SHARE et al., 1964). The capacity for cardiac glycosides to produce emesis appears to differ from species to species. It is, for instance, more prominent in cats and dogs than in rats (FÖRSTER et al., 1965). A recent meal as well as prolonged starvation tends to diminish the emetic effects of digitalis (BOYD et al., 1959).

In terms of the direct effects of cardiac glycosides on the gastrointestinal tract, these drugs enhance muscle activity in vivo in guinea pigs and rats at high concentrations, and if the concentration is increased even more, paralysis occurs (HOFFMANN and LENDLE, 1953). This effect is not affected by atropine or acetylcholine. The concentrations involved, however, are so high that it is doubtful if these findings have any relevance in clinical situations. Digitalis therefore appears to produce emesis by a direct effect on the chemoemetic trigger zone in the medulla oblongata. It can increase gastrointestinal motility directly and at very high concentrations cause paralysis of gastrointestinal smooth muscle.

IV. Respiratory Effects

Almost 50 years ago, it was found that the lethal dose of digitoxin and strophanthin G in rats was increased substantially by artificial ventilation of the animals. This was in keeping with later findings (HOFFMANN and LENDLE, 1951) that in guinea pigs toxic doses of digitoxigenin produced respiratory arrest before cardiac standstill occurred. Similarly, strophanthidin caused respiratory paralysis in rats before the onset of cardiotoxicity (LENDLE and OLDENBURG, 1950), but as pointed out by these authors, these doses were far in excess of doses that would be cardiotoxic in more sensitive animals. In dogs, artificial respiration failed to change the lethal dose of ouabain (STICKNEY and MEYERS, 1973), probably because cardiotoxic effects occur before significant respiratory depression is elicited.

Respiratory stimulation has been described after administration of more modest doses of cardiac glycosides. Increased phrenic nerve discharge and hyperventilation have been reported after intravenous digoxin administration in cats (VIANA, 1977; PACE and GILLIS, 1976; GILLIS et al., 1972).

As far as humans are concerned, therapeutic doses of digitoxin have no effect on minute volume but appear to depress the stimulatory effect of inhaled CO_2 (RIGGERT and SCHWAB, 1955). In a later study, using strophanthin K, these findings could not be substantiated (SCHWAB and WAGNER, 1957). It therefore appears as if the respiratory response to cardiac glycosides is determined by the species, the type of glycoside, and the dose used.

Two mechanisms have been proposed for respiratory paralysis caused by the cardiac glycosides. Ouabain administered directly into certain medullary areas in cats produced respiratory depression (BASU-RAY et al., 1972), suggesting a central site of action. On the other hand, experiments with intact animals as well as isolated phrenic nerve-diagphragm preparations suggested that potassium loss from respiratory and other muscles, with resultant hyperkalemia and disturbance of the sodium-potassium gradient across cell membranes, could be responsible for direct paralysis (GREEFF and WESTERMANN, 1955). Evidence for both mechanisms has been obtained when studying the effect of intravenous digitoxin on phrenic nerve impulses and direct recordings of action potentials from the diaphragm in decerebrate cats (DAL-RI and SCHMIDT, 1960). Rapid injection of the cardiac glycoside resulted mainly in peripheral muscle paralysis. Slow infusion primarily caused central depression.

Studying phrenic nerve stimulation by digitalis, PACE and GILLIS (1976) found that digoxin caused increased vagal, phrenic, and sympathetic nerve activity. Denervation of reflexogenic areas abolished the vagal effect, diminished the effect on the phrenic nerve, and failed to influence sympathetic nerve activity. They concluded that the vagus was activated peripherally, the sympathetic nerves centrally, and the phrenic nerve by both a peripheral and central mechanism.

It is therefore reasonable to conclude that moderate doses of cardiac glycosides can increase phrenic nerve activity by an action on peripheral reflexogenic as well as central medullary sites. Very high doses can depress respiration centrally or cause peripheral paralysis of respiratory muscles. This respiratory depression can only be elicited in vivo in animals such as rats, which are resistant to digitalis cardiotoxicity. In animals with more sensitive hearts, this will only be seen if the drug is applied centrally or the experiment is done in vitro. The clinical relevance of the respiratory effects of cardiac glycosides remains debatable.

B. Endocrine Effects of Cardiac Glycosides

The steroid nature of the cardiac glycosides have prompted several investigations into the effects of cardiac glycosides on endocrine glands and the interaction with various hormones. Particular attention has been given to hormones with a cyclopentanoperhydrophenanthrene nucleus. In addition, certain side-effects of the glycosides could have a hormonal basis and have induced certain workers to examine them.

I. Estrogenic Effects and Gynecomastia

The steroid structure of the cardiac glycosides and the occurrence of gynecomastia as a side-effect has led several workers to examine possible estrogenic effects. Several reports of mammary hypertrophy during digitalis therapy in both males (LE WINN, 1953) and females (CALOV and WHYTE, 1954; BLOCH, 1961) have been published. Furthermore, it was found that menopausal women receiving cardiac glycosides showed a greater degree of vaginal epithelium proliferation than a control group (TATRA and NISSEL, 1969). These workers also made the same observation in ovariectomized mice. Postmenopausal woman on digoxin therapy for longer than 2 years showed estrogenic changes in the vaginal epithelium and had significantly less postmenopausal symptoms than control patients (KITZES-COHEN et al., 1978).

Few workers have examined the possible estrogenic effects of the cardiac glycosides objectively. In an extensive investigation in young mice (RICKEN, 1975), it was found that although certain cardiac glycosides alone had no effect, they potentiated the increase in weight of the uterus produced by estrogen administration. The cardiac glycosides increased the estrogen content of various organs including the uterus while decreasing urinary excretion. Studying liver homogenates, a decrease in hydrophilic estrogen metabolites were found, and it was concluded that the cardiac glycosides inhibited metabolism of the estrogens studied (β-estradiol and diethylstilbestrol). The increase of serum levels of estrogens found in humans on long-term digoxin therapy (STOFFER et al., 1973) adds substance to these animal experiments. STOFFER et al. (1973) also showed a decrease in luteinizing hormone and testosterone levels, which the authors interpreted as being secondary to the raised estrogen values. In conclusion, it is suggested that a rise in the total estrogen content of the body rather than direct estrogenic effect could be the basis of digitalis-induced estrogen-like effects, such as gynecomastia and changes in the vaginal cytology.

II. Adrenocortical Hormones

The steroid nature of the cardiac glycosides as well as the effects on the movement of potassium and sodium ions across membrane have made the adrenal cortex the most extensively investigated endocrine gland in terms of digitalis effects.

Various experimental approaches have been used. Workers have looked at the effect of cardiac glycosides on adrenal weight and histology, the effect on the production and secretion of cortical hormones, the effect on the production of ACTH, and the interaction between adrenal homones and digitalis in terms of their effects on various target organs and processes.

Strophanthin K administration (METZLER and GREEFF, 1954; METZLER, 1951), ouabain administration (DEANE and GARDENER, 1951; GOMOLL, 1967), and digitoxin administration (GVOZDJÁK et al., 1961) produce an increase in weight of the adrenal glands and the heart in guinea pigs and rats. Histologically, this is due mainly to a proliferation of the zona glomerulosa (DEANE and GARDNER, 1951; GARDNER et al., 1954; GOMOLL, 1967).

In rats it was found that adrenal hypertrophy preceded cardiac hypertrophy (METZLER and GREEFF, 1954). Hypophysectomy did not prevent adrenal hypertrophy, but cardiac hypertrophy could be prevented by adrenalectomy. Cortisone given alone caused adrenal atrophy with cardiac hypertrophy, whereas DOCA caused adrenal atrophy without cardiac hypertrophy. The authors interpreted this as indicating that digitalis has a direct effect on the adrenal gland independent of pituitary function and that cardiac hypertrophy was secondary to glucocorticoid release from the adrenals.

It is interesting that chronic administration of cardiac glycosides appears to partially antagonize the adrenal and cardiac hypertrophy induced by swimming (METZLER, 1951; MOSKOPF and SARRE, 1954).

In 1951 BAZZI and NICOLISI reported decreased excretion of 11-oxysteroids in two patients treated with digitalis. In a study of 23 patients, decreased excretion of both 11-oxy and 17-ketosteroids was found (LASCHÉ et al., 1951). Intravenous administration of acetylstrophanthin to anesthetized dogs produced a significant decrease in aldosterone concentration in adrenal vein blood without affecting hydrocortisone or corticosterone secretion (GOMOLL and SHERROD, 1964). These authors suggested that this effect was therefore not due to ACTH elaboration or an alteration in blood flow. Similar observations were made in a later in vitro study (CUSHMAN, 1969). In rats high doses of digitoxin and digitoxigenin produced a transient increase in plasma corticosterone levels (NAKAGAWA and WANATABA, 1966), but it appears as if chronic administration of ouabain to rats depresses the ability of the adrenals to secrete corticosterone in vitro (GOMOLL, 1967). In a study on adrenocortical mitochondria, it was found that ouabain enhanced and other glycosides inhibited the 11-hydroxylation system (TSAO and KING, 1967).

The occurrence of secondary hyperaldosteronism in patients with cardiac failure is well known (DAVIS, 1960), and digitalis administration in the context of this condition leads to reversal of this phenomenon (CARPENTER et al., 1962) due to the improvement in hemodynamics and renal blood flow.

The exclusion of pituitary involvement in effects of cardiac glycosides on the adrenal cortex on the basis of indirect evidence has been discussed (METZLER and GREEFF, 1954). Although chronic ouabain administration depresses the ability of the adrenal cortex to secrete corticosterone in vitro, the response to ACTH is markedly enhanced (GOMOLL, 1967). In a study on rats, it was reported that digitoxin administration for 30 days produced a significant fall in the ACTH content of the hypophysis (VERNIKOS-DANELLIS and MARKS, 1962); these workers postulated a direct effect on the anterior pituitary. The suggestion of others (METZLER and GREEFF, 1954) that cardiac glycosides in the rat can cause corticosteroid release can, however, also explain these findings.

A large number of studies have looked at corticosteroid-cardiac glycoside interactions. Glucocorticoid administration appears to protect against digitalis toxicity in vivo (METZLER and HERGOTT, 1951). Aldosterone in the cat appears to antagonize the inotropic response to ouabain in vitro (LEFER and SAYERS, 1965). Corticosterone did not antagonize the inotropic effects of strophanthin G in the rat in vitro (REITER and NOÉ, 1960). Adrenalectomy greatly potentiated the toxic effects of ouabain (UNTERMAN et al., 1955; GREEFF, 1957), although it was found that adrenalectomy does not change the responsiveness of myocardial Na^+, K^+-ATPase to strophanthin K (BORSCH-GALETKE et al., 1972). Adrenalectomy or metyrapone administration alone decreased the specific activity of the enzyme. Cardiac glycosides appear to antagonize the effects of corticosteroids on urinary electrolytes (GREEFF, 1957; SULSER et al., 1959).

III. Thyroid Hormones

Several workers have reported that thyroid disease can alter the pharmacokinetics (CROXSON and IBBERTSON, 1975; VARADI and FÖLDES, 1976) and the pharmacodynamics (MORROW et al., 1963) of cardiac glycosides, but few studies have shown that these drugs can affect the thyroid gland. A standard reference on drug side-effects makes no mention of effects on the thyroid gland (ROBINSON, 1975).

HALONEN et al. (1952), however, reported that strophanthin retards the metamorphosis of tadpoles whereas it is enhanced by thyroxin and suggested that cardiac glycosides inhibit thyroid function. In agreement with this notion are the findings that cardiac glycosides histologically produced a resting phase in guinea pig thyroid glands (KUUSITO et al., 1953) and an inhibition of ^{131}I uptake by sheep thyroid slices (WOLFF and MAUREY, 1958). There is no evidence to suggest that digitalis will influence the thyroid gland in clinical situations, but thyroid function may influence cardiac glycoside pharmacokinetics and pharmacodynamics (see this Handbook, Vol. 56/II, Chaps. 11 and 14).

IV. Other Endocrine Glands

Apart from the endocrine glands discussed, there does not appear to be evidence that the cardiac glycosides can influence other endocrine glands. Strophanthin in man can, however, elicit a rise or a fall in blood glucose concentrations (WOLFF, 1955), depending on the functional state of the autonomic nervous system. In rats (KYPSON et al., 1968) and in dogs (KIEN et al., 1960), cardiac glycosides increase glucose uptake and metabolism by myocardium and skeletal muscle, an action similar to that of insulin. It is therefore conceivable that cardiac glycosides could secondarily influence the endocrine functions of the pancreas, but this appears to be of little clinical significance.

References

Abiko, Y., Mukahira, K., Tanabe, T.: On the role of vagi and sinus nerves in the reflexogenic inhibition of sympathetic discharges induced by strospeside in cats. Jpn. J. Pharmac. *15*, 143–148 (1965)

Anagnoste, B., Goldstein, M: The effects of ouabain on cathecolamine biosynthesis in different areas of rats' brains. Pharmacologist *9*, 210 (1967)

Angelucci, L., Lorentz, G., Baldieri, M.: The relation between noradrenaline content of rabbit heart muscle and the amount of k-strophanthin needed to produce arrhythmias. J. Pharm. Pharmacol. *18*, 775–782 (1966)

Attree, T., Sawyer, P., Turnbull, M.J.: Interaction between digoxin and tricyclic anti-depressants in the rat. Eur. J. Pharmacol. *19*, 294–296 (1972)

Basu-Ray, B.N., Dutta, S.N., Pradhan, S.N.: Effects of microinjections of ouabain into certain medullary areas in cats. J. Pharmacol. Exp. Ther. *181/2* 357–361 (1972)

Bazzi, U., Nicolisi, G.: Modificazioni indotte dalla somministrazione a dosi massive di sostanze a nucleo sterolico sulla eliminazione di steroidi ormonici dell' urina (11-ossicorticosteroidi, 17-chetosteroidi). II. Azione della digitale. Boll. Soc. Ital. Biol. Sper. *27*, 1307–1308 (1951)

Bloch, K.: Zur Pathogenese der Mamma-Hypertrophie bei der Digitalis-Applikation. Z. Kreislaufforsch. *50*, 591–595 (1961)

Borison, H.L., Brizzee, K.R.: Morphology of emetic chemoreceptor trigger zone in the cat medulla oblongata. Proc. Soc. Exp. Biol. Med. *77*, 38–42 (1951)

Borison, H.L., Wang, S.C.: The vomiting center: Its destruction by radon implantation in dog medulla oblongata. Am. J. Physiol. *166*, 712–717 (1951)

Borison, H.L., Wang, S.C.: Physiology and pharmacology of vomitting. Pharmacol. Rev. *5*, 193–230 (1953)

Borsch-Galetke, E., Dransfeld, H., Greeff, K.: Specific activity and sensitivity to strophanthin of the $Na^{+}+K^{+}$-activated ATP-ase in rats and guinea-pigs with hypoadrenalism. Naunyn Schmiedebergs Arch. Pharmacol.*274*, 74–80 (1972)

Boyajy, L.D., Nash, C.B.: Influence of reserpine on arrhythmias, inotropic effects, and myocardial potassium balance induced by digitalis materials. J. Pharmacol. Exp. Ther. *148/2*, 193–201 (1965)

Boyd, E.M., Brown, M.D., Cassell, W.A.: Delayed vomiting induced in dogs by intramuscular digoxin. J. Pharm. Pharmacol. *11*, 742–746 (1959)

Busse, W. von, Loennecken, S.J.: Pharmakologische Beeinflußbarkeit reflektorisch-toxischer Digitaliswirkungen (Bradykardie und Erbrechen) über die Herznerven. Naunyn Schmiedebergs Arch. Pharmacol. *220*, 232–244 (1953)

Busse, W. von, Göing, H., Lendle, L.: Einfluß des erhöhten Vagustonus auf die Strophanthintoxizität am Herzen. Z. Kreislaufforsch. *44*, 604–608 (1955)

Calove, W.L., Whyte, H.M.: Oedema and mammary hypertrophy: a toxic effect of digitalis leaf. Med. J. Aust. *1*, 556–557 (1954)

Carpenter, C.C.J., Davis, J.O., Wallace, C.R., Hamilton, W.F.: Acute effects of cardiac glycosides on aldosterone secretion in dogs with hyperaldosteronism secondary to chronic right hart failure. Circ. Res. *10*, 178–187 (1962)
Chai, C.Y., Wang, H.H., Hoffmann, B.F., Wang, S.C.: Mechanisms of bradycardia induced by digitalis substances. Am. J. Physiol. *212*, 26–34 (1967)
Croxson, M.S., Ibbertson, H.K.: Serum digoxin in patients with thyroid disease. Br. Med. J. *3*, 566–568 (1975)
Cushman, P.: Inhibition of abdosterone secretion by ouabain in dog adrenal cortical tissue. Endocrinology *84*, 808–813 (1969)
Dagget, W.M., Weisfeldt, M.L.: Influence of the sympathetic nervous system on the response of the normal heart to digitalis. Am. J. Cardiol. *16*, 394–405 (1965)
Dal-Ri, H., Schmidt, G.: Central and peripheral alteration of respiration by toxic cardiac glycosides in the rat. Naunyn Schmiedebergs Arch. Pharmacol. *239*, 158–169 (1960)
Davis, J.O.: Mechanism of salt and water retention in congestive heart failure. Am. J. Med. *29*, 486 (1960)
Deane, H.W., Gardener, L.I.: Notes and comments: Ouabain and the adrenal cortex. Endocrinology *48*, 237–238 (1951)
Dengler, J.H., Michaelson, I.A., Spiegel, H.E., Titus, E.: The uptake of labelled norepinephrine by isolated brain and other tissues of the cat. Intern. J. Neuropharmacol. *1*, 23–28 (1962)
Doggett, N.S., Case, G.: Some observation on the interaction between cardiac glycosides and reserpine in the heart and central nervous system. Toxicol. Appl. Pharmacol. *33*, 87–93 (1975)
Doggett, N.S., Spencer, P.S.J.: Pharmacological properties of centrally administered ouabain and their modification by other drugs. Br. J. Pharmacol. *42*, 242–253 (1971)
Dresbach, M.: Additional experiments relative to the origin of glycoside emesis using cats and dogs. Am. J. Physiol. *126*, 480 (1939)
Dresbach, M.: Further experiments in attempt to locate site of emetic action of digitalis glycosides. J. Pharmacol. Exp. Ther. *91*, 307–316 (1947)
Dresbach, M., Waddell, K.C.: Question of reflex vomiting from heart induced by digitalis bodies. J. Lab. Clin. Med. *14*, 625–630 (1929)
Dutta, S., Marks, B.H., Schoener, E.P.: Accumulation of radioactive cardiac glycosides by various brain regions in relation to the dysrhythmogenic effect. Br. J. Pharmacol. *59*, 101–106 (1977)
Eckstein, J.W., Abboud, F.M., Pereda, S.A.: Hemodynamic responses to administration of acetyl strophanthidin before and after ganglionic blockade in normal dogs and in dogs treated with reserpine. J. Lab. Clin. Med. *58*, 814–815 (1961)
Eggleston, C.: The antagonism between atropine and certain central emetics. J. Pharmacol. Exp. Ther. *9*, 11–25 (1916)
Eikenburg, D.C., Stickney, J.L.: Inhibition of sympathetic neuronal transport and ouabain-induced cardiac arrhythmias. Res. Commun. Chem. Pathol. Pharmacol. *18/4*, 587–599 (1977)
Farges, J.P., Ollanger, M., Lievre, M., Faucon, G.: Influence of cardiac cholinergic impregnation on the effects of ouabain on atrial and ventricular myocardium. J. Pharmacol. (Paris) *9/1*, 65–75 (1978)
Fawaz, G.: Effect of reserpine and pronethalol on the therapeutic and toxic actions of digitalis in the dog heart-lung preparation. Br. J. Chemother. *29*, 302–308 (1967)
Förster, W., Kalsow, H.: Über den unterschiedlichen Einfluß von Dichlorisoproterenol auf den positiv inotropen Effekt verschiedener Digitaliskörper am isolierten Vorhof-Aurikelpräparat des Meerschweinchen. Acta. Biol. Med. Ger. *15*, 71–78 (1965)
Förster, W., Sziegoleit, W., Guhlke, I.: Vergleichende Untersuchung einiger extra-cardialer Effekte von herzwirksamen Glykosiden. Arch. Int. Pharmacodyn. *155/1*, 165–182 (1965)
Gaffney, T.E., Kahn, J.B., van Maanen, E.F.: A mechanism of the vagal effect of cardiac glycosides. J. Pharmacol. Exp. Ther. *122*, 423–429 (1958)
Gaitondé, B.B., Joglekar, S.N.: Role of catecholamines in the central mechanism of emetic response induced by peruvoside and ouabain in cats. Br. J. Pharmacol. *54/2*, 157–162 (1975)

Gaitondé, B.B., McCarthy, L.E., Borison, H.L.: Central emetic action and toxic effects of digitalis in cats. J. Pharmacol. Exp. Ther. *147/3*, 409–415 (1965)
Gardner, L.I., Berman, H., Deane, H.W.: Metabolic ompetition between descoxycorticosterone acetate and strophanthin-g in the rat. Endocrinology *55*, 417–427 (1954)
Gillis, R.A.: Cardiac sympathetic nerve activity: Changes induced by ouabain and propranolol. Science *166*, 508–510 (1969)
Gillis, R.A., Raines, A., Sohn, Y.J., Levitt, B., Standaert, F.G.: Neuroexcitatory effects of digitalis and their role in the development of cardiac arrhythmias. J. Pharmacol. Exp. Ther. *183/1*, 154–158 (1972)
Gillis, R.A., Helke, C.J., Kellar, K.J., Quest, J.A.: Autonomic nervous system actions of cardiac glycosides. Biochem. Pharmacol. *27*, 849–856 (1978)
Göthert, M.: Der Einfluß von Digitoxin auf die Noradrenalinkonzentration im Myocard. Arzneim. Forsch. *21*, 1333–1334 (1971)
Gold, H., Kwit, N.T., Otto, H., Fox, T.: On vagal and extravagal factors in cardiac slowing by digitalis in patients with auricular fibrillation. J. Clin. Invest. *18*, 429–437 (1939)
Gold, H., Greiner, T., Cattell, M., Modell, W., Gluck, J., Marsh, R., Mathes, S., Hudson, D., Robertson, D., Warshaw, L., Otto, H., Kwit, N., Kramer, M.: Difference in the relation of cardiac to emetic actions in oral and parenteral digitalization. Am. J. Med. *13*, 124–144 (1952)
Goldstein, M., Ohi, Y., Backstrom, T.: The effect of ouabain on catecholamine biosynthesis in rat brain cortex slices. J. Pharmacol. Exp. Ther. *174*, 77–82 (1970)
Gomoll, A.W.: Influence of ouabain on the formation of corticosterone in the rat adrenal gland. J. Pharmacol. Exp. Ther. *156/2*, 300–309 (1967)
Gomoll, A.W., Sherrod, T.R.: Influence of acetylstrophanthin on aldosterone secretion in the intact anesthetized dog. Arch. Int. Pharmacodyn. *150/3–4*, 469–482 (1964)
Goodman, D.J., Rossen, R.M., Cannom, D.S., Rider, A.K., Harrison, D.C.: Effect of digoxin and atrioventricular conduction: Studies in patients with and without cardiac autonomic innervation. Circulation *51/2*, 251–256 (1975)
Greeff, K.: Mineralstoffwechselwirkung und Toxicität des Strophanthins bei nebennierenlosen Ratten. Arch. Exp. Pathol. Pharmakol. *231*, 391–400 (1957)
Greeff, K., Kasperat, H.: Vergleich der neurotoxischen Wirkung von Digitalisglykosiden und Geninen bei intra-cerebraler und intravenöser Injektion an Mäusen, Ratten und Meerschweinchen. Naunyn Schmiedebergs Arch. Pharmacol. *242*, 76–89 (1961)
Greeff, K., Westermann, E.: Untersuchung über die muskellähmende Wirkung des Strophanthins. Naunyn Schmiedebergs Arch. Pharmacol. *226*, 103–113 (1955)
Greenspan, K., Lord, T.J.: Digitalis and vagal stimulation during atrial fibrillation: effects on atrioventricular conduction and ventricular arrhythmias. Cardiovasc. Res. *7*, 241–246 (1973)
Guozdjak, J., Niederland, T.R., Honanova, J., Kovacikova, B.: Dynamic investigation of the biochemical changes in the adrenal glands of rats after the administration of digitoxin. Arch. Int. Pharmacodyn. *81/3–4*, 390–396 (1961)
Halonen, P.I., Kuusisto, A.N., Koskelo, P.: On the effect of strophanthin on thyroid activity. Cardiologia *21*, 727–734 (1952)
Harvey, S.C.: The effects of ouabain and phenytoin on myocardial noradrenaline. Arch. Int. Pharmacodyn. *213*, 222–234 (1975)
Hatcher, R.A., Eggleston, C.: The emetic action of digitalis bodies. J. Pharmacol. Exp. Ther. *4*, 113–134 (1912)
Hatcher, R.A., Weiss, S.: Studies on vomiting. J. Pharmacol. Exp. Ther. *22*, 139–193 (1923)
Hatcher, R.A., Weiss, S.: Seat of emetic action of digitalis bodies. J. Pharmacol. Exp. Ther. *32*, 37–53 (1927)
Heim, F., Bänder, A.: Der Einfluß von Strophanthin auf die Vagus-Erregbarkeit des normalen und geschädigten Warmblüterherzens. Arch. Exp. Pathol. Pharmakol. *212*, 456–462 (1951)
Helke, C.J., Souza, J.D., Hamilton, B.L., Morgenroth, V.H., Gillis, R.A.: Evidence for a role of central serotonergic neurones in digitalis-induced cardiac arrhythmias. Nature *263*, 246–248 (1976)

Hoffmann, G., Lendle, L.: Kritisches zur Digitalisauswertung am Meerschweinchen. Naunyn Schmiedebergs Arch. Pharmacol. *212*, 376–385 (1951)
Hoffmann, G., Lendle, L.: Nachweis extrakardialer Digitaliswirkungen an der Ratte. Arch. Exp. Pathol. Pharmakol. *217*, 184–193 (1953)
Hohensee, T., Lendle, L.: Über die Frage einer Vaguswirkung von Digitalisglykosiden. Naunyn Schmiedebergs Arch. Pharmacol. *207*, 388–408 (1949)
Joubert, P.H.: Digitalis in clinical practice. S. Afr. Med. J. *50*, 146–152 (1976)
Joubert, P.H., van der Meer, L.: Digitalis preparations show pharmacodynamic differences with important clinical implications. In: IUPHAR 7th International Congress of Pharmacology, Paris 16–21 July, 1978, Abstracts. Boissien, J.R., Ledrat, P. (eds.), p. 917. Oxford: Pergamon 1978
Joubert, P.H., van der Meer, L.: A specific cardiac glycoside for cardiac failure and another for atrial fibrillation. S. Afr. Med. J. *56*, 1040–1042
Kelliher, G.J., Roberts, J.: Effect of 6-hydroxydopamine on ouabain induced arrhythmia. Clin. Res. *20*, 857 (1972)
Kien, G.A., Gomoll, A.W., Sherrod, T.R.: Action of digoxin and insulin on transport of glucose through myocardial cell membrane. Proc. Soc. Exp. Biol. Med. *103*, 682–685 (1960)
Kirpekar, S.M., Prat, J.C., Yamamoto, H.: Effects of metabolic inhibitors on norepinephrine release from the perfused spleen of the cat. J. Pharmacol. Exp. Ther. *172*, 342–350 (1970)
Kitzes-Cohen, R., Neri, A., Schehter, A., Rosenfeld, J.: Estrogen like activity of digoxin, its correlation to vaginal cytology, to digoxin serum levels and Kupperman index. 7th International Congress of Pharmacology, Paris, 16–21 Juli 1978, Abstracts. Boissier, J.R., Lechat, P. (eds.), p. 675. Oxford: Permagon 1978
Koch-Weser, J.: Beta-receptor blockade and myocardial effects of cardiac glycosides. Circ. Res. *28*, 109–118 (1971)
Korth, C., Marx, H., Weinberg, S.: Über die Wirkung des Strophanthins auf das Zentralnervensystem. Naunyn Schmiedebergs Arch. Pharmacol. *185*, 42–56 (1937)
Kuusisto, A.N., Koskelo, P., Halonen, P.I.: Über die Wirkung von Digitalis und Methyl-Thiourazil auf die Schilddrüse bei gleichzeitiger Zufuhr. Ann. Med. Exp. Fenn. *31*, 74–81 (1953)
Kypson, J., Triner, L., Nahas, G.G.: The effects of cardiac glycosides and their interaction with catecholamines on glycolysis and glycogenolysis in skeletal muscle. J. Pharmacol. Exp. Ther. *164/1*, 22–30 (1968)
Lasché, E.M., Perloff, W.H., Durant, Th. M.: Some aspects of adrenocortical function in cardiac decompensation. Am. J. Med. Sci. *222*, 459–467 (1951)
Lefer, A.M., Sayers, G.: Antagonism of the inotropic action of ouabain by aldosterone. Am. J. Physiol. *208*, 649–654 (1965)
Leitz, F.H., Stefano, F.J.E.: Effect of ouabain and desimipramine on the uptake and storage of norepinephrine and metaraminol. Eur. J. Pharmacol. *11*, 278–285 (1970)
Lendle, L., Mercker, H.: Extrakardiale Digitaliswirkungen. Ergeb. Physiol. *51*, 200–298 (1961)
Lendle, L., Mercker, H., Rohr, H.: Über die Herzvaguswirkung unter dem Einfluß von Digitalglykosiden. Naunyn Schmiedebergs Arch. Exp. Pathol. Pharmak. *219*, 352–361 (1953)
Lendle, L., Oldenburg, D.: Prüfung extrakardialer Wirkung an der Digitalis-unterempfindlichen Ratte. Arch. Exp. Pathol. Pharmakol. *211*, 243–263 (1950)
Lendle, L., Wienke, H.: Zur Frage der Sensibilisierung von Vaguswirkungen auf die Herzfrequenz durch Digitalis. Arch. Exp. Pathol. Pharmakol. *213*, 373–386 (1951)
Levitt, B., Roberts, J.: Effect of quinidine and pronethalol on acetylstrophanthidin-induced ventricular arrhythmia in cats treated with reserpine. Circ. Res. *19*, 622–631 (1966)
Levy, J.V., Richards, V.: Inotropic effects of ouabain on rabbit left atria in presence of *beta*-adrenergic blocking drugs. Proc. Soc. Exp. Biol. Med. *119*, 278–281 (1965a)
Levy, J.V., Richards, V.: The influence of reserpine pretreatment on the contractile and metabolic effects produced by ouabain on isolated rabbit left atria. J. Pharmacol. Exp. Ther. *147/2*, 205–211 (1965b)

Le Winn, E.B.: Gynaecomasta during digitalis therapy. N. Engl. J. Med. *248*, 316–319 (1953)
Le Winter, M.M., Crawford, M.H., O'Rourke, R.A., Kaliner, J.S.: The effects of oral propranolol, digoxin, and combination therapy on the resting and exercise electrocardiogram. Am. Heart J. *93/2*, 202–209 (1977)
Massaro, G.D., Finnerty, F.A., Ryan, M.: The effect of reserpine and digitalis on the heart rate during nor-adrenaline infusion. Am. Heart J. *59*, 401–403 (1960)
Matheny, J.L., Ahlquist, R.P.: Effect of digitalis on carotid sinus reflex in the rabbit. Fed. Exp. Proc. *32*, 717 (1973)
McLain, P.L.: Effects of cardiac glycosides on spontaneous efferent activity in vagus and sympathetic nerves of cats. Int. J. Neuropharm. *8*, 379–387 (1969)
McLain, P.L., Kruse, T.K., Redick, T.F.: The effect of atropine on digitoxin bradycardia in cats. J. Pharmacol. Exp. Ther. *126*, 76–81 (1959)
Melville, K.I., Shister, H.E.: General systemic effect and electrocardiographic changes following injections of digitalis glycosides into the lateral ventricle of the brain. Am. Heart J. *53*, 425–438 (1957)
Metzler, A. von: Über die Wirkung von Strophanthin, Percorten, akuter und chronischer Ammoniakvergiftung auf Herz und Nebennieren von ruhenden und arbeitenden Meerschweinchen. Arch. Kreislaufforsch. *17*, 56–71 (1951)
Metzler, A. von, Greeff, K.: Über den Mechanismus der durch K-Strophanthin verursachten Nebennieren- und Herzhypertrophie. Arch. Exp. Pathol. Pharmakol. *222*, 352–359 (1954)
Metzler, A. von, Hergott, J.: Über den Einfluß des Percortens auf toxische Strophanthindosen am Froschherzen in situ. Klin. Wochenschr. *29/5–6*, 91–92 (1951)
Morrow, D.H., Gaffney, T.E., Braunwald, E.: Studies on digitalis. VII. Influence of hyper- and hypothyroidism on the myocardial response to ouabain. J. Pharmacol. Exp. Ther. *140*, 324–328 (1963)
Moskopf, E., Sarre, H.: Der Einfluß von Digitoxin auf die Nebenniere des Meerschweinchens im Schwimmversuch. Klin. Wochenschr. *32*, 327 (1954)
Nadeau, R., de Champlain, J.: Comparative effects of 6-hydroxy-dopamine and of reserpine on ouabain toxicity in the rat. Life Sci. *13*, 1753–1761 (1973)
Nadeau, R.A., James, T.N.: Antagonistic effects on the sinus node of acetylstrophanthidin and adrenergic stimulation. Circ. Res. *13*, 388–391 (1963)
Nakgawa, A., Watanabe, M.: Digitaliswirkungen auf die Cortocosteron-Bildung in der Nebennierenrinde. Naunyn Schmiedebergs Arch. Pharmacol. *254*, 355–363 (1966)
Oberg, B., Thorèn, P.: Studies on left ventricular receptors, signalling in non-medullated vagal afferents. Acta Physiol. Scand. *85*, 145–163 (1972)
Östling, G.: Effect of large digitalis doses on the empty stomach of man. Acta Pharmacol. Toxicol. (Kbn.) *6*, 165–168 (1950)
Pace, D.G., Gillis, R.A.: Neuroexcitatory effects of digoxin in the cat. J. Pharmacol. Exp. Ther. *199/3*, 583–600 (1976)
Pace, D.G., Martin, P.M.: Interactions between digoxin and brief vagal bursts influencing atrioventricular conduction. J. Pharmacol. Exp. Ther. *205/3*, 657–665 (1978)
Pace, D.G., Thibodeaux, H.: Effect of digoxin on carotid sinus and sympathetic nerve activity. Fed. Proc. *32*, 717 (1973)
Pinschmidt, N.W.: Relation of carotid and aortic mechanisms to digitalis emesis. Proc. Soc. Exp. Biol. Med. *61*, 7–9 (1946)
Popov, N., Forster, W.: Über den Einfluß verschiedener Digitaliskörper auf die Monaminoxydaseaktivität in Ratten-, Meerschweinchen- und Katzengehirn. Acta Biol. Med. Ger. *17*, 221–231 (1966)
Quest, J.A., Gillis, R.A.: Carotid sinus reflex changes produced by digitalis. J. Pharmacol. Exp. Ther. *177/2*, 650–661 (1971)
Quest, J.A., Thibodeaux, H., Clancy, M.M., Evans, D.E.: Neural mechanisms involved in digitalis-induced bradycardia. Fed. Proc. *32*, 717 (1973)
Reiter, M., Noè, J.: Besteht ein Antagonismus zwischen Corticosteron und Strophanthin bezüglich der inotropen Wirkung am Herzmuskel? Naunyn Schmiedebergs Arch. Pharmacol. *238*, 75–76 (1960)

Ricken, K.: Interaktionen von Digitalisglykosiden mit östrogenen Hormonen. Inaugural-Dissertation, Universität Düsseldorf 1975
Riggert, H., Schwab, M.: Der Einfluß von Digitoxin auf die Atmung des herzgesunden Menschen. Z. Ges. Exp. Med. *125*, 80–92 (1955)
Robinson, B.F.: Meyler's side effects for drugs. In: A survey of unwanted effects of drugs reported in 1972–1975, Dukes, M.N. (ed.), pp. 428–460, Vol. VIII. Amsterdam, Oxford: Excerpta Medica, New York: American Elsevier 1975
Roy, A.R., Chatterjee, M.L.: Effect of ouabain on the catecholamine content of heart and adrenal gland of rabbit. Life Sci. *9*, 395–401 (1970)
Saxena, P.R., Bhargava, K.P.: The importance of a central adrenergic mechanism in the cardiovascular responses to ouabain. Eur. J. Pharmacol. *31*, 332–346 (1975)
Schwab, M., Wagner, P.H.: Der Einfluß von K-Strophanthin auf die Atmung des herzgesunden Menschen. Z. Ges. Exp. Med. *129*, 28–32 (1957)
Seifen, E.: Evidence for participation of catecholamines in cardiac action of ouabain: positive chronotropic effect. Br. J. Pharmacol. *51/4*, 481–490 (1974)
Share, N.N., Chai, C.Y., Wang, S.C.: Emesis induced by intracerebroventricular injections of apomorphine and deslanoside in the dog. Fed. Proc. *23*, 349 (1964)
Spann, J.F., Sonnenblick, E.H., Cooper, T., Chidsey, C.A., Willman, V.L., Braunwald, E.: Studies on digitalis XIV: Influence of cardiac norepinephrine stores on the response of isolated heart muscle to digitalis. Circ. Res. *19*, 326–331 (1968)
Stickney, K.L.: Differential species sensitivity to the inhibitory effect of cardiac glycosides on 3-H-1-noradrenaline accumulation by tissue slices. Arch. Int. Pharmacodyn. Ther. *224*, 215–229 (1976)
Stickney, J.L., Lucchesi, B.R.: The effect of sympatholytic agents on the cardiovascular responses produced by the injection of acetylstrophanthidin into the cerebral ventricles. Eur. J. Pharmacol. *6*, 1–7 (1969)
Stickney, J.L., Meyers, F.H.: Digitalis toxicity: Development of cardiac arrhythmias in spontaneously breathing vs. artificially respired dogs. Am. Heart J. *85/4*, 501–505 (1973)
Stoffer, S.S., Hynes, K.M., Jiang, N-S., Ryan, R.J.: Digoxin and abnormal serum hormone levels. J. Am. Med. Assoc. *225*, 1643–1644 (1973)
Sulser, F., Kunz, H.A., Gantenbein, R., Wilbrandt, W.: Zur Frage einer antagonistischen Wirkung zwischen Herzglykosiden und Corticosteroiden auf die Elektrolytausscheidung der Niere. Naunyn Schmiedebergs Arch. Pharmacol. *235*, 400–411 (1959)
Swamy, V.C., Hamlin, R.L., Wolf, H.H.: Influence of myocardial catecholamines on the cardiac action of ouabain. J. Pharm. Sci. *54/10*, 1505–1507 (1965)
Takagi, M., Zanuttini, D., Khalil, E., Bellet, S.: Tolerance of reserpinized dogs to digitalis. Am. J. Cardiol. *15*, 203–205 (1965)
Tanz, R.D.: The action of ouabain on cardiac muscle treated with reserpine and dichloroisoproterenol. J. Pharmacol. Exp. Ther. *144*, 205–213 (1964)
Tanz, T.D., Coram, W.M., Brining, C., Cavaliere, T.: The inotropic action of ouabain in relation to ventricular norepinephirine content. Arch. Int. Pharmacodyn. Ther. *173*, 294–305 (1968)
Tatra, G. von, Nissel, W.: Zur Frage der östrogenen Wirksamkeit von Herzglykosiden. Wien. Med. Wochenschr. *24*, 460–463 (1969)
Toda, N., West, T.O.: The influence of ouabain on cholinergic responses in the sinoatrial node. J. Pharmacol. Exp. Ther. *153/1*, 104–113 (1966)
Tsao, D.P.N., King, T.E.: Communications: Effect of cardiac glycosides on 11-hyroxylation of the adrenal system. Arch. Biochem. Biophys. *118*, 259–260 (1967)
Unterman, D., DeGraff, A.C., Kuppermann, H.S.: Effect of hypoadrenalism and excessive doses of desoxycorticosterone acetate upon response of the rat to ouabain. Circ. Res. *3*, 280–284 (1955)
Varadi, A., Földes, J.: Serum digoxin in patients with thyroid disease. Br. Med. J. *1976 I*, 175
Vernikos-Danellis, J., Marks, B.H.: Pituitary inhibitory effects of digitoxin and hydrocortisone. Proc. Soc. Exp. Biol. Med. *109*, 10–14 (1962)
Viana, A.P.: Role of adrenergic mechanisms in cardiac and respiratory effects of digoxin. Pharmacology *15*, 436–444 (1977)

Wang, S.C., Borison, H.L.: A new concept of organization of the central emetic mechanism: recent studies on the sites of action of apomorphine, copper sulfate, and cardiac glycosides. Gastroenterology *22/1*, 1–12 (1952)
Weaver, L.C., Akera, T., Brody, T.M.: Digoxin toxicity: Primary sites of drug action on the sympathetic nervous system. J. Pharmacol. Exp. Ther. *197/1*, 1–9 (1976)
Weinberg, S.J., Haley, T.J.: Centrally mediated effects of cardiac drugs: Strophanthin-K, quinidine, and procainamide. Circ. Res. *3*, 103–109 (1955)
Wolff, G.: Über das Verhalten des Blutzuckers nach Strophanthin-Injektionen. Med. Monatsschr. *8*, 514–518 (1955)
Wolff, J., Maurey, J.R.: Cardiac glycosides and thyroidal iodide transport. Nature *4640*, 957 (1958)
Zinnith, F.: Die Vaguswirkung auf den Blutdruck unter Rescinnamin und Strophanthin. Acta. Biol. Med. Ger. *3*, 82–89 (1959)

CHAPTER 25

Effects of Cardiac Glycosides on Kidneys

O. HEIDENREICH and H. OSSWALD

A. Introduction

Since 1785 when WITHERING introduced digitalis as a diuretic, it has been questionable whether its diuretic effect was due to a direct action on the kidney or to changes in extrarenal hemodynamics. Today, it can be stated that cardiac glycosides, though acting mainly on the cardiovascular system, exert direct renal tubular effects that can be demonstrated even in healthy volunteers. Inhibition of tubular electrolyte reabsorption by up to 50% of the filtered load has been reported in animal experiments where high intrarenal concentrations of the glycosides were achieved. Despite the existence of marked species differences in sensitivity, the decrease of tubular electrolyte reabsorption by a specific inhibition of Na^+, K^+-stimulated ATPase is such a constant phenomenon that cardiac glycosides are used as a tool to investigate renal electrolyte transport processes. Comprehensive reviews of the older literature were written by STRAUB (1924), LENDLE (1935), ROTHLIN and BIRCHER (1954), and LENDLE and MERCKER (1961).

B. Effects on Renal Hemodynamics

The effects of cardiac glycosides on renal hemodynamics are small and rather uncharacteristic at *therapeutic* doses. Higher concentrations administered directly into the renal artery or added to the perfusion fluid in the isolated perfused kidney led to renal vasoconstriction, a decrease in glomerular filtration rate, and finally to anuria.

FARBER et al. (1951) observed in *human volunteers* no or only small changes in glomerular filtration rate (GFR) and renal plasma flow (RPF) following i.v. injection of 1–1.5 mg digoxin. The same holds true for patients with congestive cardiac failure, despite a reduction in venous pressure in this group. Changes in renal venous hydrostatic pressure seem to be of importance since BLAKE et al. (1949) have demonstrated a decrease in electrolyte excretion following elevation of renal venous pressure in the dog. No changes in GFR were found after digitoxin administration to patients with congestive cardiac failure (GÖLTNER et al., 1956). A small decrease in GFR accompanied by an increase in renal vascular resistance was reported by DOERING et al. (1953) in healthy subjects after i.v. injection of 0.35 mg strophanthin K.

In the *rat* GFR was unchanged (RAMSEY and SACHS, 1967) or slightly decreased after infusion of diuretic doses of ouabain into the renal artery (STRIEDER et al., 1974). In the isolated perfused rat kidney 1.2–2.4 × 10^{-6} *M* ouabain induced initial vasoconstriction together with a decrease in GFR (ROSS et al., 1974). Prevention of vasoconstriction by the use of the calcium antagonist verapamil (4.4×10^{-6} *M*) allowed the application of higher doses of ouabain, which then enhanced sodium excretion further (SCHUREK et al., 1976). Ouabain also increased renal vascular resistance and reduced GFR in the isolated heart-lung whole blood perfused rabbit kidney (ROSENFELD et al., 1969). NECHAY and PARDEE (1965) did not, however, observe changes in inulin excretion after unilateral infusion of ouabain into the renal portal circulation of the *chicken*.

In *dogs* application of 0.06–0.25 mg digoxin into one renal artery did not affect the excretion of endogenous creatinine (HYMAN et al., 1956). Similar findings were reported by HOFMANN and SHERROD (1967) who failed to observe a change in GFR after $0.033 \text{ mg} \times \text{kg}^{-1}$ acetylstrophanthidin administered intravenously. The majority of investigators have, however, reported a decrease in GFR (measured as inulin or creatinine clearance), in renal blood flow (RBF) (measured by means of a bubble flow or electromagnetic flowmeter), and in RPF (estimated as PAH clearance) when higher doses of cardiac glycosides were injected or infused into the renal artery (CADE et al., 1961; STRICKLER and KESSLER, 1961; TANABE et al., 1961; NECHAY and CHINOY, 1968; SEJERSTED et al., 1971; TORETTI et al., 1972). Addition of KCl ($75 \text{ µmol} \cdot \text{min}^{-1}$) or $CaCl_2$ ($33 \text{ µmol} \cdot \text{min}^{-1}$) to the glycoside infusions potentiated the decrease of creatinine and PAH clearance to 60% of the value measured in their absence. Potassium and calcium chloride alone did not influence renal hemodynamics (HEIDENREICH et al., 1965, 1966). Rapid injection of $85 \text{ µg} \cdot \text{kg}^{-1}$ ouabain into one renal artery induced an immediate fall in RBF (electromagnetic flowmeter), which could be prevented by prior administration of acetylcholine (LIE et al., 1974). In the same study, autoregulation of RBF was found to be markedly reduced by diuretic doses of ouabain. In dogs in which the heart was replaced by a mechanical intrathoracic pump, very high doses of ouabain ($0.25 \text{ mg} \cdot \text{kg}^{-1}$ or more i.v.) could be applied and led to a decrease in urine flow resulting in a complete renal shut down (CSÁKY et al., 1965).

C. Effects on Renal Excretion of Solutes and Water

I. Excretion of Sodium

Even therapeutic doses of cardiac glycosides reduce tubular sodium reabsorption. FARBER et al. (1951) injected 1.5 mg digoxin i.v. in normal subjects and observed a distinct increase in sodium excretion in 16 of 21. The mean chloride excretion and urine volume increased simultaneously whereas the mean potassium excretion remained unchanged. Since the renal hemodynamics were unaffected, these results provided the first evidence for a direct tubular effect of cardiac glycosides in man.

Figure 1 shows the renal excretion of sodium after i.v. injection of 1.5 mg digoxin in healthy subjects and in patients in congestive cardiac failure compared to untreated controls. It may be seen that sodium excretion increased after digoxin in both normal subjects and patients in congestive cardiac failure, the increase being much greater in the latter group. Similar results were obtained in healthy subjects with 0.35 mg i.v. strophanthin K (DÖRRIE et al., 1954) and oral digitoxin (GÖLTNER and SCHWAB, 1954) as well as in 29 patients with congestive heart failure

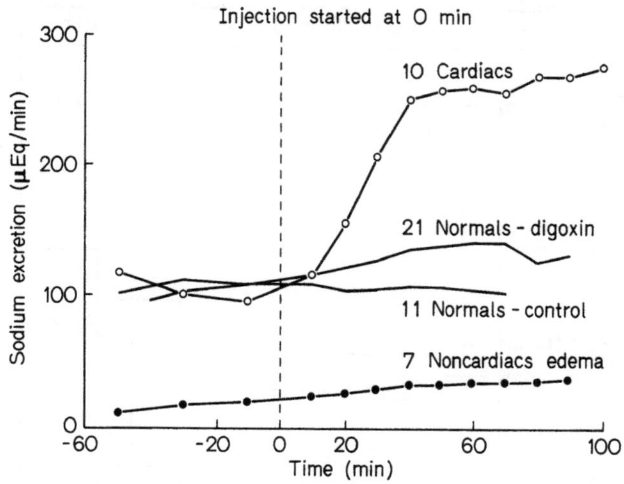

Fig. 1. Effect of intravenously administered digoxin on sodium excretion in normal subjects and in patients with edema due to congestive cardiac failure and other causes. FARBER et al., 1951

after digitoxin (GÖLTNER et al., 1956). The last three papers report an increase in plasma potassium and a decrease in plasma sodium levels.

Intravenous injections of cardiac glycosides also induce natriuresis in the *Necturus maculosus* (SCHATZMANN et al., 1958) and in the dog (HOFMANN and SHERROD, 1967). The rat differs from other species in its relative insensitivity to cardiac glycosides. This difference results from the very low sensitivity of rat Na^+, K^+-dependent ATPase to ouabain and most other glycosides (REPKE et al., 1965). In addition, the binding of cardiac glycosides to rat kidney Na^+, K^+-ATPase is less stable than in other species (ALLEN and SCHWARTZ, 1969). Effective doses of glycosides in the rat (0.5–5 $mg \cdot kg^{-1}$) are thus two orders of magnitude higher than those for man or other species. Such high doses lead initially to extrarenal electrolyte disturbances, namely, a marked hyperkalemia and hyponatremia. This electrolyte shift is assumed to result from a loss of potassium from the intracellular compartment, which in turn is replaced by sodium (GREEFF, 1956, 1957). The above high doses usually also result in an increased sodium excretion, probably due to inhibition of tubular sodium reabsorption. However, the concomitant increase in potassium excretion has to be considered, at least in part, as a consequence of the hyperkalemia (BODE and GREEFF, 1956; VOGEL, 1961; VOGEL and KLUGE, 1961; HELLER and TATA, 1967). Natriuresis as a regular effect of 5 $mg \cdot kg^{-1}$ ouabain i.v. could only be demonstrated in rats when kept on a low potassium diet (DUARTE et al., 1971). Administration of 0.5–5 $mg \cdot kg^{-1}$ meproscillaridin or proscillaridin by a gastric tube failed to produce natriuresis although kaliuresis occurred (FRIEDRICH et al., 1978). In adrenalectomized rats, strophanthin K exhibits natriuretic activity at the low dose of 0.25 $mg \cdot kg^{-1}$ (GREEFF, 1957). The latter effect can be abolished by administration of cortexone (SULSER et al., 1959).

To avoid extrarenal influences, many investigators have used isolated perfused kidney preparations. In these preparations, cardiac glycosides inhibited tubular sodium reabsorption in all species studied, including poikilothermic animals. A na-

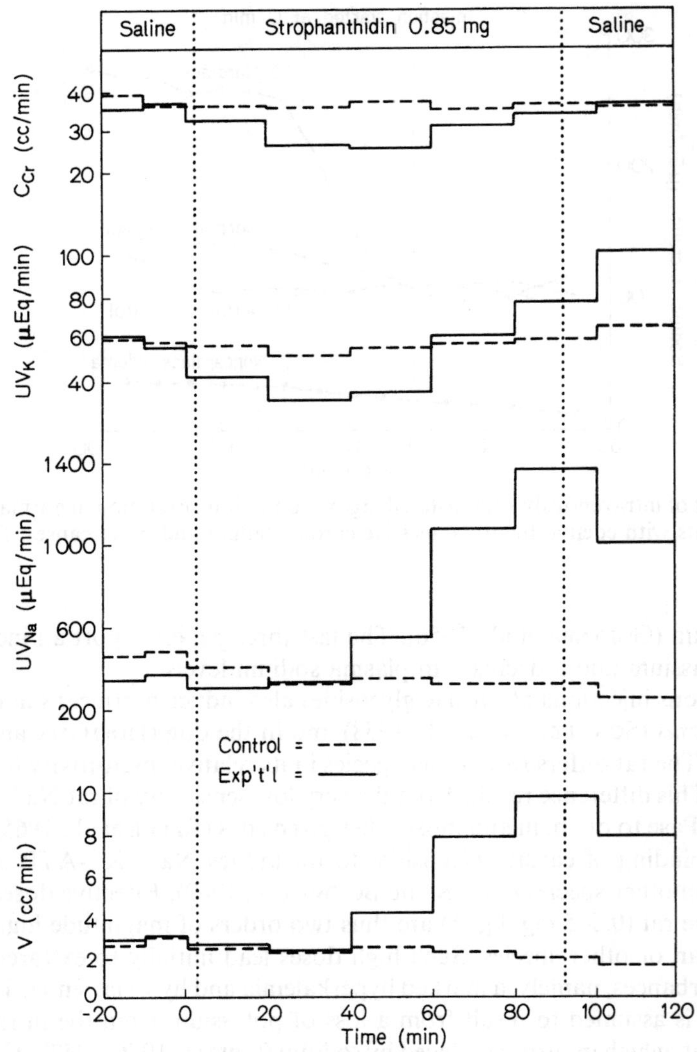

Fig. 2. Effects of an infusion of 0.85 mg strophanthidin over 92 min into one renal artery of a dog on urine volume, sodium and potassium excretion, and creatinine clearance. STRICKLER and KESSLER, 1961

triuretic effect following ouabain administration was observed in the perfused *Necturus* kidney (GIEBISCH et al., 1973; WHITTEMBURY et al., 1975), in the perfused toad kidney (WHITTEMBURY and FISHMAN, 1969), and in the isolated perfused kidney of *Rana ridibunda* (VOGEL, 1962; VOGEL et al., 1962). Convallatoxin ($3.6 \cdot 10^{-6}$ *M*) induced natriuresis in the isolated frog kidney added to the aortal or renoportal perfusion fluid (VOGEL and BUCHHEIM, 1962).

High doses of ouabain or strophanthidin ($250\ \mu g \cdot min^{-1}$) infused into the renoportal circulation of the chicken inhibit sodium reabsorption (ORLOFF and BURG, 1960), whereas lower doses ($1\ \mu g \cdot min^{-1}$) increase sodium and fluid reabsorption

(PALMER and NECHAY, 1964). Ouabain reduced tubular sodium reabsorption by up to 50% of the filtered load in the whole blood perfused rabbit kidney (ROSENFELD et al., 1969) and the isolated perfused rat kidney (RUEDAS and WEISS, 1967; BOWMAN et al., 1973; ROSS et al., 1974; SCHUREK et al., 1976). The renal effects of cardiac glycosides are best demonstrated in the intact animal by direct injection or infusion into one renal artery. If suitable doses are chosen, the natriuretic effect occurs in the ipsilateral kidney. The contralateral kidney responded with natriuresis only when higher doses were applied. STRIEDER et al. (1974) observed natriuresis following ouabain infusion into the renal artery of normal or potassium-depleted rats. The excreted sodium amounted to only 3% of the filtered load. RAMSEY and SACHS (1967) reported an increase in sodium excretion in only 18 of 33 experiments, whereas potassium excretion increased markedly in each case.

The dog is a suitable experimental animal to study the renal action of cardiac glycosides due to the high sensitivity of its Na^+, K^+-ATPase to cardiac glycosides. The first report of the diuretic effect of digoxin in the dog following its infusion into one renal artery is that from HYMAN et al. (1956). A typical experiment from STRICKLER and KESSLER (1961) in which 0.85 mg strophanthidin was infused over 92 min into one renal artery is depicted in Fig. 2.

A unilateral diuresis was evident within 60 min. Sodium excretion increased from a control value of 360 $\mu Eq \cdot min^{-1}$ to a maximum of 1,360 $\mu Eq \cdot min^{-1}$. At maximal diuresis 26% of the filtered sodium was excreted. Potassium excretion was reduced within the first 60 min and increased thereafter.

Figure 3 shows a representative experiment demonstrating the decline in the tubular action of ouabain after a single injection of 50 $\mu g \cdot kg^{-1}$ into the renal artery. The effect of ouabain was maximal within 10 min and decayed exponentially at a rate of 60%/h (WILDE and HOWARD, 1960). The diuretic and natriuretic effect of cardiac glycosides following infusions into the renal artery of dogs was confirmed in numerous papers (KOCH, 1959; WILDE and HOWARD, 1960; CADE et al., 1961; TANABE et al., 1961; GROLLMAN et al., 1962; VOGEL, 1963; VOGEL and LAUTERBACH, 1963; VOGEL et al., 1963; HEIDENREICH et al., 1965, 1966, 1967, 1971; NECHAY and CHINOY, 1968; HENDLER et al., 1969; HOOK, 1969; MARTINEZ-MALDONADO et al., 1969, 1970, 1972; NECHAY and NELSON, 1970; NELSON and NECHAY, 1970, 1971; ALLEN et al., 1971; SEJERSTED et al., 1971; TORETTI et al., 1972; BRADY and NECHAY, 1974; LIE et al., 1974; NECHAY, 1977; ROBINSON et al., 1977).

II. Excretion of Potassium

The effect of cardiac glycosides on potassium excretion depends on the experimental conditions and is subject to marked species differences. In the isolated perfused *frog kidney*, natriuretic glycoside concentrations do not seem to influence the excretion of potassium, even when potassium concentration in the perfusate was elevated fourfold (VOGEL, 1962; VOGEL et al., 1962, 1963). In the *chicken*, low doses of ouabain (1 $\mu g \cdot min^{-1}$) stimulate potassium reabsorption, whereas higher doses (80 $\mu g \cdot min^{-1}$) are clearly kaliuretic (PALMER and NECHAY, 1964). High doses of strophanthidin (250 $\mu g \cdot min^{-1}$) reduce potassium excretion by the chicken only when potassium excretion was elevated by infusions of KCl or K_2SO_4 (ORLOFF and BURG, 1960).

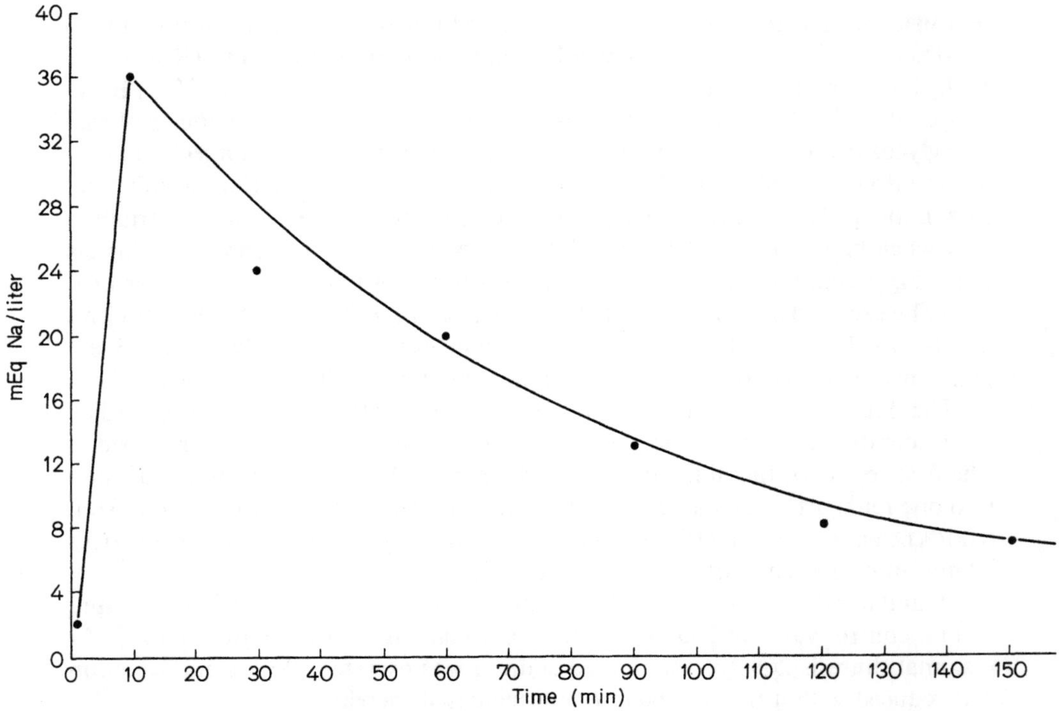

Fig. 3. Time course of the decline of renal tubular action of ouabain in a dog weighing 23 kg after rapid injection of 50 µg · kg^{-1} into the renal artery. The *ordinate* value of each single point on the plot is the minimum sodium concentration taken from one stop-flow curve. Each stop-flow run was sampled at the time indicated on the *abscissa* commencing at the time of ouabain injection. Plasma sodium concentration remained constant throughout the successive sampling periods. WILDE and HOWARD, 1960

Detection of a direct effect of cardiac glycosides on tubular handling of potassium in the *rat* is complicated by the concomitant hyperkalemia after intravenous application of glycosides (see Chap. 23). Direct administration of glycosides into the renal artery inhibit the reabsorption of potassium more markedly than that of sodium (RAMSEY and SACHS, 1967). In recollection micropuncture experiments, STRIEDER et al. (1974) observed a direct and specific effect of ouabain on the distal tubular transport of potassium. Similar observation were made by DUARTE et al. (1971). Figure 4 provides data from distal tubular recollection experiments of STRIEDER et al. (1974) in which the effects on distal tubular sodium and potassium transport are summarized.

Both sodium and potassium concentrations in distal tubular fluid samples were elevated, and fractional excretion rates increased significantly. GIEBISCH (1976) failed to find a decrease of potassium excretion by ouabain in the rat even when renal potassium excretion had been maximally stimulated by the administration of acetazolamide and hypertonic potassium sulfate solution. He interpreted these findings, which seem to be unique for the rat, as an inhibitory effect of cardiac glycosides on an active transport step at the luminal cell membrane of the distal tubule. The isolated perfused rat kidney, however, responded to 0.2 m*M* ouabain in

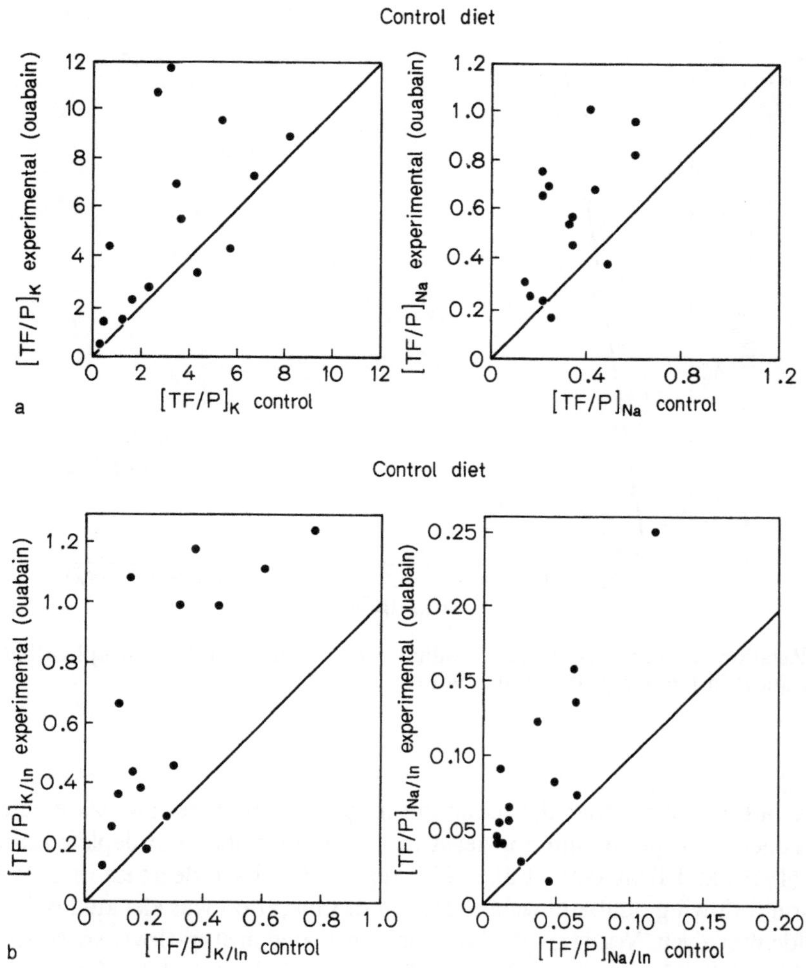

Fig. 4. Distal tubular potassium and sodium tubular fluid:plasma (*TF/P*) concentration ratios and fractional excretion rates before and after administration of ouabain into the left renal artery of the rat. Animals had been on a normal potassium intake. STRIEDER et al., 1974

the perfusion fluid with reduced potassium excretion after excretion rates had been elevated initially by furosemide, which itself caused net secretion of potassium (BOWMAN et al., 1973).

Infusion of natriuretic doses of cardiac glycosides into the renal artery of the *dog* (references see Sect. C.I) induce an elevated potassium excretion, which was lower and more variable than that seen in the rat. Potassium secretion by the distal tubule usually depends on the sodium delivery and is therefore increased by infusion of saline or administration of furosemide (SUKI et al., 1965). If one compares the relationship between potassium and sodium excretion as depicted in Fig. 5 it is apparent that the strophanthidin-infused kidney excretes far less potassium than the contralateral control kidney for each increment in sodium load (CADE et al., 1961).

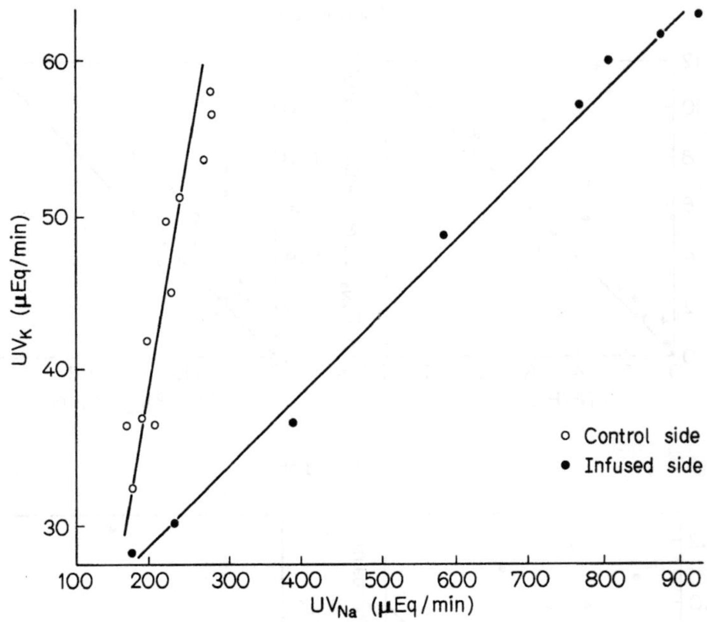

Fig. 5. Relation between potassium and sodium excretion in control and strophanthidin-infused kidneys in the dog. CADE et al., 1961

The authors claim that the reduction in potassium excretion was greatest in dogs loaded with potassium and least in a severely potassium-depleted animal. KOCH (1959) and ROBINSON et al. (1977) reported a sharp decrease in potassium excretion following ouabain when excretion of this cation was elevated before the glycoside was given. VOGEL et al. (1963) demonstrated in stop-flow experiments an inhibition of distal potassium secretion if it was initially stimulated by infusion of potassium chloride. The few data available for man reveal a variable and in general small increase in potassium excretion following administration of cardiac glycosides (FARBER et al., 1951; GÖLTNER and SCHWAB, 1954; DÖRRIE et al., 1954; GÖLTNER et al., 1956).

III. Excretion of Other Electrolytes and Nonelectrolytes

1. Calcium and Magnesium Reabsorption

Cardiac glycosides inhibit calcium and sodium reabsorption by the frog kidney to about the same extent (VOGEL, 1962; VOGEL et al., 1962, 1963; VOGEL and TERVOOREN, 1964). An increase in calcium excretion has also been reported in the dog (VOGEL and LAUTERBACH, 1963; VOGEL et al., 1963; HEIDENREICH et al., 1966).

KUPPFER and KOSOVSKY (1965) compared the clearances of Na, Ca, and Mg after infusions of strophanthin K or digoxin into the renal artery of dogs and observed a more than proportional increase in Ca and Mg clearance.

2. Hydrogen Ion Excretion

Injection of strophanthidin into the renoportal circulation of the *chicken* inhibits hydrogen ion excretion resulting in alkalinization of the urine (ORLOFF and BURG, 1960). In contrast, the *rat* responds to infusion of ouabain into the renal artery by progressive urine acidification of the urine with a twofold increase of H^+ concentration (RAMSEY and SACHS, 1967). In the *dog*, infusions of cardiac glycosides only slightly alter urine pH (HEIDENREICH et al., 1967). CADE et al. (1961) studied H^+ transport during the elaboration of both acid and alkaline urine. The bulk H^+ transport, as measured by the reabsorption of bicarbonate from an alkaline urine, was not significantly affected by the infusion of strophanthidin. Experiments with an acid urine, however, revealed that the ability of the kidney to maintain a steep transtubular H^+ concentration gradient was severely impaired.

3. Chloride and Phosphate Reabsorption

Chloride is the main anion to accompany the cardiac glycoside induced increase in urinary sodium excretion in man and dog (literature for humans: FARBER et al., 1951; GÖLTNER and SCHWAB, 1954; DÖRRIE et al., 1954; GÖLTNER et al., 1956). Phosphate reabsorption is also inhibited in the dog (KUPPFER and KOSOVSKY, 1965), though only to a slight degree (CADE et al., 1961).

4. Glucose Reabsorption

In the isolated perfused *frog* kidney, convallatoxin 6.4×10^{-6} *M* lowers the glucose reabsorption by about 30% inducing marked glucosuria (VOGEL and TERVOOREN, 1964). In the isolated perfused *rat kidney*, ouabain similarly leads to a decrease in net glucose reabsorption by about 30% (RUEDAS and WEISS, 1967). SCHUREK et al. (1976), using a similar preparation but counteracting vasoconstriction by addition of verapamil, found that glucose reabsorption remained at a high level (−7%), despite complete inhibition of the Na^+, K^+-ATPase. They suggested that the difference between their results and those of RUEDAS and WEISS (1967) might be due to renal hypoxia secondary to ouabain-induced vasoconstriction occurring in the absence of verapamil.

Ouabain infusion of $72\ \mu g \cdot kg^{-1} \cdot min^{-1}$ into a renal artery of normal *dogs* induced marked glucosuria (ROBINSON et al., 1977). HOFMANN and SHERROD (1967) determined the maximal tubular glucose transport *(Tm_G)* in unanesthetized dogs and found a reduction of Tm_G by only 7% after i.v. injection of $33\ \mu g \cdot kg^{-1}$ acetylstrophanthidin. In a dog preparation where cardiac function was maintained by a mechanical intrathoracic pump, higher doses of ouabain or acetylstrophanthidin administered intravenously depressed Tm_G up to 90% (CSÁKY et al., 1965). KUPPFER and KOSOVSKY (1965) did not, however, find any change in Tm_G when they infused digoxin or strophanthin K in doses of 50–80 μg/kg · h into the left renal artery of dogs.

5. Excretion of Organic Anions

Renal handling of uric acid by rabbit kidney has been analyzed using stop-flow techniques. Application of $40\ \mu g \cdot kg^{-1}$ ouabain produced a marked increase in the proximal U/P urate : U/P inulin ratio. No influence on distal tubular urate concentration was detected. In free flow experiments urate clearance was usually increased after ouabain (BERNDT and BEECHWOOD, 1964). Excretion of p-aminohippurate (PAH) by the isolated perfused frog kidney was only slightly reduced by convallatoxin and cassaine (VOGEL et al., 1966). In the dog, ouabain infusions of 0.7–1.3 $\mu g \cdot kg^{-1} \cdot min^{-1}$ for 60–80 min reduced ipsilateral renal extraction of PAH at a low plasma PAH concentration (2 mg-%). At high plasma PAH concentrations of 30–48 mg-%, the early response to ouabain was characterized by a depression of Tm_{PAH} but to a lesser degree than the inhibition of sodium reabsorption. PAH secretion gradually returned to control levels, whereas the depression of sodium reabsorption was persistent (NECHAY and CHINOY, 1968). In rabbit-kidney slices, strophanthidin interferes with cation transport and PAH accumulation (BURG and ORLOFF, 1962). These authors assume, however, that the interference of ouabain with PAH accumulation is secondary to the associated potassium depletion and does not represent a specific effect on PAH transport.

6. Excretion of Organic Cations

Numerous organic cations are secreted by a specific renal tubular transport mechanism, which is independent of the secretory mechanism for organic acids (for references, see PETERS, 1960). Renal tubular secretion of the cation *N'*-methylnicotinamide was reduced ipsilaterally after ouabain (40–80 $\mu \cdot min^{-1}$) was infused into the renoportal circulation of the chicken. Potassium reversed the inhibitory effects of ouabain on the transport of both *N'*-methylnicotinamide and sodium (NECHAY and PARDEE, 1965).

IV. Excretion of Water

A review of renal excretion data after application of cardiac glycosides reveals that they increase the urinary flow rate and electrolyte excretion in a roughly parallel manner. The increase in urine volume seems to be less pronounced during water diuresis as compared to the hydropenic state. Evidence for a direct action of cardiac glycosides on tubular water permeability is not available in the literature. The relation between solute excretion, i.e., osmolar clearance (C_{osm}) and urine volume ($\dot{V}$), is generally described as free water clearance ($C_{H_2O} = \dot{V} - C_{osm}$) or as free water reabsorption ($T^C_{H_2O} = C_{osm} - \dot{V}$), respectively. Infusions of cardiac glycosides into the renal arteries of dogs pretreated by an infusion of isotonic saline reduce the $T^C_{H_2O}$ to zero (HEIDENREICH et al., 1967; HENDLER et al., 1969). The same holds true for hydropenic dogs receiving vasopressin (TORETTI et al., 1972). Impairment of the concentration capacity of the kidney by cardiac glycosides has also been reported by MARTINEZ-MALDONADO et al. (1969, 1970). During water diureses, induced by a 2.5% glucose infusion, free water clearance decreased markedly, indicating an impaired diluting ability of the kidneys. It is generally assumed that diuretics that reduce both free water clearance and free water reabsorption act by inhibition of sodium chloride reabsorption in the ascending limb of Henle's loop (SELDIN et al., 1966), thereby diminishing the normally existing corticomedullary concentration gradient. Unfortunately, there are no data available on the ouabain effects on the concentration gradients in the dog kidney. In the rat, the corticomedullary gradients of sodium and urea were unaffected by application of 1.5–4.5 mg·kg^{-1} ouabain (HELLER and TATA, 1967). However, as discussed in Sect. C.I, ouabain has only a weak natriuretic action in this species.

V. Quantitative Differences in the Diuretic Activity of Cardiac Glycosides

There is little doubt that *no qualitative* differences exist in the renal effects of different cardiac glycosides. There are, however, some *quantitative* distinctions in the dose-response curves for various glycosides. VOGEL (1962) and VOGEL et al. (1962) investigated the diuretic activity of five cardiac glycosides in the isolated perfused *frog* kidney. The rank potency for inhibition of sodium transport at a concentration of $3.6 \cdot 10^{-6}$ *M* was convallatoxin > digitoxin > strophanthin K > ouabain > digoxin. In the *rat*, the same investigators found a slightly different rank order after intravenous application: convallatoxin again had the most potent diuretic effect, on both an absolute basis and on comparison of the median lethal doses. The diuretic activities of cymarin, strophanthin K, digoxin, lanatosid C, digitoxin, and scillaren A were lower than that of convallatoxin. Digitoxigenin was ineffective (VOGEL, 1961; VOGEL and KLUGE, 1961). BODE and GREEFF (1956) observed the following rank potency assessed by the degree of potassium excretion after subcutaneous application of the individual cardiac glycosides in doses from 0.5 to 2 mg·kg^{-1}: strophanthin K > convallatoxin > ouabain > scillaren > strophanthidin K. In a more recent study, GYÖRY et al. (1972) found scilliroside, a glycoside from red squill, to be the most potent diuretic in the rat. The in vitro sensitivity of rat Na^+, K^+-ATPase to scilliroside was 100-fold higher than to other cardiac glycosides and comparable to that shown for ouabain by a similar ATPase from kidneys of more glycoside-sensitive animals like the guinea pig or rabbit.

STRICKLER and KESSLER (1961) studied the diuretic properties of nine digitalis-like steroids following their infusion into one renal artery of the *dog*. Only those five steroids (digitoxin, ouabain, strophanthidin, emicymarin, and scillaren A) known to affect cardiac activity and transepithelial cation movement in vitro increased sodium and water excretion from the infused kidney. Scillaren A had a delayed action but was the most potent diuretic, being effective at a dose of 0.5 mg/dog. GROLLMAN et al. (1962) also came to the conclusion that scillaren has a more potent diuretic activity than digoxin or lanatoside C. VOGEL (1963) and VOGEL and LAUTERBACH (1963) infused $2.3 \cdot 10^{-8}$ mol·min^{-1} of each of three cardiac glycosides for 120 min into one renal artery of dogs and found convallatoxin and digoxin to have potent natriuretic activity, whereas digitoxin was without effect. They concluded that the natriuretic potency of cardiac glycosides is inversely correlated to the extent of their plasma protein binding. This assumption, however, is not supported by the results of STRICKLER and KESSLER (1961) who observed a good diuretic effect of digitoxin in the dog. In addition, HEIDENREICH et al. (1965, 1967) determined in dogs diuretic threshold doses of ouabain, scillaren A, and proscillaridin at 0.2 $\mu g \cdot kg^{-1} \cdot min^{-1}$, 0.09 $\mu g \cdot kg^{-1} \cdot min^{-1}$, and 0.065 $\mu g \cdot kg^{-1}$ min^{-1}, respectively. Since ouabain has the lowest plasma protein binding of these three glycosides, the correlation between plasma protein binding and natriuretic potency as mentioned above is not confirmed.

Differences in the diuretic threshold doses of individual glycosides could depend on a variety of factors including differences in tissue binding and accumulation. Measurements of renal tissue concentration of ^3H-ouabain and ^3H-proscillaridin following infusion into the renal artery of dogs show that the latter had significantly lower tissue levels when compared with ouabain: 51% lower in the cortex and 26% in the medulla (HEIDENREICH et al., 1971), indicating that diuretic threshold doses do not correlate with tissue levels. The binding of cardiac glycosides to renal plasma membranes and Na^+, K^+-ATPase are discussed in Sect. E.I (see also Chaps. 14 and 15).

D. Site of Renal Action

Experiments on different isolated perfused tubular segments of the *rabbit* nephron revealed that ouabain or strophanthidin 10^{-5} *M* added to the bath completely inhibited sodium and fluid reabsorption by the proximal convoluted tubule (BURG and ORLOFF, 1968), the thick ascending limb of Henle's loop (BURG and GREEN, 1973), and the cortical collecting tubule (GRANTHAM et al., 1970). Fluid absorption by the proximal tubule of *Necturus maculosus* was inhibited by 50% after ouabain (10 μg/animal i.v.) (SCHATZMANN et al., 1958). In perfusion experiments in the same species applying the split-oil droplet and other micropuncture techniques, GIEBISCH et al. (1973) and WHITTEMBURY et al. (1975) demonstrated that ouabain partially inhibited total renal fluid and net sodium transport including that occurring in the proximal tubule.

Cardiac glycosides (10^{-6}–10^{-3} *M*) applied by peritubular capillary perfusion inhibit transepithelial sodium transport in the *rat* proximal tubule by up to 50% as measured by the split-oil droplet technique (GYÖRY et al., 1972). Such local inhibition of sodium and fluid reabsorption by the proximal convoluted tubule seen under stop-flow conditions may be compensated for under free flow conditions in the kidney in vivo by a prolonged contact time, increase in tubular pressure, etc., as known from numerous studies (RECTOR et al., 1966). As illustrated in Fig. 4, ouabain directly inhibits fluid and electrolyte reabsorption in the distal tubule of the rat (STRIEDER et al., 1974). Similar results were reported by DUARTE et al. (1971).

Cardiac glycosides inhibit glucose reabsorption and PAH secretion, markers of proximal tubular function, in the *dog* (see Sect. C.III). The experiments on water excretion in which both C_{H_2O} and $T^C_{H_2O}$ decreased (Sect. C.IV) revealed an inhibition of sodium and chloride transport in the ascending limb of Henle's loop, the so-called diluting segment. There must also exist some distal tubular effect of cardiac glycosides on potassium transport in the dog since they inhibited the initial increase in potassium secretion. An inhibition of distal sodium and potassium reabsorption was also detectable in stop-flow experiments (WILDE and HOWARD, 1960; VOGEL et al., 1963).

E. Mechanism of Tubular Action of Cardiac Glycosides

I. Distribution and Localization of Na^+, K^+-ATPase in the Kidney

The rates of transport of solutes and water differ markedly in different regions of the nephron, as does the distribution of Na^+, K^+-ATPase. Since cardiac glycosides bind to Na^+, K^+-ATPase in the kidney, as in other tissues, we shall consider the localization of this enzyme along the nephron before discussing the effects of the cardiac glycosides.

The first report of a ouabain-sensitive Na^+, K^+-ATPase in the kidney (guinea pig) was that of POST et al. (1960). The enzyme was subsequently found in rabbit kidney (WHITTAM and WHEELER, 1961) where it had a heterogenous distribution (BONTING et al., 1961). These investigators found a twofold higher enzyme activity in homogenates of cat renal medulla when compared to those from the renal cortex. This finding was confirmed in several species including: dog (BONTING et al., 1962; HOOK, 1969; MARTINEZ-MALDONADO et al., 1969; HENDLER et al., 1971; ALLEN et al., 1971; NELSON and NECHAY, 1971; SAITO et al., 1976), rabbit (BEYTH and GUTMAN, 1969), rat (JØRGENSEN, 1969; HENDLER et al., 1971), guinea pig, squirrel monkey, and partridge (HENDLER et al., 1971). The higher Na^+, K^+-ATPase activity in the outer medulla (red medulla) coincides with the predominance of this enzyme in the thick ascending limbs of the loop of Henle.

The enzyme preparations derived from both regions of the kidney appear to be similar in that their K_m for ATPase is almost identical, as is the K_m for sodium. In addition, the capacity of the enzyme to bind ouabain in vitro is exactly the same in preparations derived from both cortex and medulla suggesting that the high activity in the medulla is the result of higher concentrations of the enzyme (HENDLER et al., 1971). The authors suggest that the cortical and medullary Na^+, K^+-ATPase are the same enzyme.

Microdissection methods have allowed greater precision in the demonstration of Na^+, K^+-ATPase distribution along the nephron (SCHMIDT and DUBACH, 1969). This work shows that the thick ascending limb of the loop of Henle has a four to eight times higher Na^+, K^+-ATPase activity compared to the proximal convoluted tubule. Also, the distal tubule has a four to five times higher enzyme activity compared to that of the proximal tubule (Fig. 6).

The intracellular localization of the Na^+, K^+-ATPase was investigated by two different methods. Quantitative histochemistry by ultramicrodissection revealed that Na^+, K^+-ATPase activity is limited to the basal area of the tubular epithelium (i.e., contraluminal side). The luminal brush border fragments lack Na^+, K^+-ATPase activity (SCHMIDT and DUBACH, 1971). This was confirmed by HEIDRICH et al. (1972) using crude membrane preparations of renal tissue from rats subjected to free flow electrophoresis. Na^+, K^+-ATPase was exclusively found in basolateral membrane fractions whereas the apical membrane fractions (brush border) did not contain significant Na^+, K^+-ATPase activity. The functional implication of this cell polarity is discussed below (Sect. E.III). Essentially, the same pattern of ouabain-sensitive renal Na^+, K^+-ATPase distribution was obtained in rats by ERNST (1975) and in rabbit and human kidney by BEEUWKES and ROSEN (1975) using a cytochemical method and subsequent electron probe analysis. KYTE (1976a, b) determined Na^+, K^+-ATPase distribution in the kidney by means of immunoferritin

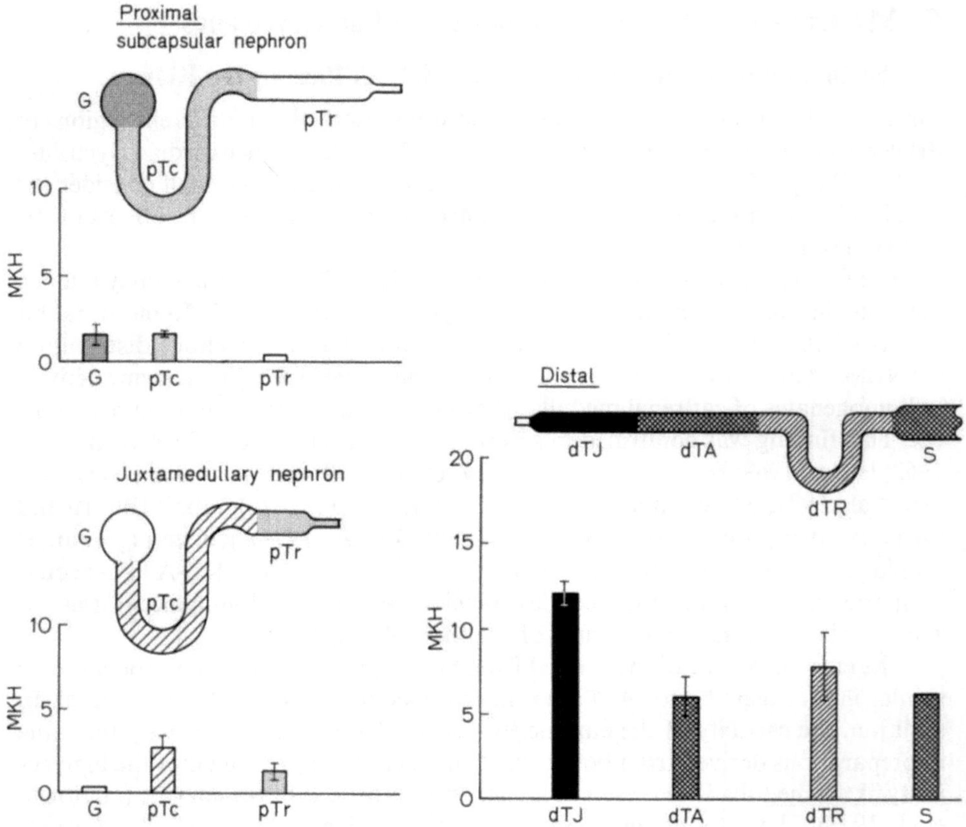

Fig. 6. Na^+, K^+-ATPase activity in MKH (mol $P_i \cdot kg^{-1}$ dry weight $\cdot h^{-1}$ incubation) in various segments of the rat nephron. G = glomerulus; pTc = proximal tubule convoluted portion; pTr = proximal tubule straight portion; dTJ = distal tubule inner stripe; dTA = distal tubule outer stripe; dTR = distal tubule cortex; S = collecting tubule. SCHMIDT and DUBACH, 1969

staining. Although confirming the results mentioned above, his findings suggest the presence of a very low but significant Na^+, K^+-ATPase activity in the brush border membrane of the proximal tubules.

II. Binding of Cardiac Glycosides to Renal Tissue

It is generally assumed that binding of cardiac glycosides to Na^+, K^+-ATPase is directly related to the inhibition of the enzyme. The binding of ouabain to purified Na^+, K^+-ATPase from the outer medulla of dog kidney is saturable and reversible (TOBIN and SEN, 1970; LANE et al., 1973). It is influenced by the concentration of sodium, potassium, ATP, and magnesium in the incubation medium (SCHWARTZ et al., 1975). The maximal amount of ^3H-ouabain bound specifically to purified renal Na^+, K^+-ATPase lies between 3.0–5.7 nmol/mg protein (KYTE, 1972; LANE et al., 1973; MATSUI et al., 1977). Ouabain seems to interfere with the high affinity binding sites of the enzyme to potassium and sodium (INAGAKI et al., 1974; KANIIKE et al., 1976; MATSUI et al., 1977). Ouabain-induced diuresis in the dog kidneys seems to be antagonized by an increase in plasma potassium concentration (CADE et al., 1961; HEIDENREICH et al., 1965). There are species differences in the binding of cardiac glycoside to renal tissue preparations. The binding of ouabain to intact cultured kidney cells of guinea pigs was about three times higher than to those of rabbits when incubated with the glycoside (10^{-5}–10^{-4} *M*) in the presence of 6 m*M* potassium (BAKER and WILLIS, 1972). However, half of the Na^+, K^+-ATPase ~ouabain complex prepared from guinea pig kidney dissociates within 2.5 min whereas the half maximal dissociation of rabbit kidney ATPase ~ouabain complex is 10 min under the same conditions (TOBIN and SEN, 1970; TOBIN and BRODY, 1972). It is difficult to compare the rate constant of half maximal association and dissociation of the ouabain Na^+, K^+-ATPase complex when different enzyme preparations are used since ouabain binds to sites other than Na^+, K^+-ATPase in renal tissue (ALMENDARES and KLEINZELLER, 1971; BAKER and WILLIS, 1972). The rat is known to be relatively insensitive to cardiac glycosides (REPKE et al., 1965). A detailed study of the interaction of ouabain with renal Na^+, K^+-ATPase of the rat was reported by ALLEN and SCHWARTZ (1969). It was found that the inhibition of Na^+, K^+-ATPase by ouabain is independent of temperature and time and that the enzyme binds ouabain only loosely with a different stoichiometry compared to other species. For detailed discussions of interactions between cardiac glycosides and Na^+, K^+-ATPase see Chap. 14.

III. Correlation Between Inhibition of Na^+, K^+-ATPase and Tubular Reabsorption of Solutes and Fluid

The inhibitory effects of cardiac glycosides on tubular fluid and electrolyte reabsorption determined for the whole kidney has been confirmed in both functionally isolated tubules in vivo and isolated perfused tubules in vitro. The reduction in renal electrolyte reabsorption by ouabain in vivo correlates well with a similar reduction in Na^+, K^+-ATPase activity suggesting that the Na^+, K^+-ATPase is the diuretic receptor of cardiac glycosides (PALMER and NECHAY, 1964; MARTINEZ-MALDONADO et al., 1969; HOOK, 1969; NELSON and NECHAY, 1970, 1971; NECHAY and NELSON, 1970; ALLEN et al., 1970; GYÖRY et al., 1972; TORETTI et al., 1972; MARTINEZ-MALDONADO et al., 1972; NECHAY, 1974; BRADY and NECHAY, 1974; SCHUREK et al., 1976; SAITO et al., 1976; SEJERSTED et al., 1977). Exposure of slices of renal cortex, isolated tubules, or whole kidney to ouabain or strophanthidin increased the intracellular content of sodium, chloride, and calcium while the intracellular content of potassium and magnesium were decreased, a finding consistent with the active role of Na^+, K^+-ATPase in generating the sodium gradient across the plasma membrane of the cells of tubules (BURG and ORLOFF, 1962; NAHMOD and WALSER, 1966; ABRAMOW et al., 1967; MACKNIGHT, 1968; WILLIS, 1968a, b; MAUDE, 1969; HOSKINS and HOLLAND, 1970; WEINSCHELBAUM et al., 1972; FRIEDRICHS and SCHONER, 1973; NOÉ and CRABBÉ; LAW, 1976; ROBINSON et al., 1977). Additional evidence to support the role of Na^+, K^+-ATPase in tubular electrolyte and water transport was the demonstration by SCHMIDT et al. (1975) of a reduction in Na^+, K^+-ATPase activity in distal tubules and thick ascending limbs of the loop of Henle in adrenalectomized rats, a pathophysiologic state associated with severe deficiency of tubular sodium and water reabsorption (for references, see Hierholzer and Lange, 1974). Aldosterone replacement was associated with a return of the Na^+, K^+-ATPase activity to control values (Schmidt et al., 1975). During maturation of the kidney, the tubular electrolyte reabsorption capacity parallels the ouabain-sensitive Na^+, K^+-ATPase activity along the nephron (Schmidt and Horster, 1977). Thus, it appears to be established that the Na^+, K^+-ATPase is the essential enzyme promoting tubular sodium reabsorption.

If the Na^+, K^+-ATPase were the only enzyme responsible for transtubular sodium transport, a complete inhibition of sodium reabsorption would be expected after complete inhibition of the Na^+, K^+-ATPase by ouabain. This assumes that the only source of energy for sodium transport is ATP splitting. Györy and Kinne (1971) have calculated that in rat proximal tubule 80% of the sodium transport is ATP dependent, the remaining energy requirement being provided by non-ATP high-energy intermediates.It is beyond the scope of this chapter to discuss the energy requirements for sodium transport, but it is noteworthy that at least 20% of the sodium transport can be achieved without splitting ATP.

Additional limitations of the concept that the cardiac glycoside-sensitive Na^+, K^+-ATPase is the only principle for active sodium transport in the kidney can be seen in a number of reports where a reduced but significant rate of sodium and fluid reabsorption in whole kidney and in renal tubules was still maintained despite almost complete inhibition of Na^+, K^+-ATPase by cardiac glycosides (Martinez-Maldonado et al., 1972; Györy et al., 1972; Brady and Nechay, 1974; Ross et al., 1974; Besarab et al., 1976; Schurek et al., 1976; Ross and Bullock, 1976; Imai et al., 1977; Ullrich et al., 1977; Robinson et al., 1977). In fact, it was concluded from the above results that there must be a "second sodium pump" independent from Na^+, K^+-ATPase (Whittembury and Proverbio, 1970; Weinschelbaum et al., 1972; Brady and Nechay, 1974; Whittembury and Proverbio, 1976). Several ouabain-insensitive ATPases have been found to be present in basolateral or brush border membrane preparations of tubular cells (Kinne-Saffran and Kinne, 1974a, b); Liang and Sacktor, 1976; Amelsvoort et al., 1977).

The effects of cardiac glycosides on tubular transport of calcium, glucose, amino acids, phosphate, and other organic anions can be explained on the basis of the secondary active cotransport of these solutes with sodium (Fig. 7). The steep sodium gradient across the plasma membrane generated by the Na^+, K^+-ATPase provides the driving force for the transmembranal movement of the solutes, which use specific carriers for translocation. The different permeabilities of the luminal and basolateral plasma membranes for the solutes control the direction of transepithelial transport. This concept is discussed in detail by Ullrich et al. (1977), Ullrich and Frömter (1978), and Kinne and Murer (1978)

IV. Transepithelial Electric Potential Difference

The effect of ouabain on transepithelial electric potential difference (PD) was investigated in proximal tubule, thin and thick ascending limbs of the loop of Henle, and cortical collecting ducts. The PD across the proximal tubular epithelium is de-

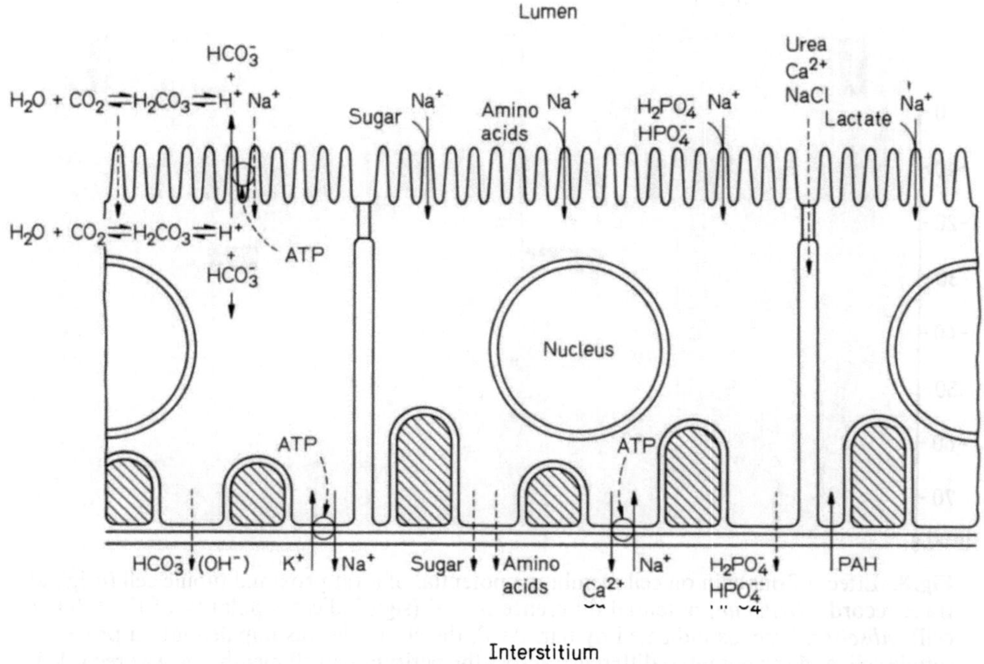

Fig. 7. Proximal tubule of rat kidney. The location of the three ATPases, the co- and countertransport processes, and the exit steps at the contraluminal cell side are shown as are the contraluminal PAH uptake and the paracellular route for the passive fluxes of urea, Ca^{2+}, Na^+, and Cl^-. The sodium-amino acid uptake and amino acid exit arrows stand for several systems. The sugar and amino acid entrace systems have different specificity from the respective exit systems. ULLRICH and FRÖMTER, 1978

pendent on the fluid composition in the lumen. Luminal perfusates containing glucose, bicarbonate, and amino acids or artificial ultrafiltrate of the plasma produce a negative transepithelial PD of 2.0–5.8 mV (lumen negative) in the rabbit proximal tubule in vitro. This is reduced to zero following contraluminal administration of ouabain (10^{-5} *M*) (BURG and ORLOFF, 1970; KOKKO and RECTOR, 1971; KOKKO, 1973; LUTZ et al., 1973; CARDINAL et al., 1975). However, with a perfusate mimicking the composition as normally present in the late proximal tubular fluid (lacking glucose, amino acids, lactate, bicarbonate), the PD was zero and ouabain was without effect. Ouabain increases the electric resistance of the isolated perfused proximal tubule of rabbits (LUTZ et al., 1973). The above findings are consistent with the concept of an active sodium transport dependent on Na^+, K^+-ATPase activity in the proximal tubule.

The origin of the lumen-negative PD in the in vivo rat proximal tubule has been considered in detail by FRÖMTER et al. (1973), FRÖMTER (1974), and FRÖMTER and GESSNER (1975). Ouabain increased the lumen-positive active transport potential in the midproximal convoluted tubule by an average of 0.2 mV and diminished the potential response to luminal perfusion with glucose (FRÖMTER and GESSNER, 1975). In addition, ouabain caused a depolarization of the peritubular cell membrane as shown in Fig. 8; again this is consistent with the concept that ouabain acts

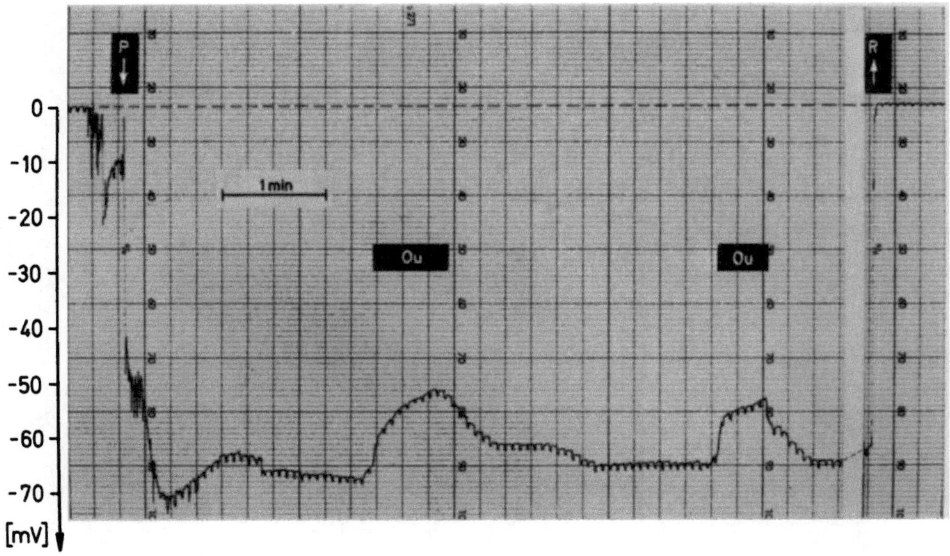

Fig. 8. Effect of ouabain on cell membrane potential of a rat proximal tubule cell (original trace record). *Ordinate*, potential difference in mV (sign indicates polarity of the tubular cell). *Abscissa*, time, as indicated by bar. At *P*, the electrode was impaled into a proximal tubule cell, and the potential difference across the peritubular cell membrane was recorded. *Ou* indicates application of 18 mmol/liter ouabain in HCO_3-Ringer's solution to the peritubular side. In the interval between ouabain applications, the peritubular perfusion fluid contained 18 mmol/liter raffinose. The pulses indicate electrode resistance, 1 mV = 15 $M\Omega$, pulse interval 3.3 s. Note rapid onset of the ouabain effect. Courtesy of FRÖMTER and GESSNER, 1975

on active ion transport (FRÖMTER and GESSNER, 1975). MARSH and MARTIN (1977) found that ouabain could lower the PD in the thin ascending limb of the loop of Henle of the hamster kidney by about 0.6 mV lumen positive.

The thick ascending limb of the loop of Henle exhibits a transepithelial PD that is lumen positive. The PD varies 30–5 mV depending on the luminal NaCl concentration (50–150 mEq/liter). Ouabain (10^{-5} *M*) reduced the PD almost to zero (BURG and GREEN, 1973; ROCHA and KOKKO, 1973). Although the lumen-positive potential is evidence for active chloride transport, the effect of ouabain does not conflict with a role of Na^+, K^+-ATPase in the chloride transport of this tubular segment. Ouabain also reduced unidirectional sodium fluxes from lumen to bath (ROCHA and KOKKO, 1973). Although attempts have been made to explain active chloride transport in the thick ascending limb on the basis of Na^+, K^+-ATPase activity (KILL, 1977; JØRGENSEN, 1978), this question needs further elucidation. In comparison, active chloride transport (secretion) in the shark rectal gland was also inhibited by ouabain, an effect that could be explained on the basis of an inhibition of Na^+, K^+-ATPase (EPSTEIN et al., 1978). The isolated perfused cortical collecting duct of the rabbit has a lumen-negative transepithelial PD of about 20–70 mV (BURG et al., 1968; GRANTHAM et al., 1970). This PD seems to result from active sodium and potassium transport and can be reduced almost to zero by addition of ouabain (10^{-5} *M*) to the bathing medium (GRANTHAM et al., 1970). All studies on transepithelial PD in various segments of the nephron agree that ouabain reduces the electrogenic activity of the Na^+, K^+-ATPase.

V. Effects on Oxygen Consumption and Renal Metabolism

The effects of ouabain on renal metabolism and oxygen consumption (Q_{O_2}) were first investigated in slices of renal cortex and medulla. GORDON (1965) reported a decrease in Q_{O_2}, CO_2 production, and $^{14}CO_2$ formation from glucose ^{14}C in rat kidney slices when incubated in the presence of 10^{-4} *M* ouabain. The fall in Q_{O_2} due to ouabain has been subsequently confirmed in kidney slices from dogs (WEINSCHELBAUM DE JAIRALA et al., 1969, 1972), ground squirrels, hamsters (WILLIS, 1968a, b; WILLIS and FANG, 1970), and for kidneys from rats and guinea pigs (MACKNIGHT, 1968; FRIEDRICH and SCHONER, 1973; NOÉ and CRABBÉ, 1975). This effect can be antagonized by increasing the concentration of potassium in the medium (WILLIS, 1968a, b; DE JAIRALA et al., 1969).

Ouabain stimulated gluconeogenesis in rat kidney cortex slices but did not affect adenylate cyclase and phosphodiesterase activity or glycolysis (FRIEDRICHS and SCHONER, 1973). This enhancement of gluconeogenesis by ouabain (0.1–4 m*M*) was additive to its stimulation by L-epinephrine, dibutyryl-cAMP, and acetoacetate. Concomitant with the increase in glucose production by the slices, CO_2 generation from free fatty acids was diminished and tissue levels of malate, lactate, and α-ketoglutarate were elevated (FRIEDRICHS and SCHONER, 1973). Ouabain increased tissue levels of ATP and prevented the inhibition of gluconeogenesis induced by high potassium concentration in the medium. From these results, the authors assumed that the higher energy state of the cells in the presence of ouabain may have been the cause for the increased gluconeogenesis. Accordingly, the increase of the ATP:ADP ratio in dog kidney cortex in vivo by ouabain (URBAITIS and KESSLER, 1971) was interpreted as a reduced utilization of high-energy substrates by the kidney cells when the Na^+, K^+-ATPase was inhibited.

The oxygen consumption of the kidney in vivo parallels the tubular reabsorption of sodium (DEETJEN and KRAMER, 1961; LASSEN et al., 1961; KIIL et al., 1961). FÜLGRAFF et al. (1970) reported a 32% reduction in Q_{O_2} in dog kidney after infusion of ouabain ($0.5\ \mu g \cdot kg^{-1} \cdot min^{-1}$) into one renal artery. The ratio of transported sodium to Q_{O_2} remained unchanged during ouabain infusion due to a proportional decrease in sodium reabsorption. SEJERSTED et al. (1971) have measured the metabolic rate of renal tissue in the dog in vivo by means of implanted thermocouples sensing local heat production. They found that ouabain reduced the metabolic rate in the cortex and outer medulla by 28% and 52%, respectively. Tubular sodium reabsorption was reduced by 22% of the filtered load. Similar results were reported for the dog (LIE et al., 1974) and for the isolated perfused rat kidney (SCHUREK et al., 1976). To prevent the vasoconstrictive effect of high doses of ouabain, SEJERSTED et al. (1977) have simultaneously infused mannitol and acetylcholine. Under these conditions the infusion of $0.1\ mg \cdot kg^{-1}$ of the glycoside over 3–5 min into one renal artery of the dog reduced Q_{O_2} by 45% and reduced the renal extraction of lactate to zero.

The conclusions from the studies measuring renal oxygen consumption after administration of ouabain are similar in regard to the relation between transported sodium and Na^+, K^+-ATPase as discussed in Sect. E.III–IV:

1) The Na^+, K^+-ATPase is the diuretic receptor for ouabain.
2) Sodium transport by renal tubular epithelium is mainly dependent on the Na^+, K^+-ATPase activity.
3) Inhibition of renal Na^+, K^+-ATPase reduces the oxygen consumption of the kidney.

References

Abramow, M., Burg, M.B., Orloff, J.: Chloride flux in rabbit kidney tubules in vitro. Am. J. Physiol. *213*, 1249–1253 (1967)

Allen, J.C., Schwartz, A.: A possible biochemical explanation for the insensitivity of the rat to cardiac glycosides. J. Pharmacol. Exp. Ther. *168*, 42–46 (1969)

Allen, J.C., Lindenmayer, G.E., Schwartz, A.: An allosteric explanation for ouabain-induced time-dependent inhibition of sodium, potassium-adenosine triphosphatase. Arch. Biochem. Biophys. *141*, 322–328 (1970)

Allen, J.C., Martinez-Maldonado, M., Eknoyan, G., Suki, W.N., Schwartz, A.: Relation between digitalis binding in vivo and inhibition of sodium, potassiumadenosine triphosphatase in canine kidney. Biochem. Pharmacol. *20*, 73–80 (1971)

Almendares, J.A., Kleinzeller, A.: The ouabain inhibition of sugar transport in kidney cortex cells. Arch. Biochem. Biophys. *145*, 511–519 (1971)

Amselvoort, J.M.M. van, Pont, J.J.H.H.M. de, Stols, A.L.H., Bonting, S.L.: Is there a plasmamembrane located anion-sensitive ATPase? II. Further studies on rabbit kidney. Biochim. Biophys. Acta *471*, 79–91 (1977)

Baker, P.F., Willis, J.S.: Binding of the cardiac glycoside ouabain to intact cells. J. Physiol. (Lond.) *224*, 441–462 (1972)

Beeuwkes, R., Rosen, S.: Renal sodium-potassium adenosine triphosphatase optical localization and X-ray analysis. J. Histochem. Cytochem. *23*, 828–839 (1975)

Berndt, W.O., Beechwood, E.C.: Effect of ouabain on uric acid transport in the rabbit. Pharmacologist *6*, 191 (1964)

Besarab, A., Silva, P., Epstein, F.H.: Multiple pumps for sodium reabsorption by the perfused kidney. Kidney Int. *10*, 147–153 (1976)

Beyth, Y., Gutman, Y.: Ontogenesis of microsomal ATPase in the rabbit kidney. Biochim. Biophys. Acta *191*, 195–197 (1969)

Blake, W.D., Wegria, R., Keating, R.P., Ward, H.P.: Effect of increased renal venous pressure on renal function. Am. J. Physiol. *157*, 1–13 (1949)

Bode, H., Greeff, K.: Über die Wirkung des Strophanthins auf die renale Kalium- und Natriumausscheidung bei Ratten. Naunyn-Schmiedebergs Arch. Pharmacol. *227*, 436–443 (1956)

Bonting, S.L., Simon, K.A., Hawkins, N.M.: Studies on sodium-potassium-activated adenosine triphosphatase. I. Quantitative distribution in serval tissues of the cat. Arch. Biochem. Biophys. *95*, 416–423 (1961)

Bonting, S.L., Caravaggio, L.L., Hawkins, N.M.: Studies on sodium-potassium-activated adenosine-triphosphatase. IV. Correlation with cation transport sensitive to cardiac glycosides. Arch. Biochem. Biophys. *98*, 413–419 (1962)

Bowman, R.H., Dolgin, J., Coulson, R.: Interaction between ouabain and furosemide on Na and K excretion in perfused rat kidney. Am. J. Physiol. *224*, 1200–1205 (1973)

Brady, J.M., Nechay, B.R.: Maximal effects of ouabain on renal sodium reabsorption and ouabain-sensitive adenosine triphosphatase activity in the dog. J. Pharmacol. Exp. Ther. *190*, 346–351 (1974)

Burg, M.B., Green, N.: Function of the thick ascending limb of Henle's loop. Am. J. Physiol. *224*, 659–668 (1973)
Burg, M.B., Orloff, J.: Effect of strophanthidin on electrolyte content and PAH accumulation of rabbit kidney slices. Am. J. Physiol. *202*, 565–571 (1962)
Burg, M.B., Orloff, J.: Electrical potential difference across proximal convoluted tubules. Am. J. Physiol. *219*, 1714–1716 (1970)
Burg, M.B., Orloff, J.: Control of fluid absorption in the renal proximal tubule. J. Clin. Invest. *47* 2016–2024 (1968)
Burg, M.B., Issaacson, L., Grantham, J., Orloff, J.: Electrical properties of isolated perfused rabbit renal tubules. Am. J. Physiol. *215*, 788–794 (1968)
Cade, J.R., Shalhoub, R.J., Canessa-Fischer, M., Pitts, R.F.: Effect of strophanthidin on the renal tubules of dogs. Am. J. Physiol. *200*, 373–379 (1961)
Cardinal, J., Lutz, M.D., Burg, M.B., Orloff, J.: Lack of relationship of potential difference to fluid absorption in the proximal renal tubule. Kidney Int. *7*, 94–102 (1975)
Csáky, T.Z., Prachuabmoh, K., Eiseman, B., Ho, P.M.: The effect of digitalis on the renal tubular transport of glucose in normal and in heartless dogs. J. Pharmacol. Exp. Ther. *150*, 275–278 (1965)
Deetjen, P., Kramer, K.: Die Abhängigkeit des O_2-Verbrauches der Niere von der Na-Rückresorption. Pflügers Arch. *273*, 639–650 (1961)
Doering, P., Sancken, H., Schubert, W., Schwab, M.: Der Einfluß von Digitalisglykosiden auf das Glomerulusfiltrat, den renalen Plasmafluß und die intrarenale Hämodynamik beim gesunden Menschen. Dtsch. Arch. Klin. Med. *200*, 814–820 (1953)
Dörrie, H., Göltner, E., Schwab, M.: Der Einfluß von Strophanthin auf die Plasmaelektrolyte und die Wasser- und Elektrolytausscheidung der Niere beim herzgesunden Menschen. Klin. Wochenschr. *32*, 165–170 (1954)
Duarte, C.G., Chomety, F., Giebisch, G.: Effect of amiloride, ouabain, and furosemide on distal tubular function in the rat. Am. J. Physiol. *221*, 632–640 (1971)
Epstein, F.H., Silva, P., Stoff, J.: Active chloride transport powered by Na-K-ATPase in shark rectal gland. In: Current problems in clinical biochemistry. Vol. 6: Biochemical nephrology. Guder, G., Schmidt, U. (eds.), pp. 107–122. Bern, Stuttgart, Vienna: Huber 1978
Ernst, S.A.: Transport ATPase cytochemistry: ultrastructural localization of potassium dependent and potassium independent phosphatase activities in rat kidney cortex. J. Cell. Biol. *66*, 586–608 (1975)
Farber, S.J., Alexander, J.D., Pellegrino, E.D., Earle, D.P.: The effect of intravenously administered digoxin on water and electrolyte excretion and on renal functions. Circulation *4*, 378–386 (1951)
Friedrich, L., Hofmann, H.P., Kreiskott, H., Raschack, M.: Tierexperimentelle Untersuchungen zu extrakardialen Wirkungen des Herzglykosides Meproscillarin. Arzneim. Forsch. *28*, 503–506 (1978)
Friedrichs, D., Schoner, W.: Stimulation of renal gluconeogenesis by inhibition of the sodium pump. Biochim. Biophys. Acta *304*, 142–160 (1973)
Frömter, E.: Electrophysiology and isotonic fluid absorption of proximal tubules of mammalian kidney. In: MTP International review of science. Physiology Series One, Vol. 6: Kidney and urinary tract physiology. Thurau, K. (ed.), pp. 1–38. London: Butterworths, Baltimore: University Park Press 1974
Frömter, E., Gessner, K.: Effect of inhibitors and diuretics on electrical potential differences in rat kidney proximal tubule. Pflügers Arch. *357*, 209–224 (1975)
Frömter, E., Rumrich, G., Ullrich, K.J.: Phenomenologic description of Na^+, Cl^-, and HCO_3 absorption from proximal tubules of the rat kidney. Pflügers Arch. *343*, 189–220 (1973)
Fülgraff, G., Bieg, A., Wolf, K.: Der renale Sauerstoffverbrauch nach Strophanthin und 6-aminonicotinamid. Naunyn-Schmiedebergs Arch. Pharmacol. *266*, 43–49 (1970)
Giebisch, G.: Effects of diuretics on renal transport of potassium. Methods Pharmacol. *4A*, 121–164 (1976)
Giebisch, G., Sullivan, L.P., Whittembury, G.: Relationship between tubular net sodium reabsorption and peritubular potassium uptake in the perfused necturus kidney. J. Physiol. (Lond.) *230*, 51–74 (1973)

Göltner, E., Schwab, M.: Der Einfluß von Digitoxin auf die Plasmaelektrolyte und die Wasser- und Elektrolytausscheidung der Niere beim herzgesunden Menschen. Klin. Wochenschr. *32*, 542–546 (1954)

Göltner, E., Koch, R., Schwab, M.: Der Einfluß von Digitalisglykosiden auf das Glomerulusfiltrat, die Wasser- und Elektrolytausscheidung der Niere und die Plasmaelektrolyte bei Herzinsuffizienz. Naunyn-Schmiedebergs Arch. Pharmacol. *228*, 251–270 (1956)

Gordon, E.E.: Influence of ouabain on metabolism of rat kidney. Biochim. Biophys. Acta *104*, 606–608 (1965)

Grantham, J.J., Burg, M.B., Orloff, J.: The nature of transtubular Na and K transport in isolated rabbit renal collecting tubules. J. Clin. Invest. *49*, 1815–1826 (1970)

Greeff, K.: Über die Wirkung des Strophanthins auf den Elektrolythaushalt. Dtsch. Med. Wochenschr. *81*, 666–668 (1956)

Greeff, K.: Mineralstoffwechselwirkung und Toxizität des Strophanthins bei nebennierenlosen Ratten. Arch. Exp. Pathol. Pharmakol. *231*, 391–400 (1957)

Grollman, A., Suki, W., Ghavamian, M.: The direct action of squill on the kidney as compared to certain other digitalis bodies. Arch. Int. Pharmacodyn. Ther. *140*, 55–60 (1962)

Györy, A.Z., Kinne, R.: Energy source for transepithelial sodium transport in rat renal proximal tubules. Pflügers Arch. *327*, 234–260 (1971)

Györy, A.Z., Brendel, U., Kinne, R.: Effect of cardiac glycosides and sodium ethacrynate on transepithelial sodium transport in in vivo micropuncture experiments and on isolated plasma membrane Na-K ATPase in vitro of the rat. Pflügers Arch. *335*, 287–296 (1972)

Heidenreich, O., Fülgraff, G., Laaff, H.: Die diuretische Wirkung einiger Herzglykoside und ihre Beeinflussung durch Calcium- und Kaliumionen. Naunyn-Schmiedebergs Arch. Pharmacol. *251*, 169 (1965)

Heidenreich, O., Laaff, H., Fülgraff, G.: Der Einfluß von Calcium- und Kaliumionen auf die diuretische Wirkung von Herzglykosiden beim Hund. Naunyn-Schmiedebergs Arch. Pharmacol. *255*, 317–327 (1966)

Heidenreich, O., Baumeister, L., Fülgraff, G., Hahnege, V., Laaff, H.: Vergleichende Untersuchungen über die diuretische Wirkung und die akute Toxizität von g-Strophanthin, Scillaren A und Proscillaridin an Hunden. Arch. Int. Pharmacodyn. Ther. *166*, 1–10 (1967)

Heidenreich, O., Graf, R., Dierkesmann, R.-H.: Determination of diuretic effective concentrations of g-strophanthin-H^3 and proscillaridin-H^3 in renal tissue. Naunyn-Schmiedebergs Arch. Pharmacol. [Suppl.] *270*, R59 (1971)

Heidrich, H.G., Kinne, R., Kinne-Saffran, E., Hannig, K.: The polarity of the proximal tubule cell in rat kidney. J. Cell. Biol. *54*, 232–245 (1972)

Heller, J., Tata, P.S.: Effect of some metabolic inhibitors on the electrolyte and urea concentration gradients in rat kidney. Physiol. Bohemoslov. *16*, 428–440 (1967)

Hendler, E.K., Toretti, J., Weinstein, E., Epstein, F.H.: Functional significance of the distribution of Na-K-ATPase within the kidney. J. Clin. Invest. *48*, 37a (1969)

Hendler, E.K., Torretti, J., Epstein, F.H.: The distribution of sodium-potassium-activated adenosine triphosphatase in medulla and cortex of the kidney. J. Clin. Invest. *50*, 1329–1337 (1971)

Hierholzer, K., Lange, S.: The effects of adrenal steroids on renal function. In: MTP International review of science. Physiology Series One, Vol. 6: Kidney and urinary tract physiology. Thurau, K. (ed.), pp. 273–333. London: Butterworths, Baltimore: University Park Press 1974

Hofmann, L.M., Sherrod, T.R.: Effect of acetyl strophanthidin on glucose Tm and electrolyte excretion in the unanesthetized dog. Arch. Int. Pharmacodyn. Ther. *165*, 14–24 (1967)

Hook, J.B.: A positive correlation between natriuresis and inhibition of renal Na-K adenosine triphosphatase by ouabain. Proc. Soc. Exp. Biol. Med. *131*, 731–734 (1969)

Hoskins, B., Holland, W.C.: The effect of ouabain on calcium exchange in renal tubules of the rabbit. Arch. Int. Pharmacodyn. Ther *187* 37–45 (1970)

Hyman, A.L., Jaques, W.E., Hyman, E.S.: Observation on the direct effect of digoxin on renal excretion of sodium and water. Am. Heart J. *52*, 592–608 (1956)

Imai, M., Seldin, D.W., Kokko, J.P.: Effect of perfusion rate on the fluxes of water, sodium, chloride, and urea across the proximal convoluted tubule. Kidney Int. *11*, 18–27 (1977)
Inagaki, C., Lindenmayer, G.E., Schwartz, A.: Effects of sodium and potassium on binding of ouabain to the transport adenosine triphosphatase. J. Biol. Chem. *249*, 5135–5140 (1974)
Jørgensen, P.L.: Regulation of the Na^+-K^+)-activated ATP hydrolyzing enzyme system in rat kidney. II. The effect of aldosterone on the activity in kidneys. Biochem. Biophys. Acta *192*, 326–334 (1969)
Jørgensen, P.L.: Structure and function of Na-K-ion pump or Na, K-ATPase in mammalian kidney. In: Current problems in clinical, biochemistry, Vol. 8: Biochemical nephrology. Guder, G., Schmidt, U. (eds.), pp. 133–143. Bern, Stuttgart, Vienna: Huber 1978
Kaniike, K., Lindenmayer, G.E., Wallick, E.T., Lane, L.K., Schwartz, A.: Specific sodium-22 binding to a purified sodium and potassium adenosine triphosphatase. Inhibition by ouabain. J. Biol. Chem. *251*, 4794–4795 (1976)
Kiil, F.: Renal energy metabolism and regulation of sodium reabsorption. Kidney Int. *11*, 153–160 (1977)
Kiil, F., Aukland, K., Refsum, H.E.: Renal sodium transport and oxygen consumption. Am. J. Physiol. *20*, 511–516 (1961)
Kinne, R., Murer, H.: Recent advances in the understanding of renal amino acid and sugar transport. In: Proc. 7th. Int. Congr. of Nephrology, pp. 601–608. Basel: Karger 1978
Kinne-Saffran, E., Kinne, R.: Presence of bicarbonate stimulated ATPase in the brushborder microvillus membranes of the proximal tubule. Proc. Soc. Exp. Biol. Med. *146*, 751–753 (1974a)
Kinne-Saffran, E., Kinne, R.: Localization of a calcium-stimulated ATPase in the basal-lateral plasma membranes of the proximal tubule of rat kidney. J. Membr. Biol. *17*, 263–274 (1974b)
Koch, A.: Effect of ouabain on renal potassium secretion. Physiologist *2*, 72 (1959)
Kokko, J.P.: Proximal tubule potential difference, dependence on glucose on glucose, HCO_3, and amino acids. J. Clin. Invest. *52*, 1362–1367 (1973)
Kokko, J.P., Rector, F.C.: Flow dependence of transtubular potential difference in isolated perfused segments of rabbit proximal convoluted tubule. J. Clin. Invest. *50*, 2745–2750 (1971)
Kuppfer, S., Kosovsky, J.D.: Effects of cardiac glycosides on renal tubular transport of calcium, magnesium, inorganic phosphate, and glucose in the dog. J. Clin. Invest. *44*, 1132–1143 (1965)
Kyte, J.: The titration of cardiac glycoside binding site of the ($Na^+ + K^+$)-adenosine triphosphatase. J. Biol. Chem. *247*, 7634–7641 (1972)
Kyte, J.: Immunoferritin determination of distribution of ($Na^+ + K^+$) ATPase over the plasma membranes of renal convoluted tubules. I. Distal segment. J. Cell. Biol. *68*, 287–303 (1976a)
Kyte, J.: Immunoferritin determination of distribution of ($Na^+ + K^+$) ATPase over the plasma membranes of renal convoluted tubules. II. Proximal segment. J. Cell. Biol. *68*, 304–318 (1976b)
Lane, L.K., Copenhaver, J.H., Lindenmayer, J.R.G.F., Schwartz, A.: Purification and characterization of and (H^3)-ouabain binding to the transport adenosine-triphosphatase from outer medulla of canine kidney. J. Biol. Chem. *248*, 7197–7200 (1973)
Lassen, N.A., Munck, O., Thaysen, J.H.: Oxygen consumption and sodium reabsorption in the kidney. Acta Physiol. Scand. *51*, 371–375 (1961)
Law, R.O.: The effects of ouabain and ethacrynic acid on the intracellular sodium and potassium concentrations in renal medulary slices incubated in cold potassium-free Ringer solution and reincubated at 37 degrees C in the presence of external potassium. J. Physiol. (Lond.) *254*, 743–758 (1976)
Lendle, L.: Digitaliskörper und verwandte herzwirksame Glykoside (Digitaloide). In: Handbuch der experimentellen Pharmakologie. Heffter-Heubner (ed.), Vol. 1, pp. 11–241. Berlin: Springer 1935
Lendle, L., Mercker, H.: Extrakardiale Digitaliswirkungen. Ergeb. Physiol. *51*, 199–298 (1961)

Liang, C.T., Sacktor, B.: Bicarbonate-stimulated ATPase in the renal proximal tubule luminal (brush border) membrane. Arch. Biochem. Biophys. *176*, 285–297 (1976)
Lie, M., Sejersted, O.M., Raeder, M., Kiil, E.: Comparison of renal responses to ouabain and ethacrynic acid. Am. J. Physiol. *226*, 1221–1226 (1974)
Lutz, M.D., Cardinal, J., Burg, M.B.: Electrical resistance of renal proximal tubule perfused in vitro. Am. J. Physiol. *225*, 729–734 (1973)
Macknight, A.D.C.: Water and electrolyte contents of rat renal cortical slices incubated in medium containing p-chloromercuribeneoic acid or p-chloromercuribenzoic acid and ouabain. Biochim. Biophys. Acta *163*, 500–505 (1968)
Marsh, D.J., Martin, C.M.: Origin of electrical Pd's in hamster thin ascending limbs of Henle's loop. Am. J. Physiol. *232*, F348–F357 (1977)
Martinez-Maldonado, M., Allen, J.C., Eknoyan, G.E., Suki, W., Schwartz, A.: Renal concentrating mechanism: Possible role for sodium-potassium activated adenosine triphosphatase. Science *165*, 807–808 (1969)
Martinez-Maldonado, M., Eknoyan, G., Allen, J.C., Suki, W.N., Schwartz, A.: Urine dilution and concentration after digoxin infusion into the renal artery of dogs. Proc. Soc. Exp. Biol. Med. *134*, 855–860 (1970)
Martinez-Maldonado, M., Allen, J.C., Inagaki, C., Tsaparas, N., Schwartz, A.: Renal sodium-potassium-activated adenosine triphosphatase and sodium reabsorption. J. Clin. Invest. *51*, 2544–2551 (1972)
Matsui, H., Hayashi, Y., Homareda, H., Kimimura, M.: Ouabain-sensitive 42K binding to Na^+, K^+-ATPase purified from canine kidney outer medulla. Biochem. Biophys. Res. Commun. *75*, 373–380 (1977)
Maude, D.L.: Effects of K and ouabain on fluid transport and cell Na in proximal tubule in vitro. Am. J. Physiol. *216*, 1199–1206 (1969)
Nahmod, V.E., Walser, M.: The effect of ouabain on renal tubular reabsorption and cortical concentrations of several cations and on their association with subcellular particles. Mol. Pharmacol. *2*, 22–36 (1966)
Neachy, B.R.: Relationship between inhibition of renal $Na^+ + K^+$-ATPase and natriuresis. Ann. N.Y. Acad. Sci. *242*, 501–518 (1974)
Nechay, B.R.: Biochemical basis of diuretic action. J. Clin. Pharmacol. *17*, 626–641 (1977)
Nechay, B.R., Chinoy, D.A.: Effect of ouabain on renal transport of p-aminohippuric acid (PAH) and blood flow in the dog. Eur. J. Pharmacol. *3*, 322–329 (1968)
Nechay, B.R., Nelson, J.A.: Renal ouabain-sensitive adenosine triphosphatase activity and Na^+ reabsorption. J. Pharmacol. Exp. Ther. *175*, 717–726 (1970)
Nechay, B.R., Pardee, L.M.: Inhibition of N'-methylnicotinamide secretion by ouabain in the chicken kidney. J. Pharmacol. Exp. Ther. *147*, 270–276 (1965)
Nelson, J.A., Nechay, B.R.: Effects of cardiac glycosides on renal adenosine triphosphatase activity and Na^+ reabsorption in dogs. J. Pharmacol. Exp. Ther. *175*, 727–740 (1970)
Nelson, J.A., Nechay, B.R.: Interaction of ouabain and K^+ in vivo with respect to renal adenosine triphosphatase activity and Na^+ reabsorption. J. Pharmacol. Exp. Ther. *176*, 558–562 (1971)
Noé, G., Crabbé, E.J.: Imcomplete inhibition of sodium transport-related arobic metabolism upon exposure of guinea pig renal cortex slices to ouabain. Arch. Int. Physiol. Biochim. *83*, 343–345 (1975)
Orloff, J., Burg, M.: Effect of strophanthidin on electrolyte excretion in the chicken. Am. J. Physiol. *199*, 49–54 (1960)
Palmer, R.F., Nechay, B.R.: Biphasic renal effects of ouabain in the chicken: Correlation with a microsomal Na^+-K^+ stimulated ATP-ase. J. Pharmacol. Exp. Ther. *146*, 92–98 (1964)
Peters, L.: Renal tubular excretion of organic bases. Pharmacol. Rev. *12*, 1–35 (1960)
Post, R.L., Merrit, C.R., Kinsolving, C.R., Albright, C.D.: Membrane adenosine triphosphatase as a participant in the active transport of sodium and potassium in the human erythrocyte. J. Biol. Chem. *235*, 1796–1802 (1960)
Ramsey, A.G., Sachs, G.: Effect of ouabain on Na^+ and K^+ excretion in the rat. Proc. Soc. Exp. Biol. Med. *126*, 294–298 (1967)
Rector, F.C., Brunner, F.P., Sellman, J.C., Seldin, D.W.: Pitfalls in the use of micropuncture for the localization of diuretic action. Ann. N. Y. Acad. Sci. *139*, 400–407 (1966)

Repke, K., Est, M., Portius, H.J.: Über die Ursache der Speciesunterschiede in der Digitalisempfindlichkeit. Biochem. Pharmacol. *14*, 1785–1802 (1965)
Robinson, J.W.L., Mirkovitch, V., Sepulveda, F.V.: A comparison of the effects of ouabain and ethacrynic acid on the dog kidney in vivo and vitro. Pflügers Arch. *371*, 9–18 (1977)
Rocha, A.S., Kokko, J.P.: Sodium chloride and water transport in the medullary thick ascending limb of HENLE. Evidence for active chloride transport. J. Clin. Invest. *52*, 612–623 (1973)
Rosenfeld, S., Kraus, L., McCullen, A., Low, W., Morales. J.: Effect of ouabain and potassium on isolated perfused rabbit kidney. Proc. Soc. Exp. Biol. Med. *130*, 65–71 (1969)
Ross, B.D., Bullock, S.: Energy requirements for metabolic and excretory activities of perfused rat kidney. Curr. Probl. Clin. Biochem. *6*, 86–98 (1976)
Ross, B.D., Leaf, A., Silva, P., Epstein, F.H.: Na-K-ATPase in sodium transport by the perfused rat kidney. Am. J. Physiol. *226*, 624–629 (1974)
Rothlin, E., Bircher, R.: Pharmacodynamische Grundlagen der Therapie mit herzwirksamen Glykosiden. Ergeb. Inn. Med. Kinderheilkd. *5*, 457–552 (1954)
Ruedas, G., Weiss, Ch.: Die Wirkung von Änderungen der Na-Konzentration im Perfusionsmedium und von Strophanthin auf die Glucoseresorption der isolierten Rattenniere. Pflügers Arch. *298*, 12–22 (1967)
Saito, H., Nishida, H., Monma, Y., Tanabe, T.: Intrarenal distribution and ATPase inhibiting activity of ouabain in dogs. Jpn. J. Pharmacol. *26*, 171–178 (1976)
Schatzmann, H.J., Windhager, E.E., Solomon, A.K.: Single proximal tubules of the necturus kidney. II. Effect of 2,4-dinitrophenol and ouabain on water reabsorption. Am. J. Physiol. *195*, 570–574 (1958)
Schmidt, U., Dubach, U.C.: Activity of (Na^+-K^+) stimulated adenosine triphosphatase in the rat nephron. Pflügers Arch. *306*, 219–226 (1969)
Schmidt, U., Dubach, U.C.: Na-K stimulated adenosine-triphosphatase: Intracellular localisation within the proximal tubule of the rat nephron. Pflügers Arch. *330*, 265–270 (1971)
Schmidt, U., Horster, M.: Na-K-activated ATPase: activity maturation in rabbit nephron segments dissected in vitro. Am. J. Physiol. *233*, F55–F60 (1977)
Schmidt, U., Schmid, J., Schmid, H., Dubach, U.C.: Sodium- and potassium-activated ATPase. A possible target of aldosterone. J. Clin. Invest. *55*, 655–660 (1975)
Schurek, H.J., Aulbert, E., Ebel, H.: The effect of Ca ion antagonist verapamil on ouabain inhibition of renal sodium reabsorption. Studies in the isolated perfused rat kidney. In: Current problems in clinical biochemistry, Vol. 6: Renal metabolism in relation to renal function. Schmidt, U., Dubach, U.C. (eds.), pp. 281–288. Bern, Stuttgart, Vienna: Huber 1976
Schwartz, A., Lindenmayer, G.E., Allen, J.C.: The sodium-potassium adenosine triphosphatase pharmacological, physiological, and biochemical aspects. Pharmacol. Rev. *27*, 3–134 (1975)
Sejersted, O.M., Lie, M., Kiil, F.: Effect of ouabain on metabolic rate in renal cortex and medulla. Am. J. Physiol. *220*, 1488–1493 (1971)
Sejerstedt, O.M., Mathiesen, Ø, Kiil, F.: Oxygen requirement of renal Na, K-ATPase dependent sodium reabsorption. Am. J. Physiol. *232*, F152–F158 (1977)
Seldin, D.W., Eknoyan, G., Suki, W.N., Rector, F.C.: Localization of diuretic action from the pattern of water and electrolyte excretion. Ann. N.Y. Acad. Sci. *139*, 328–343 (1966)
Straub, W.: Die Digitalisgruppe. In: Handbuch der experimentellen Pharmakologie. Heffter-Heubner (ed.), Vol. II/2, pp. 1355–1452. Berlin: Springer 1924
Strickler, J.C., Kessler, R.H.: Direct renal action of some digitalis steroids. J. Clin. Invest. *40*, 311–316 (1961)
Strieder, N., Khuri, R., Wiederholt, W., Giebisch, G.: Studies on the renal action of ouabain in the rat. Effects in the non-diuretic state. Pflügers Arch. *349*, 91–107 (1974)
Suki, W., Rector, F.C., jr., Seldin, D.W.: The site of action of furosemide and other sulfonamide diuretics in the dog. J. Clin. Invest. *44*, 1458–1469 (1965)
Sulser, F., Kunz, H.A., Gantenbein, R., Wilbrandt, W.: Zur Frage einer antagonistischen Wirkung zwischen Herzglykosiden und Corticosteroiden auf die Elektrolytausscheidung der Niere. Naunyn-Schmiedebergs Arch. Pharmacol. *235*, 400–411 (1959)

Tanabe, T., Tsunemi, I., Abiko, Y., Dazai, H.: On the diuresis from the unilateral kidney produced by ouabain injected directly into the renal artery. Arch. Int. Pharmacodyn. Ther. *133*, 452–462 (1961)
Tobin, T., Brody, T.M.: Rates of dissociation of enzyme-ouabain complexes and K 0.5 values in ($Na^+ + K^+$) adenosine triphosphatase from different species. Biochem. Pharmacol. *21*, 1553–1560 (1972)
Tobin, T., Sen, A.K.: Stability and ligand sensitivity of ^{3}H-ouabain binding to (Na^+-K^+)-ATPase. Biochim. Biophys. Acta *198*, 120–131 (1970)
Toretti, J., Hendler, E., Weinstein, E., Longnecker, R.E., Epstein, F.H.: Functional significance of Na-K-ATPase in the kidney: Effects of ouabain inhibition. Am. J. Physiol. *222*, 1398–1405 (1972)
Ullrich, K.J., Frömter, E.: Active and passive transtubular transport in the proximal convolution. Proc. 7th. Int. Congr. Nephrology, pp. 147–154. Basel: Karger 1978
Ullrich, K.J., Capasso, G., Rumrich, G., Papavassilou, F., Klöss, S.: Coupling between proximal tubular transport processes. Studies ouabain, SITS and HCO_3-free solutions. Pflügers Arch. *368*, 245–252 (1977)
Urbaitis, B.K., Kessler, R.H.: Actions of inhibitor compounds on adenine nucleotides of renal cortex and sodium excretion. Am. J. Physiol. *220*, 1116–1123 (1971)
Vogel, G.: Vergleichende Untersuchungen zur diuretischen Aktivität verschiedener herzwirksamer Glykoside. Naunyn-Schmiedebergs Arch. Pharmacol. *241*, 553 (1961)
Vogel, G.: Die Beeinflussung des tubulären Natrium-, Kalium- und Calcium-Transportes der isolierten, künstlich perfundierten Froschniere durch verschiedene herzwirksame Steroide. Naunyn-Schmiedebergs Arch. Pharmacol. *243*, 354–355 (1962)
Vogel, G.: Über die renale Manipulierung von Herzglykosiden als Ursache ihrer unterschiedlichen diuretischen Aktivität. Arch. Exp. Pathol. Pharmakol. *245*, 69 (1963)
Vogel, G., Buchheim, S.: Über die Abhängigkeit der natriuretischen Wirkung kardiotoner Steroide von der Darbietungsrichtung. Pflügers Arch. *276*, 312–316 (1962)
Vogel, G., Kluge, E.: Vergleichende Untersuchungen zur diuretischen Aktivität verschiedener herzwirksamer Steroide. Arzneim. Forsch. *11*, 848–850 (1961)
Vogel, G., Lauterbach, F.: Über das Verhalten von Herzglykosiden in der Niere als Ursache ihrer unterschiedlichen diuretischen Aktivität. Arch. Exp. Pathol. Pharmakol. *244*, 334–350 (1963)
Vogel, G., Tervooren, U.: Zur Lokalisation der Wirkung kardiotoner Steroide auf verschiedene Transporte in der Niere. Pflügers Arch. *280*, 46–49 (1964)
Vogel, G., Buchheim, S., Lehmann, H.D.: Über die Beeinflussung des tubulären Na-, K- und Ca-Transportes der isolierten künstlich perfundierten Froschniere durch herzwirksame Steroide verschiedener Polarität. Pflügers Arch. *275*, 12–22 (1962)
Vogel, G., Kröger, W., Tervooren, U.: Untersuchungen zur Wirkung kardiotoner Steroide und ihrer Lokalisation auf verschiedene tubuläre Ionentransporte in der Niere mit besonderer Berücksichtigung der Kalium-Transportmechanismen. Pflügers Arch. *277*, 502–512 (1963)
Vogel, G., Stoeckert, I., Tervooren, U.: Hemmung renal tubulärer Substanztransporte durch Diuretica. Naunyn-Schmiedebergs Arch. Pharmacol. *255*, 245–253 (1966)
Weinschelbaum de Jairala, S.E., Goldman, L., Vieyra, A. Garcia, A.P., Rasia, M.L.: Effect of calcium, potassium, and ouabain on the oxygen consumption of external medulla slices from dog kidney. Biochim. Biophys. Acta *183*, 137–143 (1969)
Weinschelbaum de Jairala, S., Vieyra, A., MacLaughlin, M.: Influence of ethacrynic acid and ouabain on the oxygen consumption and potassium and sodium content of the kidney external medulla of the dog. Biochim. Biophys. Acta *279*, 320–330 (1972)
Whittam, R., Wheeler, K.P.: The sensitivity of a kidney ATPase to ouabain and to sodium and potassium. Biochim. Biophys. Acta *51*, 622–624 (1961)
Whittembury, G., Fishman, J.: Relation between cell Na extrusion and transtubular absorption in the perfused toad kidney: The effect of K, ouabain, and ethacrynic acid. Pflügers Arch. *307*, 138–153 (1969)
Whittembury, G., Proverbio, F.: Two modes of Na extrusion in cells from guinea pig kidney cortex slices. Pflügers Arch. *316*, 1–25 (1970)

Whittembury, G., Proverbio, F.: Dual sodium pumps in the kidney. In: Intestinal ion transport. Robinson, J.W.L. (ed.), pp. 1–11. Lancaster: Medical and Technical Publishing Company 1976
Whittembury, G., Diezi, F., Diezi, J., Spring, K., Giebisch, G.: Some aspects of proximal tubular sodium chloride reabsorption in necturus kidney. Kidney Int. 7, 293–303 (1975)
Wilde, W.S., Howard, P.J.: Renal tubular action of ouabain on Na and K Transport during stop-flow and slow-flow technique. J. Pharmacol. Exp. Ther. *130*, 232–238 (1960)
Willis, J.S.: Ouabain inhibition of ion transport and respiration in renal cortical slices of ground squirrels and hamsters. Biochim. Biophys. Acta *163*, 506–515 (1968a)
Willis, J.S.: The interaction of K^+, ouabain, and Na^+ on the cation transport and respiration of renal cortical cells of hamsters and ground squirrels. Biochim. Biophys. Acta *163*, 516–530 (1968b)
Willis, J.S., Fang, L.S.: Li^+-stimulation of ouabain-sensitive respiration and (Na^+-K^+)-ATPase of kidney cortex of ground squirrels. Biochim. Biophys. Acta *219*, 486–489 (1970)
Withering, W.: An account of the foxglove and some of its medical uses: with practical remarks on dropsy and other diseases. London: Robinson 1785 (Reprinted in Medical classics, Vol. 2, pp. 295–443. Baltimore: Williams & Wilkins 1937–1938)

Author Index

Page numbers in *italics* refer to bibliography

Subject Index

Handbook of Experimental Pharmacology

Continuation of "Handbuch der experimentellen Pharmakologie"

Springer-Verlag
Berlin
Heidelberg
New York

Handbook of Experimental Pharmacology

Continuation of "Handbuch der experimentellen Pharmakologie"

Springer-Verlag
Berlin
Heidelberg
New York

MIX
Papier aus verantwortungsvollen Quellen
Paper from responsible sources
FSC® C105338

If you have any concerns about our products,
you can contact us on
ProductSafety@springernature.com

In case Publisher is established outside the EU,
the EU authorized representative is:
Springer Nature Customer Service Center GmbH
Europaplatz 3, 69115 Heidelberg, Germany

Printed by Libri Plureos GmbH
in Hamburg, Germany